AF745602

Springer Geography

The Springer Geography series seeks to publish a broad portfolio of scientific books, aiming at researchers, students, and everyone interested in geographical research. The series includes peer-reviewed monographs, edited volumes, textbooks, and conference proceedings. It covers the entire research area of geography including, but not limited to, Economic Geography, Physical Geography, Quantitative Geography, and Regional/Urban Planning.

More information about this series at http://www.springer.com/series/10180

Assefa M. Melesse · Wossenu Abtew
Editors

Landscape Dynamics, Soils and Hydrological Processes in Varied Climates

Springer

Editors
Assefa M. Melesse
Department of Earth and Environment
Florida International University
Miami, FL
USA

Wossenu Abtew
South Florida Water Management District
West Palm Beach, FL
USA

ISSN 2194-315X ISSN 2194-3168 (electronic)
Springer Geography
ISBN 978-3-319-18786-0 ISBN 978-3-319-18787-7 (eBook)
DOI 10.1007/978-3-319-18787-7

Library of Congress Control Number: 2015942474

Springer Cham Heidelberg New York Dordrecht London
© Springer International Publishing Switzerland 2016
This work is subject to copyright. All rights are reserved by the Publisher, whether the whole or part of the material is concerned, specifically the rights of translation, reprinting, reuse of illustrations, recitation, broadcasting, reproduction on microfilms or in any other physical way, and transmission or information storage and retrieval, electronic adaptation, computer software, or by similar or dissimilar methodology now known or hereafter developed.
The use of general descriptive names, registered names, trademarks, service marks, etc. in this publication does not imply, even in the absence of a specific statement, that such names are exempt from the relevant protective laws and regulations and therefore free for general use.
The publisher, the authors and the editors are safe to assume that the advice and information in this book are believed to be true and accurate at the date of publication. Neither the publisher nor the authors or the editors give a warranty, express or implied, with respect to the material contained herein or for any errors or omissions that may have been made.

Printed on acid-free paper

Springer International Publishing AG Switzerland is part of Springer Science+Business Media (www.springer.com)

Preface

Landscape dynamics and variability have a close linkage to ecosystem dynamics affecting fluxes of water, energy and mass. The spatiotemporal variability of energy, water and mass budget controls the functional response and behaviour of landscapes, and this degree of response is different in different climates. Landscape changes are linked to climate change through energy partitioning on the surface, evapotranspiration, rainfall and carbon release or sequestration. Critical resources loss as forest cover, and soil has direct link with landscape. The availability of soil moisture which is related to topography and vegetation cover controls surface runoff and the subsurface flow and hence groundwater recharge.

The book, *Landscape Dynamics, Soils and Hydrological Processes in Varied Climates*, presents the results of various studies on the dynamics and processes governing the spatiotemporal variability of the most important natural resources, landscapes\land use, soil and water across the spectra of varied climates. These processes are believed to be highly variable across regions, and their linkages to environmental variability in different climatic regions of the world are an important aspect worth understanding and documenting. What makes this book unique are as follows: (1) it integrates and presents the basic processes driving these variability as well as the case studies in various regions; (2) varied climatic zones from arid to semi-arid to humid and wet are represented; and (3) the results from various scales, laboratory, field, watershed, basin and region are included.

The book presents the results of scientific studies and analysis of processes governing the dynamics of land/landscapes, soils and water across varied climatic regions. The linkage and interaction of these processes with ecosystem dynamics in various environmental settings and their effect on the fluxes of water, soil/sediment and energy are presented. The book discusses spatial as well as temporal variability of land cover, biophysical variables, sediment/soils, water, energy and contaminant in varied climatic zones. Topics on physical characteristics of soils and soil moisture and tillage operations, surface and groundwater dynamics and management, soil and geomorphological processes, soil detachment and sediment deposition, impact of climate change on vegetation, and water and sediment dynamics are presented. The book presents landscape mapping and analysis, soil erosion,

erosion control method, water resources stress, floods, groundwater flow assessment, climate change impacts, applications of soil and hydrological models, climate models and remote sensing for resource evaluation. The stress created by population growth on forest cover, wildlife, soils and water resources is sufficiently covered in several chapters.

The book has 36 chapters in seven parts: (1) Landscape and Land Cover Dynamics, (2) Rainfall–Runoff Processes, (3) Floods and Hydrological Processes, (4) Groundwater Flow and Aquifer Management, (5) Sediment Dynamics and Soil Management, (6) Climate Change Impact on Sediment and Water Dynamics and (7) Water and Watershed Management.

This book is beneficial for resource managers, agricultural and environmental professionals, policy-makers, students and teachers. It is a valuable source of information for graduate and undergraduate students at college level as well as for different users who are involved in research. It is a useful reference and text for courses such as ecohydrology, arid zone hydrology, landscape dynamics, soil processes and similar courses.

Assefa M. Melesse
Wossenu Abtew

Contents

Part II Rainfall–Runoff Processes

Part III Floods and Hydrological Processes

Part V Sediment Dynamics and Soil Management

Part VI Climate Change Impact on Sediment and Water Dynamics

Chapter 1
Introduction

Assefa M. Melesse and Wossenu Abtew

Abstract Dynamics of landscape features can significantly affect ecohydrological processes with strong connection to the fluxes of water, energy, and mass (pollutants and sediment). Knowledge of the spatial variability of these landscape variables is useful information in understanding how landscape patterns are related to hydrological variables including soil moisture, runoff, evapotranspiration, and groundwater flow. Among other environment parameters, topography is a determinant for magnitudes and spatial distributions of water and energy fluxes over natural landscapes.

Keywords Landscape dynamics · Ecohydrological processes · Spatial variability · Fluxes

1.1 Overview

Landscape pattern is important to the range of issues of environmental processes, including hydrological connectivity processes, the temporal storage of runoff and soil moisture, fluxes of energy and sediment delivery (Popp et al. 2009; Hou et al. 2014). The spatial variability of landscape features such as topography, soils, and vegetation defines the spatial pattern of hydrological state variables like soil moisture, runoff, evapotranspiration, and groundwater flow. This spatial variability thereby controls the functional behavior of the landscape in terms of its hydrologic response impacting the patterns of vegetation and hence local climate. Fluxes of

A.M. Melesse (✉)
Department of Earth and Environment, Florida International University, Modesto a. Maidique Campus, Miami, FL 33199, USA
e-mail: melessea@fiu.edu

W. Abtew
South Florida Water Management District, 3301 Gun Club Road, West Palm Beach, FL 33416, USA
e-mail: wabtew@sfwmd.gov

© Springer International Publishing Switzerland 2016
A.M. Melesse and W. Abtew (eds.), *Landscape Dynamics, Soils and Hydrological Processes in Varied Climates*, Springer Geography,
DOI 10.1007/978-3-319-18787-7_1

water, energy, and nutrient and their transport in landscapes are highly variant and influenced by topography and other physical features of watersheds. Land use is closely related to the characteristics of human activities, which in turn determine the anthropogenic substances carried into erosion systems through soil detachment, runoff process, and sediment transport (Shi et al. 2013). The most important landscape variables and factors that require knowledge of their variability and hence impact in dictating the fluxes of energy, water, contaminants, and sediment include soil properties, land use, and topography.

1.2 Soils and Soil Erosion

Soil characteristics of various landscapes are different and are the result of geological and other natural processes impacting water resources availability. Efforts to characterize the high spatial and temporal variability of soils' hydrological properties (infiltration capacity, bulk density, soil moisture content and hydraulic conductivity, and others), with reasonable accuracy are limited by the need to take a large number of samples, which is laborious, expensive, and time-consuming. Understanding soil's characteristics and how they impact hydrological processes of landscapes is key to watershed management. Soil, the medium of food production, has a slow accretion rate. But as a result of changes in landscape and land use, soil erosion in many regions is a serious problem. Population increase and the need for more food production, in regions as the Blue Nile basin, have resulted in deforestation and expansion of arable land to steep slope terrains. As a result, severe soil erosion is observed and attempt to control or reduce erosion is usually not successful. Land cover change impacts soils, water storage, and runoff generation with wider implication of streamflow and water quality.

1.3 Landscape and Land Use

Watersheds exhibit high variability in their topography, geology, land cover, soils, and land management. This variability in turn controls the extent of water and energy redistribution affecting partitioning of precipitation into soil moisture, runoff, evapotranspiration, infiltration, and groundwater flow. This partitioning and redistribution dictate the availability of water in the different storages and water movement among the various storages.

One of the most important landscape variables that play an important role in influencing hydrological regimes is land use/land cover. Land cover impacts the partitioning of precipitation into runoff and infiltration affecting available soil moisture and groundwater recharge. Land cover also plays an important role in impacting local and large-scale hydrometeorological processes and climate dynamics through the exchange of moisture and energy between the earth and atmosphere.

Accurate and up-to-date information on land cover and the state of the environment are critical components of flood management, environmental planning, and management. Land-cover information is used in watershed modeling to estimate the value of surface roughness or friction as it affects the velocity of the overland flow of water. It may also be used to determine the amount of rainfall infiltration on a surface. Surface energy fluxes are highly related to the land cover though the albedo and emissivity characteristics of the land cover surfaces.

Energy (i.e., latent heat, sensible heat, and soil heat) fluxes and land surface environment variables (e.g., management practices and topography) determine land surface characteristics and availability of moisture, and thus are commonly used to understand the land surface–atmosphere interactions. The partitioning of net radiation into latent, sensible, and soil heat fluxes is mainly controlled by vegetation cover, temperature gradient, vapor pressure difference, and wind speed. These energy fluxes have been used for estimating irrigation water demand (e.g., Kustas 1990; Bastiaanssen 2000; Kustas et al. 2004; Melesse and Nangia 2005), evaluating wetland restoration (e.g., Loiselle et al. 2001; Mohamed et al. 2004; Melesse et al. 2006, 2007; Oberg and Melesse 2006), and understanding the effects of land management on vegetation cover (e.g., Kustas et al. 1994, 2004; Kustas and Norman 1999; French et al. 2000; Hemakumara et al. 2003; Melesse et al. 2008).

Deforestation is the most anthropogenic-induced landscape change on earth. According to Porter and Brown (1991), the rate of deforestation in tropical forests was 15.4 million ha per year. Globally, 60 % of deforestation is for subsistence and commercial agriculture, 20–25 % is for commercial logging and 15–20 % is for cattle ranching cash crop plantation. Land use change, mainly global deforestation, contributes 12.5 % of greenhouse gas emissions (Herzog 2009).

Urbanization is another anthropogenic landscape change with local environment change as heat island effect. Heat, moisture, and momentum fluxes are significantly altered by urban landscape in contrast to rural areas. In the northern hemisphere, urban areas on the average have 12 % less solar radiation, 14 % more rainfall, 8 % more clouds, 10 % more rainfall, 15 % more thunderstorms, 10 times higher pollutant concentration, and are 2 °C warmer (Taha 1997).

Drainage of wetlands and floodplains are other anthropogenic major landscape changes with changes in energy and water distribution. According to the US EPA, in the lower 48 states there were 89 million ha of wetlands in 1600 which are reduced by half currently (http://water.epa.gov/type/wetlands/vital_status.cfm Accessed 25 February 2015).

1.4 Topography

Topography plays an important role in the distribution and flux of water and energy within the natural landscape. Surface runoff, evaporation, and infiltration are hydrologic processes that take place at the ground–atmosphere interface. The quantitative assessment of these processes depends on topographic configuration of the landscape, which is one of several controlling boundary conditions.

The topographic configuration of a landscape is a control boundary condition for the hydrologic processes of surface runoff, evaporation, and infiltration, which take place at the ground–atmosphere interface. For example, wetness index (WI) provides a description of the spatial distribution of soil moisture in terms of topographic information. WI is computed as

$$\mathrm{WI} = \ln\frac{A}{S} \tag{1.1}$$

where A and S are the specific drainage (i.e., flow accumulation) area and slope, respectively.

As A increases and/or S decreases, WI becomes larger, indicating that soil moisture content will increase. Because WI takes into account local slope variations, it has proven to be a reasonable indicator for soil wetness, flow accumulation, saturation dynamic, water table fluctuation, evapotranspiration, soil horizon thickness, organic matter content, pH, silt and sand content, and plant cover density.

The microtopography and latent heat flux are found to be well correlated. Areas with higher WI values are identified as the areas receiving more overland flows (i.e., with greater flow accumulations) and having a smaller gradient. These areas have higher soil moisture but a higher evaporation rate than the areas with lower WI values. The correlation between WI and soil moisture is further verified by the observation that when water is a limiting factor of an agricultural field, the crop in the areas with higher WI values tends to grow better than the crop in the areas with lower WI values. This can be attributed to more water availability for transpiration (i.e., latent heat demand) in areas with higher WI values. The latent heat flux increases at a greater rate when the WI is lower, which is partially due to the proportional relationship between available water for evapotranspiration and WI.

References

Bastiaanssen WGM (2000) SEBAL-based sensible and latent heat fluxes in the irrigated ediz Basin, Turkey. J Hydrol 229:87–100

French AN, Schmugge TJ, Kustas WP (2000) Estimating surface fluxes over the SGP site with remotely sensed data. Phys Chem Earth 25(2):167–172

Hemakumara HM, Chandrapala L, Moene AF (2003) Evapotranspiration fluxes over mixed vegetation areas measured from large aperture scintillometer. Agric Water Manag 58(2): 109–122

Herzog T (2009) World greenhouse gas emissions in 2005. World Resources Institute

Hou J, Fu B, Liu Y, Lu N, Gao G, Zhou J (2014) Ecological and hydrological response of farmlands abandoned for different lengths of time: evidence from the Loess Hill Slope of China. Global Planet Change 113:59–67

Kustas WP, Li F, Jackson TJ, Prueger JH, MacPherson JI, Wolde M (2004) Effects of remote sensing pixel resolution on modeled energy flux variability of croplands in Iowa. Remote Sens Environ 92(4):535–547

Kustas WP (1990) Estimates of evapotranspiration with a one-and two-layer model of eat transfer over partial canopy cover. J Appl Meteorol 29:704–715

Kustas WP, Perry EM, Doraiswamy PC, Moran MS (1994) Using satellite remote sensing to extrapolate evapotranspiration estimates in time and space over a semiarid Rangeland basin. Remote Sens Environ 49(3):275–286

Kustas WP, Norman J (1999) Evaluation of soil and vegetation heat flux predictions using simple two-source model with radiometric temperatures for partial canopy cover. Agric For Meteorol 94:13–29

Loiselle S, Bracchini L, Bonechi C, Rossi C (2001) Modeling energy fluxes in remote wetland ecosystems with the help of remote sensing. Ecol Model 45(2):243–261

Melesse A, Nangia V (2005) Spatially distributed surface energy flux estimation using remotely-sensed data from agricultural fields. Hydrol Process 19(14):2653–2670

Melesse A, Oberg J, Beeri O, Nangia V, Baumgartner D (2006) Spatiotemporal dynamics of evapotranspiration and vegetation at the Glacial Ridge prairie restoration. Hydrol Process 20 (7):1451–1464

Melesse A, Nangia V, Wang X, McClain M (2007) Wetland restoration response analysis using MODIS and groundwater data. Special Issue: Remote Sens Nat Resour Environ Sens 7:1916–1933

Melesse AM, Frank A, Nangia V, Liebig M, Hanson J (2008) Analysis of energy fluxes and land surface parameters in grassland ecosystem: remote sensing perspective. Int J Remote Sens 29 (11): 3325–3341

Mohamed YA, Bastiaanssen WGM, Savenije HHG (2004) Spatial variability of evaporation and moisture storage in the swamps of the upper Nile studied by remote sensing techniques. J Hydrol 289:145–164

Oberg J, Melesse AM (2006) Evapotranspiration dynamics at an ecohydrological restoration site: an energy balance and remote sensing approach. J Am Water Resour Assoc 42(3): 565–582

Popp A, Vogel M, et al (2009) Scaling up ecohydrological processes: role of surface water flow inwater-limited landscapes. J Geophys Res Biogeosci:114

Porter G, Brown JW (1991) Global environmental politics. Westview Press, Boulder

Shi ZH, Ai L et al (2013) Partial least-squares regression for linking land-cover patterns to soil erosion and sediment yield in watersheds. J Hydrol 498:165–176

Taha H (1997) urban climates and heat islands: albedo, evapotranspiration and anthropogenic heat. Energy Build 25:99–103

Part I
Landscape and Land Cover Dynamics

Chapter 2
GIS and Remote Sensing-Based Forest Resource Assessment, Quantification, and Mapping in Amhara Region, Ethiopia

Mulatie Mekonnen, Tsegaye Sewunet, Mulu Gebeyehu, Bayleyegn Azene and Assefa M. Melesse

Abstract World forest resources are continually depleting. Assessing and quantifying the current forest resources status is a prerequisite for forest resources improvement planning and implementation. The objectives of this study are to assess, quantify, and map forest resources in the Amhara National Regional State, Ethiopia. GIS, GPS, and Remote Sensing technologies were applied for the study. As a result, forest distribution map is prepared. Most of the forest covers were found along the lowland belt of Mirab Gojam, Awi, and Semen Gonder zones bordering the neighboring country, Sudan and the Tigray and Benishangul-Gumz regions. The total forest cover of the region is 12,884 km^2, that is, about 8.2 % of the total land area. Including bushlands, it is about 21,783 km^2, which is about 13.85 %. Woodlands, natural dense forest, riverine forest, bushlands, and plantations are 740,808, 463,950, 20,653, 889,912, and 62,973 ha in area with percentage coverage of 4.71, 2.95, 0.13, 5.66, and 0.40 respectively. GIS, GPS, and Remote Sensing were found to be important tools for forest resource assessment and mapping.

M. Mekonnen (✉) · M. Gebeyehu · B. Azene
Amhara National Regional State, Bureau of Agriculture, Natural Resource Conservation and Management Department, PO Box 1188, Bahir Dar, Ethiopia
e-mail: mulatiemekonneng@gmail.com; mulatie.mekonnen@yahoo.com

M. Gebeyehu
e-mail: mulu_gebeyehu@yahoo.com

B. Azene
e-mail: aze_bayleyegn@yahoo.com

T. Sewunet
Amhara National Regional State, Bureau of Finance and Economic Development, Bahir Dar, Ethiopia
e-mail: tsegayesewinet@gmail.com

A.M. Melesse
Department of Earth & Environment, Florida International University, 11200 SW 8th Street, Miami, USA
e-mail: melessea@fiu.edu

© Springer International Publishing Switzerland 2016

A.M. Melesse and W. Abtew (eds.), *Landscape Dynamics, Soils and Hydrological Processes in Varied Climates*, Springer Geography,
DOI 10.1007/978-3-319-18787-7_2

Keywords Forest resource · Deforestation · GIS · Remote sensing · Amhara region · Ethiopia · Blue Nile basin

2.1 Introduction

The Amhara National Regional State (ANRS) is located in the northwestern part of the country, Ethiopia. It is situated between 8° 45′–13° 45′N latitude and 35° 15′–40° 20′E longitude. The region covers about 157,127 km^2. It has common boundaries with four national regional states of the country, Oromiya in the south, Afar in the east, Tigray in the north, and Benishangul-Gumuz in the west. It also shares a common boundary with the neighboring county, Sudan, in the west. According to the 2007/08 census, the region has about 20,650,419 people. About 87.3 % of the population lives in rural areas and 12.7 % lives in urban areas. Agriculture and livestock production are the main farming systems of the region. Forestry is the other product of the region that the population depends on for domestic firewood, construction, and utility pole contributing to the country's economy.

The region has a wide biodiversity of flora and fauna. However, increase in the population, livestock pressure, and increased demand for arable land are causing a significant depletion of forest resources. Loss of cover results in high rate of soil erosion, loss of soil fertility, and degradation of water resources. These factors in turn, adversely affect agricultural productivity. Soil fertility is further depressed where animal dung and crop residue are diverted for fuel to compensate for the shortage of wood. The cumulative effect of this chain of events is reflected in the prevailing land degradation, poor economic performance, and accelerated poverty.

The regional government with the Bureau of Agriculture (BoA) has made great efforts to manage the remnant forests as regular and priority state forests, community or privately owned forests. Studies conducted to identify and quantify the forest cover of the region are limited. Even if there are studies, they were conducted a decade before and they could not show the current status of forest cover. Hence, this study was initiated to assess, quantify, and map the current forest cover status of the region using GIS and remote sensing techniques together with secondary and primary data at ground level.

Land use/land cover dynamics is an important landscape process capable of altering the fluxes of water, sediment, contaminants, and energy. Mainly caused by humans, impact of land use on water resources availability is high. Degraded watersheds tend to accelerate overland flow reducing soil moisture and baseflow recharge and increases sediment detachment and transport. Various studies used land cover mapping tools and methods to understand land use changes, inventory of forest and natural resources, as well as understand the changes in the hydrologic behavior of watersheds (Getachew and Melesse 2012; Mango et al. 2011a, b; Wondie et al. 2011, 2012; Melesse et al. 2008; Melesse and Jordan 2002, 2003; Mohamed et al. 2013; Heinen et al. 1989).

The specific objectives of the study reported in this chapter are to: (1) identify and quantify the forest resources of the region, (2) assess the spatial distribution of the regional forest resource and map it, (3) develop regional forest resource database and access information for decision makers, researchers, development practitioners, professionals, and others, and (4) assess the incense and bamboo resources of the region.

2.1.1 Scope of Study

The scope of the study was spatially limited in the Amhara National Regional State, which covers about 157,127 km^2. Objectively, the study is aimed to identify, quantify, and map the spatial distribution of forest resources in the region. The study is also charged with identifying incense potential areas in the region. Moreover, identification of bamboo resources and its potential sites is another area of concentration for the study. To conduct the study, field observations, collection of secondary information from the offices of agriculture at zonal and district level, ground level primarily GPS-based data, and satellite imagery were used.

2.2 Forest Resources in Ethiopia

The definition of forest is still ambiguous. According to FAO (2001) forest is defined as "land with a tree crown cover of more than 10 % and an area of more than 0.5 ha; the trees should be able to reach a minimum height of 5 m at maturity." UNFCCC (2006) also defined forest as a minimum land area of 0.05–1 ha, with tree crown cover more than 10–30 % and tree height of 2–5 m at maturity.

Ethiopia was rich in natural forests. Several authors and national or subnational inventory projects have carried out assessments and documented forest resources of Ethiopia. Close to 40 % of Ethiopia might have been covered by high forests as recently as the sixteenth century as historical sources indicate (EFAP 1994a). Another estimate put the original forest cover as a percentage of the total land area at 25 % compared with 48 % for the world (Earth Trends 2007). EFAP (1994a) notes that about 16 % of the land area was estimated to have been covered by high forests in the early 1950s, which declined to 3.6 % in the early 1980s and further declined to 2.7 % in 1989.

In the recent forest proclamation (No. 542/2007), high forests, woodlands, bamboo forests are recognized as forests. Following the definition of FAO (2001), the vegetation of Ethiopia that may qualify as 'forests' is natural high forests, woodlands, plantations, and bamboo forests, with an estimated area of 30.8 % (35.13 million ha). If the shrublands are added, the estimated cover is over 50 % (61.62 million ha). On the other hand, the recent data on forest resources of

Ethiopia reported in FAO (2010) puts Ethiopia among countries of the world with forest cover of 10–30 %. According to this report Ethiopia's forest cover (FAO definition) is 12.2 million ha (11 %). It further indicated that the forest cover shows a decline from 15.11 million ha in 1990 to 12.2 million ha in 2010, during which 2.65 % of the forest cover was deforested.

The growing need for fuelwood and agricultural land and overgrazing by livestock, coupled with improper forest and land tenure policies, are believed to be the major causes of forest degradation (Mulugeta and Melaku 2007). Extensive forest fires have also resulted in further losses. Dramatic deforestation has been associated with political transitions from the Imperial to the Derg regime, and especially from the latter to the Ethiopian People's Revolutionary Democratic Front (EPRDF) (Birhan 2009). Although such challenges had occurred at different times, it has been suggested that the northern Ethiopian Highlands had little forest in the past (Melaku 2003). Considerable afforestation has also been carried out notably in population centers. The government has also introduced a national tree planting campaign every year all over the country in which every person is encouraged to plant trees, with significant influence on the attitude of the population toward trees.

2.2.1 Forest Resource Base in the Amhara Region

At the national level, in Amhara region, modern tree planting using introduced tree species (mainly Australian Eucalyptus) started in 1895 when Emperor Menelik II (1888–1892) looked into solutions for alleviating shortage of firewood and construction wood in the capital, Addis Ababa. However, the historic rapid expansion of large-scale and community plantations occurred during the Derg regime, which resulted in the establishment of large-scale plantations. Several fuelwood projects funded by NGOs spread over the country with marked concentrations around big cities such as Bahir Dar, Dessie, Gondar, and Debre-Berhan. These plantations have often been established for supplying the huge demand for wood products in Ethiopia. But forest resources in the region have experienced so much pressure due to increasing need for wood products, conversion to agriculture, agricultural investments, incense investments, and settlements.

Few studies have been conducted to identify the forest resource of the region. Proceedings of the Ethiopian national forest cover workshop estimated the regional forest resource cover at about 5.91 % of the total area of the region. The assessment of the BoA was 0.48 % natural high forest, 4.2 % woodland, and 1.23 % plantation forests (Bane et al. 2008). Such studies proved that the regional forest resource is dwindling as the population growth and overgrazing pressure increased from day-to-day.

2.2.2 *Forest Ownership*

Federal Policy is now governed by the Forestry Development Conservation and Utilization Proclamation, issued in September 2007 (542/2007), which repealed Proclamation 94/1994. The policy recognizes two types of forests, state and private. State forests are any protected or productive forests owned by the federal or a regional state. Private forests are forests other than state forests that are developed by any private person and include forests developed by members of a peasant association or by any association organized by private individuals, investors, and governmental and nongovernmental organizations. In reality, no official maps exist, no management plans have been prepared, and not much has been done except budget allocation for guards. Regional and federal resources to delineate state forests and prepare management plans are extremely limited. As a consequence, considerable uncertainty exists as to whether the remaining forests in different parts of the country would be allocated for agricultural investment.

2.2.2.1 State Forests

State forests comprise the priority state forests, regular state forests, and project forests. The priority forests are 17, which include Wof-washa in North Shoa; Erkie in Oromiya; Yegof in Dessie Zuria; Denkoro Chaka in South Wollo; Woinye in North Wollo; Guna, Alemsaga, and Tara Gedam in South Gondar; Matebia, Angereb and Gundo Gordim in North Gondar; Kinbaba and Sekela Mariam in West Gojam; Yeraba and Abafelase in East Gojam; Kahtasa, and Elala Guangua in the Awi zone. There are some project forests which are found in the region as Gondar fuelwood in North Gondar, Dessie fuelwood plantation in South Wollo, and Debre Birhan fuelwood plantation in North Shoa. There are also about 124 regular state forests in different districts. Regional national parks, such as, Alatish and Semen parks in North Gondar, Denkoro Chaka from priority state forests in South Wollo are the other most important forest resources which are habitat for wild life.

2.2.2.2 Community and Private Forests

Private or community forests are forests that are developed by any private person or forests developed by members of a peasant association or by any association organized by private individuals, investors, and governmental and nongovernmental organizations. In Amhara region, the community forest is considered as public forestry developed and conserved by the Kebele (community) in the communal lands for the Kebele development purposes and conservation of degraded areas. There are also many forests that were developed and conserved by the government and have now been transferred to the community at Kebele level to be managed and used as community forests. Community and privately owned forests are major sources of forest resources in the region. The availability of many communal lands in the rural areas can be used for tree plantation to diversify the forest cover. The severe shortage of wood for fuel, construction, and industries calls for huge forest development programs in the region, which creates an income opportunity for the rural poor.

2.2.3 *Incense and Bamboo Potential*

2.2.3.1 Incense Potential

Boswellia papyrifera is a deciduous multipurpose tree species known for its commercial product called frankincense or gum *olibanum*. It is mostly found in the lowland areas of the Amhara region which accounts for the lion share incense production in Ethiopia. Ethiopia is also one of the world's leading producers of incense, notably frankincense (product of Boswellia spp.) and myrrh and myrrh-like resins, products of Commiphora spp. (Lemenih and Teketay 2003). The production and trade volumes of gums and resins in Ethiopia have been increasing since the 1990s. Between 1998 and 2007, Ethiopia exported about 25,192 tons, an annual average of 2519 tons, of natural gums and resins with a value of 307,248,000 Eth. Birr equivalents to 34,138,670 USD (Lemenih and Kassa 2008).

In the western, northern, and northwestern lowlands of Ethiopia, the principal gum and resin producing species is *B. papyrifera*. In Metema District of the Amhara region, *B. papyrifera* accounts for 51 % of the woody plant density on average (Eshete 2002). Variations occur from site-to-site as a function of local climatic and anthropogenic factors. The density of *B. papyrifera* in the Metema area ranges from 64 to 225 stems per ha (Gebrehiwot 2003).

According to Tadesse et al. (2002) in the Metema district, *B. papyrifera* provide an annual yield of 6.7–451.4 g per tree, and others reported an annual yield of 207–352 g per tree. These variations in incense yield are attributed to tree size and tapping intensity. Generally, trees with bigger diameters at breast height (DBH) yield more incense than trees with smaller DBH. Similarly, increased tapping intensity increases incense yield per tree, although this has also been shown to affect tree vitality and reproductive biology.

Ecologically, the species is important since it can grow in areas where other trees fail to grow. The wood of Boswellia is used for poles and timber and for industrial manufacturing of matchboxes and boards. The leaves provide dry-season fodder, and the flowers are a good source of nectar for bees. Leaves, bark, and roots are also used in traditional medicines. The species is recommended for economic development and desertification control. Figure 2.1 shows woodlands dominated by *B. papyrifera* Species in Metema, Quara, and Belesa districts.

Fig. 2.1 Woodlands dominated by *Boswellia papyrifera* species in Metema, Quara, and Belesa districts, (*Photo* by Mulatie Mekonnen, December, 2011)

2.2.3.2 Bamboo Resource

The other most important species in the Amhara region is bamboo. There are two types of bamboo species in the region: highland bamboo (in Amharic, kerkeha) and lowland bamboo (in Amharic, shimel). Environmental and ecological differences of zones create conditions for growing these two different bamboo species. As their names indicate lowland bamboo grows in lowland areas of the region at altitude below 1800 m above sea level following the Abay basin and the Sudan border. The highland bamboo grows in the highland at altitude above 1800 m above sea level (Bereket 2008).

Planting bamboo and using it as a source of income has a very long history. The farmers are well aware of the local value of the resource. Especially, the highland bamboo in Awi, East and West Gojam zones is used for almost every utensil and furniture used in homes, fencing, construction, and agricultural equipment. The available resources are at good locations and near to potential markets such as the cities of Bahir Dar and Addis Ababa. The farmers have experience in propagation, harvesting, collection, and marketing the resource.

The lowland bamboo is also a type of important indigenous species. It grows well in the three zones (North Gondar, Awi, and West Gojam) and some in East Gojam. There was an estimated large lowland bamboo but due to lack of awareness of its economic value, it is being cleared and set on fire for farmland clearing. The current size of lowland bamboo resource is expected to be further reduced due to land clearing for resettlement of farmers from less productive areas. Figure 2.2 shows the highland bamboo species in Awi and West Gojam zones of the region.

Fig. 2.2 Sample highland bamboo species in Awi and West Gojam zones (*Photo* by Mulatie Mekonnen, February, 2012)

2.2.4 Opportunities and Challenges

2.2.4.1 Opportunities and Potentials

The Amhara National Regional State has wide agro-ecological or agro climatic zones, which are favorable for the growth of diversified flora species. The climatic variations help the region to have different and well-developed natural and planted forest resources; owned as state, community, and private forests. Currently, there is a huge local and export market potential for the forest products which can motivate the producers. Additionally, the attention of ecotourism in the region, the world-wide attention on global warming, and the presence of economically important plant species create a better chance for the forest resources of the region.

2.2.4.2 Challenges and Constraints

Although there are favorable agro-ecological zones in the region to develop and conserve forest resources; the region faces many problems. Currently, the prevailing population growth needs to satisfy wood resource and crop demand by encroachment of the forest lands. The rearing of large number of cattle and the grazing effect retards the growth of newly growing forest seedlings and saplings. Livestock trampling also leads to aggravated soil erosion which is unfavorable to tree and other vegetation growth. The problems are clear and understandable to every concerned body, but design of a system and community involved approach is lacking to alleviate the problem. There is limited strong forest protection and development policy and strategy. However, forest policies, action programs, strategies, and proclamation that can support and guide the forestry development have already been approved at the federal level in 2007. There is still no regional forest policy, strategy, and proclamation to control deforestation and illegal forest product movement and encroachments. Figures 2.3, 2.4, and 2.5 show deforestation

Fig. 2.3 Deforestation of economically important forest species in North Gonder Zone (*Boswellia papyrifera*)

Fig. 2.4 Forest deforestation and agricultural expansion (North Gonder Zone)

Fig. 2.5 Contribution of overgrazing for forest destruction (North Gonder Zone)

of economically important forests species (*B. papyrifera*) and other forests due to investment agricultural practices.

In general, the major challenges for the development and conservation of forest resources include: deforestation, poor forest management and utilization, agricultural investment on forest products (like incense), unbalanced gap between demand and supply of wood resources, poor survival rate of planted forest seedlings, weak institutional capacity, turnover of skilled manpower and lack of awareness of the community in forestry importance, and poor infrastructures.

2.2.5 Current Forest Management Situations and Mitigation Measures

Remnant natural large forests and woodland species are mostly found in the lowland areas of the region. They host diverse tree species of economic importance, the most important of which is *B. papyrifera* and other forest types like natural forests, woodlands, riverine, church, and lowland bamboo forests. However, due to agricultural expansion, overgrazing, fire, resettlements, and investment activities, these natural forests are becoming more vulnerable for destruction. Hence, it demands great care and proper forest management.

The regional BoA has promoted technology extensions through awareness creation and capacity building in the development and conservation of forest resources. Planting two trees per person at the end of the second millennium and three trees per person at the beginning of the third millennium have been taking place and planned to create awareness and promote plantation for forest recovery. Agroforestry practices such as woodlots, home gardens, trees on cropland, and farm boundary plantations are the most common practices on farmland. Currently, the most common tree species for community woodlots and private tree investments in the region are Eucalyptus spps., Acacia *decurrence,* and *Cupressus lusitanica.* Eucalyptus woodlots used to be extensively planted on farmland and increasing numbers of farmers are being encouraged to plant small on-farm woodlots in semi-urban areas where the returns from the sale of firewood and poles are attractive (EFAP 1994b; Zebene and Hulten 2003).

Although the region lacks its own forest policy, strategies, and proclamation the federal government forest policies, strategies, and proclamation are implemented in the management of the available forest resources. The regional government has also shown concern and allocated a budget for forest management and nursery extensions. Different development projects are assigned and oriented in supporting the development and sustainable utilization of the natural resources. But more than this should be done by BoA and other concerned bodies to conserve the remnant forest resource of the region.

2.3 Methodology

2.3.1 Materials and Methods

The materials used for this study were GPS, satellite image, GIS software, and topographic maps. Secondary data were also used from each district Agricultural Offices in addition to information from the experts. ArcGIS software was used to digitize the forest cover from the satellite imagery to analyze the results and to map the spatial distribution of the forest cover. GPS was used to collect GCP (ground control points) points and digital camera was used to take photographs that help for ground verification.

To carry out the study, on-screen digitizing from SPOT (5 m resolution) satellite imagery was used. Intensive GPS data collection and field observation were conducted for ground truthing. About 4800 GCPs and 820 pictures were taken from different locations for image interpretation. Field observations, discussions with experts, and secondary data assessments were done in more than 97 % of the regional woredas or districts.

Direct field observation and discussion with zonal and district experts were done to identify forest species. Secondary data from the district offices were also collected to crosscheck with the satellite image and was used as input for this chapter. Woreda experts, who are familiar with their forest resources distribution and type, helped in identification.

2.3.2 Forest Classifications Methods

There are different forest classification methods. Sebsebe and Edwards (2006) identify vegetation cover types as afroalpine and sub-afroalpine, dry evergreen montane forests and grassland complex, moist evergreen montane forests, evergreen scrub, Combretum-terminalia woodland and savannah, Acacia-commiphora woodland, lowland semi–evergreen forest, desert and semidesert scrubland. FAO (2000) identifies forest types as shrubland, grassland, savannah, deciduous broadleaf forests, deciduous needle leaf forests, evergreen broadleaf forests, evergreen needle leaf forests, and mixed forests. EFAP (1994b) broadly classify forest types as: natural high forest (both disturbed and undisturbed), lowland woodlands, bushlands, shrublands, wooded grasslands, plantations, and trees around farms. In this study, forest cover types were classified as natural dense forest, dense woodland, open woodland, dense bushland, open bushland, riverine, and plantation forests. This classification is based on previous classifications like EFAP (1994a, b) and WBISPP (2002). Moreover, expert knowledge, the existing forest situation of the region, and work easiness were taken into account. Each classification type is described as follows.

Fig. 2.6 Sample natural dense forests in the South Gonder, North Gonder, Awi, and East Gojam zones (*Photo* by Mulatie Mekonnen, December, 2011)

2.3.2.1 Natural Dense Forest

Natural dense forest is a type of forest cover in which the forest comprises of diverse tree and shrub species with full canopy cover (greater than 80 % canopy cover). Tree height of the dense forest is from 5 to 12 m and higher. Such forest types are observed in the priority and protected state forests, parks, churches, and the riparian forests. The species composition of this forest type is high and rich. Figure 2.6 shows some of such natural forest in the region.

2.3.2.2 Woodlands (Dense and Open Woodland)

The woodlands occupy more areas in the lowland areas of the elevated plains and plateau-escarpments between altitudes of 500 and 1900 m and correspond to a wide range of ecological conditions ranging from the semiarid to humid. The woodlands are characterized by an upper story of 5–12 m with high trees forming a somewhat closed canopy. However, it is sparse enough to let sufficient light penetrate to the dense thicket-like lower story of 1–3 m high shrubs. Open woodlands are lands covered with mainly sparsely distributed tree/shrub species and grasses with 20–50 % tree cover, while dense woodlands are woodlands that have relatively

Fig. 2.7 Sample dense woodland in the North Gonder zone of Amhara region (*Photo* by Mulatie Mekonnen, December, 2011)

Fig. 2.8 Sample open woodland in the North Gonder zone of Amhara region (*Photo* by Mekonnen, December, 2011)

closely populated trees/shrubs from 50 to 80 % tree cover. Figures 2.7 and 2.8 show dense and open woodlands observed and taken at the ground, respectively.

2.3.2.3 Bushlands/Shrublands

Dense and Open Bushlands: Bushland/shrublands is a land on which there is vegetation which is either a remnant of the natural forest or altered that is still representative of the structure and floristic of the natural forest. Bushlands are lands covered with bushy type plant species and a woody cover of more than 20 %. It can be classified as dense (Fig. 2.9) and open (Fig. 2.10) based on the closeness of the

Fig. 2.9 Sample dense bushland in the North and South Gonder Zones of Amhara region (*Photo* by Mulatie Mekonnen, December, 2011)

Fig. 2.10 Sample open bushland in Waghimra zone of the Amhara region (*Photo* by Mulatie Mekonnen, December, 2011)

bushes. The canopy cover of the dense bushlands is more than 50 % and open bushlands are from 20 to 50 %.

2.3.2.4 Riverine or Riparian Forest

Riverine or riparian forests are located along sides of rivers and are part of a highly integrated system that includes the stream channel (Markm and Hunter 2000). In the study area, only a small area of most landscapes is occupied by riverine forests (Fig. 2.11) while its environmental contribution is profound.

2.3.2.5 Plantation Forest

A plantation is regularly harvested planted trees (Fig. 2.12). Plantation forests that are developed by different programs and individuals fulfill the gap between demand and supply of wood resources. Plantation forests have a wide agro-ecological

Fig. 2.11 Sample riverine forests in West Gojam, North Gonder and Awi zones of Amhara region (*Photo* by Mulatie Mekonnen, December, 2012)

Fig. 2.12 Sample plantation forest in West Gojam and North Shewa zones of the Amhara region (*Photo* by Mulatie Mekonnen, December, 2011)

coverage from mid to highland areas and consist mainly of Eucalyptus spps., *C. lusitanica,* and Acacia *decurrence*. These species have extraordinary quick growth rate where the indigenous species take a longer time to mature. This factor helped the spread over a large area through private initiatives as well as communities. At first, the plantations were limited to the surroundings of towns. But now, it is widely expanded to the rural areas where the demand for wood resource has increased at an alarming rate. Moreover, it is becoming a main source of income as cash forest.

Plantation forests are mainly found in Awi, North Shewa, South Gonder, South Wollo, East and West Gojam zones. These plantation forests range from large scale to woodlots and homesteads as clearly seen on the field and from satellite imagery. Sometimes, plantations might be mixed with the naturally grown species. In such cases, if the forests are dominated by the naturally grown species, they are considered as natural forests. But, if the planted species coverage is more than the naturally grown forest species, it is considered as plantation forest.

2.4 Results and Discussions

2.4.1 Results

Based on the collected data and field observation, the forest cover of the region is identified and quantified through screening and digitizing from SPOT satellite imagery. As a result, forest distribution map is prepared (Fig. 2.13). Most of the forest covers were found along the lowland belt of Mirab Gojam, Awi, and Semen Gonder zones bordering the neighboring country of Sudan and Tigray and Benishangul-Gumz regions. The total forest covers of the region 1,288,383 ha or 12,884 km^2 is about 8.2 % of the total land area (Table 2.1). Including bushlands it is about 2,178,295 ha or 21,783 km^2, which is about 13.85 %. Woodlands, natural dense forest, riverine forest, bushlands, and plantations coverage is shown in Table 2.1.

Each forest cover type including bushlands is described in detail in Tables 2.2 and 2.3. Accordingly, the dominant cover is bushlands that cover 889,912 ha, followed by woodlands and natural dense forest, which covers 740,808 and 463,950 ha, respectively. The riverine forest, which is part of natural forest, covers about 20,653 ha. Plantation forest has 62,973 ha coverage in the region.

As indicated in Table 2.3, North Gonder zone has the largest forest cover, about 23.96 %. Awi zone (8.86 %) is the second zone of the region with forest resources. North Shewa, West Gojam, Awi, South Gonder, and East Gojam have good coverage of plantation forest that could serve as a good example for other zones of the region. Awi zone is the best example for Acacia *decurrence* coverage, which serves as an important income source through charcoal production in addition to forest ecology.

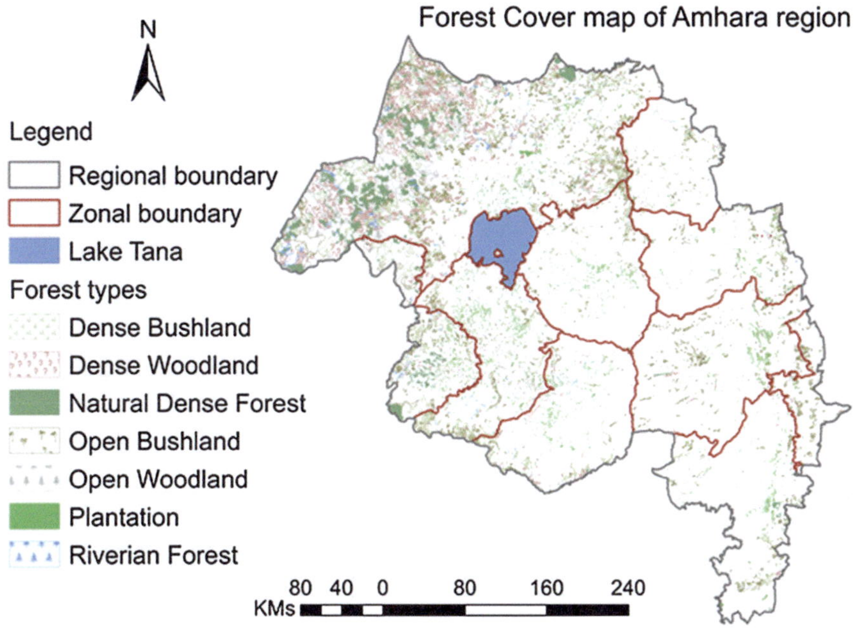

Fig. 2.13 Forest cover distribution map of the Amhara region

Table 2.1 Forest types and area coverage in the Amhara region (in ha)

No.	Forest type	Area (ha)	Forest cover (%)
1	Woodlands	740,808	4.71
2	Natural dense forest	463,950	2.95
3	Plantation	62,973	0.40
4	Riverine forest	20,652	0.13
	Sum	1,288,383	

Table 2.2 Amhara region forest cover by forest type (in ha)

No.	Forest type	Area (ha)	Cover (%)
1	Dense woodland	415,380	2.64
2	Natural dense forest	463,950	2.95
3	Open woodland	325,428	2.07
4	Riverian forest	20,653	0.13
5	Dense bushland	482,643	3.07
6	Open bushland	407,269	2.59
7	Plantation	62,973	0.40
	Sum	2,178,295	13.85

Even if the different agroclimatic zones of the region create a good opportunity for incense and bamboo resources development, man-made factors are affecting the resources tremendously. Severe degradation of lowland bamboo is observed in the

Table 2.3 Percentage of forest cover in the Amhara region by zone

Zone/forest type	Natural forest	Woodlands	Plantation	Riverine	Sum
Awi	5.53	2.76	0.49	0.08	8.86
N/Gonder	8.49	14.83	0.24	0.4	23.96
N/Shewa	0.35	0.76	0.62	0.003	1.73
W/Gojam	0.71	0.72	0.57	0.05	2.05
E/Gojam	0.87	0.95	0.36	0.02	2.2
S/Gonder	0.19	0.15	0.38	0.003	0.73
Waghimra	0.12	0.29	0.08	0.004	0.49
N/Wollo	0.21	0.49	0.46	0.02	1.18
S/Wollo	0.13	0.25	0.7	0.02	1.1
Oromia	–	0.63	0.16	0.03	0.82

Excluding bushlands

lowland parts of North Gonder, Awi, and West Gojam zones although different studies report availability in these areas. In this study, attempt was made to quantify and map the incense and bamboo resources of the region, but satellite imagery resolution was low. Therefore, further studies shall be conducted using high resolution imagery which can distinguish species clearly.

2.4.2 Discussions

GIS and satellite image-based forest resource data lacks in the Amhara region in particular and in Ethiopia in general. Therefore, it is difficult to provide adequate discussion of the results. FAO (2010) puts Ethiopia among the countries of the world with forest cover of 10–30 %. This chapter indicates that Ethiopia's forest cover (FAO definition) is 12.2 million ha (11 %). It also further indicates that the forest cover shows a decline from 15.11 million ha in 1990 to 12.2 million ha in 2010, during which 2.65 % of the forest cover was deforested. In this study the forest cover of Amhara region is about 8.2 % (4.71, 2.95, 0.4, and 0.13 % for woodlands, natural dense forest, plantation, and riverine forests, respectively) from the total land area of the region. Including bushlands (5.66 %), it is about 13.85 %.

2.5 Conclusions, Limitations, and Recommendations

2.5.1 Conclusions

The forest cover of Amhara Region is 1,288,383 ha (12,884 km^2), that is, about 8.2 % of the total land area of the region with woodlands (4.71 %), natural dense forest (2.95 %), plantation forest (0.4 %), and ravine forest (0.13 %). Including bushlands (5.66 %), the forest cover is about 13.85 %. Natural forest cover is better in the lowland parts of the region bordering Benishangul and Tigray regions and the neighboring country, Sudan. This forest cover can contribute to slow down the rapidly expanding desertification south of the Sahara desert. The highland parts of the region have little natural forest cover, instead there is better coverage of man-made forest plantation.

Area closures are playing an important role in increasing the bushland coverage of the region; North Wollo, South Wollo, and Oromia zones are the best examples. North Gonder Zone has a very good incense resource. Quara, Metema, West Armacho, Tegede, Adiarkay, East Belesa, and Tach Armachiho are incense potential woredas of the zone. Districts like Jawi from Awi zone, Shindi from West Gojam, and Ebinat from South Gonder also have good incense potential. Awi, West Gojam, and South Gonder zones have better highland bamboo resource in the region.

Inaccessibility and hot temperature played a vital role in conserving the remaining forest resources of the region. Satellite imagery, GIS, and GPS technologies are found to be essential tools in identifying, quantifying, and mapping forest resources. Such technologies are vitally important to reach and assess resources otherwise inaccessible and remote.

2.5.2 *Limitations of the Study*

Since the satellite imagery used in this study was taken in 2008, recent plantations, closed area regenerations, and deforestation are not included. Lack of studies on the forest resources of the region limited reference resources. Agroforestry practices were not considered or included.

2.5.3 *Recommendations*

Further forest species inventory and proper management are vitally important. Higher resolution satellite imagery taken at different times, that is, satellite image taken in November for the western and northern lowlands, and in January for other parts of the region, should be used for further study. Investors are creating a bad shadow on the natural forest cover, in the name of agriculture and incense investments. Resettlements are also causing deforestation. Therefore, care should be taken.

The forest management guidelines formulated by the ANRS, BOA should be strengthened and implemented to conserve forest resources of the region, particularly the lowland forests bordering the Benishangul and Tigray regions and the neighboring country, Sudan. Forest cover can contribute to slow down the rapidly expanding desertification south of the Sahara desert. Promoting agricultural practices like honey bee, controlled livestock rearing, ecotourism, and others is more preferred for sustainable use of the forest resource of the region.

Detail studies should be carried out on farm forestry or agroforestry resource using higher resolution and better coverage satellite imagery. Forest density, diversity, and population can be further areas of study. Regional forest resource inventory and database are vital for forest resource management.

Acknowledgment The Amhara National Regional State (ANRS), Bureau of Agriculture (BoA) and Bureau of Finance and Economic Development (BOFED) are acknowledged for giving attention and for organizing the technical team for the study. We are also grateful to the Sustainable Water Harvesting and Institutional Strengthening in Amhara (SWISHA) and the North Gondar Zone Sustainable Natural Resource Management Projects, which have assisted in providing financial support. We appreciate Belachew, Tesfaye and Hailu for their dedicated support in driving a long distance day and night. Our thanks also go to the zonal and district officials and experts who assisted in providing secondary data as well as stayed with us in the field supervisions and data collections. Finally, our deep appreciation and respect extends to the small-scale farmers for conserving the forest resources of the region and who are struggling all their lives for a better future.

References

Bane J, Nune S, Mekonen A, Bluffstone R (2008) Policies to increase forest cover in Ethiopia. In: Proceedings of a policy workshop September 2007. Ethiopian Development Research institute, Addis Ababa, Jan 2008

Bereket H (2008) Study on establishment of bamboo processing plants in Amhara regional state. MSc thesis, Addis Ababa University, Ethiopia

Birhan A (2009) Impact of community based forestry on forest status and local people. The case of gora community forest, Bededo PA, Tehuledere District, South Wello, Ethiopia. MSc thesis 2009, p 1

Earth Trends (2007) Country profiles: forests, grasslands, and dry lands—Ethiopia, Earth trends. http://www.idp-uk.org/OurProjects/Environment/Forests,grasslands,%20drylands (FAO)%20%20for_cou_231.pdf. Accessed 14 Sept 2014

EFAP (1994a) Ethiopian forestry action program. The challenge for development volume II. Addis Ababa, Ethiopia

EFAP (1994b) Ethiopian forestry action program: the challenge for development. Final report, volume II, Ministry of natural resources development and environmental projection, Addis Ababa

Eshete A (2002) Regeneration status, soil seed banks and socio-economic importance of *B. papyrifera* in two woredas of North Gonder Zone, Northern Ethiopia. Master's thesis, Swedish University of Agricultural Sciences, Skinnskatteberg, Sweden

FAO (2000) State of the world's forests. Food and Agricultural Organization of the United Nations Development Programme, Rome

FAO (2001) Trees outside forests: towards rural and urban integrated resources management. Rome, Italy. ftp://www.fao.org/docrep/fao/005/y1785e/y1785e00.pdf. Accessed 14 Sept 2014

FAO (2010) Global forest resources assessment 2010—country report Ethiopia. Food and Agriculture Organisation (FAO), Rome, Italy. www.fao.org/forestry/fra/fra2010/en/. Accessed 14 Sept 2014

Gebrehiwot K (2003) Ecology and management of *Boswellia papyrifera* (Del.) Hochst dry forests in Tigray, Northern Ethiopia. PhD thesis, Georg-August University of Göttingen, Gottingen, Germany

Getachew HE, Melesse AM (2012) Impact of land use/land cover change on the hydrology of angereb watershed, Ethiopia. Int J Water Sci 1(4):1–7. doi:10.5772/56266

Lemenih M, Kassa H (2008) Management guidelines for *Boswellia papyrifera* and its frankincense in Ethiopia. Center for International Forestry Research–Ethiopia, Addis Ababa, Ethiopia

Heinen JT, Lyon JG (1989) The effects of changing weighting factors on the calculation of wildlife habitat index values: a sensitivity analysis. Photogram Eng Remote Sens 55(10):1445–1447
Lemenih M, Teketay D (2003) Frankincense and myrrh resources of Ethiopia. I. Distribution, production, opportunities for dry land economic development and research needs. Ethiop J Sci SINET 26:63–72
Mango L, Melesse AM, McClain ME, Gann D, Setegn SG (2011a) Land use and climate change impacts on the hydrology of the upper Mara River Basin, Kenya: results of a modeling study to support better resource management. Hydrol Earth Syst Sci 15 2245–2258. doi:10.5194/hess-15-2245-2011 (Special issue: Climate, weather and hydrology of East African Highlands)
Mango L, Melesse AM, McClain ME, Gann D, Setegn SG (2011b) Hydro-meteorology and water budget of Mara River basin, Kenya: a land use change scenarios analysis. In: Melesse A (ed) Nile River Basin: hydrology, climate and water use. Springer Science Publisher, Berlin, Chapter 2, pp 39–68. doi:10.1007/978-94-007-0689-7_2
Markm L, Hunter JR (2000) Maintaining biodiversity in forest ecosystems. University of Cambridge, UK
Melaku B (2003) Forest property rights, the role of the state, and institutional exigency: the Ethiopian experience. Doctor's dissertation, Swedish University of agricultural sciences, Sweden. ISSN 1401–6249, ISBN 91-576-6429-3
Melesse AM, Jordan JD (2002) A comparison of fuzzy vs. augmented-ISODATA classification algorithm for cloud and cloud-shadow discrimination in landsat imagery. Photogram Eng Remote Sens 68(9):905–911
Melesse AM, Jordan JD (2003) Spatially distributed watershed mapping and modeling: land cover and microclimate mapping using landsat imagery part 1. J Spat Hydrol (e-journal) 3(2)
Melesse A, Weng Q, Thenkabail P, Senay G (2008) Remote sensing sensors and applications in environmental resources mapping and modeling. Sensors 7:3209–3241 (Special issue: Remote Sensing of Natural Resources and the Environment)
Mohammed H, Alamirew A, Assen M, Melesse AM (2013) Spatiotemporal mapping of land cover in Lake Hardibo Drainage Basin, Northeast Ethiopia: 1957–2007. Water conservation: practices, challenges and future implications. Nova Publishers, Hauppauge, pp 147–164
Mulugeta L, Melaku B (2007) Best practices, lesson learnt and challenges encountered the Ethiopian and Tanzanian Experiences, FARMAfrica/ SOS-Sahel
Sebsebe D, Edwards S (2006) Diversity of vegetation types, agriculture systems and crops in Ethiopia. In: Facilitating the implementation and adaptation of integrated pest management (IPM) in Ethiopia. Planning workshop from 13–15 October 2003, Melkasa Agricultural Research Center, EARO. Jointly organised by the association for advancement of IPM (ASAI) and the Ethiopian agricultural research organization (EARO). DCG proceeding, Feb 2006
Tadesse W, Teketay D, Lemenih M, Fitwi G (2002) Review and synthesis on the state of knowledge of Boswellia species and commercialization of frankincense in the dry lands of eastern Africa. In: Chikamai BN (ed) Country report for Ethiopia. Kenya Forestry Research Institute, Nairobi, Kenya, pp 11–35
UNFCCC (2006) Definition, modalities, rules and guidelines relating to landuse, land-use change and forestry activities under the Kyoto Protocol, p 3. http://unfccc.int/resource/docs/2005/cmp1/eng/08a03.pdf
WBISPP (2002) Manual for woody biomass inventory and strategic planning project. Ministry of Agriculture, Addis Ababa, Ethiopia
Wondie M, Schneider W, Melesse AM, Teketay D (2011) Spatial and temporal land cover changes in the Simen Mountains National Park, a world heritage site in Northwestern Ethiopia. Remote Sens 3:752–766. doi:10.3390/rs3040752
Wondie M, Schneider W, Melesse AM, Teketay D (2012) Relationship among environmental variables and land cover in the Simen Mountains National Park, a world heritage site in Northern Ethiopia. Int J Remote Sens Appl (IJRSA) 2(2):36–43
Zebene A, Hulten H (2003) Tree diversity management in the traditional agroforestry land-use of Sidama, southern Ethiopia. Acta Universitatis Agriculturae, SLU. SILVESTRIA 263(1):1–28

Chapter 3
Landscape Changes Impact on Regional Hydrology and Climate

Wossenu Abtew and Assefa M. Melesse

Abstract There is enough evidence that landscape change has impact on regional energy and water balance resulting in climate change. Climate change also changes landscape. The expansion of the Sahara Desert corresponds to land degradation where the role of vegetation in keeping surface energy fractionation in balance and initiation of the rainfall process is diminished. Non-vegetated dry areas appropriate more solar energy to sensible heat (surface temperature increase) than latent heat (evapotranspiration). Evapotranspiration has cooling effect and is part of the hydrologic cycle. Solar energy reflectance (albedo) is dependent on surface characteristics. In this chapter, cases of landscape change and climate change in Africa, China, and the United States are presented. Application of remote sensing in observation of land cover, surface temperature, and energy partitioning is presented for monitoring wetland and dryland landscapes.

Keywords Landscape change · Climate change · Evapotranspiration · Surface temperature · Rainfall · Climate change · Albedo · Energy flux · Remote sensing

3.1 Introduction

Landscape or type of surface cover on earth is part of the global energy and mass circulation and distribution. Land cover is also related to soils, surface, and groundwater. Erosion of bare soils, surface runoff, and infiltration of water into the

W. Abtew (✉)
South Florida Water Management District, 3301 Gun Club Road, West Palm Beach, FL 33416, USA
e-mail: wabtew@sfwmd.gov

A.M. Melesse
Department of Earth and Environment, Florida International University, Miami, FL 33199, USA
e-mail: melessea@fiu.edu

© Springer International Publishing Switzerland 2016

A.M. Melesse and W. Abtew (eds.), *Landscape Dynamics, Soils and Hydrological Processes in Varied Climates*, Springer Geography,
DOI 10.1007/978-3-319-18787-7_3

ground are related to land cover. Through surface heating and evapotranspiration, land cover and regional climate are connected. Adaptation to global warming-related climate change also includes land use/land cover changes. Land use change is widely accepted to alter local response to precipitation in the form of surface runoff and infiltration. In a modeling study in UK, it was shown that land use change impact hydrology of spatially larger areas as catchments as the sum of smaller scale local effects and the impact can be separated from other factors such as climate variation (Rust et al. 2014). Dale (1997) stated that land use influences the flux of mass and energy and land use change has more impact on ecological variables than climate change.

The major source of energy that drives the climate of the surface of the earth is solar energy. Solar radiation receiving surface characteristics include reflectance (albedo), fraction of solar energy reflected by the surface; and partitioning of the received energy into sensible and latent heat. The role of the oceans, which cover 70 % of the earth, in energy and mass transfer, is so critical that an earth would not be imagined without such a cover. The infrared reflectance of water is low. Water heat storage capacity is higher than soil and rock and the bulk of solar energy storage is in the oceans which are 3800 m deep on average. As a result, the large thermal inertia of the ocean is a determining factor for the earth's climate. Both spatial and vertical temperature, density, and salinity gradients create oceanic circulation and determine climate patterns coupled with temperature and pressure gradient-driven atmospheric circulation.

This chapter deals with only terrestrial landscapes. Energy and water are interchangeable through the evapotranspiration and condensation processes. The estimated solar energy input at normal direction to earth is 1350 W m^{-2} where 450 W m^{-2} is reflected back by the atmosphere. Further, solar energy reaching the earth is partly reflected back with the amount depending on surface or landscape characteristics. Snow surfaces reflect back more than vegetated surfaces. The net solar radiation that is not reflected back becomes the energy for maintaining the earth's environment in several ways. Changing the landscape changes the energy retained on the surface and its fractionation; this in turn affects the regional temperature, rainfall, and runoff.

Heat, moisture, and momentum fluxes are significantly altered by urban landscape in contrast to rural areas. In the northern hemisphere, urban areas on the average have 12 % less solar radiation, 14 % more rainfall, 8 % more clouds, 10 % more snowfall, 15 % more thunderstorms, 10 times higher pollutant concentration, and are 2 °C warmer (Taha 1997). Anthropogenic changes in land use and land cover and their impact on rainfall and concerns for the continuing changes are documented by Pielke et al. (2007). Also, climate change influences landscape change. Climate change in the western United States of America is expected to impact forest ecology through changes in fire regime with wildfire implications for urban-wildland interfaces. A need for adaptive wildland management is suggested to manage risks of wildfire that could impact urban areas on the fringe (Keeton et al. 2007).

3.2 Surface Energy Budget

Incoming solar radiation (R_S) is measured as energy flux density of both direct and diffuse sky radiation passing through a horizontal plane of given area (1 m^2). Solar radiation varies diurnally, seasonally, by atmospheric conditions such as cloud cover, dust, smog, etc., and location. Net solar radiation (R_n) is net shortwave radiation, which is the balance of incoming radiation and reflected back radiation. The reflected back radiation amount is determined by the characteristics of the surface and angle of incidence of incoming radiation. The surface characteristic is represented by albedo, fraction of energy reflected by the surface. Different surfaces have varying albedo under similar conditions (Table 3.1). Urban landscape has lower albedo than rural areas and it impacts mainly local climate as the total urban area on the planet is 0.44 % (Spangmyr 2010).

Change in surface characteristics has a lot of importance. Pigmented active photosynthesising micro algae and cyanobacteria on the surface of Greenland ice reduce reflectance and as a result the additional energy retained causes melting (Yallop et al. 2012). An experimental study in comparison to spectral reflectance from clear water and algae-laden water with varying suspended solids concentrations was reported. One of the conclusions was, between 400 and 700 nm wavelengths, clear water has more reflectance than algae-laden water with the same amount of suspended sediment (Han 1997).

The practical importance of surface reflectance change is illustrated by the following recent experience. The Florida Power and Light's Turkey point nuclear plant is cooled by water from a 2389 ha of meandering canal cooling pond. The power plant has regulatory safety limitation of 100 °F (37.8 °C) maximum cooling water temperature. In the summer of 2014, rising temperature and algae bloom in the cooling ponds threatened to force the shutdown of two nuclear reactors. The power company asked to control the algae with herbicides and for a permit to use alternative cooling water source, groundwater (The Miami Herald, July 17, 2014). The company also requested to increase the cooling water temperature regulatory limit.

Table 3.1 Albedo of various surfaces of different landscapes (Oke 1992; Ahrens 2006)

Surface	Range of Albedo
Water	0.03–0.10
Coniferous forest	0.05–0.15
Deciduous forest	0.15–0.20
Soil (dark and wet)	0.05–0.40
Soil (light and dry)	0.15–0.45
Grass	0.16–0.26
Deseret sand	0.4
Tundra	0.2
Fresh snow	0.80–0.90

In August 2014, the company obtained approval to withdraw water from one of the regional canals (L-31E Canal) and obtained temporary approval from the Nuclear Regulatory Commission to deviate from the water temperature limit. The company's preferred maximum temperature is 104 °F or 40 °C (The Miami Herald, September 5, 2014).

The net radiation (R_n) absorbed by the surface is the sum of the net short (solar) and long (thermal) wave radiations as given by Eq. 3.1.

$$R_n = (R_S \downarrow - R_S \uparrow) + (R_L \downarrow - R_L \uparrow) \quad (3.1)$$

where $R_S\downarrow$ and $R_S\uparrow$ are the incoming and outgoing, or reflected shortwave radiations, and $R_L\downarrow$ and $R_L\uparrow$ are incoming and outgoing long wave radiations, respectively. Shortwave radiation, the net shortwave radiation (R_n), in Eq. 3.1, is the difference between $R_S\downarrow$ and $R_S\uparrow$ (Abtew and Melesse 2013).

The net radiation (R_n) is partitioned into three fluxes at the earth's surface: surface heat storage (G), surface temperature change or sensible heat (H), and latent heat or evapotranspiration (E) as shown in Eq. 3.2 and Fig. 3.1. Figure 3.1 depicts incoming radiation, outgoing radiation from the atmosphere and the surface, surface heat storage, evaporation, condensation, momentum, mass, and energy transfer processes.

$$R_n = G + H - E \quad (3.2)$$

where R_n is net radiation at the surface, LE is latent heat or moisture flux (ET in energy units), H is sensible heat flux to the air, and G is soil heat flux. Energy flux models solve Eq. (3.2) by estimating the different components separately.

The ratio of sensible heat to latent heat is the Bowen ratio (β). If β is less than unity, a great proportion of the available energy at the surface becomes latent heat

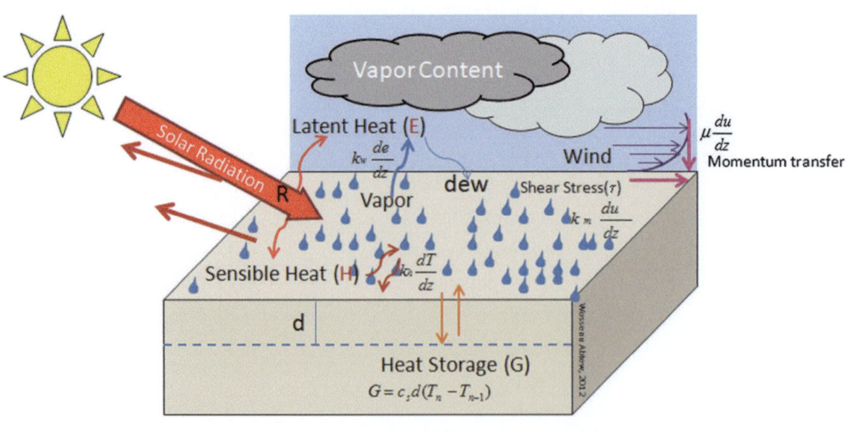

Fig. 3.1 Mass and energy fluxes and momentum transfer on a surface

than sensible heat. This will be the case with open water, wet, and vegetated surfaces. If the ratio is greater than unity, the proportion of energy allocated to sensible energy is higher than latent heat energy. This is characteristic of dry surfaces.

The interchangeability of energy and water is shown through latent heat of vaporization and latent heat of condensation. A kilogram of water uses 2.45 MJ of energy to convert from liquid into vapor and gives out about the same energy when the same amount of vapor condenses back to liquid water. Open water, wetlands, and vegetated surfaces proportionally use most of the energy as latent heat or evapotranspiration resulting in cooling effect "green space" while paved and dry surfaces use most of the energy as sensible heat raising the local temperature "urban heat island." The link between landscape, hydrology, and climate comes through the evapotranspiration process. The process of evapotranspiration is instrumental in temperature and water distribution in time and space and there is ever-growing evidence that human interference with vegetation cover and water flows has resulted in major changes in temperature distribution (Eiseltova et al. 2012). An illustration of surface temperature differences with land cover is depicted in Fig. 3.2, as published by Eiseltova et al. (2012). Evaporative forest surface has lower temperature (25 °C) than drained meadow and road surface (45 °C).

In recognition of the relationship of landscape and surface temperature, programs as urban greening are recognized to reduce the urban heat island effect. A study on the recognition of evapotranspiration links to the energy and water budget of urban areas and mitigation of urban heat through evaporative cooling was reported (Jacobs et al. 2014). Urban green spaces as parks, golf courses, gardens, residential lawns, urban trees, green roofs and undeveloped areas, pools, lakes, and water bodies contribute to cooling. Urban green spaces improve the environment by modifying air temperature, air quality, building cooling cost, micro-climate, biodiversity, and social stabilization (Nouri et al. 2013).

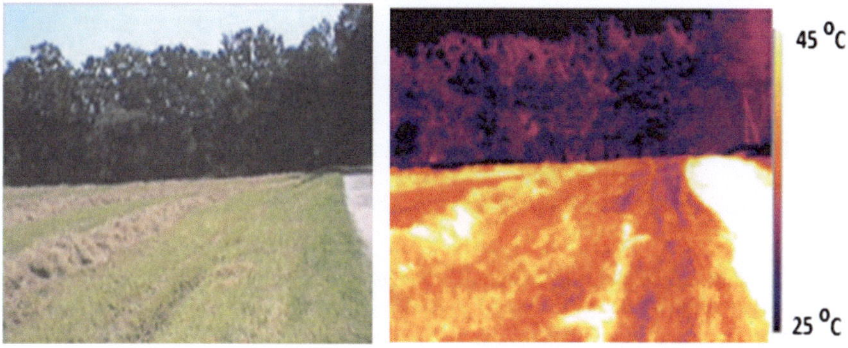

Fig. 3.2 Surface temperature variation between forest and drained meadow with road on the side land covers (*source* Eiseltova et al. 2012, figure orientation modified)

3.3 Landscape Changes Impact on Climate

3.3.1 Landscape and Climate Change in Africa

Major landscape changes with documented climate change impacts are deforestation and drainage of wetlands. Deforestation and desertification relationships have been widely reported. Satellite map study of the Mau Forest in Western Kenya show extreme surface temperature increases following deforestation (Eiseltova et al. 2012). The Sahara is advancing south by more than 1.5 km a year and rainfall has decreased by 30 % in the last 40 years (Chineke et al. 2011). The same study cited Darfur in Sudan as a grim lesson and also concluded that climate change and unsustainable ecosystem management could result in food shortage in Nigeria. In their modeling of the interaction between vegetation and climate, Zheng and Eltahir (1997) concluded that deforestation along the southern coast of West Africa may result in complete collapse of monsoon circulation resulting in a significant rainfall reduction.

Rainfall in the Sahel region of Africa has been declining since the 1970s. Rainfall amount is positively correlated to normalized difference vegetation index (NDVI) where higher NDVI values indicate a greener landscape (Tucker et al. 1991). Climate change and ecological degradation are linked in the Darfur region of Sudan where conflict is observed between herders and farmers. In another study of the persistent drought in Sahel, it is postulated that a strong positive feedback mechanism is operating reflecting changes in surface properties mainly albedo and soil moisture through the amount of energy retained and latent flux (Courel et al. 1984). A dry season albedo change from 0.30 in 1973 to 0.20 in 1979 was found reflecting changes in vegetation cover.

North Central Ethiopian rainfall decline in the twentieth century has been reported with mainly decline in summer rainfall (Sileshi and Zanke 2004). Deforestation of Ethiopian highlands has changed the landscape from 40 % of forest cover a century ago to a mere 3 % cover currently. As a result, a billion cubic meter of fertile top soil is lost by erosion every year (Bishaw 2001). A study in the semiarid North Kordofan region of Sudan shows signs of desertification related to changes in land use/land cover (Dafalla et al. 2014).

3.3.2 Landscape and Climate Change in Other Regions

Like most parts of the world, China also has gone through drastic landscape change with current estimate of 14 % forest area. The link between landscape change and climate is widely reported. Modeling results on climate effects of land use change over China showed regionally varied impacts. The current human-induced land scape changes result in reduced precipitation and decrease in temperature south of the Yangtze, increased precipitation north of the Yangtze, summer climate changes in south China with increase in precipitation and temperature, reduction in precipitation in northern China, and temperature rise in northwest China (Xuejie et al. 2007).

The impact of anthropogenic land-cover changes on the climate of Florida peninsula is anecdotally mentioned as a change in the rain machine. In the last 100 years, Central and South Florida have gone through landscape change with urbanization in the coasts and agricultural production in the interior. The transformation resulted in drainage of wetlands, channelization of rivers, and clearing of vegetation covers. Sea breeze-related summer convective rainfall events over the peninsula are primarily driven by thermal properties of land surfaces and the ocean. Since land use change alters surface energy flux and temperature, it is hypothesized that these changes affect convective rainfall processes. A regional atmospheric system model was applied to simulate warm season convective rainfall and surface sensible weather using pre-1900 and 1993 land use for the months of July and August. The analysis showed that sensible heat increased and latent heat decreased from pre-1900 land use to 1993 (Marshall and Pielke 2003). Trend analysis of July and August observed convective rainfall from 1924 to 2000 showed a declining trend. Summer time maximum temperature increased in the region. The model simulation results confirmed the observations and anecdotal evidences.

Three-dimensional meteorological simulation of downtown Los Angeles by increasing albedo by 0.14 and the entire basin by 0.08 decreased peak summer time temperature by 1.5 °C. Results indicate that surface modification strategies such as vegetative cover increase can reduce energy demand and atmospheric pollution by 5–10 % (Sailor 1995).

3.4 Landscape Change Monitoring with Remote Sensing

3.4.1 Wetland Hydropattern and Vegetation Changes

Vegetation cover and groundwater level changes over the period of restoration are the two most important indicators of the level of success in wetland ecohydrological restoration. In surface energy fractionation, increase in water and vegetated surfaces increases latent heat (evapotranspiration) part of the surface energy. Change in wetland hydropattern and vegetation cover can be evaluated from fractional vegetation cover (FVC) changes and latent heat flux using Moderate Resolution Imaging Spectroradiometer (MODIS) data. Therefore, remote sensing supplemented with surface hydrology observations can be applied to evaluate changes in large-scale wetlands.

Application of this method has shown positive results as demonstrated in evaluating changes in a regional river and wetland system restoration of the Kissimmee River in Florida, USA. Based on remote sensing energy budget and on-site groundwater level monitoring, changes were detected in a five 5-year period.

3.4.2 Remote Sensing

Remote sensing uses measurements of the electromagnetic radiation, usually sunlight reflected in various bands, to characterize the landscape, infer surface properties, or in some cases actually estimate hydrologic state variables. Measurements of the reflected solar radiation (visible and short wave infrared sensors) give information on land cover, extent of surface imperviousness, and albedo. Thermal radiation (thermal-infrared sensors) gives estimates of surface temperature and surface energy fluxes.

Researchers have conducted studies using vegetation indices to derive the relationship between remotely sensed radiance and biophysical properties of forests (Boyd et al. 1996; Curran et al. 1992). Multi-temporal NDVI data derived from Landsat sensors (Spanner et al. 1990; Danson and Curran 1993) and the Advanced Very High Resolution Radiometer (AVHRR) have been used for land-cover mapping and land use change studies (Stone et al. 1994; Tucker et al. 1991; Lambin and Strahler 1994). For land-cover mapping, the radiance recorded in the middle-infrared (MIR) (1.3–3 μm) and long wave thermal-infrared (TIR) (8–14 μm) wave bands provide important additional and supplementary information to that provided by the reflectance data measured in visible (0.4–0.7 μm) and near-infrared (NIR) (0.7–1.3 μm) bands. Data acquired in the MIR and TIR wave bands can discriminate among vegetation types and assess changes in land use (Baret et al. 1988; Panigrahy and Parohar 1992; Melesse and Jordan 2002).

Remote sensing-based energy flux and surface parameters from different vegetated and non-vegetated surfaces are studied by various researchers. Energy flux from agricultural field (Kustas 1990; Bastiaanssen 2000; Kustas et al. 2004; Melesse and Nangia 2005; Senay et al. 2008, 2007), wetlands (Loiselle et al. 2001; Mohamed et al. 2004; Oberg and Melesse 2005; Melesse et al. 2006, 2007; Lagomasino et al. 2015) rangeland and other vegetated surfaces (Kustas et al. 2003; Kustas and Norman 1999; French et al. 2000; Hemakumara et al. 2003; Melesse et al. 2008), lakes (Melesse et al. 2009) and desert (Wang et al. 1998) are studied. These studies have shown the application of remote sensing in spatial mapping of flux and surface parameter to characterize the response of land surface to vegetation dynamics.

3.4.3 Remotely Sensed Data

Remote sensing application to landscape changes considers assessing the spatio-temporal changes of vegetation cover and latent heat flux (evapotranspiration in energy units) of the study area. Remotely sensed images from MODIS aboard Terra sensor were used in the evaluation of Kissimmee River basin restoration efforts study. Images for the months of April, September, and December from 2000 to 2004 were acquired and processed. Daily surface temperature, NDVI and albedo were also acquired from the Land Processes Distributed Active Archive Center (LP DAAC) (http://lpdaac.usgs.gov/modis/dataproducts.asp Accessed January 19, 2015) and used in the surface energy balance computation. Micrometeorological data acquired from National Climatic Data Center (NCDC) include air temperature and wind speed.

The study found that remote sensing-based ET estimation before and after restoration was useful in identifying areas with higher ET, mainly due to restoration of hydrology.

3.4.3.1 Fractional Vegetation Cover (FVC) Mapping

The NDVI (Rouse et al. 1974) is a measure of the degree of greenness in the vegetation cover of a watershed. It is the ratio of the difference to the sum of the reflectance values of NIR and red bands. In highly vegetated areas, the NDVI typically ranges from 0.1 to 0.6, in proportion to the density and greenness of the plant vegetation. Clouds, water, and snow, which have larger visible reflectance than NIR reflectance, will yield negative NDVI values. Rock and bare soil areas have similar reflectance in the two bands and result in NDVI values near zero.

To understand the change in the FVC for images of different scenes and dates, the scaled NDVI ($NDVI_S$) has been used by many researchers (Price 1987; Che and Price 1992; Carlson and Arthur 2000) as expressed by Eq. 3.3.

$$NDVI_S = \frac{NDVI - NDVI_{low}}{NDVI_{high} - NDVI_{low}} \tag{3.3}$$

where $NDVI_{low}$ and $NDVI_{high}$ are values for bare soil and dense vegetation, respectively. Carlson and Ripley (1997) found the relationship between FVC and scaled NDVI as expressed by Eq. 3.4.

$$FVC \approx (NDVI_S)^2 \tag{3.4}$$

where FVC ranges between 0 and 1. The FVC is an indicator of the level of vegetation cover at a pixel level, which is a very good estimate of the percent of pixel area covered by vegetation.

3.4.3.2 Latent Heat Mapping

Remote sensing-based evapotranspiration (ET) estimations using the surface energy budget equation are proving to be one of the most recently accepted techniques for areal ET estimation (Morse et al. 2000). Surface Energy Balance Algorithms for Land (SEBAL) is one of such models utilizing Landsat images and images from other sensors with a TIR band to solve Eq. (3.3) and hence generate areal maps of ET (Bastiaanssen et al. 1998a, b; Morse et al. 2000).

SEBAL requires weather data such as solar radiation, wind speed, precipitation, air temperature, and relative humidity in addition to satellite imagery with visible, NIR, and thermal bands. SEBAL uses the model routine of ERDAS Imagine in order to solve the different components of the energy budget equations. In the absence of horizontally advective energy, the surface energy budget of land surface satisfies the law of conservation of energy as expressed by Eq. 3.2.

3.4.3.3 Net Radiation

Net radiation (R_n) is estimated based on the relationship given by Bastiaanssen et al. (1998a) as expressed by Eq. 3.5.

$$R_n = R_{S\downarrow}(1 - \alpha) + R_{L\downarrow} - R_{L\uparrow} - R_{L\downarrow}(1 - \varepsilon_s)\ (\text{W/m}^2) \tag{3.5}$$

where $R_S\downarrow$ (W m^{-2}) is the incoming direct and diffuse shortwave solar radiation that reaches the surface; α is the surface albedo, the dimensionless ratio of reflected radiation to the incident shortwave radiation; $R_L\downarrow$ is the incoming longwave thermal radiation flux from the atmosphere (W m^{-2}); $R_L\uparrow$ is the outgoing longwave thermal radiation flux emitted from the surface to the atmosphere (W m^{-2}), ε_s is the surface emissivity, the (dimensionless) ratio of the radiant emittance from a gray body to the emittance of a black body.

3.4.3.4 Soil Heat Flux

The soil heat flux (*G*) is the rate of heat storage in the ground by conduction. Studying irrigated agricultural regions in Turkey, Bastiaanssen (2000) suggested an empirical relationship for *G* given as Eq. 3.6.

$$G/R_n = 0.2(1 - 0.98\text{NDVI}^4)\ (\text{W m}^{-2}) \tag{3.6}$$

where NDVI is the normalized difference vegetation index (dimensionless).

3.4.3.5 Sensible Heat Flux (*H*)

Sensible heat flux is the rate of heat loss to the air by convection and conduction due to a temperature difference. Using the equation for heat transport, sensible heat flux can be calculated by Eq. 3.7.

$$H = \frac{\rho C_p(T_a - T_s)}{r_{ah}}\ (\text{Wm}^{-2}) \tag{3.7}$$

where ρ is the density of air (kg m^{-3}), C_p is the specific heat of air (1004 J kg^{-1} K^{-1}), T_a is the air temperature (K), T_s is surface temperature (K) derived from the thermal band of Landsat images, and r_{ah} is the aerodynamic resistance (s m^{-1}).

3.4.3.6 Latent Heat Flux (LE)

With R_n, G, and H known, the latent heat flux is the remaining component of the surface energy balance to be calculated by SEBAL. Rearranging Eq. (3.4) gives the latent heat flux as shown in Eq. 3.8.

$$\text{LE} = R_n - G - H \ (\text{W}\,\text{m}^{-2}) \tag{3.8}$$

The detailed technique for estimating latent and sensible heat fluxes using remotely sensed data from Landsat and other sensors is documented and was tested in Europe, Asia, Africa, and in Idaho in the US and proved to provide good results (Bastiaanssen et al. 1998a, b; Wang et al. 1998; Bastiaanssen 2000; Morse et al. 2000).

3.5 NDVI-TS-Albedo Relationships

From MODIS data, monthly values of NDVI, TS and albedo were generated for the months of April, September, and December from 2000 to 2004. The selection of the months was designed to represent the different times of the year. Using these values as layers, unsupervised classification was run using Iterative Self-organizing Data Analysis (ISODATA) algorithm (ERDAS 1999). This classification yielded 30 classes for each month. Combining the resulting land-cover classes from each run (3 months × 5 years) gave a scattergram of NDVI–TS–Albedo (Fig. 3.3). Figure 3.4 shows the scattergram of TS and NDVI and Albedo and NDVI correlation. It is shown that, albedo and NDVI and surface temperature and NDVI have negative relationship, especially for NDVI > 0.5 with R^2 value of 0.61 and 0.15, respectively (Fig. 3.4). Higher latent heat losses from the vegetated surface lead to a cooler surface and lower surface temperature compared to bare ground. This relationship is not clearly defined in less vegetated surface (water bodies and bare ground) as shown on the left-hand side of the graph (Fig. 3.4).

3.5.1 Fractional Vegetation Cover Changes

FVC for the month of April from 2000, 2002, and 2004 was generated and comparisons were made (Fig. 3.5). It is shown that FVC has shown changes along the

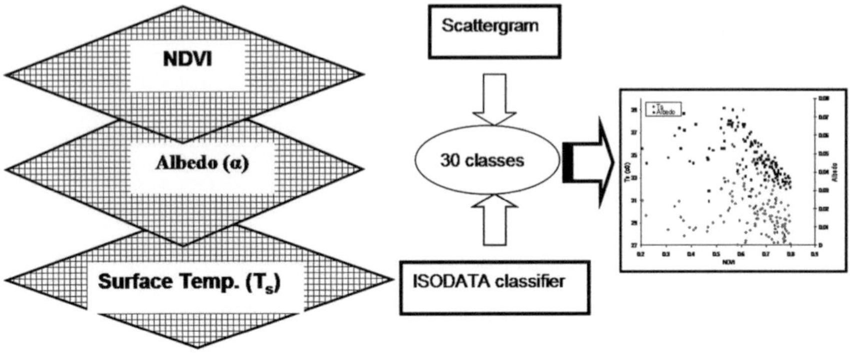

Fig. 3.3 Scattergram of NDVI–TS–Albedo

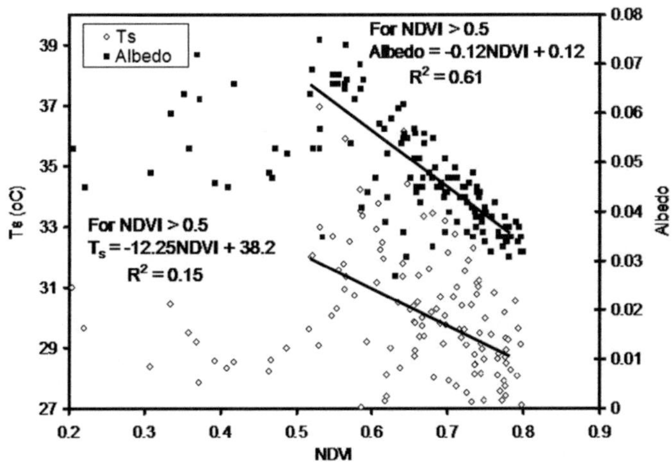

Fig. 3.4 Scattergram of NDVI–TS–Albedo with linear correlations

river, especially in the middle portion of the watershed. Although changes are not significant, mean April FVC of 0.15. 0.16, and 0.17 for 2000, 2002, and 2004, respectively, the trend is an indicator of some response of the vegetation along the river to the restoration work. The actual changes in the FVC will require a field sampling and ground observation. This study does not identify the type of vegetation and if this response is a desirable one.

3.5.2 *Latent Heat Flux Dynamics*

Latent heat grids were generated from Landsat imagery for the month of April (2000, 2002, and 2004). Figure 3.6 show maps of latent heat in watts per square

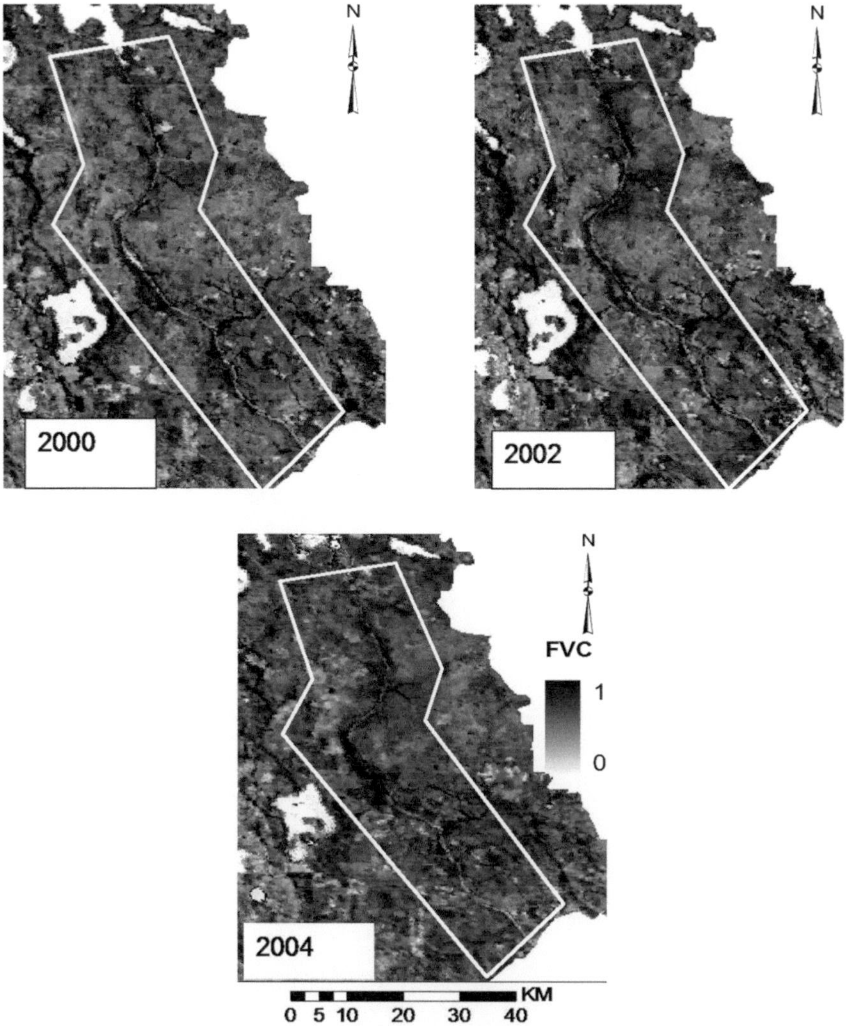

Fig. 3.5 Comparison of fractional vegetation cover (*FVC*) of Kissimmee River basin for 3 years

meter. As depicted in Fig. 3.6, latent heat values were higher in 2002 and 2004 than 2000 on areas along the river. The average April LE for 2000, 2002, and 2004 were 128, 135, and 139 W m^{-2}, respectively. The removal of flood control structures and re-channelization of the river to its natural course will increase the floodplain area and in turn latent heat flux will increase. It is shown that higher latent heat flux along the river can be attributed to the increased floodplain areas and vegetation cover. The rainfall volume for the month of April (2000, 2002, and 2004) was 40, 10, and 35 mm, respectively.

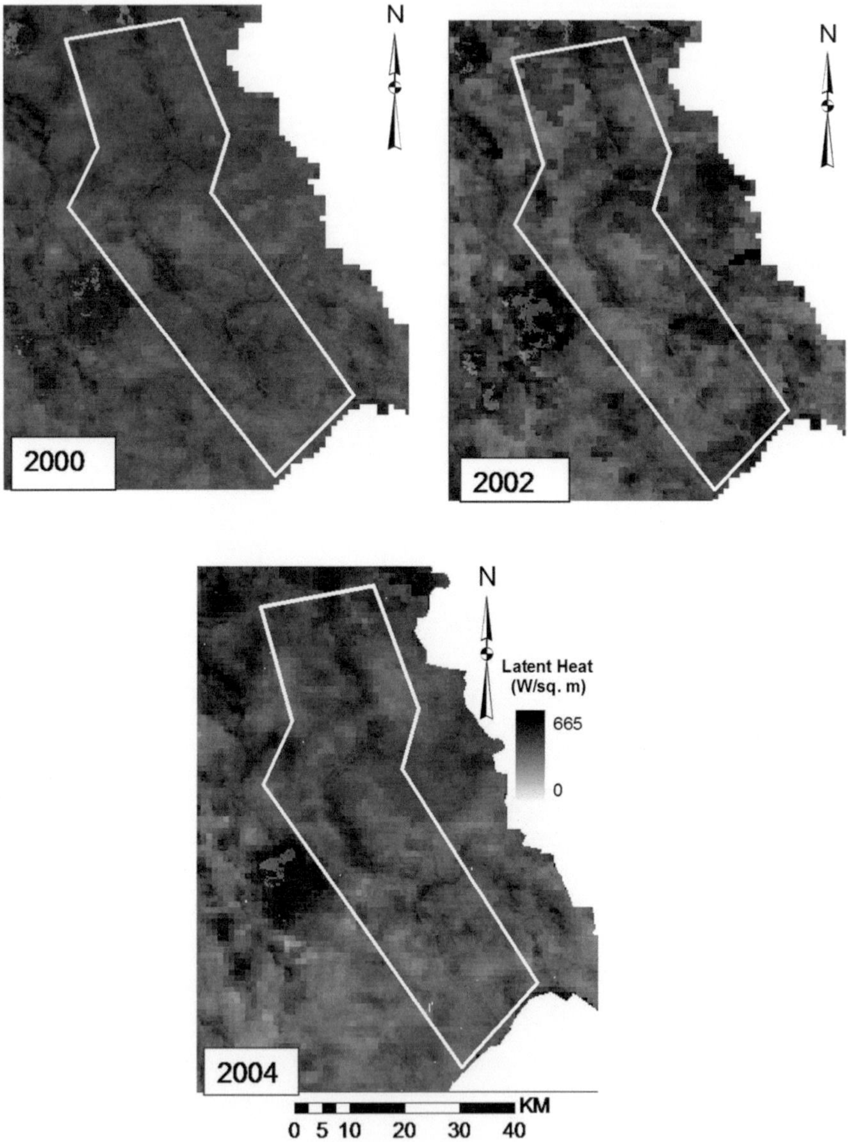

Fig. 3.6 Latent heat variation from year-to-year for Kissimmee River basin

3.5.3 Evaporative Flux (EF)

The MODIS remote sensing data were used to compute the EF (Eq. 3.9) values in April for the years 2002–2004 (Fig. 3.7). The results indicate that the EF values vary spatiotemporally and are higher in the lower part of the basin along the

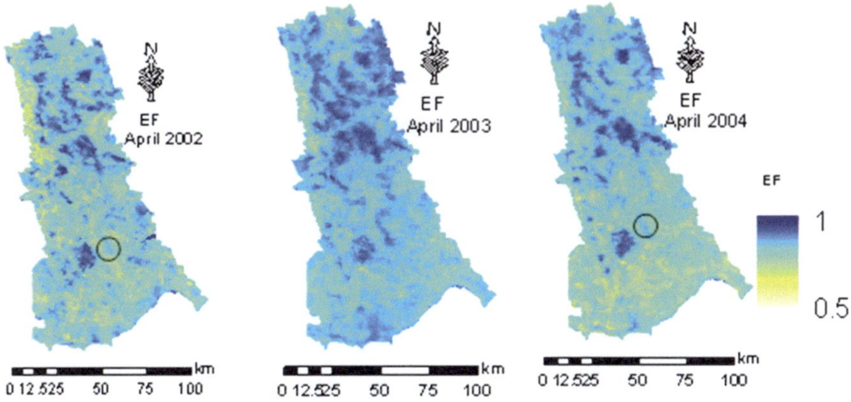

Fig. 3.7 Evaporative flux map of Kissimmee River and floodplain

restored areas in 2003 and 2004. The removal of flood control structures and rechannelization of the river to its natural course will increase the floodplain area and in turn higher latent heat flux and EF. It is shown that higher latent heat flux along the river can be attributed to the increased floodplain areas and vegetation cover.

$$\mathrm{EF} = \frac{\mathrm{LE}}{\mathrm{LE} + H} = \frac{\mathrm{LE}}{R_{\mathrm{n}} - G} \tag{3.9}$$

Response of the Kissimmee basin's hydrology and vegetation to the recent restoration was evaluated using data from MODIS-based FVC, spatial latent heat flux, and groundwater records. The NDVI–TS–albedo relationship was also analyzed for the 2000–2004 period. Using NDVI, TS and albedo values for the month of April, unsupervised classification was conducted and scattergram was generated. Results show that for the highly vegetated portion, a negative correlation between NVDI–TS and NDVI-albedo was observed. It was also indicated that for the less vegetated (lower NDVI) part, the NDVI–TS–albedo relationship was not clearly defined.

The FVC was increased for 2002 and 2004 than 2000 for areas along the Kissimmee River indicating response to the floodplain restoration. The spatial latent heat flux, which is evapotranspiration in energy units, also showed an increase in 2002 and 2004 compared to 2000, which can be attributed to large areas of vegetated surface. This change was mainly seen along the river where most of the restoration work is going and changes in the hydrology are expected.

Groundwater level records from selected monitoring wells were also used to compare spatiotemporal variations in the groundwater levels. Analysis of groundwater level data (2000–2004) from eight monitoring wells showed that, the average monthly level of groundwater increased by 20 and 34 cm between 2000 and 2004, and 2000 and 2003, respectively. Taking into account the amount of rainfall, this observation is valid and reasonable.

3.5.4 Non-evaporative Energy Flux (NEF)

Evaporative energy flux (EF) is complementary with NEF, which is computed as 1 − EF (Eq. 3.10) as function of latent heat (LE), sensible heat (H), net radiation (R_n), and soil heat flux (G). EF indicates the availability of evaporable water; a high EF value indicates the presence of wet surface and green vegetative cover. In contrast, a high NEF value indicates water deficit and drought.

$$\text{NEF} = 1 - \text{EF} = \frac{H}{\text{LE} + H} = \frac{H}{R_n - G} \tag{3.10}$$

To understand the onset of droughts in the Mara River basin (1°23′38″S, 35°11′24″E) in Kenya and Tanzania, the spatially distributed NEF was estimated using remotely sensed data acquired in February, May, August, and October of 2004 from

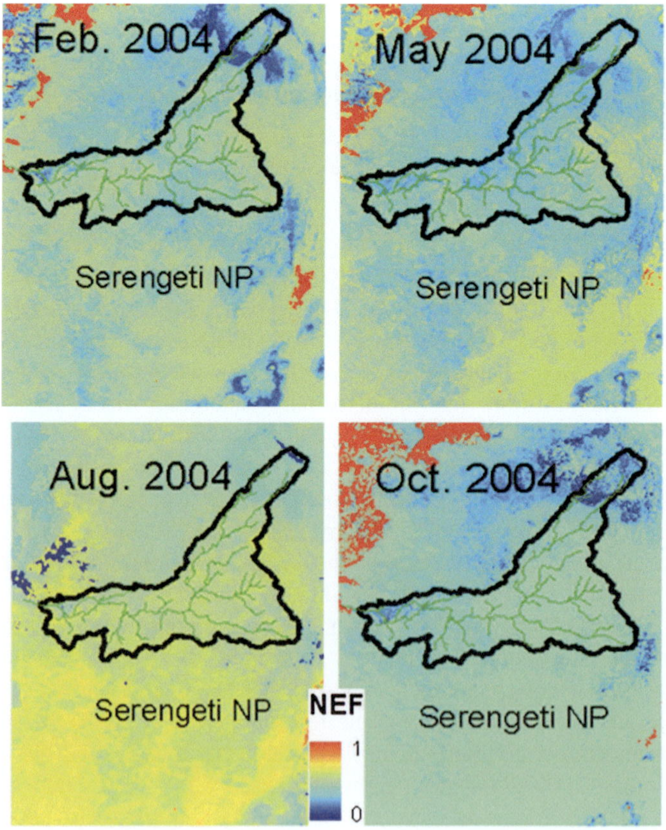

Fig. 3.8 The non-evaporative flux (*NEF*) maps for the Mara River basin and the Serengeti National Park (*SNP*)

a MODIS sensor aboard Terra. In addition, data on daily surface temperature, NDVI, and albedo were obtained from the LP DAAC and used in the surface energy balance computation. Further, data on air temperature and wind speed were collected from the Kenyan and Tanzanian meteorological offices. The dryness of the region can be evaluated as shown in Fig. 3.8 for the months of February, May, August, and September 2004.

3.6 Summary

Landscapes and topography influence the fluxes of moisture, energy, nutrient, and chemicals. Human caused landscape change has been documented as the result of land use change to meet material and other needs. These changes bring about alterations in thermalscapes, hence influencing local weather and regional climates. The impacts of these changes on climate will also affect the magnitude, distribution, and pattern of precipitation and vegetation cover causing hydrological and ecological alterations. The impacts of large-scale climate processes on local and watershed scale hydrometeorological and ecological processes are better understood using climate models. Climate models are capable of predicting climate shift and impact models are capable of assessing potential watershed-scale ecohydrological changes. The role of remote sensing in capturing land use changes, energy partitioning, precipitation monitoring, and other ecohydrological changes contribute input parameters for large-scale model applications and studies.

References

Abtew W, Melesse A (2013) Evaporation and evapotranspiration measurements and estimations. Springer, New York

Ahrens CD (2006) Meteorology today. An introduction to weather, climate and the environment, 8th edn. Thomson Brooks, Cole

Baret F, Guyot G, Begue A, Maurel P, Podaire A (1988) Complimentarily of middle-infrared reflectance for monitoring wheat canopies. Remote Sens Environ 26:213–215

Bastiaanssen WGM (2000) SEBAL-based sensible and latent heat fluxes in the irrigated ediz Basin, Turkey. J Hydrol 229:87–100

Bastiaanssen WGM, Menenti M, Feddes RA, Holtslag AM (1998a) The surface energy balance algorithm for land (SEBAL): part 1 formulation. J Hydrol 212–213:198–212

Bastiaanssen WGM, Pelgrum H, Wang J, Ma Y, Moreno J, Roerink GJ, van der Wal T (1998b) The surface energy balance algorithm for land (SEBAL): part 2 validation. J Hydrol 212–213:213–229

Bishaw B (2001) Deforestation and land degradation in the Ethiopian highlands: a strategy for physical recovery. Northeast African Stud 8(1):7–26

Boyd DS, Foody GM, Curran PJ, Lucas RM, Honzaks M (1996) An assessment of radiance in Landsat TM middle and thermal infrared wave bands for the detection of tropical regeneration. Int J Remote Sens 17:249–261

Carlson TN, Arthur ST (2000) The impact of land use-land cover changes due to urbanization on surface microclimate and hydrology: a satellite perspective. Glob Planet Change 25:49–65
Carlson TN, Ripley AJ (1997) On the relationship between fractional vegetation cover, leaf area Index and NDVI. Remote Sens Environ 62:241–252
Che N, Price JC (1992) Survey of radiometric calibration results and methods for visible and near-infrared channels of NOAA-7,-9 and –11 AVHRRs. Remote Sens Environ 41:19–27
Chineke TC, Idinoba ME, Ajayi OC (2011) Seasonal evapotranspiration signature under a changing landscape and ecosystem management in Nigeria: implications for agriculture and food security. Am J Sci Ind Res. doi:10.5251/ajsir.2011.2.2.191.204
Courel MF, Kandel KS, Rasool SI (1984) Surface albedo and Sahel drought. Lett Nat 307:528–531
Curran PJ, Dungan JL, Gholz HL (1992) Seasonal LAI in slash pine estimated with Landsat TM. Remote Sens Environ 39:3–13
Dafalla MS, Abdel-Rahman EM, Siddig KHA, Ibrahim IS, Csaplovics E (2014) Land use land cover changes in Northern Kordofan State of Sudan: a remotely sensed data analysis (chapter 15). In: Melesse A et al (eds) Nile River basin ecological challenges, climate change and hydropolitics. Springer, New York
Dale VH (1997) The relationship between land-use change and climate change. Ecol Appl 7 (3):753–769
Danson FM, Curran PJ (1993) Factors affecting the remotely sensed response of coniferous forest plantations. Remote Sens Environ 43:55–65
Earth Resources Data Analysis System (ERDAS) (1999) ERDAS field guide. ERDAS Inc, Atlanta
Eiseltova M, Pokorny J, Hesslerova P, Ripl W (2012) Evapotranspiration—a driving force in landscape sustainability (chapter 14). In: Irmak A (ed) Evapotranspiration—remote sensing and modeling. InTech, Croatia
French AN, Schmugge TJ, Kustas WP (2000) Estimating surface fluxes over the SGP site with remotely sensed data. Phys Chem Earth 25(2):167–172
Han L (1997) Spectral reflectance with varying suspended sediment concentrations in clear and algae-laden waters. Photogram Eng Remote Sens 63(6):701–705
Hemakumara HM, Chandrapala L, Moene AF (2003) Evapotranspiration fluxes over mixed vegetation areas measured from large aperture scintillometer. Agric Water Manag 58(2): 109–122
Jacobs CM, Elbers J, Brolsma R, Moors O, Rodreguez-Carretero M, van Hove BM (2014) Assessment of urban evapotranspiration in the Netherlands. In: An internal symposium on evapotranspiration: challenges in measurement and modelling from leaf to landscape scale and beyond, Raleigh, North Carolina, 7–14 April 2014. ASABE, USA
Keeton WS, Mote PW, Franklin JF (2007) Climate variability, climate change, and western wildfire with implication for the urban-wildland interface. Adv Econ Environ Res 6:225–253
Kustas WP (1990) Estimates of evapotranspiration with a one-and two-layer model of heat transfer over partial canopy cover. J Appl Meteorol 29:704–715
Kustas WP, Norman JM (1999) Evaluation of soil and vegetation heat flux predictions using simple two-source model with radiometric temperatures for partial canopy cover. Agric Forest Meteorol 94:13–29
Kustas WP, Norman JM, Anderson MC, French AN (2003) Estimating sub-pixel surface temperatures and energy fluxes from the vegetation index–radiometric temperature relationship. Remote Sens Environ 85(4):429–440
Kustas WP, Li F, Jackson TJ, Prueger JH, MacPherson JI, Wolde M (2004) Effects of remote sensing pixel resolution on modeled energy flux variability of croplands in Iowa. Remote Sens Environ 92(4):535–547
Lagomasino D, Price RM, Whitman D, Melesse AM, Oberbauer S (2015) Spatial and temporal variability in spectral-based evapotranspiration measured from Landsat 5TM across two mangrove ecotones. Agric Forest Meteorol doi:10.1016/j.agronet.2014.11.017
Lambin EF, Strahler AH (1994) Indicators of land cover change—vector analysis in multi-temporal space at coarse spatial scale. Int J Remote Sens 15:2099–2119

Loiselle S, Bracchini L, Bonechi C, Rossi C (2001) Modeling energy fluxes in remote wetland ecosystems with the help of remote sensing. Ecol Model 45(2):243–261
Marshall CH, Pielke RA (2003) The impact of anthropogenic land-cover change on the Florida peninsula sea breezes and warm season sensible weather. Mon Weather Rev 132:28–52
Melesse AM, Jordan JD (2002) A comparison of fuzzy vs. augmented-ISODATA classification algorithm for cloud and cloud-shadow discrimination in Landsat imagery. Photogram Eng Remote Sens 68(9):905–911
Melesse A, Nangia V (2005) Spatially distributed surface energy flux estimation using remotely-sensed data from agricultural fields. Hydrol Process 19(14):2653–2670
Melesse AM, Oberg J, Beeri O, Nangia V, Baumgartner D (2006) Spatiotemporal dynamics of evapotranspiration and vegetation at the Glacial Ridge Prairie restoration. Hydrol Process 20 (7):1451–1464
Melesse A, Nangia V, Wang X, McClain M (2007) Wetland restoration response analysis using MODIS and groundwater data. Spec Issue Remote Sens Nat Res Environ Sens 7:1916–1933
Melesse A, Frank A, Nangia V, Liebig M, Hanson J (2008) Analysis of energy fluxes and land surface parameters in grassland ecosystem: remote sensing perspective. Int J Remote Sens 29 (11):3325–3341
Melesse A, Abtew W, Desalegne T (2009) Evaporation estimation of Rift Valley Lakes in ethiopia comparison of models. Sensors 9(12):9603–9615. doi:10.3390/s91209603
Mohamed YA, Bastiaanssen WGM, Savenije HHG (2004) Spatial variability of evaporation and moisture storage in the swamps of the upper Nile studied by remote sensing techniques. J Hydrol 289:145–164
Morse A, Tasumi M, Allen RG, Kramber W (2000) Application of the SEBAL methodology for estimating consumptive use of water and streamflow depletion in the Bear River basin of Idaho through remote sensing. Final report submitted to the Raytheon Systems Company, Earth Observation System Data and Information system Project, by Idaho Department of Water Resources and University of Idaho, 107 pp
Nouri H, Beecham S, Kazemi F, Hassanli AM (2013) A review of ET measurement techniques for estimating the water requirements of urban landscape vegetation. Urban Water J 10(4): 247–259
Oberg J, Melesse AM (2005) Wetland evapotranspiration dynamics vs. ecohydrological restoration: an energy balance and remote sensing approach. J Am Water Res Assoc 42 (3):565–582
Oke TR (1992) Boundary layer climates, 2nd edn. Routledge, New York
Panigrahy S, Parohar JS (1992) Role of middle-infrared bands of Landsat thematic mapper in determining the classification accuracy of rice. Int J Remote Sens 13:2943–2949
Pielke RA, Adegoke J, Beltran-Prezkurat A, Hiemstra CA, Lin J, Nair US, Niyogi D, Nobis TE (2007) An overview of regional land-use and land-cover impacts on rainfall. Tellus B 59 (3):587–601
Price JC (1987) Calibration of satellite radiometers and the comparison of vegetation indices. Remote Sens Environ 21:15–27
Rouse JW, Haas RH, Schell JA, Deering DW (1974) Monitoring vegetation systems in the Great Plains with ERTS. In: Proceedings of third earth resources technology satellite-1 symposium, vol 351. NASA SP, Greenbelt, pp 3010–3017
Rust W, Corstanje R, Holman IP, Milne AE (2014) Detecting land use and land management influences on catchment hydrology by modelling and wavelets. J Hydrol 517:378–389
Sailor DJ (1995) Simulated urban climate response to modification in surface albedo and vegetative cover. J Appl Meteorol 34:1694–1704
Senay GB, Budde M, Verdin JP, Melesse AM (2007) A coupled remote sensing and simplified energy balance approach to estimate actual evapotranspiration from irrigated fields. Spec Issue Remote Sens Nat Res Environ Sens 7:979–1000
Senay GB, Verdin JP, Lietzow R, Melesse AM (2008) Global daily reference evapotranspiration modeling and validation. J Am Water Res Assoc (JAWRA) 44(4):969–979

Sileshi Y, Zanke U (2004) Recent changes in rainfall and rainy days in Ethiopia. Int J Climatol 24:973–983

Spangmyr M (2010) Global effects of albedo change due to urbanization. In: Seminar series no. 180. Department of Earth and Ecosystem Sciences, Lund University

Spanner MA, Pierce LL, Running SW, Peterson DL (1990) The seasonality of AVHRR data of temperate coniferous forests: relationship with LAI. Remote Sens Environ 33:97–112

Stone TA, Schleeinger P, Houghton RA, Woodwell GM (1994) A map of the vegetation of South America based on satellite imagery. Photogram Eng Remote Sens 60:541–551

Taha H (1997) Urban climates and heat islands: albedo, evapotranspiration and anthropogenic heat. Energy Build 25:99–103

Tucker CJ, Dregne HE, Newcomb WW (1991) Expansion and contraction of the Sahara Desert from 1980 to 1990. Science 253(5017):299–301 (new series)

Wang J, Bastiaanssen WGM, Ma Y, Pelgrum H (1998) Aggregation of land surface parameters in the oasis-desert systems of Northwest China. Hydrol Process 12:2133–2147

Xuejie G, Zhang D, Chen Z, Pal JS, Giorgi F (2007) Land use effects on climate in China as simulated by a regional climate model. Sci Chin D Earth Sci 50(4):620–628

Yallop ML, Ansio AM, Perkin RG, Cook J, Telling J, Fagan D, MacFarlane J, Stibal M, Barker G, Bellas C, Hodson A, Tranter M, Wdham J, Roberts NW (2012) Photophysiology and albedo-changing potential of the ice algal community on the surface of the Greenland Ice. ISME J 6(12):2302–2313

Zheng X, Eltahir EAB (1997) The response to deforestation and desertification in a model of West African monsoon. Geophys Res Lett 24(2):155–158

Chapter 4
Multitemporal Land Use/Land Cover Change Detection for the Batena Watershed, Rift Valley Lakes Basin, Ethiopia

Gebiaw T. Ayele, Solomon S. Demessie, Kassa T. Mengistu, Seifu A. Tilahun and Assefa M. Melesse

Abstract A majority of the rural population in Ethiopia depends on agriculture. Land use changes during the past couple of decades are mostly linked to agricultural development attributed to factors such as population pressure and environmental changes. Mapping land use/land cover (LULC) to analyze the type, rate, and extent of changes in land use patterns has far reaching significance for policy/decision makers and resource managers to provoke the wide range of applications at regional scales for erosion, landslide, land planning, forest management, and ecosystem conservation. The focus of this chapter is to depict quick and practical approaches to generate spatially and temporally quantified information on land cover dynamics using high-resolution satellite images for the years (1973–2008) in Batena watershed and its environs in southwestern Ethiopia. To quantify the magnitude of LULC change, supervised classification technique was applied using Landsat Thematic Mapper (TM) and Enhanced Thematic Mapper Plus (ETM+) images employing Bayesian maximum likelihood classifier (MLC) with the aid of ground truth training sites. A majority/minority analysis was used for smoothing the classification results

G.T. Ayele (✉) · S.A. Tilahun
Faculty of Civil and Water Resource Engineering, Bahir Dar Institute of Technology, Bahir Dar University, 252, Bahir Dar, Ethiopia
e-mail: gebeyaw21@gmail.com

G.T. Ayele · S.A. Tilahun
Department of Hydraulic and Water Resources Engineering, Blue Nile Water Institute, Bahir Dar University, 252, Bahir Dar, Ethiopia

S.S. Demessie
Department of Civil & Environmental Engineering, University of California, Los Angeles, Los Angeles 90095, CA, USA

K.T. Mengistu
Department of Irrigation and Water Resources Engineering, Arba Minch University, Arba Minch, Ethiopia

A.M. Melesse
Department of Earth and Environment, Florida International University, Miami, USA

© Springer International Publishing Switzerland 2016

A.M. Melesse and W. Abtew (eds.), *Landscape Dynamics, Soils and Hydrological Processes in Varied Climates*, Springer Geography,
DOI 10.1007/978-3-319-18787-7_4

and the accuracy of image classification was carried out by means of a confusion matrix generated through geographic information system (GIS) overlay of the classified maps and the test samples. The classification accuracy was further verified by the strong kappa statistical estimate of more than 90 % as a measure of overall agreement between image and reference data. The final output of remote sensing imagery revealed five land cover classes: Grazing land, bush land, mixed forest, dominantly cultivated agricultural land, and water body. It has been discovered that, there were more active LULC change processes in the area in the first study period (1973–1984) than the second study period (1984–1995) and the third study period (1995–2003). On the other hand, areal extent of cultivated and uncultivated agricultural land has been on a steady decline from 39.7 % in 1995 to 41.4 % in 2003 and a mere 50.1 % in 2008. In the first period, nearly half of the landscape underwent land cover change with more than 17 % of the entire landscape experiencing agricultural expansion. In the second period, the extent of the changes was limited to less than 1/3 of the total area with a smaller amount of agricultural area expansion than before. Though the rate of land cover change was observed to vary across the three periods of study, a general decline of forest cover and amplified increase of agricultural lands of more than 41.7 % was found in the area.

Keywords Change detection · Land cover dynamics · Maximum likelihood classifier · Landsat imagery · Remote sensing · GIS · Batena watershed · Ethiopia

4.1 Introduction

Land use/land cover (LULC) change has increasingly become a key research priority for national and international research programs examining global environmental change and impact analysis of the changes, which is a standard requirement for land use planning and sustainable management of natural resources as highlighted by many researchers (Petit et al. 2001). Researchers in their finding have argued that more focused management intervention requires information on the rates and impacts of LULC change as well as the distribution of these changes in space and time as a central component in present strategies for managing natural resource and monitoring environmental changes (Tiwari and Saxena 2011). Land use/cover change can affect biodiversity, biogeochemical cycles, soil fertility, hydrological cycles, energy balance, land productivity, and the sustainability of environmental services (Lupo et al. 2001). In most parts of the world, agriculture is the primary driver of land use change. But few comprehensive studies have been undertaken at a global level on long-term historical changes of land cover due to land use practice (Goldewijk and Ramankutty 2004). However, land cover change has recently become a major concern for research on global warming and global change; it draws attention and has emerged as a research agenda for many researchers (Lambin et al. 2003). Wondie et al. (2011) analyzed LULC change of

Simen Mountains National Park, a World Heritage Site in northwestern Ethiopia and linked to human interference and encroachment to the park.

LULC dynamics is an important landscape process capable of altering the fluxes of water, sediment, contaminants, and energy. Mainly caused by human, impact of land use on water resources availability is high. Degraded watersheds tend to accelerate overland flow reducing soil moisture and baseflow recharge and increases sediment detachment and transport. Various studies used land cover mapping tools and methods to understand land use changes, inventory of forest, and natural resources as well as understand the changes in the hydrologic behavior of watersheds (Getachew and Melesse 2012; Mango et al. 2011a, b; Wondie et al. 2011, 2012; Melesse and Jordan 2003; Mohamed et al. 2013).

The changes of land use are the result of the interaction between the socio-economic conditions of the society, institutional and environmental factors added with population pressure, land use type, and climatic conditions. LULC analysis is also devoted to the relations between land use and the socioeconomic and bio-physical variables that act as the driving forces of LULC change, associated with the intensification of agriculture, cattle production, and urbanization, could have a profound impact on the hydrological processes in small watersheds and at the regional level (Mendoza et al. 2002). Proper change detection in land cover types requires the spatial resolution of satellite images (Swinne and Veroustaete 2008; Yu et al. 2011).

Since LULC changes are products of prevailing interacting natural and anthropogenic factors (Fashona and Omojola 2005) and their utilization by man in time and space (Clevers 2004), the need to track the change pattern and areal extent of LULC in small-scale watersheds with Landsat imagery and to quantify the rate of change is a critical input for natural resource management policy decisions. Geospatial techniques are used to monitor the LULC dynamics and have an important role in natural resources conservation, management, monitoring, and assessment of catchment characteristics, for the study of hydrologic response and flow regime. GIS and remote sensing tools are used to understand the rate and magnitude of LULC change and to derive accurate and time-valued spatial distribution information (Carlson and Azofeifa 1999; Guerschman et al. 2003; Rogana and Chen 2004; Zsuzsanna et al. 2005).

In this chapter, detailed space–time scale LULC map is produced and the rate and magnitude of change is quantified. To this effect adequate amount of Earth's surface data are collected using remote sensing tools which provide an excellent source-updated LULC information and changes that can be extracted and analyzed efficiently (Bauer et al. 2003). Since the launch of the first remote sensing satellite (1972), multiresolution and multitemporal satellite data available in various data archives have been used as a base for various environmental studies including LULC change analysis and to present a reliable database for long-term change detection using Landsat imageries recorded in the last four decades using Multispectral Scanner (MSS), Thematic Mapper (TM), and Enhanced Thematic Mapper Plus (ETM+) sensors.

As part of change detection analysis, image differencing with the principle of subtraction of images of two different time periods of the same location to evaluate the change pattern in different temporal levels (Dimyati 1995) is important to planners in monitoring the consequences of the change on the area to plan and assess the pattern and extent of the change, to model and predict the future level of change, and to analyze the driving forces of changes (Moshen 1999). Classified images are associated with errors, to reduce these errors, post-classification refinements are done using an error matrix approach. The error matrix is computed as the total number of correct class predictions divided by total number of cells (Verbyla 1986). Accuracy assessment of classified images is undertaken to verify the extent to which classified imagery is accurate, by using producers and users' accuracies assessment and Kappa statistics (Sexton et al. 2013). These assessments make it possible to correct conservative and optimistic biases in image classification due to misclassification of land cover classes. For this study, accuracy was assessed using the error matrix and the result validated using Kappa coefficient values. Congalton (1996) indicated that Kappa coefficient values greater than 0.80 show strong agreement, and between 0.40 and 0.80 as moderate agreement, and below 0.40 considered poor agreement.

4.1.1 Land Use Change Issues

Land is becoming a scarce resource due to immense agricultural and demographic pressure, which requires extraction of information on LULC and possibilities for their optimal use for selection, planning, and implementation of land use schemes to meet the increasing demands for basic human needs and welfare. It is also noted that LULC change studies are proven to be essential for the qualification and quantification of central environmental processes, environmental change and influence of environmental management on biodiversity, water budget, radiation budget, trace gas emissions, carbon cycling, livelihood (Verburg et al. 2002, 2004), urban and rural agricultural land loss (Lambin et al. 2003; Muzein 2008), and a wide range of socioeconomic and ecological processes (Ozbakir et al. 2007), which on the aggregate affects global environmental change and the biosphere (Fashona and Omojola 2005).

Land use and land cover are dynamic in nature and understanding the interaction and relationship of anthropogenic activities with the environment is essential (Prakasam 2010). The finding of many researches across abroad and particularly in the study basin about LULC change and driving forces in small and larger watersheds fails to explain either the length of time series of records or quality of the analysis while viewing the Earth from space, which is crucial to study the situations of rapid and often unrecorded land use change by which observations of the Earth from space provide objective information on human utilization of the landscape.

There are different studies in the Rift valley basin and Bilate watershed, (the study area, Batena is the subwatershed of Bilate), southern Ethiopia; the effect of LULC changes and management practices in watershed hydrological responses (Kassa 2009), assessment of LULC dynamics and its impact on soil loss (Loppiso 2010), analysis of biomass degradation as an indicator of environmental challenge (Degelo 2007), and watershed modeling to understand catchment flow responses and the impact of catchment dynamics on runoff generation (Negash 2014). However, there is no detailed fine time scale and well-trained time series of LULC change study in small watersheds like Batena signifying the anthropogenic impact on the trend and rate of LULC change.

4.1.2 Objectives

The overall objective of the study is to quantify the LULC change trajectories with the available multitemporal Landsat imagery and to detect the rate and magnitude of change for a small rural area, Batena, in the Rift Valley Lakes Basin, Ethiopia.

The specific objectives of this study are to

1. identify the decadal LULC change at different spatial and temporal scales;
2. produce an LULC map of the area and identify the rate, nature, trend, and magnitude of change; and
3. assess the accuracy of the classification technique.

4.2 Study Area and Datasets

4.2.1 Study Area Description

The study watershed, Batena, is a 116.7 km^2 rural watershed located in the mountainous part of Rift Valley lakes basin, southwestern part of Ethiopia (Fig. 4.1). The region is typified by steep to moderate hill slopes of dominantly clayey permeable soils, with insignificant areal coverage of Leptosols and Regosols.

The topography of Batena watershed signifies two distinct features; the highlands, ragged mountainous areas in the north most and northwest part of the watershed and lowlands in the eastern most part of the watershed. The altitude in the watershed ranges from 2063 m in the low lands up to 2947 m in the highlands. Topography of the watershed is mountainous, with an average slope exceeding 25 % (4 % in the flat areas and more than 30 % in the mountainous part of the area). The area has a bimodal rainfall pattern with high spatiotemporal variability. According to the FAO-UNESCO soil classification, the dominant soils in the watershed includes the high base status, high activity, and clay-enriched Chromic

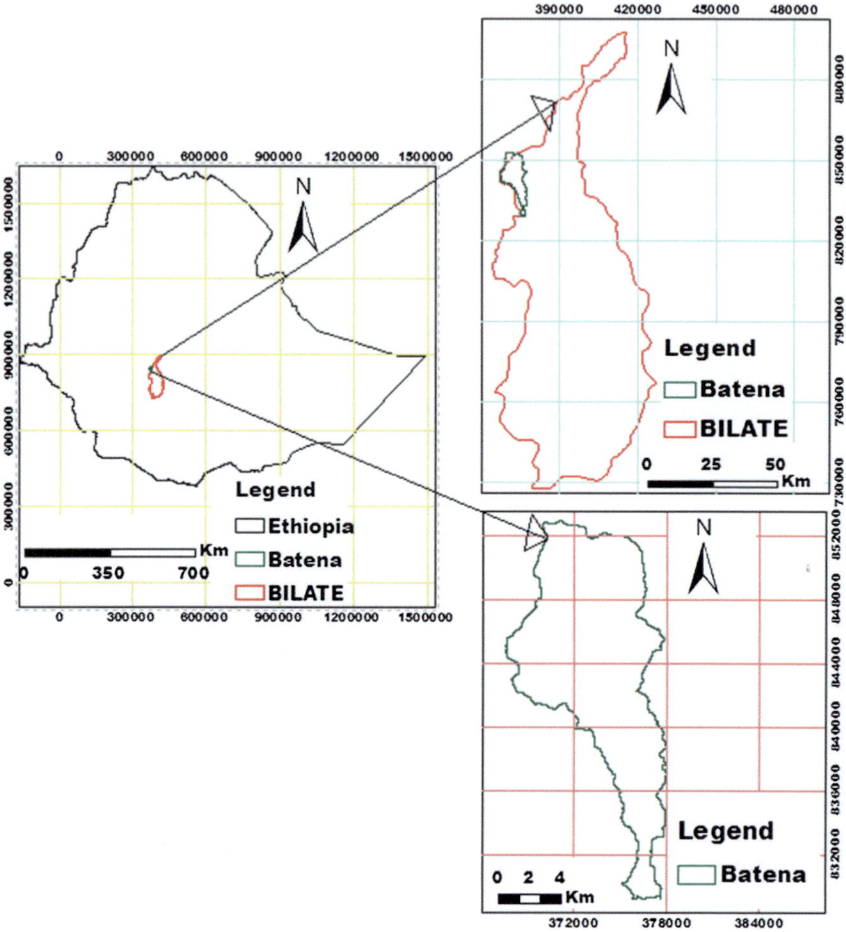

Fig. 4.1 Location of the study area

Luvisols with about 94.11 % areal coverage and the deep, well-drained, low-activity, P fixation, and strongly structured, tropical soils with diffuse horizon boundaries (Humic Nitisols).

4.2.2 Field Data Sampling and Design

The field survey includes reconnaissance of the area and ground truthing to produce decadal LULC maps, spatial extent, and neighborhood features. The LULC change is analyzed by using WGS 1984 datum referenced remotely sensed medium resolution Landsat satellite from MSS, TM, and ETM+ sensors digital data.

The Landsat satellites acquire imagery in a regular, tiled fashion, following the World Reference System (WRS1 for MSS, WRS2 for TM and ETM+). The MSS sensor provides the oldest and lowest quality Landsat data, from 1972 to present.

4.3 Methods

4.3.1 Image Classification

For proper image classification and change detection MSS, TM, and ETM+ images have passed through different pre- and post-classification and change detection steps including digital image processing for manipulation and interpretation of digital images, image resampling, image enhancement, spatial enhancement, radiometric, enhancement, spectral enhancement, and image transformation.

Image preprocessing: A series of sequential preprocessing operations including atmospheric correction or normalization, image registration, geometric correction, and masking (e.g., for clouds, water, irrelevant features) is important before image classification. The normalization of satellite imagery takes into account the combined, measurable reflectance of the atmosphere, aerosol scattering and absorption, and the Earth's surface. Geometric rectification of the imagery resamples or changes the pixel grid to fit that of a map projection or another reference image. This becomes especially important when scene to scene comparisons of individual pixels in applications such as change detection are being sought. Due to availability of longer date of acquisition of satellite images, the Landsat images are preferred, though their lower resolution may affect the level of information that can be extracted from the images.

Image resampling: MSS images have low spatial resolution but TM and ETM+ images have relatively higher resolution. For LULC change analysis with the different years of satellite imagery, date of data acquisition and resolution of images must be similar as much as possible. Therefore, the bands of MSS images were resampled to the higher spatial resolution before use for analysis.

Spatial Image Enhancement: Image enhancement process involves solely improving the appearance of imagery to assist in visual interpretation and analysis. Spatial Image enhancement method helps to increase spatial resolution of an image using the spatial enhancement techniques of resolution merge which enables to integrate imagery of different spatial resolution or pixels. Since a higher resolution imagery is a single band (panchromatic band with resolution of 15 m in ETM+ images) while multispectral imagery generally have a lower resolution (TM 30 m resolution). These techniques are often used to produce high resolution, multispectral imagery to improve the interpretability of the data by having color resolution information.

Radiometric Image Enhancement: As part of radiometric corrections, stripping and banding corrections have been made, which occurs if a detector goes out of adjustment; it provides readings consistently greater than or less than the other detectors for the same band over the same ground cover (Crippen 1989a). The effects of the atmosphere upon remotely sensed data are not considered errors, since they are part of the signal received by the sensing device (Bernstein 1983). This defect was corrected by conversion of radiance to reflectance. For the conversion of radiance to reflectance, the characteristics of the satellite images used have been known from header files of the image.

Spectral Image Enhancement: Principal Component analysis is used as a method of data compression, to produce uncorrelated output bands, to segregate noise components, and to reduce the dimensionality of data sets. It allows a redundant data to be compacted into none correlated and independent fewer bands of more interpretability than the source data. Principal Component bands produce more colorful color composite images than spectral color composite images because the data is uncorrelated (Richards 1999).

Image Transformation: Unlike to image enhancement operations which are applied only to single channel of data at a time, image transformation usually involves combined process of data from multiple spectral bands. Arithmetic operations are performed to combine and transform the original band to "new" image which better displays or highlights certain features in the scene. Tasseled cap transformation is used in the research for enhancing spectral information content of Landsat TM data. Tasseled cap transformation, especially optimizes data viewing for vegetation studies, is calculated from data of the related six TM bands. Three of the six tasseled cap transform bands are often used: band 1 (brightness, measure of soil), band 2 (greenness, measure of vegetation), and band 3 (wetness, interrelationship of soil, and canopy moisture). The tasseled cap transformed bands are layer stacked for analysis of LULC change. The layer stacked landsat images of 1973, 1984, 2003, and 2008 are radiometric enhanced for haze and noise reduction.

4.3.2 Land Cover Mapping

Field Sampling Design and Image Classification: Field sampling is made to validate land cover interpretation results from satellite images, for qualitative characteristic descriptions of each land use and land cover classes. For better accuracy image classification, it is found important to convert the Landsat radiance values to reflectance. To eliminate the effects of atmospheric scattering and absorption; and to increase the accuracy of surface-type classification, the Landsat digital number (DN) values were converted to top-of-atmosphere (TOA) reflectance. After converting Landsat radiance values to reflectance, their Normalized Difference Vegetation Index (NDVI) is calculated for both radiance values and TOA reflectance to evaluate the effects of converting processes. This is performed in a two-step procedure, first, by converting DN values to spectral radiance and second, by transferring the sensor-detected radiance into TOA reflectance. In the first stage, DN values of the sensor measurements are converted into spectral radiance measured by satellite sensors. In the second stage, the sensor-detected radiance is transferred into the ground surface reflectance.

To evaluate the effects of converting processes, NDVI is calculated for both radiance values and TOA reflectance of Landsat using the following formula:

$$\mathrm{NDVI} = \frac{(\mathrm{Band}\ \ 4 - \mathrm{Band}\ \ 3)}{\mathrm{Band}\ \ 4 + \mathrm{Band}\ \ 3} \tag{4.1}$$

where Band 4 is near-infrared and Band 3 is visible red reflectance or the NDVI can also calculate with this expanded form as:

$$\mathrm{NDVI} = \frac{\mathrm{NIR}_{\mathrm{ref}} - \mathrm{RED}_{\mathrm{ref}}}{\mathrm{NIR}_{\mathrm{ref}} + \mathrm{RED}_{\mathrm{ref}}} \tag{4.2}$$

where NDVI is a simple numerical indicator that can be used to analyze remote sensing measurement, $\mathrm{NIR}_{\mathrm{ref}}$ is reflectance for vegetation at near-infrared spectral band, and $\mathrm{RED}_{\mathrm{ref}}$ is reflectance for vegetation at red spectral band. Figure 4.2 shows the NDVI classes for the different land cover types.

Unsupervised Classification: Classification is the process of sorting pixels into a finite number of individual classes, or categories of data, based on their data file values using both unsupervised and supervised classification. Prior to supervised classification, unsupervised classification is done by taking the NDVI values using the Iterative Self-Organizing Data analysis (ISODATA) algorithm. Figure 4.3 shows results of the unsupervised classification. This technique repeatedly performs an entire classification with a standard deviation or distance threshold and recalculates statistics with minimum user inputs for locating clusters and also is relatively simple and has considerable intuitive appeal. The classic supervised and unsupervised classification techniques have been used successfully on medium to lower resolution imagery (imagery with pixel sizes of 5 m or larger), however, one of their significant disadvantages is that their statistical assumptions generally preclude their application to high resolution imagery. They are also hampered by the necessity for multiple bands to increase the accuracy of the classification.

Supervised Classification: Following the computer-automated unsupervised classification; supervised classification is performed on the selected classification scheme employing Bayesian maximum likelihood classifier (MLC). LULC maps are produced for the years 1973, 1984, 2003, and 2008 to investigate changes that occurred between these periods. The image classification accuracy depends on factors such as number of spectral bands in the imagery, target/background contrast, signature quality, ground truth information, and image quality.

A majority/minority analysis is used for smoothing the classification results. The accuracy of classification is carried out by means of a confusion matrix generated through geographic information system (GIS) overlay of the classified maps and the test samples. The image classification accuracy is further assessed by calculating the Kappa coefficient '$\hat{k}$'. The kappa statistics is an estimate of measure of overall agreement between image data and the reference (ground truth) data. Essentially,

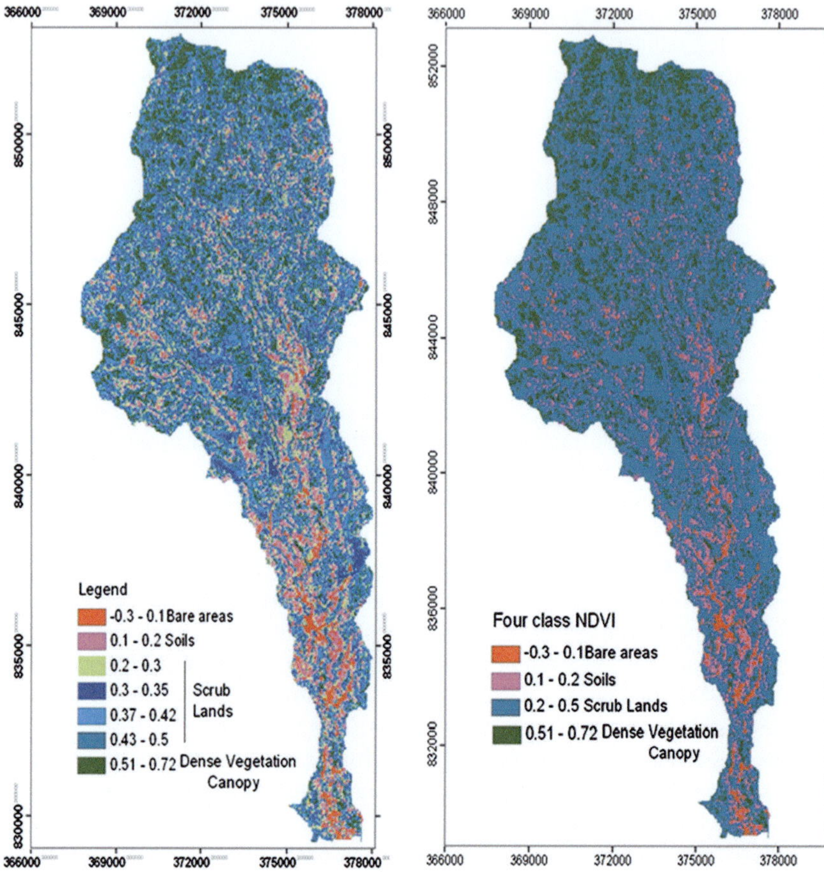

Fig. 4.2 Unsupervised classification with NDVI values (conversion of landsat radiance values to reflectance)

land cover change detection involves the ability to quantify temporal effects using multitemporal data sets. Remotely sensed data obtained from Earth orbiting satellites is an important data source for land cover change detection because of repetitive coverage at short intervals and consistent image quality (Singh 1989).

Land Use/Land Cover Change Detection: Change detection involves the use of multispectral data sets to discriminate area of land cover change between dates of imaging. Ideally, change detection procedures should involve data acquired by the same or similar sensor and be recorded using the same spatial resolution, viewing geometry, spectral bands, radiometric resolution, and the time of the day. Some aspects of change detection application using remote sensing are LULC change analysis, forest and vegetation change analysis, wetland change analysis, forest fire, and landscape change. Change detection techniques based on multispectral and multitemporal remotely sensed data have demonstrated a great potential as a means

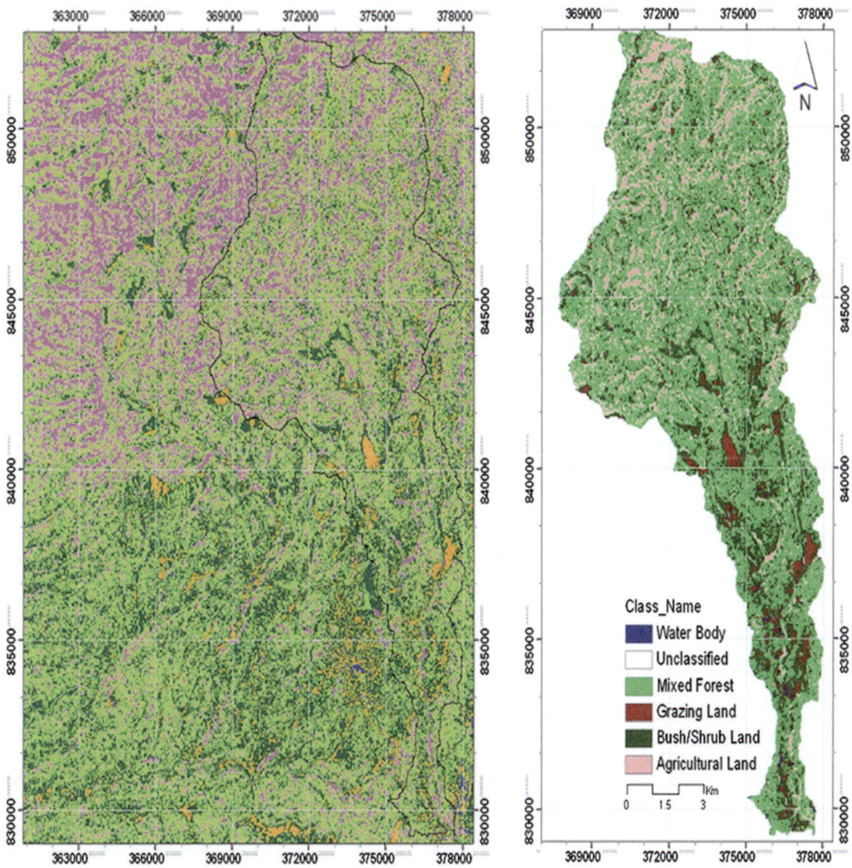

Fig. 4.3 Unsupervised classification false color composite of Batena (2008)

to understand landscape dynamics and to monitor differences in land use and land cover patterns over time, irrespective of the causal factors (Jensen 1996). Therefore, attempt is made in this study to quantify the magnitude of change and change rate, spatial distribution of change types, and change trajectories of land cover types. Accuracy assessment of change detection results are performed over the study periods from 1973 to 2008. The most widely used change detection algorithm is the Post-Classification Comparison (PCC) which detects changes between hand-labeled region classes (Currit 2005; Petit et al. 2001). This technique provides detailed change trajectories between the two images. In addition, the independent classification processes reduce the impact of multitemporal effects due to atmosphere or sensor differences (Lu et al. 2004). The changed pixels extracted between the study period are used to define the "from-to" LULC class trend of change, the area coverage and their trajectories. Hienen and Lyon (1989) conducted a sensitivity analysis to calculate habitat index.

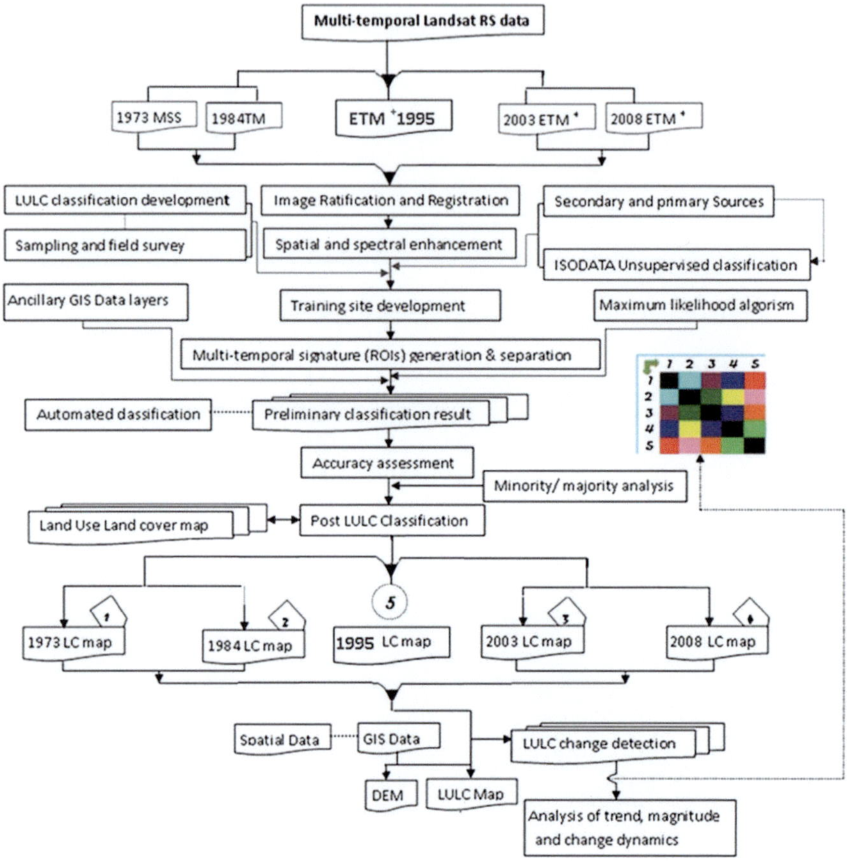

Fig. 4.4 General working conceptual framework

The general working step (Fig. 4.4) and LULC change quantification are described in a flowchart with high level of detail from the scratch; year of Landsat imagery and mapping false color unsupervised and supervised classification verified by accuracy assessment to land cover mapping and finally LULC change detection.

4.4 Results and Discussions

As per the field survey and output of Landsat imagery classification, the land cover for the watershed is mainly characterized by Grazing land, bush/Shrub land, sparsely populated semi-mixed forest on the mountainous part of the area and in line with the streams, dominantly cultivated agricultural land, and an insignificant areal coverage of water body. Supervised classification with the selected Bayesian

Table 4.1 Description of the land use classes in the study area

Class	Description
Agricultural land	It is the land cover under the crop cultivation of annual crops. Scattered settlements surrounded by agricultural lands are classified as agricultural lands, since the low spatial resolution landsat imagery fails to separate the scattered rural settlements with agricultural lands
Grazing land	It is the area covered with both communal and private pasture lands which retain the grass cover for a year and above. Here it includes the fallow lands which stay without tillage more than a year in the area
Scrub lands	Land covered by shrubs, bushes, and young regeneration. It is the area of land covered with short to dwarf tree species with spares to densely populated land status. This includes the Accasia species: Accasia (Abysinica, Albida, Ducerence, Melanoxylone, Senegal, Seal, and Saligna), Eucalyptus species, and miscellaneous species
Mixed forest	It includes the area of land covered with sparsely populated forest, riverine trees and artificially planted indigenous and nonindigenous groups of trees like eucalyptus globules (Nech Bahiezaf), Eucalptus Comanduleses (Key Bahirzaf), Eucalyptus Saligna (Girar Bahirzaf), Cordia Africana (Wanza), Sasbania Sesban (Yemeno Zaf), etc

maximum likelihood classification scheme as a parametric decision rule and well-developed method from statistical decision theory has proved the presence of four land cover classes with their detailed property and water body (Table 4.1).

Maximum likelihood algorithm based image classification technique gave good results for digital change detection in land use/land cover classes. As per the field survey and output of Landsat imagery, the land cover for the watershed is mainly characterized by grazing land, bush land, semi-mixed forest, dominantly cultivated agricultural land, and an insignificant areal coverage of water body (Table 4.2).

Land Cover Change Detection: The images are acquired in one swath/scene 169/055 (path/row) covering the whole area of the Batena catchment. Images from the Landsat satellites have been acquired since 1972, with a variety of characteristics to consider and available through the GLCF/Global Land Cover Facility.

Satellite images of 1973, 1984, 1995, 2003, and 2005 were used to analyze the size of the land use classes through time. A decreasing trend is shown for the mixed forest and grazing land year after year and shrub lands have shown a changeable and fluctuating trend throughout the analysis periods. The spatial extent of agricultural areas has revealed an amplified response mainly at the expense of the mixed forest and grazing land. The degree of expansion for agricultural land is shown with 0.984 strong positive correlation coefficient (Fig. 4.5).

The change detection statistics table (Table 4.3) provides a detailed tabulation of changes between two classified images. The statistics tables list the initial state classes in the columns and the final state classes in the rows. This is required for a complete accounting of the distribution of pixels that changed classes. For each initial state class (that is, each column), the table indicates how these pixels were classified in the final state image. The Class Total row indicates the total number of pixels in each initial state class, and the Class Total column indicates the total

Table 4.2 Comparative numeric figure for response of land use in time

Class		Variation of land use area share with time									
Land use		Year									
		1973		1984		1995		2003		2008	
		km^2	%	km^2	%	km^2	%	km^2	%	km^2	%
Grazing land		15.95	13.67	10.201	8.74	11.69	10.02	12.220	10.48	5.022	4.30
Agriculture land		11.1	9.51	30.967	26.55	46.32	39.71	48.298	41.40	58.431	50.09
Bush/shrub land		71.11	60.93	51.418	44.08	34.45	29.53	34.927	29.94	52.315	44.85
Mixed forest		19.77	16.94	23.861	20.45	24.09	20.65	21.193	18.17	0.873	0.75
Water body		0	0.00	0.209	0.18	0.11	0.09	0.019	0.02	0.015	0.01
%	Kappa coefficient	92.33		98.38		93.30		99.95		99.96	
	Overall accuracy	95.13		99.16		95.19		99.97		99.98	
Total		116.7		116.66		116.66		116.66		116.656	

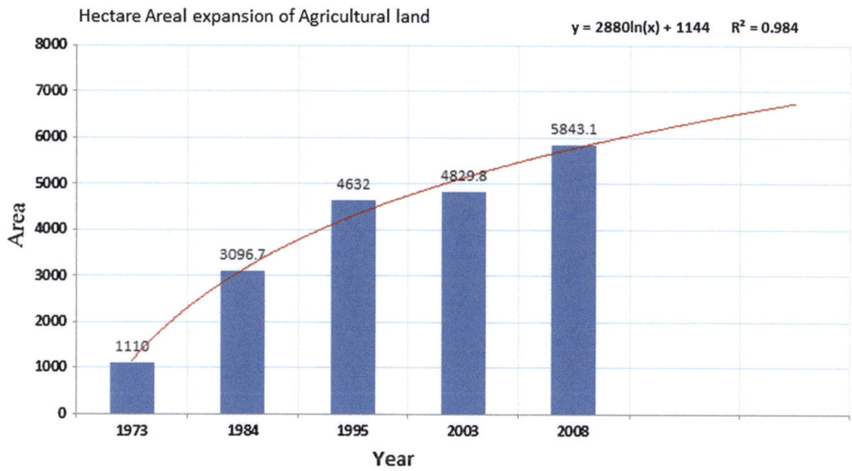

Fig. 4.5 Time response areal expansion of agricultural land (ha)

number of pixels in each final state class. The Row Total column is a class-by-class summation of all final state pixels that fell into the selected initial state classes. The Class Changes row indicates the total number of initial state pixels that changed classes. The Image Difference row is the difference in the total number of equivalently classed pixels in the two images, computed by subtracting the Initial State Class Totals from the Final State Class Totals. An Image Difference that is positive indicates that the class size increased and a negative indicates decrease in class size. Unchanged areas are shown in diagonal cells. In general, columns describe at what expense the new cover classes are established.

The Image Difference change detection statistics in Table 4.3 provides a detailed tabulation of changes between two classified images. The Image Difference is the

Table 4.3 Change detection statistics between final state (1984) and initial state image (1973) in hectare area measure

	Final state image (1984)							
Initial state image (1973)	Class	Grazing land	Agricultural land	Bush/shrub land	Mixed forest	Water body	Row total	Class total
	Grazing land	190	190	640	10	0	1030	1030
	Agricultural land	300	530	1690	680	0	3200	3200
	Bush/shrub land	980	300	3390	480	0	5150	5150
	Mixed forest	130	80	1380	800	0	2390	2390
	Water body	0	0	20	0	0	20	20
	Class total	1590	1110	7110	1980	0	0	0
	Class changes	1400	580	3720	1180	0	0	0
	Image difference	−560	2090	−1960	410	20	0	0

difference in the total number of equivalently classed pixels in the two images, computed by subtracting the Initial State Class Totals from the Final State Class Totals. An Image Difference that is positive indicates that the class size increased and a negative indicates decrease in class size. Agricultural land (+2090 ha), water body (+20 ha), and mixed forest (+410 ha) have shown an increase in spatial neighborhood at the expense of scrub land (−1960 ha) and grazing land (−560 ha). Figure 4.6 shows the spatial map of the land cover for the study area in 1973 and 2008, a 35-year range.

A more extended and time series analysis of the change statistics is shown for all the years considered from 1973 to 2008. The areal expansion in hectare is tabulated (Table 4.4) for detailed change detection statistical analysis for years between final state image (2008) and initial state image (1973, 1984, 1995, and 2003).

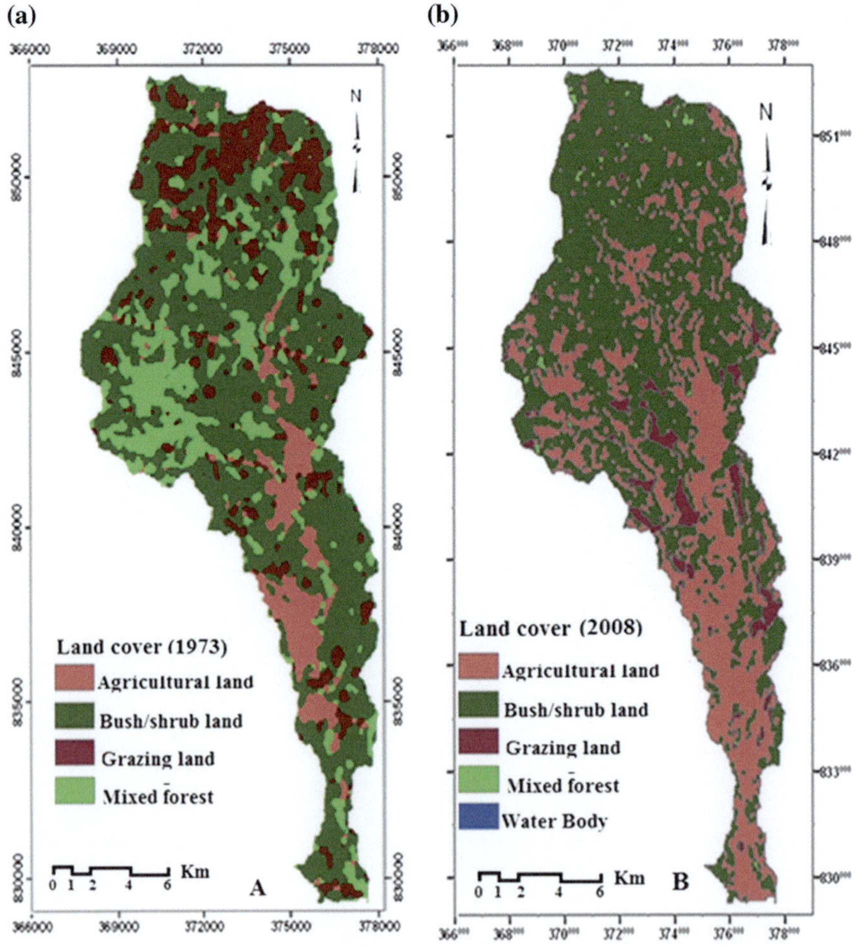

Fig. 4.6 Comparison of landsat land use/land cover map class and areal coverage for the years: **a** 1973 and **b** 2008

Table 4.4 Change detection (ha) statistics between final state image (2008) and initial state image (1973, 1984, 1995, and 2003) area (hectare)

Final state image (2008)									
	Year	Class	Grazing land	Agricultural land	Bush/shrub land	Mixed forest	Water body	Row total	Class total
Initial state images for the years (1973, 1984, 1995, and 2003)	*Final state image (2008)*								
	1973	Grazing land	73	45	373	10	0	501	501
		Agricultural land	446	2308	2432	657	0	5843	5843
		Bush/shrub land	1048	246	2648	1289	0	5231	5231
		Mixed forest	27	0	40	22	0	89	89
		Water body	0	0	1	0	0	1	1
		Class total	1595	1110	7111	1977	0	0	0
		Class changes	1522	291	2845	1955	0	0	0
		Image difference	−1093	3244	−263	−1889	1	0	0
	Final state image (2008)								
	1984	Grazing land	398.43	65.25	19.17	6.03	7.11	495.99	495.99
		Agricultural land	348.21	3230.31	1955.07	302.49	7.02	5843.1	5843.1
		Bush/shrub land	273.60	1387.17	1504.51	2060.19	6.03	5231.5	5231.5
		Mixed forest	2.16	3.33	38.34	46.17	0.00	90.00	90.00
		Water body	0.45	0.00	0.00	0.00	1.17	1.62	1.62
		Class total	1022.85	3032.75	5173.74	2414.88	21.33	0.00	0.00
		Class changes	624.42	1455.75	2012.58	2368.71	20.16	0.00	0.00
		Image difference	−526.86	1157.04	1714.41	−2324.88	−19.71	0.00	0.00

(continued)

Table 4.4 (continued)

Final state image (2008)

Year	Class	Grazing land	Agricultural land	Bush/shrub land	Mixed forest	Water body	Row total	Class total
Final state image (2008)								
1995	Grazing land	392	15	66	24	0	496	496
	Agricultural land	403	2446	902.6	431	7	4189.6	4189.6
	Bush/shrub land	371	2156	2449	1909	3	6888	6888
	Mixed forest	3	14	27	45	0	90	90
	Water body	0	0	0	0	1	2	2
	Class total	1169	4632	3444.6	2409	11	0	0
	Class changes	777	2185	995.6	2364	10	0	0
	Image difference	−673	−442.4	3443.4	−2319	−9	0	0
Final state image (2008)								
2003	Grazing land	437.64	41.99	18.52	5.85	0.00	504.00	504.00
	Agricultural land	491.57	3789.26	855.87	706.40	0.00	5843.1	5843.1
	Bush/shrub land	287.94	1033.55	2655.65	1254.36	0.00	5231.5	5231.5
	Mixed forest	2.03	26.72	17.22	42.72	0.00	88.70	88.70
	Water body	1.46	0.00	0.00	0.00	0.00	1.46	1.46
	Class total	1220.65	4888.36	3547.26	2009.30	0.00	0.00	0.00
	Class changes	783.01	1102.26	891.61	1966.61	0.00	0.00	0.00
	Image difference	−716.65	951.58	1684.24	−1920.64	1.46	0.00	0.00

As shown in Table 4.4, for the 1973–2008 period, the land cover changes were grazing land (−1093 ha), scrub land (−263 ha), and mixed forest (−1889 ha) to agricultural land (+3244 ha) and water Body (+1 ha). The areal extent of the agricultural land (2008) has increased to 3244 ha at the expense of 1093 ha of grazing land, 262 ha from scrub land, 1889 ha from mixed forest. Water body (2008) has increased to 1 hectare at the expense of the Scrub land.

4.5 Conclusions and Recommendations

4.5.1 Conclusions

As a key technical challenge, it is reasonable to expect the variation in LULC change in the river catchment of Batena in time due to which more robust projections can be made to alleviate the probable impacts of land use change on local climate and agricultural response pattern. To provide a road map of how to address the impact of this change, this chapter has addressed a number of specific research questions.

This study indicated that, Batena land cover from 1973 to 2008 showed rapid change with high growth in agricultural areas at the expense of the sparse and densely vegetated lands while grazing land has reduced marginally and water body is showing almost stagnant over time. Generally, the decadal analysis of LULC change shows a general decreasing trend for the mixed forest and grazing land year after year and scrub lands have shown a changeable trends throughout the analysis periods.

The magnitude of land use change varies considerably over space and time, reflecting the differences of landscape characteristics between locations in the catchment. To provide relevant conclusions and recommendations for policy makers, it is important to properly capture the major driving force to changes in LULC. Over the period of this analysis, it is shown that LULC change is the result of immediate and often radical human activities, natural effects over longer period of time, and complex interactions between several biophysical and socioeconomic conditions. Understanding the types and impacts of LULC change is an essential indicator for resource base analysis and development of effective and appropriate response strategies for sustainable management of natural resources in study area.

4.5.2 Recommendation

In this study, mapping of different types of land cover and change detection is carried out using digital image processing techniques. The spatial pattern and change detection in LULC could serve as guiding tool, in biodiversity conservation and environmental development.

The satellite imagery used for classifying land cover types and detecting land cover conditions should undergo atmospheric corrections by preprocessing satellite sensor imagery since the electromagnetic radiation signals received by the satellite sensors can be scattered and absorbed by the atmospheric gases and aerosols. The land use/cover type classification should be based from the conversion of Landsat sensor imagery into TOA reflectance.

There are some problems in getting highly accurate land use data, particularly in the mountainous landscapes of tropical climates. One of the main problems when generating land use maps from digital images is the confusion of spectral responses from different features. Discrimination of land cover types, including vegetation types, through the use of remote sensing techniques in the mountainous areas of Batena is a very difficult task because of the complex structure and composition of vegetation communities added with low spatial resolution of Landsat imagery. In addition, the mountain topography leads to a significant shadowing effect, which becomes a particular problem in the digital image processing. It is recommended to use one of the techniques to achieve improvement in digital classification by incorporation of ancillary data, such as a digital elevation model (DEM), geomorphometric variables (relief, convexity, slope, aspects, and incidence). DEM integration in image classification will increase the classification accuracy of digital data by describing the distribution of terrain components which contribute to spectral response, identify sites for fieldwork, and geographically stratify training areas or homogeneous regions.

Land use induced changes in the surface energy budget can affect climate across all scales: local, regional, and global. The combined impacts of land use and climate change are likely to dramatically affect natural resources and ecosystems. In view of such uncertainties, the coupling impact of land use and climate change should be seen through a simple feedback mechanism signifying the effect of their interaction to the rainfall–runoff generation and to characterize the impact of this change on small catchment hydrology.

Environmental problems have no boundaries and are interrelated. This chapter recommends further works to quantify the local vegetation cover change and its cumulative impact on regional and global climate changes.

Acknowledgments The authors acknowledge the International Water Management Institute (IWMI) "Nile Basin Development Challenge of the Consultative Group on International Agricultural Research program for water and food (NBDC-CGIAR-CPWF)", Horn of Africa Regional Environment Center and Network, Demand Driven Action Research Program (HoA-REC/N-DDAR), and the Ethiopian Ministry of Education for their financial support to conduct this research work.

References

Bauer M, Yuan F, Saway K (2003) Multi-temporal landsat image classification and change analysis of land cover in the twin cities (Minnesota) metropolitan area. Workshop on the analysis of multi-temporal remote sensing images, Italy

Bernstein R (1983) Image geometry and rectification. Chapter 21 in manual of remote sensing. In: Colwell RN (eds) Falls church. American Society of Photogrammetry, Virginia

Carlson T, Azofeifa S (1999) Satellite remote sensing of land use changes in and around San Jose, Costa Rica. Remote Sens Environ 70:247–256

Clevers JB (2004) Land cover classification with the medium resolution imaging spectrometer (MERIS). In: EARSeLeProceedings 3

Congalton R (1996) A review of assessing the accuracy of classification of remotely sensed data. Remote Sens Environ 37:35–46

Crippen R (1989a) A simple spatial filtering routine for the cosmetic removal of scan-line noise from landsat tm p-tape imagery. Photogram Eng Remote Sens 55(3):327–331

Currit N (2005) Development of a remotely sensed, historical land-cover change data base forrural Chihuahua, Mexico. Int J Appl Earth Obs Geoinf 7:232–247

Degelo S (2007) Analysis of biomass degradation as an indicator of environmental challenge of Bilate watershed

Dimyati (1995) An analysis of landuse/landcover change using the combination of MSS landsat and land use map—case of Yogyakarta, Indonesia. Int J Remote Sens 17(5):913–944

Fashona M, Omojola A (2005) Climate change, human security and communal clashes in Nigeria. In: Int'l workshop on human security and climate change. Oslo, Norway

Getachew HE, Melesse AM (2012) Impact of land use /land cover change on the hydrology of Angereb Watershed, Ethiopia. Int J Water Sci 1(4):1–7. doi:10.5772/56266

Goldewijk K, Ramankutty NK (2004) Land cover change over the last three centuries due to human activities: the availability of new global data sets. Geojournal 61:335–344. Kluwer Academic Publishers, The Netherlands

Guerschman J, Paruelo J, Bela C, Giallorenzi M, Pacin F (2003) Land cover classification in the Argentine Pampas using multi-temporal Landsat TM data. Int J Remote Sens 24:3381–3402

Heinen JT, Lyon JG (1989) The effects of changing weighting factors on the calculation of wildlife habitat index values: a sensitivity analysis. Photogram Eng Remote Sens 55(10):1445–1447

Jensen JR (1996) Introductory digital processing: a remote sensing perspective, 2nd edn. Prentice-Hall, Upper Saddle River

Kassa T (2009) Watershed hydrological responses to changes in land use and land cover, and management practices at Hare Watershed, Ethiopia

Lambin EF, Geist HJ, Lepers E (2003) Dynamics of land-use and land-cover change in tropical regions. Ann Rev Environ Res 28:205–241

Loppiso S (2010) Assessment of land use land cover dynamics and its impact on soil loss: using GIS and remote sensing, in Shashogo Woreda, Southern Ethiopia

Lu D, Mausel P, Brondízio E, Moran E (2004) Change detection techniques. Int J Remote Sens 25:2365–2401

Lupo F, Reginster I, Lambin EF (2001) Monitoring land-cover changes in West Africa with SPOT vegetation: impact of natural disasters in 1998–1999. Int J Remote Sens 22:2633–2639

Mango L, Melesse AM, McClain ME, Gann D, Setegn SG (2011a) Land use and climate change impacts on the hydrology of the upper Mara River Basin, Kenya: results of a modeling study to support better resource management, special issue: climate, weather and hydrology of East African highlands. Hydrol Earth Syst Sci 15:2245–2258. doi:10.5194/hess-15-2245-2011

Mango L, Melesse AM, McClain ME, Gann D, Setegn SG (2011b) Hydro-meteorology and water budget of Mara River basin, Kenya: a land use change scenarios analysis, In: Melesse A (ed) Nile River basin: hydrology, climate and water use. Springer Science Publisher, Chapter 2, pp 39–68. doi:10.1007/978-94-007-0689-7_2

Melesse AM, Jordan JD (2003) Spatially distributed watershed mapping and modeling: land cover and microclimate mapping using landsat imagery part 1. J Spat Hydrol (e-journal) 3(2):1–29

Mendoza M, Bocco G, Bravo B (2002) Spatial prediction in hydrology: status and implication in the estimation of hydrological processes for applied research. Prog Phys Geogr 26(3):319–338

Mohammed H, Alamirew A, Assen M, Melesse AM (2013) Spatiotemporal mapping of land cover in Lake Hardibo Drainage Basin, Northeast Ethiopia: 1957–2007. Water conservation: practices, challenges and future implications. Nova Publishers, New York, pp 147–164

Moshen A (1999) Environmental landuse change detection and assessment using multi-temporal satelitte imageries. Zanjan University

Muzein B (2008) Remote sensing and gis for landcover/landuse change detection and analysis in the semi-natural ecosystem and agriculture landscapes of the Central Ethiopian Rift Valley. Fakultät Forst- Geo-und Hydrowissenschaften Institut Fernerkundung

Negash W (2014) Catchment dynamics and its impact on runoff generation: coupling watershed modelling and statistical analysis to detect catchment responses. Int J stat Anal Detect Catchment Responses 6:73–87

Ozbakir B, Bayram B, Acar U, Uzar M, Baz I, Karaz I (2007) Synergy between shoreline change detection and social profile of waterfront zones: a case study in Istanbul. In: Conference paper at the international conference for photogrammetry and remote sensing, Istanbul, Turkey 16–18 May

Petit C, Scudder T, Lambin E (2001) Quantifying processes of land-cover change by remote sensing: resettlement and rapid land-cover changes in southeastern Zambia. Int J Remote Sens 22:3435–3456

Prakasam C (2010) Land use and land cover change detection through remote sensing approach: a case study of Kodaikanal taluk, Tamilnadu. Int J Geomatics Geosci 1(2):189–206

Richards J (1999) Remote sensing digital image analysis: an introduction. Springer, Berlin, p 240

Rogana J, Chen D (2004) Remote sensing technology for mapping and monitoring land-cover and landuse change. Prog Plann 61:301–325

Sexton JO, Urban DL, Donohue MJ, Song C (2013) Long-term land cover dynamics by multi-temporal classification across the landsat-5 record. J Remote Sens Environ 128:246–258

Singh A (1989) Digital change detection techniques using remotely-sensed data. Int J Remote Sens 10(6):989–1003

Swinne E, Veroustaete F (2008) Extending the SPOT-VEGETATION NDV timeseries (1998–2006) back in time with NOAA-AVHRR data (1985–1998) for southern Africa. IEEE Trans Geosci Remote Sens 46(2):558–572

Tiwari MK, Saxena A (2011) Change detection of land use/landcover pattern in an around Mandideep and Obedullaganj area, using remote sensing and GIS. Int J Technol Eng Syst 2 (3):398–402

Verburg P, Schot P, Dijst M, Veldkamp (2004) Lan duse change modelling: current practices and research priorities. Geo Journal 51(4):309–324

Verburg P, Soepboer W, Veldkamp A, Limpiada R, Espaldon V, Mastura S (2002) Modelling the spatial dynamics of land use: the CLUE-S model. Environ Manage 30(3):391–405

Verbyla (1986) Potential prediction bias in regression and discriminate analysis. Can J Forest Res 16:1255–1257

Wondie M, Scrhneider W, Melesse AM, Teketay D (2011) Spatial and temporal land cover changes in the Simen mountains national park, a world heritage site in Northwestern Ethiopia. Remote Sens 3:752–766. doi:10.3390/rs3040752

Wondie M, Schneider W, Melesse AM, Teketay D (2012) Relationship among environmental variables and land cover in the Simen Mountains national park, a world heritage site in Northern Ethiopia. Int J Remote Sens Appl (IJRSA) 2(2):36–43

Yu W, Gu S, Zhao XQ, Xiao J, Tang Y, Fang J, Jiang S (2011) High positive correlation between soil temperature and NDVI from 1982 to 2006 in alpine meadow of the three river sources region of Qinghai-Tibetan plateau. Int J Appl Earth Obs Geoinf 13(4):528–535

Zsuzsanna D, Bartholy J, Pongracz R, Barcza Z (2005) Analysis of landuse/land-cover change in the Carpathian region based on remote sensing techniques. Phys Chem Earth 30:109–115

Chapter 5
Analyses of Land Use/Land Cover Change Dynamics in the Upland Watersheds of Upper Blue Nile Basin

Rahel S. Asres, Seifu A. Tilahun, Gebiaw T. Ayele and Assefa M. Melesse

Abstract Investigating land use/land cover (LULC) change is important for effective sustainable land resource management and for understanding the changes in hydrological processes. In this chapter, we investigated LULC dynamics in three experimental watersheds: Mizewa (27 km^2), Debre Mawi (5.23 km^2) and Enchilala (4.14 km^2) of upper Blue Nile Basin, Ethiopia. These watersheds are experimental watersheds for investigating runoff processes, erosion, and soil conservation practices. The LULC changes were measured through interpretation of Landsat images of 1973, 1986, 2000, and 2013 supported by repeated field visits. Based on the image analysis and field survey, cultivated, forest, shrub/bush land, and grazing land are the major LULC classes during the study periods. The result showed an increase in cultivated land mainly at the expense of shrub/bush, forest, and grazing land for all three watersheds. The cultivated land increased from 30 to 68 % in Mizewa, from 67 to 80 % in Debre Mawi, and from 55 to 86 % in Enchilala within the past 40 years. This is likely associated with the population that has been steadily increasing at a growth rate of 2–3 % per year during the past five decades. However,

R.S. Asres (✉)
Department of Water Resources and HydraulicEngineering, Gondar University, 196, Gondar, Ethiopia
e-mail: rahelseifu3@yahoo.com

S.A. Tilahun
Faculty of Civil and Water Resource Engineering, Bahir Dar Institute of Technology, Bahir Dar University, 26, Bahir Dar, Ethiopia
e-mail: satadm86@gmail.com

G.T. Ayele
Faculty of Civil and Water Resource Engineering, Bahir Dar Institute of Technology, Bahir Dar University, 252, Bahir Dar, Ethiopia
e-mail: gebeyaw21@gmail.com

G.T. Ayele
Department of Hydraulic and Water Resources Engineering, Blue Nile Water Institute, Bahir Dar University, 252, Bahir Dar, Ethiopia

A.M. Melesse
Department of Earth and Environment, Florida International University, Miami, USA
e-mail: melessea@fiu.edu

© Springer International Publishing Switzerland 2016
A.M. Melesse and W. Abtew (eds.), *Landscape Dynamics, Soils and Hydrological Processes in Varied Climates*, Springer Geography,
DOI 10.1007/978-3-319-18787-7_5

the rate of change of cultivated land of Mizewa (27 ha/year) is greater than Debre Mawi (1.5 ha/year) and Enchilala (3.2 ha/year) watersheds. The higher rate of change is the same for the other land uses. This is likely because the rate of expansion of cultivated land into steep slope, degraded, and marginal lands was found to be much more in Mizewa, which is located on the main road from Bahir Dar to Debre Tabor. The expansion of cultivated land and the decrease of vegetative covers through forest will cause a decrease in evapotranspiration and increase in overland and subsurface flow and will increase soil erosion. This leads to growth of numerous gullies as observed in Debre Mawi and Enchilala. Therefore, the current trends in LULC must be improved, toward the resource management and conservation of the existing vegetation and other natural resources in all the three watersheds. These should be done in collaboration with all stakeholders including local communities, government, and NGOs for effective management of natural resources.

Keywords Upper Blue Nile · Land use/land cover · Upland · Cultivated land · Remote sensing

5.1 Introduction

Land use change is any physical or biological change attributable to management including conversion of grazing to cropping, pollution and land degradation, vegetation removal, and conversion to nonagricultural uses (Quentin 2006). Though the extent and rates of change in land cover and land uses are known with some certainty (Turuner and Meyer 1994), the scientific community has now come to recognize diverse roles of land use and land cover (LULC) change (Geist and Lambin 2002). According to Turner (1993), most of the earth's surface is already modified, except those areas that are peripheral in location or are fairly inaccessible. One of the most significant global challenges in this century relates to management of the transformation of the earth's surface occurring through changes in LULC (Mustared 2004).

Land cover classification has recently been a hot research topic for a variety of applications (Liang 2002). A great deal of research has to understand major shifts in LULC and relate them to changing environmental conditions. During the next decades, land use dynamics will play a major role in driving the changes in the global environment (Szejwach and Baulies 1998). Generally, agriculture is found to be the major driver of land cover change in tropical regions. Over the past 50 years in East Africa, there has been expansion of agriculture at the expense of grazing land (Olson and Matima 2006; Yitaferu 2007). Before 1950, semiarid and subhumid areas were predominantly pastoral with scattered settlement and cultivation but from then onwards, there has been a significant transformation of grazing land to mixed crop-livestock agriculture. Understanding the mechanisms leading to LULC changes in the past is crucial to understand the current changes and predict future alterations. These changes occurred at different time periods, places, and degrees of magnitude and with diverse biophysical implications (Szejwach and Baulies 1998). Therefore, land use and land cover change (LULCC) research needs to deal with the identification, qualitative description, and parameterization of factors which drive changes in LULC, as well as the integration of their consequences and feedbacks (Szejwach and Baulies 1998). However, one of the major challenges in LULCC analysis is to link the behavior of people to biophysical information in the appropriate spatial and temporal scales (Codjoe 2007).

LULC dynamics is an important landscape process capable of altering the fluxes of water, sediment, contaminants, and energy. Mainly caused by humans, impact of land use on water resources availability is high. Degraded watersheds tend to accelerate overland flow reducing soil moisture and baseflow recharge and increasing sediment detachment and transport. Various studies used land cover mapping tools and methods to understand land use changes, inventory of forest and natural resources, as well as understand the changes in the hydrologic behavior of watersheds (Getachew and Melesse 2012; Mango et al. 2011a, b; Wondie et al. 2011, 2012; Melesse and Jordan 2003; Mohamed et al. 2013).

The hydrology of the Nile River basin has been studied by various researchers. These studies encompass various areas including stream flow modeling, sediment dynamics, teleconnections and river flow, land use dynamics, climate change impact, groundwater flow modeling, hydrodynamics of Lake Tana, water allocation, and demand analysis (Melesse et al. 2009a, b, 2011, 2014; Abtew et al. 2009a, b; Abtew and Melesse 2014a, b, c; Yitayew and Melesse 2011; Chebud and Melesse 2009a, b, 2013; Setegn et al. 2009a, b, 2010; Melesse 2011; Dessu and Melesse 2012, 2013; Dessu et al. 2014).

Land use and land cover mapping is one of the most important and typical applications of remote sensing data (Chrysoulakis 2004). Remotely sensed data are useful to study human and environment interactions, especially LULC changes. Few studies have been conducted to understand LULC change and other related issues in the proposed study areas. Yitaferu (2007) has done satellite image analysis of the Lake Tana basin between 1985/86 and 2001/03 and found that croplands increased by about 4.2 %, which largely occurred at the expense of grasslands and shrublands. Furthermore, forest cover in the basin was found to have increased by about 0.23 % in the same time frame.

In many parts of the highlands of Ethiopia, agriculture has gradually expanded from gently sloping land into the steeper slopes of the neighboring mountains. According to many literatures, population that has been steadily increasing at a growth rate of 2.67 % per year during the past five decades is the major cause for this expansion. In some areas, expansions of cultivation, commonly into steeper slopes and marginal areas, may have been done without appropriate soil and water conservation measures. Despite this increase, the agricultural productivity is lagging behind the population growth rate.

The impact of population growth on the environment and poverty is not simple and one directional (Solomon 1994). For instance, in the past four decades, areas in the Blue Nile basin have undergone dramatic LULC changes, with the result that almost all land units have been converted into cultivated land (Hurni and Zeleke 2001). The results from Hurni and Zeleke (2001) in Dembecha area, for example, show that the natural forest cover declined from 27 % in 1957 to 2 % in 1982 and 0.3 % in 1995, whereas cultivated land increased from 39 % in 1957 to 70 % in 1982 and 77 % in 1995.

Previous studies indicate depleted forest cover at the national level, major land cover changes at the local level, loss of biodiversity, soil and environmental degradation in the country. LULC change also modifies the hydrological cycle of a watershed area by altering both the balance between rainfall and evaporation and the runoff response of the area, and subsequently affects water resources. Hence, vegetation removal results in an increase in surface runoff and a decrease in evapotranspiration that may also in turn lead to lower rainfall in semiarid areas.

By the same reasoning, a land cover change study by Bewket (2002) in the Chemoga watershed reported the sudden appearance of a pond in 1998, using satellite image analysis, presumably created by the increased water yield due to a decrease in vegetation high up in the area. In the above two studies at Umbulo and Chemoga watersheds, the clearing of forests was assumed to bring high surface runoff and less evaporation that leads to higher water yield induced by land cover change and later creates a temporary water reservoir/pond. The appearance of temporary water ponds in the above two watershed studies was correlated with the decline of vegetation cover, clearly indicating the land cover change effect on the hydrological flow of the watershed.

Similarly, studies have reported the drying of Lake Cheleleka in Southern Ethiopia as a result of long-term land use changes and subsequent sediment deposition in the lake (Ayenew 2004; Dessie and Kleman 2007; Gebreegziabher 2005). This disappearance of the lake directly correlates with the deforestation record of the Awassa watershed from 16 % forest cover in 1972 to 2.8 % in 2000 (Dessie and Kleman 2007). As a result, 44 million m^3 of sediment load was deposited in the lake within 35 years.

A similar case has been found in the eastern Ethiopian highlands where Lake Haramaya reportedly disappeared, due to water abstraction, siltation, and clearing of land for farming on its watershed. The expansion of farming around the lake catchment resulted in increased siltation of the lake that decreased the lake's volume and surface area, which in turn increased the rate of evaporation. Since land cover determines the rate of soil loss, the removal of vegetation by conversion of land to cultivation reduces the protection of soil cover, minimizes the regrowth capacity of vegetation, and speeds up sheet and gully erosion (Zerihun and Mesfin 1990).

Based on the estimates of the severity and extent of erosion in the mid-1980s, FAO (1986) concluded that about half (about 27 million ha) of the highland land area of Ethiopia was "significantly eroded" and over one-fourth (14 million ha) was "seriously eroded." It also concluded that over 2 million ha of farm lands had reached the "point of no return" in the sense that they were unlikely to sustain economic crop production in the future.

Land cover change, especially deforestation, not only facilitates the physical removal of soil but also accelerates the deterioration of the basic soil properties (Gebresamuel et al. 2010). The formation and advancement of gully erosion are common effects of soil erosion. Gully development in the Umbulo catchment was extended from upslope to the middle and lower slopes at the same pace as the rate of forest reduction from the catchment, indicating the influence of land cover change for the formation of soil erosion, since vegetation was providing soil protection (Moges and Holden 2009).

LULC change has also implications for Nile River flow and for downstream countries, Sudan and Egypt, because of the water and sediment carried by the Blue Nile from Ethiopia. Due to continuous cultivation in the past centuries, most agricultural land in the Ethiopian highlands is degraded and has become shallower, resulting in more overland flow and less base flow (Hurni and Zeleke 2001). This might result in a greater amount of surface runoff flowing to downstream countries of Sudan and Egypt. However, retaining rain water has become increasingly difficult due to loss of land cover which will reduce the volume of reservoirs by sedimentation. It is widely accepted that sustainable utilization and conservation of land and water resources can be achieved through sustainable development and management of the river basin. An understanding of the variability in time and space of land cover and land use provides favorable foundations for effective management of land and other natural resources. Except for a few studies (Hurni and Zeleke 2001; Solomon 1994) around Chemoga watershed, a limited number of studies on LULCC were documented at the upper Blue Nile River basin.

In this chapter, the spatial statistics of land use/land cover change in the upland watersheds of Mizewa, Debre Mawi and Enchilala, Upper Blue Nile basin is identified over the period of four decades. This chapter also entails the importance of analyzing the LULCC of the area for spatial planning, management, environmental protection, and economic production. Furthermore, documentation of the LULCC provides information for better understanding of historical land use practices, current land use patterns, and future land use trajectories. The main objective of the study reported in the chapter is to quantify and analyze major land use/land cover changes in the three watersheds in the past 40 years, and also present its implication on land and water resources degradation and hydrologic alterations to downstream areas.

5.2 Study Areas and Methodology

5.2.1 Study Area Description

This section describes three experimental watersheds in the upper Blue Nile (Fig. 5.1), Lake Tana Subbasin. One of the watersheds is Mizewa, which is found at about 65 km from Bahir Dar town with 27 km^2 area coverage. Mizewa has varying

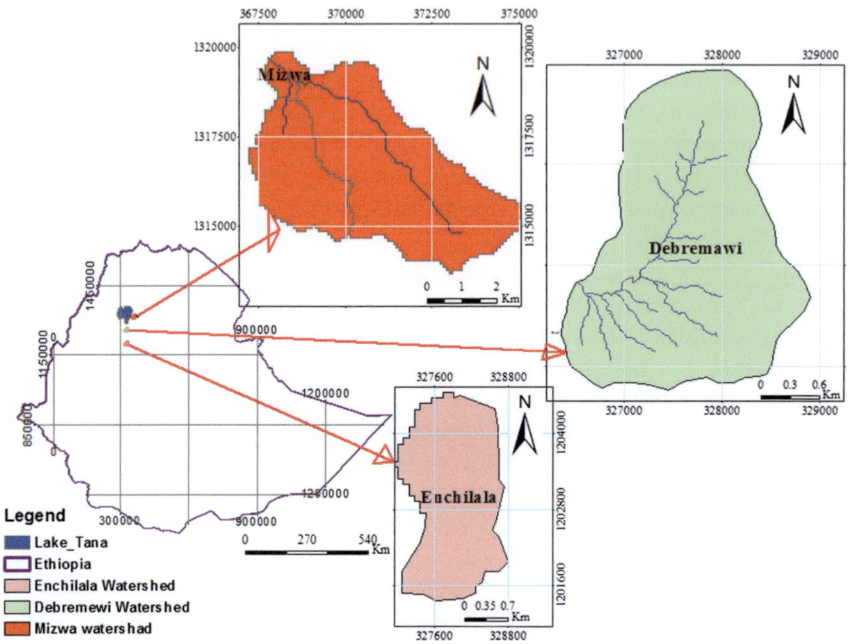

Fig. 5.1 Location map of the three experimental watersheds

elevation ranges from 1774 to 2410 m above mean sea level. The second experimental watershed is Debre Mawi with 523 ha areal coverage and 30 km south of Bahir Dar. The third experimental watershed is Enchilala having 414 ha total area and found 220 km south of Bahir Dar.

5.2.2 *Land Cover Classification*

There are two general approaches for image classification: Supervised and Unsupervised. In the case of supervised classification, the classifier delineates specific land cover types based on training sites but in the case of unsupervised classification no training sites are needed and land cover classes are generated based on the number of classes requested. Effective land use classification by use of remote sensing image depends on separating land cover type of interest into sets of different signatures that represent the data in a form suited to the particular algorithm used (Rechardes and Kelly 1984).

In this study, the selected satellite images are classified using supervised classification. The final land use/land cover map and legend were built using interpretations of digital images, field data, and the general observation knowledge.

5.2.3 *Change Detection*

The method used for land use land/cover change detection in this chapter is the comparison statistics. Using the post-image classification procedure, the percentage area for each land cover classes was derived from the classified image for each year (1973, 1986, 2000, and 2013) separately, using ERDAS imagine, an image processing software. Finally, the areal coverage for each land cover over time is compared.

5.3 Results and Discussion

5.3.1 *Accuracy Assessment*

To assess the classification accuracy of Mizewa watershed, 50 ground truth control GPS points were used and the overall accuracy of 76 % and Kappa statistics of 67 % were achieved.

For Debre Mawi, with 38 ground control points for land cover classification validation, an overall accuracy and kappa coefficient of 74 and 65 % were recorded, respectively. The same is true for Enchilala watershed with 35 ground control GPS points and overall accuracy of 71 % and the Kappa statistics of 62 % were found. Statistics Kappa values can be categorized as <40 % poor, 40–55 % fair, 55–70 % good, 70–85 % very good, and greater than 85 % as excellent (Monserved 1990). According to this classification scale, the Kappa statistical value for all watersheds lies in a fair category.

5.3.2 *Land Use and Land Cover Map*

Band combinations of red, blue, and green are used to display the raw stacked images in standard color composite. The spectral band combination for displaying images often varies (Trotter 1998). False color composition (FCC) for bands 2, 3, and 4 is used for the analysis.

The study identified similar four land use cover classes for the three experimental watersheds; Mizewa, Debre Mawi, and Enchilala. Agriculture has expanded at the expense of forests, savannas, and steppes in all parts of the world to meet the demand for food and fiber (Lambin et al. 2003). The same is true in the three experimental watersheds in which the agriculture area goes on increasing for all the analysis periods. The four identified LULC classes are cultivated, forest, shrub/bush, and grazing lands. Figures 5.2, 5.3 and 5.4 show the land cover maps of the three study sites.

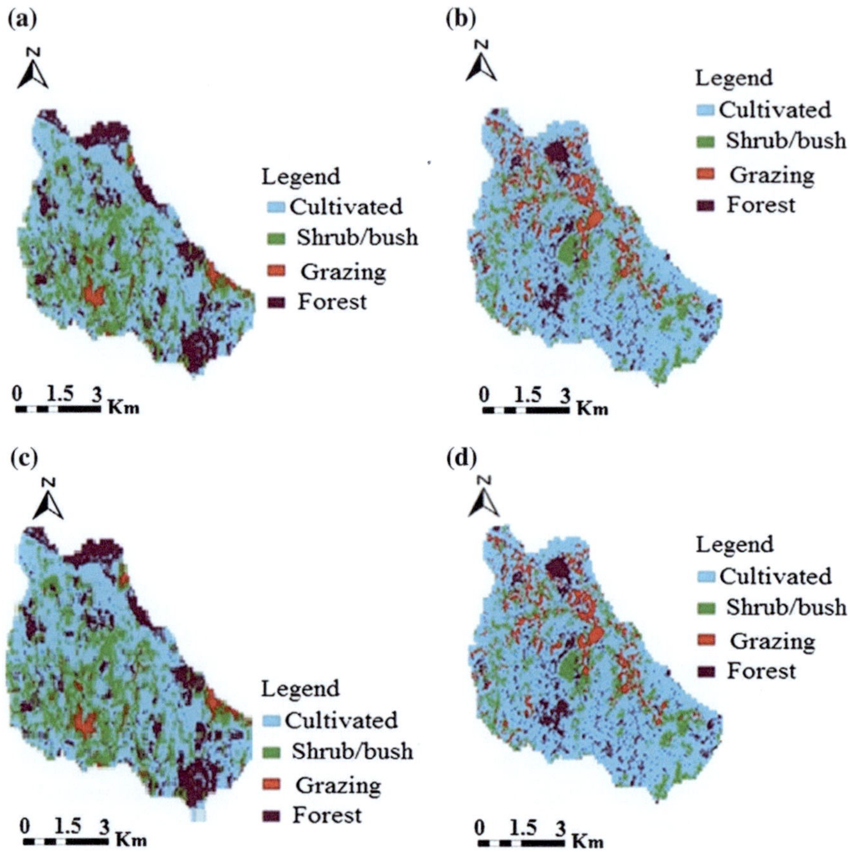

Fig. 5.2 Land cover map of **a** 1973, **b** 1986, **c** 2000, and **d** 2013 of Mizewa watershed

5.3.3 Land Use and Land Cover Dynamics

The magnitude and rate of land cover change are displayed in Figs. 5.5, 5.6, 5.7, 5.8, 5.9 and 5.10 for all experimental watersheds. The agricultural area is expanding at the expense of grazing lands (Figs. 5.5, 5.6, 5.7, 5.8, 5.9 and 5.10).

In 1973, the greatest share of LULC from all classes is shrub/bushland, which covers an area of 1005 ha and contributing 37 % of the total area (Fig. 5.5). Forest and grazing lands cover an area of 781 (29 %) and 163 ha (6 %), respectively, whereas the areal coverage of the cultivated land is 751 ha (28 %) from the total area of the Mizewa watershed. This result shows that 72 % of the total area of the watershed was covered by shrub, forest, and grazing land in 1973 and the remaining 28 % was covered by cultivated land, which indicates that much of the area was

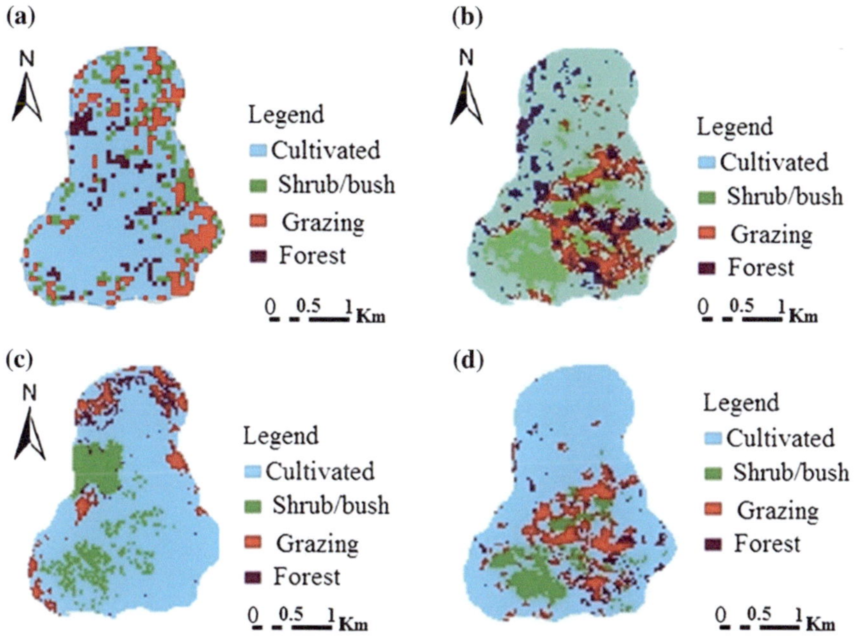

Fig. 5.3 Land cover map of **a** 1973, **b** 1986, **c** 2000, and **d** 2013 of Debre Mawi watershed

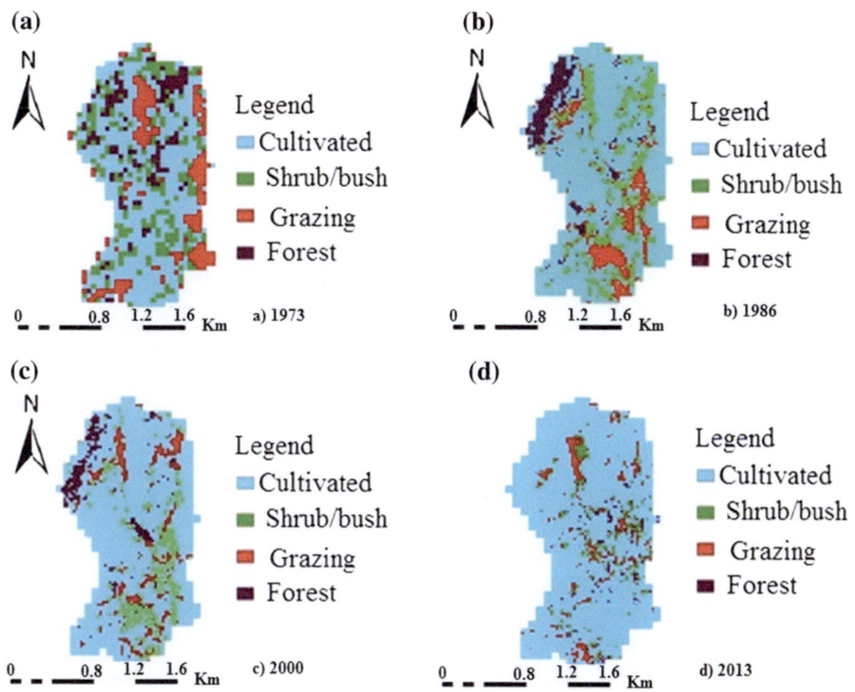

Fig. 5.4 Land cover map of **a** 1973, **b** 1986, **c** 2000, and **d** 2013 of Enchilala watershed

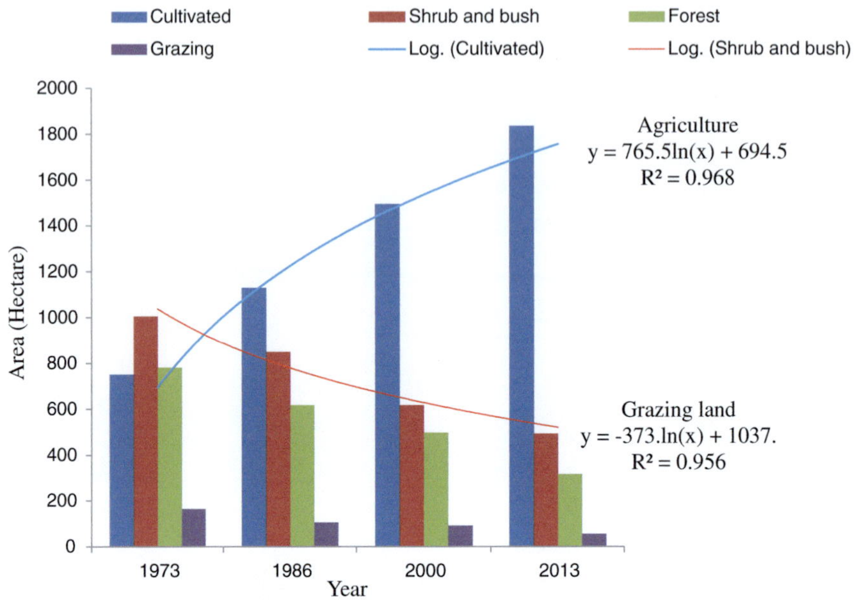

Fig. 5.5 Land use/land cover types and area coverage of Mizewa watershed for the years: 1973, 1986, 2000, and 2013

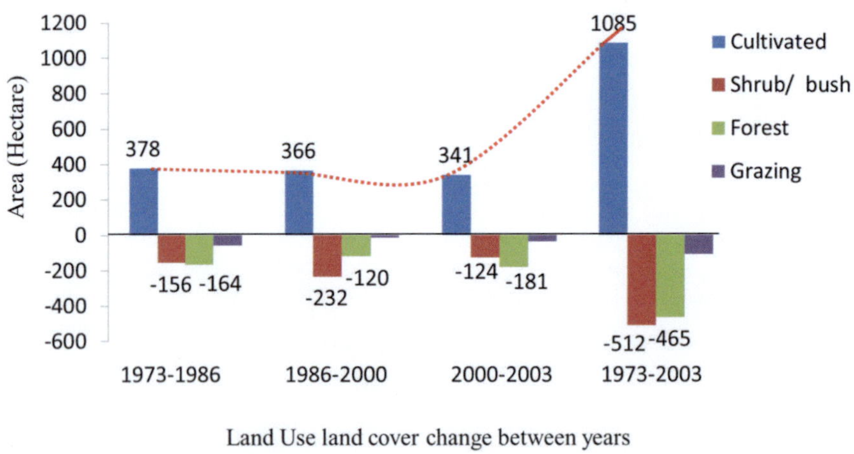

Fig. 5.6 Land use/land cover change statistics in the Mizewa watershed; 1973–1986, 1986–2000, 2000–2013, and 1973–2013

covered by green vegetation in 1973. But the change dynamics is different for the rest of the years as depicted in Fig. 5.5. Figure 5.6 shows the land cover areal extent change for the years selected for Mizewa watershed.

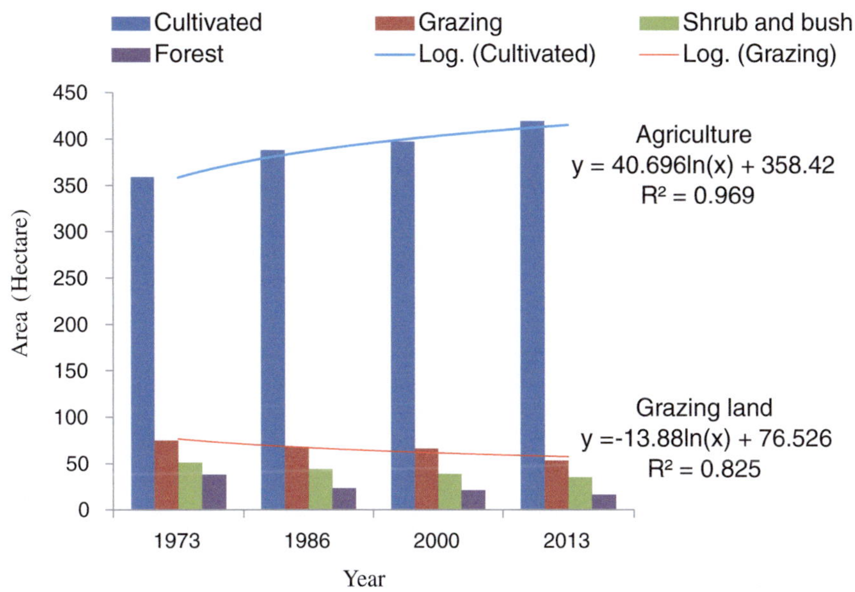

Fig. 5.7 Land use/land cover types and area coverage of Debre Mawi watershed for the years 1973, 1986, 2000, and 2013

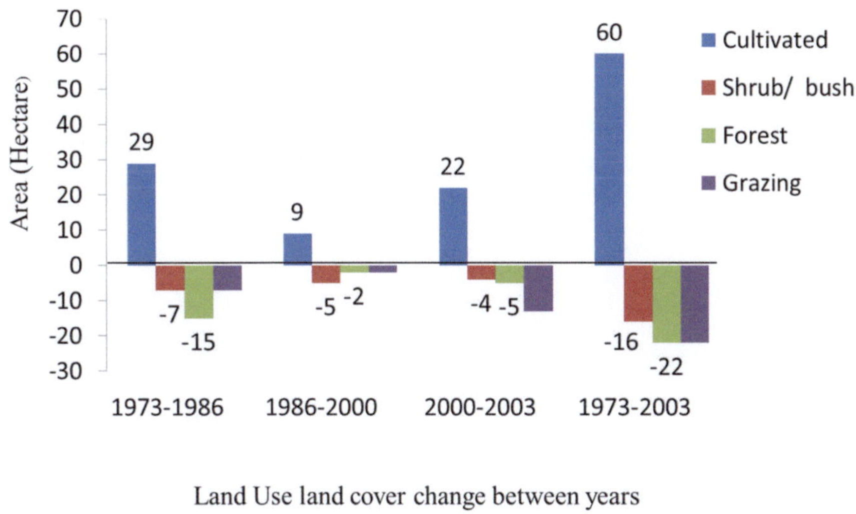

Fig. 5.8 Land use/land cover change statistics in the Debre Mawi watershed; 1973–1986, 1986–2000, 2000–2013, and 1973–2013

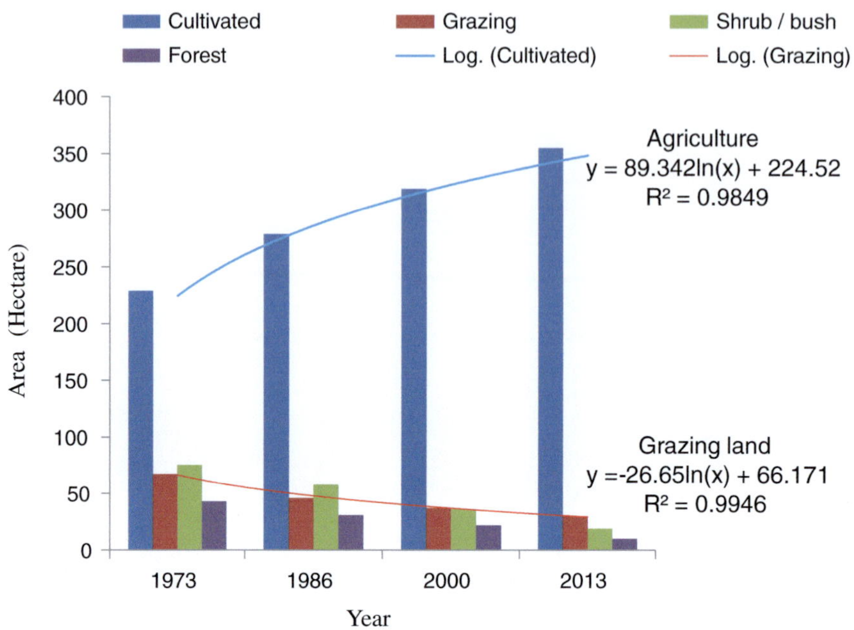

Fig. 5.9 Land use/land cover types and area coverage of Enchilala watershed in four different periods (1973, 1986, 2000, and 2013)

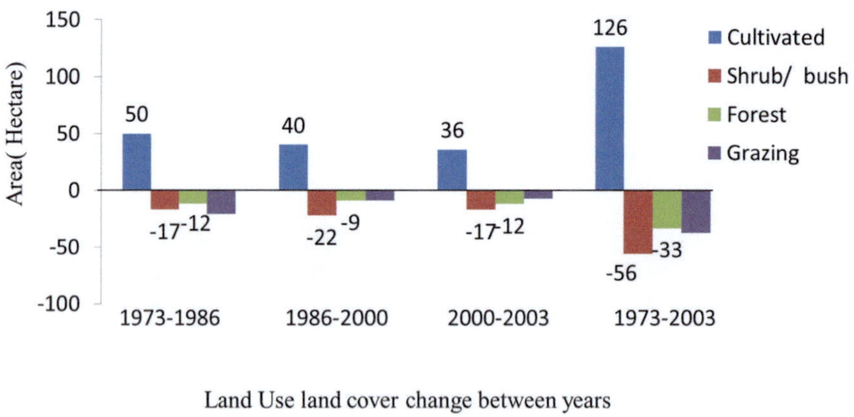

Fig. 5.10 Land use/land cover changes in the Enchilala watershed for the periods 1973–1986, 1986–2000, 2000–2013, and 1973–2013

The year 2000 is used for illustration (Fig. 5.7) and the greatest share of LULC from all classes is cultivated land with 397 ha (76 %). Shrub/bushland and grazing land cover an area of 39 ha (7.4 %) and 66 ha (12.6 %), respectively. Forest covers only 21 ha (4 %) of the watershed.

As shown in Figs. 5.8 and 5.10, agriculture increased in 2003 at the expanse of forest, brush, and grazing lands for Debre Mawi, and Enchilala watersheds. The change dynamics is different for the different analysis periods (Fig. 5.9). For the Enchilala watershed in 2013, the greatest share of LULC from all classes is cultivated land, which covers 355 ha (86 %). Shrub/bushland and grazing land cover 19 ha (5 %) and 30 ha (7 %) respectively. The least area is also covered by forest land, which is 10 ha (2 %). This shows that only 14 % of the total area of the watershed was covered by shrub, forest, and grazing land in 2013 and the remaining 86 % was covered by cultivated land. From the above results, it is clearly shown that cultivated land increases at the expense of forest, grazing land, and others leading to use marginal lands and accelerate land degradation.

5.3.4 *Rate of Land Use/Land Cover Change*

Table 5.1 shows the rate of change of cultivated, shrub/bush; grazing and forest cover for Mizewa, Debre Mawi and Enchilala watersheds. This result showed that the rate of change of land covers of Mizewa is greater than that of Debre Mawi and Enchilala. With the same context, the rate of change of land cover of Debre Mawi is slightly less than Enchilala.

5.4 Discussion

In 1973, the dominant land cover of Mizewa was shrub/bushland, however, in Debre Mawi and Enchilala watersheds, cultivated land was the dominant. This indicates that there was expansion of cultivated land in Debre Mawi and Enchilala than Mizewa in 1973.

However, in 1986, 2000, and 2013 the dominant land cover of all three observed watersheds was cultivated land, whereas shrub/bushland and grazing land were the

Table 5.1 Rate of land use/land cover change of all studied watersheds

Land use/land cover type	1973–2013 Mizewa		1973–2013 Debre Mawi		1973–2013 Enchilala	
	Area change (ha)	Rate of change (ha/year)	Area change (ha)	Rate of change (ha/year)	Area change (ha)	Rate of change (ha/year)
Cultivated	+1085	+27.1	+60	+1.5	+126	+3.1
Grazing	−108	−2.7	−22	−0.5	−37	−0.9
Shrub and bush	−512	−12.8	−16	−0.4	−56	−1.4
Forest	−465	−11.6	−22	−0.5	−33	−0.8

second dominant land cover of Mizewa and Debre Mawi watersheds, respectively. In 1973 and 1986 the second dominant land cover of Enchilala watershed was shrub/bushland; however, in 2000 and 2013 the second dominant land cover was grazing land. Hence in 2000 and 2013 grazing land was the second dominant land cover of Debre Mawi and Enchilala watersheds. Hence from the LULC analysis of three watersheds, there was the expansion of cultivated land and decrease of grazing, shrub/bush and forest land in all periods.

The analysis of LULC change for 40 years between 1973 and 2013 of this study showed that cultivated land increased by 60 ha and the shrinkage of grazing, shrub/bush and forestland by 22, 16, and 22 ha, respectively.

Changes in land cover in the Chemoga watershed in the Blue Nile basin within 1957–1998, farmland and settlements gained area between 1957 and 1982 was 2246 ha (13 % increase) but lost around 586 ha (2 % decrease) between 1982 and 1998. The increase between 1957 and 1982 corresponds to the population growth. The decrease in the latter 16-year period, though very small, might seem to contradict expectations, given the increase in population. But this decrease is attributable to increased tree planting at the household level (Woldeamlak 2002).

Land cover changes interfere with the land phase of the hydrological cycle. As is well known, land under little vegetative cover is subject to high surface runoff and low water retention. The increased runoff causes sheet erosion to intensify and rills and gullies to widen and deepen. The masses of sedimentary materials removed from hill slopes accumulate in low-lying areas downstream, where they create

Fig. 5.11 Typical gully on Ethiopian highlands, picture taken from Debre Mawi watershed (Photo by Seifu A. Tilahun 2014)

Fig. 5.12 Typical gully at in Enchilalawatershed on 2014 (**a**), gullies initiated by failure of soil bunds in saturated bottomlands of Debre Mawi watershed on 2013 (**b**), Initiation of gully erosion in saturated areas in the year 2012 (**c** and **d**). (*Source* Seifu A. Tilahun 2012, 2014)

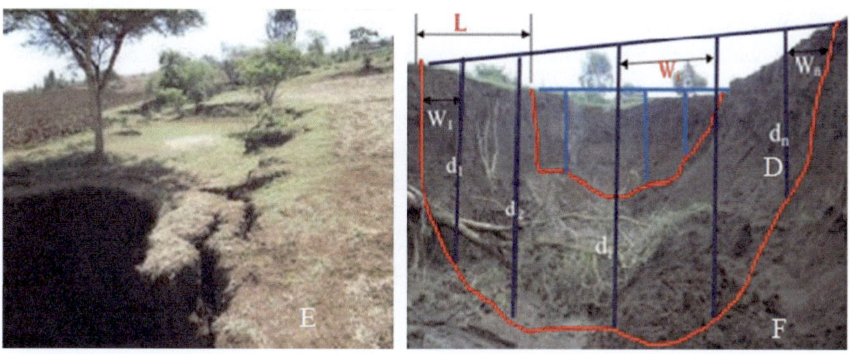

Fig. 5.13 Gully picture on the same site taken on different times. On 17th April, 2010 (**a**), and gully spread for the same site on 17 August, 2010 (*D*). It is shown that the tree is shown to be collapsed after 4 months from the first photo shot. (*Source* Seifu A. Tilahun 2012)

problems of water pollution, reservoir siltation, and sediment deposition on important agricultural lands. These problems have already emerged in the Chemoga watershed (Woldeamlak 2002).

Discharge and precipitation analysis over a 40-year period (1964–2003) for Bahir Dar and Kessie, representing the upper part of Blue Nile and El Diem at the border between Sudan and Ethiopia, shows that discharge during the long rainy season (June–September) increased in all the three stations. This could be explained by erosion of hillside lands that stored some of the water before they became eroded and were transformed into contributing areas of direct runoff. However, the precipitation did not change significantly over the entire basin (Zelalem et al. 2010).

Debre Mawi watershed is significantly characterized by intensive agriculture. Agricultural practice was even more prominent in 1973 and shows an increasing trend at the expense of grazing, forest, and bushlands. Time series reduction in the vegetative cover has resulted in rapid soil erosion, which in turn decrease in infiltration and facilitate runoff generation capacity of land during the rainy season. This finally leads to growth of numerous gullies in Debre Mawi and Enchilala. Seifu (2012) has assured this dynamic process and described the severity of soil erosion in Debre Mawi watershed detailing how intensive agricultural expansion hazards soil erosion and formation of gullies. Figures 5.11, 5.12 and 5.13 entail how gullies are spreading from time-to-time because of an immense increase in agricultural area at the expense of the different land covers in Debre Mawi.

5.5 Conclusion and Recommendation

LULC change is analyzed for three watersheds in the upper Blue Nile River basin using Landsat images of 1973, 1986, 2000, and 2013, supported by focus group discussions and repeated field visits. Cultivated, forest, shrub/bush, and grazing land were the major LULC classes identified in the studied watersheds.

Comparison of the LULC dynamics between three watersheds indicated that Mizewa watershed has shown the greatest rate of change in land covers within 40 years than Debre Mawi and Enchilala watersheds. This is because the rate of expansion of cultivated land into steep slope, degraded, and marginal lands was found to be much more in Mizewa, as farmers cut and clear the shrubs and forests and convert these land uses into agriculture. Observation and discussion with the farmers also proved that even at 2410 masl, it was found that forests and bushlands were converted for cultivation in Mizewa watershed.

Population growth is the greatest driving force for LULC change in the observed three watersheds. This is due to the growing demand for land for cultivation and settlement, forests for fuel, while construction purposes would become greater which would cause cutting of trees and converting of these lands into cultivated land to meet the growing food demand. Hence, environmental consequences of this population growth lead to cultivation of shallow soils and steep slopes followed by accelerating erosion and overexploitation of forest land. The effect of population pressure is often considered to be an important driver of deforestation and recognized by many studies (Marai and Pahari 1999).

In Debre Mawi and Enchilala watersheds, deforestation and inappropriate farming practices caused high soil erosion leading to rapid growth of numerous gullies forcing farmers to abandon their cultivated and grazing lands and clearing other areas for farming. Therefore, the current trends in LULC must be checked by protecting the remaining vegetation and avoiding further deforestation.

Although settlements have direct relationships to land cover change, they were not successfully classified due to the limited spatial resolution of the Landsat images and the pattern of settlements of all three watersheds.

Creating awareness among the communities concerning optimal use of natural resources, conservation practices and their benefits would play a significant role in rehabilitation of the environment. Conservation measures through afforestation programs are immediate requirements for natural resources management to prevent the wide spreading of gullies over the experimental watersheds and for the Blue Nile basin in general.

References

Abtew W, Melesse AM, Desalegn T (2009a) Spatial, inter and intra-annual variability of the Blue Nile River Basin Rainfall. Hydrol Process 23(21):3075–3082

Abtew W, Melesse AM, Desalegn T (2009b) El Niño southern oscillation link to the Blue Nile river basin hydrology. Hydrol Process Spec Issue Nile Hydrol 23(26):3653–3660

Abtew W, Melesse AM (2014a) Nile river basin hydology. In: Melesse AM, Abtew W, Setegn S (eds) Nile river basin: ecohydrological challenges, climate change and hydropolitics, pp 7–22

Abtew W, Melesse AM (2014b) Climate teleconnections and water management. In: Nile river basin. Springer International Publishing, pp 685–705

Abtew W, Melesse AM (2014c) Transboundary rivers and the Nile. In: Nile river basin, Springer International Publishing, pp 565–579

Ayenew T (2004) Environmental implications of changes in the levels of lakes in the Ethiopian Rift since 1970. Reg Environ Change 4:192–204

Bewket W (2002) Land cover dynamics since the 1950s in Chemoga watershed, Blue Nile Basin, Ethiopia. Mt Res Dev 22(3):263–269

Chebud Y, Melesse AM (2013) Stage level, volume, and time-frequency change information content of lake Tana using stochastic approaches. Hydrol Process 27(10):1475–1483. doi:10.1002/hyp.9291

Chebud YA, Melesse AM (2009a) Numerical modeling of the groundwater flow system of the gumera sub-basin in Lake Tana basin, Ethiopia. Hydrol Process Spec Issue Nile Hydrol 23 (26):3694–3704

Chebud YA, Melesse AM (2009b) Modeling Lake Stage and Water Balance of Lake Tana, Ethiopia. Hydrol Process 23(25):3534–3544

Chrysoulakis N (2004) Combining satellite and socio economic data for land use models estimation. In: Proceedings of third workshop of EARSEL special interest group on remote sensing for developing countries (in press)

Codjoe SNA (2007) Integrating remote sensing, GIS, census, and socioeconomic data in studying the population-land use/cover Nexus in Ghana. Lit Update Afr Dev xxxll:197–212

Dessu SB, Melesse AM (2012) Modeling the rainfall-runoff process of the Mara river basin using SWAT. Hydrol Process 26(26):4038–4049

Dessu SB, Melesse AM (2013) Impact and uncertainties of climate change on the hydrology of the Mara river basin. Hydrol Process 27(20):2973–2986

Dessu SB, Melesse AM, Bhat M, McClain M (2014) Assessment of water resources availability and demand in the Mara river basin. CATENA 115:104–114
Dessie G, Kleman J (2007) Pattern and magnitude of deforestation in the South Central Rift Valley region of Ethiopia. Mt Res Dev 27:162–168
Gebreegziabher Y (2005) Assessment of the water balance of Lake Awassa Catchment, Ethiopia (Unpublished MSc Thesis). ITC, Enschede, The Netherlands
Gebresamuel G, Singh BR, Dick Ø (2010) Land-use changes and their impacts on soil degradation and surface runoff of two catchments of Northern Ethiopia. Acta Agric Scand Sect B: Plant Soil Sci 60:211–226
Geist HJ, Lambin EF (2002) Proximate causes and underlying driving forces of tropical deforestation. Bioscience 52:143
Getachew HE, Melesse AM (2012) Impact of land use/land cover change on the hydrology of Angereb watershed, Ethiopia. Int J Water Sci 1(4):1–7. doi:10.5772/56266
Hurni H, Zeleke G (2001) Implication and land use/land cover dynamics for mountain resources degradation. In the Northwe Chrysoulakis stern Ethiopian highlands. Mt Res Dev 21:184–191
Lambin EF, Geist HJ, Lepers E (2003) Dynamics of land use and land cover change in tropical regions. Annu Rev Environ Resour 28:205–241
Liang S (2002) Atmospheric of landsat ETM + land surface imagery:validation and applications. IEEE Trans Geosci Remote Sens 41:1260
Mango L, Melesse AM, McClain ME, Gann D, Setegn SG (2011a) Land use and climate change impacts on the hydrology of the upper Mara river basin, Kenya: results of a modeling study to support better resource management, special issue: climate, weather and hydrology of East African highlands. Hydrol Earth Syst Sci 15:2245–2258. doi:10.5194/hess-15-2245-2011
Mango L, Melesse AM, McClain ME, Gann D, Setegn SG (2011b) Hydro-meteorology and water budget of Mara river basin, Kenya: a land use change scenarios analysis. In: Melesse A (ed) Nile river basin: hydrology, climate and water use. Springer Science Publisher, Chapter 2, pp 39–68. doi:10.1007/978-94-007-0689-7_2
Melesse AM, Loukas AG, Senay G, Yitayew M (2009a) Climate change, land-cover dynamics and ecohydrology of the Nile river basin. Hydrol Process Spec Issue Nile Hydrol 23 (26):3651–3652
Melesse AM, Abtew W, Desalegne T, Wang X (2009b) Low and high flow analysis and wavelet application for characterization of the Blue Nile river system. Hydrol Process 24(3):241–252
Melesse AM, Abtew W, Setegn S, Dessalegne T (2011) Hydrological variability and climate of the upper Blue Nile river basin In: Melesse A (ed) Nile river basin: hydrology, climate and water use, Springer Science Publisher Chapter 1, pp 3–37. doi:10.1007/978-94-007-0689-7_1
Melesse AM (2011) Nile river basin: hydrology, Climate and Water Use. Springer Science & Business Media, New York
Melesse AM, Abtew W, Setegn SG (2014) Nile river basin: ecohydrological challenges, climate change and hydropolitics. Springer Science & Business Media, New York
Melesse AM, Jordan JD (2003) Spatially distributed watershed mapping and modeling: land cover and microclimate mapping using landsat imagery part 1. J Spat Hydrol (e-journal) 3(2)
Moges A, Holden NM (2009) Land cover change and gully development between 1965 and 2000 in Umbulo Catchment, Ethiopia. Mt Res Dev 29:265–276
Mohammed H, Alamirew A, Assen M, Melesse AM (2013) Spatiotemporal mapping of land cover in Lake Hardibo Drainage basin, Northeast Ethiopia: 1957–2007. Water conservation: practices, challenges and future implications. Nova Publishers, New York, pp 147–164
Monserved RA (1990) Method for comparing global vegitation maps. WP-90, ll ASA
Mustared JF (2004) Land use and land cover change pathways and impacts, land change science observing, monitoring and understanding trajectories of change on the earth's surface. Kluwer, Dordrecht pp 411–429
Olson JM, Matima JW (2006) Sustainable intesification of mixed crop-livestock systems, land use change impacts and dynamics policy brief international livestock research institute, Nairobi, Kenya, vol 1

Quentin FB (2006) Drivers of land use change, final report mathing opportunities to motivations. Royal Melbourne Institute of Technology, Australia, Department of sustainability and environmental and primary industries, ESAI project

Rechardes JA, Kelly DJ (1984) On the concept of spectral class. Int J Remote Sens 5:987–991

Seifu TA (2012) Observations and modeling of erosion from spatially and temporally distributed sources in the (semi) humid ethiopian highlands

Setegn SG, Srinivasan R, Dargahi B, Melesse AM (2009a) Spatial delineation of soil erosion prone areas: application of SWAT and MCE approaches in the Lake Tana basin, Ethiopia. Hydrol Process Spec Issue Nile Hydrol 23(26):3738–3750

Setegn SG, Srinivasan R, Melesse AM, Dargahi B (2009b) SWAT model application and prediction uncertainty analysis in the Lake Tana basin. Ethiop Hydrol Process 24(3):357–367

Setegn SG, Bijan Dargahi B, Srinivasan R, Melesse AM (2010) Modelling of sediment yield from Anjeni Gauged watershed, Ethiopia using SWAT. JAWRA 46(3):514–526

Solomon A (1994) Land use dynamic, soil conservation and potential for use in mute area geographical bernensis, Berne, Switzerland: Illubabor Region, Ethiopia, Africa studies A13

Szejwach G, Baulies X (1998) LULCC data requirements workshop survey of needs, gaps and priorities on data for land use land cover change research. In: GBP/IHDP-LULCC and IGBP-DIS (ed) LULCC report

Trotter CM (1998) Characterising the tropic effect at red wavelengths using juvenile contier canopies. Int J Remote Sens 19:2215–2221

Turner BL (1993) Relating land use and global land cover change: a proposal for an IGBP—HDP No core project Report from the IGBP-HDP working group on land use land cover change. Swedish Academy of sciences, Stockholm: Joint publication of the IGBP, HDP No. 5

Turuner BL, Meyer WB (1994) Change in land use and land cover. A global perspective. Cambridge University Press, New York

Wondie M, Schneider W, Melesse AM, Teketay D (2011) Spatial and temporal land cover changes in the Simen Mountains national park, a world heritage site in Northwestern Ethiopia. Remote Sens 3:752–766. doi:10.3390/rs3040752

Wondie M, Schneider W, Melesse AM, Teketay D (2012) Relationship among environmental variables and land cover in the Simen Mountains national park, a world heritage site in Northern Ethiopia. Int J Remote Sens Appl (IJRSA) 2(2):36–43

Yitaferu B (2007) Land degradation and options for sustainable land management in the lake tana basin. Ph.D. University of Bern, Switzerland

Yitayew M, Melesse AM (2011) Critical water resources management issues in nile river basin, In: Melesse A (ed) Nile river basin: Hydrology, Climate and Water Use. Springer Science Publisher, Chapter 20, 401–416, doi:10.1007/978-94-007-0689-7_20

Zelalem TK, Mohamed YA, Steenhuis TS (2010) Hydrological process, trends in rainfall and runoff in the Blue Nile basin: 1964–2003. Online in wiley online library (wileyonlinelibrrary.com)

Zerihun W, Mesfin T (1990) The status of the vegetation in the Lake Region of Rift Valley of Ethiopia and the possibilities of its recovery. Sinet: Ethiop. J. Sci. 13:97–120

Chapter 6
Land Use and Land Cover Change Impact on Groundwater Recharge: The Case of Lake Haramaya Watershed, Ethiopia

Shimelis B. Gebere, Tena Alamirew, Broder J. Merkel and Assefa M. Melesse

Abstract Anthropogenic actions have been dramatically changing the land cover of the earth with a substantial impact on the soil, water, and atmosphere. Haramaya watershed is located in the eastern part of Ethiopia which encompasses the dry Lake Haramaya, primarily used for agricultural production with burgeoning population, dramatic changes in land use land cover have been observed over the past few decades. The land cover changes have impacted the water balance of the watershed by changing groundwater level. This study focuses on assessing the impact of land use land cover changes on groundwater recharge potential of the watershed. Future land use change was simulated using CLUE-S (Conversion of Land Use and its Effects at Small regional extent) land use change model. The result showed an increase in chat cultivation from 6276 ha in 2011 to 7282 and 7000 ha in 2028 under current conditions (scenario-1) and good watershed management (scenario-2), respectively. Chat (*Catha edulis*) also referred to as Khat, is a stimulant plant chewed as a tradition but labeled as drug by the World Health Organization (WHO). Cultivated land declined from 4975 ha in 2011 to 3999 and 4013 ha in 2028 under both scenarios 1 and 2, respectively. The simulated result of the WetSpass water balance model showed that the groundwater recharge in the watershed is strongly influenced by land use land cover change. The annual groundwater recharge in the year 2011 ranged from 0 to 90 mm. A land use land

S.B. Gebere (✉) · B.J. Merkel
Department for Geology, Technische Universität Bergakademie Freiberg, Freiberg, Germany
e-mail: shimelisberhanu@yahoo.com

B.J. Merkel
e-mail: merkel@geo.tu-freiberg.de

T. Alamirew
Water and Land Resource Centre, Addis Ababa, Ethiopia
e-mail: tena.a@wlrc-eth.org

A.M. Melesse
Department of Earth and Environment, Florida International University, Modesto A. Maidique Campus, Miami, FL 33199, USA
e-mail: melessea@fiu.edu

© Springer International Publishing Switzerland 2016

A.M. Melesse and W. Abtew (eds.), *Landscape Dynamics, Soils and Hydrological Processes in Varied Climates*, Springer Geography,
DOI 10.1007/978-3-319-18787-7_6

cover projection to 2028 with baseline and good management scenarios showed the range of recharge values decreased to 0–83 and 0–87 mm, respectively. At the same time, groundwater level will continue declining due to increased abstraction. Therefore, it is recommended that the concerned authorities should consider the impact of land use change on the water resources of the watershed in order to optimally utilize the available water resources and to find alternative water sources to fill the deficit resulting from groundwater table decline.

Keywords Groundwater recharge · Land use land cover · CLUE-S · WetSpass · Climate change · Haramaya watershed · Eastern ethiopia

6.1 Introduction

Land use land cover changes are known to have impact on groundwater–surface water interaction. However, the effects of such changes are poorly understood (Scanlon et al. 2005; Stonestrom et al. 2009). Anthropogenic activities and natural factors are the sources of the changes and cause significant impact on the environment (Mulungu and Kashaigili 2012). Anthropogenic impact is immediate while natural impact is long term. Over the past few decades, human actions are gradually changing the natural landscapes in various forms (Yanda and Munishi 2007). Land resources are intensively utilized by mankind and have resulted in significant changes of land use land cover (Bronstert 2004). Forest clearance for firewood, subsistence agriculture, commercial agricultural expansion, urbanization, and over grazing by livestock are some of the changes that have huge impact on water resources, productivity of land, and damage on ecosystems (Chhabra et al. 2006).

In developing countries, most importantly in Ethiopia, the resource bases such as forest, land, and water, are declining considerably. In developing countries, information regarding rate of depletion is often lacking (FAO 2009). Rapid human population growth and agriculture based economies are the major driving forces for the changes (Warra et al. 2013). DeFries and Eshleman (2004) and Mulungu and Kashaigili (2012) underline the importance of understanding the impact of land use land cover changes on water resources. The drying of Lake Haramaya and its wetland system (Assen 2011) and the heavy decline of the Awash Valley swampy areas (EIB 2014) are some of the examples that exhibit decline in water resources associated with the changes in land use and land cover. Lake Haramaya was once a beautiful lake, 228 ha large and 10–12 m deep (Zinabu 2012). It completely dried in 2011. Currently, the groundwater of the dry Lake Haramaya watershed is overexploited to supply water to the city of Harar, to the towns of Awoday and Haramaya and local communities, for domestic water supply and irrigation.

Land use/land cover dynamics is an important landscape process capable of altering the fluxes of water, sediment, contaminants, and energy. Mainly caused by human, impact of land use on water resources availability is high. Degraded watersheds tend to accelerate overland flow reducing soil moisture and base flow recharge and increases sediment detachment and transport. Various studies used land cover mapping tools and methods to understand land use changes, inventory of forest, and natural resources as well as understand the changes in the hydrologic behavior of watersheds (Getachew and Melesse 2012; Mango et al. 2011a, b; Wondie et al. 2011, 2012; Melesse and Jordan 2002, 2003; Melesse et al. 2008; Mohamed et al. 2013).

Many researchers have worked to quantify the potential impact of land use/land cover changes on hydrological processes. Dams et al. (2008) showed the impact of land use change on the groundwater system of the Kleine Nete catchment, Belgium using the CLUE-S (the Conversion of Land Use and its Effects at Small regional extent) model along with WetSpass and MODFLOW. They found large changes in groundwater near the major cities of their study area. Mkaya et al. (2013) evaluated the impact of land use change on catchment hydrology in Taita Hills, Kenya using SWAT model and their study revealed an increase in surface runoff and sediment yield within the catchment as a result of the impact of land use change. In Dill catchment of south-east Germany, Eckhardt and Ulbrich (2003) used the SWAT-G model and showed up to 50 % decrease of mean groundwater recharge and streamflow.

Water resource engineers and hydrologist use hydrological modeling in investigation of hydrological systems. There are many models available for land use simulation (Mas et al. 2014), the CLUE-S model is one among many models. It is used in this study together with geographic information systems for its dynamic and spatially explicit raster based procedure. It is a multiscale land use change model (Verburg et al. 2002). The model is used to simulate land use change and its environmental effects at a small scale based on an empirical analysis (DeKoning et al. 1999). The model can also help to understand the complex system of biophysical and socioeconomic factors.

The study area is the dry Lake Haramaya watershed which is located in the eastern highlands of Ethiopia. Groundwater is the only source of freshwater available for various uses in the watershed. Yet, little has been done to quantify the potential use of groundwater in the watershed. Recently, the groundwater level is gradually declining due to high abstraction rate as a result of rapid increase in population, agricultural intensification, and expansion in the area. According to Dams et al. (2008) current research has mostly neglected to evaluate changes in quantity of groundwater but focus on groundwater qualities.

Groundwater recharge in the study area is susceptible to land use land cover changes. Over the past few decades, a rapid population growth has led to the clearing of forest covers for expansion of agricultural lands which caused significant lowering of the groundwater table. Little is known about the factors that affect groundwater recharge of the watershed. Most researches in Ethiopia only focused toward determining the scale of land use land cover change and there are few studies that focus on its impact on water resources availability. Hence, the aim of this study is to evaluate the impact of land use land cover change on groundwater recharge in the eastern highlands of Ethiopia using CLUE-S land use change model coupled with WetSpass water balance model.

6.2 The Study Area

Haramaya watershed is located in the Eastern highlands of Ethiopia (9° 22′ 58″–9° 31′ 22″N and 41° 53′ 44″–42° 12′ 43″E) and covers an area of 15,271 ha (Fig. 6.1). The watershed consists of a vast depression area bounded by adjacent highlands. The elevation of the area ranges from 2460 m above sea level at extreme north eastern parts to 2006 m above sea level near the central depression area. The watershed contains no perennial river; the only source of water available to recharge groundwater is rainfall. Groundwater is heavily exploited during the dry season for domestic, industrial, and irrigation purpose. Geologically, it is generally constituted by rocks ranging in age from Precambrian to recent deposition. Stratigraphy classification are from bottom to top, granite (Precambrian), sandstone, and limestone (Mesozoic sedimentary rocks), and recent sediments, Quaternary sediments (Tadesse et al. 2010). The predominant soil types of the watershed are characterized by medium- to fine-textured materials.

The climate of the study area is semiarid with annual average rainfall of 743 mm. Rainfall is bimodal, with peaks in April and May (98 mm for each month) and in August (140 mm). January is the driest month (8.5 mm) followed by December, November, and February (Fig. 6.2). The elevation and latitude of the catchment has resulted in moderate temperature with an average of 16.4 °C, varying from 12.6 °C in December to 19.0 °C in June.

Figure 6.3 shows mean annual rainfall and temperature temporal variation in the watershed. The mean annual temperature between 1970 and 2010 ranged from 12.4 to 19.8 °C while annual rainfall during same time period ranged from 469 to 1104 mm.

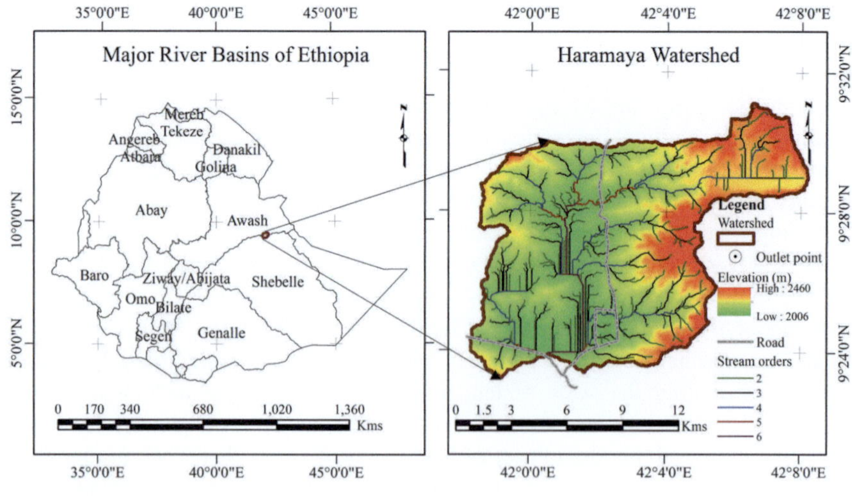

Fig. 6.1 The Haramaya watershed in eastern Ethiopia

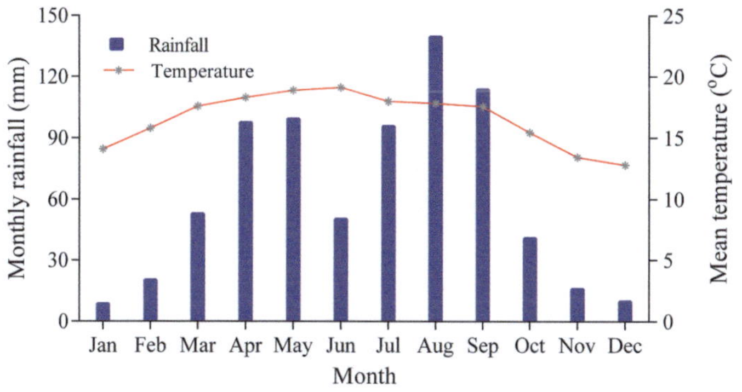

Fig. 6.2 Haramaya watershed mean monthly rainfall and temperature variation

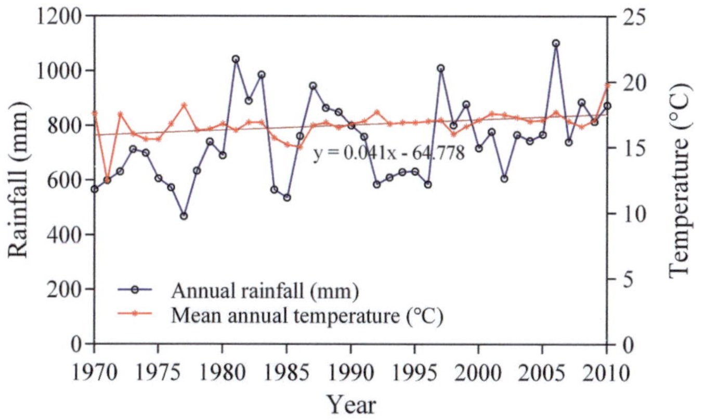

Fig. 6.3 Annual rainfall and mean annual temperature variability of Haramaya watershed

Despite its undulating topography, the land is intensively cultivated and agricultural activities in the watershed always have maintained priority in economic activities. In particular, cultivation of cereals, vegetables, and the cash crop chat (*Catha edulis*) and livestock rearing has great importance in the area. According to Mesfin et al. (2011), the average land holding size of a household in the area is about 0.74 ha. Activities other than agriculture are limited.

The increasing population size is believed to aggravate the stress to the environment (Grimm et al. 2008). A rapid population growth and unplanned human settlement are the major driving forces to the need of increase food production and environmental stresses such as soil, forests, water, and fisheries (Zhou et al. 2008). The population living around the watershed is estimated at 166,597 in 1994 and 271,018 in 2007, with 63 % increase (CSA 1994, 2007).

6.3 Methodology

6.3.1 Data Collection and Preparation

Socioeconomic and biophysical data that have impact on the rate and spatial pattern of land use change of the watershed such as digital elevation models (DEM), landsat images, climatic data, soil map, population density, and groundwater level data were used in the study.

ArcGIS 9.3 software was used to delineate and calculate the area of the watershed using DEM data. ArcView 3.2 was used to convert all the required data such as elevation, soil, slope, distance to road, distance to town, and population into raster format to enable analysis and modeling in the CLUE-S modeling framework. WetSpass model is an ArcView 3.2 extension which was used to estimate groundwater recharge of the study area. The raster grids were exported directly to ASCII file format. The precipitation, temperature, wind speed, potential evapotranspiration, soil, topography, slope, and groundwater level maps for the WetSpass model were prepared in ArcView 3.2.

Landsat images of path/row 166/54 of 30 m spatial resolution for the year 1995 and 2011 was obtained from USGS website and was processed using ERDAS IMAGINE 9.2 software. The images were classified in to eight land use types using supervised image classification technique of maximum likelihood method. Compared with other classification methods, supervised maximum likelihood classifier gives higher classification accuracy and it is the most commonly and widely used technique (Mengistu and Salami 2007; Reis 2008; Solaimani et al. 2010; Dewan and Yamaguchi 2009; Yacouba et al. 2009; Chandola and Vastavai 2010). This technique is based on the decision rule that unknown class pixels belong to particular class with the highest likelihood among several classes (Franklin et al. 2003; Solaimani et al. 2010).

6.3.2 The CLUE-S Model

The CLUE-S is an improved model for small-scale regional application (Verburg et al. 1999). The model dynamically allocates land use changes to simulate the future land use change based on a combination of empirical analysis of location suitability and spatial analyses (Verburg et al. 1999; Zheng et al. 2012). Compared with other empirical models, it simulates multiple land use types simultaneously. The criteria to select the CLUE-S model for this study were based on its flexibility and performance on small watershed, and the possibility of linking the output to the WetSpass model to estimate groundwater recharge of the watershed. Also, input data for this model are available in the watershed.

The CLUE-S model works on annual time step. It comprises two distinct modules, a nonspatial demand module, which calculates the total area of all land use types at each time step; and a spatially explicit allocation module. The basis of the spatially explicit allocation module is a statistical model describing the suitability of locations for each land use type under consideration, which is assumed to be a function of spatially explicit driving factors.

Some land use covers are very unlikely to be converted to specific types of land use types but some land use changes are possible. For example, forest plantation and settlement areas cannot be converted into water bodies. Most land use changes follow a certain cycle. The conversion matrix can be used to specify the possible and not possible land use conversion. The conversion matrix indicates to what other land use types the present land use type can be possibly be converted or not during the next time step. For example, forest can directly be converted to either agricultural or grass land while it is not possible to get a direct conversion of the reverse. Therefore for each land use type, a value needs to be specified that represents the relative elasticity to change, 0 for conversion is not possible and 1 for conversion is possible. A more detailed description of the model is given in Verburg et al. (2002).

6.3.3 Land Use Scenarios

Two land use demand scenarios have been tested in this study based on past and current land use condition and from the socioeconomic condition of the area. The current effort made by Haramaya University to rehabilitate the watershed environment of the study area has also been considered. The near future land use change trends were extrapolated based on the land use changes of the recent past trends and the trends were corrected for changes in population growth to prepare the land use demand scenarios. The future scenarios focused on the following land use and land cover classes, cultivated land, grass land, bare land, "chat"/shrubs, forest plantation, settlement, water body, and shallow water body.

The recent past land use scenario was obtained by processing land use maps of 1995 and 2011 Landsat images. The two scenarios in this study are baseline and good management scenarios. In the baseline scenario, the growth in population is assumed to follow the current trend and the future land use requirement is expected to change linearly and extrapolated based on the trend in the time period from 1995 to 2011. The farmers in the watershed are expected to continue doing as they have in the past. In this scenario, the historical evolution of the major land use covers such as cultivated land and chat/shrubs are extrapolated till 2028 using regression equations. This is because the water resource may not be available beyond 2028 if it continues to be utilized at the current trend.

The second scenario, the good management scenario, is directly related with the watershed rehabilitation activities started by Haramaya University. From what was observed in the watershed, it is impossible to take the land back from the farmers to reforest and create buffer zone in the watershed. However, the best alternative is to work together with the farmers and create awareness on the consequences of land use change in the watershed resources. Hence, this scenario is developed with the assumption that Haramaya University can manage to conserve the remaining forest, wetland, and water body. As a result, forest plantation, water body, and shallow water are expected to remain the same in this scenario.

6.3.4 *Spatial Analysis*

The land use map of 2011 was used in a logistic regression analysis to understand the spatial relationship between current land use/land cover and driving factors which are represented by beta value (Table 6.1). For each types of land use, a separate regression model is used. SPSS software was used to extract the beta coefficients. To simulate future land use changes, it is necessary to know the factors which drive the spatial distribution of land use. Hence, in this study, the following driving factors were used, population density, soil type, soil texture, altitude, distance to the nearest roads, and distance to the nearest towns (Fig. 6.4). The beta values calculated for each land use classes by logistic regression were used as inputs to the CLUE-S model.

The relationship between land use change and its driving factors of the study area was established through logistic regression for each land cover type given a set of defined driving factors as shown in Eq. 6.1 (Verburg et al. 2002):

$$\text{Log}\left(\frac{P_i}{1-P_i}\right) = \beta_0 + \beta_1 X_{1,i} + \beta_2 X_{2,i} + \cdots + \beta_n X_{n,i} \quad (6.1)$$

where P_i is the probability of the occurrence of a specific land use type, X_i is the driving factors, and β_n is the regression coefficients.

Table 6.1 Regression analysis results between land uses and driving factors in the watershed

Driving factors	Bare land	Chat/shrubs	Cultivated land	Forest	Grass land	Settlement	Shallow water	Water body
Topography					−0.0028			0.5819
Population density			0.0539				−0.4803	−1.0065
Slope	−0.04	0.042	−0.0334	−0.1275			0.4011	
Distance to road		−0.0002			0.0002	0.0004	−0.0024	
Distance to the nearest town			0.0003					−0.0089
Soil type			−0.5176					
Soil texture			1.4196					
Constant	4.323	0.541	−29.808	7.794	6.705	2.931	274.895	−583.94

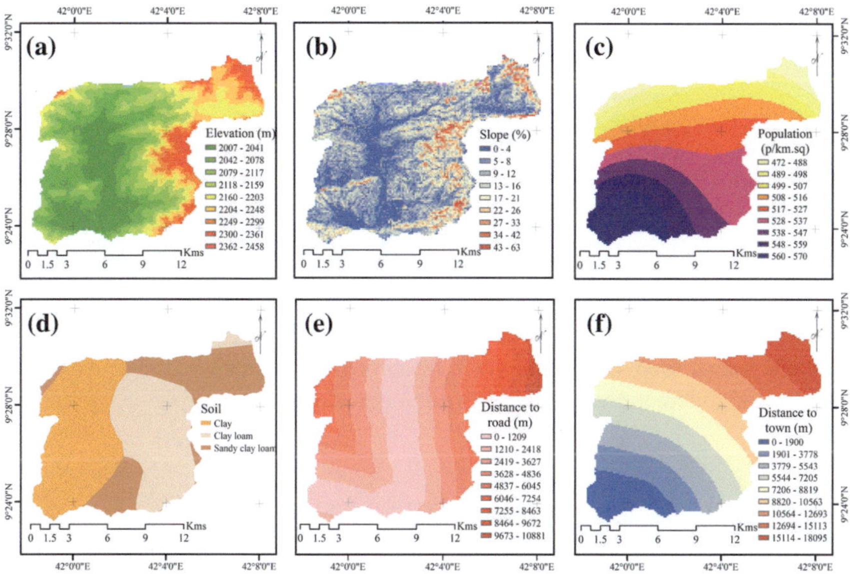

Fig. 6.4 Land use change drivers map. **a** Elevation, **b** slope, **c** population density, **d** soil, **e** distance to road, and **f** distance to the nearest town

6.3.5 The WetSpass Model

Water and energy transfer between soil, plants, and atmosphere under quasi-steady state conditions (WetSpass) is especially suited for studying long-term effects of land use changes on the water regime in a watershed (Kuisi and El-Naqa 2013). The model was originally developed for temperate regions in Europe by Batelaan and DeSmedt (2001) based on WetSpa model, which is a GIS-based distributed hydrological model for flood prediction and water balance simulation on catchment scale (DeSmedt et al. 2000).

The model is capable of simulating spatially distributed seasonal and yearly averages of evapotranspiration, surface runoff, groundwater recharge, soil evaporation, transpiration and interception using physical and empirical relationships. The WetSpass model accounts for the spatial variation of groundwater recharge that result from land use, climatic parameters, groundwater depth, soil, slope, and topography. The water balance components of vegetated, bare soil, open water, and impervious surfaces were performed to calculate the water balance at raster cell level and added up to calculate the total water balance of a raster cell using Eqs. 6.2–6.4 (Batelaan and DeSmedt 2007).

$$\mathrm{ET}_{\mathrm{raster}} = a_{\mathrm{v}}\mathrm{ET}_{\mathrm{v}} + a_{\mathrm{s}}E_{\mathrm{s}} + a_{\mathrm{o}}E_{\mathrm{o}} + a_{\mathrm{i}}E_{\mathrm{i}} \tag{6.2}$$

$$S_{raster} = a_v S_v + a_s S_s + a_o S_o + a_i S_i \tag{6.3}$$

$$R_{raster} = a_v R_v + a_s R_s + a_o R_o + a_i R_i, \tag{6.4}$$

where ET_{raster}, S_{raster}, R_{raster} are total evapotranspiration, surface runoff, and groundwater recharge of a raster cell, respectively, and E is evaporation. Each has an area component denoted by a_v, a_s, a_o, and a_i for vegetated, bare soil, open water, and impervious surfaces, respectively. For calculation of the water balance of each of the above-mentioned components of a raster cell, precipitation is taken as the starting point and the rest of the processes, which are interception, runoff, evapotranspiration, and recharge, follow in an orderly manner. This order is a prerequisite for the seasonal time scale with which the processes is quantified. The water balance for the different components is treated hereafter.

6.3.6 *Groundwater Recharge Calculation*

WetSpass physical quasi-steady-state model was used to quantify spatially varying groundwater recharge taking into account the soil texture, land-use, slope and meteorological conditions of the study area. Groundwater recharge is calculated as a residual term of the water balance using Eq. 6.5 (Batelaan and DeSmedt 2007).

$$R_v = P - S_v - ET_v - I, \tag{6.5}$$

where ET_v is actual evapotranspiration (LT^{-1}) (sum of transpiration and evaporation from bare soil found in between the vegetation), S_v is surface runoff (LT^{-1}), P is average seasonal precipitation (LT^{-1}), and I is interception by vegetation (LT^{-1}).

6.4 Results and Discussion

6.4.1 *Predicted Land Use and Land Cover Changes*

Continuous groundwater abstraction, deforestation, and expansion of agricultural land along with rapid population growth rate in the watershed have led to a significant decline of groundwater level and shrinkage of surface water. Two future land use scenarios were assumed, baseline (scenario-1) and good management scenario (scenario-2). Using these scenarios, two future land use land cover maps were predicted for 2028 using the CLUE-S model as shown in Fig. 6.5. Under certain possible future scenarios assumptions, linking the land use changes with relevant driving factors, the model decreases land use allocation uncertainty for

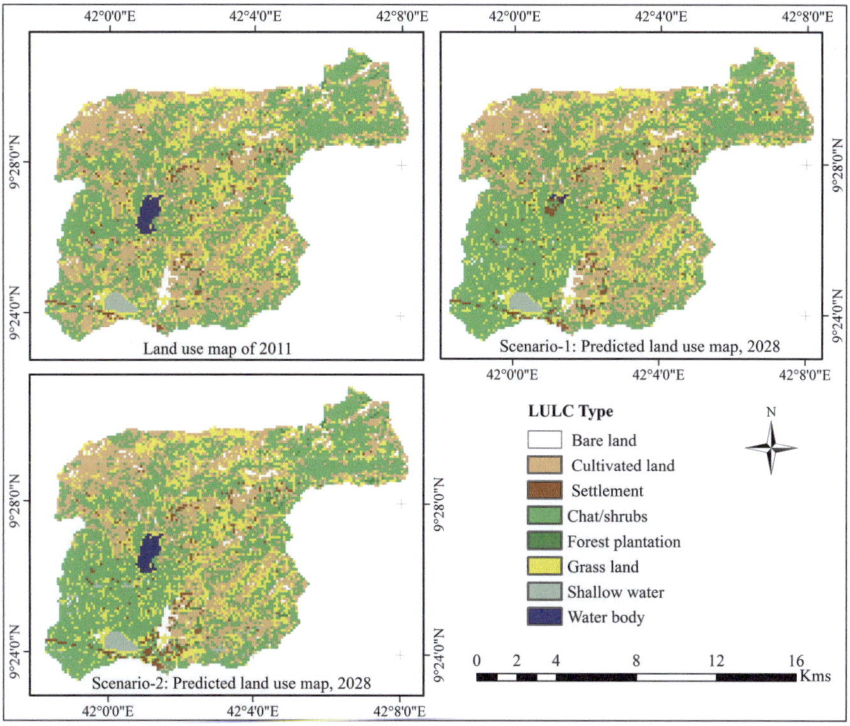

Fig. 6.5 Predicted land use maps using CLUE-S model

spatially explicit and multi scale simulation of land use change (Dams et al. 2008; Ganasri et al. 2013).

Under baseline scenario, scenario-1, the land use and land cover changes are expected to occur at the same rate as the land use change trend between 1995 and 2011. Since bare land and shallow water showed very small changes, in this analysis, their change is neglected and kept same. Extrapolation was made for the two main land use and land cover types, cultivated land and chat/shrubs. Both land use types cover more than 70 % of the total area of the watershed (Table 6.2). Chat/shrubs land was the largest land cover type, about 41.1 % of the total area in 2011 and steadily increased throughout the entire simulation period while cultivated land declined. This is due to the economic benefit of producing chat over other crops. Producing and selling of chat in the area is more profitable to the local farmers than cultivation of cereals and vegetables.

It is apparent in this scenario that water body cover declined from 1 % in the current cover to 0.1 % cover by 2028. This result is associated with the high demand of water for irrigation, municipal use, and industrial consumption due to a rapid growth of population in the watershed and the nearby cities. Loss of forest plantation was also noticed, from 0.7 % cover in 2011 to 0.3 % in 2028. In the

Table 6.2 Simulated land use and land cover changes between 2011 and 2028

LULC classes	2011		2028			
			Scenario-1		Scenario-2	
	ha	%	ha	%	ha	%
Bare land	358.6	2.3	358.6	2.3	358.6	2.3
Chat/shrubs	6275.5	41.1	7282.1	47.7	6999.8	45.8
Cultivated land	4975.2	32.6	3998.9	26.2	4013.3	26.3
Forest plantation	105.1	0.7	44.6	0.3	105.1	0.7
Grass land	3024	19.8	3109	20.4	3109	20.4
Settlement	254.9	1.7	339.8	2.2	339.8	2.2
Shallow water	121	0.8	121	0.8	188.6	1.2
Water body	157	1	17.3	0.1	157	1
Total	15271.2	100	15271.2	100	15271.2	100

watershed, forest cover is cleared for various purposes as fire wood, house construction, and marketing. Due to rise in population in the watershed, growth in construction of houses is expected in the near future.

In the good management scenario, scenario-2, the watershed rehabilitation activity started by Haramaya University was taken under consideration. In the land use planning for the watershed development, it was assumed that it will be unlikely to stop the farmers from producing chat at the expense of other land use types. Therefore, it is assumed that the University will manage to protect the remaining forest and water body with motivation of farmers to produce more agricultural crops, which demand less water compared to chat production and through afforestation.

But, farmers' resistance to control chat production will result in a modest increase in area, 41.1 % in 2011 to 45.8 % in 2028. However, compared with the assumption in scenario-1 there is less production of chat in scenario-2. While cultivated land showed a decline from 32.6 % in 2011 to 26.3 % in 2028, it has increased marginally compared with the assumption in scenario-1. Assuming Haramaya University rehabilitation project will succeed, forest plantation, bare land, and water body are expected to remain relatively stable between 2011 and 2028. As a result, shallow water body shows an increase. Since agriculture is the mainstay for the farmers in the watershed, reduction of the total area of production of both cultivated land and chat production is not expected to happen in both scenarios.

6.4.2 WetSpass Model Outputs

The main focus of this study is to estimate the impact of land use land cover changes on groundwater recharge of the watershed. Figure 6.6 shows the simulated groundwater recharge of the watershed. In the two future land use scenarios, there is a decreasing trend in groundwater recharge compared with groundwater recharge in 2011. In all the maps, there is higher groundwater recharge rate toward the southern part of the watershed. The southern part of the watershed promotes high groundwater recharge rate due to its geomorphological features. Having lower elevation and gentle slope, it promotes high rate of recharge to groundwater by retarding runoff to increase infiltration time while the eastern, western, and north eastern parts of the watershed are less likely to promote high rate of groundwater recharge due to their steep topographic nature.

The annual groundwater recharge rate in 2011 ranged between 0 and 90 mm but for both future land use scenarios it shows a decrease in the recharge rate. For scenario-1, the rate of groundwater recharge lowered to 0–83 and 0–87 mm for scenario-2. The scenario-2 rate is higher because with the assumption of watershed rehabilitation project initiated by Haramaya University, most land use types are expected to remain the same. Hence, more positive result might come from watershed rehabilitation initiative.

Table 6.3 shows the percentage of groundwater recharge contribution from different land use and land cover types. Under recent condition and in both scenarios, grass land contribution toward groundwater recharge is higher than the remaining land covers. In 2011, 34.1 % of groundwater recharge was through grass land while for both scenario-1 and scenario-2 contribution is 25.7 and 35.1 %, respectively. Like grass land, chat/shrubs, and cultivated land nearly contribute the same amount of recharge to groundwater in each simulation periods. This shows the importance of covering the surface with vegetation to facilitate groundwater recharge. Unlike grass land, chat/shrubs, and cultivated land, groundwater recharge contribution from bare land is very little. This is due to the fact that in the absence of vegetation cover, the surface cannot hold water for longer time and also most of the bare lands are located on the steepest part of the watershed.

The already depleted forest plantation has very low contribution toward groundwater recharge in current and both future land use scenarios. This result implies that there is much work to be done to protect existing forest cover and

Table 6.3 Groundwater recharge contribution of different land use and land cover type in 2011 and 2028

LULC classes	2011 (%)	2028	
		Scenario-1 (%)	Scenario-2 (%)
Bare land	4.7	8.4	2.8
Chat/shrubs	27.4	21.8	26.2
Cultivated land	29.8	22.5	24.3
Forest plantation	1.2	3.5	2.3
Grass land	34.1	25.7	35.1
Settlement	2.7	16.8	8.3
Shallow water	0.1	1.2	0.8
Water body	0	0.1	0.2

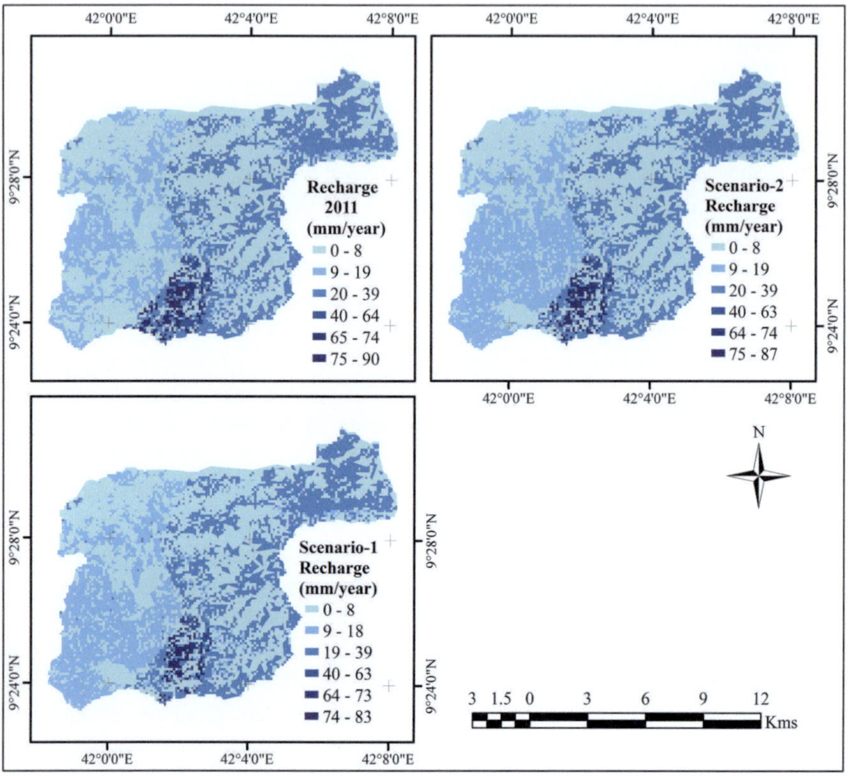

Fig. 6.6 WetSpass simulated yearly recharge maps in Haramaya watershed

encourage reforestation to increase the depleted forest of the watershed. Observing the result from scenario-1, baseline scenario, if the land use and land cover modifications continues at the present rate, recharge contribution toward groundwater from chat/shrubs, cultivated land, and grass land will significantly decrease. The impact of this scenario can easily be observed in Fig. 6.6. The maximum ground recharge potential in 2011 was 90 mm but it decreased to 83 mm under the assumption of scenario-1. Therefore, protecting the remaining water body as well as forest cover must be given attention in order to reduce further damage to the existing environment.

As shown in Table 6.4, soil type has impact on groundwater recharge contribution. Three different types of soils are found in the watershed, clay, clay loam, and sandy clay loam (Fig. 6.4d). The sandy clay loam soil is located in the southern low land area and north eastern part of the watershed. In the current and both future land use scenarios, sandy clay loam soil contributes almost half of the groundwater recharge while clay and clay loam soils each contribute around quarter to groundwater recharge. For example in 2011, the groundwater recharge contribution from sandy clay loam soil was 50.5 % while clay and clay loam soils contributes

Table 6.4 Groundwater recharge contribution of different soil types

LULC classes	2011 (%)	2028	
		Scenario 1 (%)	Scenario 2 (%)
Clay	22.7	25.2	24.6
Clay loam	26.8	28.5	27.1
Sandy clay loam	50.5	46.3	48.3

22.7 and 26.8 %, respectively. Looking at the results from scenario-1 and scenario-2, the groundwater recharge contribution from the three soil types is not much different from the result obtained in 2011.

In general, comparing the result of simulated groundwater recharge under baseline scenario with that of good management and recent land use scenarios, decrease in groundwater recharge is observed. This is mainly due to the decline in forest cover. Cultivated land, chat/shrubs, and grass land with sandy clay loam soil have higher amount of groundwater recharge in the watershed. This is as a result of good permeability of the soil with its location in the gentle topography. This result implies that the low groundwater recharge contributions from decreasing forest plantation need to be improved.

6.5 Conclusions

This chapter presents investigation on the impact of land use and land cover changes on groundwater recharge system of Lake Haramaya watershed in Ethiopia. Analysis of historical trends of climatic data (1970–2010) in the watershed projected a 0.7 °C increase in temperature by 2028, which consequently affect the groundwater recharge rate. Rainfall data shows no specific trends. In this study, the CLUE-S was simulated using CLUE-S model to project the impact of land use changes under current land management (baseline) and good land management scenarios. The result from the simulation revealed an increase in chat production and decline in cultivated land. Chat production in the watershed is a very important practice due to sociocultural and economic factors.

In order to understand the impact of land use and land cover changes on groundwater recharge of the watershed, WetSpass water balance model was used. The model uses the output from CLUE-S model for simulation of recharge. The simulation result revealed a decreasing trend in groundwater recharge in the watershed under both scenarios. Compared with recent groundwater recharge rate, model results showed decrease in groundwater recharge rate. This is due to the fact that forest is cleared continuously. In this study, the impact of groundwater abstraction on groundwater recharge is not taken into account due to the limited availability of abstraction data. Thus, there is a need for more hydrogeological studies in order to evaluate the impact of both reduced recharge and increased abstraction on available groundwater. According to this study, the watershed will experience groundwater level decline as a result of land use changes.

Unregulated abstraction of groundwater in the watershed is continuing with no due regards to its sustainability. Besides, there is a drastic change in the land use land cover of the watershed. This will negatively affect the ecosystem service of the watershed. Forest cover decline and loss of water body in the watershed are anticipated results of land use land cover changes. Therefore, it is important to evaluate future groundwater recharge improvements with better water resources management. In this study, the application of modeling tools is demonstrated to show consequences of land use change on groundwater under baseline and good land management scenarios. Watershed and water management with enforcement is a necessity for sustainable groundwater water use.

Acknowledgments This study is funded by the Engineering Capacity Building Program of Ethiopia (ECBP) of the Ministry of Education, Ethiopia. The authors acknowledge the support rendered by the National Meteorological Agency (NMA).

References

Al Kuisi M, El-Naqa A (2013) GIS based spatial groundwater recharge estimation in the Jafr basin, Jordan—application of WetSpass models for arid regions. Rev Mex Cienc Geol 30 (1):96–109

Assen M (2011) Land use/cover dynamics and its implications in the dried Lake Alemaya watershed, eastern Ethiopia. J Sustain Dev Afr 13(4):267–284

Batelaan O, DeSmedt F (2007) GIS-based recharge estimation by coupling surface–subsurface water balances. J Hydrol 337:337–355

Batelaan O, DeSmedt F (2001) WetSpass: a flexible, GIS based, distributed recharge methodology for regional groundwater modelling. In Impact of Human Activity on Groundwater Dynamics, Maastricht

Bronstert A (2004) Rainfall-runoff modeling for assessing impacts of climate and land-use change. Hydrol Process 18:567–570

Chandola V, Vastavai RR (2010) Multi-temporal remote sensing image classification—a multi-view approach. In: The proceedings of the 2010 conference on intelligent data understanding

Chhabra A, Geist H, Houghton RA, Haberl H, Braimoh AK, Vlek PLG, Patz J, Xu J, Ramankutty N, Coomes O (2006) Multiple impacts of land-use/cover change. In: Land-use and land-cover change: local processes and global impacts. Springer, Berlin

CSA (1994) The 1994 population and housing census of Harari region Volume II analytical Report. Central Statistical Agency, Addis Ababa, Ethiopia

CSA (2007) Population and housing census report-Harari region – 2007. Central Statistical Agency, Addis Ababa, Ethiopia

Dams J, Woldeamlak ST, Batelaan O (2008) Predicting land-use change and its impact on the groundwater system of the Kleine Nete catchment, Belgium. Hydrol Earth Syst Sci 12:1369–1385

DeFries R, Eshleman KN (2004) Land-use change and hydrologic processes: a major focus for the future. Hydrol Process 18:2183–2186

DeKoning GHJ, Verburg PH, Veldkamp A, Fresco LO (1999) Multi-scale modelling of land use change dynamics in Ecuador. Agric Syst 61(2):77–93

DeSmedt F, Liu YB, Gebremeskel S (2000) Hydrologic modeling on a catchment scale using GIS and remote sensed land use information. Risk Anal 2:295–304

Dewan AM, Yamaguchi Y (2009) Using remote sensing and GIS to detect and monitor land use and land cover change in Dhaka metropolitan of Bangladesh during 1960–2005. Environ Monit Assess 150(1–4):237–249

Eckhardt K, Ulbrich U (2003) Potential impacts of climate change on groundwater recharge and streamflow in a central European low mountain range. J Hydrol 284(1):244–252

EIB (Ethiopian Institute of Biodiversity) (2014) Ecosystems of Ethiopia. Retrieved from http://www.ibc.gov.et/ibc/ecosm/

FAO (2009) Sustaining communities, livestock and wildlife. A decision support tool. Project implemented by Food and Agriculture Organization of the United Nations in collaboration with African Wildlife Foundation. ILRI and Government of Tanzania on a GEF/WB fund in Monduli and Simanjiro Districts, Northern Tanzania

Franklin J, Phinn S, Woodcock C, Rogan J (2003) Rationale and conceptual framework for classification approaches to assess forest resources and properties. In: Wulder M, Franklin S (eds) Methods and applications for remote sensing of forests: concepts and case studies. Kluwer Academic Publishers, Dordrecht

Ganasri BP, Raju A, Dwarakish GS (2013) Different approaches for land use land cover change detection: a review. J Eng Technol 2(3):44–48

Getachew HE, Melesse AM (2012) Impact of land use /land cover change on the Hydrology of Angereb watershed, Ethiopia. Int J Water Sci 1(4):1–7. doi:10.5772/56266

Grimm NB, Faeth SH, Golubiewski NE, Redman CL, Wu J, Bai X, Briggs JM (2008) Global change and the ecology of cities. Science 319(5864):756–760

Mango L, Melesse AM, McClain ME, Gann D, Setegn SG (2011a) Land use and climate change impacts on the hydrology of the upper Mara River Basin, Kenya: results of a modeling study to support better resource management. Hydrol Earth Syst Sci 15:2245–2258 (Special issue: climate, weather and hydrology of East African Highlands). doi:10.5194/hess-15-2245-2011

Mango L, Melesse AM, McClain ME, Gann D, Setegn SG (2011b) Hydro-meteorology and water budget of Mara River basin, Kenya: a land use change scenarios analysis, In: Melesse A (ed) Nile River Basin: hydrology, climate and water use, Chap. 2. Springer Science Publisher, 39–68. doi:10.1007/978-94-007-0689-7_2

Mas J-F, Kolb M, Paegelow M, Teresa M, Olmedo C, Houet T (2014) Inductive pattern-based land use/cover change models: a comparison of four software packages. Environ Model Softw 51:94–111

Melesse AM, Jordan JD (2003) Spatially distributed watershed mapping and modeling: land cover and microclimate mapping using landsat imagery part 1. J Spat Hydrol (e-journal) 3(2)

Melesse AM, Jordan JD (2002) A comparison of fuzzy vs. augmented-ISODATA classification algorithm for cloud and cloud-shadow discrimination in landsat imagery. Photogram Eng Remote Sens 68(9):905–911

Melesse AM, Weng Q, Thenkabail P, Senay G (2008) Remote sensing sensors and applications in environmental resources mapping and modeling. Spec Issue Remote Sens Nat Resour Environ Sens 7:3209–3241

Mengistu DA, Salami AT (2007) Application of remote sensing and GIS inland use/land cover mapping and change detection in a part of south western Nigeria. Afr J Environ Sci Technol 1(5):99–109

Mesfin W, Fufa B, Haji J (2011) Pattern, trend and determinants of crop diversification: empirical evidence from smallholders in eastern Ethiopia. J Econ Sustain Dev 2(8):78–89

Mkaya DM, Mutua BM, Kundu PM (2013) Evaluation of the impact of land use change on catchment hydrology: the case of Wundanyi River catchment in Taita hills, Kenya. Res J Agric Environm Manage 2(5):92–98

Mohammed H, Alamirew A, Assen M, Melesse AM (2013) Spatiotemporal mapping of land cover in Lake Hardibo Drainage Basin, Northeast Ethiopia: 1957–2007. Water conservation: practices, challenges and future implications. Nova Publishers, New York, pp 147–164

Mulungu DMM, Kashaigili JJ (2012) Dynamics of land use and land cover changes and implications on river flows in Simiyu River catchment, Lake Victoria Basin in Tanzania. Nile Basin Water Sci Eng J 5(2):23–35

Reis S (2008) Analyzing land use/land cover changes using remote sensing and GIS in rize, North-East Turkey. Sensors 8(10):6188–6202
Scanlon BR, Reedy RC, Stonestrom DA, Prudic DE, Dennehy KF (2005) Impact of land use and land cover change on groundwater recharge and quality in the southwestern US. Glob Change Biol 11(10):1577–1593
Solaimani K, Arekhi M, Tamartash R, Miryaghobzadeh M (2010) Land use/cover change detection based on remote sensing data (a case study; Neka Basin). Agric Biol J North Am 1 (6):1148–1157
Stonestrom DA, Scanlon BR, Zhang L (2009) Introduction to special section on impacts of land use change on water resources. Water Resour Res 45:1–3
Tadesse N, Bheemalingeswara K, Abdulaziz M (2010) Hydrogeological investigation and groundwater potential assessment in Haromaya watershed. Eastern Ethiopia 2(1):26–48
Verburg PH, Soepboer W, Veldkamp A, Limpiada R, Espaldon V, Mastura S (2002) Modeling the spatial dynamics of regional land use: the CLUE-S model. Environ Manage 30(3):391–405
Verburg PH, Veldkamp T, Bouma J (1999) Land use change under conditions of high population pressure: the case of Java. Glob Environ Change 9(4):303–312
Warra HH, Mohammed AA, Nicolau MD (2013) Spatio-temporal impact of socio-economic practices on land use/ land cover in the kasso catchment, bale mountains, ethiopia. Geogr ser 59(1):95–120
Wondie M, Schneider W, Melesse AM, Teketay D (2011) Spatial and temporal land cover changes in the Simen Mountains National Park, a world heritage site in Northwestern Ethiopia. Remote Sens 3:752–766. doi:10.3390/rs3040752
Wondie M, Schneider W, Melesse AM, Teketay D (2012) Relationship among environmental variables and land cover in the Simen Mountains National Park, a world heritage site in Northern Ethiopia. Int J Remote Sens Appl (IJRSA) 2(2):36–43
Yacouba D, Guangdao H, Xingping W (2009) Assessment of land use cover changes using NDVI and DEM in Puer and Simao counties, Yunnan Province, China. World Rural Observations 1 (2):1–11
Yanda P, Munishi P (2007) Hydrologic and land use/cover change analysis for the Ruvu river (uluguru) and Sigi river (east usambara) watersheds. Dar es Salaam, Tanzania
Zheng X, Zhao L, Xiang W, Li N, Lv L, Yang X (2012) A coupled model for simulating spatio-temporal dynamics of land-use change: a case study in Changqing, Jinan, China. Landscape Urban Plann 106:51–61
Zhou Q, Li B, Sun B (2008) Modelling spatio-remporal pattern of landuse change using multitemporal remotely sensed imagery. The International Archives of the Photogrammetry, Remote Sens Spat Inform Sci XXXVII(B7):729–734
Zinabu TZ (2012) Ground water quality determination of former Lake Haramaya, Haramaya District, Eastern Haranghe Zone, Oromoia Regional State, Ethiopia. J Appl Sci Environ Manage 16(3):245–252

Part II
Rainfall–Runoff Processes

Chapter 7
Runoff Estimation and Water Demand Analysis for Holetta River, Awash Subbasin, Ethiopia Using SWAT and CropWat Models

Mahtsente Tibebe, Assefa M. Melesse and Birhanu Zemadim

Abstract This chapter discusses the hydrology of Holetta River, Ethiopia, its seasonal variability and water management in the watershed. Soil and Water Assessment Tool (SWAT) modeled the rainfall–runoff process of the watershed. Statistical [coefficient of determination (R^2), Nash–Sutcliffe efficiency coefficient (NSE), and index of volumetric fit (IVF)] and graphical methods were used to evaluate the performance of the model. The result showed that R^2, NSE and IVF were 0.85, 0.84 and 102.8, respectively, for monthly calibration and 0.73, 0.67 and 108.9, respectively, for monthly validation. These indicated that SWAT model performed well for simulation of the hydrology of the watershed. After modeling the rainfall runoff relation, the water demand of the area was assessed. CropWat model was applied and survey analyses were performed to calculate the water demand in the area. The total water demand for the three major users was 0.313, 0.583, 1.004, 0.873, and 0.341 million cubic meters (MCM) from January to May, respectively. The average flow obtained from SWAT simulation was 0.749, 0.419, 0.829, 0.623, and 0.471 MCM from January to May, respectively. From 5 months, the demand and the supply showed a gap during February, March, and April with 0.59 MCM. To solve the gap created by the demand, alternative source of water supply should be studied and integrated water management systems should be implemented.

Keywords Holetta River · SWAT · Water demand · CropWat · Runoff

M. Tibebe (✉)
Ethiopian Institute of Agricultural Research, P.O. Box 252, Addis Ababa, Ethiopia
e-mail: mahtsenti@yahoo.com

A.M. Melesse
Departments of Earth and Environment, Florida International University, AHC 5-390, 11200 SW 8th Street, Miami, FL 33199, USA
e-mail: melessea@fiu.edu

B. Zemadim
International Crops Research Institute for the Semi-Arid Tropics (ICRISAT) (Regional hub-WCA), BP 320 Bamako, Mali

© Springer International Publishing Switzerland 2016

A.M. Melesse and W. Abtew (eds.), *Landscape Dynamics, Soils and Hydrological Processes in Varied Climates*, Springer Geography,
DOI 10.1007/978-3-319-18787-7_7

7.1 Introduction

Ethiopia is endowed with a huge surface and ground water resources. Many perennial and annual rivers exist in the country. A number of lakes, dams, and reservoirs exist in various parts of the country. Holetta River is one of the rivers found in the upper part of Awash River basin and facing challenges of runoff variability and scarcity of water availability during the dry season. The Holetta River is the main source of surface water in the study area and it is a perennial river having three major users; Holetta Agricultural Research Center (HARC), Tesdey Farm, and Village Farmers. In addition to increasing water demand in the area, there is no facility to store the water during the rainy season for future use in the dry season. Therefore, the competition for water is increasing due to scarcity of water and increasing pressure by the expanding population and increasing irrigation. In order to alleviate this challenge, integrated water resources management (IWRM) is essential. Therefore, this chapter discusses the water availability of Holetta River and the water management in the watershed using GIS, statistical methods, and hydrological model.

7.2 Theoretical Background

7.2.1 Global Water Management and Allocation Issues

IWRM is a way of analyzing the change in demand and operation of water institutions that evaluate a variety of supply side and demand side management measures to determine the optimal way of providing water services. Demand side management includes any measure or initiative that will result in the reduction in the expected water usage or water demand. Supply side management includes any measure or initiative that will increase the capacity of a water resource or water supply system to supply water (Buyelwa 2004).

7.2.2 Hydrological Models

A hydrological model is a simplified representation of a real-world system, and consists of a set of simultaneous equations or a logical set of operations contained within a computer program. Models have parameters, which are numerical measures of a property or characteristics that are constant under specified conditions. Computer modeling offers a methodology to investigate hydrological processes and make predictions on what the flow might be in a river given a certain amount of rainfall. There are different types of models, with different amounts of complexity, but all are a simplification of reality and aim to either make a prediction or improve our understanding of biophysical processes (Davie 2008). Figure 7.1 shows different types of models (Chow et al. 1988).

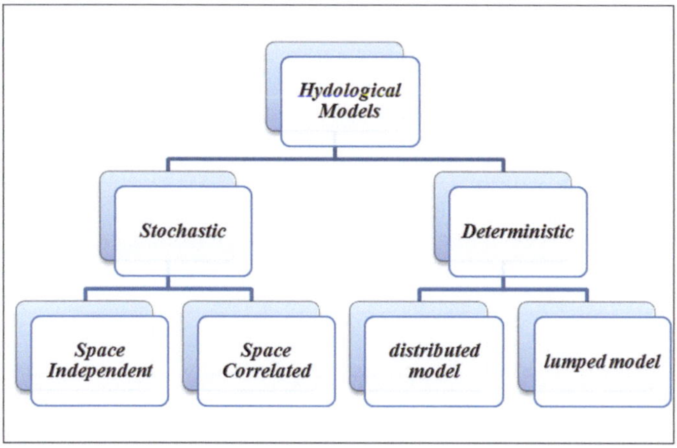

Fig. 7.1 Hydrological model classifications (Chow et al. 1988)

For this study, SWAT model was selected because it has the following capabilities:

- It is physically based and distributed model
- Was capable of operating on a watershed scale with several subbasins
- Allowed topographical, land use, and soil differences
- Was capable of simulating several management practices
- Could simulate long periods of time.

7.2.3 Description of SWAT Model

Soil and Water Assessment Tool (SWAT) is a river basin, or watershed, scale model developed by Jeff Arnold for the US Department of Agriculture (USDA)—Agricultural Research Service (ARS) (Neitsch et al. 2005). The model predicts the impact of land management practices on water, sediment, and agricultural chemical yields in large, complex watersheds with varying soils, land use, and management conditions over long periods. The model is physically based and distributed requiring specific information about soil, topography, weather, and land management practices within the watershed. The physical processes associated with water movement, sediment movement, crop growth and nutrient cycling are directly modeled by SWAT using readily available input data (Arnold et al. 1998). For modeling purposes, the watershed can be divided into a number of sub-watersheds or subbasins. Input information for each subbasin is organized into the following categories: climate, hydrological response units (HRUs), ponds/wetlands, groundwater, and the main channel or reach.

HRUs are portion of a subbasin that possesses unique land use, management, and soil attributes. A subbasin will contain at least one HRU, a tributary channel, and a main channel or reach. HRUs are used in most SWAT runs because they simplify a run lumping all similar soil and land use areas into a single response unit and it will increase the accuracy (Neitsch et al. 2004).

Simulation of the hydrology of a watershed can be separated into two major divisions. The first division is the land phase of the hydrologic cycle which controls the amount of water, sediment, nutrient, and pesticide loadings to the main channel in each subbasin. The second division is the water or routing phase of the hydrologic cycle, which refers to the movement of water, sediments, etc., through the channel network of the watershed to the outlet (Neitsch et al. 2005).

The application of SWAT in predicting stream flow and sediment as well as evaluation of the impact of land use and climate change on the hydrology of watersheds has been documented by various studies (Dessu and Melesse, 2012, 2013; Dessu et al. 2014; Wang and Melesse 2005, 2006; Wang et al. 2006, 2008a, b, c; Behulu et al. 2013, 2014; Setegn et al. 2014; Mango and Melesse 2011a, b; Getachew and Melesse 2012; Assefa et al. 2014; Grey et al. 2013; Mohammed et al. 2015).

7.2.4 Description of CropWat Model

CropWat is a decision support system developed by the Land and Water Development Division of Food and Agriculture Organization (FAO) for planning and management of irrigation (Derek et al. 1998). CropWat is a practical tool to carry out standard calculations for reference evapotranspiration, crop water requirements (CWRs), and crop irrigation requirements, and more specifically the design and management of irrigation schemes. For this study, CropWat 8.0 was used. CropWat 8.0 is a computer program for the calculation of CWRs and irrigation requirements from existing or new climatic and crop data. Furthermore, the program allows the development of irrigation schedules for different management conditions and the calculation of scheme water supply for varying crop patterns. In CropWat 8.0, the calculation of CWRs is carried out per decade.

7.3 Holetta River SubBasin Features

The following data were collected in order to characterize the features of Holetta River subbasin. All meteorological data (rainfall, temperature, relative humidity, wind speed, and sunshine hour) were collected from National Meteorology Agency (NMA) and HARC. Flow data and GIS data (topographic, land use/cover data and map, soil map) were collected from Ministry of Water and Energy (MoWE). Primary data of crop type and area coverage were collected from major water users of Holetta River (HARC, Tsedey Farm, and Farmers). These data were collected

Table 7.1 Geographical locations of meteorological stations

No.	Station	Record period	Coordinate		Elevation (m amsl)	Data collected
			XPR	YPR		
1	Addis Alem	1994–2004	475,810.95	981,592.52	2100	Rainfall
2	Holetta	1994–2004	447,252.34	1,003,731.64	2395	Climate and flow
3	Kimoye	1994–2004	423,058.00	998,462.26	2260	Rainfall
4	Welenkomi	1994–2004	423,058.00	996,021.93	2160	Rainfall

from literature, field survey, and questionnaire. One of the meteorological stations (Holetta) is found inside the watershed. The other meteorological stations, found outside the watershed are Addis Alem, Kimoye, and Welenkomi. Table 7.1 shows the geographical location of these stations. The meteorological data measured from Holetta station are Rainfall, Maximum and Minimum temperature, Relative humidity, Wind speed, and Sunshine hour. All the other meteorological stations were used only for rainfall data. The consistency, homogeneity, and outlier test for the data were performed using Excel software and XLSTAT software. The percentage of missing data for rainfall is 14 % at Addis Alem station, 13 % at Kimoye station, 1 % at Holetta station, and 18 % at Welenkomi station. Missing data were filled from observations at the three nearby stations using the normal ratio method.

7.3.1 Location and Topography of Holetta Watershed

The Holetta watershed is located in the upper part of Awash River basin, Ethiopia. It lies at an altitude of 2069–3378 m above mean sea level (amsl) and located at a latitude range of 80° 56′N to 9° 13′N and longitude range of 38° 24′E–38° 36′E. It is a watershed with drainage area of 403.47 km^2. Figure 7.2 shows the location of Holetta Watershed.

According to Food and Agricultural Organization (FAO/UNESCO 1974) slope classification, most slope in watershed (54.01 %) are flat to gently undulating with a dominant slopes ranging between 0 and 8, 41.9 % of the area is rolling to hilly with a dominant slopes ranging between 8 and 30 % and only 4.09 % are steeply dissected to mountainous with dominant slopes over 30 %. The slope classification of Holetta watershed is shown in Fig. 7.3.

7.3.2 Climate

The central and most of the eastern part of the country have two rainy periods and one dry period. These seasons are known locally as the main *Kiremt* rains from June to September, small *Belg* rains, from February to May, and dry *Bega* season from October to January. The annual rainfall of the Holetta watershed ranges between

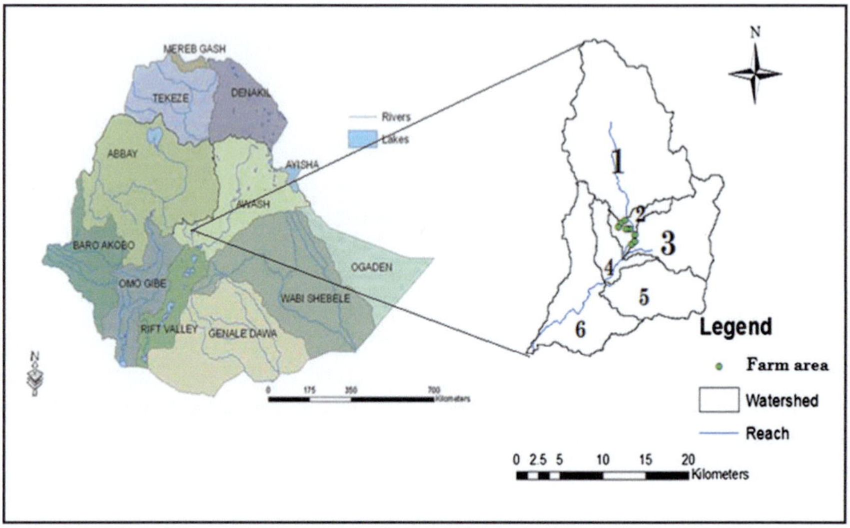

Fig. 7.2 Location of Holetta watershed

Fig. 7.3 Graphical distributions of Holetta watershed slopes

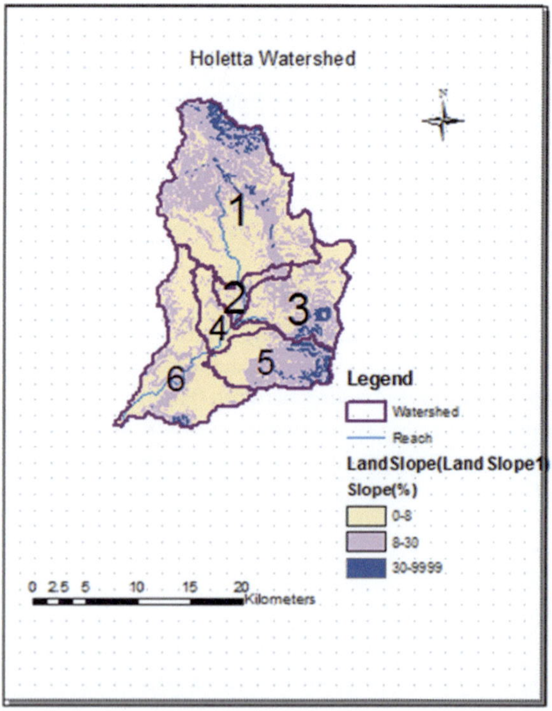

818 and 1226 mm, with a bimodal pattern of main rainy season during June to September and short rainy season during January to May. There is relatively intensive rainfall during June to August with the highest mean monthly rainfall recorded in July (243 mm). The months with the lowest rainfall are November and December.

The climate data obtained from Holetta station showed that the air temperature in the area ranges from 6 to 23 °C. The mean maximum temperature was 25 °C. Based on meteorological data from 1994 to 2004, the mean monthly relative humidity value varies from 45 to 85 %. Figure 7.4 shows the average rainfall, temperature, and relative humidity of Holetta watershed.

7.3.3 Land Use/Land Cover

The land use map of Awash River basin clipped and dissolved into Holetta River watershed. Then, the clipped land use map was used for SWAT land use reclassification. According to SWAT land use classification (Fig. 7.5), the watershed has five categories. These are, Agricultural Land-Row Crops (AGRR) with an area of 13.54 %, Agricultural Land-Close-Grown (AGRC)—0.17 %, Wetlands-Mixed (WETL)—0.14 %, Forest-Deciduous (FRSD)—57.26 %, and Forest-Mixed (FRST)—28.9 %.

7.3.4 Soil Classification

The soil map of Awash River basin clipped and dissolved into Holetta River watershed. Then, the clipped soil map was used for SWAT soil reclassification. Based on SWAT soil reclassification (Fig. 7.6), the watershed has four soil categories. These are Chromic Luvisols (Chluvisols) with an area of 33.26 %, Humic

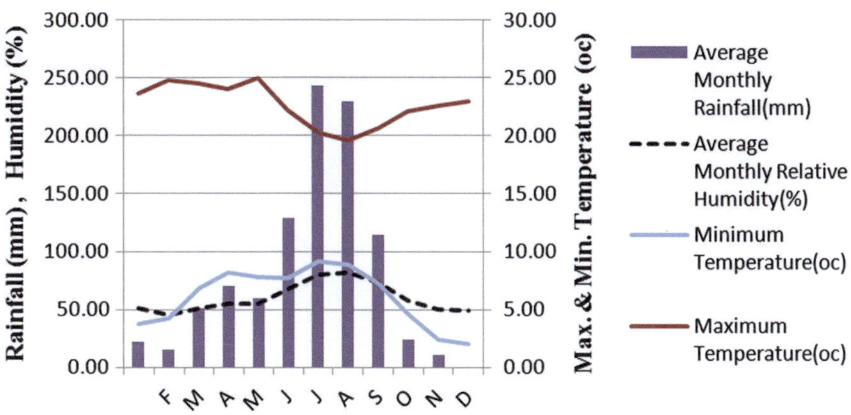

Fig. 7.4 Average rainfall, temperature, and relative humidity of Holetta watershed (1994–2004)

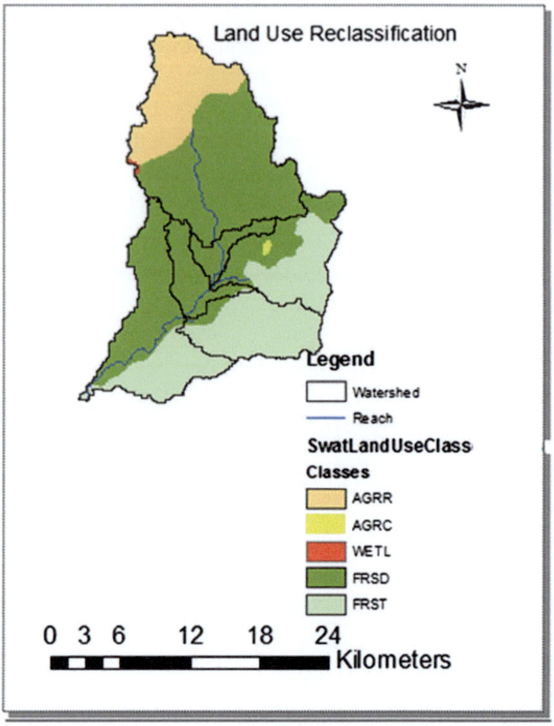

Fig. 7.5 Land use classification of SWAT model for Holetta watershed

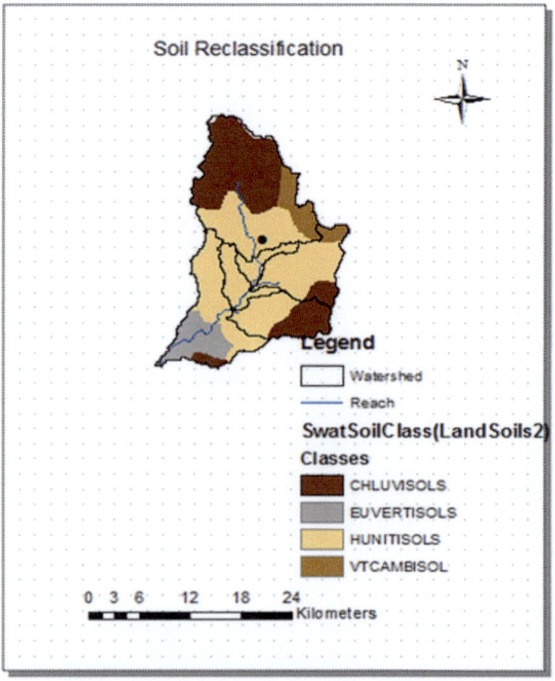

Fig. 7.6 Soil classification of SWAT model for Holetta watershed

Nitisols (Huntisols)—56.57 %, Vertic Cambisols (Vtcambisol)—1.71 %, and Eutric Vertisols (Euvertisols)—8.27 %. Based on their texture, Vtcambisol and Euvertisol are classified as clay, whereas Chluvisols and Huntisols are classified as loam (Berhanu et al. 2013).

7.3.5 Flow Data

The Holetta River is a tributary of the larger Awash River, which joins it after traveling about 25 km downstream of the gauging station. The Holetta River is the main source of surface water in the study area. The River is gauged since 1975 and for this study, the 1994–2004 time series of the river discharge data was used. The daily flow data from gauging station was used for sensitivity analysis, model calibration (1994–1999) and validation (2000–2004).

The average annual flow at Holetta River was 44 million cubic meter (MCM). The flow was low from January to May and it started to increase in June. The peak flow volume was 17 MCM, which occurred in August, and the minimum was 0.524 MCM in February. Figures 7.7 and 7.8 show the average monthly flows at Holetta River and the monthly rainfall runoff relations for Holetta subbasin, respectively.

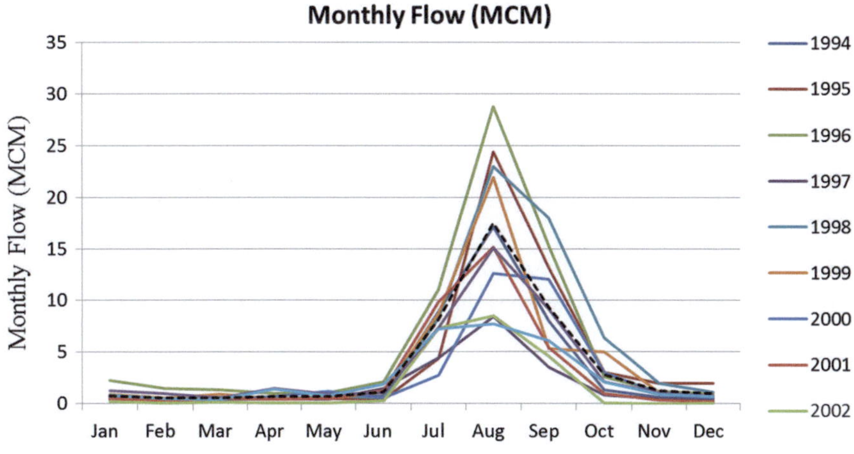

Fig. 7.7 Average monthly flows at Holetta River (1994–2004)

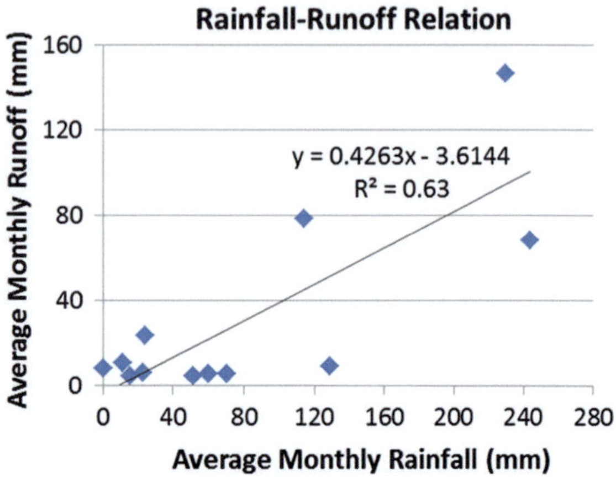

Fig. 7.8 Monthly rainfall runoff relations for Holetta subbasin (1994–2004)

7.4 Hydrological Analysis

Watershed delineation and determination of HRUs were the first step in SWAT model analysis. Then, weather station and all the necessary data were fitted. After setting and running SWAT model, sensitivity analysis, calibration, and validation were performed. In this study, the calibration and validation were performed at subbasin 1, which is found in the upper part of the watershed (Fig. 7.2). A long-term data was required for the analysis and the results are highly dependent on the accuracy of the data.

7.4.1 Watershed Delineation and Determination of HRUs

The Holetta River watershed was delineated by SWAT model and it has six subbasins. Then, the subbasins were divided into HRUs. The HRUs can be determined either by assigning only one HRU for each subbasin considering the dominant soil/land use combinations, or by assigning multiple HRUs for each subbasin considering the sensitivity of the hydrologic processes based on a certain threshold values of soil/land use combinations. In this study, a multiple HRU definition with a threshold value of 15 % for land use, 20 % for soil class, 5 % for slope were given and as a result, 33 HRUs were identified.

7.4.2 Sensitivity Analysis

Sensitivity analysis was performed for the entire period (1994–2004). About 270 iterations have been done by SWAT sensitivity analysis for flow calibration with

Table 7.2 Result of sensitivity analysis of flow at Holetta subbasin

Rank	Parameter	Description	Mean
1	Canmx	Maximum canopy storage (mm)	0.18
2	Alpha_Bf	Base flow alpha factor (days)	0.15
3	Revapmn	Threshold water depth in the shallow aquifer for "revap" (mm)	0.15
4	Gwqmn	Threshold water depth in the shallow aquifer for flow (mm)	0.06
5	Gw_Revap	Groundwater "revap" coefficient	0.06
6	Esco	Soil evaporation compensation factor	0.04
7	Cn2	Initial SCS CN II value	0.01
8	Sol_K	Saturated hydraulic conductivity (mm/hr)	0.00

the output of 26 parameters were reported as sensitive in different degree of sensitivity for flow. Among these 26 parameters, eight of them have more effect on the simulated result when changed. Based on the result of sensitivity analysis, Table 7.2 shows the most sensitive parameters for the watershed. Then, these parameters were used for calibration.

7.4.3 Model Calibration

After sensitivity analysis has been carried out, the calibration of SWAT model was done manually. The calibration was carried out using the output of the sensitivity analysis of the model and by changing the more sensitive parameter at a time, while keeping the rest of the parameters constant. The analysis of simulated result and observed flow data comparison was considered daily and monthly. The calibration was performed until the best-fit curve of simulated versus measured flow was obtained.

The sensitive parameters were adjusted based on the allowable range until the best fitting value was found. Table 7.3 showed the initial/default and finally adjusted parameter values.

The SWAT model performance was evaluated using statistical and graphical methods of comparing simulated with observed flow data. The goodness-of-fit statistics was used in describing the model's performance relative to the observed data. These statistical measures were the coefficient of determination (R^2), NSE, and index of volumetric fit (IVF) between the observations and the final best simulations. Figures 7.9 and 7.10 show the daily and monthly graphical performance evaluation of SWAT model during calibration period, respectively. Both the daily and monthly graphs implied that the model simulation is best fitted with the observed flow measurement.

The daily calibration result showed that the regression coefficient (R^2) was 0.57, NSE was 0.55, and IVF was 102.62 %. In addition, based on monthly calibration,

Table 7.3 Initial and final adjusted value of calibrated flow parameters at Holetta subbasin

No.	Parameter	Default	Range (lower–upper limit)	Final calibrated value
1	Canmx	0	0–10	10
2	Alpha_Bf	0.048	0–1	0.4
3	Revapmn	1	0–1	0.01
4	Gwqmn	0	0–5000	70
5	Gw_Revap	0.02	0.02–0.2	0.2
6	Esco	0	0–1	0.01
7	Cn2	72	±50 %	+12 %
8	Soil_K	18	0–2000	120

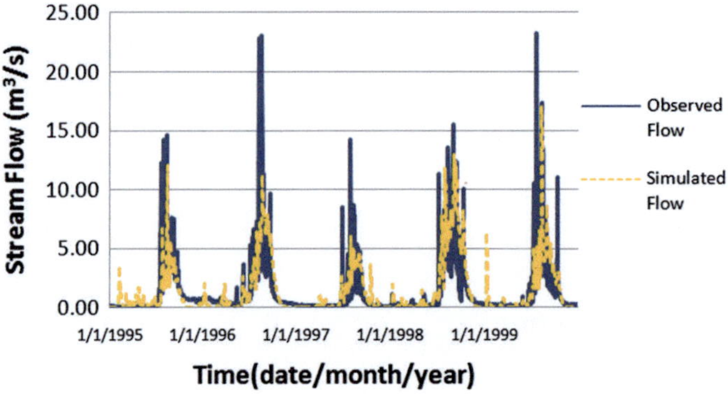

Fig. 7.9 Observed and simulated hydrograph after daily calibration

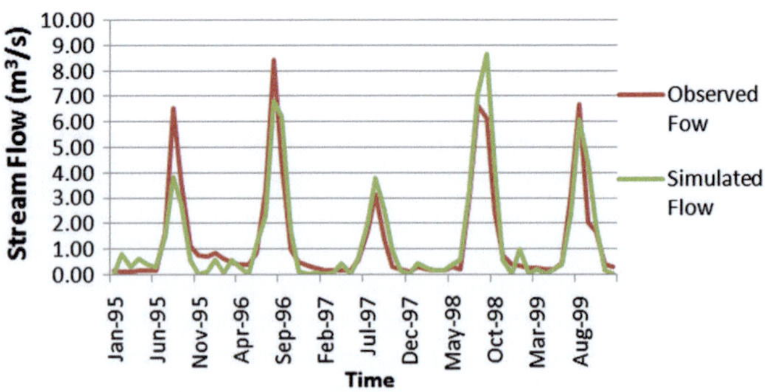

Fig. 7.10 Observed and simulated hydrograph after monthly calibration

the result showed that the regression coefficient (R^2) was 0.85, NSE was 0.84, and IVF was 102.8 % (Fig. 7.11). These indicated that the model performance was very good and highly acceptable.

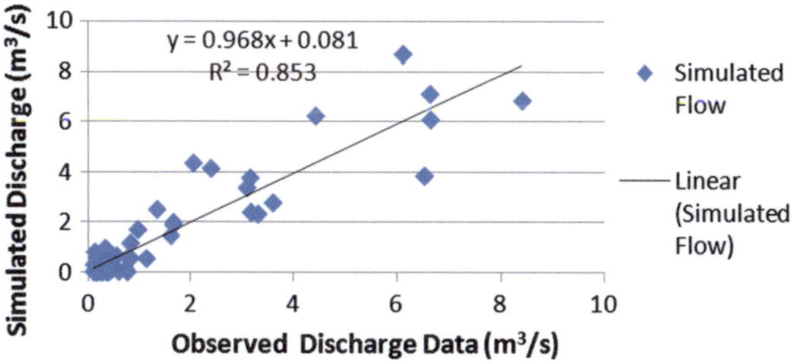

Fig. 7.11 Scattered plot and correlation between simulated and observed monthly flow during calibration

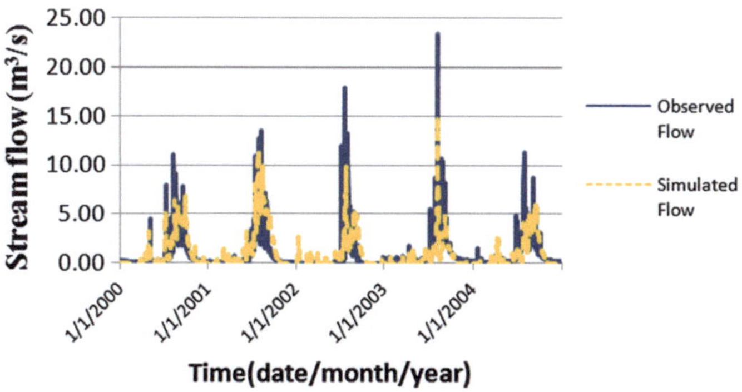

Fig. 7.12 Observed and simulated hydrograph during daily model validation

7.4.4 Model Validation

The validation process was performed by running the model for different time periods outside the calibration using the previously calibrated input parameters. Figures 7.12 and 7.13 show the daily and monthly graphical performance evaluation of SWAT model during validation period, respectively. Both the daily and monthly graphs implied that the model simulation is best fitted with the observed flow measurement.

The three goodness-of-fit measures were also calculated for the validation period. The daily calibration result showed that the regression coefficient (R^2) was 0.44, NSE was 0.4, and IVF was 108.9 %.

In addition, based on the result of monthly validation, the regression coefficient was 0.73, NSE was 0.67, and IVF was 108.9 % (Fig. 7.14). These results indicated that the model performance was good in the acceptable limit.

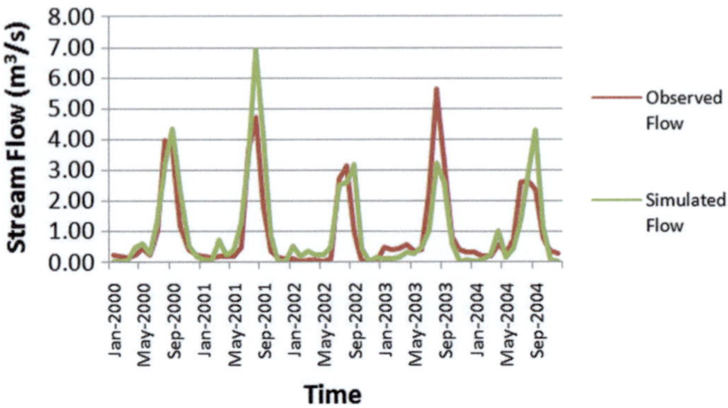

Fig. 7.13 Observed and simulated hydrograph during monthly model validation

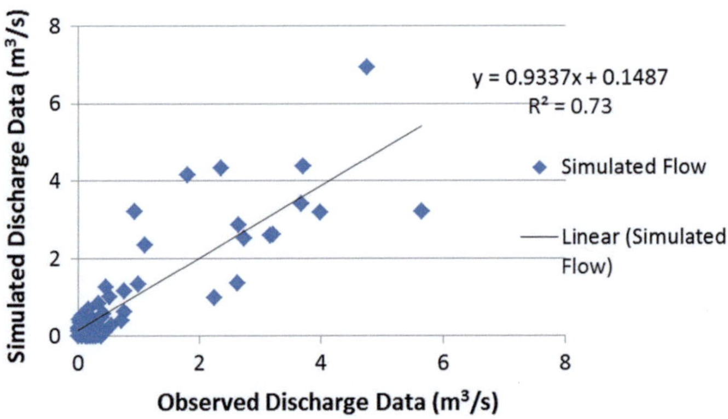

Fig. 7.14 Scattered plot and correlation between simulated and observed monthly flow during validation

7.4.5 Runoff Estimation for Holetta Watershed

The Holetta watershed was divided into six subbasins. Only one of the subbasin found in the upper part of the watershed was gauged. Calibration and validation of SWAT model were performed at subbasin 1. Then, regionalization approach was used to estimate runoff for the ungauged subbasins of the watershed.

In this study, Spatial Proximity method was used to estimate runoff at subbasins 2, 3, 4, and 5 where majority of the users were located. Figures 7.15 and 7.16 show the monthly simulation result of SWAT model at the subbasins, respectively. The mean flow (m^3/s) at the subbasin 2, 3, 4, and 5 was shown in Table 7.4.

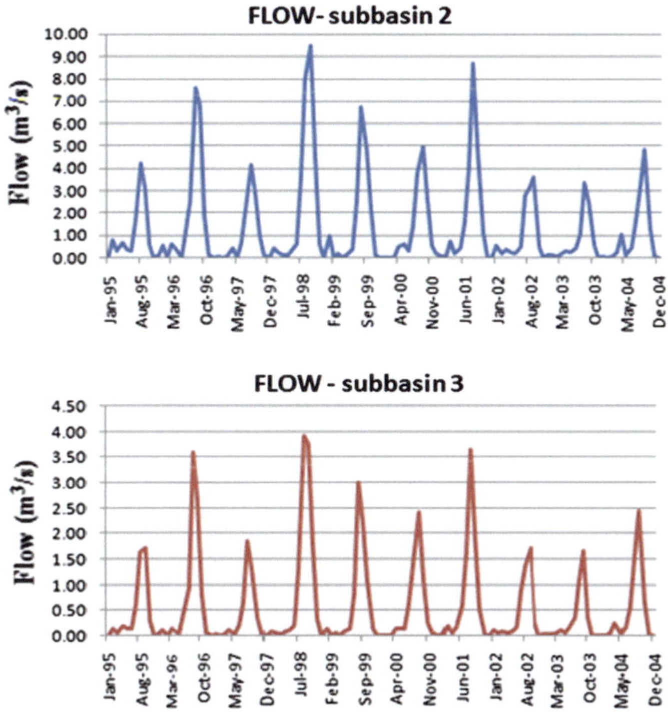

Fig. 7.15 Monthly SWAT simulation result at subbasins 2 and 3

7.5 Questionnaire Analysis

The survey form was used to identify information which includes the number of Holetta River consumers, major crops grown by irrigation, the total area coverage, conflict between users and water management system in the watershed. Overall 100 respondents were interviewed, 60 of them were from farmers, 10 from HARC, 10 from Tsedey Farm, 10 from *Kebele,* and 10 from Agricultural office. Then, the questionnaire was analyzed with Excel software and simple statistical description method was used. The majority of downstream users of Holetta River were from four *Kebele*s. These are *Medi Gudina, Dewana Lafto, Tulu Wato Dalecha,* and *Hamus Gebeya*. For detailed questionnaire survey only one *kebele* was selected which is *Medi Gudina*. *Tsedey* Farm is located at subbasins 2 and 3; HARC and *Medi Gudina kebele* are located at subbasin 2, whereas *Dewana Lafto*, T*ulu Wato Dalecha,* and *Hamus Gebeya* are located at subbasins 3, 4, and 5 (Fig. 7.17). *Kebele* is the smallest administrative unit of Ethiopia similar to a ward, a neighborhood, or a localized and delimited group of people.

According to the collected data, the majority of users have been using the river for more than 10 years and 51.67 % of the users use the river for 30–50 years. All

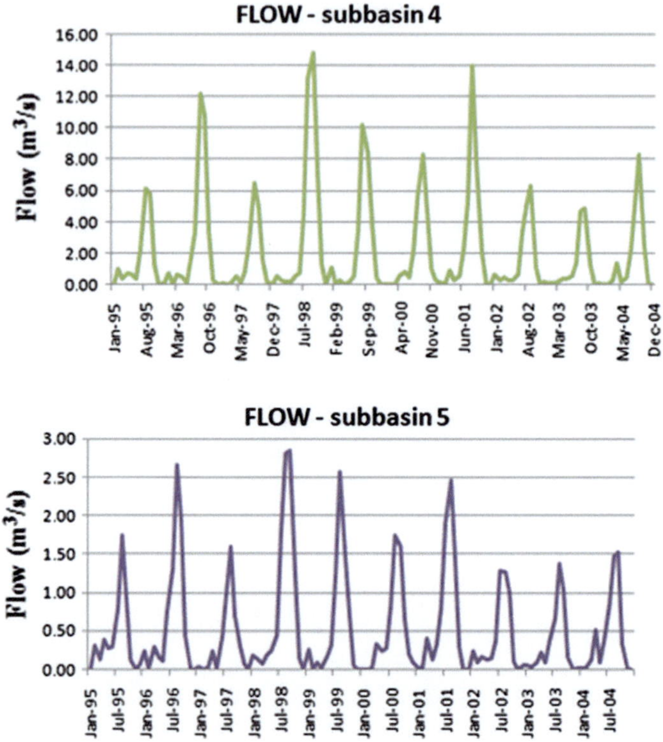

Fig. 7.16 Monthly SWAT simulation result at subbasins 4 and 5

Table 7.4 Summary of mean flow (m^3/s) at the subbasins

Subbasin	Mean daily flow (m^3/s)	Mean monthly flow (m^3/s)	Mean annual flow (m^3/s)
2	1.358	1.351	1.358
3	0.564	0.561	0.564
4	2.109	2.099	2.109
5	0.525	0.522	0.525

the farmers responded that they use the Holetta River for irrigation, livestock, and human consumption but the main use of the river is for irrigation. HARC and *Tsedey* Farm use the river only for irrigation purpose.

In the survey, it was planned to determine the major crops grown in the study area. The major crops grown are potato, tomato, cabbage, carrot, onion, and lettuce. The farmers' response showed that the three major crops are potato with 96.67 %, cabbage with 91.67 %, and tomato with 56.67 %. They use furrow irrigation to grow these crops during the off-season mainly from January to June. The area of irrigated land for each crops was about 0.25 ha. The survey also indicated that the major crops for HARC are potato, cabbage, barely, and apple. Potato, tomato, and

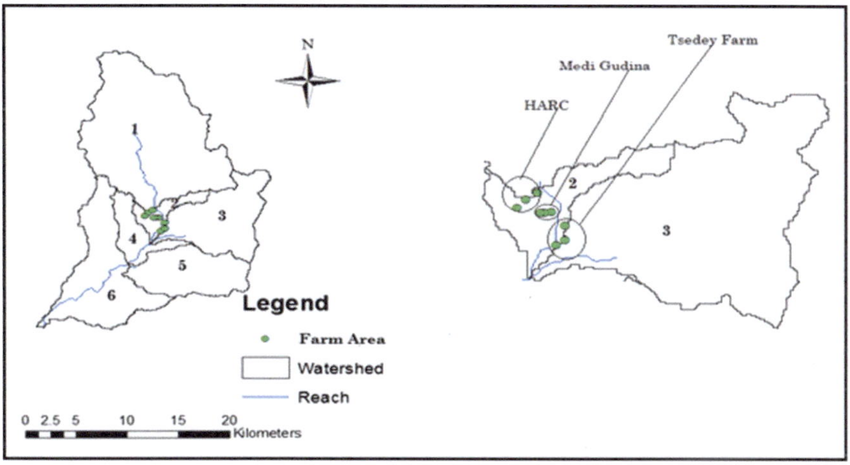

Fig. 7.17 Location of users of Holetta River

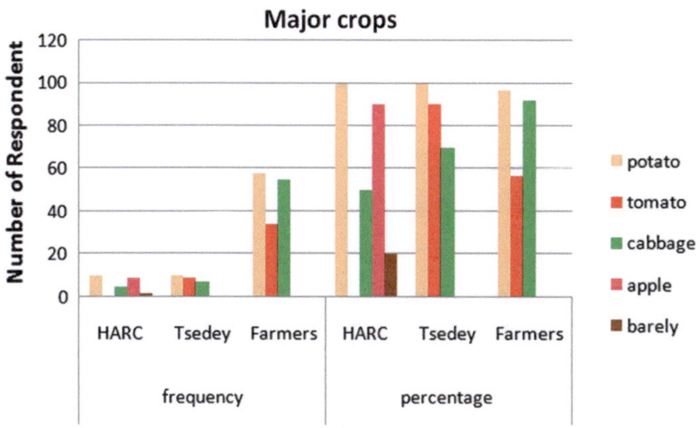

Fig. 7.18 Summary of major crops for the three users of Holetta River

cabbage are the major crops for *Tsedey* Farm. Figure 7.18 explained the major crops for the three users of Holetta River.

All the farmers responded that the only source of water for irrigation is the river and there is no alternative means, but there are springs and wells for human consumption. About 63.33 % of the farmers agreed that there is conflict between the users. On contrary, 36.67 % of the farmers replied that there is no conflict. HARC and *Tsedey* Farm respondents believed that there is a conflict between users of Holetta River. They also mentioned that this conflict mostly occurs at the turning points and during allocation of the water. Even though it is not well established, there is an irrigation committee, which settles these conflicts.

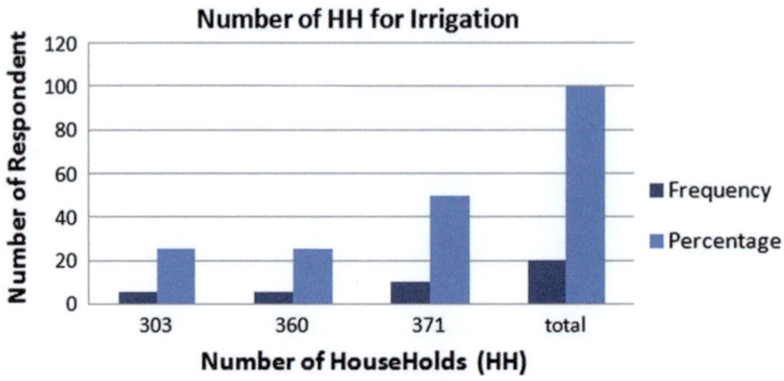

Fig. 7.19 Summary of irrigation users of Holetta River

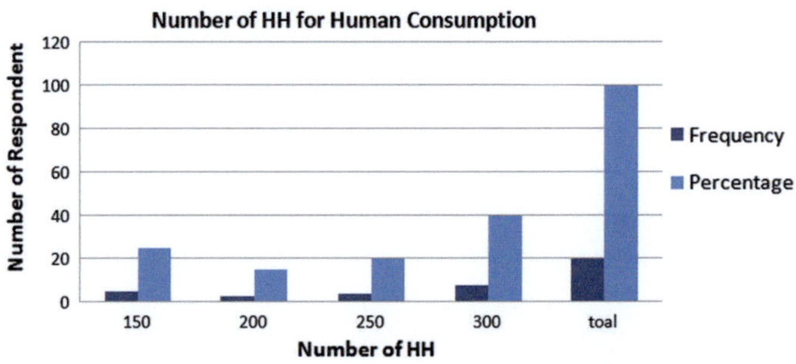

Fig. 7.20 Summary of human consumption users of Holetta River

During the survey, attempts were made to collect information about the number of households and livestock that use Holetta River at subbasin 2. According to the survey from Agricultural office and *kebele*, about 371 households use the river for irrigation purpose and 300 households use for human consumption (Figs. 7.19 and 7.20).

The collected data indicated that some of the livestock exist in the subbasin 2 were ox, cow, sheep, goat, horse, and donkey. According to the survey, the approximate number of livestock summarized in Table 7.5.

Table 7.5 Summary of livestock which uses Holetta River

Type of livestock	Number
Ox	154
Cow	250
Sheep	500
Goat	200
Horse	33
Donkey	34

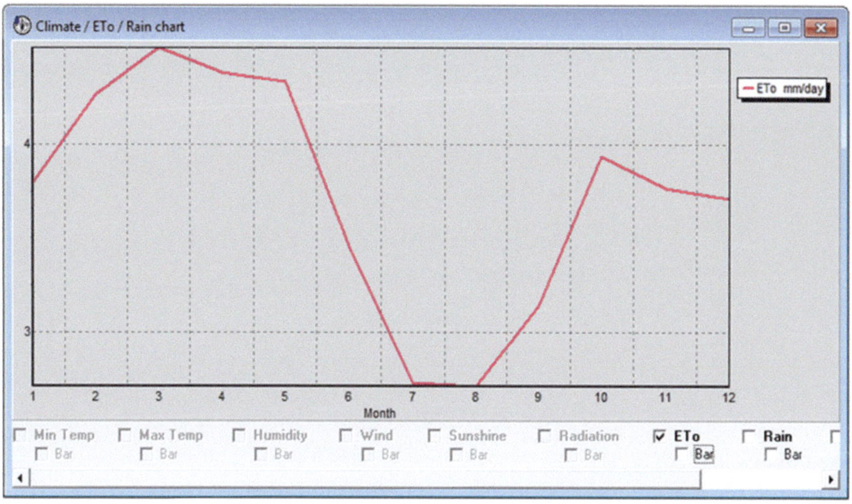

Fig. 7.21 Reference Evapotranspiration (ETo) used by CropWat 8.0

7.6 CropWat Model Analysis

Reference evapotranspiration, effective rainfall, crop pattern data, and soil data were used for CropWat model analysis. The major crops identified from the survey analysis were used in the calculation of CWR.

7.6.1 Reference Evapotranspiration

First, monthly maximum and minimum temperatures, relative humidity, sunshine hour, and wind speed data (1994–2004) were fitted in CropWat model. Then, the model calculated crop evapotranspiration values based on the FAO Penman–Montieth equation. Figure 7.21 showed the calculated reference evapotranspiration.

7.6.2 Effective Rainfall

To account for the losses due to runoff or percolation, a choice was made from the four methods given in CropWat 8.0 (Fixed percentage, dependable rain empirical formula, USDA Soil Conservation Service). Rainfall data from 1994 to 2004 was taken to calculate effective rainfall and dependable rain empirical formula has been used (Fig. 7.22).

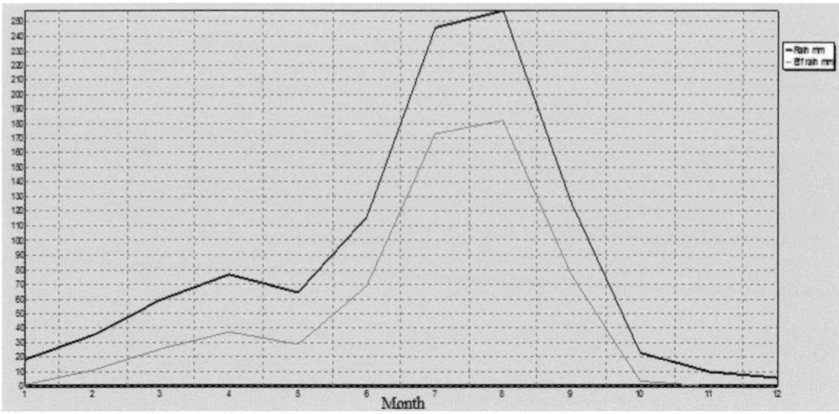

Fig. 7.22 Rainfall versus effective rain calculated by CropWat 8.0

7.6.3 Crop and Soil Data

CWR and irrigation requirements were calculated only for the major crops in the study area. The major crops are potato, cabbage, apple, and barely for HARC; potato, cabbage, and tomato for Tsedey Farm and farmers. The development stages, Kc factor, and root depth of each crop was taken from FAO-24 (Doorenbos and Pruitt 1992) and FAO-33 (Doorenbos and Kassam 1986).

The soil data required by the CropWat model includes total available soil moisture, maximum rain infiltration rate, maximum root depth, initial soil moisture depletion, and initial available soil moisture. The soil data used in the model was the same for all crops except the maximum root depth.

7.6.4 Crop Water Requirement and Irrigation Requirement

In order to estimate the water demand for agricultural use/irrigation for each crop, evapotranspiration, effective rainfall, data of crop type, area coverage, and soil data were fitted in CropWat model. The water demand of irrigation was assumed to occur during the growing season. All calculation procedures as used in CropWat 8.0 are based on the FAO-56 guidelines (Allen et al. 1998). The CWR and irrigation requirement (IR) of each crop for the entire growing period was summarized below. Table 7.6 describes the total CWR and irrigation requirement for each crop and Table 7.7 shows the irrigation requirement for the month of January to May.

Table 7.6 Estimation of total crop water requirement and irrigation requirement

Crop	CWR (mm)	Effective rain (mm)	Net IR (mm)
Potato	440.1	78.3	360.9
Cabbage	425.4	73.5	350.6
Tomato	600.8	116.8	480.8
Apple	668.7	103.6	565.0
Barely	466.2	86.7	378.7

Table 7.7 Estimation of irrigation water requirement (mm/month) for each crop

Month	Potato	Cabbage	Tomato	Barely	Apple
January	32	45.3	38.7	19.1	125
February	69.7	82.50	68	95	114.5
March	138.1	122.70	122.5	144.3	121.7
April	110.8	100.30	122.7	104.9	102.5
May	10.2		118.3	15.4	101.3

7.7 Water Demand Analysis

The result of CropWat model and survey analysis was used as an input for the calculation of water demand. The CropWat calculated the irrigation water requirement of the major crops in the area. The survey analysis indicated the area coverage and number of users of Holetta River.

Based on the result of CropWat model and survey analysis, the irrigation water demand for the three major users of Holetta River was calculated. The period was taken only for the dry seasons, from January to May. Tables 7.8, 7.9 and 7.10 show the monthly irrigation requirement of major crops in MCM for HARC, Tsedey Farm and farmers, respectively.

The three other *kebele* farmers only differ based on the area of irrigated land. *Dewana Lafto Kebele* has 94 ha of irrigated land, *Tulu Wato Dalecha* has 150 ha, and *Hamus Gebeya* has 218 ha. Therefore, the irrigation requirement for these kebeles was summarized in Table 7.11.

Table 7.8 Monthly irrigation requirement (MCM) for each major crop of HARC

Crop type	Area (ha)	Total IR (MCM)				
		January	February	March	April	May
Potato	6	0.00192	0.004182	0.008286	0.006648	0.000612
Cabbage	3	0.001359	0.002475	0.003681	0.003009	
Apple	6	0.0075	0.00687	0.007302	0.00615	0.006078
Barely	5	0.000955	0.00475	0.007215	0.005245	0.00077
Total	20	0.01173	0.01828	0.02648	0.02105	0.00746

Table 7.9 Monthly irrigation requirement (MCM) for each major crop of Tsedey Farm

Crop type	Area (ha)	Total IR (MCM)				
		January	February	March	April	May
Potato	7	0.00224	0.004879	0.009667	0.007756	0.000714
Cabbage	5	0.002265	0.004125	0.006135	0.005015	
Tomato	6	0.002322	0.00408	0.00735	0.007362	0.007098
Total	18	0.006827	0.013084	0.023152	0.020133	0.007812

Table 7.10 Monthly irrigation requirement (MCM) for each major crop of farmers

Crop type	Area (ha)	Total IR (MCM)				
		January	February	March	April	May
Potato	92.75	0.02968	0.064647	0.128088	0.102767	0.009461
Cabbage	92.75	0.042016	0.076519	0.113804	0.093028	
Tomato	92.75	0.035894	0.06307	0.113619	0.113804	0.109723
Total	278.25	0.10759	0.20424	0.35551	0.3096	0.11918

Table 7.11 Total monthly irrigation requirement (MCM) for the four Kebele farmers

Kebele	Total IR (MCM)				
	January	February	March	April	May
Medi Gudina	0.10759	0.204236	0.355511	0.3096	0.119184
Dewana Lafto	0.03633	0.068987	0.120127	0.10459	0.040182
Tulu wato Dalecha	0.058	0.1101	0.19165	0.1669	0.06425
Hamus Gebeya	0.084293	0.160026	0.278545	0.242533	0.093188
Total	0.286213	0.543348	0.945832	0.823622	0.316804

Then, the total monthly irrigation requirement (IR) for all the three major users was added and summarized (Table 7.12). Based on the analysis, the total irrigation water demand of all the three users was 0.305, 0.575, 0.995, 0.865, and 0.332 MCM for January, February, March, April and May, respectively.

Tsedey Farm and HARC use the river only for irrigation purpose but the farmers further use the river for human consumption and livestock. Therefore, the water demand for human consumption and livestock was calculated for the farmers.

Table 7.12 Total monthly irrigation requirement (MCM) for all major users of Holetta River

Total IR for the three (MCM)				
January	February	March	April	May
0.304774	0.5747088	0.99546775	0.8648068	0.33207550

Water demand for livestock and human consumption was estimated by multiplying the number of user/consumer by standard consumption

$$\mathrm{CR} = \frac{N * q * t}{1000} \quad (7.1)$$

where CR is human and livestock consumptive requirement (m^3);
N is the consumer size (number);
q is the consumptive rate (liter/day) and, t is the number of days.

Based on the above formula, the monthly human consumption at *Medi Gudina Kebele* was calculated and showed in Tables 7.13, 7.14 and 7.15. The monthly livestock consumption at the same *Kebele* was calculated and showed in Tables 7.16, 7.17 and 7.18. The total human consumptive requirement was 0.00279, 0.0025, 0.00279, 0.0027, and 0.0279 MCM for January, February, March, April and May, respectively. According to the result, total livestock consumptive requirement was 0.0059, 0.0053, 0.0059, 0.0057, and 0.0059 MCM for January, February, March, April and May, respectively (Table 7.19).

Monthly value of irrigation requirement, human consumptive requirement, and livestock consumptive requirement was added in order to get the overall water demand of the three major users of Holetta River. Finally, Table 7.19 summarizes the total water demand requirement of each month for all the three users.

The total water demand of all three major users was 0.313, 0.583, 1.004, 0.873 and 0.341 MCM for January, February, March, April and May, respectively. The available river flow from January to May was taken from the result of SWAT simulation at subbasins 2, 3, 4, and 5. The flow taken is the inflow (m^3/s) at each subbasin. The average flow was 0.749, 0.419, 0.829, 0.623 and 0.471 MCM for January, February, March, April and May, respectively. From 5 months, the

Table 7.13 Human consumptive requirement for January, March, and May

Description	Quantity	t (days)	N (number)	q (liter/day)	Consumptive requirement CR(m^3) = $N*q*t/1000$	Consumptive requirement CR (MCM)
# of HH	300	31	1500	15	**697.5**	**0.0006975**
# of members	5					
Liters/day	15					

Table 7.14 Human consumptive requirement for February

Description	Quantity	t (days)	N (number)	q (liter/day)	Consumptive requirement CR(m^3) = $N*q*t/1000$	Consumptive requirement CR (MCM)
# of HH	300	28	1500	15	**630.0**	**0.00063**
# of members	5					
Liters/day	15					

Table 7.15 Human consumptive requirement for April

Description	Quantity	*t* (days)	*N* (number)	*q* (liter/day)	Consumptive requirement CR (m^3) = *N***q***t*/1000	Consumptive requirement CR (MCM)
# of HH	300	30	1500	15	**675.0**	**0.000675**
# of members	5					
Liters/day	15					

Table 7.16 Livestock consumptive requirement for January, March, and May

Type of livestock	*N* (number)	*q* (liter/head/day)	*t* (days)	Consumptive requirement CR (m^3) = *N***q***t*/1000	Consumptive requirement CR (MCM)
Ox	154	45	31	214.83	
Cow	250	130	31	1007.5	
Sheep	500	7.5	31	116.25	
Goat	200	7.5	31	46.5	
Horse	33	45	31	46.035	
Donkey	34	45	31	47.43	
Total				1478.545	0.001478545

Table 7.17 Livestock consumptive requirement for February

Type of livestock	*N* (number)	*q* (liter/head/day)	*t* (days)	Consumptive requirement CR (m^3) = *N***q***t*/1000	Consumptive requirement CR (MCM)
Ox	154	45	28	194.04	
Cow	250	130	28	910	
Sheep	500	7.5	28	105	
Goat	200	7.5	28	42	
Horse	33	45	28	41.58	
Donkey	34	45	28	42.84	
Total				1335.46	0.001335460

Table 7.18 Livestock consumptive requirement for April

Type of livestock	*N* (number)	*q* (liter/head/day)	*t* (days)	Consumptive requirement CR (m^3) = *N***q***t*/1000	Consumptive requirement CR (MCM)
Ox	154	45	30	207.9	
Cow	250	130	30	975	
Sheep	500	7.5	30	112.5	
Goat	200	7.5	30	45	
Horse	33	45	30	44.55	
Donkey	34	45	30	45.9	
Total				1430.85	0.001430850

Table 7.19 Overall summary of total water demand and supply at Holetta watershed

	January	February	March	April	May
Total IR for the three (MCM)	0.30477425	0.5747088	0.99546775	0.86480675	0.3320755
Human consumptive requirement CR (MCM)	0.0027900	0.002520	0.0027900	0.0027	0.0027900
Livestock consumptive requirement CR (MCM)	0.00591418	0.00534184	0.00591418	0.0057234	0.00591418
Total (MCM)	0.313	0.583	1.004	0.873	0.341

Table 7.20 The summary of available flow and water demand in the study area

	January	February	March	April	May
Flow (MCM)	0.749	0.419	0.829	0.623	0.471
Total water demand (MCM)	0.313	0.583	1.004	0.873	0.341
Difference	0.436	−0.164	−0.175	−0.25	0.13

demand and the supply showed a gap during February, March, and April. This indicated that there is shortage of supply during these months with 0.59 MCM (Table 7.20).

7.8 Summary

The study was conducted to estimate runoff at Holetta watershed and to model rainfall runoff relation in the area. The study also analyzed the water demand and the gap between the river water supply and demand.

The rainfall runoff process of the watershed was modeled by SWAT. According to SWAT classification, the watershed was divided into 6 subbasins and 33 HRUs. Only one subbasin at the upstream side was gauged. Therefore, sensitivity analysis, calibration, and validation of the model were performed at this subbasin and then the calibrated model was used to estimate runoff for the ungauged part of the watershed. The result of sensitive analysis showed that 26 parameters were sensitive; out of 26, eight of them are the most sensitive ones. These parameters were used for model calibration.

The performance of the model was evaluated by statistical and graphical methods. The statistical methods used were coefficient of determination (R^2), NSE, and IVF. The result showed that R^2, NSE, and IVF were 0.85, 0.84 and 102.8, respectively, for monthly calibration and 0.73, 0.67 and 108.9, respectively, for monthly validation. Therefore, this indicated that SWAT model performed well for simulation of the hydrology of the watershed.

After modeling the rainfall runoff relation and studying availability of water at the Holetta River, the water demand in the area was assessed. CropWat model was used to calculate the irrigation water requirement for major crops and the area coverage was determined from the questionnaire. The study identified the three major users of Holetta River that is Holetta Agriculture Research Center, Tsedey Farm, and village farmers. Based on the analysis, the total irrigation water demand of all the three users was 0.305, 0.575, 0.995, 0.865, and 0.332 MCM for January, February, March, April, and May, respectively. In addition to irrigation, the farmers use the river for livestock and human consumption. Therefore, the study also included the water demand for livestock and human's use. According to the result, livestock consumptive requirement was 0.0059, 0.0053, 0.0059, 0.0057 and 0.0059 MCM for January, February, March, April and May, respectively. The human consumptive requirement was 0.00279, 0.0025, 0.00279, 0.0027 and 0.00279 MCM for January, February, March, April and May, respectively. Overall, the water demand in the area was 0.313, 0.583, 1.004, 0.873, and 0.341 for January, February, March, April and May, respectively. The available river flow from January to May was taken from the result of SWAT simulation at subbasins 2, 3, 4, and 5. The average flow was 0.749, 0.419, 0.829, 0.623, and 0.471 MCM for January, February, March, April and May, respectively. From 5 months, the demand and supply showed a gap during February, March, and April. Therefore, it is possible to conclude that there is shortage of river water supply during February, March, and April comparing the water demand with the available river flow at the same months. The total shortage of supply during these months was 0.59 MCM.

In addition to shortage of water supply, the analysis of the questionnaire indicated that there is a conflict between users at diversion points and during water allocation. There is an irrigation committee to settle this conflict but the conflict has become more and more a concerning issue in the area.

The water demand analysis showed that there was shortage of river water supply for February, March, and April. During these months, there was also conflict between users at diversion and water allocation. Therefore, in order to solve water shortage, alternative source of water supply like groundwater and water harvesting technologies should be studied and integrated water management system should be implemented. In addition to this, to improve the efficiency of irrigation water, different irrigation methods like drip irrigation should be improved in the area.

In order to minimize the conflict, well-established irrigation committee including all the users with a clear guide and management rules is required and water allocation system should be developed. In addition to this, water management and irrigation training should be improved in the area in order to establish river management system and to properly use the scarce water resource.

Acknowledgments The authors would like to express their deepest thanks for Holetta Agriculture Research Center, Ministry of Water and Energy, Ethiopian Institute of Water Resource and United States Agency for International Development (USAID)/HED for all their support.

The chapter is part of the MSc thesis work by the first author.

References

Allen GR, Pereira SL, Raes D, Smith M (1998) Crop evapotranspiration—guidelines for computing crop water requirements. FAO irrigation and drainage paper no 56, Rome

Arnold JG, Sirinivansan R, Muttiah RS, Williams JR (1998) Large area hydrologic modeling and assessment, part 1: model development. J Am Water Resour Assoc 34(1):73–89

Assefa A, Melesse AM, Admasu S (2014) Climate change in upper Gilgel Abay river catchment, Blue Nile Basin Ethiopia, In: Melesse AM, Abtew W, Setegn S (eds) Nile river basin: ecohydrological challenges, climate change and hydropolitics, pp 363–388

Behulu F, Setegn S, Melesse AM, Fiori A (2013) Hydrological analysis of the upper tiber basin: a watershed modeling approach. Hydrol Process 27(16):2339–2351

Behulu F, Setegn S, Melesse AM, Romano E, Fiori A (2014) Impact of climate change on the hydrology of upper Tiber river basin using bias corrected regional climate model. Water Resour Manage: 1–17

Berhanu B, Melesse AM, Yilma Seleshi Y (2013) GIS-based hydrological zones and soil geo-database of Ethiopia. Catena 104:21–31

Buyelwa S (2004) Water conservation and water demand management strategies for the water services sector. Ministry of Water Affairs and Forestry, South Africa

Chow VT, Maidment DR, Mays LW (1988) Applied hydrology. McGraw-Hill, USA

Davie T (2008) Fundamentals of hydrology. Taylor and Francis e-Library, UK

Derek C, Martin S, El Khaled (1998) CropWat for Windows: user guide. University of Southampton, Southampton

Dessu SB, Melesse AM (2012) Modeling the rainfall-runoff process of the mara river basin using SWAT. Hydrol Process 26(26):4038–4049

Dessu SB, Melesse AM (2013) Impact and uncertainties of climate change on the hydrology of the mara river basin. Hydrol Process 27(20):2973–2986

Dessu SB, Melesse AM, Bhat M, McClain M (2014) Assessment of water rcsources availability and demand in the mara river basin. CATENA 115:104–114

Doorenbos J, Kassam AH, Bentvelsen CL, Branscheid V, Plusje JM, Smith M, Uittenbogaard GO, VanDerWal HK (1986) Yield response to water. FAO irrigation and drainage paper 33, FAO, Rome

Doorenbos J, Pruitt WO, Aboukhaled A, Damagnez J, Dastane NG, Van Den Berg C, Rijtema PE, Ashford OM, Frere M, FAO field staff (1992) Crop water requirement. FAO irrigation and drainage paper 24, FAO, Rome

FAO/UNESCO (1974) Soil map of the world, vol 1. Food and Agriculture Organization of the United Nations and UNESCO, Paris

Getachew HE, Melesse AM (2012) Impact of land use/land cover change on the hydrology of Angereb watershed, Ethiopia. Int J Water Sci 1(4):1–7. doi:10.5772/56266

Grey OP, Webber DG, Setegn SG, Melesse AM (2013) Application of the soil and water assessment tool (SWAT model) on a small tropical island state (Great River Watershed, Jamaica) as a tool in integrated watershed and coastal zone management. Int J Trop Biol Conserv 62(3):293–305

Mango L, Melesse AM, McClain ME, Gann D, Setegn SG (2011a). Land use and climate change impacts on the hydrology of the upper Mara River Basin, Kenya: results of a modeling study to support better resource management. (Special issue: climate, weather and hydrology of East African Highlands). Hydrol Earth Syst Sci 15: 2245–2258. doi:10.5194/hess-15-2245-2011

Mango L, Melesse AM, McClain ME, Gann D, Setegn SG (2011b) Hydro-meteorology and water budget of Mara river basin, Kenya: a land use change scenarios analysis, In: Melesse A (ed) Nile river basin: hydrology, climate and water use, Chapter 2, 39–68. Springer Science Publisher. doi: 10.1007/978-94-007-0689-7_2

Mohammed H, Alamirew T, Assen M, Melesse AM (2015) Modeling of sediment yield in Maybar gauged watershed using SWAT, northeast Ethiopia. CATENA 127:191–205

Neitsch SL, Arnold JG, Kiniry JR, Srinivasan R, Williams JR (2004) Soil and water assessment tool input/output file documentation. Texas Water Resources Institute, Collage Station, Texas

Neitsch SL, Arnold JG, Kiniry JR, Srinivasan R, Williams JR (2005) Soil and water assessment tool theoretical documentation. Texas Water Resources Institute, Collage Station, Texas

Setegn SG, Melesse AM, Haiduk A, Webber D, Wang X, McClain M (2014) Spatiotemporal distribution of fresh water availability in the Rio Cobre watershed, Jamaica. CATENA 120:81–90

Wang X, Melesse AM (2005) Evaluations of the SWAT model's snowmelt hydrology in a Northwestern Minnesota watershed. Trans ASAE 48(4):1359–1376

Wang X, Melesse AM (2006) Effects of STATSGO and SSURGO as Inputs on SWAT model's snowmelt simulation. J Am Water Resour Assoc 42(5):1217–1236

Wang X, Melesse AM, Yang W (2006) Influences of potential evapotranspiration estimation methods on SWAT's hydrologic simulation in a Northwestern Minnesota watershed. Trans ASAE 49(6):1755–1771

Wang X, Shang S, Yang W, Melesse AM (2008a) Simulation of an agricultural watershed using an improved curve number method in SWAT. Trans Am Soc Agri Bio Eng 51(4): 1323–1339

Wang X, Yang W, Melesse AM (2008b) Using hydrologic equivalent wetland concept within SWAT to estimate streamflow in watersheds with numerous wetlands. Trans Am Soc Agri Bio Eng 51(1):55–72

Wang X, Garza J, Whitney M, Melesse AM, Yang W (2008c) Prediction of sediment source areas within watersheds as affected by soil data resolution. In: Findley PN (ed) Environmental modelling: new research, Chap 7, pp 151–185, Nova Science Publishers, Inc., Hauppauge, NY 11788. ISBN: 978-1-60692-034-3

Chapter 8
Spatiotemporal Variability of Hydrological Variables of Dapo Watershed, Upper Blue Nile Basin, Ethiopia

Mulatu L. Berihun, Assefa M. Melesse and Birhanu Zemadim

Abstract This chapter discusses the results of an experimental hydrological study in Dapo watershed located at the head of the upper Blue Nile Basin in Ethiopia. Detailed spatial measurements of rainfall and soil moisture were conducted at five rain gauge stations and fifteen soil moisture stations of varying depths. It was observed that there is relatively higher consistency of rainfall data in four manual rain gauge stations. Spatial and temporal distribution of soil moisture (%) shows uniform pattern of distribution at different sample depths and most of the stations showed high degree of correlation between the station values. However, unlike the soil moisture, the groundwater level at three automatic monitoring stations showed relatively differing pattern of distribution as indicated by the low degree of correlation among station values. Measurements also showed that the baseflow contribution is greater than the direct runoff contribution at the outlet of the watershed indicating streamflow generation in the watershed dominated by subsurface flow.

Keywords Hydrological processes · Dapo watershed · Blue Nile River basin · Rainfall · Soil moisture · Groundwater · Baseflow

M.L. Berihun (✉)
Faculty of Civil and Water Resources Engineering, Bahir Dar Institute of Technology, Bahir Dar University, P.O. Box 26, Bahir Dar, Ethiopia
e-mail: mulatuliyew@yahoo.com

A.M. Melesse
Department of Earth and Environment, Florida International University, Florida, USA

B. Zemadim
International Crops Research Institute for the Semi- Arid Tropics (ICRISAT), Bamako, Mali

© Springer International Publishing Switzerland 2016
A.M. Melesse and W. Abtew (eds.), *Landscape Dynamics, Soils and Hydrological Processes in Varied Climates*, Springer Geography,
DOI 10.1007/978-3-319-18787-7_8

8.1 Introduction

In many parts of the world, catchments are ungauged and this is particularly true in the developing world, where gauging networks are in decline as a result of lack of resources to install and maintain and human resources to operate (Mazvimavi 2003). This severely hampers runoff prediction in these catchments. Analysis of land use changes and the impact on runoff prediction in the research catchment is difficult without the existence of long-term hydrometeorological data. According to Gelfan (2005), a large body of hydrometeorological observations and measurements has been accumulated for decades and allowed hydrologists to form general understanding of the main processes which control runoff generation under different physiographic and climatic conditions. Moreover, understanding of the hydrological processes in the area required additional information beyond conventional data such as rainfall and discharge data, which are still invaluable. Hence, the International Water Management Institute (IWMI) worked on three watershed research areas in the highlands of Blue Nile basin to study rainwater management strategies in an effort to improve rural livelihoods. Dapo watershed is one of the IWMI research areas located in Diga District, west of Addis Ababa, capital city of Ethiopia. In this watershed, IWMI established hydrological and meteorological data monitoring instruments. These instruments are intended to provide information on watershed hydrological processes (i.e., water budgets and partitioning of rainfall between soil moisture, groundwater, and runoff) and provide baseline data for hydrological modeling. The IWMI led establishment of monitoring networks in Dapo watershed is presented in the work of Zemadim et al. (2013). Hydrological processes within a watershed define how precipitation reaches the watershed outlet, how long water is stored in the surface water, soil water, and groundwater systems (Jochen et al. 2008).

Rainfall and runoff do not generally provide sufficient information for a single determination of hydrological response through solution of the model inverse problem. Therefore, the identification of hydrological processes requires further observations or investigations within the catchment to characterize dominant water flow pathways (Latron and Gallart 2008).

Nevertheless, the characterization of dominant runoff processes is not an easy task, especially when such processes occur below the soil surface. In consequence, even if surface runoff processes have often been well identified, as infiltration or saturation excess overland flow, processes occurring below the surface are still not properly understood (Beven 1989). Therefore, it is very important to understand the hydrological process of the watershed to characterize the water fluxes.

Soil moisture is a key state variable that controls hydrological process at various spatial scales below the surface. The spatial distribution of soil moisture has been studied in both ground-based and remote sensing measurements. Wendroth et al. (1999) studied the spatiotemporal patterns and covariance structures of soil water status, and revealed that plot spatial patterns of soil water status were persistent in time. Ridolfi et al. (2003) investigated the soil moisture dynamics along a hill slope and showed that the stochastic dynamics are the result of complex and nonlocal interactions between climate, soil, vegetation, and hill slope shape. Soil in the valley bottom rapidly becomes saturated during a storm (Dubreuil 1985), sometimes expanding the saturated valley bottom area and rainfall on the saturated area results in saturation overland flow (Frankenberger et al. 1999). Identification of patterns of soil moisture response to rainfall and especially the vertical dynamics of soil moisture at the hill slope or plot scale can be useful for the investigation of runoff generation processes in a previously ungauged or data scarce catchment (Blume et al. 2007). Therefore, it is very important to understand the hydrological process of the watershed to characterize the water fluxes under consideration of different hydrological data below the surface.

Hydrology of the Nile River basin has been studied by various researchers. These studies encompass various areas including streamflow modeling, sediment dynamics, teleconnections and river flow, land-use dynamics, climate change impact, groundwater flow modeling, hydrodynamics of Lake Tana, water allocation and demand analysis (Melesse et al. 2009a, b, 2011, 2014; Abtew et al. 2009a, b; Yitayew and Melesse 2011; Chebud and Melesse 2013, 2009a, b; Setegn et al. 2009a, b, 2010; Getachew and Melesse 2012; Melesse 2011; Abtew and Melesse 2014a, b, c; Dessu and Melesse 2012, 2013; Dessu et al. 2014).

In Dapo watershed, hydrological research to identify the watershed behavior and hydrological process has not been conducted despite its importance and location in the headwater of the Blue Nile River.

The objective of this research reported in this chapter is, therefore, to understand hydrological processes in the Dapo watershed, using limited continuous data on rainfall, groundwater table, and soil moisture dynamics. The analysis concentrates on rainfall, runoff, soil moisture, and groundwater level and their relationships at a daily scale to define some general characteristics of the response of the catchment.

8.2 Description of the Study Area

8.2.1 Location

Dapo watershed (Fig. 8.1) is located in East Wollega in Diga District 28 km from Nekemte town, southwest of Addis Ababa and it covers an area of 17 km^2. The Diga District is located in the southwest of the Blue Nile (Abbay) basin and is bordered in the northeast by Guto Gida District, and in the west by the Didessa River, one of the major tributaries to Blue Nile River, on the north by Sasiga District and on the south and southeast by Jimma Arjo and Leka Dulecha Districts. The area is one of the highest rainfall receiving regions of the Ethiopian highlands. The altitude in the study area varies from 1352 to 2009 m above mean sea level (a. m.s.l). The Dapo River is a tributary of the Dimtu River and it is perennial. Point flow monitoring instrument is installed at a bridge located on the main road from

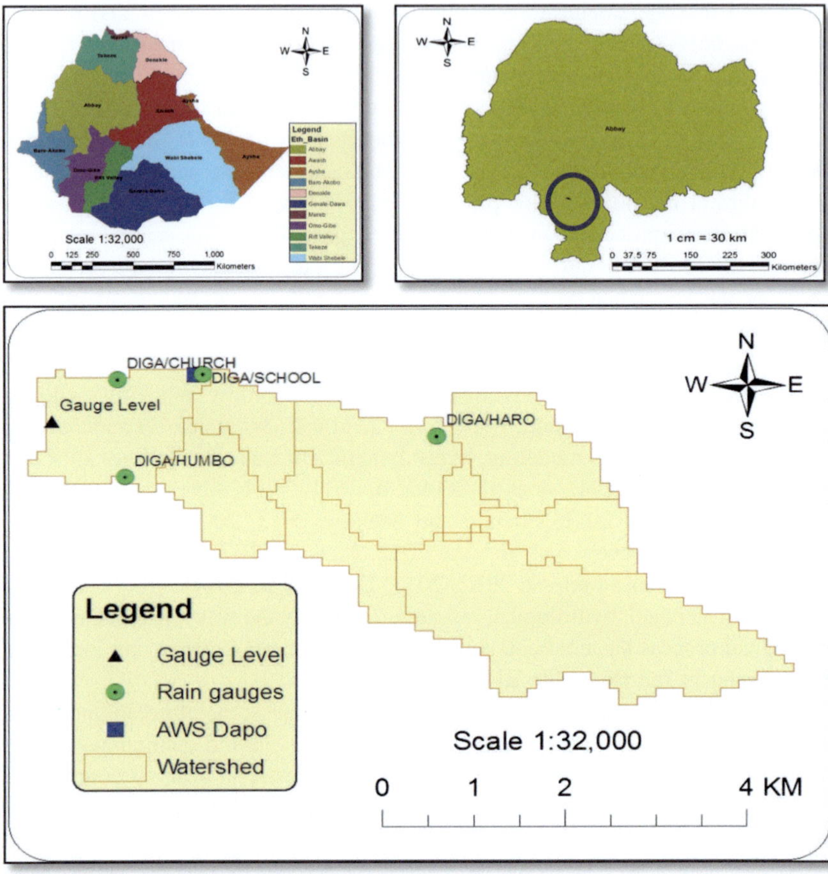

Fig. 8.1 Location of Dapo watershed

Diga to Arjo (09°03.141′N, 36°17.650′E, altitude 1347 m.a.s.l) (Zemadim et al. 2010; Zemadim et al. 2011, 2013).

Dapo watershed is one of the research sites established by the IWMI where hydrological and meteorological data monitoring instruments are operational since August 2011 (Zemadim et al. 2013). Currently, IWMI handed over the established monitoring station in Dapo watershed to the Ministry of Water Irrigation and Energy (MoWIE).

8.2.2 Climatic Characteristics

The area is one of the highest rainfall receiving regions of the Ethiopian highlands. In the watershed, there are four ordinary rain gauges and one automatic weather station (AWS). The watershed lies under a warm, humid climate, with typical total annual rainfall of 4300 mm; rainfall mostly occurring between May and October in a monomodal distribution. The highest rainfall occurs between July and August and similarly, the minimum rainfall is between November and May. The major rainfall during the rainy season, between June and September, accounted for 70 % of the total annual rainfall. Total annual potential evaporation, precipitation, discharge, and average temperature are approximately 1670 mm, 4300 mm, 1200 mm, and 23.35 °C, respectively.

8.2.3 Topography, Soils, and Land-Use/Cover

Topography is defined by a digital elevation model (DEM), which describes the elevation of any point in a given geographic area at a specific spatial resolution in a digital file. DEM is one of the essential inputs required to delineate the watershed. DEM is used to analyze the slope, flow direction and accumulation, and stream orders of the watershed. The DEM used in this study was obtained from Shuttle Radar Topography Mapping (SRTM) with a resolution of 90 m × 90 m. The watershed has a shape of open book and the river flows nearly at the center of the watershed from southern to western direction (Fig. 8.1).

The watershed is in the slope category from 0 to 60 % according to the Food and Agriculture Organization (FAO) slope classification (Table 8.1). The slope range of 8–15 % covers the largest area of the watershed (39.2 %) followed by a 32 % areal coverage with a slope of 15–30 % (Table 8.1).

The major type of soil in the watershed has 100 % Haplic Alisols. Agro-silviculture and Agropastoral are the most dominant land covers in the watershed. Currently, the watershed is used for agricultural production, mostly maize. The land from the middle to downstream of the watershed and around river edge is difficult for agricultural production because of the presence of large trees. Crops are planted once a year but there is small plateau area at the outlet of the watershed which is planted twice a year.

Table 8.1 Slope classification and area coverage of Dapo watershed

Slope range (%)	Area (km^2)	Area coverage (%)
0–2	0.15	0.93
2–5	0.97	5.95
5–8	2.07	12.69
8–15	6.39	39.24
15–30	5.23	32.12
30–60	1.4	8.62
>60	0.07	0.44

8.3 Hydrometeorology

The primary data of rainfall, runoff, soil moisture, and groundwater level were collected from August 2011 to November 2012 from the networks of monitoring stations. Several factors were considered in designing the networks in the watershed, including land use and land cover, topography, drainage pattern, agronomic conditions, local climate patterns (based on indigenous knowledge), and equipment safety and accessibility (Zemadim et al. 2013). The presence of varied land cover and land use necessitated the establishment of soil moisture probes and groundwater monitoring devices in rain-fed farmland, irrigated farmland, grazing area, and inside or close to eucalyptus plantations. Transects of soil moisture probes, groundwater monitoring devices, and rain gauges were installed to cover a range of elevations. Both manual stream-level gauges (i.e., stage boards) and automatic stream-level recording gauges were installed at the watershed outlets and upstream on selected tributaries. Data capturing resolution from the different instruments ranges from hourly to daily.

8.3.1 Rainfall

Rainfall data was collected and measured by means of four manual rain gauges and one automatic rain gauge which are located at different physiographic locations of the watershed. The period of data was selected for each station taking into consideration the availability of continuous observations.

Continuous daily data of rainfall, air temperature, relative humidity, net radiation, soil temperature, and wind direction and wind speed at 2 m height were collected from AWS.

Rainfall data screening from the five rain gauge stations was first done by visual inspection of the daily data. A comparative analysis of the available data series for all stations is made to determine how the stations can be combined and used in the estimation of runoff coefficient and in the model simulation. Comparison of temporal and spatial distribution of rainfall data at four manual and one automatic rain gauge stations was done on daily and monthly time steps. Correlation coefficient was used to analyze the spatial relationships among the five rain gauge stations at daily and monthly time steps. Areal rainfall data was determined using Thiessen polygon method to account for the spatial variability.

Based on density and mode of rainfall data, rainfall events in the study area were grouped into five classes: light rainfall, moderate rainfall, heavy rainfall, rain storm, and heavy storm.

Rainfall Characteristics: The trend in daily rainfall for August 08, 2011–November 30, 2012 is plotted in Fig. 8.2. In all five rainfall stations, the rainfall pattern is similar. In the watershed, low rainfall occurs during dry season (November–April) and high rainfall occurs in the wet season (June–August). It is readily noticeable in Fig. 8.2 that much of the rainfall occurs from May to

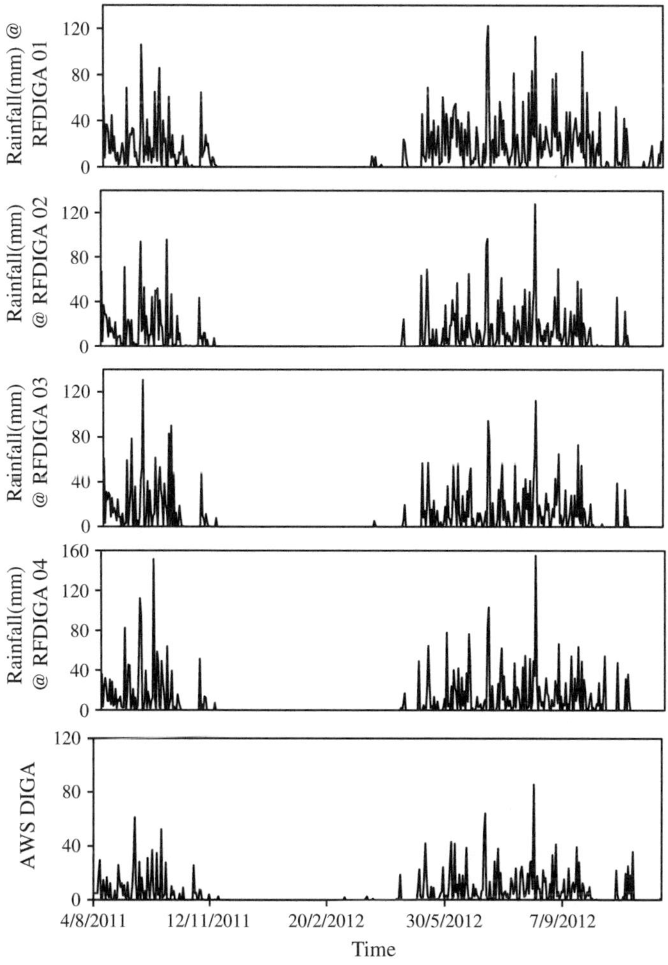

Fig. 8.2 Daily temporal and spatial pattern of rainfall at five rain gauge station and from the AWS in the Dapo watershed

September in the wet season. Wet season accounts for 61.78, 62.55, 61.01, 61.70, and 63.05 % of the annual rainfall in station RG DIGA 01, RG DIGA 02, RG DIGA 03, RG DIGA 04, and AWS DIGA, respectively. From five rain gauge stations RG DIGA 01 (DIGA/HARO) shows larger recorded total rainfall value than the other rain gauge stations because the rain gauge station is located at the highest elevation compared to the other stations (Table 8.2). The rainfall record from the weather station (AWS DIGA) shows the lowest rainfall record.

Besides RG DIGA 02 and AWS are located in the same place, however, the poor correlation coefficient between the rainfall records of the two stations (Table 8.3) implied that the rainfall records of the AWS are questionable, which probably require a local calibration.

Table 8.2 Description and location of ordinary rain gauge and automatic weather station

Station code	Location name	*X* (m)	*Y* (m)	*Z* (m)
AWS DIGA	AWS	204,065	1,002,326	1423
RG DIGA 01	DIGA/HARO	206,711	1,001,565	1548
RG DIGA 02	DIGA/SCHOOL	204,153	1,002,337	1427
RG DIGA 03	DIGA/CHURCH	203,207	1,002,277	1414
RG DIGA 04	DIGA/HUMBO	203,286	1,001,070	1456

Table 8.3 Rainfall characteristics for August 04, 2011–November 10, 2012, including the number of occurrence of precipitation event/group, amount and type of rainfall, and total rainfall

RF station name	Precipitation statistics	Light rain	Moderate rain	Heavy rain	Rain storm	Heavy storm
RG DIGA 01	Occurrence	81	80	62	21	5
	Total (mm)	378.2	1228.7	2146.6	1367.6	548.6
RG DIGA 02	Occurrence	88	58	32	19	1
	Total (mm)	366.4	968.7	1146.2	1271.5	127.6
RG DIGA 03	Occurrence	74	53	34	21	2
	Total (mm)	347.1	863.1	1189.9	1355.6	243
RG DIGA 04	Occurrence	81	47	36	17	4
	Total (mm)	376	769.1	1347.1	1111.4	553.6
AWS DIGA	Occurrence	133	50	24	5	–
	Total (mm)	443.39	807.1	788.14	313.68	–

Detailed statistics of the rainfall characteristics are given in Table 8.3. The China National Weather Bureau Classification System defines 0.1–9.9 mm rainfall as light rain, 10.0–24.9 mm as moderate rain, 25.0–49.9 mm as heavy rain, 50.0–99.9 mm as rainstorm, and 100–249.9 mm as heavy storm (Shumin 2011). While single light rains (≤10 mm rain events) are common, rainstorms-to-heavy storm (≥100 mm rain events) are rare in the study area.

Also the classification of areal rainfall of the watershed shows most of rainfall categories under light rain and only twelve events under the range of heavy storm (≥100 mm rainfall events).

Beside the visual and graphical data comparison, assessment of spatial homogeneity of the daily rainfall data indicated poor correlation between records of the five rain gauge stations. The lowest and the highest correlation coefficient (r) for a daily time step are 0.19 and 0.79, respectively (Table 8.4). Similarly, the lowest and the highest correlation coefficient at monthly time step are 0.96 and 0.99, respectively (Table 8.5).

The correlation coefficient (r) between rainfall stations RG DIGA 02 and RG DIGA 03 are 0.79 and 0.99 at daily and monthly time steps, respectively. This shows that there is consistency of data between both stations when compared to the other stations.

Table 8.4 Correlation coefficient matrix of the five rain gauge stations for the period (August 2011–November 2012) based on daily data

Correlation coefficient (*r*)	RF DIGA 01	RF DIGA 02	RF DIGA 03	RF DIGA 04	AWS DIGA
RF DIGA 01	1.00	0.67	0.60	0.67	0.29
RF DIGA 02	0.67	1.00	0.79	0.58	0.32
RF DIGA 03	0.60	0.79	1.00	0.59	0.41
RF DIGA 04	0.67	0.58	0.59	1.00	0.19
AWS DIGA	0.29	0.32	0.41	0.19	1.00

Table 8.5 Correlation coefficient matrix of the five rain gauge stations for the period (August 2011–November 2012) based on monthly data

Correlation coefficient (*r*)	RF DIGA 01	RF DIGA 02	RF DIGA 03	RF DIGA 04	AWS DIGA
RF DIGA 01	1.00	0.98	0.96	0.96	0.98
RF DIGA 02	0.98	1.00	0.99	0.99	0.98
RF DIGA 03	0.96	0.99	1.00	0.99	0.96
RF DIGA 04	0.96	0.99	0.99	1.00	0.96
AWS DIGA	0.98	0.98	0.96	0.96	1.00

Temporal distribution of areal rainfall in Dapo watershed is similar to other five rain gauges stations. Figure 8.3 shows the total annual rainfall of five rain gauge stations and its total annual areal average rainfall.

8.3.2 *Streamflow*

Streamflow Data: The flow data in the watershed was obtained from daily stage readings from August 4, 2011 to November 30, 2012 and current meter readings during wet and dry seasons by using manual gauge level. This station is located at the outlet of the watershed and has a contributing area of about 17 km^2.

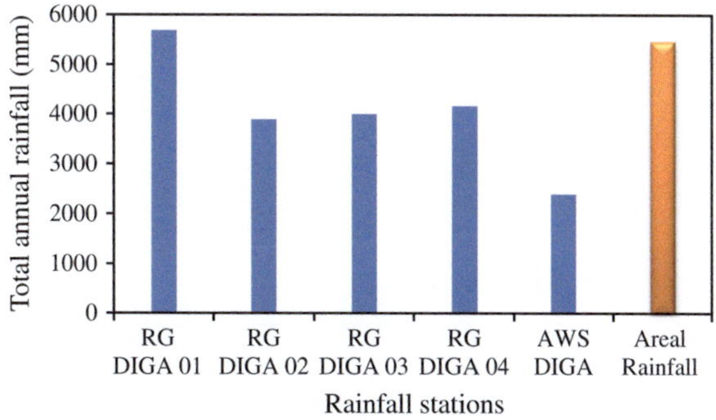

Fig. 8.3 Total annual rainfall and areal rainfall of Dapo watershed

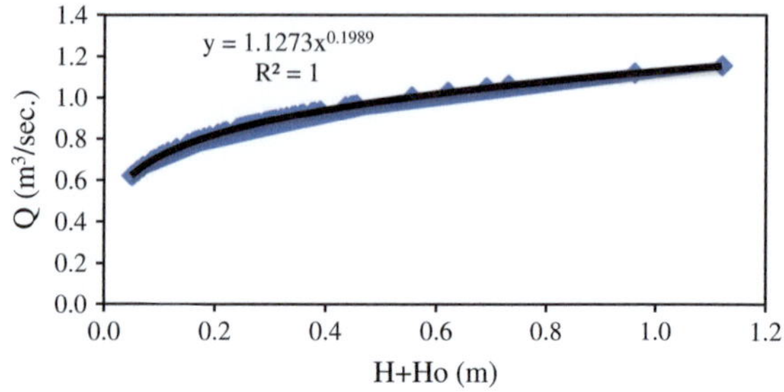

Fig. 8.4 Rating curve of Dapo watershed at the bridge measurement station

The flow station had been calibrated and a rating curve (Fig. 8.4) for stage height with corresponding runoff was established by least square solution method from which a relationship for the calculation of flow in cubic meter per second from the daily stage measurements was derived. Baseflow was subtracted from total flow to compute the event runoff coefficient. The runoff coefficient in the watershed was calculated directly by dividing direct runoff to rainfall at event bases.

Streamflow and Event Runoff Coefficient: The flow is obtained by developing a rating curve equation of $Q = 1.127(H + H_0)^{0.198}$ (Q—discharge, H—river water level, and H_0—initial river water level) by stage discharge method and R^2 = 1, the value of H_0 is 20 cm.

The river in the watershed is perennial and flow is continuous. The peak runoff occurs between August and September and decrease gradually from October up to May then the volume again increases (Fig. 8.5). The flow decreases gradually from October through May.

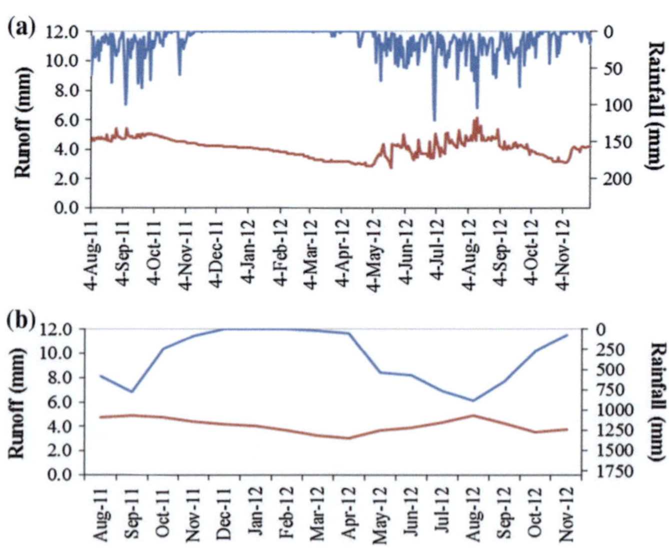

Fig. 8.5 **a** Daily and **b** Monthly temporal pattern of runoff with respect to rainfall in the watershed

The cumulative quantity of the streamflow in the watershed is 371.82 m^3/s (1972.25 mm) during the period between August 04, 2011 and November 30, 2012.

Runoff coefficients (*C*) were highly variable (0.001–0.98) with a coefficient of variation of 0.61. Most of the runoff coefficient values are smaller than 0.5 indicating that the contribution of rainfall is small to direct runoff rather it infiltrates and contributes to the subsurface flow.

8.3.3 Soil Moisture

Soil Moisture data: In Dapo watershed, four transect lines were laid to collect soil moisture data from fifteen soil moisture (SM) monitoring stations (Fig. 8.6). The transects were laid considering variations in physiographic; slope and land use/land cover conditions. The first transect line that consists of five soil moisture monitoring sites was installed at the lower portion of the watershed that neighbors the watershed outlet. The second transect line, consisting of four soil moisture monitoring stations, represents the middle part of Dapo watershed. The third transect with three soil moisture monitoring stations is located in the midway between the middle and upper part of the watershed. The fourth transect line is located in the upper part of the watershed. These monitoring stations are located with groundwater monitoring stations together at different transects of the watershed. Soil moisture data at 10, 20,

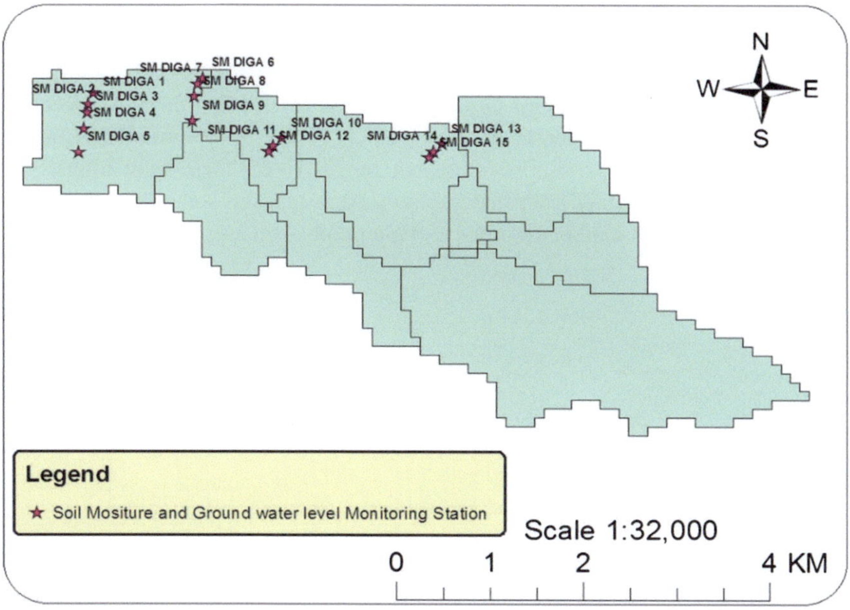

Fig. 8.6 Soil moisture and groundwater level monitoring station in the watershed

30, 40, 60, and 100 cm was measured during the field campaigns at the same sampling points using a profile probe-type PR2 with access tubes at daily time step. Access tubes are fiberglass tubes inserted into pre-augured holes in the soil.

Temporal and spatial distribution of daily and monthly average soil moisture (%) at 15 soil moisture stations of different sample depths is done in the analysis. Correlation among soil moisture stations was evaluated. The relation and threshold value of average soil moisture is made with the relation to runoff coefficient, runoff (mm), and average groundwater level (mm) at daily and monthly time steps with linear regression method.

Spatial and Temporal Variability of Soil Moisture: Soil moisture values over the watershed at 15 soil moisture stations show consistent patterns within 10–100 cm soil moisture sample depth during the study period. Similarly, the soil moisture (%) mostly increases with increasing of sample depth at all monitoring stations. This indicates the corresponding patterns for soil moisture within 10–100 cm showing less topographic influence on wetness, although low-lying flats were consistently wetter than hill slopes.

The temporal distribution of soil moisture is not smooth at daily time steps rather it shows sudden peak and sudden low values during the study period (Fig. 8.7). This is because of the daily effect of rainfall in the watershed. Similarly, at monthly time steps, temporal distribution graph of soil moisture from August 2011 to March 2012 (Fig. 8.8) shows a gradual decreasing pattern, with increasing records with the depth of the soil.

Pearson correlation coefficient of average soil moisture in 15 soil moisture stations shows spatial relation between the stations. At sample depth of 10 and 20 cm, the correlation coefficient range >0.5 accounting 96.44 and 99.11 %, respectively. At sample depth 40 and 60 cm the correlation coefficient (<0.5) is high relative to other sample depth with values of 39.56 and 39.11 %, respectively. This shows that the high variability or the relationships was not statistically significant between soil moisture station at sample depths of 40 and 60 cm (Table 8.6). Similarly, at sample depth of 100 cm the correlation coefficient ranges (0.85–1) account higher percentage compared to other sample depth indicating that the soil moisture is more associated within the station when the sample depth increases.

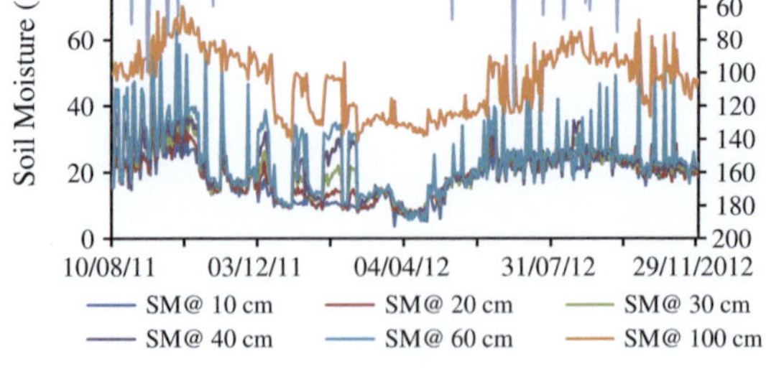

Fig. 8.7 Daily temporal distribution of average soil moisture (%) at different sample depth

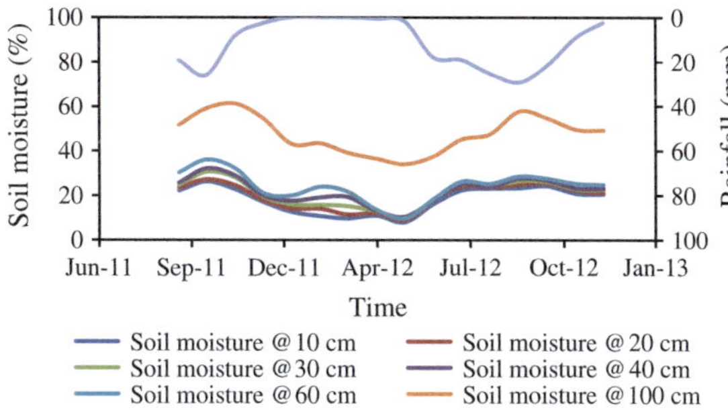

Fig. 8.8 Monthly temporal distribution of average soil moisture (%) at different sample depth

Table 8.6 Pearson correlation coefficients' ranges (%) for the relation between the average of the fifteen soil moisture measurements at 0–100 cm

Depth	% of pearson correlation coefficient range (*r*)					
	0–0.5 (%)	0.5–0.65 (%)	0.66–0.75 (%)	0.76–0.85 (%)	0.85–1.00 (%)	>0.5 (%)
10 cm	3.56	11.56	40.44	28.00	16.44	96.44
20 cm	0.89	15.11	33.78	35.11	15.11	99.11
30 cm	13.33	35.11	16.89	20.89	13.78	86.67
40 cm	39.56	11.11	8.00	23.11	18.22	60.44
60 cm	39.11	17.33	7.56	18.67	17.33	60.89
100 cm	19.56	19.11	28.89	13.33	19.11	80.44

The spatial relationships were also analyzed by linear regression analysis among four soil moisture stations' transects at daily and monthly time steps (Tables 8.7 and 8.8).

In general, the spatial relationships between each soil moisture transects were positive and highly strong at all soil moisture sample depths, especially on monthly time steps. This shows the consistency or low variability of soil moisture value between soil moisture transect.

Table 8.7 Daily spatial relationships between transects average soil moisture (%) at different sample depth

Transect 1 and 2						
Depth (cm)	10	20	30	40	60	100
R^2	0.79	0.88	0.87	0.84	0.88	0.80
Transect 3 and 4						
Depth (cm)	10	20	30	40	60	100
R^2	0.76	0.88	0.88	0.79	0.90	0.60

Table 8.8 Monthly spatial relationships between transects average soil moisture (%) at different sample depth

Transect 1 and 2						
Depth (cm)	10	20	30	40	60	100
R^2	0.90	0.98	0.97	0.92	0.97	0.91
Transect 3 and 4						
Depth (cm)	10	20	30	40	60	100
R^2	0.85	0.98	0.92	0.89	0.98	0.94

8.3.4 Groundwater

Groundwater Level (GWL) data: All groundwater monitoring stations (tube wells) were installed near soil moisture monitoring stations at each transect line during August 01–03, 2011. Totally, 15 tube wells were augured having the same geographical coordinates with soil moisture monitoring sites. Groundwater level measurement was done using dip meter on August 05, 2011. In the location where groundwater level record was observed, pressure transducers (PT) were inserted in tube wells for continuous groundwater level recording. Hence, four tube wells were set to record water level readings automatically using PT.

Spatial and Temporal Variability of Groundwater Level: Groundwater changes followed a clear temporal pattern that was repeated each water year. This involved (i) a wetting-up period, corresponding to a rise in the water table, which started with the end of spring rainfall events and lasted until regeneration of the water reserves was complete; (ii) a saturation period over summer and half of autumn, when the water table was close to the soil surface, and (iii) a drying-down period involving a progressive decline in the water table levels from the end of autumn or the beginning of winter to the beginning of summer. This is illustrated in Fig. 8.9, which shows the seasonal evolution of the daily water table depth measured at the four groundwater monitoring stations from August 2011 to November 2012. The daily rainfall and runoff is also plotted in the upper part of the figure. It is noteworthy that, in the water year 2011–2012, the rise of the water table did not start until late spring, and thus the saturation period was very short. The ground surface elevations of the three shallow wells are 1384, 1427, and 1411 m, respectively, for GWL DIGA 9, GWL DIGA 11, and GWL DIGA 12.

The decline of groundwater reserves was closely associated with the increasing water deficit observed (Fig. 8.9) from the end of autumn (November) until the end of spring (May), which was triggered by a marked increase in evapotranspiration and a decrease in rainfall. The low evapotranspiration during half of autumn and summer ensures a high water table level, even with low rainfall levels. Similarly, high runoff only occurred when the water table was close to the soil surface, highlighting the importance of catchment moisture conditions on the streamflow response under the consideration of lag time. As a result of inter-seasonal climatic variability, the length of the saturation period was variable among seasons. The

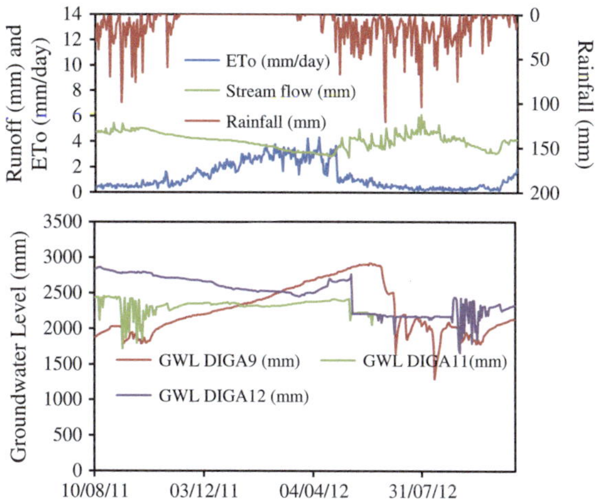

Fig. 8.9 Daily rainfall, runoff, and depth to the water table reference to original ground level at the study location from August 10, 2011 to November 10, 2012

variability of this length of the saturation period influences the annual water yield of the catchment. Among groundwater level monitoring stations, GW DIGA 9 daily and monthly water table data are more associated with runoff compared to other monitoring stations.

In addition, Fig. 8.9 shows spatial and temporal variability of the water table. This spatial variability is indicative of the degree of variability among locations. The high spatial variability of water table fluctuation was greater during the wetting-up and drying-down transition periods, especially during the wet periods, seems to be a more characteristic feature of the watershed.

Generally, during the rainy season, water table at all locations rises to soil surface and shows peaks with respect to rainfall peaks. Likewise, the runoff increases in this season and also more of contribution from baseflow.

To further investigate the spatial variability of the water table, the relationships between daily water table depths at different groundwater level monitoring locations were analyzed.

Table 8.9 shows that there was a weak correlation among the water level monitoring locations. This implies that the data is not consistent between each location. Similarly, the relationships between daily water table, runoff, and rainfall

Table 8.9 Relationship matrix of daily water table depth at different groundwater monitoring locations

R^2	GWL DIGA 9	GWL DIGA 11	GWL DIGA 12
GWL DIGA 2	0.12	0.04	0.00
GWL DIGA 9	1.00	0.05	0.00
GWL DIGA 11	0.05	1.00	0.32
GWL DIGA 12	0.00	0.32	1.00

Table 8.10 Relationship between rainfall and water table depth at different groundwater level monitoring locations

Location	GWL DIGA 9	GWL DIGA 11	GWL DIGA 12
R^2 @ daily	0.04	0.25	0.26
R^2 @ monthly	0.16	0.70	0.82

Table 8.11 Relationship between runoff and water table depth at different groundwater level monitoring locations

Location	GWL DIGA 9	GWL DIGA 11	GWL DIGA 12
R^2 @ daily	0.43	0.09	0.23
R^2 @ monthly	0.57	0.24	0.29

are not strong at daily and monthly time steps (Tables 8.10 and 8.11) except at monthly time step for DIGA11 and DIGA12 where rainfall and GWL have higher correlation.

Generally, the relationship at monthly water table with rainfall and runoff is better than the daily relationship.

8.3.5 *Relation Between Rainfall, Runoff, and Soil Moisture*

The relationship of daily average soil moisture, rainfall, and runoff analysis is done by linear regression method. The relationship between daily rainfall and runoff has an R^2 of 0.55 but both of them are not strongly associated. This is because the watershed runoff volume gets large contribution from baseflow and found that the baseflow is more than direct runoff in the watershed.

The relationship between daily and monthly average soil moisture and rainfall is shown in Table 8.12. In both cases, the relationship decrease when the soil moisture sample depth increases. Similarly, the relationship between monthly average soil moisture and rainfall is more significant when compared to daily relationships and the relationship is high at sample depth of 10 cm ($R^2 = 0.64$) and this is because the top soil attains saturation quickly responding to rainfall.

Table 8.12 Relationship between average soil moisture (%) and rainfall at different soil moisture sample depth

Soil moisture sample depth	10 cm	20 cm	30 cm	40 cm	60 cm	100 cm
R^2 @ daily	0.19	0.16	0.09	0.07	0.04	0.06
R^2 @ monthly	0.64	0.63	0.56	0.47	0.33	0.25

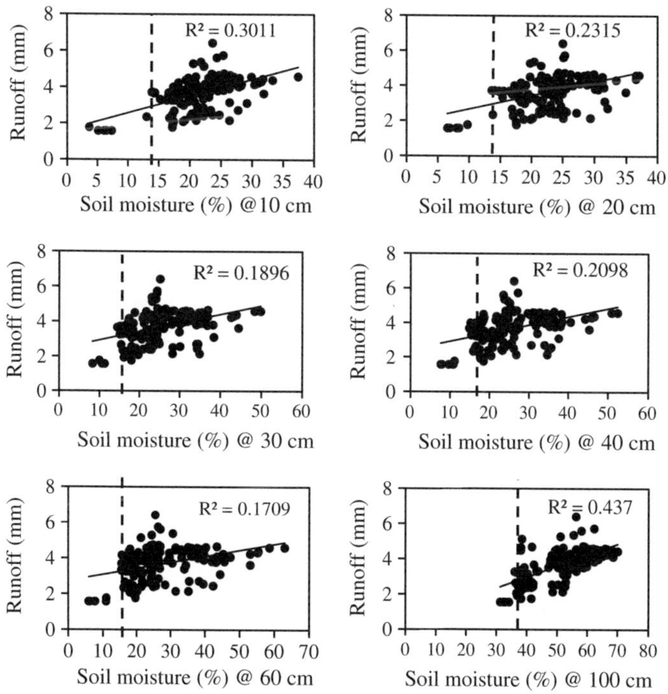

Fig. 8.10 Relationships and threshold behavior in relation between daily average soil moisture prior to the event and runoff, the vertical line highlights the soil moisture threshold

The relationship between average soil moisture (%) at 10–100 cm (defined as the mean of the 15 soil moisture measurements) and runoff coefficient for the rainfall–runoff events during the study period was strongly nonlinear and allowed the identification of a soil moisture threshold value (approximately 14 and 35 % at 10–60 and 100 cm sample depth, respectively) above which runoff significantly increased (Fig. 8.10). A clear threshold behavior was also observed in the soil moisture at 10–100 cm and groundwater relationship (Fig. 8.11). Discharge and water table level were low during dry condition and a gradual increase occurred when the 14 and 35 % at 10–60 and 100 cm sample depth, respectively, moisture threshold were exceeded.

Like relationships between average soil moisture and rainfall, the relationships between daily average soil moisture, runoff, and average groundwater level in the watershed decrease with increase of sample depth (Fig. 8.11). As the soil sample depth increases, its moisture nears to groundwater table and saturation of the soil significantly associated with rising of groundwater level. Similarly, Fig. 8.11 shows the relationships between daily average soil moisture and groundwater level. The scatter plot shows negative slope between them.

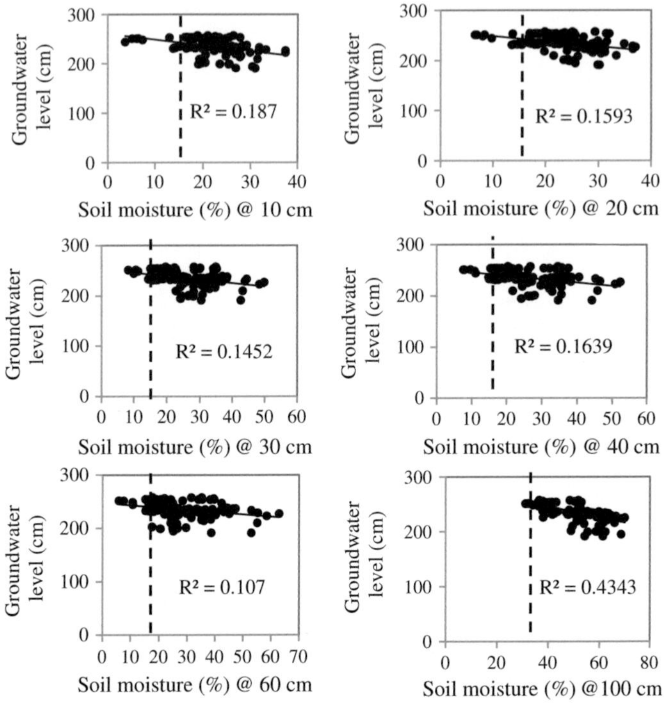

Fig. 8.11 Relationships and threshold behavior in relation between daily average soil moisture prior to the event and average groundwater level, the *vertical line* highlights the soil moisture threshold

8.4 Summary

The study area is one of the highest rainfall receiving regions of the Ethiopian highlands with a unimodal rainfall characteristics. Generally, the spatial relationships between rain gauge stations with rainfall data are more significant at monthly than daily time steps in the watershed. The AWS rainfall data was shown to be consistently less than the ordinary rain gauge values, meaning that there need to be a local calibration of the AWS rainfall record.

The river in the watershed is perennial and during the dry season, the rainfall is null and low flow occurs due to the contribution of subsurface water. The base flow is greater than direct runoff in the watershed accounting more than 85 % of streamflow.

The spatial correlation coefficient between soil moisture monitoring stations at different sample depths shows good relationships. Also the correlation of average soil moisture (%) between each soil moistures transect is very high compared to individual stations, especially during monthly time steps. Generally, the spatial and temporal variability of the soil moisture in the watershed was very low and it shows the consistency of data among the stations and transects. But the variability of soil moisture increases with increasing of soil moisture sample depth.

The temporal and spatial dynamics of the water table was investigated in the watershed. The water table followed a marked seasonal evolution involving a drying-down period of the water table from the end of autumn (associated with increasing evapotranspiration), a wetting-up period commencing with the summer rainfalls, and a wet period during the end of spring up to first of autumn, when the water table was close to the soil surface at all study locations. The water table fluctuations also varied substantially among locations and the spatial and temporal relationship between the stations was also poor, this is because of different watershed behavior on the station.

The relationship between water table levels and runoff was quite weak, especially for the daily data, suggesting little connection between groundwater and the runoff response.

Daily and monthly relationships and degree of correlation between rainfall and soil moisture decrease with increasing of soil moisture sample depth and the degree of correlation is high at monthly time steps than for the daily data. This is because the soil surface does not respond quickly to rainfall at daily time step and also the response is high as the soil moisture sample depth decreases in both daily and monthly time steps. A clear response of soil moisture and runoff to rainfall input was observed, whereas the watershed water table was less reactive, especially during the dry season.

Acknowledgments We would like to express our appreciation to the International Water Management Institute for providing the primary data and Diga research field data observers. Special thanks go to Diga catchment coordinator Mr. Tolera Megersa for providing the necessary material for the study. This chapter is part of the thesis work by the first author.

References

Abtew W, Melesse AM (2014a) Nile River basin hydrology. In: Melesse AM, Abtew W, Setegn S (eds) Nile River basin: ecohydrological challenges, climate change and hydropolitics, pp 7–22

Abtew W, Melesse AM (2014b) Climate teleconnections and water management. In: Nile River basin. Springer International Publishing, Switzerland, pp 685–705

Abtew W, Melesse AM (2014c) Transboundary rivers and the Nile. In: Nile River basin. Springer International Publishing, Switzerland, pp. 565-579

Abtew W, Melesse AM, Desalegn T (2009a) Spatial, inter and intra-annual variability of the Blue Nile River basin Rainfall. Hydrol Process 23(21):3075–3082

Abtew W, Melesse AM, Desalegn T (2009b) El Niño southern oscillation link to the Blue Nile River basin hydrology. Hydrol Process Spec Issue Nile Hydrol 23(26):3653–3660

Beven K (1989) Interflow. In: Morel-Seytoux HJ (ed) Unsaturated flow in hydrologic modeling—theory and practice, NATO ASI series C, vol 275. Kluwer, pp 191–219

Blume T, Zehe E, Bronstert A (2007) Use of soil moisture dynamics and patterns for the investigation of runoff generation processes with emphasis on preferential flow. Hydrol Earth Syst Sci Discuss 4:2587–2624

Chebud YA, Melesse AM (2009a) Numerical modeling of the groundwater flow system of the Gumera Sub-Basin in Lake Tana Basin, Ethiopia. Hydrol Process, Spec Issue Nile Hydrol 23 (26):3694–3704

Chebud YA, Melesse AM (2009b) Modeling lake stage and water balance of Lake Tana, Ethiopia. Hydrol Process 23(25):3534–3544

Chebud Y, Melesse AM (2013) Stage level, volume, and time-frequency change information content of Lake Tana using stochastic approaches. Hydrol Process 27(10):1475–1483. doi: 10.1002/hyp.9291

Dessu SB, Melesse AM (2012) Modeling the rainfall-runoff process of the Mara River basin using SWAT. Hydrol Process 26(26):4038–4049

Dessu SB, Melesse AM (2013) Impact and uncertainties of climate change on the hydrology of the Mara River basin. Hydrol Process 27(20):2973–2986

Dessu SB, Melesse AM, Bhat M, McClain M (2014) Assessment of water resources availability and demand in the Mara River basin. CATENA 115:104–114

Dubreuil PL (1985) Review of field observations of runoff generation in the tropics. J Hydrol 80:237–264

Frankenberger JR, Brooks ES, Walter MT, Walter MF, Steenhuis TS (1999) A GIS-based variable source area hydrology model. Hydrol Process 13:805–822

Gelfan AN (2005) Dynamic-stochastic models of river runoff generation. Hydrological systems modeling—vol II. Water Problems Institute of Russian Academy of Sciences, Moscow

Getachew HE, Melesse AM (2012) Impact of land use/land cover change on the hydrology of Angereb Watershed, Ethiopia. Int J Water Sci 1, 4:1–7. doi: 10.5772/56266

Jochen W, Uhlenbrook S, Lorentz S, Christian L (2008) Identification of runoff generation processes using combined hydrometric, tracer and geophysical methods in a headwater catchment in South Africa. Hydrol Sci–J–des Sci Hydrol 53(1):65–80

Latron AJ, Gallart BF (2008) Runoff generation processes in a small Mediterranean research catchment. J Hydrol 358:206–220

Mazvimavi D (2003) Estimation of flow characteristics of ungauged catchments, case study in Zimbabwe. PhD thesis, Wageningen University, Wageningen

Melesse AM (2011) Nile River basin: hydrology, climate and water use. Springer Science & Business Media, Berlin

Melesse AM, Loukas Athanasios G, Senay Gabriel, Yitayew Muluneh (2009a) Climate change, land-cover dynamics and ecohydrology of the Nile River basin. Hydrol Process, Spec Issue Nile Hydrology 23(26):3651–3652

Melesse A, Abtew W, Desalegne T, Wang X (2009b) Low and high flow analysis and wavelet application for characterization of the Blue Nile River System. Hydrol Process 24(3):241–252

Melesse A, Abtew W, Setegn S, Dessalegne T (2011) Hydrological variability and climate of the Upper Blue Nile River basin. In: Melesse A (ed) Nile River basin: hydrology, climate and water use, Chap. 1. Springer Science Publisher, Berlin, pp 3–37. doi: 10.1007/978-94-007-0689-7_1

Melesse A, Abtew W, Setegn SG (2014) Nile River basin: ecohydrological challenges, climate change and hydropolitics. Springer Science & Business Media, Berlin

Ridolfi L, D'Odoricoc P, Porporato A, Rodriguez II (2003) Stochastic soil moisture dynamics along a hillslope. J Hydrol 272:264–275

Setegn SG, Srinivasan R, Dargahil B, Melesse AM (2009a) Spatial delineation of soil erosion prone areas: application of SWAT and MCE approaches in the Lake Tana Basin, Ethiopia. Hydrol Process 23(26):3738–3750

Setegn SG, Srinivasan R, Melesse AM, Dargahi B (2009b) SWAT model application and prediction uncertainty analysis in the Lake Tana Basin, Ethiopia. Hydrol Process 24(3):357–367

Setegn SG, Dargahi B, Srinivasan R, Melesse AM (2010) Modelling of sediment yield from Anjeni Gauged watershed, Ethiopia using SWAT. J Amer Water Resour Assoc 46(3):514–526

Shumin H, Yonghui Y, Tong F, Dengpan X, Juana PM (2011) Precipitation-runoff processes in Shimen hillslope micro-catchment of Taihang Mountain, north China. Hydrol Process 26 (9):1332–1341

Wendroth O, Pohla W, Koszinski S, Rogasik H, Ritsema CJ, Nielsen DR (1999) Spatial-temporal patterns and covariance structures of soil water status in two northeast-German field sites. J Hydrol 215:38–58

Yitayew M, Melesse AM (2011) Critical water resources management issues in Nile River basin. In: Melesse AM (ed) Nile River basin: hydrology, climate and water use, Chap. 20. Springer Science Publisher, Berlin, pp 401–416. doi: 10.1007/978-94-007-0689-7_20

Zemadim B, Matthew MC, Gerba L (2010) Hydrology reconnaissance. Report CPWF Nile Project 2. International Water Management Institute, Ethiopia

Zemadim B, Matthew MC, Bharat SM, Wale Abeyou (2011) Integrated rainwater management strategies in the Blue Nile basin of the Ethiopian highlands. Int J Water Resour Environ Eng 3 (10):220–232

Zemadim B, McCartney M, Langan S, Sharma B (2013) A participatory approach for hydrometeorological monitoring in the Blue Nile River basin of Ethiopia. Colombo, Sri Lanka: International Water Management Institute (IWMI). (IWMI research report 155), p 32. doi: 10.5337/2014.200. Available online http://www.iwmi.cgiar.org/publications/iwmi-research-reports/iwmi-research-report-155/

Chapter 9
Runoff and Soil Loss Estimation Using N-SPECT in the Rio Grande de Anasco Watershed, Puerto Rico

Matilde Duque and Assefa M. Melesse

Abstract Over the last decades, the Rio Grande de Añasco watershed in the western part of Puerto Rico (PR) has been experiencing changes in land use due to conversion of agricultural lands into suburban use. The conversion contributed to sediment movements and pollutant loads to rivers and other water bodies. Agricultural practices contributed to nutrients to rivers via surface runoff and erosion. According to the US Environmental Protection Agency (EPA), concentrations of sediments from the uplands of the watershed are the main non-point sources of runoff entering the Mayagüez Bay, PR. The Non-point Source Pollution and Erosion Comparison Tool (N-SPECT) was used to study the relationships between land cover, soil characteristics, topography, and precipitation to assess spatial patterns of surface water runoff, non-point source pollution, and erosion. This paper uses N-SPECT to calculate runoff and erosion in the Rio Grande de Añasco watershed. Results show the most permeable soils are located in the northwest side of the Rio Grande de Añasco; while the highest probability of soil loss is in areas located in the west side of the watershed. The event-based runoff depth patterns coincide with the precipitation spatial patterns where the south part of the watershed, which lies in the Maricao Municipality, is expected to have major runoff events.

Keywords Rio grande de añasco watershed · N-SPECT · Event-based runoff depth · Soil loss · Phosphorus and nitrogen loads

---

M. Duque · A.M. Melesse (✉)
Department of Earth and Environment, Florida International University, 11200 SW 8th Street, Miami, FL 33199, USA
e-mail: melessea@fiu.edu

M. Duque
e-mail: mduqu001@fiu.edu

© Springer International Publishing Switzerland 2016

A.M. Melesse and W. Abtew (eds.), *Landscape Dynamics, Soils and Hydrological Processes in Varied Climates*, Springer Geography,
DOI 10.1007/978-3-319-18787-7_9

9.1 Introduction

One of the major sources of water pollution in Puerto Rico is nutrient over-enrichment concentrations and the presence of coliform bacteria in watersheds due to the lack of water quality standards and nutrient identification in impaired waters (JCA 2003). This lack of standards impedes the implementation of adequate control over nutrient and sediment concentrations on water bodies and streams needed to assess the problem (JCA 1990). According to Torres and Francisco (2009), in 2004 some sectors of Rio Grande de Añasco (RGA) watersheds were classified by the Environmental Protection Agency (EPA) into categories 1, 3, and 5. Category 1 corresponds to waters that meet the standards, category 3 refers to insufficient data to determine if the waters meet the standards, and category 5 waters that do not meet the standards.

A number of studies used laboratory, field scales, and modeling studies to understand soil erosion and sediment dynamics in various regions (Defersha and Melesse 2011, 2012; Defersha et al. 2011, 2012; Maalim and Melesse 2013; Maalim et al. 2013; Setegn et al. 2010; Melesse et al. 2011; Msagahaa et al. 2014; Wang et al. 2008; Mekonnen and Melesse 2011; Setegn et al. 2009; Mohammed et al. 2015).

Previous studies have been mainly focused on quantifying non-point source contamination by nutrients and also estimating the concentration of coliforms and fecal bacteria that may be transported by runoff. The transport of nutrients such as nitrogen (N) and phosphorous (P) to surface waters can occur via surface runoff and erosion from urban and rural agricultural lands. Sotomayor-Ramirez et al. (2004) suggest that rural watersheds may experience sediment movement associated with agricultural preparation and land construction. In addition, the major change in the watershed was caused by conversion of 10 % of the land area from agricultural to suburban use, which increased the concentrations of total phosphorous (TP) and dissolved phosphorous (DP) by 60 %. Agricultural activities, waste from ranching and crops, grass, and feedlots are frequently a non-point source of pollution of soils and waters with N.

According to the EPA, the volume of sediment coming from the upland portion of the watershed is the main non-point source runoff that enters Mayaguez Bay impacting fisheries and coral reef in the bay. The sediments in the Mayaguez Bay come from the influx of different sources affecting marine life.

The Puerto Rico Water Resource and Environmental Research Institute (PRWARERI) points out that the main contributors of nutrients loads are non-point source such as unsewered communities in urban, suburban, and rural areas, which cause eutrophication in fresh waters. Research by Ramon-Gines (1997) quantified the mean TP concentration greater than 0.1 mg P/L as the factor causing eutrophication of the lakes in central Puerto Rico. In 2001, Sotomayor et al. (2004) found that in the major rivers of Puerto Rico a trend of total phosphorus (TP) concentration fluctuated from 0.04 mg P/L (mg poured per L) to 0.29 mg P/L, and identified strong correlation between the TP and the presence of *Fecal coliform bacteria* and *Streptococcal bacteria*. They also studied the relation of land use and amount of phosphorus (P) loads in five sub-watersheds within the RGA watershed. The study quantified the TP from 0.01 to 0.180 mg/L with no significant difference among watersheds but with a positive correlation between land use such as agriculture and pasture versus urban or forest and rangeland versus urban. Data collected at USGS' station during 2007–2008 showed a 6.32 mg/L of TP discharged and 0.87 mg/L of N discharged. In 2005, Warne et al. (2005) calculated a mean annual runoff of 910 mm during 10 water-years (1900–2000) for the entire isle and also a mean annual suspended-sediment discharge between 2.7 and 9 million metric tons. In addition, they calculated the mean annual runoff for the RGA watershed in 1170 mm. This chapter estimates the annual runoff and soil loss in any place within the watershed, allowing the identification of areas with potential risk for these events to occur. Also, the study delineated the nutrient pollutions patterns.

The specific objectives of the study are to estimate (1) the soil loss within in the Rio Grande Añasco (RGA) watershed, (2) the runoff in the RGA watershed, and (3) the nutrients loading into the Mayagüez Bay, Puerto Rico.

9.2 Study Area and Gis Dataset

9.2.1 Study Area

The Mayagüez Bay Watershed, also called Rio Grande de Añasco (RGA) watershed, is one of the largest watersheds in Puerto Rico (Fig. 9.1). The main river is Rio Grande de Añasco, which begins at 1204 m above sea level and flows westward for 74 km to discharge into the Añasco/Mayagüez Bay in the West Coast of Puerto Rico (Sotomayor-Ramirez et al. 2004). The main tributaries are Guilarte, Lemani, Guayo, Prieto, Toro, Blanco, Canas, Lajas, Guabo, Bucarabones, Mayaguecilla, Arenas, and Casei. Small streams in the mountains are intermittent, while those in the lowland are perennial, and almost all are channeled for agricultural purposes.

Its predominant weather is characterized with the Caribbean patterns: the climate is warm with wet summers (from April to November), dry winters, and cooler temperatures in the mountain regions. According to the United States Geological Survey (USGS), the weighted average annual rainfall at Mayagüez is approximately 2200 mm. Mountain regions have an elevation of between 300 and 340 m above sea level, while upland areas have rugged steep topography with frequent slopes from 70 to 100 %. According to Perez-Alegria et al. (2005), during the hurricane season (June–November) intense rainfall precipitation generates soil detachment, which is transported to streams and coastal valleys around the island.

The Rio Grande de Añasco serves to drain nearly 50 000 hectares of the watersheds, which is an important socioeconomic natural resource with scenic attraction and an important ecological area in the Mayaguez Bay (Sotomayor-Ramirez et al. 2004). In the watershed, agriculture activities and natural forests are the most abundant type of land use, with agriculture being the main economic activity. Crops grown in the area are coffee, plantain, fruits, and citrus.

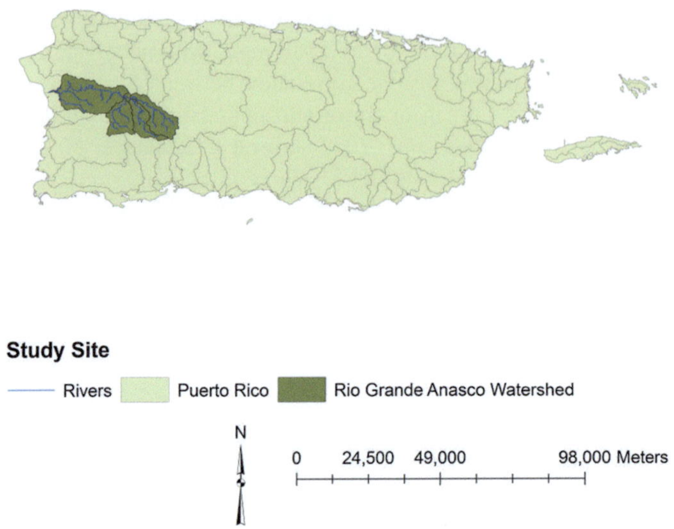

Fig. 9.1 Rio Grande de Añasco watershed, Puerto Rico

9.2.2 *GIS Dataset*

For this study, spatial and aspatial data in the format of raster, vectors, and tabular data were obtained. Raster datasets were masked to the RGA watershed, setting the snap extent to the DEM to align the grids with different sizes. Obtained data were projected to WGS_1984_UTM_Zone20 N. Vector files, such as rivers, lakes, water management units, counties, soil, and watershed shapefiles were clipped to the RGA watershed. Tabular data have been modified and converted into ASCII files prior to loading to N-SPECT.

Digital Elevation Model (*DEM*): DEM for Puerto Rico was downloaded from http://seamless.usgs.gov, processed in decimal degrees using North America Datum (NAD 83), and projected in WGS_1984_UTM_Zone20 N. Elevation values were provided in units of meters, resolution of 1 arc-second, and equivalent to approximately 30 m.

Land Cover Data: Information was obtained from the USGS (2009) and the National Land Cover Database Commonwealth of Puerto Rico. Remote sensing image with land cover layer was downloaded from www.csc.noaa.gov/crs/lca/data.html. The projection used was USA_Contiguos_Albers_ Equal_Area_Conic_ USGS_version1. Curve Number data was converted into ASCII file and land cover image was exported as a grid before loading into N-SPECT.

Precipitation: This raster describes the mean annual precipitation (mm) over Puerto Rico, derived from climate data collected at weather stations in the period 1990–2000. Precipitation was projected in WGS_1984_ UTM_Zone20N and downloaded from http://www.ncdc.noaa.gov/oa/climate/climatedata.html. Precipitation

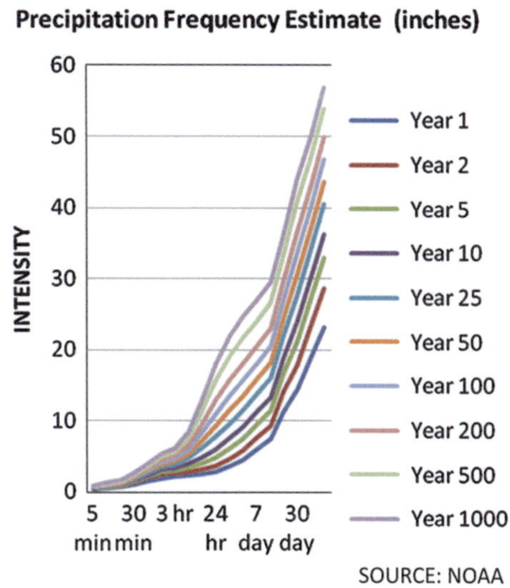

Fig. 9.2 Precipitation frequency estimate for Rio Grande de Añasco

grid was converted into inches before loading into N-SPECT. Figure 9.2 shows the Precipitation Frequency Estimate in inches for the RGA.

R factor: Rainfall-runoff erosive factor is a TIFF raster with data of the average annual summation values in a normal year's rain. Its projection was WGS-1984_UTM_ Zone_20N. Values varied from 15 (for forest) to 220 (for barren land).

Soil Data Layer: Sources of this information were the U.S. Department of Agriculture, Natural Resources Conservation, and the Service Soil Survey Geographic (SSURGO) database published in 2006. Survey areas for the RGA watershed were soil_pr684, soil_pr688, and soil_pr787. Soil shapefiles' projection was NAD_1983_UTM_Zone19N. The soil attribute table was combined with the SSURGO table containing K factor values and hydrogroup classifications. Several modifications, such as assignment of an appropriate hydrological value to all polygons and conversion to a numeric classification scheme, have been made before loading into N-SPECT.

Stream Discharge and Gage Locations: Data from Station 50144000, named Rio Grande de Añasco near San Sebastian, Rio Grande Añasco Basin, located at latitude 18°17′05″, longitude −67°03′05″, referenced to North American Datum of 1927 was downloaded from USGS Water Resource and the National Hydrography Data set (NHD), where Puerto Rico is No. 21 with 5 subregions.

Nitrogen Phosphorous Concentration: Source of these tabular data was USGS Water Resource, and the data were obtained from the stations located in the Hydrologic Unit Code 21010003. The USGS 50147900 Quebrada Las Marias at Marias, located at latitude 18°21′50.8″, longitude −67°06′40.8″ NAD27; station USGS 50144000 Rio Grande de Añasco at San Sebastian located at 18°17′05″

longitude −67°03′05″ NAD27; and station USGS 5014600 Rio Grande de Añasco near Añasco, located at latitude 18°16′31″, longitude −67°7′37″ NAD27.

9.3 Methodology

9.3.1 Non-point Source Pollution and Erosion Comparison Tool (N-SPECT)

N-SPECT (Eslinger et al. 2005) was developed by the U.S. National Oceanic and Atmospheric Administration (NOAA) as public-domain software which runs as an extension to Environmental System Research Institute (ESRI) ArcGIS software and requires the ESRI's Spatial Analyst extension. N-SPECT is designed for use with any watershed, and it examines the relationships between land cover, soil characteristics, topography, and precipitation to assess spatial patterns of surface water runoff, non-point source of pollution, and erosion. N-SPECT evaluates annual or event-based runoff, erosion, and sediment delivery. Limitations of the tools are the omission of some processes such as atmospheric deposition, storm drainage, stream diversion, and landslides among others.

9.3.2 Drainage Area Identification and Watershed Delineation

Watershed of the Río Grande de Añasco was delineated in ArcHydro from USGS Digital Elevation Model (DEM) maps (1:20,000 scale) to determine the drainage point for each catchment. N-SPECT was used to create a hillshade raster to visualize and understand the topographic influence on runoff and pollutant production patterns across the landscape. The watershed delineation is a tool that allows resource managers and land use planners to explore the physical processes associated with drainage patterns, targeting those catchments that are more susceptible to erosion (N-SPECT Technical Guide 2004).

9.3.3 Soil Loss Modeling Using Universal Soil Loss Equation (USLE)

Universal Soil Loss Equation (USLE) was developed in 1978 as tool for soil conservation. USLE estimates average soil loss resulting from splash, sheet, and rill erosion from agricultural plots. Data required for the USLE calculation is public and available in GIS format assigning factor values for soil prediction in a grid cell. USLE has been widely used because of its relative simplicity and robustness, although it has many limitations (Kinnell 2001). USLE model can be expressed by the following equation:

$$A = R * K * LS * C * P \tag{9.1}$$

where: is expressed in MJ mm ha^{-1} h^{-1} year^{-1}. *R* factor calculation includes total precipitation and kinetic energy of raindrops when falling onto the soil, and is affected by rainfall intensity and raindrop size. *R* factor does not consider the current land use; it provides an overall indicator and highlights erosion-prone areas.. Technical Guide: Default *C* Factor Values.

A mean annual soil loss, expressed in t ha^{-1} year^{-1}.

R rainfall erosivity *f*actor is the potential ability of the rainfall to cause erosion in a soil with no protection; R factor is expressed in MJ mm ha^{-1} h^{-1} yr^{-1}. R factor calculation includes total precipitation and kinetic energy of raindrops when fall onto the soil, and is affected by rainfall intensity and raindrop size. R factor does not consider the current land use; it provides an overall indicator and highlights erosion-prone areas.

K Soil erodibility factor refers to the vulnerability of soil to detachment and transport caused by raindrops and runoff. *K* factor depends on a number of mineralogical, chemical, morphological, and physical attributes (Perez-Rodriguez et al. 2007). The units are t h MJ^{-1} mm^{-1}. *K* factor was taken from SSURGO dataset.

LS Topographic factor is the effect of topography on erosion, and it depends on the slope steepness factor (*S*) and the slop length factor (*L*). LS represents the influence of the surface on runoff speed. Grid was created by N-SPECT.

C Cover management factor is related to the land use. It is a reduction factor to soil erosion vulnerability. *C* factor is the ratio of soil loss from an area according to its cover and management; thus, it represents the conditions that can be changed to reduce the erosion (Beskow et al. 2009). This *C* factor was taken from the N-SPECT Technical Guide: Default *C* Factor Values.

P Supporting practice factor highlights the relationship between soil loss in cropped soil and its support practice (Pandey et al. 2007). If there is no supporting practice in the area, $P = 1$. The *P* factor used for this analysis was 1.

9.3.4 Determination of Rainfall-Runoff Relationship Using N-SPECT

According to the N-SPECT Technical Guide, steps to calculating runoff are as follows:

Create the Curve Number Grid: Curve Number (CN) is used to calculate runoff depth estimation, which represents the infiltration capacity of the soil in a range from 0 to 100, with 0 being no runoff and 100 indicating no infiltration. Each land cover classification is related with hydrological soil types, and a corresponding CN is calculated for each type. When a soil has a dual hydrological classification (e.g., A/D, B/D, etc.), the highest curve number is used. N-SPECT uses the runoff curve numbers developed by NRCS that represent the overall permeability. The CN grid is generated from the combination of land cover and hydrological soil group at each cell.
Maximum Potential Retention (R): Retention is the ability of the soil to absorb or retain moisture. Retention is expressed in inches.

$$\text{Retention} = (1000/\text{CN}) - 10 \tag{9.2}$$

Initial Abstraction (I): Abstraction is the losses that occur before runoff begins. Abstraction is the water stored by surface depressions, and the water intercepted by vegetation, evaporation, and infiltration.

$$I = 0.2 * \text{Retention} \tag{9.3}$$

Precipitation Grid: Grid must be in inches. According to NOAA, Puerto Rico falls in the Type II Distribution developed by NRCS, which is the most intense short duration rainfall. Type II distribution is common for most countries in the Atlantic Ocean region.
Runoff Calculation: N-SPECT estimates the event-based runoff depth according to the equation taken from Urban Hydrology of Small Watersheds:

$$\text{Event-based Runoff depth} = \frac{(P - I)^2}{(P - I) + R} \tag{9.4}$$

In places where abstraction is greater than rainfall, N-SPECT sets runoff to zero. This prevents the reintroduction of artificial sinks to the runoff analysis.

The annual runoff depth is estimated using the following equation, which accounts for the average number of days it rains per year. For this analysis 7 rain days were used.

$$Q = \frac{P - (I * R_\mathrm{d})^2}{[(P - (I * R_\mathrm{d})) + (R * R_\mathrm{d})]} \tag{9.5}$$

where

Q annual runoff depth,
I initial abstraction,
R_d rainy days, and
R maximum potential retention.

The runoff volume is calculated as the product of land cover, soils, and precipitation data sets. The output raster data describe the runoff volume produced at each cell, the upslope contributions are not considered.

9.3.5 Modeling and Estimating Pollution Loads by Simple Method

The Simple Method introduced by Schueler (1987) provides a reasonable water quality and pollutant load estimates. The method requires a small amount of information such as annual precipitation, stormwater runoff pollutant concentration and impervious cover (NHDES 2010).

The pollutants were calculated according to the equation,

$$L = 0.226 * R * C * A \tag{9.6}$$

where:

L Annual load (pounds),
R Annual runoff (inches),
R $P * Pj$,
P annual rainfall in inches,
Pj fraction of annual rainfall events produce runoff, i.e., 0.9,
C Pollutant concentration (mg/l),
A Area (acres) and Unit conversion factor = 0.226.

9.4 Results and Discussions

9.4.1 Drainage Area Identification and Watershed Delineation

Watershed delineation is essential for hydrological analysis, as it serves to link a plot of land with its stream and river network and point of discharge to the sea. Drainage patterns allow identifying different levels of sediment transportation in streams and also reflect the ability of landscapes to flush suspended material through their drainage network. Figure 9.3 shows the drainage points in RGA watershed that allows calculating the drainage density based on the sum of all channel lengths in every catchments divided by the catchment area. Drainage density is strongly related to the effectiveness of surface runoff and erosion. High drainage density depends on high flood peaks, high sediment production, and steep hill slopes making this area unsuitable for agriculture due to high development costs (Dunne and Leopold 1978). To delineate a watershed, several steps are necessary, including to remove artificial sinks and to generate the flow direction grid, which is determined by evaluating the relative elevation of the eight neighbor cells. Then, it is necessary to create the flow accumulation grid, which estimation is based on the flow direction grid. Flow accumulation grid is used to derive a stream network. 41 catchments areas and 21 drainage points were created and identified with the watershed delineation.

Fig. 9.3 Catchments and drainage points in Rio Grande de Añasco watershed

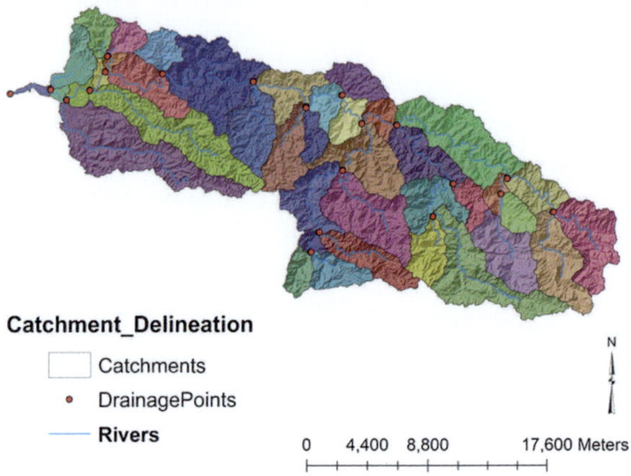

9.4.2 *Determination of Rainfall-Runoff Relationship*

Most of the soil in the RGA watershed corresponds to Hydrologic Soil Group (HSG) of C and D (Fig. 9.4) characterized for its slow and very slow rate of water infiltration. HSG of C are soil with moderately fine to fine texture or soils with a layer that impedes downward movement of water and D corresponds to soils with a clay layer at or near the surface, and shallow soils over nearly impervious material (NSPECT 2004).

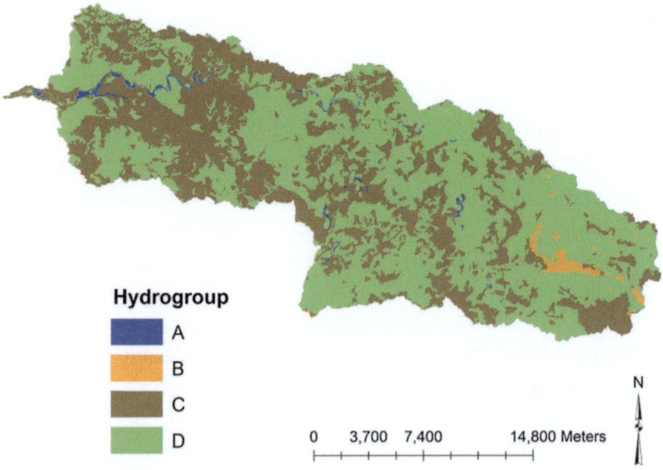

Fig. 9.4 Hydrologic soils group in the Rio Grande de Añasco watershed

Values for Curve Numbers (CN) were taken from the National Engineering Handbook and then assigned to every land use before loading them into N-SPECT. Figure 9.5 illustrates the CN in the watershed. Permeable soils are located in the northwest side of the RGA with less probabilities of runoff. The highest values of CN, greater than 91, correspond to urban and suburban areas in the watershed.

Input data sets for USLE calculation are rainfall erodibility factor, soil erodibility factor, and LS grid. Figure 9.6, shows that rainfall erodibility factor (R factor) for the RGA watershed has a range of 15–220 in./ha^{-1}/year^{-1}, with 89.48 % of the watershed having the lowest value; thus the erosive rainfall potential is low.

It should be noted for the soil erodibility factor (K factor) that the considerable area of the watershed has a K factor lower than 0.2. It indicates a low susceptibility to water erosion. According to Beskow et al. (2009), in USLE application, the topographic factor (LS factor) is important since this parameter characterizes runoff speed, indicating the soil erosion risk.

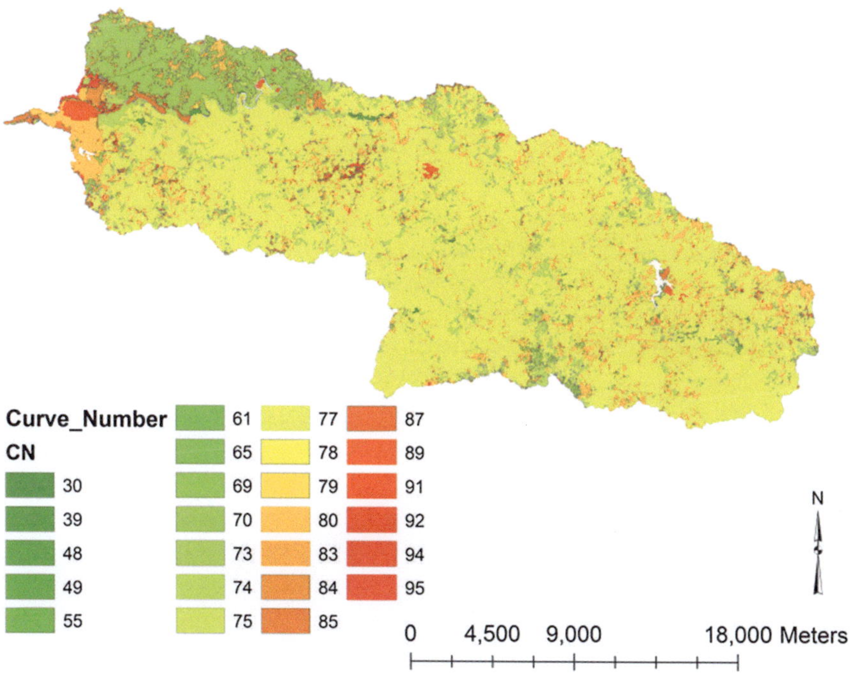

Fig. 9.5 Runoff curve numbers in Rio Grande de Añasco watershed

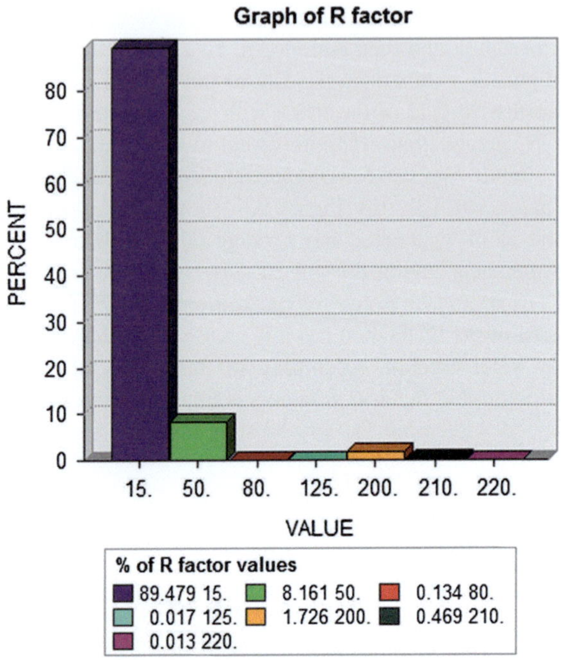

Fig. 9.6 Graph showing the R factor

The distribution of the annual soil loss is presented in Fig. 9.7. The map shows that the major part of the watershed is predicted to have low soil losses; the most significant area with medium values is the land located in the western side where the land cover is crops and urban use (city of Añasco). The high value of soil loss is 56.98 Tons/ha/year are presented scattered within the watershed.

Fig. 9.7 Soil loss in Ton/ha/year

9.4.3 *Runoff Calculation*

Results of the event-based runoff depth estimated with N-SPECT are shown in Fig. 9.8, where the runoff spatial patterns variability are similar to the rainfall spatial patterns in the watershed. Range values are from 1152 to 2313 mm that agree with the mean annual runoff of 1170 mm reported by Warne et al. (2005) for the RGA watershed. The highest runoff value is located in a protected forest land area with high steep topography in the municipality of Maricopa.

The major land cover in the RGA watershed is comprised of 74.89 % forestland, followed by 20.52 % grassland; urban areas represent 3.1 %, agriculture 0.98 %, water 0.43 %, wetland 0.06 %, and barren land 0.02 %. As shown in Fig. 9.9, grasslands are the major contributor to runoff, even though they do not make up the largest percentage of land cover in the watershed. This result is related to the cover factor, since soils in forestlands could be better protected from erosion.

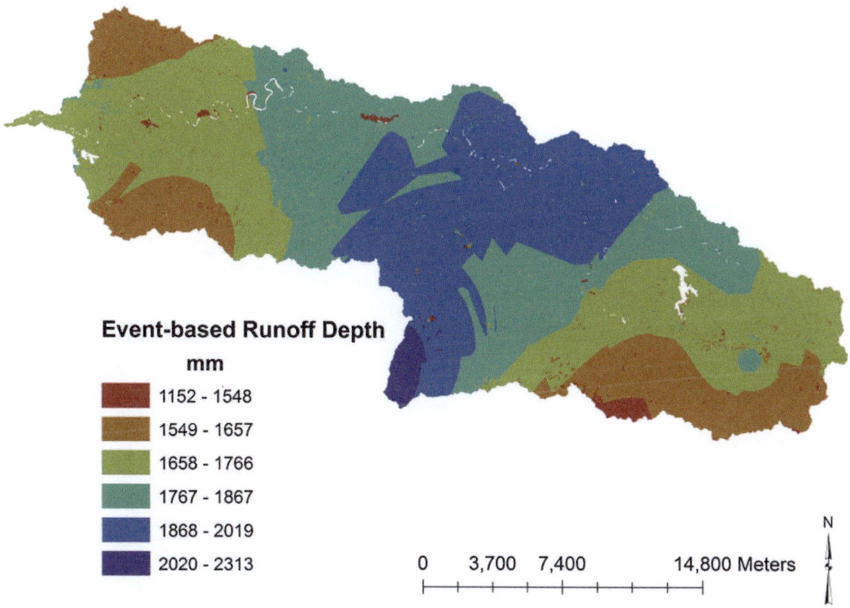

Fig. 9.8 Event-based runoff depth (mm)

Annual runoff depth estimation has range values from 151 mm that correspond to areas in the northwest of the RGA, to 1926 mm in the south-central part of the watershed. Medium-low values correspond to a wide area in the central part of the RGA watershed. The result of the estimation of the annual runoff depth is presented in Fig. 9.10.

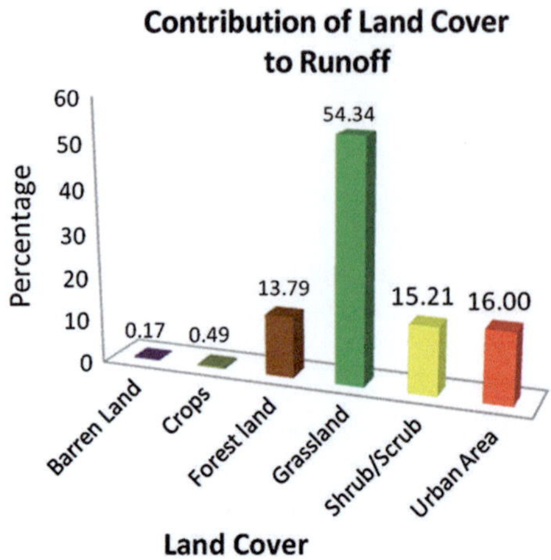

Fig. 9.9 Percentage of runoff by land cover

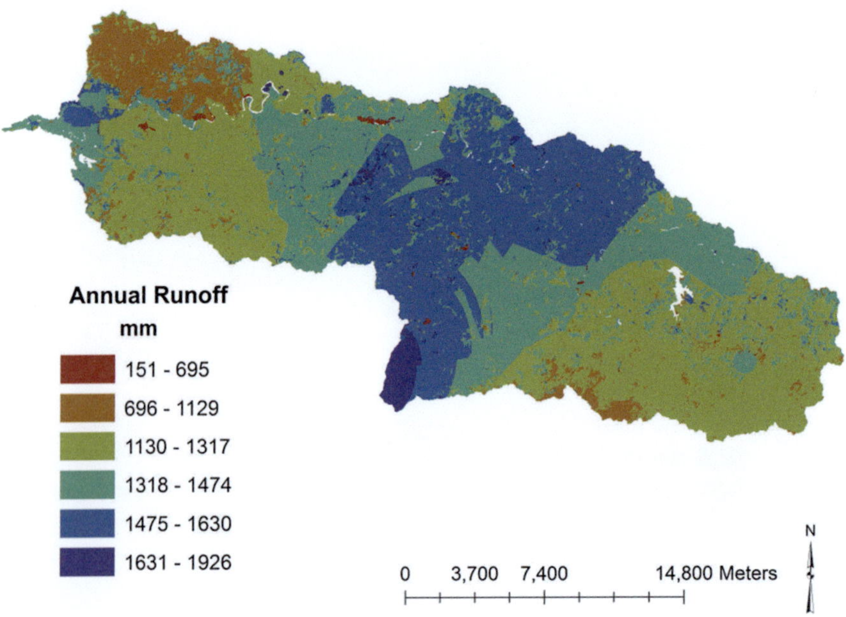

Fig. 9.10 Annual runoff depth of the study area

The total annual runoff volume was 21,811,931 mm and it was estimated by sub-watershed (see Fig. 9.11). Runoff volume takes into account soil type, land cover, and the rainfall duration and intensity.

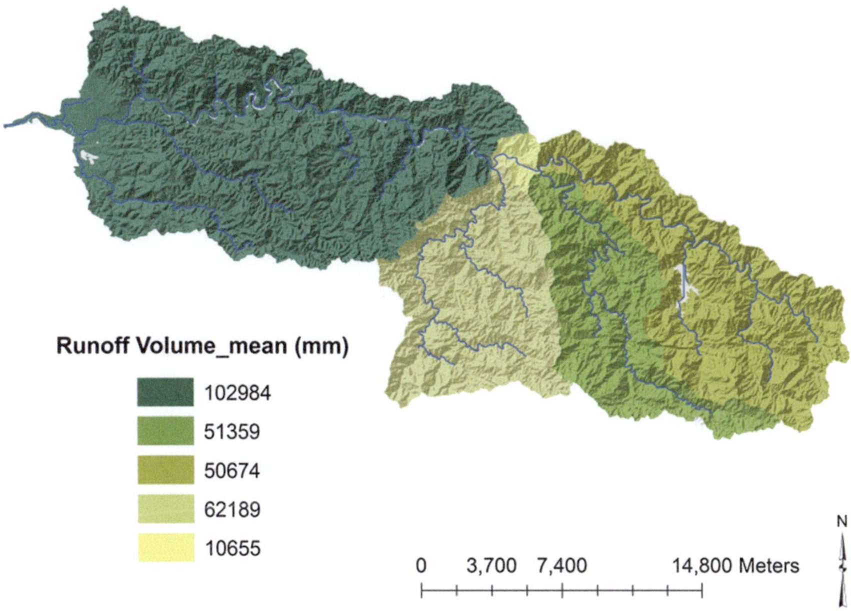

Fig. 9.11 Runoff volume by sub-watershed

9.4.4 Modeling and Estimating Pollution Loads

Pollutant concentrations are strongly correlated with land uses in upstream contributing areas. Storm events increase non-point source contributions by transporting pollutants in solid organic form and absorbed on soil particles. Table 9.1 shows the data spreadsheet with the nutrient discharge measurement in three stations along the RGA watershed.

Pollutant loads were calculated using the mean of the total discharge for each nutrient. TP mean is 0.07 mg/L, and 0.518 mg/L for N. Figures 9.12 (TP) and 9.13 (N) show the expected concentration of TP with range values from 0 to 26,349 kg/ha. Expected N loads vary from 0 to 174, 588 kg/ha that can be exported.

Table 9.1 Water quality spreadsheet

AGENCY	Station ID	Date	Total phosphorus (mg/L)	Nitrogen (mg/L)
USGS	50146000	2007-03-07	0.037	0.34
USGS	50144000	2007-03-08	0.042	0.41
USGS	50147900	2007-03-08	0.053	0.14
USGS	50144000	2007-08-28	0.14	1.4
USGS	50147900	2007-08-28	0.049	0.12
USGS	50146000	2007-08-29	0.13	1.13
USGS	50147900	2009-10-27	0.042	0.09

Source USGS Water Resources

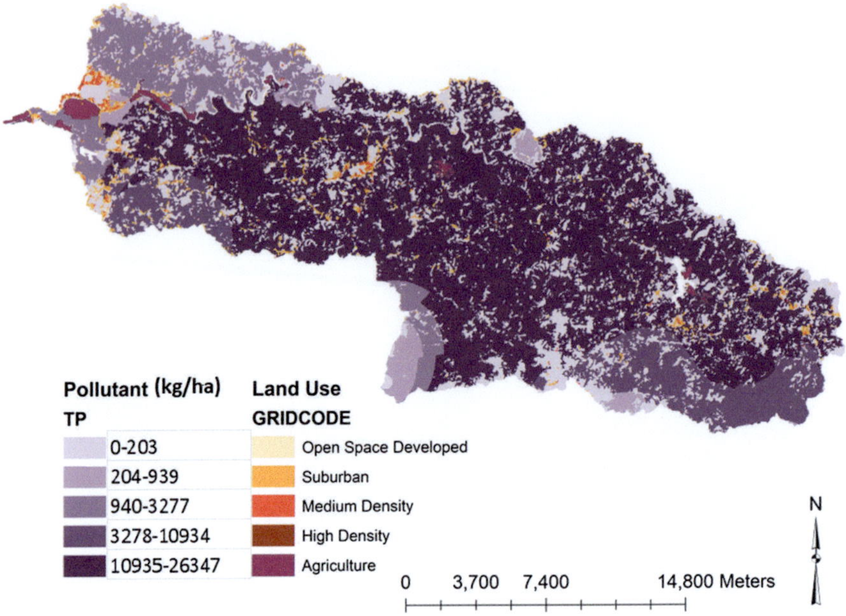

Fig. 9.12 Total phosphorous estimates

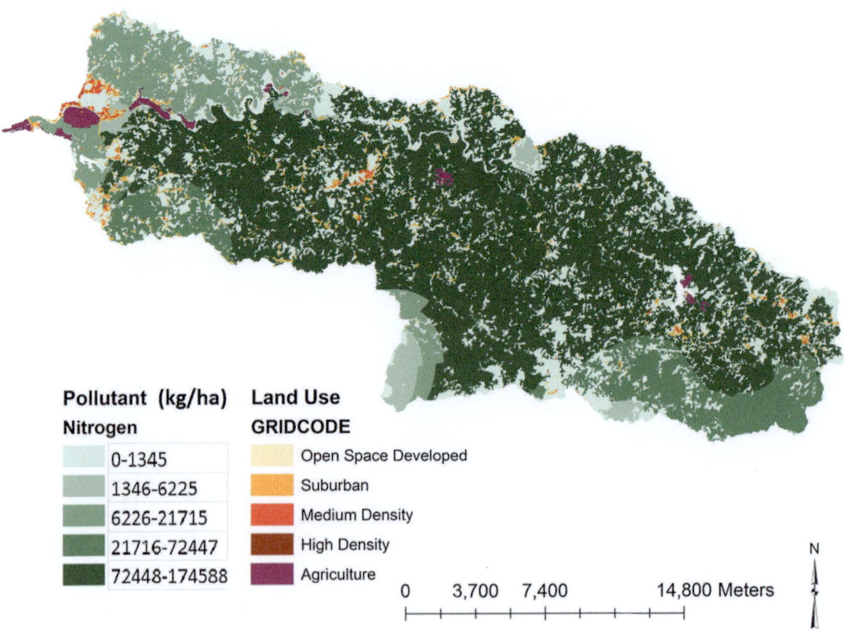

Fig. 9.13 Nitrogen loads estimates

9.5 Conclusions

Results from USLE helped spatial identification of the most susceptible areas to soil erosion which are located around the city of Añasco, with values of 29 Ton/ha/year. The largest part of the watershed presents low potential of erosion and low rainfall vulnerability. Highest value of event-based runoff depth was 2313 mm located in the south side of the watershed in a protected forested area. Medium-high values of runoff from 1767 to 1867 mm are likely to occur around the cities of Añasco and Las Marias, in the central part of the watershed, where the land cover corresponds to grasslands. The results from the annual runoff depth shows the highest values, 1471–1926 mm in the central-south area of the watershed which also has more precipitation events. Spatial patterns of runoff are very similar to the precipitation patterns. N-SPECT was found to be an effective computer tool to quantify runoff with the consideration of topography, climate, soil, and land use.

Acknowledgments The authors thank the U.S. Geological Survey (USGS) for the data set used in this chapter. Furthermore, we would like to thank Jaime H. Arbelaez for his comments and suggestions which surely improved this chapter.

References

Beskow S, Mello CR, Norton LD, Curi N, Viola MR, Avanzi JC (2009) Soil erosion prediction in the Grande River basin, Brazil using distributed modeling. CATENA 79:49–59

Defersha MB, Melesse AM (2011) Field-scale investigation of the effect of land use on sediment yield and surface runoff using runoff plot data and models in the Mara River basin, Kenya. CATENA 89:54–64

Defersha MB, Melesse AM (2012) Effect of rainfall intensity, slope and antecedent moisture content on sediment concentration and sediment enrichment ratio. CATENA 90:47–52

Defersha MB, Quraishi S, Melesse AM (2011) Interrill erosion, runoff and sediment size distribution as affected by slope steepness and antecedent moisture content. Hydrol Proc 7 (4):6447–6489

Defersha MB, Melesse AM, McClain M (2012) Watershed scale application of WEPP and EROSION 3D models for assessment of potential sediment source areas and runoff flux in the Mara River basin, Kenya. CATENA 95:63–72

Dunne T, Leopold LB (1978) Water in environmental planning. W.H. Freeman and Company, New York

Eslinger DL, Jamieson Carter H, Dempsey Ed, VanderWilt M, Wilson B, Meredith A (2005) The nonpoint-source pollution and erosion comparison tool. NOAA Coastal Services Center, Charleston, SC. http://csc.noaa.gov/nspect/. Accessed 11 Feb 2010

Junta de Calidad Ambiental (JCA) (1990) Estándares de Calidad de Agua de Puerto Rico. Estado Libre Asociado de Puerto Rico, Hato Rey, PR. http://www2.pr.gov/agencias/jca/Documents/Leyes%20y%20Reglamentos/Reglamentos/Reglamentos/Reglamento%20Est%C3%A1dares%20Calidad%20de%20Agua%202010.pdf. Accessed 15 Jan 2010

Junta de Calidad Ambiental (JCA) (2003) Informe sobre el Estado y Condición del Ambiente en Puerto Rico (2003). Capitulo 2. http://www2.pr.gov/agencias/jca/Documents/Publ icaciones%20de%20Inter%C3%A9s/Informes%20Ambientales/Informe%20Ambiental%202002/Primeras%20P%C3%A1ginas.pdf. Accessed 2 Feb 2010

Kinnell PIA (2001) Slope length factor for applying the USLE-M to erosion grid cells. Soil Tillage Res 58:11–17
Maalim FK, Melesse AM (2013) Modeling the impacts of subsurface drainage systems on runoff and sediment yield in the Le Sueur watershed, Minnesota. Hydrol Sci J 58(3):1–17
Maalim FK, Melesse AM, Belmont P, Gran K (2013) Modeling the impact of land use changes on runoff and sediment yield in the Le Sueur watershed, Minnesota using GeoWEPP. CATENA 107:35–45
Mekonnen M, Melesse AM (2011) Soil erosion mapping and hotspot area identification using GIS and remote sensing in northwest Ethiopian highlands, near Lake Tana, In: Melesse A (ed) Nile River basin: hydrology, climate and water use, Chapter 10. Springer, Berlin, pp 207–224. doi:10.1007/978-94-007-0689-7_10
Melesse AM, Ahmad S, McClain M, Wang X, Lim H (2011) Sediment load prediction in large rivers: ANN approach. Agric Water Manag 98:855–886
Mohammed H, Alamirew T, Assen M, Melesse AM (2015) Modeling of sediment yield in Maybar gauged watershed using SWAT, northeast Ethiopia. CATENA 127:191–205
Msagahaa J, Ndomba Melesse AM (2014) Modeling sediment dynamics: effect of land use, topography and land management. In: Melesse AM, Abtew W, Setegn S (eds) Nile River basin: ecohydrological challenges, climate change and hydropolitics. Springer, Berlin, pp 165–192
New Hampshire Department of Environmental Services (NHDES) (2010) Pollutant loading calculations, Chapter 8. Guidance for estimating pre- and post-development stormwater pollutant loads. WD-10-11. http://des.nh.gov/organization/divisions/water/stormwater/documents/wd-08-20a_ch8.pdf
NOAA Coastal Service Center (NSPECT 2004) Nonpoint-source pollution and erosion comparison tool. Technical guide. Digital coast. http://www.csc.noaa.gov/digitalcoast/tools/nspect/TechnicalGuide. Accessed 10 Jan 2010
NOAA Coastal Service Center (2009) Coastal remote sensing. Land cover analysis. Coastal NLCD classification scheme. http://www.csc.noaa.gov/crs/lca/tech_cls.html#8
Pandey A, Chowdary VM, Mal BC (2007) Identification of critical erosion prone areas in the small agricultural watershed using USLE, GIS and remote sensing. Water Res Manage 21:729–746
Perez-Alegria L, Olivieri L, Rivera F (2005) GIS-linked soil erosion model for sustainable management of the Rio Grande de Arecibo watershed. Project no. PR-MS-00011. Nifa PR. USDA-Universidad de Puerto Rico
Perez-Rodriguez R, Marquez MJ, Bienes R (2007) Spatial variability of the soil erodibility parameters and their relation with the soil map at subgroup level. Sci Total Environ 378:166–173
Ramos–Ginés O (1997) Water balance and quantification of total phosphorus and total nitrogen loads entering and leaving Lago de Cidra, Central Puerto Rico. U.S. Geological Survey. Water Resour Invest Rep 96–4222
Schueler T (1987) Controlling urban runoff: a practical manual for planning and designing urban BMPs. Metropolitan Washington Council of Governments. Washington, DC.
Setegn SG, Srinivasan R, Dargahil B, Melesse AM (2009) Spatial delineation of soil erosion prone areas: application of SWAT and MCE approaches in the Lake Tana basin, Ethiopia. Hydrol Process (Special Issue): Nile Hydrol 23(26):3738–3750
Setegn SG, Bijan Dargahi B, Srinivasan R, Melesse AM (2010) Modelling of sediment yield from Anjeni Gauged watershed, Ethiopia using SWAT. JAWRA 46(3):514–526
Sotomayor-Ramirez D, Martinez G, Perez-Alegria L (2004) Nutrient discharge from Mayaguez Bay watershed. Final progress report. Project no. CIMP-002. University of Puerto Rico, Department of Agricultural and Biosystems Engineering
Torres V, Francisco J (2009) Desarrollo y Aplicación de un Índice de Calidad de Agua para ríos en Puerto Rico, Universidad de Puerto Rico. http://prwreri.uprm.edu/publications/PR_2009_ 01. pdf. Accessed 9 Feb 2010

U.S. Department of Agriculture, Natural Resources Conservation Service (2006) Soil Survey Geographic (SSURGO) database for Arecibo Area, Puerto Rico northern part. pr682. Available online at: http://SoilDataMart.nrcs.usda.gov/

U.S. Geological Survey (USGS) (2009) National elevation dataset (NED), 2nd edn. Available online at: http://seamless.usgs.gov

Wang X, Garza J, Whitney M, Melesse AM, Yang W (2008) Prediction of sediment source areas within watersheds as affected by soil data resolution. In: Findley PN (ed) Environmental modelling: new research, Chapter 7. Nova Science Publishers, Inc., Hauppauge, pp 151–185. ISBN 978-1-60692-034-3

Warne AG, Webb RMT, Larsen MC (2005) Water, sediments and nutrient discharge characteristics of river in Puerto Rico, and their potential influence on coral reefs: U.S. Geological Survey. Scientific investigations report 2005-5206, 58 p

Chapter 10
Rainfall–Runoff Processes and Modeling: The Case of Meja Watershed in the Upper Blue Nile Basin of Ethiopia

Solomon Berhane, Birhanu Zemadim and Assefa M. Melesse

Abstract Understanding the basic relationships among rainfall, runoff, soil moisture, and groundwater level are vital for effective and sustainable water resources planning and management. The current study was conducted to understand the dynamics of the hydrological processes and model rainfall–runoff relationship in Meja watershed in the Upper Blue Nile River Basin of Ethiopia. Meja watershed is part of the three research sites of the International Water Management Institute (IWMI) developed in early 2010. The study utilized primary data of soil moisture, shallow groundwater level, rainfall and runoff collected from the hydrological monitoring networks in the watershed. Hydrological models, Hydrologiska Byråns Vattenbalansavdelning (HBV) and Rainfall Runoff Library Soil Moisture Accounting and Routing (RRL SMAR), were configured to understand the relationship between rainfall and runoff in the watershed. Relationships among rainfall, runoff, soil moisture, and groundwater level were developed to understand the dynamics of hydrological processes in the watershed.

Keywords Hydrological processes · Rainfall–runoff process · Meja watershed · HBV model · RRL SMAR model · Upper Blue Nile Basin

S. Berhane
Ethiopian Institute of Water Resources, Addis Ababa University, Addis Ababa, Ethiopia
e-mail: solber200315@yahoo.com

B. Zemadim (✉)
International Crops Research Institute for the Semi-Arid Tropics (ICRISAT), West and Central Africa, Bamako, Mali
e-mail: z.birhanu@cigar.org

A.M. Melesse
Department of Earth and Environment, Florida International University, Miami, FL, USA

© Springer International Publishing Switzerland 2016

A.M. Melesse and W. Abtew (eds.), *Landscape Dynamics, Soils and Hydrological Processes in Varied Climates*, Springer Geography,
DOI 10.1007/978-3-319-18787-7_10

10.1 Introduction

Watershed-based planning and management requires a thorough understanding of the hydrological processes and accurate estimation of runoff and other hydrological variables. The determination of runoff is essential to address soil and water conservation practices in the watershed. The information pertaining to occurrence of runoff further helps in integrated soil and water management practices such as prioritizing watersheds, erosion control, and selection of sites for conservation measures.

Design of effective soil and water conservation practices should take into account understanding of biophysical conditions in the area. Conversely, soil and water conservation practices affect the runoff processes of the watershed. Understanding runoff generation processes is, therefore, paramount importance for land and water resources management (Zemadim et al. 2011).

Complete and reliable hydrological and meteorological data is important for an effective and sustainable water resources planning and management. Due to the absence of reliable and gauged data, design of hydraulic structures and other water resource planning activities have been a challenge (Woinishet 2009). To get such data, there is a need to develop and maintain hydrometric stations capable of collecting these data continuously. Currently in Ethiopia, most of the available gauging stations are located nearby access roads. Because of this, most of the rivers which are inaccessible to roads are not gauged.

Rainfall is the most important hydrologic parameter which is used as an input for different water resource management activities and hydrological modeling. In order to achieve good rural and urban water management strategies, occurrence, distribution, characteristics and patterns of rainfall, which is highly variable in space time must be studied (Buytaert et al. 2000). Understanding the rainfall process is critical for the solution of several regional environmental problems of integrated water resources management at regional scales, with implications for agriculture, climate change, and natural hazards such as floods and droughts (Manfreda et al. 2003).

In mountain regions, in addition to the stochastic nature of rainfall, the precipitation pattern may be influenced by the irregular topography. The large variability in altitude, slope, and aspect may increase rainfall variability by means of processes such as rain shading and strong winds. The best method to improve the quality of spatial rainfall estimation is to increase the density of the monitoring network (Goovaerts 2000).

Spatial variability of rainfall can be affected by different factors such as geographical and morphological factors, for example, area exposure to the direction of wind and the characteristics of its surface (roughness, vegetative canopy). Elevation difference can also affect the spatial variability of rainfall. Therefore, quantification and a good knowledge of the characteristics of hydrological input data are essential for a correct interpretation of modeling results and planning of water management activities (Jakeman and Hornberger 1993).

The rainfall–runoff process in a catchment is a complex phenomenon governed by large number of known and unknown physiographic factors that vary both in space and time. The rain falling on a catchment undergoes a number of transformations and abstractions through various processes such as interception, detention, evapotranspiration, overland flow, infiltration, interflow, percolation, subsurface flow, and baseflow and emerges as runoff at catchment outlet (David 2003).

The relation between rainfall intensity and the discharge is not linear. However, at the catchment scale, due to the uncertainty of all the hydrological parameters, it might be assumed that the rainfall–runoff relation follows a linear relationship. Accurate understanding of the hydrological functioning of a catchment is not possible, if only rainfall (input) and discharge (output) data are available, as many different processes or process combinations may lead to similar hydrographs. Indeed, rainfall and discharge do not generally provide sufficient information for a single determination of hydrological response through solution of the model inverse problems (Wheater et al. 1991). Therefore, the identification of runoff generation processes requires further observations or investigations of soil moisture status within the catchment to characterize dominant water flow pathways.

Blume et al. (2007) used soil moisture data to understand the dynamics of soil moisture and its influence on investigation of runoff generation processes using dye experiment and soil moisture profile probes. Based on their study, the combination of high temporal resolution but spatially scarce soil moisture data with episodic additional measurements are proved to be useful for the investigation of runoff generation processes, especially with respect to preferential flow. Thus soil moisture/flow patterns were shown to be persistent in time and highly variable in space. The most likely explanation for the observed flow patterns is a combination of hydrophobicity with strong gradients in unsaturated conductivities, where flow paths are caused either by the presence of roots or the highly heterogeneous distribution of through fall and thus water input (Blume et al. 2007).

Field observations were conducted at Bukit Tarek Experimental Watershed in Peninsular Malaysia to investigate the relationship between rainfall–runoff responses and variation in soil moisture in a tropical rain forest (Noguchi et al. 1997). Storm flow depended strongly on the antecedent wetness as represented by the initial runoff rate. Though heavy rains fell in almost every month, the soil moisture decreased when fair weather was sustained.

Water level fluctuation measurement in observation wells is an important aspect of groundwater studies. Water level fluctuations are mostly influenced by hydrological, hydrometeorological, and hydrogeological phenomenon such as groundwater recharge, artificial recharge, groundwater pumpage, and return flows from irrigation. In many cases, there may be more than one phenomena/process operating simultaneously (Sklash and Farvolden 1979).

Under undisturbed natural conditions, hydrographs do not show any change in tendency with time because the recharge balances with the discharge. Aquifer response to recharge or discharge is reflected in water level fluctuations measured at different time periods. At any specific point, the change in water level below ground surface depends not only on rates of pumping and recharge, but also on the intrinsic characteristics of the geological formations. Long and steady low rainfall on a loamy saturated soil with a highly permeable geological section and deep water table condition can result in a significant rise in the water table. Whereas, an intense rainfall event of shorter duration on a dry clayey soil with a shallow water table may not raise the water table a considerable amount (Sklash and Farvolden 1979).

Latron and Gallart (2008) used limited and continuous data of rainfall, soil moisture, groundwater, and stream flow to analyze the runoff generation process in a small Mediterranean catchment, called Can Villa catchment. According to their study, the relationship between runoff and the depth to water table showed much more scatter than is usually observed under more humid conditions. Likewise, water table variations (rise or fall) were on some occasions not in phase with runoff changes, suggesting somewhat more intricate hydrological behavior.

Spatial and temporal variability and complexity of hydrological processes and limited availability of spatially and temporally distributed hydrologic, climatologic, geologic, and land use/land cover data challenge the ability to forecast hydrological data. Hydrological models are useful tools to solve such practical problems of forecasting hydrological data. From operational water resources management point of view, hydrological models are developed to guide the formulation of water resource management strategies by understanding spatial and temporal distribution of water resources.

Hydrology of the Nile River basin has been studied by various researchers. These studies encompass various areas including stream flow modeling, sediment dynamics, teleconnections and river flow, land use dynamics, climate change impact, groundwater flow modeling, hydrodynamics of Lake Tana, water allocation and demand analysis (Melesse et al. 2010, 2011, 2014; Abtew et al. 2009a, b; Yitayew and Melesse 2011; Chebud and Melesse 2009a; b, 2013; Assefa et al. 2014; Melesse 2011; Abtew and Melesse 2014a, b, c; Dessu and Melesse 2012, 2013; Dessu et al. 2014).

In the Ethiopian Blue Nile Basin, different studies were conducted using HBV model at a catchment level. Ayalew (2007) studied variations of climate change impacts in different hydroclimatologic regimes and indicated that HBV model performs well with the coefficient of determination during calibration and validation periods of 0.79 and 0.82, respectively. Similarly, Kumela (2011) reported a coefficient of determination during calibration and validation of 0.89 and 0.87, respectively.

The current study utilized the hydrological and meteorological data from monitoring stations established by the International Water Management Institute (IWMI) from May to August 2011. Primary data collected include rainfall, soil moisture, shallow groundwater level, and runoff. Detailed description of the monitoring network establishment was presented in the work of Zemadim et al. (2012, 2013).

The purpose of the study reported in this chapter was to understand runoff generation processes and model rainfall runoff relationship and get an alternative mechanism for estimating runoff, soil moisture, and groundwater level by using statistical analysis and rainfall–runoff models using Hydrologiska Byråns Vattenbalansavdelning (HBV) and Rainfall Runoff Library Soil Moisture Accounting and Routing (RRL SMAR) models.

10.2 Description of the Study Area

The Jeldu area is located in the south of the Upper Blue Nile (Abay) Basin at geographical coordinates of 9°1′N, 37°40′E with altitudes ranging from 1328 to 3200 m above sea levels (masl). It is predominantly a highland area and rainfall varies from 900 mm in the lower parts of the area to 1350 mm at higher altitudes. Mean daily temperature ranges from 3 °C to 24 °C and mean annual minimum temperatures is 8.5 °C and mean annual maximum temperature is 19 °C. The mean daily evapotranspiration is 4 mm (Zemadim et al. 2011).

10.2.1 Drainage Network

The Meja River originates at high altitude just outside Jeldu in the Ginchi Woreda. The headwater is in a flat wide valley, which is a wetland heavily utilized for livestock grazing. It then drops steeply and flows through a relatively narrow deeply incised valley. Numerous tributaries drain into the Meja from both the east and west. These are also deeply incised mountain streams with relatively small catchments with 3–4 km^2 drainage area (Zemadim et al. 2011). The drainage network together with monitoring stations established by IWMI are presented in Fig. 10.1.

10.2.2 Physiographic, Land Use, and Soil Condition

The physiographic conditions in the headwater of the catchment near Galessa consist of cultivated hills, eucalyptus plantations, and flat grazing areas which become swampy during the rainy season. The land use of the study area can be categorized mainly as agricultural and small part is agro-pastoral. Eucalyptus globules are the main tree planted in the area for construction and income generation purposes. The gauging station was established in a narrow gorge after identifying a stable river bank that consists of rock which is not susceptible for landsliding and bank erosion. The sides of the bank at the gauge location consist of slopes of greater than 60 %, cultivated and are eroded.

The major soil group in Meja catchment is Haplic Alisols which can form in wide variation of parent materials having high activity clay minerals such as

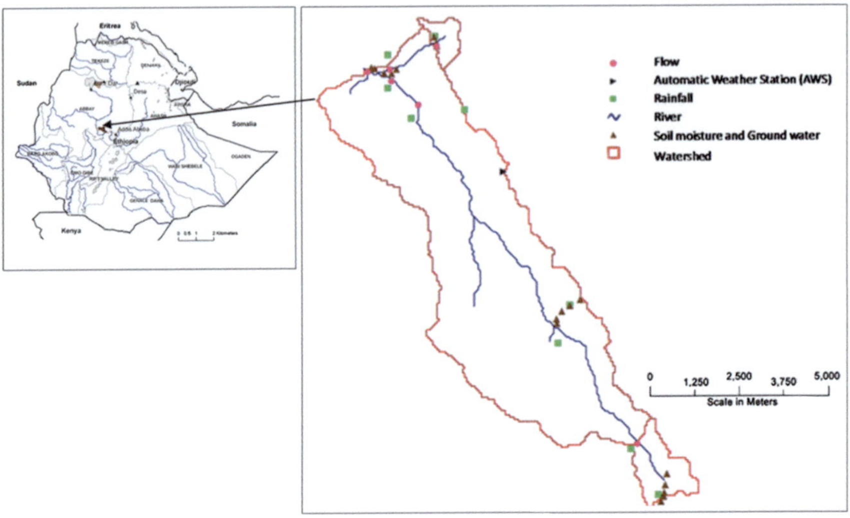

Fig. 10.1 Hydrological and meteorological monitoring stations in the Meja watershed. *Source* Zemadim et al. (2013)

vermiculite or smectate. The other soil types occurred in Meja catchment are Eutric Leptosols, Eutric Vertisols, Haplic Nitisols, Chromic Luvisols, and Eutric Nitisols. Upper part of Meja watershed is dominated by Eutric Nitisols and Haplic Alisols.

10.3 Dataset and Methods

10.3.1 Dataset

Rainfall data was collected from the field using nine ordinary rain gauges and one automatic weather station. Similarly, flow records were collected using current meter and water level recording instruments called stage boards (Fig. 10.2). Flow data were collected from four gauging stations. Soil moisture data was collected using profile probes PR2/6 and with access tubes and data was collected from 6 depths down to 100 cm from the ground surface (i.e., at 10, 20, 30, 40, 60, and 100 cm). Groundwater data was collected using dip meter called SEBA Electric Contact Meter type KLL, which is used to measure the water level in all tube wells manually. In addition, readings of weather conditions were recorded using Campbell Scientific automatic weather station. All of these data were collected from July 15, 2011 to September 30, 2012. Detailed description of the monitoring stations and geographic coordinates is presented in Table 10.1.

Galessa Legajaba Meja at bridge

Fig. 10.2 Images of the water level measuring instruments located in Meja watershed. *Photo credit* Zemadim (2011) and Langan (2011)

10.3.2 Data Analysis

All data were evaluated visually and using charts for errors in recording and missed information. After screening and factual error correction, various techniques were used to generate information from the collected data. For example, areal rainfall was determined using Thiessen polygon from the rain gauges to account for the spatial variability. Similarly, a rating curve was established to relate flow level measurements with stream discharge.

There are 6 layers of soil moisture readings taken from depths of 100, 200, 300, 400, and 1000 mm for the 18 stations. Groundwater level data were analyzed by plotting the time series of groundwater level over the rainfall and soil moisture and evapotranspiration was calculated using FAO Penman–Monteith equation. Further relationships of rainfall, runoff, soil moisture, and groundwater level were established to understand runoff generation mechanisms in the watershed.

10.3.3 Rainfall–Runoff Modeling

Researchers have used different hydrological models to simulate watershed hydrology at various scales. The most commonly used hydrological model capable of simulating river discharge in ungauged watershed is Soil Water Assessment Tool (SWAT). The application of SWAT in predicting stream flow and sediment as well as evaluation of the impact of land use and climate change on the hydrology of watersheds has been documented by various studies (Dessu and Melesse 2012, 2013; Dessu et al. 2014; Wang et al. 2008a, b, c; Wang et al. 2006; Wang and Melesse 2005, 2006; Behulu et al. 2013, 2014; Setegn et al. 2014; Mango et al. 2011a, b; Getachew and Melesse 2012; Assefa et al. 2014; Grey et al. 2013; Mohamed et al. 2015).

In this study, conceptual hydrological models of HBV and RRL SMAR were used stream flow modeling. Calibration is a major aspect of hydrological modeling

Table 10.1 Hydrological and meteorological stations installed in the Meja watershed, Jeldu District

Parameter	Station location	Station code	Period of data availability	Coordinates		Altitude (m)
				Northing	Easting	
Water level	Galesa	Galesa FG[a]	From June 2011	09°09′03.7″	038°09′03.7″	2806
	Kolu	LagaJaba FG 1	From June 2011	09°18′03.3″	038°03′27.1″	2732
		LagaJaba FG 2	From September 2012	09°15′00.0″	038°36′	2512
		Meja FG 1	From June 2011	09°17′29.1	038°01′49.9″	2409
		Meja FG 2	From June 2012	09°14′14.4″	038°30′	2444
		Meja FG 3	From September 2012	09°17′29.1	038°01′49.9″	2477
Precipitation	Galesa	Galesa RG[a] 1	From July 2011	09°07′55.0″	038°08′37.5″	2960
		Galesa RG 2	From June 2012	09°08′00.4″	038°08′	3035
	Serity	Serity RG 1	From July 2011	09°12′12.3″	038°06′33.2″	2946
		Serity RG 2	From June 2012	09° 11.34′.0″	038°2.31′	2903
	Kolu	Kolu RG 1	From July 2011	09°18′17.0″	038°03′24.9″	2786
		Kolu RG 2	From July 2011	09°17′52.2″	038°02′18.6″	2531
		Kolu RG 3	From June 2012	09°17.11 0″	038°02.31′ 0″	2522
		Kolu RG 4	From June 2012	09°16.42 0″	038°02.8′ 0″	2578
		EdensaGelan RG	From July 2011	09°16′37.1″	038°04′06.6″	2862

(continued)

Table 10.1 (continued)

Parameter	Station location	Station code	Period of data availability	Coordinates		Altitude (m)
				Northing	Easting	
Weather	Gojjo Town	Jeldu AWS	From August 2011	09°15′13.4″	038°05′00.9″	2942
Groundwater and soil moisture	Galesa	SM[a] Galesa 1	From July 2011	09°07′46.0″	038°08′41.1″	3020
		SM Galesa 2	From July 2011	09°07′52.9″	038°08′45.1″	2990
		SM Galesa 3	From July 2011	09°07′58.0″	038°08′45.8″	2973
		SM Galesa 4	From July 2011	09°08′08.9″	038°08′47.5″	2964
		SM Galesa 5	From July 2011	09°08′23.5″	038°08′49.3″	2990
	Serity	SM Serity 1	From July 2011	09°12′19.6″	038°06′48.1″	3007
		SM Serity 2	From July 2011	09°12′11.8″	038°06′33.7″	2944
		SM Serity 3	From July 2011	09°12′04.4″	038°06′22.5″	2933
		SM Serity 4	From July 2011	09°11′53.8″	038°06′14.5″	2904
		SM Serity 5	From July 2011	09°11′47.8″	038°06′15.3″	2846
	Kolu	SM Kolu 1	From July 2011	09°18′16.2″	038°03′23.7″	2792
		SM Kolu 2	From July 2011	09°18′08.1″	038°03′26.7″	2740
		SM Kolu 3	From July 2011	09°18′04.1″	038°03′27.1″	2710
		SM Kolu 4	From July 2011	09°17′35.2″	038°01′55.2″	2480
		SM Kolu 5	From July 2011	09°17′33.8″	038°02′00.0″	2463
		SM Kolu 6	From July 2011	09°17′27.2″	038°02′13.4″	2488
		SM Kolu 7	From July 2011	09°17′26.5″	038°02′24.0″	2494
		SM Kolu 8	From July 2011	09°17′33.1″	038°02′30.7″	2518

Source Zemadim et al. (2013)
[a]*FG*—flow gauge, *RG*—rain gauge, *SM*—soil moisture

Table 10.2 Efficiency criteria for evaluating model performance

Objective function	Definition	Value for perfect fit
Nash–satcliffe efficiency (Reff)	$1-\frac{\sum(Q_{obs}-Q_{sim})^2}{\sum(Q_{obs}-\overline{Q_{sim}})^2}$	1
Efficiency using ln(Q) (logReff)	$1-\frac{\sum(\text{In}Q_{obs}-\text{In}Q_{sim})^2}{\sum(\text{In}Q_{obs}-\overline{\text{In}Q_{sim}})^2}$	1
Coefficient of determination (R^2)	$\frac{\left(\sum(Q_{obs}-\overline{Q_{obs}})(Q_{sim}-\overline{Q_{sim}})\right)^2}{\sum(Q_{obs}-\overline{Q_{obs}})^2\sum(Q_{sim}-\overline{Q_{sim}})^2}$	1

Note: Q_{obs} and Q_{sim} are simulated and observed flows, respectively

and is aimed at fitting simulated versus measured discharge with minimal residuals. To do this, the available dataset was split into two. Two-thirds of the data, i.e. (July 15, 2011 to May 5, 2012) were used for model calibration and the remaining dataset (May 6, 2012 and September 30, 2012) was used for model validation. Performance of the model was determined using coefficient of determination and Nash–Sutcliff efficiency criteria. The evaluation criteria's are presented in Table 10.2.

10.4 Results and Discussion

10.4.1 Areal Rainfall Estimation

Areal rainfall was estimated from 11 rain gauges (Table 10.3). Based on topographic locations of rain gauges, the watershed was divided into three zones; Galessa represented the upper part of the watershed, Serity represented the middle part of the watershed and Kolu represents the lower part of the watershed.

Table 10.3 Rain gauge stations with their annual sum rainfall

Station No.	Station name	Elevation (m)	Mean annual rainfall (mm)
1	Kolu 1near Melka	2522	1370
2	Kolu Elementary	2531	1400
3	Kolu 2	2578	1310
4	Kolu Church	2786	1417
5	Edensa Gelan	2862	1241
6	Jeldu Metreology	2867	1642
7	Hintodale	2903	1362
8	Aws	2942	1218
9	Serity	2946	1397
10	Galessa 1	2960	1460
11	Galessa 2	3035	1464

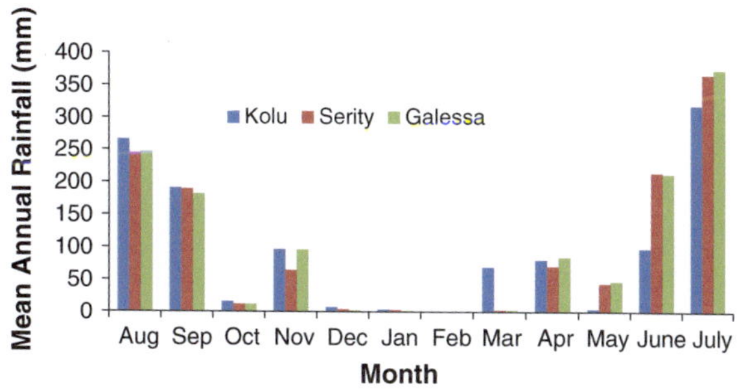

Fig. 10.3 Monthly patterns of areal rainfall for Kolu, Serity, and Galessa

The result indicated that there is not much difference in the maximum records of daily rainfall from the three zones. However, the occurrence period of the maximum rainfall in the three sub-watersheds is different. Overall, the average daily rainfall in the watershed was found to be 4 mm/day.

Figure 10.3 shows monthly patterns of areal rainfalls of Galessa, Serity, and Kolu. The pattern of monthly rainfall is similar for the three sites. Monthly average of rainfall of Kolu, Serity, and Galessa is 107, 117 and 115 mm, respectively. In Kolu, maximum total monthly rainfall was observed during August 2011, while in Serity and Galessa the maximum total monthly rainfall was observed during July 2012.

The monthly rainfall correlation at the three sites is stronger than the daily time step. Coefficient of determination of monthly rainfall between Galessa and Kolu, Serity and Kolu, Serity and Galessa are 0.83, 0.85 and 0.95, respectively.

10.4.2 Rainfall and Runoff Relationship

Figure 10.4 shows the pattern of rainfall and runoff in the Meja watershed. It is shown that high rainfall causes high runoff, but occurrence of runoff does not mean there was high rainfall. When there was no rainfall for four months in a year (October, November, December and February), runoff varies from 0.16 to 0.83 mm/day. During this time, runoff was affected by other factors such as soil

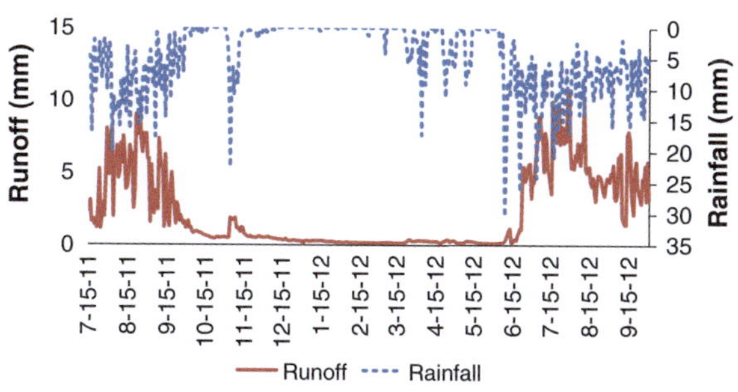

Fig. 10.4 Daily rainfall runoff pattern of Meja watershed

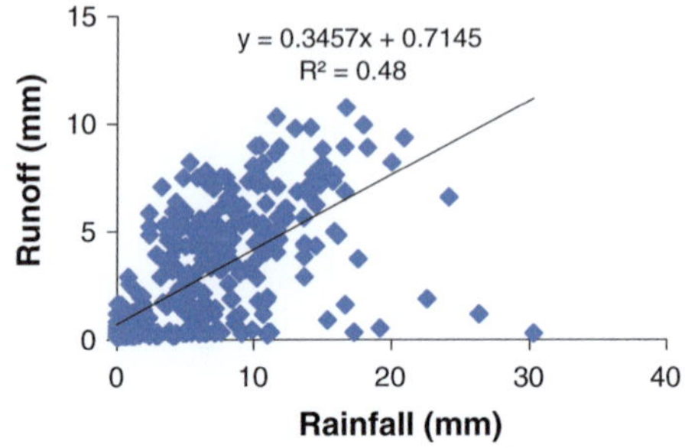

Fig. 10.5 Daily rainfall–runoff relationship of Meja watershed

moisture and groundwater flow or baseflow. Meja watershed requires only few storms at the beginning of wet season to satisfy the watershed and begin producing runoff but stream flow did not immediately return to dry season levels instead it steadily decreases. There is moderate relationship between daily rainfall and runoff. The computed flows in dry season (October to May) and wet season (June to September) were 96 and 571 mm, respectively.

Coefficient of determination of rainfall and runoff in Meja watershed on a daily basis is 0.48 and correlation coefficient is 0.7 indicating the presence of moderate relationship between rainfall and runoff (Fig. 10.5).

10.4.3 Rainfall and Groundwater Level Relationship

The rainfall and groundwater level relationship for the middle part of the catchment (Serity) showed an increase in groundwater level due to the summer rainfall and small or no response during the winter season (Fig. 10.6). There is moderate relationship between monthly rainfall and groundwater in this case, with a coefficient of determination of 0.56 and correlation coefficient of 0.75. Figure 10.6 depicts that the occurrence of instantaneous high rainfall does not automatically increase the shallow

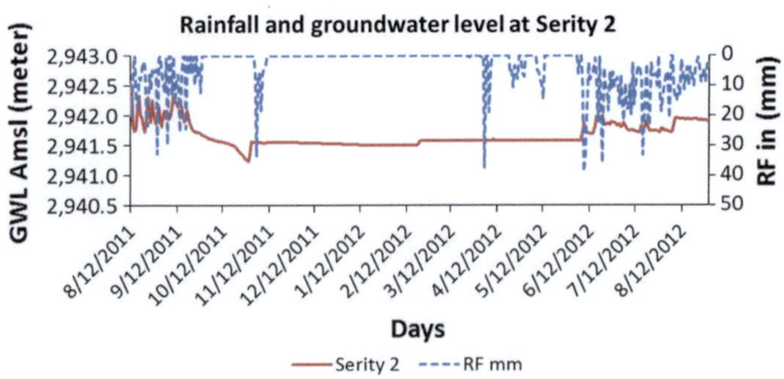

Fig. 10.6 Daily rainfall and groundwater level pattern of Serity

groundwater level, signifying the slow response of the subsurface water movement or rapid surface runoff condition due to poor drainage capacity.

Rainfall and groundwater level relationship in Kolu nested sub-watershed was analyzed using data from two tube wells called Kolu 2 and Kolu 5. Coefficient of determination between monthly groundwater level at Kolu 2 and rainfall at Kolu site is 0.80 with a correlation coefficient of 0.94. The coefficient of determination between groundwater level in Kolu 5 station and rainfall at Kolu site is 0.45 with a correlation coefficient of 0.67. This signifies strong relationship between monthly rainfall and shallow groundwater level.

Figures 10.7 and 10.8 show the daily rainfall and groundwater level patterns at Kolu 2 and Kolu 5. Accordingly, in the summer season, rainfall and groundwater level increase, but in winter season groundwater level does not change as rainfall increases. In Kolu, nested sub-watershed groundwater level relates more to flow than to rainfall, so that there may be high horizontal hydraulic permeability than the vertical hydraulic permeability. Peak groundwater level does not match with occurrence of peak rainfall; this indicates that there is delay of groundwater response to the occurrence of rainfall events.

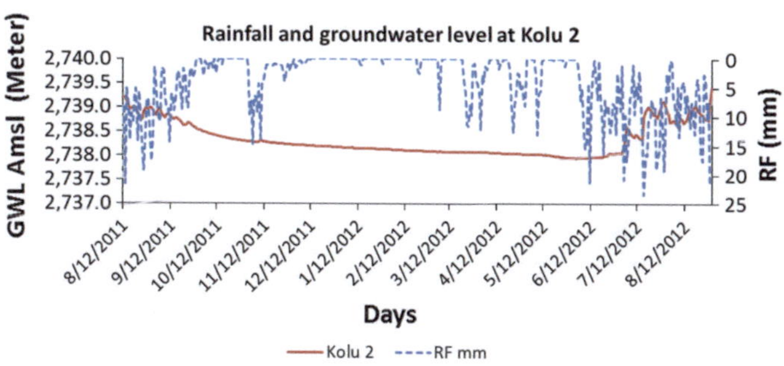

Fig. 10.7 Daily rainfall and groundwater level pattern of Kolu 2

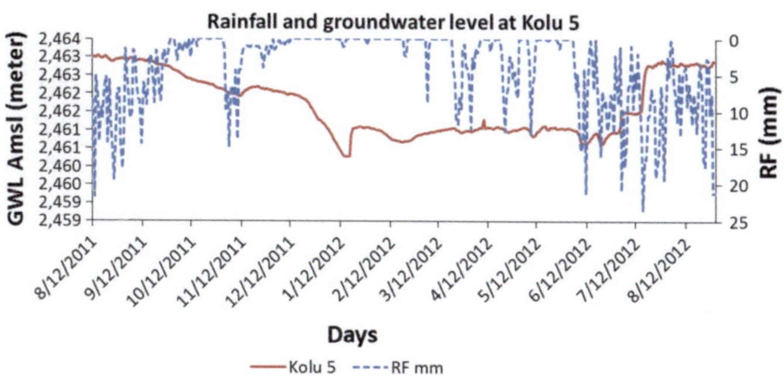

Fig. 10.8 Daily rainfall and groundwater level pattern of Kolu 5

10.4.4 Rainfall and Soil Moisture Relationship of Meja Watershed

There is strong relationship between rainfall and soil moisture at a depth of 100 mm with a coefficient of determination of 0.76 and correlation coefficient of 0.87. This means soil in the upper horizon in Meja is pervious. There is also strong relationship between rainfall and soil moisture at 300 and 400 mm layer. In 600 mm layer of soil moisture measurement, coefficient of determination between rainfall and soil moisture is 0.52 and correlation coefficient is 0.7 which indicates moderate relationship indicating soil moisture at the 600 mm layer in Meja watershed is influenced by groundwater than rainfall amount.

The analysis of soil moisture revealed presence of spatial variability of soil moisture content in the Meja watershed. This variation occurs due to heterogeneity of soil texture, vegetation cover, and topography.

Figure 10.9 shows soil moisture at different depths and rainfall variability in Meja watershed. Soil moisture increases along the layer profiles in Galessa, Serity, and Kolu sites. Generally, in Meja watershed, when the depth from the ground surface increases, volumetric monthly soil moisture also increases. In all sites, soil moisture was measured at 100, 200, 300, 400, and 600 mm layers. There was high amount of volumetric soil moisture in 600 mm layer than the 100 mm layer. This may be because of capillary rise of groundwater at 600 mm and there is more influence of rainfall at 100 mm.

10.4.5 Evapotranspiration, Rainfall, and Runoff Relationship

Evapotranspiration in Meja and its three zones Galessa, Serity, and Kolu was analyzed with rainfall and runoff pattern. The computed average daily

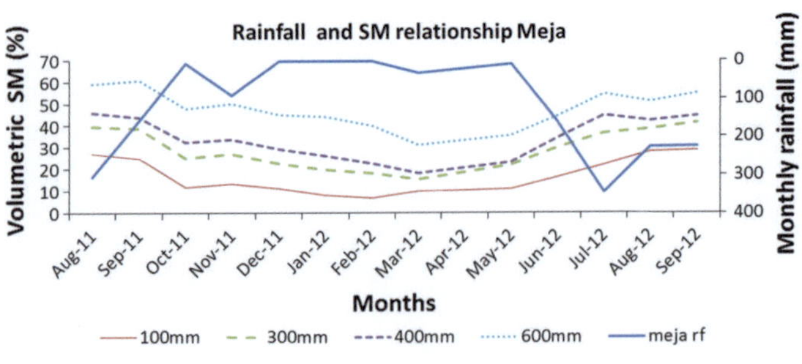

Fig. 10.9 Rainfall and volumetric soil moisture pattern of Meja watershed

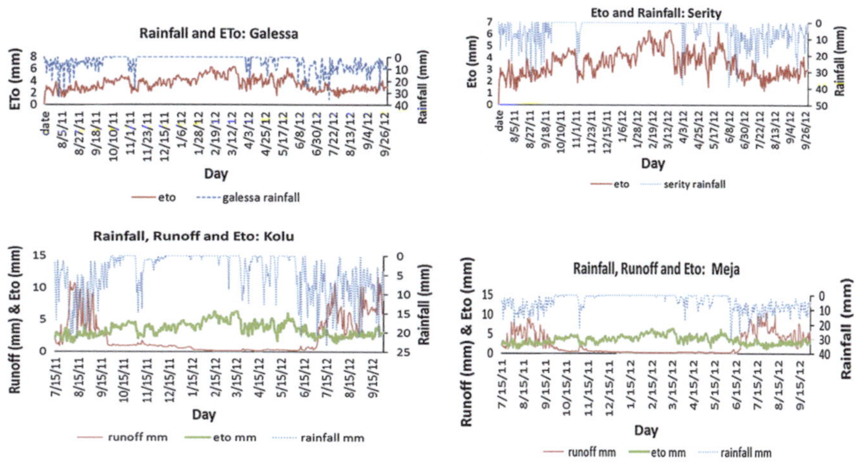

Fig. 10.10 Relationship of evapotranspiration, rainfall and runoff at Galessa, Serity, Kolu, and Meja watershed

evapotranspiration in the entire catchment was 3.9 mm/day. In the three zones, rainfall and evapotranspiration have high daily variability. Patterns of evapotranspiration, rainfall, and runoff are presented in Fig. 10.10.

Maximum evapotranspiration was observed during the driest months of the year (February, March, and April) when rainfall and runoff decreased. In the summer season, with the presence of rainfall, evapotranspiration decreases, correspondingly runoff increases as well (Fig. 10.10).

10.4.6 HBV Model Analysis

10.4.6.1 Model Description

The Hydrological Bureau Water balance-section (HBV) model is a rainfall–runoff model, which includes conceptual numerical descriptions of hydrological processes at the catchment scale. The general water balance can be described as:

$$P - E - Q = \frac{\mathrm{d}}{\mathrm{d}t}[\mathrm{SP} + \mathrm{SM} + \mathrm{UZ} + \mathrm{LZ} + \mathrm{lakes}] \qquad (10.1)$$

where P = precipitation, E = evapotranspiration, Q = runoff, SP = snowpack, SM = soil moisture, UZ = upper groundwater zone, LZ = lower groundwater zone, and lakes = lake volume.

Input data are observations of precipitation, air temperature, and estimates of potential evapotranspiration. The time step is usually one day, but it is possible to use shorter time steps. The evaporation values used are normally monthly averages, although it is possible to use daily values. Air temperature is used to adjust potential evaporation when the temperature deviates from normal values, or to calculate potential evaporation.

The model consists of subroutines for meteorological interpolation, snow accumulation and melt, evapotranspiration estimation, soil moisture accounting procedure, routines for runoff generation and finally, a simple routing procedure between subbasins and in lakes. It is possible to run the model separately for several subbasins and then add the contributions from all subbasins.

10.4.6.2 Calibration, Sensitivity Analysis, and Model Validation

Split data sampling was used to calibrate the model. Two-thirds of the data, i.e., July 15, 2011 to May 5, 2012 were used for calibration and one-third of the data, i.e., May 6, 2012 and September 30, 2012 were used to validate the model without further fine-tuning the model parameters. Sensitivity analysis was conducted during model calibration by varying the parameters and observing the simulation results. The most sensitive parameters were the response routing (PERC) and K2 which govern subsurface and baseflow contributions in Meja and Kolu. PERC defines the maximum percolation rate from the upper to the lower groundwater box. K2 is recession coefficient. Graphical display of the observed and simulated flows after calibration and verification are presented in Figs. 10.11 and 10.12. Visual inspection of the two graphs indicated that the patterns of observed and simulated flow values are reasonably closer both for rainy and dry seasons. The calibrated parameters are shown in Table 10.4 for catchment parameters and Table 10.5 depicts vegetation zone parameters.

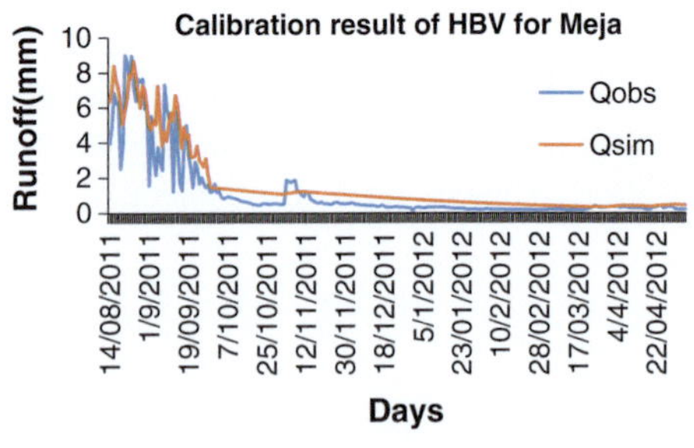

Fig. 10.11 The HBV model after calibration for Meja watershed

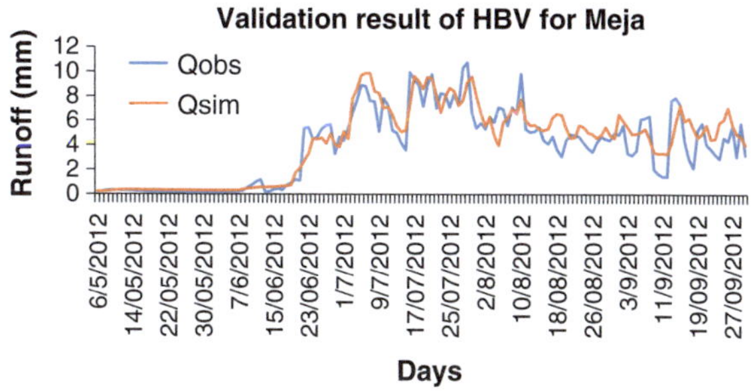

Fig. 10.12 The HBV model after validation for Meja watershed

Table 10.4 Optimal model calibration for catchment parameters

Parameter	Lower limit	Upper limit	Calibrated value
Maximum percolation to the soil lower zone (PERC) (mm/d)	0	4	1
Threshold parameter (UZL) (mm)	0	70	45
Storage or recession coefficient 0 (Ko)	0.1	0.5	0.3
Storage or recession coefficient 1 (K1)	0.01	0.2	0.01
Storage or recession coefficient 2 (K2)	0.00005	0.1	0.01
Length of triangular weighting function (MAXBAS)	1	2.5	1

Table 10.5 Optimal model calibration for vegetation parameters

Parameter	Lower limit	Upper limit	Calibrated value
Threshold temperature (TT)	−2	0.5	0
Degree-Δt factor (CFMAX)	0.5	4	3
Snowfall correction factor (SFSC)	0.5	0.9	1
Refreezing coefficient (CFR)	0.05	0.05	0.1
Water holding capacity (CWH)	0.1	0.1	0.1
Maximum soil moisture storage (FC)	100	550	150
Soil moisture value above which AET reaches PET (LP)	0.3	1	0.5
Parameter that determines the relative contribution to runoff from rain or snowmelt (BETA)	1	5	1

10.4.7 Rainfall Runoff Library Soil Moisture Accounting and Routing (RRL SMAR) Model

10.4.7.1 Model Description

The Soil Moisture Accounting and Routing model (SMAR) is a lumped conceptual rainfall runoff water balance model with soil moisture as a central theme. The model provides daily estimates of surface runoff, groundwater discharge, evapotranspiration, and leakage from the soil profile for the catchment as a whole. The surface runoff component comprises overland flow, saturation excess runoff, and saturated throughflow from perched groundwater conditions with a quick response time.

The SMAR model consists of two components in sequence, a water balance component and a routing component. The model utilizes time series of rainfall and pan evaporation data to simulate stream flow at the catchment outlet. The model is calibrated against observed daily stream flow.

The water balance component divides the soil column into horizontal layers, which contain a prescribed amount of water (usually 25 mm) at their field capacities. Evaporation from soil layers is treated in a way that reduces the soil moisture storage in an exponential manner from a given potential evapotranspiration demand. The routing component transforms the surface runoff generated from the water balance component to the catchment outlet by a gamma function model form, a parametric solution of the differential routing equation in a single input single output system. The generated groundwater runoff is routed through a single linear reservoir and provides the groundwater contribution to the stream at the catchment outlet.

The SMAR model contains five water balance parameters and four routing parameters. The water balance component uses these five parameters to describe the movement of water into and out of a generalized soil column under conditions of atmospheric forcing: C, Z, H, Y, and T

The dimensionless parameter C regulates evaporation from the soil layers. The parameter Z (mm) represents the effective moisture storage capacity of the soil contributing to the runoff generation mechanisms. Each layer holds 25 mm at field capacity. The dimensionless parameter H is used to estimate the variable H', the proportion of rainfall excess contributing to the generated runoff as saturation excess runoff or the Dunne runoff. H' is obtained as a product of H, rainfall excess and soil saturation. Soil saturation is defined as the ratio of available soil moisture in mm at time t (days) and 125 mm, representing the maximum soil moisture content of the first five layers. The parameter Y (mm/d) represents the infiltration capacity of the soil and is used for estimating the infiltration excess runoff (Hortonian runoff). The dimensionless parameter T is used to calculate the potential evaporation from pan evaporation (E). Generated surface runoff is calculated from the excess rainfall (rainfall minus potential evaporation) as saturation excess runoff (shallow subsurface flow) plus the Hortonian runoff and plus a proportion (1-G) of moisture in excess of the effective soil moisture storage capacity (g) (i.e., throughflow). The remaining proportion (G) of the latter, i.e., the deep drainage component discharged from the groundwater system to the stream, is routed through a linear reservoir and the total generated surface runoff is routed using a gamma function model form to obtain the daily total estimated discharge at the catchment outlet.

Table 10.6 Optimal model calibration parameters of RRL SMAR model for Meja

no	Parameters	Optimized	Min	Max
1	Groundwater evaporation rate (C)	0.95	0	1
2	Groundwater runoff coefficient (G)	0.05	0	1
3	Proportion direct runoff (H)	0.65	0	1
4	Storage loss coefficient	0.19	0	1
5	U.H linear routing (N)	1.44	1	6
6	U.H linear routing N * K = NK	0.98	0.01	1
7	Evaporation conversion parameter (T)	0.59	0.52	1
8	Infiltration rate (Y)	342	0	5000
9	Soil moisture total storage depth (Z)	547.84	0	5000

10.4.7.2 Calibration, Sensitivity Analysis, and Model Validation

The methodology of calibration, sensitivity analysis, and validation used in the HBV model was used for SMAR model as well. Automatic adjustment of the calibration parameters (listed in Table 10.6) resulted in a set of parameters that minimized the difference between observed and simulated discharge for the gauged catchment of Meja and its nested sub-watershed, Kolu.

As shown in Figs. 10.13 and 10.14, the patterns of observed and simulated runoff model during calibration and validation were satisfactory, however, the model couldn't capture peak and low flows.

10.4.7.3 Model Performance Evaluation

Performance of hydrological models can be assessed using Nash–Sutcliff Efficiency (NSE) criteria. According to Moriasi et al. (2007) classification of hydrologic models performance is as follows:

- Very good, if 0.75 < NSE < 1.00
- Good, if 0.65 < NSE < 0.75
- Satisfactory, if 0.50 < NSE < 0.65
- Unsatisfactory, if NSE < 0.50

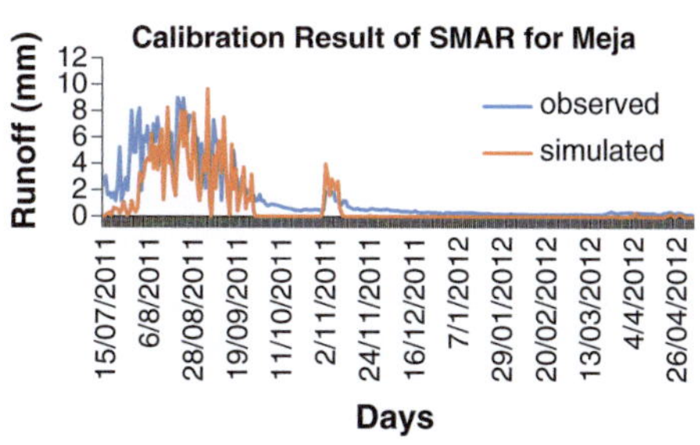

Fig. 10.13 The SMAR model after calibration for Meja watershed

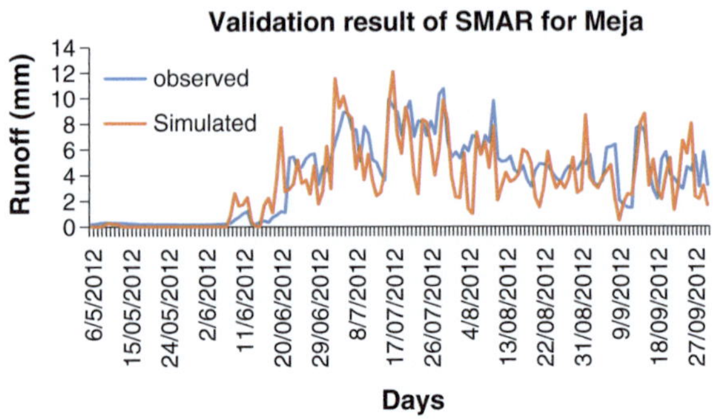

Fig. 10.14 The SMAR model verification result for Meja watershed

Table 10.7 Models performance evaluation result for calibration and validation of HBV and SMAR models for Meja watershed

Performance index	Calibration		Validation	
	HBV	SMAR	HBV	SMAR
R^2	0.85	0.72	0.84	0.64
NSE	0.78	0.65	0.81	0.57

In Table 10.7, it is shown that the performance of HBV model in Kolu during calibration and verification period was satisfactory and in Meja the model performed very good in both calibration and verification periods. SMAR model in Kolu and in Meja indicates satisfactory performance during calibration and verification periods. So that HBV model performed better than RRL SMAR in Meja and Kolu.

Meja watershed is a representative watershed of highlands in the Blue Nile Basin of Ethiopia. Our modeling result revealed that HBV model can be used in a similar highland catchments in Ethiopia after testing. Moreover, the rainfall–runoff routing of HBV model can be used to fill missed records of flow measurement, which enable the assessment of ungauged catchments.

10.5 Conclusion

Rainfall–Runoff modeling was conducted for one of the IWMI research sites in the highlands of Ethiopia using primary data collected from the field. Analyzing the relationships among rainfall, runoff, soil moisture, and groundwater through statistical techniques enabled to understand the dynamics of the hydrological processes in the Meja watershed, which is a representative of highland catchment in Ethiopia.

Analysis of rainfall data indicated that there is a weak correlation ($r^2 < 0.35$) of areal rainfall among the three zones of Meja watershed (Galessa, Serity and Kolu), however, there was a very good monthly correlation ($r^2 > 0.8$). Further, we understood that there was no strong daily rainfall and runoff relationship ($r^2 < 0.5$); this may be due to undocumented water abstractions for dry season irrigation purposes, which are common, particularly in the lower part of the watershed, Kolu.

In two parts of the watershed, Kolu and Serity, rainfall has moderate relationship with groundwater level. Occurrence of rainfall does not indicate rising of water table level immediately and absence of rainfall does not indicate falling of water level either; this indicates that factors that affect water table fluctuations is the lateral flow of water with higher hydraulic permeability than the vertical hydraulic permeability. There is strong linear relationship of rainfall and monthly averaged volumetric soil moisture in the watershed. In 600 mm layer, soil moisture has strong relationship with groundwater than rainfall but unlike this in 100 mm layer soil moisture has strong relationship with rainfall than groundwater level.

Further to the effort in understanding the runoff generation processes, hydrological models of HBV and RRL SMAR were configured to understand the relationship between rainfall and runoff in the watershed. In both models, the same input data for the same period of time were used for model calibration and verification purposes. Calibration and validation of watershed parameters were done by manual and automatic procedures. Based on the efficiency criteria such as coefficient of determination and Nash–Sutcliffe criteria, the HBV model performs better than SMAR. SMAR model could not capture low flows in Meja watershed.

Considering Meja watershed as a representative area of the highlands of Blue Nile Basin, HBV model can be recommended for use in similar types of watersheds in other parts of the country. Moreover, the tested correlation results among various variables and the HBV model configuration can also be used to predict variables like soil moisture data, and groundwater level which are always difficult and costly for measurement. Long-term data filling is another concern of modern day hydrology and models like HBV, if calibrated and validated properly with a good quality data, it is possible to generate long-term flow data, a concern in the case of ungauged catchments. Hence, we recommend the wider setup and evaluation of HBV model from smaller to larger watershed scale for efficient use of HBV model as a decision support tool.

Acknowledgment We would like to thank the International Water Management Institute for the Hydrological and Meteorological data and National Meteorological Agency of Ethiopia for the additional meteorological data used in this study. This chapter is based on the masters thesis of the first author.

References

Abtew W, Melesse AM (2014a) Nile River basin hydrology. In: Melesse AM, Abtew W, Setegn S (eds) Nile River basin: ecohydrological challenges, climate change and hydropolitics, pp 7–22

Abtew W, Melesse AM (2014b) Climate teleconnections and water management. In: Nile River basin, pp 685–705. Springer International Publishing

Abtew W, Melesse AM (2014c) Transboundary Rivers and the Nile. In: Nile River basin, pp 565–579. Springer International Publishing

Abtew W, Melesse A, Desalegn T (2009a) Spatial, inter and intra-annual variability of the Blue Nile River basin rainfall. Hydrol Process 23(21):3075–3082

Abtew W, Melesse AM, Desalegn T (2009b) El Niño Southern Oscillation link to the Blue Nile River basin hydrology. Hydrol Process 23(26):3653–3660 (Special Issue: Nile Hydrology)

Assefa A, Melesse AM, Admasu S (2014) Climate change in Upper Gilgel Abay River catchment, Ethiopia. In: Melesse AM, Abtew W, Setegn S (eds) Nile River basin: ecohydrological challenges, climate change and hydropolitics, pp 363–388

Ayalew S (2007) Variations of climate change impacts in different hydro-climatologic regimes of the Nile basin: a case study of Gilgel Abbay in the Blue Nile Sub-basin and two low land reaches (Baro and Sudd). Arba Minch University

Behulu F, Setegn S, Melesse AM, Fiori A (2013) Hydrological analysis of the Upper Tiber basin: a watershed modeling approach. Hydrol Process 27(16):2339–2351

Behulu F, Setegn S, Melesse AM, Romano E, Fiori A (2014) Impact of climate change on the hydrology of Upper Tiber River basin using bias corrected regional climate model. Water Resour Manage 28:1–17

Blume T, Zehe E, Bronstert A (2007) Use of soil moisture dynamics and patterns for the investigation of runoff generation processes with emphasis on preferential flow. Hydrol Earth Syst Sci 4:2587–2624

Buytaert W, Celleri R, Willems P, Bievre B, Guido W (2000) Spatial and temporal rainfall variability in mountainous areas: a case study from the South Ecuadorian Ande

Chebud Y, Melesse AM (2013) Stage level, volume, and time-frequency change information content of Lake Tana using Stochastic Approaches. Hydrol Process 27(10):1475–1483. doi:10.1002/hyp.9291

Chebud YA, Melesse AM (2009a) Numerical modeling of the groundwater flow system of the Gumera sub-basin in Lake Tana basin, Ethiopia. Hydrol Process 23(26):3694–3704 (Special Issue: Nile Hydrology)

Chebud YA, Melesse AM (2009b) Modeling Lake stage and water balance of Lake Tana, Ethiopia. Hydrol Process 23(25):3534–3544

David G (2003) Rainfall–runoff processes. Utah State University, Logan

Dessu SB, Melesse AM, Bhat M, McClain M (2014) Assessment of water resources availability and demand in the Mara River basin. CATENA 115:104–114

Dessu SB, Melesse AM (2012) Modeling the rainfall–runoff process of the Mara River basin using SWAT. Hydrol Process 26(26):4038–4049

Dessu SB, Melesse AM (2013) Impact and uncertainties of climate change on the hydrology of the Mara River basin. Hydrol Process 27(20):2973–2986

Getachew HE, Melesse AM (2012) Impact of land use/land cover change on the hydrology of Angereb Watershed, Ethiopia. Int J Water Sci 1(4):1–7. doi:10.5772/56266

Grey OP, Webber Dale G, Setegn SG, Melesse AM (2013) Application of the soil and water assessment tool (SWAT Model) on a small tropical island state (Great River Watershed, Jamaica) as a tool in integrated watershed and coastal zone management. Int J Trop Biol Conserv 62(3):293–305

Goovaerts P (2000) Geostatistical approaches for incorporating elevation into the spatial interpolation of rainfall. J Hydrol 228:113–129

Jakeman A, Hornberger M (1993) How much complexity is warranted in rainfall–runoff? Water Resour Res 29:2637–2649

Kumela T (2011) Performance comparison of conceptual rainfall–runoff models on Muger catchment (Abay River basin). M.Sc., theses, Addis Ababa University, Ethiopia

Latron J, Gallart F (2008) Runoff generation processes in a small Mediterranean research Catchment (Vallcebre, Eastern Pyrenees). J Hydrol 358:206–220

Manfreda S, Di Santo G, Iacobellis V, Fiorentino M (2003) A regional analysis of rainfall pattern in Southern Italy. In: Proceedings of the fourth European Graduate School Plinius conference held at Mallorca, Spain

Mango L, Melesse AM, McClain ME, Gann D, Setegn SG (2011a) Land use and climate change impacts on the hydrology of the upper Mara River basin, Kenya: results of a modeling study to support better resource management. Hydrol Earth Syst Sci 15:2245–2258 (Special Issue: Climate, weather and hydrology of East African Highlands). doi:10.5194/hess-15-2245-2011

Mango L, Melesse AM, McClain ME, Gann D, Setegn SG (2011b) Hydro-meteorology and water budget of Mara River basin, Kenya: a land use change scenarios analysis. In: Melesse A (ed) Nile River basin: hydrology, climate and water use, Chapter 2. Springer Science Publisher, Berlin, pp 39–68. doi:10.1007/978-94-007-0689-7_2

Melesse AM (2011) Nile River basin: hydrology, climate and water use. Springer Science & Business Media, Berlin

Melesse A, Abtew W, Setegn SG (2014) Nile River basin: ecohydrological challenges, climate change and hydropolitics. Springer Science & Business Media, Berlin

Melesse A, Abtew W, Setegn S, Dessalegne T (2011) Hydrological variability and climate of the Upper Blue Nile River basin. In: Melesse A (ed) Nile River basin: hydrology, climate and water use, Chapter 1. Springer Science Publisher, Berlin, pp 3–37. doi:10.1007/978-94-007-0689-7_1

Melesse A, Abtew W, Desalegne T, Wang X (2010) Low and high flow analysis and wavelet application for characterization of the Blue Nile River system. Hydrol Process 24(3):241–252

Mohammed H, Alamirew T, Assen M, Melesse AM (2015) Modeling of sediment yield in Maybar gauged watershed using SWAT, Northeast Ethiopia. CATENA 127:191–205

Moriasi D, Arnold J, Van Liew M, Bingner R, Harmel R, Veith T (2007) Model evaluation guidelines for systematic quantification of accuracy in watershed simulations. Am Soc Agric Biol Eng 50(3):885–900

Noguchi S, Nik AR, Yusop Z, Tani M, Sammori T (1997) Rainfall–runoff response and role of soil moisture variation to the response in tropical rainforest, Biket Tarek, Peninsular Malaysia. J For Res 2:125–132

Setegn SG, Melesse AM, Haiduk A, Webber D, Wang X, McClain M (2014) Spatiotemporal distribution of fresh water availability in the Rio Cobre Watershed, Jamaica. CATENA 120:81–90

Sklash M, Farvolden R (1979) The role of groundwater in storm runoff. J Hydrol 43:45–65

Wang X, Shang S, Yang W, Melesse AM (2008a) Simulation of an agricultural watershed using an improved curve number method in SWAT. Tans Am Soc Agric Bio Eng 51(4):1323–1339

Wang X, Yang W, Melesse AM (2008b) Using hydrologic equivalent wetland concept within SWAT to estimate streamflow in watersheds with numerous wetlands. Tans Am Soc Agric Bio Eng 51(1):55–72

Wang X, Melesse AM, Yang W (2006) Influences of potential evapotranspiration estimation methods on SWAT's hydrologic simulation in a Northwestern Minnesota Watershed. Trans ASAE 49(6):1755–1771

Wang X, Melesse AM (2006) Effects of STATSGO and SSURGO as inputs on SWAT model's Snowmelt simulation. J Am Water Res Assoc 42(5):1217–1236

Wang X, Melesse AM (2005) Evaluations of the SWAT model's Snowmelt hydrology in a Northwestern Minnesota Watershed. Trans ASAE 48(4):1359–1376

Wang X, Garza J, Whitney M, Melesse AM, Yang W (2008c) Prediction of sediment source areas within watersheds as affected by soil data resolution. In: Findley PN (ed) Environmental modelling: new research, Chapter 7. Nova Science Publishers, Inc., Hauppauge, NY 11788, pp 151–185. ISBN 978-1-60692-034-3

Wheater S, Langan S, Brown A, Beck B (1991) Hydrological response of the Allt Mharcaidh catchment—inferences from experimental plots. J Hydrol 123:163–199

Woinishet H (2009) Daily rainfall–runoff modeling for Beles River catchment. Masters theses, Addis Ababa University, Ethiopia

Yitayew M, Melesse AM (2011) Critical water resources management issues in Nile River basin. In: Melesse A (ed) Nile River basin: hydrology, climate and water use, Chapter 20. Springer Science Publisher, pp 401–416. doi:10.1007/978-94-007-0689-7_20

Zemadim B, Matthew MC, Bharat SM, Abeyou W (2011) Integrated rainwater management strategies in the of the Ethiopian highlands. Int J Water Res Environ Eng 3(10):220–232

Zemadim B, McCartney M, Sharma B (2012) Establishing hydrological and meteorological monitoring networks in Jeldu, Diga and Fogera Districts of the, Ethiopia report produced for challenge program on water and food Nile project 2: integrated rainwater management strategies—technologies, institutions and policies: 56 p

Zemadim B, McCartney M, Langan S, Sharma B (2013) A participatory approach for hydro meteorological monitoring in the Blue Nile River Basin of Ethiopia. Colombo, Sri Lanka: International Water Management Institute (IWMI). 32 p. (IWMI research report 155)

Chapter 11
Upstream–Downstream Linkages of Hydrological Processes in the Nile River Basin

Belete Berhanu, Yilma Seleshi, Melkamu Amare and Assefa M. Melesse

Abstract The various used of water in large transboundary river basins like the Nile River will require an understanding of the upstream–downstream hydrological linkages and impacts for better planning and management of the shared resources. Related to this understanding, the hydrological processes in the three broadly classified zones (headwaters zone, transitional zone and depositional zone) have paramount importance in the decision-making process of basin-wide water uses. Particularly, changes in the headwater zone at the Ethiopian highlands (the Blue Nile sub-basin) will have the most significant connectivity to the downstream water uses and hydrological regimes. If we compare the combination effects of the rainfall amount received by in three sub-basins (Bahr-EL-Ghazal Blue Nile and Equatorial Lakes Basin), and their larger drainage area, the two sub-basins (Bahr-El-Ghazal and Equatorial Lakes Basin) receive much greater than that of the Blue Nile sub-basin. But the contribution of flow by the western basins is comparatively low. This study uses Geographical Information System (GIS) as the base tool and 30 m SRTM Digital elevation model, high resolution mean monthly rainfall, and multi-stations (226) mean monthly potential evapotranspiration data for analysing the hydrological upstream–downstream connectivity. With these input data, the analysis has confirmed that the upstream and downstream linkages in the Nile River Basin is largely dependent on the extent of the transitional zone, in which the

B. Berhanu (✉) · Y. Seleshi
Department of Civil Engineering, Addis Ababa Institute of Technology (AAIT), Addis Ababa, Ethiopia
e-mail: beteremariam@yahoo.com

Y. Seleshi
e-mail: yilma.seleshi@aau.edu.et

M. Amare
Amare and Families Consulting Engineers P.L.C, Addis Ababa, Ethiopia
e-mail: melkamuamare@yahoo.com

A.M. Melesse
Department of Earth and Environment, Florida International University, Modesto A. Maidique Campus, Miami, FL 33199, USA
e-mail: melessea@fiu.edu

© Springer International Publishing Switzerland 2016

A.M. Melesse and W. Abtew (eds.), *Landscape Dynamics, Soils and Hydrological Processes in Varied Climates*, Springer Geography,
DOI 10.1007/978-3-319-18787-7_11

releasing function is more characterised by the evaporation process than runoff. Thus, under the current setting, the dependency of the hydrological system for the downstream reach/zone of the Nile River basin on the processes of the Blue Nile sub-basin is more significant due to the short extent of the transitional zone in this sub-basin.

Keywords Hydrological process · Upstream–downstream river linkage · Nile river basin · Headwaters zone · Transitional zone and depositional zone

11.1 Introduction

The water resources use decision-making process is constrained by our abilities to collect the required information about hydrologic systems at various spatiotemporal scales. In a river basin, hydrological events that occur in the upper stream may have a direct influence to the downstream based on the process it goes through (Nepal et al. 2014). An understanding of hydrological processes in its upstream–downstream linkages is the basis for water balance studies in the basin and will serve as an appropriate input for effective and efficient planning and management of the river basin resources. It is particularly critical in river basins of larger in size and transboundary in nature with large altitude differences, climatic features and geological settings where water use planning and management in the upstream reach will have effect on downstream uses (Blaikie and Muldavin 2004; Rasul 2014). On the other hand, studies for hydrologic processes and events occur at a wide range of scales in space and time (Klemeš 1983; Blöschl and Sivapalan 1995), and availability of data about hydrologic processes are scarce within the basin area. Therefore, investigating the upstream–downstream linkages of the hydrological process facilitates hydrologic modelling and information transferring from upstream to downstream or vice versa. This can be used to offset data shortfall problems while practicing water use planning and management activities in the river basin hydrological systems.

Upstream impacts on hydrological processes can be broadly divided into two types: (i) human-influenced activities related to land use and (ii) natural impacts related to climate (Nepal 2014). The change or the impact of these processes is largely expressed with the quantification of the fundamental components of the hydrologic cycle, such as precipitation, evapotranspiration and runoff from which water balance of a river basin is simulated. The water balance also serves as a base for the understanding of the hydrological system of the basin (Sutcliffe and Parks 1999).

Record on the Nile River goes as far back as 3600 BC where the height of the annual flood has been recorded as the most important event of Egyptians (Lyons 1906). Though, a number of efforts had been made by scientists and travellers in the investigation and documentation of the physiographic and the hydrological features of the Nile basin (Lyons 1906), the studies and publications of Sutcliffe and Parks (1999) serve as foundation for understanding of the topographical and hydrological features of the basin. Recent investigations that use different hydrological models and data sources as remote sensing, also contribute a lot to quantify hydrological processes in the basin (Senay et al. 2009; Nile 2014; Kebede and Travi 2006; Taye and Willems 2011).

Hydrology of the Nile River basin has been studied by various researchers, These studies encompass various areas including stream flow modelling, sediment dynamics, teleconnections and river flow, land-use dynamics, climate change impact, groundwater flow modelling, hydrodynamics of Lake Tana, water allocation and demand analysis (Melesse et al. 2009a, b, 2011; Abtew et al. 2009a, b; Yitayew and Melesse 2011; Chebud and Melesse 2009a, b, 2013; Dessu et al. 2012, 2013; Dessu et al. 2014; Setegn et al. 2009a, b, 2010; Melesse 2011; Melesse et al. 2014; Abtew and Melesse 2014a, b, c).

Thus, this review focuses on synthesising the available information to build understanding on the upstream–downstream linkages of hydrological processes in the Nile River basin. Particularly, it quantifies hydrological elements in the basin and sub-basins and identifies their relative impact and contributions to the whole hydrological system.

11.2 Topography and Sub-Basins in the Nile Basin

Without the good knowledge of the complex topography and sub-basin characteristics, one cannot see the hydrological process of the basin (Sutcliffe and Parks 1999). An early detailed physiographic analysis was made by Lyons (1906) that tried to address the topography, the geology and the climate of the Nile as one system. Most of the recent studies use its topographic investigation as the basis for their work. But due to limitation in technical ability to collect good topographic information, they could not address some of the details of topographic variations of the Nile basin. This review work goes into further details using the Shuttle Rader Topography Mission (SRTM) 30 m Digital Elevation Model (DEM) and spatial analysis tools of ArcGIS.

The Nile as a large river basin holds diverse topographic features as mountainous, lakes, depressions, vast wetlands, floodplains and gorges. Most of the basin area lies in the low land ranges, which has an altitude of less than 1500 m above mean sea level (amsl). The highlands in the Nile basin are the main sources of rainwater, the plateau of Ethiopia in the east is the source of the Blue Nile River and the Equatorial plateau in the south is where the White Nile originates (Fig. 11.1).

Delineating the basin boundary and the computation of the area of the basin and sub-basins are also important issues in the Nile basin hydrology. Lyons (1906) computed the basin area as 2,867,600 km^2 using the available map at scale of 1:4,000,000 and 1:2,000,000 employing grid method. Later studies approached these issues differently. Some of the studies directly refer to Lyons (1906) for their

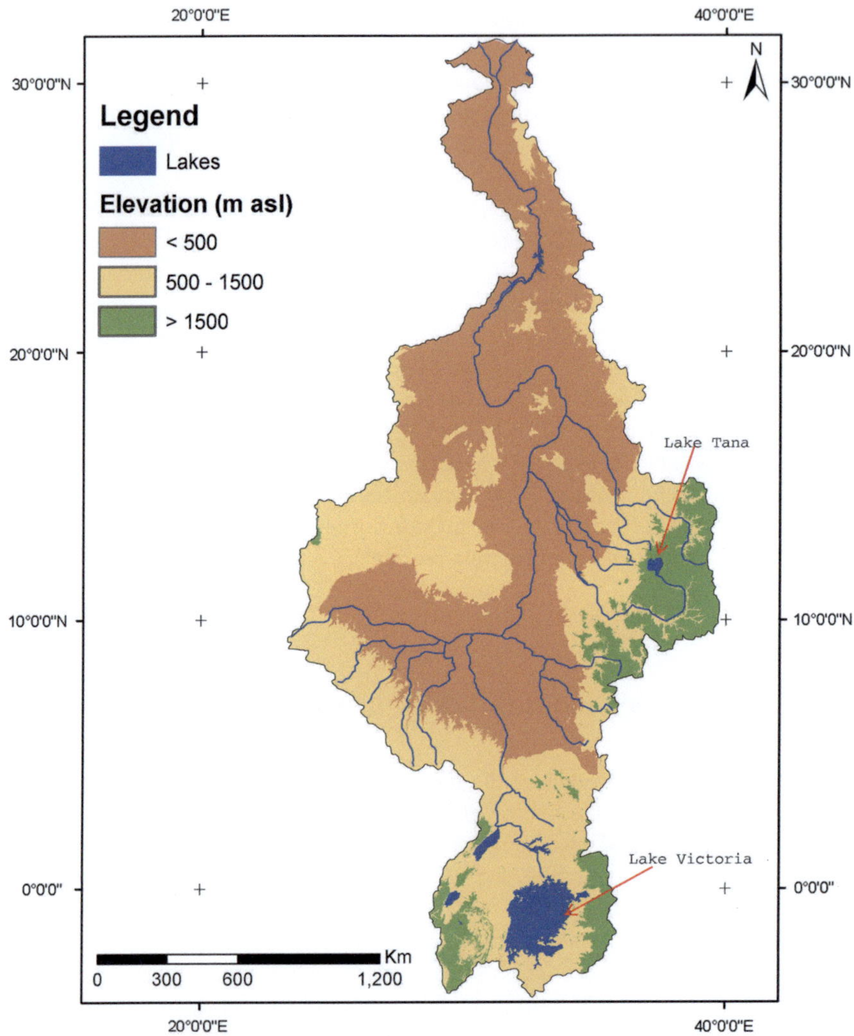

Fig. 11.1 Topographic map of the Nile basin

hydrological study (Sutcliffe and Parks 1999). Others tried to estimate the basin area differently (Zelalem 2009) as 3,112,400 km^2. Some studies only dealt with some section of the basin (Hurst and Phillips 1938; Brown et al. 1979).

Commonly, the Nile River basin is divided into three main sub-basins as White Nile, Blue Nile and Main Nile. Lyons (1906) tried to describe the Nile in six principal drainage basins; the lake plateau, the Bahr-el-Jebel, Bahr-el-Zaraf and the Bahr-el-Ghazal, the Sobat River, the White Nile and the Blue Nile and Atbara.

However, this classification also does not sufficiently describe the different topographic, climatic and other upstream–downstream linkage parameters.

Therefore, in this review work, the Nile basin is classified into nine sub-basins based on the topographic, climatic and hydrological characteristics of the respective sub-basins (Fig. 11.2). In this study, the Nile basin area is computed using GIS environment and Africa Sinusoidal projected coordinate system (Table 11.1). For large river basin like the Nile, selection of the appropriate projection system for the area computations using GIS is also essential to have good area estimation. Africa Sinusoidal projected coordinate was selected since it does not have distortion for area and distance measurement.

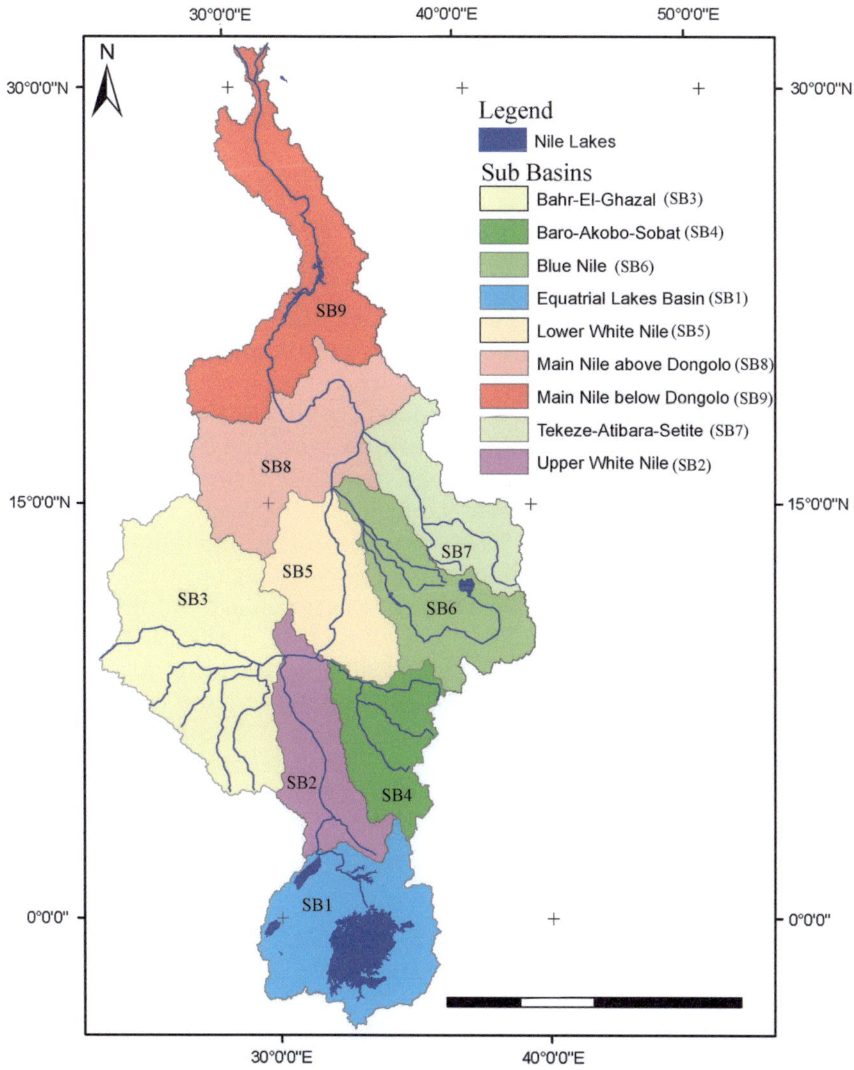

Fig. 11.2 Sub-basin map of the Nile basin

Table 11.1 Sub-basins in the Nile River basin and their drainage area

No	Sub-basin	Area (km^2)
1.	Equatorial Lakes Basin	394,147.06
2.	Upper White Nile	234,680.83
3.	Bahr-el-Ghazal	584,199.81
4.	Baro-Akobo-Pibor-Sobat	206,418.15
5.	Lower White Nile	256,040.61
6.	Blue Nile	298,382.84
7.	Tekeze-Atibara-Setite	221,685.09
8.	Main Nile upstream of Dongola	389,105.60
9.	Main Nile downstream of Dongola	443,570.58
Total basin area		3,028,230.55

11.3 River Zoning in the Nile Basin

Analysis of the longitudinal profile of streams and categorising them into different zones is the basis for upstream–downstream linkage study of a river basin. Most streams can be roughly divided into three zones (Nepal et al. 2014). Zone 1 (sources or headwaters), often has the steepest stream gradient, fast flow of water and initiation of sediment transportation. Zone 2 (transition or transfer zone,) receives some of the eroded material. It is usually characterised by wide floodplains and meandering channel patterns. Zone 3 (floodplain or depositional Zone), is primarily characterised with flatter stream bed gradient and deposition of sediments (Nepal et al. 2014).

Using the longitudinal view concept, the origin of the river channel network and the area-rainfall cumulative effect of the Nile River basin is characterised with three sources. The headwaters sections are the Ethiopian Highlands, Equatorial Lakes Plateau and the head of Bahr-El-Ghazal. These sections are the major water sources of the basin that are characterised by high altitudes and high rainfall. The middle section of the river basin is commonly known as the swamp and the Sudd area which is considered as the transitional zone of the river. Finally, part of the Nile basin around and downstream of Khartoum down to the Mediterranean Sea is grouped to be zone 3. This zone is identified as the dry zone with almost no contribution to the inflow and includes the fertile land of the delta in which maximum water use is recorded so far (Fig. 11.3). This river zoning is implemented based on the weighted overlay of annual rainfall and the gradient (limit of slope) of the land in the basin.

11.4 Rainfall in the Nile Basin

The hydrological process in downstream of a basin is highly dependent on the timing, intensity and the magnitude of rainfall in its upstream. To determine the upstream–downstream linkages of a given basin, accounting the spatial–temporal

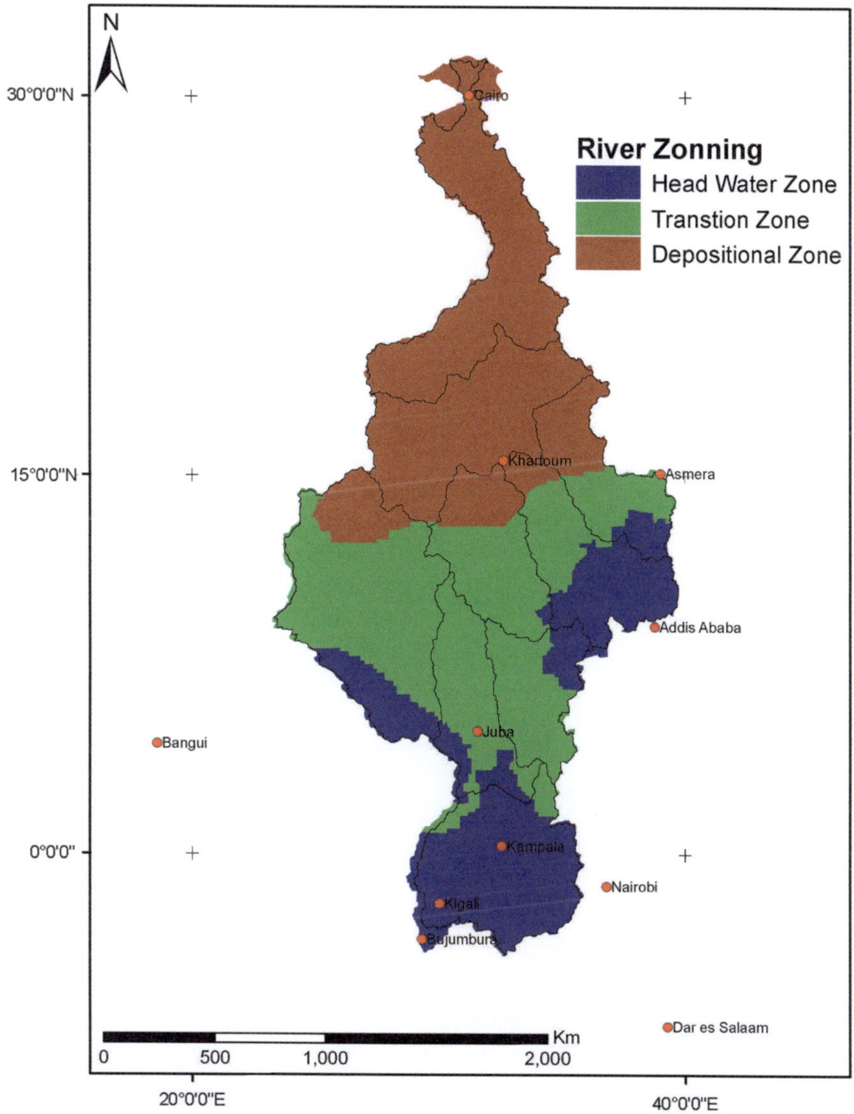

Fig. 11.3 Three broad river zones in the Nile basin based on river zoning

variation of rainfall in the basin would have a significant role. Rainfall is an important parameter for water balance analysis and inflows into the system are dependent on this parameter. The spatial and temporal distribution of rainfall can have different impact on distinct runoff generation processes (Tetzlaff and Uhlenbrook 2005). It also influences the runoff volume, peak flow and timing of hydrological response (Krajewski et al. 1991; Ogden et al. 2000).

The rainfall in the Nile River basin ranges from high rainfall in the most upstream reaches of the equatorial lakes region and the Ethiopian highlands; about 2000 mm mean annual rainfall, to arid desert condition downstream regions that receives no rainfall in a year (Batisha 2012). This climatic variability is possibly observed due to the large extent coverage of latitude (36°) and longitude (18°), large altitudinal variation (8 m below sea level to 4567 m above sea level) and the different monsoons (the longer southeasterly and shorter northeasterly monsoons) over the basin (Sutcliffe and Parks 1999).

The mean monthly and annual rainfall data over the Nile basin were extracted from very high resolution interpolated global dataset (Hijmans et al. 2005). The data set is freely available from WorldClim global climate data site (http://www.worldclim.org/). It is bias corrected and uncertainty tested dataset, which is recommended for the use in climate mapping, modelling, regional studies and understanding of climatic variations. As presented in Fig. 11.4, the spatial and temporal variability of rainfall over the Nile basin is mapped and examined using this high resolution data set.

Annual rainfall over the Nile largely decreases from the south of the basin to the North. The high rainfall area of the basin is confined to the East African lake regions and to the Ethiopian highlands (Sutcliffe and Parks 1999). The East African Lakes region receives rainfall almost throughout the year. But the rainfall in Ethiopian highlands drops out within a single season, in which the length of the wet ranges from 9 months in Baro-Akobo sub-basin in the south to 3 months in the

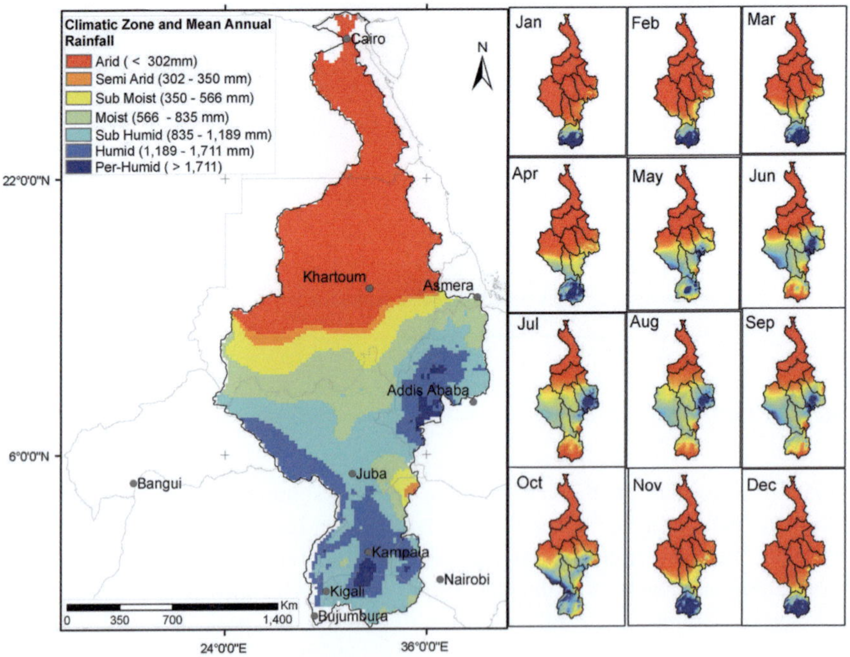

Fig. 11.4 Spatial and temporal variation of rainfall over the Nile River basin

Tekeze-Atibara sub-basin in the north. As a result of steep topography and environmentally degraded watersheds, the sub-basins in the Ethiopian highlands provide relatively quick and highly concentrated runoff to the Nile system. There is not much investigation on hydrological processes in the Bahr-El-Ghazal sub-basin. From the current review, it can be stipulated that it receives relatively considerable average annual rainfall, 835 mm. The amount of rainfall received together with the large area extent of the sub-basin, it can be anticipated that this sub-basin could have significant importance in contributing inflows to the Nile system.

Temporal variability of the wet period in the Nile basin can be categorised in three regions. The Southerly monsoon, largely located in the Equatorial Lakes region extends from October to June. The second category includes the basin area having the wet period extended from April to October. The Southern portion of the Ethiopian highlands, the Bahr-el-Ghazal sub-basin, the White Nile upstream of Malakal and the Sobat-Pibor sub-basins are likely to be included in this category. In this category, especially in the Ethiopian highlands, the wet period is limited to 3 months only towards the north direction. Large portion of the Blue Nile basin falls in this category and the Nile basin gets the largest input in terms of inflow from this sub-basin. The third category, largely located downstream of the Dongola station, is characterised as a dry spell as it receives almost no rainfall. As a result, this category of the Nile basin has no clear wet season period.

The isohyets derived from the mean annual rainfall data was used for the computation of the weighed mean annual areal rainfall of the sub-basins in the Nile system. Accordingly, the Equatorial sub-basin receives the highest mean annual rainfall (1201 mm). It is followed by the Blue Nile sub-basin (1017 mm) and the Upper White Nile sub-basin (1003 mm). These sub-basins are located on the windward side of the Ethiopian highland and East African lakes mountainous regions of the Nile system which receive high rainfall and make significant flow contributions to the Nile system. Similarly, the seasonal areal wetted rainfall for the sub-basins was also computed by the same approach. The seasonal variability of rainfall in the basin helps to compute the potential runoff in each sub-basin (Fig. 11.5).

11.5 Evapotranspiration Over the Nile

Evapotranspiration is an important part of the hydrologic cycle that describes the effect of land cover in the river basin. However, quantifying the actual evapotranspiration with space and time is challenging in water resources system analysis. Alternatively, its amount is computed with potential evapotranspiration (PET), calculated indirectly from climatic parameters and reference land covers with ample water in the area. On a global scale, evapotranspiration accounts for the loss of about 60 % of annual land precipitation and its amount increases to more than 90 % in dry-land ecosystems (Alemu et al. 2014). It is also considered as one of the largest components of the water balance of the Nile basin and accounts about 70 % of the incoming precipitation in the basin (Nile 2014).

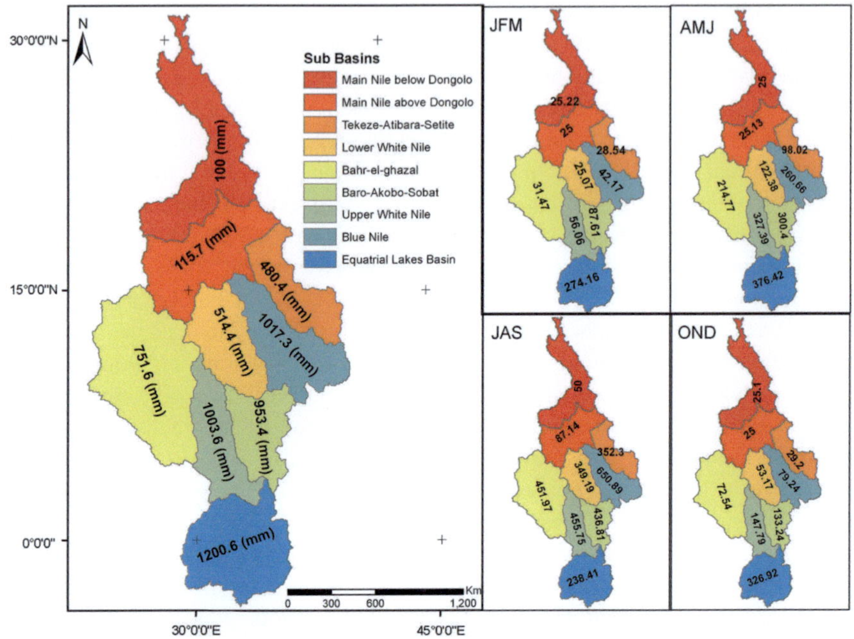

Fig. 11.5 Area weighed annual and seasonal rainfall of sub-basins in Nile River basin

For this review work, the long-term mean monthly potential evapotranspiration for 226 stations in the basin was accessed from FAO ClimWAT climate dataset (Grieser 2006). The spatial distribution of the annual, seasonal and monthly PET was interpolated using inverse distance weighted (IDW) method and shown in Fig. 11.6 to show the spatial and temporal variability of PET over the Nile basin. Accordingly, most of the Nile River basin is covered with the warm and hot thermal zone, which has high mean annual potential evapotranspiration that exceeds 1737 mm. Comparing Figs. 11.5 and 11.6, there are vast areas where PET is greater than rainfall. During period when these areas are not wet, the energy which would have been used for evapotranspiration is used to heat up the land surfaces resulting in dry and hot weather.

To understand the effect of evapotranspiration in the water balance of Nile River basin, the areal weighted annual and the seasonal potential evapotranspiration of each sub-basin was computed using the areal average of the iso-PET lines in each sub-basin. Based on analysis in this work, the downstream sub-basins, Main Nile above Dongola, Main Nile below Dongola, and the Tekezie-Atibara-Setite, show mean annual potential evapotranspiration of 2716, 2486 and 2189 mm, respectively (Fig. 11.7). In the contrary, the upstream sub-basins along the main Nile stem that constitute the main Nile above Dongola station, the mean annual potential evapotranspiration, Equatorial Lakes Basin (1434 mm) Baro-Akobo-Sobat (1603 mm) and Blue Nile (1703 mm). The annual potential evapotranspiration of

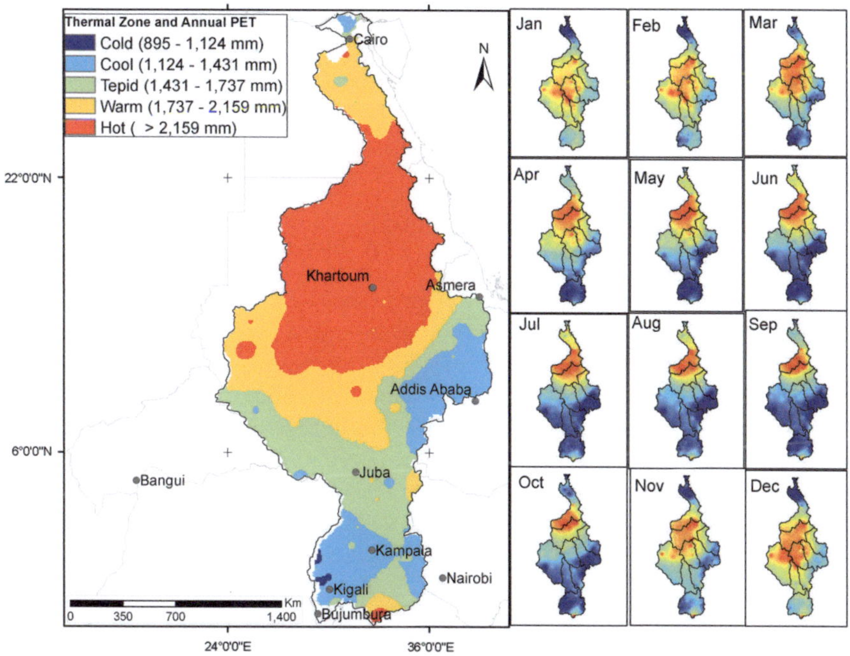

Fig. 11.6 Spatial and temporal variation of potential evapotranspiration over the Nile River basin

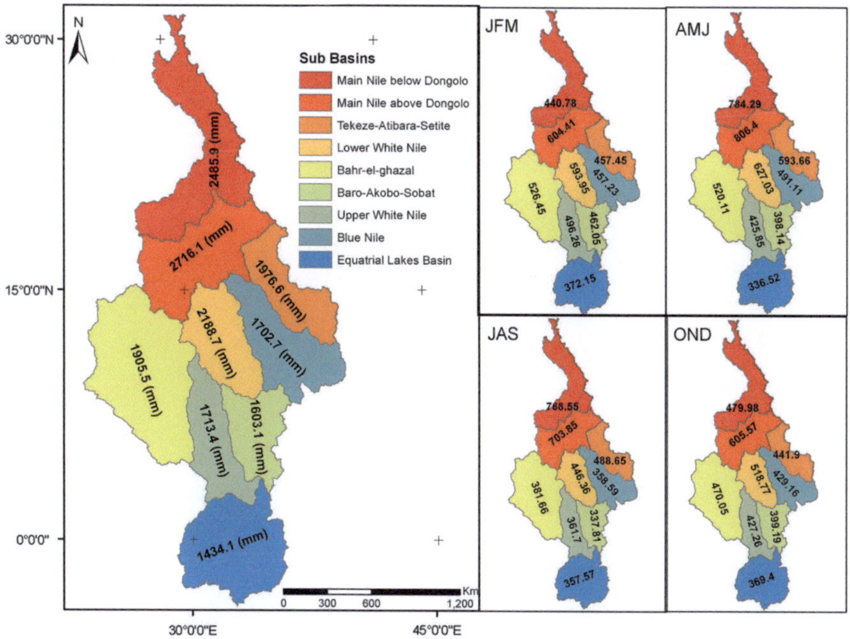

Fig. 11.7 Area weighed annual and seasonal potential evapotranspiration of sub-basins

the sub-basins is much more than the mean annual rainfall of the sub-basins which indicate the evaporation in general is the driving force for the hydrological processes of the Nile hydrosystem. But when moisture is not available for evaporation, the potential evapotranspiration would not be reached.

11.6 Runoff in the Nile River Basin

Runoff and flow characteristics of a river basin is the cumulative effect of the temporal and spatial scale changes of the hydrological process which is important for the understanding of the effect of upstream changes on the downstream system (Conway 2005). Thus, to identify the upstream–downstream linkages of the Nile River basin, basin-wide runoff was computed using the difference of the seasonal weighted areal rainfall and potential evapotranspiration over the basin. Although the water balance of a given watershed includes other variables like groundwater flow, interception, interflows, surface detention and other losses, the runoff is computed by considering the three basic hydrological elements (precipitation, evapotranspiration and runoff) as the major parts of the water balance of basin (Senay et al. 2009). As depicted in Fig. 11.8, runoff in the basin has seasonal and spatial variations. Rainfall in the upstream reach is relatively high with longer wet

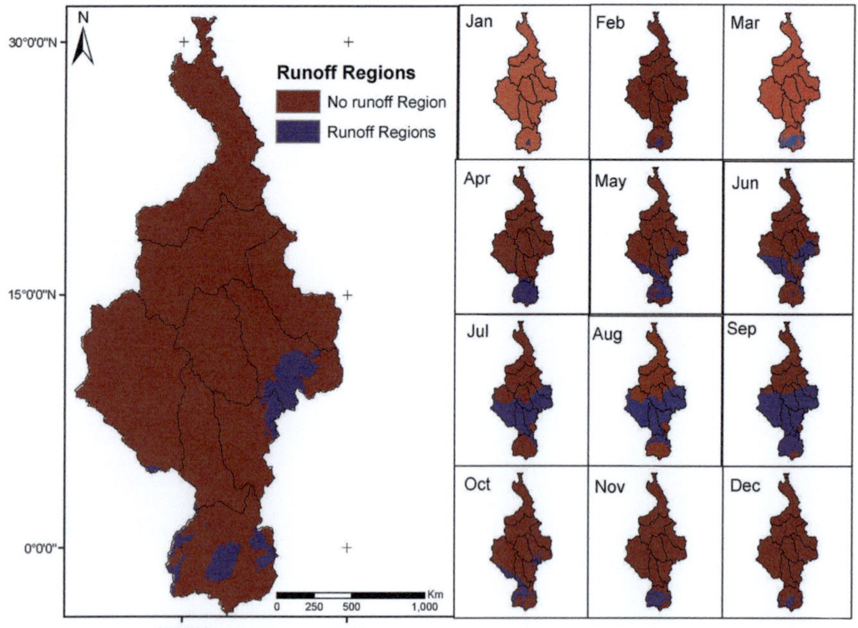

Fig. 11.8 Spatial and temporal variation of runoff over the Nile River basin

period and it is largely believed that this rainfall is the source of runoff for river flows in the basin. This further indicates the significance of the influence of hydrological process changes in the upstream to the downstream reach of the basin.

11.7 Upstream–Downstream Linkages of Processes in the Nile Basin

The different water uses in the transboundary large river basins like Nile River basin is subjected to upstream–downstream hydrological changes that might be associated with the changes in the hydrological process (Beyene et al. 2007). Despite the complex hydrologic processes in transferring flow from upstream to the downstream reach, the primary functions of a river basin can be simply characterised with three main functions: collection, storage and discharge (Black 1997). The three functions of the river basin which are responsible for the upstream–downstream linkages can be characterised by the dynamics of four elements of the hydrological cycle, precipitation, evaporation, storage and runoff (Fernandez and Sayama 2014).

The collection function describes the process of receiving precipitation from the atmosphere and channelizing the runoff supplying the storage zone. The storage zone in the river basin with its different hydrological conditions serves as the linkage between the collection and discharging zones by producing changes in the flow hydrograph base times and amount of releasing. The discharge function addresses the processes of releasing of water from the storage in the form of evaporation or runoff as surface and subsurface components.

The upstream of the basin, particularly the three headwater zones, are the major water sources for the basin river flow. The runoff generated in the headwater zones forms river inflow which leads the water to flow through the transitional zones of the basin. Due to the climatic and topographic nature of the basin, the transitional zones in the Nile River basin are mainly characterised with high rate of evapotranspiration. The high proportion of the water released into this zone is lost by evaporation. Thus the influence of the transitional zones in widening the base hydrograph time of the inflow hydrograph supplied from the headwater zones and loses of water happening in this transitional zone through evaporation describes the existing linkage of the hydrological process between the headwater and deposition zones. This further indicates the impact of changes in the hydrological process at the headwater zones to the successive transitional and deposition zones indicating existence of hydrologic connectivity between the three zones of the basin. This concept can be more illustrated by simulating water balance of the river basin at all important nodes.

The three headwater zones (Equatorial Lake basins, Blue Nile and Bahr-El Ghazal) have different levels of influences on the downstream flow system of the basin. Particularly, the changes in the headwaters zone at the Ethiopian highlands (the Blue Nile sub-basin) are contributing the most significance impacts on the

Table 11.2 Potential rainfall volumes in sub-basins of Nile River basin computed in this review work (order of list of sub-basins is as in Table 11.1)

	Sub-basin name	Sub-basin area	Mean annual rainfall	Rainfall volume in sub-basins
		(km^2)	(mm)	(km^3)
1.	Equatorial Lakes Basin	394,147	1201	473
3.	Bahr-El Ghazal	584,769	752	440
6.	Blue Nile	298,383	1017	304
2.	Upper White Nile	234,181	1004	235
4.	Baro-Akobo-Sobat	206,418	953	197
5.	Lower White Nile	256,041	514	132
7.	Tekeze-Atibara-Setite	221,685	480	107
8.	Main Nile above Dongola	389,106	116	45
9.	Main Nile below Dongola	443,580	100	44

Table 11.3 Historical flows of the Nile and contributing sub-basins (modified from Abtew and Melesse 2014; original data source Sutcliffe and Park 1999)

Reach	Annual flow (km^3)
Nile at Aswan	84.1
Atbara at Mouth	11.1
Blue Nile at Khartoum	48.3
White Nile at Khartoum	26
Sobat at Malakal	9.9

downstream water uses and hydrological system (Zachary et al. 2012). Although, they receive comparable amount of annual rainfall and covering larger area, the influence of the other headwater zones (Equatorial Lakes and Bahr-EL-Ghazal sub-basins) is comparatively low (El Bastawesy et al. 2014). Sub-basin area and mean annual rainfall combination in the Bahr-El-Ghazal and Equatorial Lake Basins is much greater than that of the Blue Nile sub-basin (Table 11.2), but the downstream effect of the changes in the Blue Nile sub-basin is much greater. This is demonstrated by historical flows at different reaches of major rivers of sub-basins and the Nile River (Table 11.3). Over 80 % of the Nile River flows are generated in the Blue Nile and Baro-Akobo-Sobat sub-basins.

11.8 Conclusions

The Nile basin is the longest river in the world that has strong upstream–downstream hydrological linkages. The Nile river flow per unit area of watershed is small (77 m^3 d^{-1} km^{-2}) compared to the Congo River (887 m^3 d^{-1} km^{-2}). Flow is significantly influenced by the process of the upstream of the basin. Particularly, the extent of the transitional zone, which is mainly characterised by evaporation, has governed the water release to the downstream. In this review work, the spatial and temporal variations of the major hydrological processes (precipitation, evapotranspiration and runoff) in the basin are mapped and used to identify the sub-basins which have significant effect on the downstream water use. Although, the three headwater zones (Equatorial Lake Basin, Blue Nile sub-basin and Bahr-El-Ghazal sub-basin) have equivalent annual rainfall volume over the basin, changes of the hydrological processes in the Blue Nile sub-basin has strong impact on the downstream of the basin. The major reason attributed to this influence is because the transitional zone of the Blue Nile sub-basin has limited storage effect; rather it serves as a hydraulic link to the lower section of the basin. Floodplains and wetlands are insignificant in the storage zone of this sub-basin. Further to this, the cause of this significant influence on the downstream reach is also clearly identified as the extent of its transitional zone is much smaller than the others. The release function in the Blue Nile sub-basin is more dependent on the runoff process than the evaporation process. Thus, upstream–downstream linkages of hydrological processes are stronger in the Blue Nile sub-basin.

References

Abtew W, Melesse AM, Desalegn T (2009a) Spatial, inter and intra-annual variability of the Blue Nile River Basin rainfall. Hydrol Process 23(21):3075–3082

Abtew W, Melesse AM, Desalegn T (2009b) El Niño Southern Oscillation link to the Blue Nile River Basin hydrology. Hydrol Process Spec Issue Nile Hydrol 23(26):3653–3660

Abtew W, Melesse AM (2014a) Chap. 2. The Nile River Basin. In: Melesse AM, Abtew W, Setegn SM (eds) Nile River Basin ecohydrological challenges, climate change and hydropolitics. Springer, New York

Abtew W, Melesse AM (2014b) Climate teleconnections and water management. In: Nile River Basin. Springer International Publishing, New York, pp 685–705

Abtew W, Melesse AM (2014c) Transboundary Rivers and the Nile. In: Nile River Basin. Springer International Publishing, New York, pp. 565–579

Alemu H, Senay GB, Kaptue AT, Kovalskyy V (2014) Evapotranspiration variability and its association with vegetation dynamics in the Nile Basin 2002–2011. Remote Sens 6(7):5885–5908

Batisha AF (2012) Hydrology of Nile River Basin in the era of climate changes. Irrig Drainage Syst Eng S5:e001. doi:10.4172/2168-9768.S5-e001

Beyene T, Dennis PL, Kabat P (2007) Hydrologic impacts of climate change on the Nile River Basin: implications of the 2007 IPCC climate scenarios. University of Washington, Seattle 98195

Black PE (1997) Watershed functions. J Am Water Resour As 33:1–11

Blaikie PM, Muldavin JS (2004) Upstream, downstream, China, India: the politics of environment in the Himalayan region. Ann Assoc Am Geogr 94(3):520–548

Bloschl G, Sivapalan M (1995) Scale issues in hydrological modelling: a review. Hydrol Process 9:251–290

Brown JAH, Ribeny FMJ, Wolanski EJ, Codner GP (1979) A summary of the Upper Nile Basin model. Snowy Mountains Engineering Corporation, Cooma (NSW 2630, Australia)

Chebud Y, Melesse AM (2013) Stage level, volume, and time-frequency change information content of Lake Tana using stochastic approaches. Hydrol Process 27(10):1475–1483. doi:10.1002/hyp.9291

Chebud YA, Melesse AM (2009a) Numerical modeling of the groundwater flow system of the Gumera Sub-basin in Lake Tana basin. Ethiop Hydrol Process Spec Issue Nile Hydrol 23(26):3694–3704

Chebud YA, Melesse AM (2009b) Modeling lake stage and water balance of lake tana. Ethiop Hydrol Process 23(25):3534–3544

Conway D (2005) From headwater tributaries to international river: observing and adapting to climate variability and change in the Nile Basin. Glob Environ Change 15(2005):99–114

Dessu SB, Melesse AM, Bhat M, McClain M (2014) Assessment of water resources availability and demand in the Mara river Basin. CATENA 115:104–114

Dessu SB, Melesse AM (2012) Modeling the rainfall-runoff process of the Mara River Basin using SWAT. Hydrol Process 26(26):4038–4049

Dessu SB, Melesse AM (2013) Impact and uncertainties of climate change on the hydrology of the Mara River Basin. Hydrol Process 27(20):2973–2986

El Bastawesy M, Safwat G, Ihab M (2014) Assessment of hydrological changes in the Nile River due to the construction of renaissance dam in Ethiopia. J Remote Sens Space Sci, Egypt. doi:10.1016/j.ejrs.2014.11.001

Fernandez R, Sayama T (2014) Hydrological recurrence as a measure for large River Basin classification and process understanding. Hydrol Earth Syst Sci Discuss 11:8191–8238. doi:10.5194/hessd-11-8191-2014

Grieser J(2006) ClIMWAT2.0. Water Resources Development and Management Service Land and Water Development Division FAO, VialedelleTerme di Caracalla, 00153 Rome, Italy

Hijmans RJ, Cameron SE, Parra J, Jones PG, Jarvis A (2005) Very high resolution interpolated climate surfaces for global land areas. Int J Climatol 25:1965–1978. doi:10.1002/joc.1276 (www.interscience.wiley.com)

Hurst HE, Phillips P (1938) The hydrology of the Lake Plateau and Bahr el Jebel. The Nile Basin, vol V. Government Press, Cairo

Kebede S, Travi Y (2006) Water balance of Lake Tana and its sensitivity to fluctuations in rainfall, Blue Nile Basin, Ethiopia. J Hydrol 316:133–247

Klemeš V (1983) Conceptualization and scale in hydrology. J Hydrol 65(1–3):1–23. doi:10.1016/0022-1694(83)90208-1

Krajewski WF, Ventakataramann L, Georgakakos KP, Jain SC (1991) A Monte Carlo study of rainfall sampling effect on a distributed catchment model. Water Resour Res 27(1):119–128

Lyons HG (1906) The physiographic of the River Nile and its Basin. Survey Department, Cairo

Melesse AM (2011) Nile River Basin: hydrology, climate and water use. Springer Science & Business Media, New York

Melesse A, Abtew W, Setegn SG (2014) Nile River Basin: ecohydrological challenges, climate change and hydropolitics. Springer Science & Business Media, New York

Melesse A, Abtew W, Setegn S, Dessalegne T (2011) Hydrological variability and climate of the Upper Blue Nile River Basin In: Melesse A (ed) Nile River Basin: hydrology, climate and water use e. Springer Science Publisher, New York Chap. 1, 3–37. doi:10.1007/978-94-007-0689-7_1

Melesse A, Athanasios GL, Senay G, Yitayew M (2009a) Climate change, land-cover dynamics and ecohydrology of the Nile River Basin. Hydrol Process Spec Issue Nile Hydrol 23(26):3651–3652

Melesse A, Abtew W, Desalegne T, Wang X (2009b) Low and high flow analysis and wavelet application for characterization of the Blue Nile River system. Hydrol Process 24(3):241–252

Nepal S, Flügel WA, Fink SAB (2014) Upstream-downstream linkages of hydrological processes in the Himalayan region. Ecol Process 3:19

Nile W (2014) Understanding of Nile Basin hydrology: mapping actual evapotranspiration over the Nile Basin. Technical Bulletin from the Nile Basin Initiative Secretariat, ISSUE: 01

Ogden FL, Sharif HO, Senarath SUS, Smith JA, Baeck ML, Richardson JR (2000) Hydrologic analysis of the Fort Collins, Colorado, flash flood of 1997. J Hydrol 228:82–100
Rasul G (2014) Why eastern Himalayan countries should cooperate in transboundary water resource management. Water Policy 16(1):19–38
Senay GB, Asante K, Artan G (2009) Water balance dynamics in the Nile Basin. Hydrol Process 23:3675–3681
Setegn SG, Srinivasan R, Dargahi B, Melesse AM (2009a) Spatial delineation of soil erosion prone areas: application of SWAT and MCE approaches in the Lake Tana Basin. Ethiop Hydrol Process Spec Issue Nile Hydrol 23(26):3738–3750
Setegn SG, Srinivasan R, Melesse AM, Dargahi B (2009b) SWAT model application and prediction uncertainty analysis in the Lake Tana Basin. Ethiop Hydrol Process 24(3):357–367
Setegn SG, Bijan Dargahi B, Srinivasan R, Melesse AM (2010) Modelling of sediment yield from Anjeni Gauged watershed. Ethiop Using SWAT JAWRA 46(3):514–526
Sutcliffe JV, Parks YP (1999) The hydrology of the Nile, IAHS Special Publication no. 5 ISBN 1-910502-75-9. IAHS Press, Institute of Hydrology, Wallingford, Oxfordshire OX10 8BB, UK
Taye MT, Willems P (2011) Influence of climate variability on representative QDF predictions of the upper Blue Nile Basin. J Hydrol 411:355–365
Tetzlaff D, Uhlenbrook S (2005) Significance of spatial variability in precipitation for process-oriented modelling. Hydrol Earth Syst Sci 9:29–41
Yitayew M, Melesse AM (2011) Critical water resources management issues in Nile River Basin. In: Melesse AM (ed) Nile River Basin: hydrology, climate and water use. Springer Science Publisher, New York, Chap. 20, 401–416. doi:10.1007/978-94-007-0689-7_20
Zachary ME, Seleshi BA, Tammo SS, Saliha AH, Birhan Z, Yilma S, Kamaleddin EB (2012) Hydrological processes in the Blue Nile, a chapter on The Nile River Basin: water, agriculture, governance and livelihoods. In: Awulachew SB et al (eds) International water management institute (IWMI), Routledge, 2 Park Square, Milton Park, Abingdon, Oxon OX14 4RN
Zelalem KT (2009) Long term hydrologic trends in the Nile Basin, a thesis presented to the faculty of the graduate school of Cornell University. In: Partial fulfilment of the requirements for the degree of Master of professional studies. Cornell University, Ithaca NY, USA

Chapter 12
Advances in Landscape Runoff Water Quality Modelling: A Review

Iqbal Hossain and Monzur Alam Imteaz

Abstract As the recognition of the concept 'Sustainable Development' is increasing throughout the world, understanding the adverse impact of water quality parameters is highly important for the protection and improvement of aquatic environments from the impact of pollution. Therefore, the measurement of water quality parameters is required to protect and improve aquatic environments from the impact of pollution. The estimation of water quality parameters from direct field measurement is costly, time-consuming and sometimes impossible. Therefore, mathematical approaches of water quality models have become prevalent in recent years for the purpose of watershed management strategies. However, water quality model parameters vary not only spatially (i.e. catchment to catchment), but also temporally (i.e. differ among different rainfall events). Because depending upon the catchment characteristics such as soil permeability and initial pollutant loads, the impact of the actual land-use and management changes. There exists a wide range of water quality models to be used in managing water quantity and quality with respect to a variety of environmental impacts. This chapter presents a rigorous literature review regarding available catchment water quality modelling techniques developed in recent years. The goal is to identify the most effective water quality modelling technique which will help in the analysis, improvement and update of best management practices (BMPs). The selected modelling technique will help watershed management authorities to enable them implement economically viable and effective management and mitigation strategies to protect aquatic environments from the impact of pollution.

Keywords Runoff · Water quality · Modelling · Pollution · Watershed

I. Hossain · M.A. Imteaz (✉)
Department of Civil and Construction Engineering, Swinburne University of Technology, Hawthorn, Melbourne, VIC 3122, Australia
e-mail: mimteaz@swin.edu.au

I. Hossain
e-mail: ihossain@swin.edu.au

© Springer International Publishing Switzerland 2016

A.M. Melesse and W. Abtew (eds.), *Landscape Dynamics, Soils and Hydrological Processes in Varied Climates*, Springer Geography,
DOI 10.1007/978-3-319-18787-7_12

12.1 Introduction

It is widely accepted that storm water pollutants are accumulated on catchment surfaces during antecedent dry days (t_d). Urban expansion, land development, agricultural activities, industrial and commercial activities, atmospheric fallout and other human activities alter the natural conditions of catchment surfaces and increase the accumulation of pollutant loads. During rainfall events, these accumulated pollutants dissolve and get transported into nearby waterways and receiving water bodies (Bannerman et al. 1993; Sartor et al. 1974).

Rainfall also brings atmospheric pollutants to catchment surfaces and dislodges dissolved and suspended pollutant particles from both impervious and pervious surfaces (Zoppou 2001). During the initial period of a rainfall event, catchment surfaces get wet and most of the soluble pollutants begin to dissolve (Kibler 1982). At the same time, some of the pollutants of a catchment surface are loosened by the energy of falling raindrops (Egodawatta et al. 2007). With an increase in rainfall, catchment surfaces become wet enough to have surface runoff which transports dissolved and suspended pollutants to downstream aquatic environments and deposit. It is well recognised that accumulation of water quality parameters originating from catchment surfaces alters the quality of receiving waterways and water bodies.

The changes in water quality parameters of an environment result in the degradation of the quality of receiving water bodies and aquatic habitats, including social, economic and environmental costs, with short and long-term consequences. Therefore, the impact of water pollution is an increasing matter of concern amongst watershed management groups. However, the severity of the deterioration of an aquatic environment depends on the amount of pollutants transported from upstream catchments and the characteristics of receiving environments. Hence, the measurement of water quality parameters is required to protect and improve aquatic environments from the impact of pollution. An accurate estimation of runoff and pollutant loads will help watershed management authorities adopt proper impact mitigation strategies. Inaccurate determination of pollutant loads can lead to the design of undersized and ineffective measures, or oversized measures with the excessive capital costs and maintenance requirements.

Many regulatory authorities from government to catchment management groups strive to implement water quality management strategies to mitigate the adverse impact of water quality parameters. However, the productivity and effectiveness of such initiatives strongly rely on the accuracy and reliability of water quality parameters measurements (Chiew and McMahon 1999). Nevertheless, the allocated resources for proper management of aquatic environments are small in relation to what is required for the remediation. The intensive monitoring, analysis and direct estimation of these pollutants on a wide scale are labour-intensive, time-consuming and prohibitively expensive with limited public funds available (Davis and Birch 2009). Well management of aquatic ecosystems is also essential with the allocated budgetary constraints. With the impact of water quality parameters on aquatic environments as a foregone conclusion, an accurate prediction of pollutant loads would enable watershed management authorities to develop more efficient impact mitigation strategies.

Therefore, to address the widespread degradation of aquatic environments, watershed management authorities need appropriate modelling techniques so that they can meet legislative requirements and societal expectations of sustainable water resources. Models are essential and powerful tools, which can influence decision-makers for the implementation of proper management strategies. Modelling techniques also help to integrate scientific understandings of the impact of management changes, and to give broader and long-term perspectives about management interventions. Water quality models are used for the prediction of waterborne pollutants from waterways and receiving water bodies.

Water quality modelling is of crucial importance for the assessment of the physical, chemical and biological changes in water bodies. Mathematical approaches of water quality models have become prevalent in recent years. Different water quality models, ranging from the detailed physical to the simplified conceptual are widely available for the simulation of various water quality parameters. However, the application of an appropriate modelling approach depends on the research goal and data availability.

Water quality models are developed primarily based on modelling approaches which replicate hydrologic, hydraulic and water quality processes (Zoppou 2001). However, proper understanding of the actual methods for the accumulation and transportation of pollutants is often lacking. The lack of knowledge on the primary pollutant processes and the lack of data make modelling approaches inherently difficult. Although numerous research studies on water quality modelling have been undertaken, unfortunately comprehensive studies on the integrated modelling approach is yet to appear in scientific literature.

The usual practices of water quality investigations are performed through separate models in isolation. This can lead to the inconsistencies and significantly biased prediction results. However, due to the lack of in-depth knowledge on the pollutant processes and the lack of data, limited attention has been given to develop integrated water quality model. On the other hand, the adoption of a continuous simulation approach is recommended in water quality modelling literature (CRC for Catchment Hydrology 2005). Although numerous water quality models have been developed since 1970, including some probabilistic models, there has been little effort in the development of a continuous modelling approach.

The primary objective of this chapter is investigation of available catchment water quality modelling techniques developed in recent years. The goal is to identify the most effective water quality modelling technique, which will help in the analysis, improvement and update of best management practices (BMPs). Consequently, this chapter focuses on the development of a detailed understanding of catchment water quality modelling for the prediction of storm water quality parameters that are transported from upstream catchments to downstream aquatic environments. Finally, the chapter concentrates on recommendation of the best modelling techniques available for the prediction of storm water quality parameters.

The investigated modelling technique will help watershed management authorities by enabling them to implement economically viable and effective management design, and mitigation strategies to protect aquatic environments from the impact of pollution.

Section 12.1 of the chapter introduced background information, aims and objectives of the chapter and provides an overview of the subsequent sections. Section 12.2 reviews literature on primary water quality parameters and their impact on aquatic environments. Sources of these pollutants are identified in Sect. 12.3. Section 12.4 describes the roles of mathematical models together with variety of models. Sections 12.5 and 12.6 of the chapter reviews literature on available water quality modelling techniques and describe water quality research and identify the existing knowledge gaps. Section 12.7 of this chapter describes data requirements of catchment water quality model. Moreover, this section demonstrates the requirements of data for calibration and validation of the developed model. Section 12.8 demonstrates the use of the calibration procedure to estimate the model parameters and description of the different sensitivity analysis approaches. Section 12.9 is discussion followed by summary in Sect. 12.10.

12.2 Primary Water Quality Parameters

Since storm water pollution leads to the significant deterioration of surrounding environments, identification of the specific characteristics and types of pollutants is critically important. To address storm water runoff water quality concerns adequately, it is important to understand types of pollutants which are present as well as their potential impact in receiving water bodies (Adams and Papa 2000). The common water quality parameters which affect the quality of an aquatic environment are litter, suspended solids, nutrients, heavy metals, hydrocarbons and total organic carbon.

12.2.1 Litter

Litter is the most obvious component of storm water pollution. Shaheen (1975) found that 20 % of the weight of pollutants which accumulate on road surfaces is litter. Over the last 30 years, pollution of the environment from the export of litter has intensified due to the production of easily disposable, non-biodegradable packaging materials, and commercial and industrial items (Wong et al. 2000). Litter contributes to the blockage of urban drains and causes the unsightly appearance of receiving waters. It interferes with the quality of water aesthetically and threatens animals, plants and fishes which habit in waterways and water bodies (MW-Web 2011).

12.2.2 Suspended Solids (SS)

It is well recognised that SS is the most prevalent component in the deterioration of the quality of water, aquatic environments and their ecosystems. Excessive levels of SS in water bodies have significant deleterious impact on physical, chemical and biological properties of receiving waterways and water bodies (Bilotta and Brazier 2008). The physical alterations of water quality due to SS are associated with the undesirable aesthetic effects of polluted water, the higher water treatment costs, reduced navigational facilities of streams, and decreased longevity of dams and reservoirs (Butcher et al. 1993). The deposition of SS can block pipes, change in the flow conditions in open channels and disrupt habitats of aquatic invertebrates and fishes (Duncan 1999). In addition, the deposition of sediment particles raises the stream bed, lakes and ponds, reduces the discharge capacity of streams, impacts navigational facilities; and hence causes floods during high rainfall events. The recreational uses of water, such as boating and swimming may be reduced due to turbidity of water.

The chemical alterations caused by SS in waterways and water bodies include release of contaminants, such as heavy metals, pesticides and nutrients (Russel et al. 1998). Furthermore, SS with a high organic content undergoes anaerobic breakdown and depletes oxygen level during decomposition. This produces a critical oxygen deficit which can lead to fish kills during low-flow conditions.

The biological effects of high-level SS on different groups of organisms are different. Increase in SS during a storm event can have the ecotoxic effects to aquatic organisms (Rossi et al. 2006). The growth rate of photosynthetic organisms is reduced due to reduction in light penetration (Secchi disc) from high concentration of SS (Akan and Houghtalen 2003). Deposition of SS can clog fish gills and reduce spawning resulting in lower fish populations or a shifting of fish species. Decline of fishery resources and serious ecological degradation of aquatic environment are results of excessive SS concentration.

Significant amounts of other pollutants are also transported as solid-bound contaminants with surface runoff. Zoppou (2001) noted that some of the main water quality parameters adhere to SS particles and are conveyed along with soluble pollutants by surface runoff. Vaze and Chiew (2004) found that particulates of TP and TN in storm water runoff are attached to the sediment particles of size 11–300 μm. Wong et al. (2000) found significant amounts of inorganic pollutants in SS. A strong correlation between SS and heavy metals, nutrients and hydrocarbons was reported (Herngren et al. 2005; Akan and Houghtalen 2003; Sartor et al. 1974). As the correlation between SS and other pollutants are strong, the presence of SS in water indicates the presence of other pollutants such as heavy metals and nutrients.

12.2.3 Nutrients

Nutrients are chemical compounds that play important roles in all forms of life. However, excessive amount of nutrients in water environment can cause detrimental effects to the health of aquatic ecosystems (Wong et al. 2000). Increased concentration of nutrients in surface waters accelerates the eutrophication process. Eutrophication is the enrichment of water bodies with dissolved nutrients resulting in excessive organic matter growth which cause the depletion of available oxygen during decomposition. Exotic species can also be favoured by high nutrient concentrations and crowd out native species. Eutrophication results in disturbance of biological equilibrium in water ecosystems. Eutrophication in surface water, as rivers, dams, lakes and estuaries, leads to the enhanced aquatic plant growth such as algal bloom. Algal bloom ultimately results in the de-oxygenation of water, reduces aesthetics and increases toxicity and contributes to the loss of diversity in aquatic environments (Drewry et al. 2006). The excessive growth of algae also alters the appearance of water bodies affecting the colour, odour, turbidity and floating matter. Decaying algae cause undesirable odours, oxygen depletion and affects the taste of drinking water.

12.2.4 Heavy Metals

The presence of heavy metals in storm water runoff is of concern because most of these metals are toxic to aquatic lives (Akan and Houghtalen 2003; Davis et al. 2001; Ball et al. 2000). Environmental toxicity is dependent largely on concentration of toxic substance. Unlike most other waterborne pollutants, heavy meals cannot be chemically destroyed or integrated into the environment (Davis et al. 2001). It may be transported attached to sediments or in a soluble or dissolved form. Most heavy metals are cumulatively accrued in an aquatic environment rather than the instantaneous flux (Ball et al. 2000). They have a strong affinity to attach to smaller particles due to the adsorption capacity of fine solids because of their relatively larger surface area (Deletic and Orr 2005). The conventional pollutant removal processes does not reduce heavy metals from storm water runoff.

12.2.5 Hydrocarbons

Hydrocarbons are crucially important in water quality study because some of their compounds are toxic to aquatic environments (Gobel et al. 2007; Akan and Houghtalen 2003). Herngren et al. (2005) found that there is a strong correlation between heavy metals and hydrocarbons. Ball et al. (2000) noted that similar to heavy metals and nutrients, hydrocarbons in storm water runoff are mostly associated with sediment particles. Wong (2006) found that many organic hydrocarbons can persist in sediment for long periods and they are dangerous at very low concentration. As a result, benthic organisms are particularly vulnerable to these organic toxicants. Therefore, an increased concentration of hydrocarbons leads to increase in toxicity in receiving waterways and water bodies.

12.2.6 Organic Carbon

Organic carbon is the measure of all carbon atoms covalently bonded in organic molecules (Wong 2006). The common impact of organic matter in water is the reduction in dissolved oxygen (DO) due to microbial oxidation. Decomposition of organic matter depletes DO into receiving water and affects aquatic environments (Akan and Houghtalen 2003). Hence, excess organic carbon means excessive oxygen demand in receiving waters and causes significant damage to aquatic lives. Moreover, Lin and Chen (1998) found a positive linear relationship between organic matter and heavy metals. Furthermore, a substantial level of organic matter leads to anaerobic conditions resulting in fish kills, foul odours and discolouration. Organic carbon in sediment solution leads to the solubility enhancement effect (Warren et al. 2003). Solubility enhancement is the reduction of solid–solution partition coefficient which reduces the total amount of sediment adsorbed, thereby increasing the soluble fraction (Goonetilleke and Thomas 2003).

12.2.7 Pathogens

Every stream contains some microorganisms, many of which are harmless. However, some of them can cause diseases to humans which are called pathogens (MW-Web 2011). The common pathogens associated with the water quality degradation include bacteria, viruses and protozoa. They are disease-causing organisms and are sometimes responsible for epidemics. Pathogens cause some fatal waterborne diseases, such as diarrhoea, dysentery, hepatitis, cholera or typhoid fever (Wong 2006).

12.3 Sources of Water Quality Parameters

Knowledge of the sources of pollutants allows decisions to be made to target the reduction of pollutants and evaluate the changes in pollutant loadings due to modification in land use and development. Proper knowledge of pollutant sources helps to estimate the effects of water system configurations on contaminant flows and their control. Bohemen and De Laak (2003) noted that the most effective way to control the flow of water quality parameters is to control them at the source. The categories of important natural pollutants are geological, hydrological and climatic. Usually, surface water is impaired by water quality parameters input from both point and non-point sources.

Pollutant loads coming from an identifiable source are called point source pollutants, while pollutants coming from many unknown sources are called non-point source pollutants. Point source pollutants can be predicted, controlled and reduced by imposing regulations. However, the prediction of the non-point source pollutants are difficult, and hence non-point source pollution, such as runoff from agricultural land, industrial and highly urbanised areas, have the profound impact on both surface runoff and groundwater. Today, after years of extensive research in this area, it is clear that non-point source pollution (also called diffuse source pollution) is a significant problem to surrounding water environments. The common non-point source pollutants are categorised as sediments, nutrients and sewage. They can originate from transportation activities, industrial/commercial activities, construction and demolition activities, corrosion of materials, vegetation, soil erosion, spills and atmospheric fallout (Goonetilleke and Thomas 2003).

12.3.1 Transportation Activities

Transportation activities are considered to be one of the major contributing sources of water quality parameters. According to Puckett (1995), 38 % of the atmospheric nitrogen emission may come from transport activities. Transport related pollutants are mainly generated on street surfaces. Therefore, street surfaces are considered to have profound impact on water quality parameters which are mainly generated from vehicles. Although highways and freeways represent only a minor portion of catchment imperviousness, they are significant sources of water quality pollution (Wong et al. 2000).

Numerous research studies have identified street surfaces as significant contributors of water quality parameters (Drapper et al. 2000; Bannerman et al. 1993; Sartor et al. 1974). However, the accumulation of water quality parameters on street surfaces varies widely depending on a wide range of factors. Gobel et al. (2007) noted that the concentration of pollutants from street surfaces vary with the traffic density, wind drift, duration and intensity of storm water events, the duration of dry weather periods, and the state of the traffic technology. Sartor et al. (1974) identified pavement conditions and pavement materials as important sources of water quality parameters on street surfaces. They also found that asphalt pavements contribute 80 % more pollutant loadings than concrete surfaces. Furthermore, the accumulation of water quality parameters on a street surface depends on the location of traffic lights, road layout, pavement surface roughness and driver's habits (Goonetilleke and Thomas 2003).

12.3.2 Industrial/Commercial Activities

Industrial and commercial activities have impact on accumulation of a wide range of water quality parameters on catchment surfaces. Due to the wide range of industrial activities and chemical use, heavy metals are especially found in urban areas. The studies of Lau and Stenstrom (2005) found that industrial and commercial areas have relatively high concentration of water quality pollutants compared to other land uses. Sartor et al. (1974) found that the accumulation of TP in a catchment surface was 14.3 g/m^2 for commercial areas and 51.8 g/m^2 for industrial areas. In addition, Bannermann et al. (1993) found that streets and parking lots are critical source areas for the generation of water pollutants in industrial and commercial areas.

The major water quality parameters from various industrial/commercial activities contain significant amount of nutrients, heavy metals and chemical toxins. The industrial sources of pollutants are exposed storage, loading and unloading, equipment, spills and leaks, industrial materials and waste products. Water quality parameters in commercial areas are generated mainly from motor fluids from parked cars, large parking lots, auto service stations, gas stations, shopping centres and restaurants.

Sartor et al. (1974) found the highest metal concentration in road sweepings from industrial areas. Kelly et al. (1996) noted that due to the extensive burning of fossil fuel, heavy metals are accumulated in industrial areas. In addition, Herngren et al. (2006) found the highest heavy metal loadings coincided with areas of the highest sediment loadings which occurred in commercial areas. Latimer et al. (1990) found that particulate pollutants in industrial sites have the highest concentration of hydrocarbons. Puckett (1995) found that 53 % of the atmospheric deposited nitrogen in the north-eastern states in the US came from large industries, such as coal and oil burning and electric utilities. Therefore, areas with industrial or commercial activities are considered to be one of the most significant sources of water quality parameters (Pitt et al. 1995).

12.3.3 Construction and Demolition Activities

Construction and demolition activities have the potential to contribute a significant amount of water quality parameters. In urban areas, construction activities are major causes for sediment generation (Jartun et al. 2008). During a relatively short period, construction sites can contribute more solids into receiving water bodies than that can be deposited naturally over several decades.

After investigating different land uses in the USA, Line et al. (2002) found that the highest sediment export rate came from a construction site which was 10 times more than from other sites. They also found the highest amount of annual TN export from construction sites. In addition, Sonzogni et al. (1980) noted that urban construction sites contribute more phosphorus than any other land use.

Wind transports on-site pollutants from construction sites which are then accumulated on roofs, front yards and roads. High traffic volume close to construction and demolition activities forces pollutants to be accumulated on curbs (Brinkmann 1985). However, pollutant loads vary considerably with the amount of construction, catchment area, management of the site and maintenance activities.

12.3.4 Vegetation

Storm water runoff contains not only inorganic pollutants but also organic vegetative pollutants. Waste vegetative matter from tree leaves and other plant materials, such as pollen, bark, twigs and grass are potential contributors of both organic pollutants and nutrients as they break down in catchments and in waterways. Novotny et al. (1985) noted that a mature tree can produce 15–25 kg of organic leaf residues which contain a significant amount of nutrients.

However, Allison et al. (1998) have questioned the importance of leaf litter as a potential source of nutrients in urban storm water. They found that potential nutrients contribution from the leaf litter is two orders of magnitude smaller than the measured total nutrient loads. Their findings were based on the outcomes of an urban catchment located in an inner city suburb of South-East Australia. Their observations confirm that water quality pollutants are dictated by site-specific factors.

12.3.5 Soil Erosion

The major contributor of SS in storm water runoff is the erosion of landscape areas either from natural or anthropogenic activities. Erosion starts with the impact of a raindrop, which can vary from 1.6 to 6.4 mm in diameter with maximum speed of 32 km h^{-1} at the ground surface. The subsequent collisions between raindrops and the earth surface break the soil particles into their components (sand, silt and clay). Surface runoff transports these smaller particles of silt and clay to downstream. The force of flowing water also further detaches soil particles and transports them into receiving water bodies.

Soil types, topography, vegetation and climatic conditions also have a significant influence on soil erosion. Any land practice that exposes soil to erosive forces of rainfall and runoff represent erosion and pollutant hazards. During the active construction period, protective vegetative cover is removed from construction sites and the unprotected soil is left exposed to rainfall (USGS 2000). The loss of vegetative cover from the ground surface allows raindrops to strike with full energy, which increases soil erosion and leads to increase the concentration of SS into storm water runoff.

12.3.6 Corrosion of Materials

In areas where metal roofs dominate, corrosion is a significant source of water quality pollution. Different metal elements, for example Copper (Cu), Zinc (Zn), Aluminium (Al), Lead (Pb) and other materials are used as roof coverings, gutters and pipes. All these materials release heavy metals in storm water runoff as a corrosion product (Gobel et al. 2007). Forster (1996) also found that the areas having appreciable coverage of metal roofs are more liable to have the accumulation of heavy metals on catchment surfaces. Similar results were obtained by Davis et al. (2001) in a comprehensive study of urban water quality parameter sources. Their study was based on a variety of building sides and roofs among other surfaces. Further study undertaken by Bannerman et al. (1993), Pitt et al. (1995) demonstrated that runoff from a galvanised roof surface contains higher heavy metals concentration than runoff from street surfaces.

Acid rain and aggressive gases contribute to a significant corrosion of roofs, gutters, paints and other metal surfaces (Brinkmann 1985). Due to the low pH value of rainwater, the corrosion process is enhanced (Gobel et al. 2007). During storm events, the corroded particles are transported with surface runoff into waterways and receiving water bodies. However, the amount of corroded pollutants on surface runoff depends on many factors.

12.3.7 Atmospheric Fallout

Airborne pollutants are deposited on the earth surface and worsen the quality of water and water environments. Wang and Li (2009) found that the main contribution to the accumulation of pollutants on a roof surface is atmospheric dry deposition rather than roof materials. The sources of atmospheric deposition include industry, traffic, refineries, power plants and waste processing companies (Bohemen and De Laak 2003). Emissions from vehicles initially contribute to pollution of atmospheric environments but return to the earth surface due to atmospheric deposition and pollute storm water runoff. Puckett (1995) noted that the atmospheric nitrogen is one of the important sources of nutrients. Davis et al. (2001) found that a significant amount of heavy metals from roof surface runoff came from atmospheric deposition.

12.3.8 Spills

This category of pollutants is difficult to define quantitatively either in terms of volume or composition (Goonetilleke and Thomas 2003), and it is even difficult to predict the occurrence of spills accurately. The major source of spills is vehicular transport. Leakages of fuel, motor oils and lubricants occur everywhere on road surfaces. The adverse impact of spills on water quality is reduced through good maintenance and management practices.

12.4 Roles of Mathematical Models and Modelling Types

Modelling is a commonly used tool in research and management of environmental systems. The ability to use a mathematical model to simulate the behaviour of a system has allowed environmental researchers and managers to predict how different scenarios and stimuli affect various environmental systems. The complex relationships between waste loads from different sources (point or non-point) and the resulting water quality responses of receiving water bodies are best described with mathematical models (Deksissa et al. 2004). Therefore, there exists a wide range of water quality models to be used in managing water quantity and quality with respect to a variety of environmental impacts. These models can predict physical, chemical and biological processes occurring during transportation of water quality parameters (Perk 2006).

12.5 Water Quality Modelling Approaches

Since 1970, a number of water quality models have been developed for the simulation of different water quality parameters of catchments and streams. Some of these models were developed specifically for individual area and geographic conditions. Each of the models was developed in a unique way with the simulation characteristics and properties that influence how its output should be interpreted. These models vary in terms of their complexity, considered approaches, the personnel and computational requirements, and data requirements for calibration and validation (Merritt et al. 2003; Charbeneau and Barrett 1998).

Modelling approaches start from the very simple conceptual type to the complex data intensive models. The simple models are unable to determine the controlling processes of pollutants because they do not employ a sufficient number of processes (Cox 2003). However, data requirements of the complex models prohibit their broader applications to the practical problems (Eatherall et al. 1998). In addition, more complex models need more computation effort and more data resources (Zoppou 2001). Snowling and Kramer (2001) also noted that complex models are generally very sensitive and therefore it is difficult to predict water quality parameters. The financial costs of a complex model further enhance the difficulty of their wider application.

According to Lindenschmidt (2006), the most complex model is not necessarily the most accurate. Increasing model complexity increases the number of degree of freedom (i.e. more parameters and variables) which increases model sensitivity. Therefore, over-parameterisation makes the calibration procedure difficult and reduces the predictive capability of models. The users often try to use the simple models rather than the complex ones because they are easier to calibrate and therefore reliable (Pearl 1978). However, oversimplification of the catchments and pollutants behaviour may leads to gross errors. On the other hand, the processes of different model parameters are interrelated, and hence increase the difficulty in pollutant prediction. Different types of models can be described as follows:

12.5.1 Empirical Model

This type of model is developed based on general observation data. The computational effort and data requirements for such models are usually less than for conceptual and physical-based models (Merritt et al. 2003). However, the empirical model type is criticised for employing unrealistic assumptions of the physics. The advantage of the empirical model is that it can be frequently used as it can be implemented in situations with limited data and parameter inputs.

12.5.2 Conceptual Model

Any model developed based on Darcy's or Newton's law is considered to be a conceptual model. The parameters of a conceptual model are usually derived by the calibration procedure against observed data. Therefore, this type of model tends to suffer from problems of identifying parameter values. Furthermore, like the empirical model type, the parameters of the conceptual model have limited physical interpretability due to the lack of unique value. However, simpler conceptual models have fewer problems than more complex models (Merritt et al. 2003).

12.5.3 Physical-Based Model

The physical-based model involves sophisticated approaches and solutions which are based on the fundamental physical equations. This model type is developed based on the known sciences to simulate the processes on the basis of systems. The standard equations used in such models are the conservation of mass and momentum for flow and the conservation of mass for sediment (Merritt et al. 2003). Theoretically, the parameters of this model are measurable and so are known. However, in practice, due to a large number of factors involved, these parameters are often calibrated against the observed data (Merritt et al. 2003). Nevertheless, there are too many parameters to be quantified, i.e. they tend to be affected by over-parameterisation. The governing processes of the physical-based model are derived at small scale and under very specific physical conditions, and hence this type of model is useful for short time interval simulation (Radwan et al. 2003).

12.5.4 Lumped and Distributed Models

Models can be considered as lumped or distributed based on application areas. Models developed without considering spatial variability are lumped models, while models reflecting spatial variability of the processes are called distributed models. Most of the urban runoff models are deterministically distributed (Nix 1994). The distributed approach is also applicable to sediment transport modelling (Merritt et al. 2003). However, data requirements of the distributed models are larger compared with the lumped model. If the estimation at the catchment outlet is sufficient, then the lumped model is sufficient.

12.5.5 Event-Based and Continuous Models

In terms of rainfall–runoff simulation, models can be categorised as event-based model and continuous model. The model used for the simulation of an individual storm event is called event-based model (Alley and Smith 1981), while the model developed for the simulation of a catchment's overall water balance for a long period is continuous model. Continuous models are more advantageous than event-based models (Tan et al. 2005). However, for the design of storm water infrastructures, event-driven models are more appropriate (Zoppou 2001).

12.5.6 Deterministic and Stochastic Models

Depending upon the nature of the prediction, models can be divided into two types, deterministic model and stochastic model. Deterministic model usually attempt to simulate the actual physical process associated with runoff quantity and quality, while stochastic model focuses on the probabilistic nature of pollutant processes (Barbe et al. 1996).

If any of the model variables is regarded as random variable having a probability distribution, then the model is called stochastic. Otherwise, the model is deterministic. For the same input, a deterministic model will produce identical results (Zoppou 2001). However, a stochastic model will always produce different model responses as one or more variables are selected from random distribution.

12.6 Catchment Water Quality Modelling Approaches

Storm water management model (SWMM) is one of the first models developed for the simulation of storm water runoff quantity and quality (Gaume et al. 1998). Since then many mathematical models have been developed by different researchers with different degrees of complexities and varying levels of accuracy based on different modelling approaches. However, most of the available models provide event-based estimates of water quality parameters from specific sites.

The basic model for pollutant processes on a particular catchment surface can be written as Eq. 12.1 (Shaw et al. 2010).

$$P_{t+\Delta t_{BW}} = P_t + \left\{ k_B \left(1 - \frac{P_t}{P_{max}} \right) - \alpha_w P_t Q_t \right\} \Delta t_{BW} \quad (12.1)$$

where P_t is the available pollutant mass (kg) accumulated on a catchment surface at time 't', P_{max} is the threshold at which additional pollutant do not accumulate on a catchment surface (kg), k_B is the pollutant build-up coefficient (kg time^{-1}), Q_t is the runoff rate (m^3 time^{-1}) at time 't', α_w is the wash-off rate constant (m^{-3}) and Δt_{BW} is the time increment for pollutant build-up or wash-off.

In catchment water quality model, the pollutant processes can be described as:

(a) Event mean concentration (EMC)
(b) Pollutant build-up
(c) Pollutant wash-off
(d) First-flush

12.6.1 Event Mean Concentration

The event mean concentration (EMC) is a method used to characterise the concentration of pollutants from storm water runoff into nearby waterways and receiving water bodies. A statistical parameter, a flow weighted average concentration of a desired water quality parameter is used during a single storm event. An EMC model assumes a single flow weighted concentration and can be used across an entire storm event. The EMC model is the simplest model to calculate storm water pollutants which is mostly useful for calculating annual pollutant loads (Charbeneau and Barrett 1998). This is the frequently used method to characterise storm water loadings.

The EMC value can be determined by calculating the cumulative mass of pollutants and dividing it by the volume of storm runoff. The classical EMC model can be demonstrated by Eq. 12.2 (Kim et al. 2006).

$$\text{EMC} = \frac{\int_0^T P_c(t)\,dt}{\int_0^T V_{Ts}(t)\,dt} \quad (12.2)$$

where EMC is the event mean concentration, $P_c(t)$ is the captured pollutant loads, V_{Ts} is the volume of runoff during the integration time interval and Ts is the duration of storm event.

However, in most cases, the monitored watershed area does not represent a single homogeneous land use. Then the observed EMC is the runoff weighted sum of all EMCs from the individual land use (Butcher 2003). For these cases, the mathematical expression of EMC can be represented by Eq. 12.3.

$$\mathrm{EMC} = \frac{\sum_{i=1}^{n} (\mathrm{EMC})_i \{(R_\mathrm{D})_i (A_\mathrm{c})_i\}}{\sum_{i=1}^{n} \{(R_\mathrm{D})_i (A_\mathrm{c})_i\}} \qquad (12.3)$$

where R_D is the depth of runoff and A_c is the contributing area.

12.6.2 Pollutant Build-up

Pollutant build-up is the accumulation of contaminants on catchment surfaces prior to rainfall events. However, during storm events, rainfall not only washes away pollutants from catchment surfaces but also deposits its own pollutants (James and Thompson 1997). This deposition is small enough to neglect in water quality modelling.

To the present, most of the build-up studies have been conducted by considering the $\mathrm{t_d}$ (antecedent dry days) as the most important parameter (Egodawatta et al. 2009; Vaze and Chiew 2002; Sartor et al. 1974). For example, the first build-up assumption justified by the work of Sartor et al. (1974) used the $\mathrm{t_d}$ as the most important parameter. Alley and Smiths (1981) also developed a pollutant build-up model that included the $\mathrm{t_d}$ as an input parameter. In addition, Rossman (2004) found that the accumulation pollutant loads on a catchment surface is a function of the number of preceding dry weather days.

On the other hand, some other researchers criticised this assumption. Based on an experimental study in Belgrade, Yugoslavia, and Lund, Sweden, on impervious surfaces, Deletic and Maksimovic (1998) found that $\mathrm{t_d}$ had only a minor effect upon the accumulation of road surface pollutants. At Aberdeen, Scotland, Deletic and Orr (2005) found that $\mathrm{t_d}$ had a weak negative influence on the accumulation of pollutant on catchment surfaces. Shaw et al. (2010) noted that the accumulation of pollutants on catchment surfaces is not deterministically related to $\mathrm{t_d}$. These contradictory results were due to the different geographical conditions and the time which had elapsed between data collections.

The common variables, such as surface type, roughness, slope, land use and weather conditions affect the redistribution of pollutant loads, and hence the build-up rate. Initially, available pollutant is another important factor in the accumulation of water quality pollutants which dominates the transport rate. Theoretically, it can be assumed that pollutants on catchment surfaces accumulate uniformly. However, the rate of pollutant build-up is significantly higher during the initial period (Ball et al. 1998; Sartor et al. 1974). Then the rate decreases as the dry days increase and eventually approaches zero due to land-use activities.

Egodawatta et al. (2007) found that pollutant build-up is significantly higher during the first two days for road surfaces and seven days for roof surfaces. Vaze and Chiew (2002) noted that pollutant build-up on road surfaces may vary along the longitudinal direction depending upon the presence of traffic signals and bottlenecks. In addition, the regional and catchment management practices influence pollutant build-up and its composition (Ball et al. 1998). Therefore, different pollutant build-up functions are used to determine the accumulation of water quality parameters.

Usually, the accumulation of water quality parameters on a catchment surface is presented by two main build-up functions, i.e. the linear and non-linear. At the very beginning, SWMM assumes that there is a linear increase of pollutants, whereby a constant amount of pollutants increase during t_d (Novotny et al. 1985). The linear pollutant build-up with t_d can be written as Eq. 12.4.

$$(B_{t_d})_D = k_L t_d \tag{12.4}$$

where $(B_{td})_D$ is the accumulation of pollutant on a particular catchment surface during t_d, k_L is the linear build-up rate constant and t_d is the antecedent dry time (days or hours).

Ball et al. (1998) tested a range of models in different forms for pollutant build-up, and they found that the power function and the hyperbolic function are the best fit pollutant accumulation model. Therefore, Alley and Smith (1981) emphasised on the non-linear build-up model.

The non-linear pollutant build-up process was first identified from field data collected by Sartor et al. (1974). From the non-linear build-up models, the power function, the exponential function and the Michaelis–Menton function are the main types found in water quality literature (Wang and Li 2009). Amongst them, the exponential function is the most widely employed build-up model (Chen and Adams 2006). Most of the available models generate accumulation of pollutants in the form of decreasing rate with increasing dry days.

The decreasing of pollutant accumulation with increasing dry days can be expressed as follows (Novotny et al. 1985):

$$\frac{dB_{t_d}}{dt} = P_i - k_w B_{t_d} \tag{12.5}$$

where P_i is the sum of all the inputs and k_w is the wash-off coefficient.

Equations 12.4 and 12.5 are based on the assumption that every storm has the capacity to remove all the available pollutants from catchment surfaces for an adequate duration of rainfall event. However, from literature review, it was easily understood that a single storm event can transport only a fraction of pollutants from a particular catchment surface (Egodawatta et al. 2007). For example, the experimental study by Vaze and Chiew (2002) showed that a significant rainfall event of 39.4 mm can remove only 35 % of pollutant loads. Therefore, in practice, the amount of accumulated pollutants on catchment surfaces has two parts; pollutants build-up during the t_d and residual pollutants not washed off by the previous storm events (Chen and Adams 2007). Hence, Charbeneau and Barrett (1998) proposed the following build-up model which accounts for the mass not washed off during the previous rainfall event.

$$B_{t_d} = (B_{t_d})_R + (B_{t_d})_D \tag{12.6}$$

where $(B_{td})_R$ is the pollutant mass not washed off during the previous rainfall event.

Based on the hypothesis provided by Charbeneau and Barrett (1998), Hossain (2012) proposed pollutant build-up model using the explicit relationship between the pollutant accumulation and t_d. The build-up model was developed based on the assumption that on a clean surface, water quality parameters accumulate until a maximum value is attained. Considering the t_d as the key variable including other hypothesises, Hossain (2012) proposed pollutant build-up by the three separate functions for impervious and pervious surfaces of a particular catchment; Power function, Exponential function and Saturation function. With the nature of the pollutant accumulation, an efficient, reliable and convenient model can be selected from the above-mentioned functions.

12.6.2.1 Power Function Build-up Model

The power function pollutant build-up equation is the best generic form for the accumulation of water quality parameters on catchment surfaces (Ball et al. 1998). The power function build-up model considers that the pollutant build-up on a catchment surface increase with the increasing number of dry days until a maximum limit is reached (Rossman 2004; Ball et al. 1998). The power function pollutant build-up model is described according to Eq. 12.7.

$$B_{td}(p,s) = \text{Min}\left[\begin{array}{c}\{A_c(s) \times F_{imp}(s) \times C_1(p,s)\}, \\ \left\{A_c(s) \times F_{imp}(s) \times C_2(p,s) \times t_d^{C_3(p,s)} + (B_{td})_D(p,s)\right\}\end{array}\right] \tag{12.7}$$

where $B_{td}(p,s)$ is the accumulation of the pollutant '*p*' (kg) on the land surface '*s*' during the t_d, $A_c(s)$ is the contributing area of the catchment (km^2), $F_{imp}(s)$ is the impervious or pervious fraction of the land surface '*s*', $C_1(p, s)$ is the maximum amount of pollutant that can be accumulated on the land surface '*s*' (kg km^2), $C_2(p, s)$ is the coefficient for the pollutant build-up parameter, $C_3(p, s)$ is the exponent for the pollutant build-up parameter and t_d is the number of antecedent dry days.

12.6.2.2 Exponential Function Build-up

The exponential pollutant build-up model assumes that the accumulation of pollutants on a catchment surface occurs as an exponential function with an upper limit asymptotically (Rossman 2004). Many catchment water quality models employ the exponential build-up representation because it is simple and can be derived as a first order process (Kim et al. 2006). In this thesis, the exponential build-up model is developed as a function of build-up rate coefficient and t_d (Eq. 12.8).

$$B_{\mathrm{td}}(p,s) = \mathrm{Min}\left[\begin{matrix}\{A_c(s)\times F_{\mathrm{imp}}(s)\times C_1(p,s)\},\\ \{A_c(s)\times F_{\mathrm{imp}}(s)\times C_1(p,s)\times\left(1-e^{-k_p\mathrm{td}}\right)+(B_{\mathrm{td}})_D(p,s)\}\end{matrix}\right] \tag{12.8}$$

where k_p is the pollutant accumulation rate coefficient (day^{-1}).

12.6.2.3 Saturation Function Build-up

The saturation build-up function assumes that the accumulation of pollutants on a particular catchment surface occurs as a linear rate and continuously declines until the saturation value is reached (Rossman 2004). The saturation build-up function is presented by Eq. 12.9.

$$B_{\mathrm{td}}(p,s) = \mathrm{Min}\left[\begin{matrix}\{A_c(s)\times F_{\mathrm{imp}}(s)\times C_1(p,s)\},\\ \left\{\frac{A_c(s)\times F_{\mathrm{imp}}(s)\times C_1(p,s)\times t_d}{t_p(p,s)+t_d}+(B_{\mathrm{td}})_D(p,s)\right\}\end{matrix}\right] \tag{12.9}$$

where $t_p(p, s)$ is the half saturation constant (the days to reach half of the maximum build-up) for the pollutant '*p*' on the land surface '*s*'.

12.6.3 Pollutant Wash-off

Pollutant wash-off is the transportation of the accumulated pollutants by surface runoff from catchment surfaces to nearby waterways and receiving water bodies (Temprano et al. 2006). The complex process of pollutant wash-off is determined by a number of factors. Numerous research studies were conducted in identifying the governing variables of pollutant wash-off. However, most of the hypotheses are based on the four influencing rainfall runoff variables; rainfall intensity, rainfall volume, flow rate and runoff volume (Egodawatta et al. 2007).

According to Akan (1987), the characteristics of raindrops, overland flow and type of pollutants affect the pollutant wash-off rate. Moreover, Bannerman et al. (1993) found that the runoff energy has a significant influence on the pollutant wash-off process. In addition, Chen and Adams (2007) noted that types and conditions of street surfaces, particle size, streets cleaning and traffic density affect the pollutant wash-off rate. Berretta et al. (2007) also noted that the dynamics of the pollutant wash-off process is affected by hydrologic parameters, the catchment characteristics and the nature of pollutants. Duncan (1999), Novotny et al. (1985) noted that the amount of pollutant washed off from catchment surfaces is influenced by the amount of pollutant accumulated during the t_d. Kibler (1982) noted that landscape modification affects pollutant wash-off rate.

However, it is difficult to find out the relative importance of each parameter on the pollutant wash-off rate (Herngren et al. 2005). The interrelationships among different variables show the difficulty in understanding the degree of influence of individual variables. Therefore, different researchers proposed different pollutant wash-off models for estimation of pollutant loads from catchment surfaces. Most of these studies focused on impervious surfaces, especially impervious road surfaces. There are only few studies of pollutant wash-off rates for residential and open land areas. Even these areas are major sources of water quality parameters in storm water runoff (Bannerman et al. 1993).

The wash-off formulations used in most of the existing models are very similar to SWMM. However, these formulations are purely empirical and contain at least one parameter which has no physical basis. SWMM's formulations are according to the algorithm of the first storm water management model developed by Metcalf and Alto (1971). They proposed that the rate of pollutant wash-off from a catchment surface is proportional to the mass of pollutant which remains on that surface (Eq. 12.10).

$$\frac{\mathrm{d}W_\mathrm{P}}{\mathrm{d}t_\mathrm{BS}} = -k_\mathrm{w} W_\mathrm{P} \tag{12.10}$$

where W_P is the pollutant wash-off from a catchment surface (mass) and t_BS is the time since the beginning of a storm event. In urban storm water quality modelling, Sartor et al. (1974) proposed a solution to this exponential equation as a function of rainfall intensity and duration. Their study concluded that the pollutant wash-off rate from an impervious surface area is proportional to the rainfall intensity and the mass of pollutant available on a particular catchment surface.

$$W_\mathrm{P} = (B_{\mathrm{t_d}})_D\left(1 - \mathrm{e}^{k_\mathrm{w} I t_\mathrm{BS}}\right) \tag{12.11}$$

However, Alley (1981) modified this exponential wash-off model and showed that pollutant wash-off is better predicted by the runoff rate instead of rainfall intensity. The expression is given by Eq. 2.12.

$$W_\mathrm{P} = (B_{\mathrm{t_d}})_D\left(1 - \mathrm{e}^{k_\mathrm{w} R_\mathrm{R} t_\mathrm{BS}}\right) \tag{12.12}$$

where R_R is the average runoff rate of a storm event (mm h^{-1}) during the time step $\mathrm{t_{BS}}$.

However, many other researchers noted that pollutant wash-off can be better predicted using the runoff volume (V), Barbe et al. (1996). The expression of pollutant wash-off relationship with runoff volume is given by Eq. 12.13.

$$W_\mathrm{P} = k_\mathrm{w} V_\mathrm{Ts}^{E_\mathrm{p}} \tag{12.13}$$

where E_p is the power of wash-off parameter.

On the other hand, Akan (1987) showed that the pollutant detachment rate at any point along a particular catchment surface is assumed to be proportional to the overland flow bottom shear stress and the amount of pollutant available on that surface. Assuming the pollutant wash-off rate is proportional to the bottom-shear stress of the overland flow and the distribution density, Akan (1987) proposed a physically-based mathematical model. The expression is shown by Eq. 12.14.

$$\frac{dP_P}{dt_{BS}} = k_W S_0 h B_{t_d,A} \tag{12.14}$$

where $B_{td,A}$ is the mass of the pollutant per unit surface area.

However, there are many storm water quality models which were developed based on the assumption that the rate of pollutant wash-off is proportional to the remaining pollutants and surface runoff (Chen and Adams 2007; Grottker 1987; Sartor et al. 1974) as shown by Eq. 12.15.

$$\frac{dW_P}{dt_{BS}} = -\alpha_w B_{t_d} q_A \tag{12.15}$$

where q_A is the runoff rate per unit catchment area.

Hossain (Hossain 2012) performed a detailed literature review of pollutant wash-off phenomenon and found that the amount of pollutant wash-off is significantly influenced by the available pollutants on a particular catchment surface. The wash-off process follows a suitable depletion law and thus influences the actual pollutant concentration that is transported through a catchment outlet into nearby waterways and receiving water bodies. The pollutant wash-off approaches proposed by Hossain (2012) are based on the model developed by Rossman (2004). Similar to the pollutant build-up model, the wash-off model is presented by three different functions; Power function wash-off, Rating curve wash-off and Exponential function wash-off.

These three wash-off functions represent the pollutant transport capacity of surface runoff from a particular catchment area (Barbe et al. 1996). These functions were developed to consider the wash-off from impervious and pervious surfaces separately. These functions consider the three important variables, i.e. the rainfall intensity, the rainfall duration and the runoff volume. Due to their high correlation with each other, the individual influence cannot be clearly discerned (Chiew and McMahon 1999).

12.6.3.1 Power Function Wash-off

The power function wash-off model assumes that the amount of pollutants removed by storm water runoff from a particular catchment surface is proportional to the product of the runoff rate per unit area raised to some power and the available pollutants on that catchment surface. In this study, the power function pollutant wash-off model was developed by assuming that the removal of storm water pollutant is proportional to the amount available at a catchment surface and directly related to surface runoff discharged from that catchment. According to this assumption, mathematical expression of the power function wash-up model is expressed by Eq. 12.16.

$$W_t(p,s) = \frac{E_1(p,s)\{q_{A,t}(s)\}^{E_2(p,s)} B_{t_d}(p,s)}{1000\, V_{Ts}(s)} \tag{12.16}$$

where $W_t(p, s)$ is the wash-off rate (mg L^{-1}) for the pollutant 'p' from the land surface 's' within time 't' during surface runoff event, $E_1(p, s)$ is the pollutant wash-off coefficient $\left(\left(\mathrm{mm\,h}^{-1}\right)^{E_2}\right)$, is the runoff rate per unit area (mm h^{-1}), $E_2(p, s)$ is the pollutant wash-off exponent (dimensionless) and $V_{Ts}(s)$ is the volume of surface runoff (m^3) within time 't'.

12.6.3.2 Rating Curve Wash-off

The rating curve wash-off model estimates the transportation of water quality parameters as a function of surface runoff discharged from a particular catchment surface. A number of previous studies found that pollutants wash-off can be better estimated by using runoff volume (Charbeneau and Barrett 1998; Barbe et al. 1996). Chiew and McMahon (1999) also established that the pollutant concentration in relation to the runoff rate provides a better prediction for some catchments. In addition, Deletic and Maksimovic (1998) found that pollutants wash-off from catchment surfaces strongly depends on the overland flow. Furthermore, Huber and Dickson (1988) noted that in the catchment water quality models, the relationship between the pollutant wash-off and the runoff volume is the most convenient and easiest formulation. Another factor is that data for water quality parameters and runoff volume are easily available.

According to the rating curve wash-off function, the amount of transported pollutant from a particular catchment surface can be expressed as proportional to the surface runoff rate raised to some power. The rating curve wash-off model employed in this study is expressed by Eq. 12.17.

$$W_t(p,s) = \frac{E_3(p,s)\{Q_t(s)\}^{E_4(p,s)}}{1000\, V_{Ts}(s)} \tag{12.17}$$

where $E_3(p, s)$ is the coefficient for the pollutant wash-off parameter, $E_4(p, s)$ is the exponent or power of the wash-off parameter (dimensionless) and $Q_t(s)$ is the runoff rate (m^3 s^{-1}) from the land surface 's' at time 't'.

12.6.3.3 Exponential Wash-off

The exponential wash-off model estimates the decreasing pollutant concentration of a given accumulated pollutant with increasing time since the start of a rainfall event. Different derivations of this equation are used in the various storm water quality models. The most common form of the exponential function assumes that the rate of pollutant wash-off from a particular catchment surface is proportional to the product of available pollutant which remains on that surface and the rainfall intensity. The expression is shown by Eq. 12.18.

$$\frac{\mathrm{d}W}{\mathrm{d}t} = -\alpha_{\mathrm{w}} W \tag{12.18}$$

where 'W' is the pollutant wash-off and α_{w} is the wash-off rate constant.

Assuming α_{w} varies as a direct proportion to the rainfall intensity, $I(s)$ i.e. $\alpha_{\mathrm{w}} = E_5(p, s)\, I(s)$, Eq. 12.17 becomes;

$$W_{\mathrm{t}}(p,s) = \frac{\left(1 - \mathrm{e}^{-E_5(p,s) \times I(s) \times t}\right) B_{t_{\mathrm{d}}}(p,s)}{1000\, V_{\mathrm{Ts}}(s)} \tag{12.19}$$

where $E_5(p, s)$ is the wash-off exponent (mm^{-1}) for the pollutant 'p' from the land surface 's' and $I(s)$ is the rainfall intensity ($\mathrm{mm\ h}^{-1}$) on the land surface 's'.

The primary boundary condition of these wash-off equations is the amount of pollutant available on a catchment surface to wash out during surface runoff event (Barbe et al. 1996). However, it is commonly recognised that pollutant wash-off is significantly greater at the beginning of a storm runoff compared to the later period, after rainfall has cleansed catchment surfaces (Novotny et al. 1985).

12.6.4 The First-Flush Phenomena

First-flush (FF) is a term used to refer to the higher concentration of pollutants in storm water runoff during the initial period of a storm event. Generally, storm water runoff containing a higher pollutant concentration is called FF. It is often noted that the highest intensity of rainfall are bursts at the initial period of a rainfall event which could cause a higher pollutant wash-off during that period of storm and increases the occurrence of FF.

Numerous researchers considered the FF phenomena as an important factor in storm water pollutant wash-off and transportation processes. They observed the highest amount of pollutant wash-off during the initial period of a storm event. Sartor et al. (1974) found that surface runoff from the first hour of a moderate-to-heavy storm would contribute more pollutant loads than sanitary wastes of the same area during the same time period. Taebi and Droste (2004) observed that a higher fraction of pollutant from catchment surfaces is transported during the initial period of a storm. Therefore, Ballo et al. (2009) noted that the effects of FF are obvious in storm water runoff. Hence, the existence of FF should be considered in storm water management strategies.

The occurrence and the nature of FF can be influenced by a range of factors. It can vary with the rainfall intensity, the runoff volume and the catchment characteristics. Different hypotheses were tested to establish the relationships between FF and rainfall, runoff, catchment area, t_d, and collection network characteristics. If there is a large quantity of available pollutant then the transport rate from catchment surfaces is higher. Therefore, FF is increased in residential areas due to storm water runoff from roof surfaces. Forster (1996) noted that roof surfaces produce a significant concentration of pollutants during the initial period of surface runoff. The imperviousness of a catchment area is also responsible for the FF phenomenon.

Although the occurrence of FF is commonly reported, the FF effects appear only to a limited number of pollutants and storm events (Deletic and Maksimovic 1998). In Melbourne, Australia Bach et al. (2010) did not observe any FF for all the catchments they studied. Therefore, the appropriateness of FF depends primarily on the nature and sources of pollution, in terms of drainage hydrology, pollutant mobility and pollutant supply. Therefore, identification of the existence or non-existence of the phenomenon is most critical.

FF is most important in storm water treatment design. Pollutant collection systems during FF are employed to capture and isolate the most polluted runoff. Storm water retention and detention basins are designed considering FF to treat the initial runoff which contains a higher concentration of pollutants. After the time of FF, the rest of the runoff is discharged without any treatment.

12.7 Data Collection

The application of any water quality model depends on calibration and validation with sufficient and reliable field data. The need for data on background or baseline conditions is an essential requirement for watershed management strategy. All of the empirical equations of any model should be calibrated and validated against adequate data. Water quality predictions through any surface runoff model are useless without local data. Lack of local data for calibration of any water quality model leads to significantly biased results for the accurate estimation of pollutant loads.

The quality of water in any drainage area depends on the land use of that particular area. According to Chen and Adams (2006), water quality model parameters vary not only spatially (i.e. catchment to catchment), but also temporally, differ among different rainfall events. Although SWMM is used in many countries, Leinster and Walden (1999) discouraged the generic application of water quality models not only from overseas countries but also from other parts of the same country. Tsihrintzis and Hamid (1998) noted that due to the absence of measured data, water quality models calibrated for other similar sites can be used only for screening purposes. Because process depends upon the catchment characteristics (permeability of soils, initial pollutant loads), the impact of the actual land use and management changes. Therefore, Puckett (1995) noted that watershed management plan needs to be developed based on individual watershed information.

According to Boorman (2003), specifying water quality parameters to be modelled without fully understanding data availability is clearly a weakness in a study that has no data collection program. Due to the complex nature of pollutants processes, significant array of data is required for calibration and validation of any water quality model. Moreover, sufficient and reliable data are essential for improving existing models and creating a new one.

However, available and reliable data collection is the major challenge in water quality study (Nicholas and Walling 1998). It was understood that water quality data are not readily available, which put potentially severe limitations to the sustainable management of the nation's water. Zoppou (2001) argued that the lack of data is the greater hindrance for the development of water quality model than the lack of suitable algorithms. Due to the scarcity of measured data, the calibration and parameters estimation are impossible even for the simplest water quality model (Gaume et al. 1998).

12.8 Calibration and Parameters Estimation

Calibration is an iterative process in which the parameters of a particular model are constantly adjusted until the deviation or standard error between the simulated and observed values are minimised to a satisfactory level. The calibration procedure includes the use of the estimated parameters in the field as well as adjustment of some parameters to match better model predictions with measured data. The calibration procedure attempts not only to identify the best set of parameters, but also helps to assess and reduce the uncertainty in parameter values (Beck 1991).

For all models, accurate prediction of pollutant concentration and loads rely on an accurate estimation of model parameters. However, the estimation of model parameters is the most critical step in any water quality model (Tsihrintzis and Hamid 1998). It is often difficult to determine the model parameters accurately because the values of many model parameters are linked and interrelated to each other. For example, in the catchment water quality models, the parameters of the pollutant wash-off model are strongly depended on the pollutant build-up. Due to the absence of available pollutants on a catchment surface, the wash-off might become zero even if there is high intensity of rainfall with significant volume of surface runoff. As a result, there is a possibility in identifying a number of alternative model parameters.

12.8.1 Sensitivity of Model Parameters

Sensitivity analysis is a formalised procedure to identify changes in model responses due to changes in the various model parameters (Snowling and Kramer 2001). This analysis forms an important part of the model validation process where model development and data gathering activities are focused (Newham 2002). The sensitivity of model parameters illuminates information on the following types of questions (Beres and Hawkins 2001):

(a) Which of the model parameters exert a significant influence on the modelling output?
(b) Which parameters are inconsequential?
(c) Do increments of any parameter produce unexpectedly large alterations in the results?

Without answering these questions, proper understanding of any mathematical model responses remains incomplete. Thus, sensitivity analysis is a significant aspect of every modeller's job.

The general purpose of sensitivity analysis is to determine which input parameters apply the most influence on the model results. This analysis also helps to acquire detailed knowledge of the controlling model parameters of any particular model. Sensitivity analysis increases the modeller's understanding about the techniques considered in the model development. Moreover, through sensitivity analysis, we can learn how to select model complexity and how to improve the quality of information derived from models for planning processes (Laenen and Dunnette 1997). In addition, sensitivity analysis is used to give insights into interactions between the different components of a mathematical model (Newham 2002). Therefore, Beres and Hawkins (2001) noted that recommendations based upon a mathematical model without an explicit sensitivity analysis lack a basic foundation.

The general procedure of sensitivity analysis is to define a model output variable that represents an important aspect of the model behaviour. The values of various input parameters are then varied and the resultant changes in the output variables are monitored. The variation of the input parameters can be done in various ways. Hence, there is a wide range of sensitivity analysis techniques described in literature.

Research on model sensitivity was undertaken to improve understanding of the behaviour of models. However, there is a general paucity of literature which reviews methods of sensitivity analysis to model components and data inputs (Newham 2002). Beres and Hawkins (2001) also noted that a well-accepted procedure in the performance of sensitivity analysis is often lacking. This reflects the difficulty in generating a general approach for sensitivity analysis across a broad range of models. Campolongo et al. (2000) identified the three main settings of sensitivity analysis of a mathematical model:

(a) Factor screening: to identify the most influential factor in a system with many factors
(b) Local sensitivity analysis: to identify the local impact of the model parameters and involves the use of partial derivatives
(c) Global sensitivity: to identify the output uncertainty from the uncertainty in the input parameters

The classification provides a useful means of structuring the current review of sensitivity analysis. This information can be used to simplify the structure and parameterisation of any mathematical model and its improvement for future application to specific problems (Newham 2002).

12.8.2 Model Selection

Although various modelling techniques are well developed, the distinction between models is not well defined (Zoppou 2001). Kronvang et al. (2009) presented the ensemble modelling results of nutrients loads for 17 European catchments. They could not find any single model to perform the best across all of the catchments. Each of the model types served a specific purpose and a particular model type cannot be recommended as appropriate for all the situations. Models are likely to contain a mix of modules from each model category. For example, while the rainfall–runoff component of a water quality model may be physical-based or conceptually, empirical relationships are used to model the pollutant processes.

The ultimate decision for each modelling exercise depends upon the needs of the users and the purpose of the modelling endeavour (Snowling and Kramer 2001). However, model should be easily applicable and usable for planning and management by water authorities (Laenen and Dunnette 1997). The choice of appropriate model depends on many factors. Tan et al. (2005) identified two main factors that define model selection; the objectives of the study and the availability of data and resources. According to Merritt et al. (2003), for the practical application, selection of the most appropriate model requires the consideration of the following factors:

(a) the intended use and objectives of the model users
(b) suitability of the model to the local conditions
(c) data requirements of the model, including spatial and temporal variation of the model inputs and the output
(d) complexity of the model structure
(e) various components of the model structure
(f) the capability of the model including the accuracy, validity and underlying assumptions

In addition, Akan and Houghtalen (2003) identified some important factors which affect the choice of a particular model for a practical application:

(a) the model is widely accepted (by engineers, consultants and regulators)
(b) the model is inexpensive (written for personal computers with good user support)
(c) the model is user-friendly (preferably using Windows environment)
(d) the model is flexible and robust
(e) the model is technologically advanced.

12.9 Discussion

Accumulation of waterborne pollutants significantly alters the aquatic ecosystem of receiving environments. General strategies and programs for watershed management always depend on the catchment water quality modelling results which involve both the runoff and the pollutants processes. Therefore, over the years, various catchment water quality models have been developed for the estimation of surface runoff and pollutant loads.

As it is not possible to achieve a strictly physically-based comprehensive operational model (Akan 1987), attempts were made to estimate the amount of transferred pollutants by using the simple event mean concentration (EMC) model. Since data requirements of the EMC model are less, it is easy to use. However, the EMC value can change between storms (Chiew and McMahon 1999; Butcher 2003), and hence the prediction of pollutant loads by the EMC model may be inaccurate for un-monitored storm events. To avoid this discrepancy arising in the EMC model, the sophisticated build-up wash-off (BUWO) models are formulated (Chen and Adams 2007). Pollutant build-up and wash-off is a continuous process which occurs on catchment surfaces during the t_d and storm events respectively. These methods are most commonly used in catchment water quality models (Obropta and Kardos 2007). However, the lack of data for the determination of the parameters of BUWO models can lead to significantly biased results in the estimation of water borne pollutants. There is no standard form of pollutant build-up and wash-off formulation (Shaw et al. 2010). However, there are minor variations amongst the available formulations and their conceptual basis is the same.

Most of the currently available storm water quality models were developed only for impervious surface. Storm water quality models for pervious area are very limited. However, water quality parameters from pervious area also play an important role in polluting aquatic environments (Deletic and Maksimovic 1998; Grottker 1987). Therefore, water quality parameters from pervious surface area should be considered in any water quality study. In this study, the catchment water quality model was developed by considering the three different types of surface areas: (i) impervious surface, (ii) pervious surface and (iii) mixed surfaces (partly impervious and partly pervious).

In water quality modelling, data is the key information enabling accurate model results. The application of any water quality model always depends on calibration and validation from adequate and reliable data. Significant array of hydrologic (precipitation) and water quality data are required to calibrate and validate any water quality model. These data may be experimental or measured or both. In addition, information about the characteristics of the physical environment of a watershed area which influence the quality of water is essential.

In order to use any mathematical model for the prediction of water quality parameters, it is essential to estimate the model parameters which are relevant to the modelling processes. The accuracy of the modelling results largely depends on the accuracy of the model parameters. Therefore, the parameters estimation is the key step in the practical application of any water quality model.

Although it is difficult to estimate model parameters, they are required for the analysis, improvement and update of existing BMPs (Tsihrintzis and Hamid 1998). Different researchers use different procedures for the estimation of model parameters. For example, Deletic and Maksimovic (1998), Kim et al. (2006) proposed indirect methods for the calculation of model parameters. An alternative approach is the estimation of the parameters by the calibration procedure using runoff quality data collected at the watershed outlet (Alley and Smith 1981) which reflects the combined effects of an entire catchment. Tsihrintzis and Hamid (1998) noted that calibrated water quality models are essential for specific regions for the prediction of the impact of different water quality parameters into receiving water bodies.

In a mathematical model, there may be several parameters which need to be determined. However, the parameters to which the modelling output is sensitive and which have significant uncertainty require special attention in their determination. Hence, it is important to know the most sensitive parameter which needs to be calculated with great care. It is also important to identify those parameters that have little influence on the behaviour of models so that they may be aggregated, modified or removed (Newham 2002). Therefore, sensitivity analysis is an integral part of any mathematical model simulation which is commonly used to examine the model behaviour, and hence sometimes influences the model formulations.

12.10 Summary

Water quality modelling techniques are extensively used for the estimation of water quality parameters from a particular catchment. The goal of a model is to reduce the complexity and effort spent on hand computation and analysis of water quality parameters. However, due to the lack of specific local information and poor understanding of the limitations of various estimation techniques and underlying physical parameters, modelling approaches are often subjected to producing gross errors. There are no adequate guidelines available to select appropriate modelling options which can be used to simulate various processes. The lack of user-friendliness and proper selection of the model parameters further hinder the application of water quality models. Before going to application of any catchment water quality model, its suitability to the particular catchment should be identified first. Also there should be adequate data for the calibration and validation of the modelling outputs.

References

Adams BJ, Papa F (2000) Urban stormwater management planning with analytical probabilistic models. Willey, New York

Akan AO (1987) Pollutant wash-off by overland flow. J Environ Eng 113(4):811–823

Akan AO, Houghtalen RJ (2003) Urban hydrology, hydraulics, and stormwater quality: engineering applications and computer modelling. Willey, New Jersey

Alley WM, Smith PE (1981) Estimation of accumulation parameters for urban runoff quality modelling. Water Resour 17(6):1657–1664

Allison RA, Chiew FHS, McMahon TA (1998) Nutrient contribution of leaf litter in urban stormwater. J Environ Manage 54(4):269–272

Bach PM, McCarthy DT, Deletic A (2010) The development of a novel approach for assessment of the first flush in urban stormwater discharges. Water Sci Technol 61(10):2681–2688

Ball IE, Jenks R, Aubourg D (1998) An assessment of the availability of pollutant constituents on road surfaces. Sci Total Environ 209(2–3):243–254

Ball JE, Wojcik A, Tilley J (2000) Stormwater quality from road surfaces: monitoring of the Hume Highway at South Strathfield. The University of New South Wales, School of Civil and Environment Engineering, Water Research Laboratory, Research Report 204

Ballo S, Liu M, Hou L, Chang J (2009) Pollutants in stormwater runoff Shanghai (China): implications for management urban runoff pollution. Prog Nat Sci 19(7):873–880

Bannerman RT, Owens DW, Dodds RB, Hornewer NJ (1993) Sources of pollutants in Wisconsin stormwater. Water Sci Technol 28(3–5):241–259

Barbe DE, Cruise JF, Mo X (1996) Modelling the build-up and washoff pollutants on urban watersheds. Water Resour Bull AWRA 32(3):511–519

Beck MB (1991) Principles of modelling. Water Sci Technol 24(6):1–8

Beres DL, Hawkins DM (2001) Plackett-burman technique for sensitivity analysis of many parameters model. Ecol Model 141(1–3):171–183

Berretta C, Gnecco I, Lanza LG, La Barbara P (2007) An investigation of wash-off controlling parameters at urban and commercial monitoring sites. Water Sci Technol 56(12):77–84

Bilotta GS, Brazier RE (2008) Understanding the influence of suspended solids on water quality and aquatic biota. Water Res 42(12):2849–2861

Bohemen HD, De Laak WHJV (2003) The influence of road infrastructure and traffic on soil, water and air quality. Environ Manage 31(1):50–68

Boorman DB (2003) LOIS in-stream water quality modelling. Part 1. Catchments and methods. Sci Total Environ 314–316:379–395

Brinkmann WLF (1985) Urban stormwater pollutants: sources and loadings. Geo J 11(3):277–283

Butcher JB (2003) Buildup, washoff, and event mean concentrations. J Am Water Resour Assoc 39(6):1521–1528

Butcher DP, Labadz JC, Potter AWR, White P (1993) Reservoir sedimentation rates in the Southern Pennine region, UK. In: McManus J, Duck RW (eds) Geomorphology and sedimentology of lakes and reservoirs. Wiley, Chichester, pp 73–93

Campolongo F, Saltelli A, Sorensen T, Tarantola S (2000) Hitchhiker's guide to sensitivity analysis. In: Saltelli A, Chan K, Scott EM (eds) Sensitivity analysis. Willey, Chichester, pp 15–50

Charbeneau RJ, Barrett ME (1998) Evaluation of methods for estimating stormwater pollutant loads. Water Environ Res 70(7):1295–1302
Chen J, Adams BJ (2006) Analytical urban storm water quality models based on pollutant buildup and washoff process. J Environ Eng 132(10):1314–1330
Chen J, Adams BJ (2007) A derived probability distribution approach to stormwater quality modelling. Adv Water Resour 30(1):80–100
Chiew FHS, McMahon TA (1999) Modelling runoff and diffuse pollution loads in urban areas. Water Sci Technol 39(12):241–248
Cox BA (2003) A review of currently available in-stream water-quality models and their applicability for simulating dissolved oxygen in lowland rivers. Sci Total Environ 314–316: 335–377
CRC for Catchment Hydrology (2005) MUSIC (model for urban stormwater improvement conceptualization), version 3, user manual. Cooperative Research Centre for Catchment Hydrology, Melbourne
Davis BS, Birch GF (2009) Catchment-wide assessment of the cost-effectiveness of stormwater remediation measures in urban areas. Environ Sci Policy 12(1):84–91
Davis AP, Shokouhian M, Ni S (2001) Loading estimates of lead, copper, cadmium and zinc in urban runoff from specific sources. Chemosphere 44(5):997–1009
Deksissa T, Meirlaen J, Ashton PJ, Vanrolleghem PA (2004) Simplifying dynamic river water quality modelling: a case study of inorganic nitrogen dynamics in the Crocodile River (South Africa). Water Air Soil Pollut 155(1–4):303–320
Deletic AB, Maksimovic CT (1998) Evaluation of water quality factors in storm runoff from paved areas. J Environ Eng 124(9):869–879
Deletic A, Orr DW (2005) Pollution build-up on road surfaces. J Environ Eng 131(1):49–59
Drapper D, Tomlinson RM, Williams P (2000) Pollutant concentrations in road runoff: Southeast Queensland case study. J Environ Eng 126(4):313–320
Drewry JJ, Newham LTH, Greene RSB, Jakeman AJ, Croke BFW (2006) A review of nitrogen and phosphorus export to waterways: context for catchment modelling. J Marine Freshw Res 58:757–774
Duncan H (1999) Urban stormwater quality: a statistical overview (Report 99/3). Cooperative Research Centre for Catchment Hydrology, Melbourne
Eatherall A, Boorman DB, Williams RJ, Kowe R (1998) Modelling in-stream water quality in LOIS. Sci Total Environ 24(210–211):499–517
Egodawatta P, Thomas E, Goonetilleke A (2007) Mathematical interpretation of pollutant wash-off from urban road surfaces using simulated rainfall. Water Res 41(13):3025–3031
Egodawatta P, Thomas E, Goonetilleke A (2009) Understanding the physical processes of pollutant build-up and wash-off on roof surfaces. Sci Total Environ 407:1834–1841
Forster J (1996) Patterns of roof runoff contamination and their potential implications on practice and regulation of treatment and local infiltration. Water Sci Technol 33(6):39–48
Gaume E, Villeneuve JP, Desbordes M (1998) Uncertainty assessment and analysis of the calibrated parameter values of an urban storm water quality model. J Hydrol 210:38–50
Gobel P, Dierkes C, Coldewey WG (2007) Stormwater runoff concentration matrix for urban areas. J Contam Hydrol 91(1–2):26–42
Goonetilleke A, Thomas E (2003) Water quality impacts of urbanisation: evaluation of current research. Unpublished Academic Exercise, Energy and Resource Management Research Program, Centre for Built Environment and Engineering Research, Queensland University of Technology, Brisbane
Grottker M (1987) Runoff quality from a street with medium traffic loading. Sci Total Environ 59:457–466
Herngren L, Goonektilleke A, Ayoko GA (2005) Understanding heavy metal and suspended solids relationships in urban stormwater using simulated rainfall. J Environ Manage 76(2):149–158
Herngren L, Goonektilleke A, Ayoko GA (2006) Analysis of heavy metals in road-deposited sediments. Anal Chim Acta 571(2):270–278

Hossain I (2012) Development of an integrated catchment-stream water quality model. PhD thesis, Faculty of Engineering and Industrial Sciences, Swinburne University of Technology, Melbourne, Australia

Huber WC, Dickson RE (1988) SWMM (stormwater management model), version 4, user manual, USEPA (US Environmental Protection Agency), Athens, GA

James W, Thompson MK (1997) Contaminants from four new pervious and impervious pavements in a parking-lot. Adv Model Manage Stormw Impacts 5:207–221

Jartun M, Ottesen RT, Steinnes E, Voldena T (2008) Runoff of particle bound pollutants from urban impervious surfaces studied by analysis of sediments from stormwater traps. Sci Total Environ 396:147–163

Kelly J, Thornton I, Simpson PR (1996) Urban geochemistry: a study of the influence of anthropogenic activity on the heavy metal content of soils in traditionally industrial and non-industrial areas of Britain. Environ Geochem 11(1–2):363–370

Kibler DF (1982) Urban stormwater hydrology. American Geophysical Union, Washington DC

Kim LH, Zoh KD, Jeong SM, Kayhanian M, Stenstrom MK (2006) Estimating pollutant mass accumulation on highways during dry periods. J Environ Eng 132(9):985–993

Kronvang B, Behrendt H, Andersen HE, Arheimer B, Barr A, Borgvang SA, Bouraoui F, Granlund K, Grizzetti B, Groenendijk P, Schwaiger E, Hejzlar J, Hoffman L, Johnsson H, Panagopoulos Y, Lo Porto A, Reisser H, Schoumans O, Anthony S, Silgram M, Venohr M, Larsen SE (2009) Ensemble modelling of nutrient loads and nutrient load partitioning in 17 European catchments. J Environ Monit 11(3):572–583

Laenen A, Dunnette DA (1997) River quality: dynamics and restoration. CRC Lewis, Boca Raton

Latimer JS, Hoffman EJ, Hoffman G, Fasching JL, Quinn JG (1990) Sources of petroleum hydrocarbons in urban water runoff. Water Air Soil Pollut 52(1–2):1–21

Lau S, Stenstrom MK (2005) Metals and PAHs adsorbed to street particles. Water Res 39(17): 4083–4092

Leinster S, Walden W (1999) The application of sophisticated stormwater quality estimation techniques in Australian catchments-A Queensland case study. In: Proceedings of water 99, joint congress, Brisbane, Australia, pp 589–595

Lin J, Chen S (1998) The relationship between adsorption of heavy metal and organic matter in river sediments. Environ Int 24(3):345–352

Lindenschmidt K (2006) The effect of complexity on parameter sensitivity and model uncertainty in river water quality modelling. Ecol Model 190(1–2):72–86

Line DE, White NM, Osmond DL, Jennings GD, Mojonnier CB (2002) Pollutant export from various land uses in the upper Neuse river basin. Water Environ Res 74(1):100–108

Merritt WS, Letcher RA, Jakeman AJ (2003) A review of erosion and sediment transport models. Environ Model Softw 18:761–799

Metcalf E, Alto P (1971) Stormwater management model. USEPA, Washington DC

MW-Web (2011) Melbourne Water. web page http://www.melbournewater.com.au. Accessed 18 June 2011

Newham LTH (2002) Catchment scale modelling of water quality and quantity. PhD thesis, The Australian National University

Nicholas AP, Walling DE (1998) Numerical modelling of floodplain hydraulics and suspended sediment transport and deposition. Hydrol Process 12(8):1339–1355

Nix SJ (1994) Urban stormwater modelling and simulation. Lewis Publishers, Boca Raton

Novotny V, Sung HM, Bannerman R, Baum B (1985) Estimating non point pollution from small urban watersheds. J Water Pollut Control Federation 57(4):339–348

Obropta CC, Kardos JS (2007) Review of urban stormwater quality models: deterministic, stochastic, and hybrid approaches. J Am Water Resour Assoc 43(6):1508–1523

Pearl J (1978) On the connection between the complexity and credibility of inferred models. Int J Gen Syst 4:255–264

Perk M (2006) Soil and water contamination: from molecular to catchment scale. Taylor and Francis Group plc, London

Pitt R, Field R, Lalor M, Brow M (1995) Urban stormwater toxic pollutants: assessment, sources and treatability. Water Environ Res 67(3):260–275
Puckett LJ (1995) Identifying the major sources of nutrient water pollution. Environ Sci Technol 29(9):408–414
Radwan M, Willems P, El-sadek A, Berlamont J (2003) Modelling of dissolved oxygen and biochemical oxygen demand in river water using a detailed and a simplified model. Int J River Basin Manage 1(2):97–103
Rossi L, Fankhauser R, Chevre N (2006) Water quality criteria for total suspended solids (TSS) in urban wet-weather discharges. Water Sci Technol 54(4–7):355–362
Rossman LA (2004) SWMM (stormwater management model), version 5, user manual. USEPA (Environmental Protection Agency), Washington DC
Russel MA, Walling DE, Webb BW, Bearne R (1998) The composition of nutrient fluxes from contrasting UK river basins. Hydrol Process 12(9):1461–1482
Sartor JD, Boyd GB, Agardy FJ (1974) Water pollutants aspects of street surface contaminants. J Water Pollut Control Fed 46(3):458–467
Shaheen DG (1975) Contribution of urban roadway usage to water pollution. Report No. EPA-600/2-75-004. US Environmental Protection Agency, Washington DC
Shaw SB, Stedinger JR, Walter MT (2010) Evaluating urban pollutant buildup/wash-off models using a Madison, Wisconsin catchment. J Environ Eng 136(2):194–203
Snowling SD, Kramer JR (2001) Evaluating model uncertainty for model selection. Ecol Model 138(1–3):17–30
Sonzogni WC, Chesters G, Coote DR, Jeffs DN, Konrad JC, Ostry RC, Robinson JB (1980) Pollution from land runoff. Environ Sci Technol 14(2):148–153
Taebi A, Droste RL (2004) First flush pollution load of urban stormwater runoff. J Environ Eng Sci 3(4):301–309
Tan KS, Chiew FHS, Grayson RB, Scanlon PJ, Siriwardena L (2005) Calibration of a daily rainfall-runoff model to estimate high daily flows. In: international congress on modelling and simulation (MODSIM), Melbourne, Australia, 2960–2966
Temprano J, Arango O, Cagiao J, Suarez J, Tejero I (2006) Stormwater quality calibration by SWMM: a case study in Northern Spain. Water SA 32(1):55–63
Tsihrintzis VA, Hamid R (1998) Runoff quality prediction from small urban catchments using SWMM. Hydrol Process 12:311–329
USGS (2000) Soil erosion from two small construction sites, Dane County, Wisconsin. USGS Fact Sheet FS-109-00, US Geological Survey
Vaze J, Chiew FHS (2002) Experimental study of pollutant accumulation on an urban road surface. Urban Water 4:379–389
Vaze J, Chiew FHS (2004) Nutrient loads associated with different sediment sizes in urban stormwater and surface pollutants. J Environ Eng 130(4):391–396
Wang B, Li T (2009) Buildup characteristics of roof pollutants in the Shanghai urban area, China. J Zhejiang Univ Sci A 10(9):1374–1382
Warren N, Allan IJ, Carter JE, House WA, Parker A (2003) Pesticides and other micro-organic contaminants in freshwater sedimentary environments—a review. Appl Geochem 18(2):159–194
Wong THF (2006) Australian runoff quality: a guide to water sensitive urban design. Engineers Media, Crows Nest
Wong T, Breen P, Lloyd S (2000) Water sensitive road design: design options for improving stormwater quality of road runoff (Technical Report No. 00/1). Cooperative Research Centre for Catchment Hydrology, Melbourne
Zoppou C (2001) Review of urban stormwater models. Environ Model Softw 16(3):195–231

Part III
Floods and Hydrological Processes

Chapter 13
Watershed Storage Dynamics in the Upper Blue Nile Basin: The Anjeni Experimental Watershed, Ethiopia

Temesgen Enku, Assefa M. Melesse, Essayas K. Ayana, Seifu A. Tilahun, Gete Zeleke and Tammo S. Steenhuis

Abstract Understanding functions of a watershed is important for implementing appropriate soil and water conservation measures and for planning and development of sustainable water resources use. Watershed storage is a significant part of a catchment water budget and its quantification provides a clue to understand the fundamental catchment hydrological processes. This study is aimed to investigate the dynamics of watershed storage of the Anjeni experimental watershed in the Upper Blue Nile basin for which a long series of rainfall and runoff data is available for this study. A daily water balance equation was used to quantify the watershed storage over the distinct rainy season. On average, 86 % of the annual rainfall occurs during distinct rainy season. The study showed that the watershed storage increases with the increase of cumulative rainfall till the watershed stores its maximum capacity. After this maximum capacity, the watershed storage remains

T. Enku (✉) · E.K. Ayana · S.A. Tilahun · T.S. Steenhuis
Faculty of Civil and Water Resources Engineering, Bahir Dar Institute of Technology, Bahir Dar University, Bahir Dar, Ethiopia
e-mail: temesgenku@gmail.com

E.K. Ayana
e-mail: essayask@gmail.com

S.A. Tilahun
e-mail: sat86@cornell.edu

T.S. Steenhuis
e-mail: tss1@cornell.edu

A.M. Melesse
Department of Earth and Environment, Florida International University, Modesto A. Maidique Campus, Miami, FL 33199, USA
e-mail: melessea@fiu.edu

G. Zeleke
Water and Land Resource Center, Centre for Development and Environment of University of Bern, Addis Ababa, Ethiopia
e-mail: gete_2004@yahoo.com

T.S. Steenhuis
Department of Biological and Environmental Engineering, Cornell University, Ithaca, USA

© Springer International Publishing Switzerland 2016

A.M. Melesse and W. Abtew (eds.), *Landscape Dynamics, Soils and Hydrological Processes in Varied Climates*, Springer Geography,
DOI 10.1007/978-3-319-18787-7_13

constant, even if rainfall continuous. The Anjeni watershed stores an average of 380 mm of water after a cumulative effective rainfall of 625 mm. Before the maximum storage was reached, about 60 % of the effective rainfall is used to wet up the watershed. Then, the remainder becomes surface runoff and interflow, during which about 40 % of the flow appeared at the outlet.

Keywords Watershed storage · Rainfall and runoff · Anjeni watershed · Blue nile basin · Ethiopia

13.1 Introduction

Understanding the functions of watersheds helps in quantifying, planning, and development of water resources. These functions are prerequisite for implementation of best management practices. Watershed storage is defined as the quantity of water that exists within a control volume. It is quantified from the simple daily water balance of the watershed that includes the soil moisture, deep ground water percolation, surface depression storage, and ground water leakage either to the neighboring watersheds or farther downstream from the outlet.

Watershed storage is one of the key functions of watersheds (Black 1997). It is a significant component of a catchment water budget and its quantification provides a fundamental understanding of catchment hydrological processes. It is a primary variable for watershed rainfall-runoff modeling (Brutsaert 2005; Kirchner 2006; Sugawara and Funiyuki 1956). The commonly applied procedure in hydrologic modeling is to calibrate the model parameters based on measured watershed inputs and measured output at the watershed outlet. However, validation of hydrologic models performance only on the basis of stream flow is misleading as models may simulate runoff while misrepresenting the hydrological processes that generate the runoff (Kirchner 2006). Thus hydrological models can provide the right answer with wrong assumptions.

Taking into account the effect of watershed storage component on rainfall-runoff model accuracy and efforts toward improving storage estimates cannot be underestimated (McNamara et al. 2011). Moreover, understanding hydrological processes by which how catchments retain and release water is central to hydrological science.

Watershed storage also serves as a measure for catchment response comparison. A review on related literature reveals that studies on storage measurements are limited. Studies toward watershed storage are slowly growing as shown by recent works (Kirchner 2009; Soulsby et al. 2009; Spence 2007, 2010). That the topic received little attention may be due to the distributed nature of watershed storage heterogeneity, difficulty of watershed characterization, and watershed scale storage measurements (McNamara et al. 2011). Despite all these importance of watershed storage, few attempts have been made to estimate the volume of subsurface water storage at the watershed scale (McDonnell 2003, 2009; Sayama et al. 2011). In addition, subsurface heterogeneity makes the storage-discharge relationship even more complicated (Beven 2006). Quantifying the total water storage is a very challenging task and previous attempts to measure watershed storage in the subsurface are hindered by ill-defined boundary conditions of ground water that are very difficult to define (Sayama et al. 2011). This study focuses on the dynamic component of total watershed storage—the amount of storage change in a system over a distinct rainy season. The main objective of the study is to understand the dynamic subsurface watershed storage in a small experimental watershed at the head water of the upper Blue Nile in the rainy season. The change in the watershed storage over three categories of wetness of the rainy season is evaluated and compared for several years.

Hydrology of the Nile River basin has been studied by various researchers. These studies encompass various areas including stream flow modeling, sediment dynamics, teleconnections and river flow, land use dynamics, climate change impact, groundwater flow modeling, hydrodynamics of Lake Tana, water allocation and demand analysis (Melesse et al. 2009a, b, c; Abtew et al. 2009a, b; Melesse et al. 2011; Dessu and Melesse 2012, 2013; Dessu et al. 2014; Yitayew and Melesse 2011; Chebud and Melesse 2009a, b, 2013 ; Setegn et al. 2009a, b, 2010; Abtew and Melesse 2014a, b, c; Melesse 2011; Melesse et al. 2014).

13.2 Study Area

Under the Soil Conservation Research Program (SCRP), seven experimental watersheds were established in Ethiopia in the 1980s. The main objective of the SCRP were to help understand land degradation processes and to combat land degradation. The program was administered jointly by the Ministry of Agriculture of Ethiopia and Centre for Development and Environment of Bern University (Bosshart 1997). The watershed for this study, Anjeni watershed (10°40′N and 37° 31′E, Fig. 13.1) is located at the head water of the Blue Nile basin southwest of the Choke Mountain (4100 m amsl). Meteorological, rainfall, steam flow, and sediment data has been collected since the establishment of this station in 1984. It is located 365 km from Addis Ababa on the main road to Bahir Dar, 15 km off road in the north direction off Dembecha town.

Anjeni watershed with a catchment area of 113.4 ha is in the moist temperate type agro–climatic zone locally named (*Weyna Dega*). It has an extended unimodal type of rainfall. The 17 years (1985–2004), except incomplete data years of 1991, 1999, and 2001, long-term mean annual rainfall is 1716 mm with standard deviation of 128 mm. The minimum, median, and maximum rainfall over the study area was 1372, 1739, and 1907 mm, respectively. About 86 % of the annual rainfall occurs during the wet seasons from mid-May to mid-October. The long-term mean monthly rainfall in the watershed is shown in Fig. 13.2. The standard deviations of

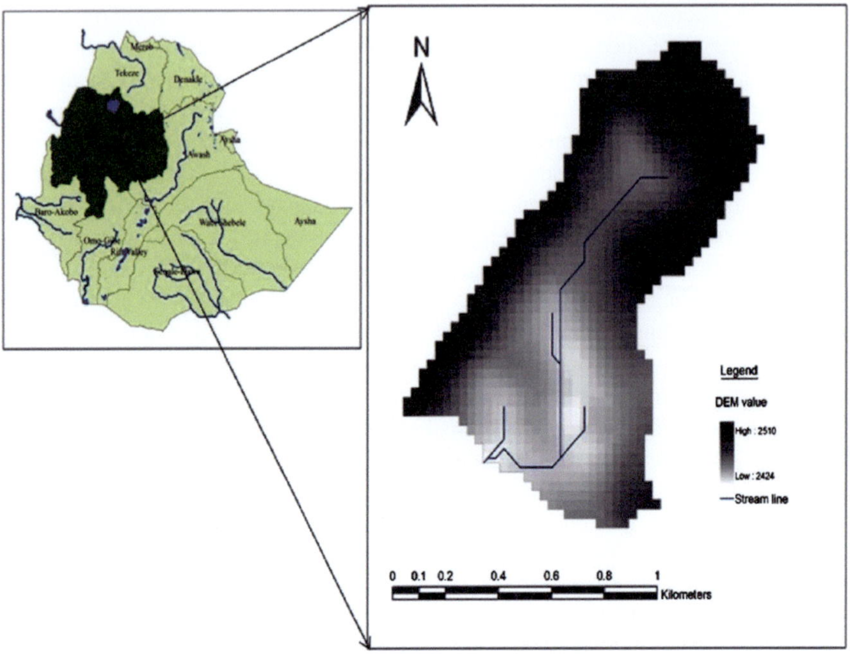

Fig. 13.1 Location map of the Anjeni watershed

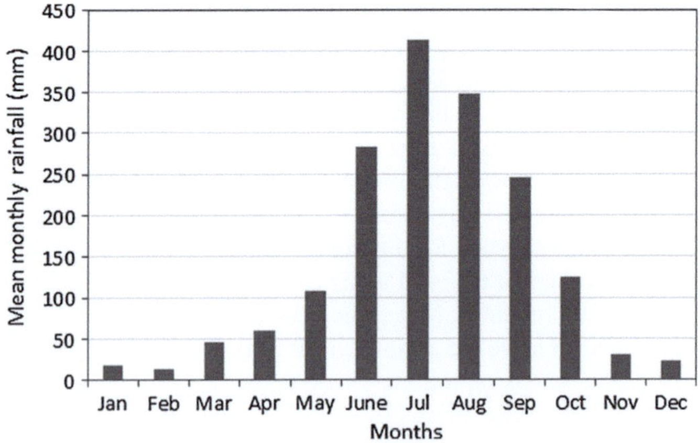

Fig. 13.2 Long-term mean monthly rainfall of the Anjeni watershed

the 17 years monthly rainfall in June, July, and August was 65 mm whereas in September it was 50 mm. The annual mean air temperature was 9.2 °C (1995–2006). The elevation of the Anjeni watershed ranges from 2405 to 2500 m.

Anjeni is a perennial stream that drains north-south to Chereka stream that joins the Birr River, which finally drains to the Blue Nile River. The watershed is among the most productive areas with cereals, beans, and oil seed as dominant crop types.

The soils of Anjeni are developed on the basalt and volcanic ash of the plateau. More than 75 % of the watershed consists of four types of soils. The southern valley bottoms of the watershed are predominantly covered with deep, well-weathered Humic Alisols (21 %); most of the northern part of the watershed is covered with the shallower Haplic Alisols (21 %). The middle area is covered with moderately deep, young Dystric Cambisols (19 %). Haplic Nitosols (17 %) is distributed mostly in the south western gently sloping part and some in the east steep upper slopes. The central northern watershed is covered with Humic Nitosols (6.6 %). Regosols and Leptosols (12 %) scattered in the northern steep slopes and in southern boundaries of the watershed with very shallow depth. Other soils like Luvisols, Leptosols, and Acrisols that adds to (3 %) are also found in small pockets in the watershed (Legesse 2009; Zeleke 2000).

13.3 Methods

On the basis of the long-term (17 years) mean rainy season rainfall, three categories of wetness are identified: *dry*, *normal,* and *wet*. A '*dry*' rainy season refers to a rainy season with total rainfall less than the long-term mean by 5 % or more. The '*normal*' season has total rainfall within 5 % of long-term mean. Season with total rainfall more than the long-term mean by 5 % or more is categorized '*wet*'. Nine rainy seasons (three in each category) were selected.

A simple water balance technique was used to estimate the water storage in the watershed. Watershed storage in distinct rainy seasons is computed as the difference between the input precipitation (P) and the outputs of stream flow and evapotranspiration (Q and ET) of the topographically defined watershed on daily basis over a distinct rainy season. Application of Eq. (13.1) should take into account how the individual components are measured or estimated, and the spatial scales over which such measurements or estimations are applicable. Rainfall measurements are basically local. Rainfall rates vary in space and time. Single rain gauge measurements do not represent areal distribution accurately. Similarly evapotranspiration estimates from a single station data are not representative of the actual areal watershed evapotranspiration. Added to this is the need to account for the accuracy of the method used to estimate ET. Stream flow is assumed as an aggregated measurement from the entire watershed, but yet the entire watershed does not contribute to stream flow. Bearing these uncertainties in mind, the point measurements of rainfall and estimates of ET from a single station data are assumed to represent the areal watershed values. It was also assumed that the entire watershed is contributing to the measured stream flow at the outlet. With all these assumptions, the analysis presented here explores the watershed storage change over the distinct rainy seasons. The storage change was computed from the daily water balance equation with Eq. 13.1.

$$S = \sum\nolimits_{t=1}^{T} (P(t) - Q(t) - ET(t)) \tag{13.1}$$

where, t (day) is the days from the beginning of the rainy season (t = 1, is the first day of the rainy season and t = T, is the last day of the rainy season, S (mm) represents the dynamic storage increase or decrease from t = 1 to t = T, P (mm) is the average daily rainfall over the watershed, Q (mm) is the daily stream flow from the entire watershed, and ET (mm) is the areal average actual evapotranspiration of the watershed. Since the absolute volume of total watershed storage cannot be quantified, using water balance method, the analysis focused exclusively on how dynamic storage changes over the distinct wet season (from the beginning to the end of the rainy season).

Rainfall was measured using pluviograph where the rain is collected in a bucket supported on a spring balance. A mechanical lever arm of the spring is connected with a pen which touches a clock mounted drum with a graph paper. The record shows the accumulation of precipitation over time. Precipitation can be computed for the required time step from the printed graph. Stream flow discharge was measured with automatic float where ink pen draws graph on a pluviograph. The automatic float readings were combined with the manual gauge readings. More detail information on the data collection and processing is found in (Hurni 1984; Bosshart 1997). During high flood events, smaller time steps (as low as 5 min) flow rates was computed and changed to volume. This smaller time step volumetric flow was summed up for the daily volumetric flow. The total daily volumetric flow was changed to depth units assuming the entire watershed contributes to runoff.

Evapotranspiration (ET) is an important hydrological variable capable of removing moisture from land surfaces and water bodies. Although there are various approaches for ET estimation, remote sensing tools area capable of providing spatial variability of energy fluxes with reasonable accuracies in different ecosystems (Senay et al. 2007, 2008; Oberg and Melesse 2006; Melesse et al. 2006, 2007, 2009a; Lagomasino et al. 2015).

The daily evapotranspiration in the wet season was estimated using Enku's temperature method given by Eq. 13.2 (Enku and Melesse 2014).

$$ET = \frac{T\max^{2.5}}{k} \tag{13.2}$$

where ET is the daily evapotranspiration (mm d^{-1}), Tmax is the daily maximum temperature (°C), k is estimated as 38 * Tmm − 63, where Tmm is the long-term mean maximum daily temperature (°C). During rainy seasons, the watershed surface is sufficiently wet so that evaporation is assumed to be at the potential rate. In analyzing the watershed storage dynamics as rainfall progresses during the rainy seasons, effective rainfall that is rainfall minus evapotranspiration (P-ET), is used instead of rainfall alone since the combined value is a more accurate estimate of the water available for movement or storage in the soil (Liu et al. 2008; Steenhuis et al. 2009). Cumulative effective rainfall during each rainy season was also calculated.

Since the starting and ending dates of the rainy season vary from year to year in the highlands of Ethiopia, a simple and consistent method was used to delineate the rainy season for each year. The start of the rainy season was determined when a cumulative rainfall exceeds cumulative evaporation for a seven day period ($P - ET > 0$) and a rainfall event day with rainfall larger than evaporation and rainfall continues afterwards. If none of the days in a subsequent 10 days period have rainfall in excess of evapotranspiration ($P - ET > 0$), then the rainy season stops.

13.4 Results

The total water storage change in the Anjeni experimental watershed during the distinct rainy seasons was evaluated in three wetness categories which were divided based on wetness. On the basis of the classification threshold, the wet seasons of 1985, 1986, and 1996 were categorized as '*dry*', that of 1987, 1990, and 1998 as '*normal*' and 1988, 1997, and 2000 as '*wet*'. Table 13.1, 13.2, 13.3 summarize the characteristic annual rainfall, rainy season duration, non-rainy days and number of rainy days. A stream flow (mm d^{-1}) vs cumulative change in storage (mm) plot helps visualizc the relationship in each of the three categories. The plot portrays how stream flow responds to rainfall after an extended dry season where watershed storage is at the minimum.

13.4.1 Dry Rainy Seasons

The rainy season in this category lasts an average of 135 days with average of 14 days of non-rainy days. The watershed storages for the 'dry' years were 335, 260, and 350 mm for a cumulative effective rainfall of 540, 400, and 670 mm, respectively (Table 13.1 and Fig. 13.3a). With the occurrence of 540 mm of average cumulative effective rainfall, (i.e., 58 % of the total effective rainfall), the average total watershed storage was 315 mm (Table 13.1). The cumulative rainfall is the cumulative difference of rainfall and ET. Figure 13.3a shows the cumulative change in storage over the rainy period vs time. The total watershed storage reached maximum after about 55 % of the total rainy season rainfall had occurred.

Stream flow (mm d^{-1}) vs cumulative change in storage (mm) plot logarithm scale (Fig. 13.3b) for the '*dry*' season shows that at the start of the rainy season, when the watershed was dry, stream flow is minimum in the order of 1 mm d^{-1} due to extended dry season. But as the rainfall progresses the watershed storage started filling up and when the cumulative watershed storage reaches 100 mm, the daily

Table 13.1 Description of rainfall, runoff, and storage in a dry rainy season category in Anjeni watershed (*Q* is runoff, *C* is runoff coefficient, Cum. is cumulative, No. is number, ΔS is change in storage, RF is rainfall)

Year	Dry rainy season			
	1985	1986	1996	Avg.
Annual Rainfall (mm)	1556	1372	1697	**1542**
Rainy season rainfall (mm)	1272	1205	1372	**1283**
Rainy season rainfall (%)	82	88	81	**83**
Rainy season duration (days)	125	129	152	**135**
No. of non- rainy days	9	13	21	**14**
No. of days with rainfall <= 5 mm	50	44	72	**55**
Cum. ΔS at threshold	335	260	350	**315**
Cum. excess RF at storage threshold	540	400	670	**537**
Cum. rainfall at storage threshold	687	553	865	**702**
Rainfall before threshold (%)	54	46	63	**54**
Total effective rainfall	945	833	984	**921**
Effective rainfall at threshold (%)	57	48	68	**58**
Q (mm)	572	510	637	**573**
Q at threshold (mm)	207	123	294	**208**
Q after threshold (mm)	365	387	343	**365**
Q before threshold (%)	36	24	46	**35**
Q after threshold (%)	64	76	54	**65**
Rainy season C	0.45	0.42	0.46	**0.45**
C before threshold	0.3	0.22	0.34	**0.29**
C after threshold	0.62	0.59	0.68	**0.63**

stream flow started increasing almost linearly with increase in cumulative change in watershed storage. After the watershed storage reached maximum threshold value, although the change in watershed storage stays almost constant, the daily stream flow continuous to increase as effective rainfall increases (Fig. 13.3b). The daily stream flow reached as high as 23 mm d^{-1} in 1996 due to a high rainfall event after storage reached maximum (Fig. 13.3b). The average maximum watershed storage (threshold) was 315 mm in the dry distinct wet season. The runoff coefficient (*C*) increased from an average of 0.29 before the watershed storage threshold to an average of 0.63 after the watershed storage threshold and 65 % of the stream flow during the rainy seasons occurred after the storage reached maximum (Table 13.1). The average runoff coefficient during the whole rainy season was 0.45 (Table 13.1).

Table 13.2 Description of rainfall, runoff, and storage in a normal wet season category in Anjeni watershed

Year	Normal rainy season			
	1987	1990	1998	Avg.
Annual Rainfall (mm)	1811	1668	1773	**1751**
Rainy season rainfall (mm)	1539	1381	1504	**1475**
Rainy season rainfall (%)	85	83	85	**84**
Rainy season duration (days)	152	130	151	**144**
No. of non- rainy days	11	8	8	**9**
No. of days with rainfall <= 5 mm	58	53	52	**54**
Cum. ΔS at threshold	380	350	416	**382**
Cum. effective RF at storage threshold	660	672	730	**687**
Cum. RF at storage threshold	937	886	942	**922**
Rainfall before threshold (%)	61	64	63	**63**
Total excess rainfall	1100	1032	1101	**1078**
Effective RF at threshold (%)	60	65	66	**64**
Q (mm)	607	598	612	**606**
Q at threshold (mm)	273	316	301	**297**
Q after threshold (mm)	334	282	311	**309**
Q before threshold (%)	45	53	49	**49**
Q after threshold (%)	55	47	51	**51**
Rainy season C	0.39	0.43	0.41	**0.41**
C before threshold	0.29	0.36	0.32	**0.32**
C after threshold	0.55	0.57	0.55	**0.56**

13.4.2 *Normal Rainy Seasons*

The rainy season in this category lasts an average of 144 days with an average 9 days of non-rainy days. It was found that, the total watershed storage in these seasons were 380 mm, 350 mm, and 416 mm at cumulative effective rainfall of 660, 670, and 730 mm, for the respective years (Table 13.2 and Fig. 13.4a). The cumulative watershed storage change in the normal wet years varies from a minimum of 350 mm in 1990 to a maximum of 416 mm in 1998. This difference could be due to the ill-defined ground water boundary condition and rainfall properties in the wet season (Table 13.2). The average total watershed storage was 380 mm with a cumulative effective rainfall of 690 mm, 64 % of the total effective rainfall during the rainy season (Table 13.2). Figure 13.4a shows the cumulative change in storage (logarithmic scale) over the wet period vs time linear scale. The total watershed storage reached maximum after about 63 % of the total rainy season rainfall had occurred.

The stream flow (mm d^{-1}) vs cumulative change in storage (mm) plot (Fig. 13.4b) for the '*Normal*' rainy season shows that at the start of the rainy season, when the watershed was dry, stream flow is minimum in the order of 1 mm d^{-1} due to extended dry season. When the cumulative watershed storage reaches 100 mm,

Table 13.3 Description of rainfall, runoff, and storage in a wet rainy season category and overall average for the three categories for Anjeni watershed

Year	Wet rainy season				Tables 13.1, 13.2, 13.3
	1988	1997	2000	Avg.	Overall average
Annual Rainfall (mm)	1855	1708	1907	**1823**	**1705**
Wet season rainfall (mm)	1649	1545	1710	**1635**	**1464**
Wet season rainfall (%)	89	90	90	**90**	**86**
Wet season duration (days)	140	158	177	**158**	**146**
No. of non-rainy days	10	12	6	**9**	**11**
No. of days with rainfall <= 5 mm	53	64	75	**64**	**58**
Cum. ΔS at threshold	430	456	415	**434**	**377**
Cum. excess RF at storage threshold	590	650	707	**649**	**624**
Cum. rainfall at storage threshold	715	900	980	**865**	**829**
Rainfall before threshold (%)	43	58	57	**53**	**57**
Total excess rainfall	1323	1090	1210	**1208**	**1069**
Rainfallexcess at threshold (%)	45	60	58	**54**	**59**
Q (mm)	631	503	678	**604**	**594**
Q at threshold (mm)	160	195	292	**216**	**240**
Q after threshold (mm)	471	308	386	**388**	**354**
Q before threshold (%)	25	39	43	**36**	**40**
Q after threshold (%)	75	61	57	**64**	**60**
Wet season C	0.38	0.33	0.4	**0.37**	**0.41**
C before threshold	0.22	0.22	0.3	**0.25**	**0.29**
C after threshold	0.5	0.48	0.53	**0.5**	**0.56**

the daily stream flow started increasing almost linearly with increase of cumulative change in watershed storage. After the watershed storage reached maximum threshold value of 380 mm, although the change in watershed storage keeps almost constant, the daily stream flow continuous to increase as effective rainfall increases (Fig. 13.4b). The daily stream flow reached as high as 28 mm d^{-1} in 1990 due to high rainfall event after watershed storage threshold (Fig. 13.4b). The runoff coefficient increased from an average of 0.32 before the watershed storage threshold to an average of 0.56 after the watershed satisfies its storage capacity. The average runoff coefficient during the normal wet season was 0.41 (Table 13.2).

13.4.3 Wet Rainy Seasons

The rainy season period in the wet category is generally prolonged, lasting on average of 158 days and 9 non-rainy days. In this category, change in storage was analyzed for three rainy seasons; 1988, 1997, and 2000. The cumulative change in storage in this category is generally larger. On average, the watershed storage was

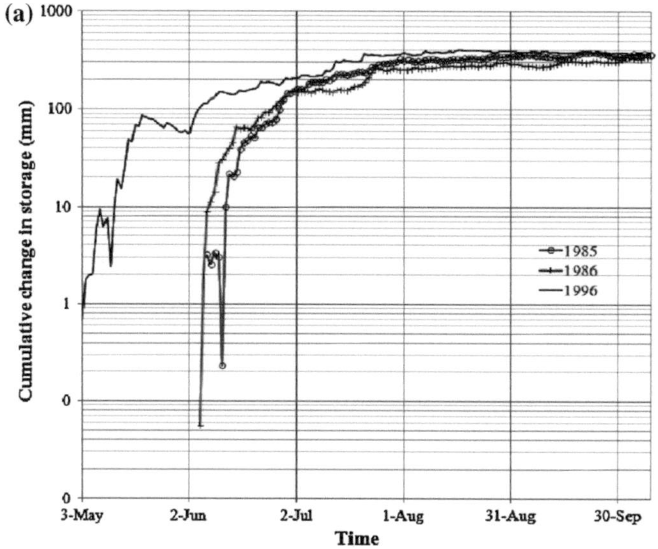

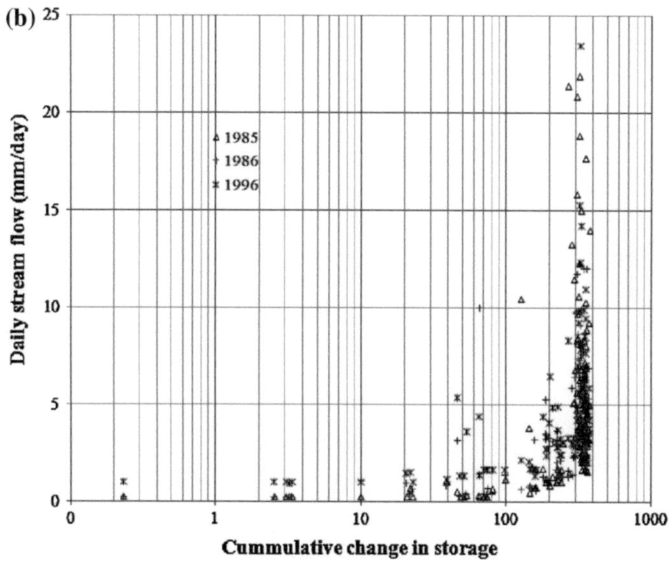

Fig. 13.3 **a** Cumulative change in storage versus time **b** Daily stream flow (mm/day) versus cumulative change in storage (mm) for the dry rainy season

435 mm after a cumulative effective rainfall of 650 mm (Table 13.3). Moreover, the watershed storage steadily keeps on increasing with increase in cumulative effective rainfall even after the watershed stores its maximum capacity (Fig. 13.5a). A possible explanation will be groundwater leakage to an adjacent watersheds and

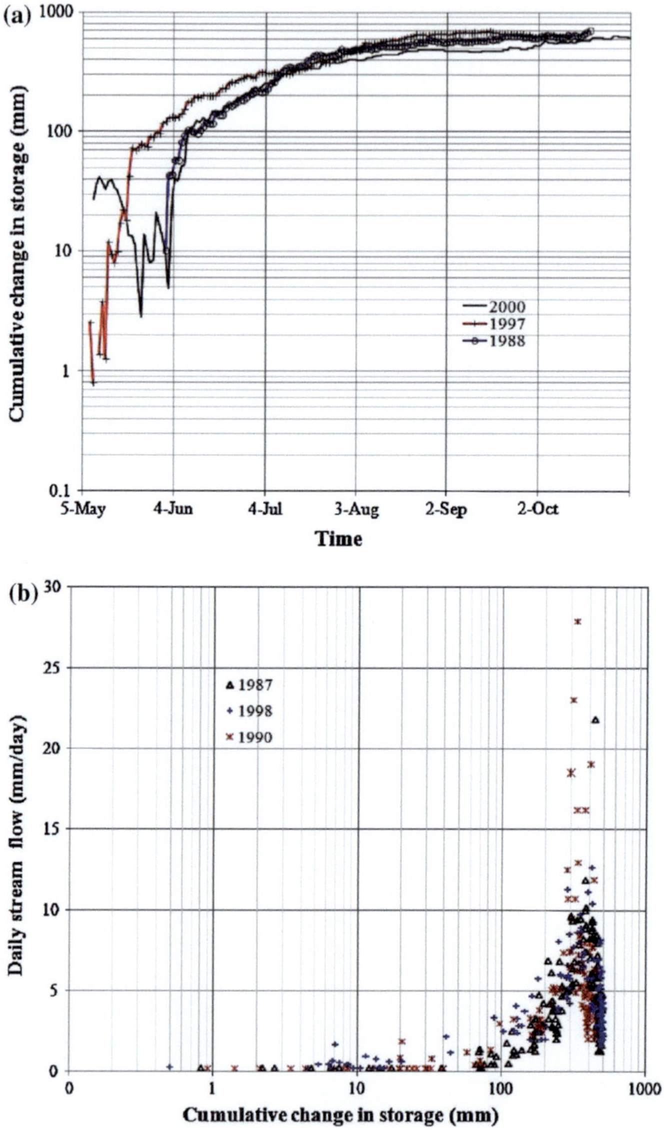

Fig. 13.4 **a** Cumulative change in storage versus time **b** Daily stream flow versus cumulative change in storage (mm) in the normal rainy seasons

farther ground water flow downstream of the gauging station and/or error of closure due to the ill-defined boundary condition of the ground water.

Stream flow (mm d^{-1}) vs cumulative change in storage (mm) plot (Fig. 13.5b) for the '*wet*' category show similar to the '*dry*' and '*normal*' rainy seasons. When the watershed is dry, stream flow is minimum in the order of 1 mm d^{-1}. Similarly, when

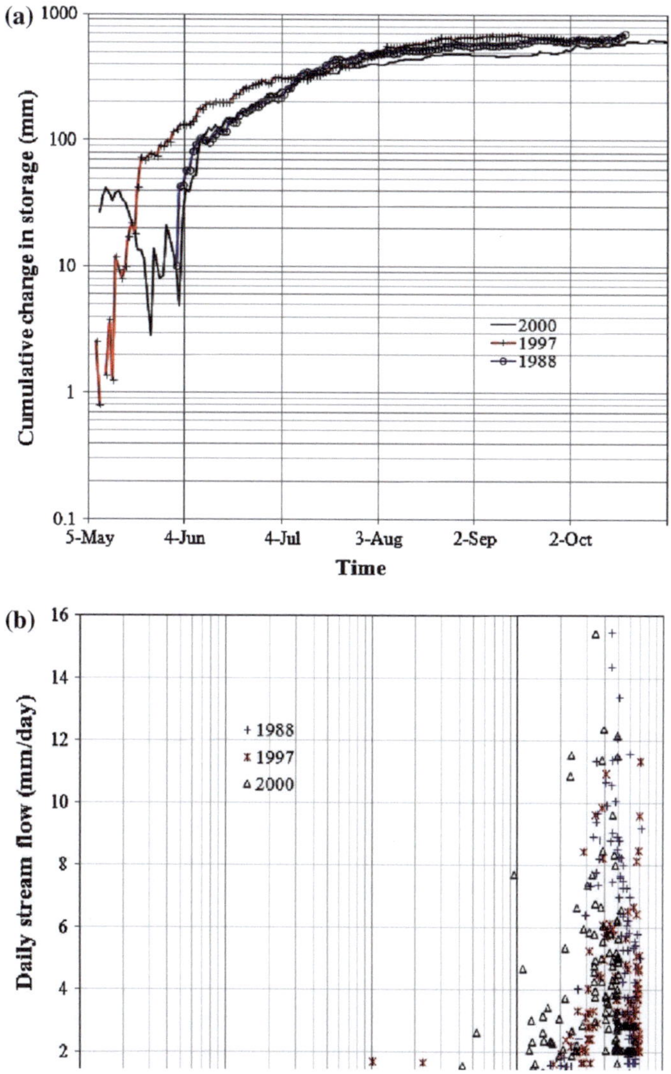

Fig. 13.5 **a** Cumulative change in storage versus time **b** Daily stream flow versus cumulative change in storage in the wet rainy season

the cumulative watershed storage reaches 100 mm, stream flow start to increase linearly with cumulative change in storage. When watershed storage reached its maximum threshold value, stream flow kept on increasing as effective rainfall increased (Fig. 13.5b). The runoff coefficient increased from an average of 0.25 before the watershed storage threshold to an average of 0.50. Sixty-four percent of the stream flow had occurred after the watershed storage reached maximum (Table 13.3). The average runoff coefficient during the wettest season was 0.37 (Table 13.3).

13.5 Discussions and Conclusions

Understanding functions of a watershed is important for appropriate soil and water conservation measures and for planning and development of water resources for sustainable use. Watershed storage is a significant part of a catchment water budget and its quantification helps to know the fundamental catchment hydrological processes. This study is aimed to understand the dynamics of watershed storages in distinct wet seasons. The daily observed rainfall and runoff data and modeled evapotranspiration (ET) estimates were used in quantifying watershed storage dynamics over distinct wet seasons using water balance approach. The dynamics of watershed storage was evaluated in three wetness categories of the distinct wet season of the Anjeni watershed. During the dry wet season, the watershed storage reached a maximum threshold average value of 315 mm after a cumulative effective rainfall of 540 mm. In the *normal* rainy season, the average total watershed storage was 380 mm after a cumulative effective rainfall of 690 mm. In the *wet* rainy season, the average watershed storage was 435 mm after a cumulative effective rainfall of 650 mm. The watershed maximum storage threshold point is generally lower in the *dry* category when compared with the *normal* and *wet* categories. The water storage of the Anjeni watershed generally ranges from 315 mm in the *dry* rainy season to 435 mm in the *wet* rainy season. The difference in the watershed storage among the rainy seasons could be due to the rainfall volume, rainfall characteristics, and the number of rainy and non-rainy days. For example, the *'dry'*, *'normal'*, and *'wet'* rainy seasons' on average last 135, 144, and 158 days, respectively (Tables 13.1, 13.2, 13.3). The *dry, normal, and wet* rainy season non-rainy days were 14, 9, and 9, respectively (Tables 13.1, 13.2, 13.3). The *dry* rainy season generally lasts for a shorter period and relatively longer non-rainy days than the *normal* and *wet* rainy seasons (Tables 13.1, 13.2, 13.3). For instance, the 1986 *dry* category rainy season lasted 129 days with 13 non-rainy days, while in 2000 *wet* category lasted an extended 177 days with only 6 non-rainy days. It is this difference that brought the difference in the watershed storage dynamics among the wetness categories.

The general increase of the watershed storage volume in the *wet* rainy seasons and the slow increase of the storage volume as rainfall progresses is most probably due to ground water flow leakage to the neighboring watersheds and ground water flow further downstream of the gauging station. The loss of water from one watershed to another through deep groundwater systems can be important in such small watersheds like Anjeni. Quantifying this ground water flux was not possible due to ill-defined boundary conditions that could over estimate the total change in storage of the watershed especially in the *wet* category. Irrespective of the wetness of the rainy season category, the watershed responds very similarly till the watershed stores 100 mm. After 100 mm of storage, stream flow increases linearly with increase in watershed storage. But once, the watershed storage reached its maximum threshold value, stream flow kept on increasing as rainfall continues while storage was almost constant at the maximum storage (Figs. 13.3, 13.4, and 13.5).

As rainfall continues to accumulate during a rainy season, the watershed eventually reaches a threshold point where it cannot store any more water. The watershed runoff response can be predicted by a linear relationship with effective rainfall indicating that the proportion of the rainfall that became runoff was constant during the remainder of the rainy season. The watershed runoff coefficient before the storage threshold was 0.29. This increased to 0.56 after the watershed storage threshold is reached. On the average, 40 % of the flow occurred before the watershed storage threshold point even though 60 % of the wet season's rain had occurred. While the remaining 60 % of the flow occurred after the storage threshold point with the remaining 40 % of the wet season rainfall (Tables 13.1, 13.2 and 13.3). The majority of the stream flow occurred after a storage threshold value with less rainfall volume. This showed that stream flow is a function of watershed storage threshold. Since the majority of the stream flow occurred after watershed storage threshold point, this helps to conclude that saturation excess runoff is the dominant runoff generation mechanism in the Anjeni watershed. This is in line with previous studies of (Collick et al. 2009; Liu et al. 2008; Steenhuis et al. 2009; Tilahun et al. 2013) which concluded that the dominant runoff generation mechanism is saturation excess. Hydrology models which consider the watershed storage dynamics as input could better model the rainfall and runoff process in watersheds in the monsoonal climates of the Upper Blue Nile. Further studies with watersheds having different sizes and data collected in sub daily time step could help better understand the functions of the watersheds in the Upper Blue Nile basin.

Acknowledgments The authors acknowledge the Amhara Region Agriculture Research Institute for providing the data free of charge, and the Ethiopian Institute of Water Resources for providing transport facilities to visit the watershed.

References

Abtew W, Melesse AM, Desalegn T (2009a) Spatial, inter and intra-annual variability of the Blue Nile River basin rainfall. Hydrol Process 23(21):3075–3082

Abtew W, Melesse AM, Desalegn T (2009b) El Niño southern oscillation link to the Blue Nile River Basin hydrology. Hydrol Process Spec Issue Nile Hydrol 23(26):3653–3660

Abtew W, Melesse AM (2014a) Nile River basin hydology. In: Melesse AM, Abtew W, Setegn S (eds) Nile River basin: ecohydrological challenges, climate change and hydropolitics, pp 7–22

Abtew W, Melesse AM (2014b) Climate teleconnections and water management. In: Nile River basin. Springer International Publishing, Berlin, pp 685–705

Abtew W, Melesse AM (2014c) Transboundary rivers and the Nile. In: Nile River basin. Springer International Publishing, Berlin, pp 565–579

Beven K (2006) Searching for the Holy Grail of scientific hydrology: as closure. Hydrol Earth Syst Sci Dis 10(5):609–618

Black PE (1997) Watershed functions. J Am Water Resour Assoc 33(1):1–11. doi:10.1111/j.1752-1688.1997.tb04077.x

Bosshart U (1997) Measurement of river discharge for the SCRP research catchments: gauging station profiles. Soil Conservation Research Programme, Research report 31, University of Berne, Switzerland

Brutsaert W (2005) Hydrology: an introduction. Cambridge University Press, UK, p 605

Chebud Y, Melesse AM (2013) Stage level, volume, and time-frequency change information content of Lake Tana using stochastic approaches, hydrological processes 27(10):1475–1483. doi: 10.1002/hyp.9291

Chebud YA, Melesse AM (2009a) Numerical modeling of the groundwater flow system of the Gumera Sub-Basin in Lake Tana Basin, Ethiopia. Hydrol Process Spec Issue Nile Hydrol 23 (26):3694–3704

Chebud YA, Melesse AM (2009b) Modeling lake stage and water balance of Lake Tana, Ethiopia. Hydrol Process 23(25):3534–3544

Collick AS, Easton ZM, Ashagrie T, Biruk B, Tilahun S, Adgo E, Awulachew SB, Zeleke G, Steenhuis TS (2009) A simple semi-distributed water balance model for the Ethiopian highlands. Hydrol Process 23(26):3718–3727

Dessu SB, Melesse AM (2012) Modeling the rainfall-runoff process of the Mara River Basin using SWAT. Hydrol Process 26(26):4038–4049

Dessu SB, Melesse AM (2013) Impact and uncertainties of climate change on the hydrology of the Mara River Basin. Hydrol Process 27(20):2973–2986

Dessu SB, Melesse AM, Bhat M, McClain M (2014) Assessment of water resources availability and demand in the Mara River Basin. CATENA 115:104–114

Enku T, Melesse AM (2014) A simple temperature method for the estimation of evapotranspiration. Hydrol Process 28(6):2945–2960

Hurni H (1984) Soil conservation research project third progress report. University of Berne and the United Nations University. Ministry of Agriculture Addis Ababa

Kirchner JW (2006) Getting the right answers for the right reasons: linking measurements, analyses, and models to advance the science of hydrology. Water Resour Res 42(3):W03S04, doi:10.1029/2005WR004362

Kirchner JW (2009) Catchments as simple dynamical systems: catchment characterization, rainfall-runoff modeling, and doing hydrology backward. Water Resour Res 45(2):W02429, doi:10.1029/2008WR006912

Lagomasino D, Price RM, Whitman D, Melesse AM, Oberbauer S (2015) Spatial and temporal variability in spectral-based evapotranspiration measured from Landsat 5TM across two mangrove ecotones, Agricultural and Forest Meteorology, in press

Legesse ES (2009) Modeling rainfall-runoff relationships for the Anjeni watershed in the Blue Nile basin. MSc thesis. Cornell University, 54 pp

Liu BM, Collick AS, Zeleke G, Adgo E, Easton ZM, Steenhuis TS (2008) Rainfall-discharge relationships for a monsoonal climate in the Ethiopian highlands. Hydrol Process 22(7):1059–1067

McDonnell JJ (2003) Where does water go when it rains? Moving beyond the variable source area concept of rainfall-runoff response. Hydrol Process 17(9):1869–1875

McDonnell JJ (2009) Classics in physical geography revisited. Hewlett JD, Hibbert AR (1967) Factors affecting the response of small watersheds to precipitation in humid areas. In Sopper WE, Lull HW (eds.) Forest hydrology. Pergamon Press, New York, pp 275–290. Prog Phys Geogr 33(2):288–293

McNamara JP, Tetzlaff D, Bishop K, Soulsby C, Seyfried M, Peters NE, Aulenbach BT, Hooper R (2011) Storage as a metric of catchment comparison. Hydrol Process 25(21):3364–3371

Melesse AM (2011) Nile River basin: hydrology, climate and water use. Springer Science & Business Media, Berlin

Melesse A, Abtew W, Setegn SG (2014) Nile River basin: ecohydrological challenges, climate change and hydropolitics. Springer Science & Business Media, Berlin

Melesse AM, Oberg J, Beeri O, Nangia V, Baumgartner D (2006) Spatiotemporal dynamics of evapotranspiration and vegetation at the glacial ridge prairie restoration. Hydrol Process 20 (7):1451–1464

Melesse A, Nangia V, Wang X, McClain M (2007) Wetland restoration response analysis using MODIS and groundwater data. Special issue: remote sensing of natural resources and the environment. Sensors 7:1916–1933

Melesse A, Abtew W, Desalegne T (2009a) Evaporation estimation of Rift Valley Lakes in Ethiopia: comparison of models. Sensors 9(12):9603–9615; doi:10.3390/s91209603

Melesse AM, Loukas Athanasios G, Senay Gabriel, Yitayew Muluneh (2009b) Climate change, land-cover dynamics and ecohydrology of the Nile River Basin. Hydrol Process Spec Issue Nile Hydrol 23(26):3651–3652

Melesse AM, Abtew W, Desalegne T, Wang X (2009c) Low and high flow analysis and wavelet application for characterization of the Blue Nile River System. Hydrol Process 24(3):241–252

Melesse AM, Abtew W, Setegn S, Dessalegne T (2011) Hydrological variability and climate of the Upper Blue Nile River Basin. In: Melesse A (ed) Nile River basin: hydrology, climate and water use, Chap. 1. Springer Science Publisher, Berlin, 3–37. doi:10.1007/978-94-007-0689-7_1

Oberg J, Melesse AM (2006) Evapotranspiration dynamics at an ecohydrological restoration site: an energy balance and remote sensing approach. J Amer Water Resour Assoc 42(3):565–582

Sayama T, McDonnell JJ, Dhakal A, Sullivan K (2011) How much water can a watershed store? Hydrol Process 25(25):3899–3908

Senay GB, Verdin JP, Lietzow R, Melesse AM (2008) Global daily reference evapotranspiration modeling and validation. J Amer Water Resour Assoc (JAWRA) 44(4):969–979

Senay GB, Budde M, Verdin JP, Melesse AM (2007) A coupled remote sensing and simplified energy balance approach to estimate actual evapotranspiration from irrigated fields. Special issue: remote sensing of natural resources and the environment., SENSORS 7:979–1000

Setegn SG, Srinivasan R, Dargahil B, Melesse AM (2009a) Spatial delineation of soil erosion prone areas: application of SWAT and MCE approaches in the Lake Tana Basin, Ethiopia. Hydrol Process Spec Issue Nile Hydrol 23(26):3738–3750

Setegn SG, Srinivasan R, Melesse AM, Dargahil B (2009b) SWAT model application and prediction uncertainty analysis in the Lake Tana Basin, Ethiopia. Hydrol Process 24(3):357–367

Setegn SG, Bijan Dargahi B, Srinivasan R, Melesse AM (2010) Modelling of sediment yield from Anjeni Gauged watershed, Ethiopia using SWAT. JAWRA 46(3):514–526

Soulsby C, Tetzlaff D, Hrachowitz M (2009) Tracers and transit times: windows for viewing catchment scale storage? Hydrol Process 23(24):3503–3507

Spence C (2007) On the relation between dynamic storage and runoff: A discussion on thresholds, efficiency, and function. Water Resour Res 43(12)

Spence C (2010) A paradigm shift in hydrology: storage thresholds across scales influence catchment runoff generation. Geogr Compass 4(7):819–833

Steenhuis TS, Collick AS, Easton ZM, Leggesse ES, Bayabil HK, White ED, Awulachew SB, Adgo E, Ahmed AA (2009) Predicting discharge and sediment for the Abay (Blue Nile) with a simple model. Hydrol Process 23(26):3728–3737

Sugawara M, Funiyuki M (1956) A method of revision of the river discharge by means of a rainfall model. Collection of research papers about forecasting hydrologic variables, The Geosphere Research Institute of Saitama University, Saitama, pp 14–18

Tilahun SA, Guzman C, Zegeye A, Engda T, Collick AS, Rimmer A, Steenhuis TS (2013) An efficient semi-distributed hillslope erosion model for the subhumid Ethiopian Highlands. Hydrol Earth Syst Sci 17(3):1051–1063

Yitayew M, Melesse AM (2011) Critical water resources management issues in Nile River Basin. In: Melesse Am (ed) Nile River Basin: hydrology, climate and water use, Chap. 20. Springer Science Publisher, Berlin, pp 401–416. doi:10.1007/978-94-007-0689-7_20

Zeleke G (2000) Landscape dynamics and soil erosion process modeling in the NorthWestern Ethiopian Highlands African studies series A 16. Geographica Bernensia Berne, Switzerland

Chapter 14
Estimation of Large to Extreme Floods Using a Regionalization Model

Khaled Haddad and Ataur Rahman

Abstract Estimation of large to extreme floods in the range of 100 years return periods to probable maximum floods (PMF) is needed in planning and designing of large water resources management projects. Due to the limited availability of observed flood data, the estimation of large to extreme floods requires significant extrapolation beyond the observed flood and rainfall data. This chapter provides a review of various techniques to estimate large to extreme floods. It also presents a case study in Australia where based on observed flood data, a large to extreme flood regionalization (LEFR) model has been developed which can be applied relatively easily as compared with rainfall runoff modeling. The LEFR model assumes that the maximum observed flood data over a large number of sites in a region can be pooled together by accounting for the at-site variation in the mean and coefficient of variation of the observed annual maximum flood data. The LEFR model has been developed and tested using data from 227 catchments in New South Wales and Victoria States in Australia. The method can easily be adapted to other Australian states and countries.

Keywords Extreme floods · Probable maximum flood · Large to extreme flood regionalization (LEFR) · Probable maximum precipitation

---

K. Haddad · A. Rahman (✉)
School of Computing, Engineering and Mathematics, University of Western Sydney, Building XB, Kingswood, Locked Bag 1797, Penrith, NSW 2751, Australia
e-mail: a.rahman@uws.edu.au

K. Haddad
e-mail: K.Haddad@uws.edu.au

© Springer International Publishing Switzerland 2016
A.M. Melesse and W. Abtew (eds.), *Landscape Dynamics, Soils and Hydrological Processes in Varied Climates*, Springer Geography,
DOI 10.1007/978-3-319-18787-7_14

14.1 Introduction

Flood estimation is needed in water resources management tasks such as planning and designing of hydraulic structures, flood control levees, flood plain management, and for various regulatory purposes. The most direct method of flood estimation is at-site flood frequency analysis, which needs reliable and relatively long flood data recorded at the site of interest. However, at many sites of interests, flood data length is either relatively short at gauged sites or completely unavailable at ungauged sites. In such case, regional flood frequency analysis method is adopted which attempts to use data recorded at the gauged sites in a region to make flood estimation at ungauged sites. This in essence attempts to substitute space for time, i.e., it uses spatial data to cover the shortfall of temporal data at the location where flood estimation is required. Regional flood frequency analysis is generally adopted to estimate design floods at ungauged catchments in the range of 2–100 years return periods.

Large and extreme floods have caused significant damage to many countries in the past. In 2011, floods were ranked in the top third position as the most common disaster after earthquake and tsunami (CRED 2012). In Australia, floods are considered to be the most common and devastating natural disaster than any other disaster (FitzGerald et al. 2010). Around 951 people were killed and another 1,326 were injured in Australia by floods from 1852 to 2011 (Carbone and Hanson 2013). The estimated annual average flood damage cost in Australia is around $400 million (Charalambous et al. 2013). However, in some years, flood damage is too high as in the 2010–2011 flood season when about 70 % of Queensland State was affected by severe flooding with the total damage to public infrastructure exceeding $5 billion (PWC 2011). A reliable estimation of large and extreme floods is important to safegurd infrastructure and to protect human lives by making improved flood estimation, design, and management.

In many applications, design floods of smaller return periods such as 20 years are needed such as for urban stormwater design. However, for large dams, probable maximum flood (PMF) estimates may be needed. Design floods have arbitrarily been categorized into various classes depending on return periods and applications (Table 14.1). For example, in Australia, 'large' floods refer to floods with 50–100 years return periods, and floods in the range of 100–2000 years return periods are referred to as 'rare' floods, while floods of 2000 years return period to the PMF are termed 'extreme' floods (Nathan and Weinmann 2001). To estimate rare floods, Australian Rainfall and Runoff (ARR) suggest rainfall-based methods, and use of regional, and paleohydrological information (Nathan and Weinmann 2001).

The recommended method in ARR is based on an annual exceedance probability (AEP) neutral approach (Nathan and Weinmann 2001). It assumes that the AEP of design rainfall event can be converted directly into the AEP of flood event through the use of representative values of other model inputs such as temporal patterns and losses. However, the AEP neutrality approach has been widely criticized as the

Table 14.1 Arbitrary classification of large to extreme floods

ARI (years)	Classification	Estimation method
50–100	Large floods	At-site flood frequency analysis, and rainfall runoff modeling, regional flood estimation methods
Beyond 100 years up to 2000 years	Rare floods	Rainfall runoff modeling and approximate methods
Beyond 2000 years and smaller than PMF	Extreme	Rainfall runoff modeling and approximate methods
PMF	Probable maximum flood (PMF)	Statistical method, rainfall runoff model using probable maximum precipitation (PMP) as model input and approximate methods

model inputs generally show a wide temporal and spatial variability and nonlinearity and selection of a representative value involves notable subjectivity (Rahman et al. 2002; Weinmann et al. 2002; Kuczera et al. 2006; Aronica and Candela 2007; Caballero and Rahman 2014).

This chapter presents a regional approach to estimate large to extreme floods by developing a large to extreme flood regionalization (LEFR) model which can be applied to both gauged and ungauged catchments. The LEFR model is relatively easy to apply in practice and offers an alternative approach to the rainfall-based estimation methods. The remainder of the chapter is organized as follows. A review of large to extreme flood estimation techniques is presented first, which is followed by a description of the adopted methodology, data preparation for the case study, results, and conclusion.

14.2 Review of Large to Extreme Flood Estimation Methods

There are a number of alternative methods to estimate large to extreme floods. For example, preliminary estimates of large to extreme floods can be made using approximate methods. These estimates can be used in the preliminary assessment of spillway and dam sites, and determination of priorities for undertaking detailed studies on selection of dam sites and large-scale water resources projects. Preliminary estimates of 50–100 years design floods can be obtained from regional flood methods where adherence to the catchment size recommended for the regional flood methods is required. Regional equations to estimate PMF are also available in different parts of the world such as Nathan et al. (1994) for Australia. In this regard, envelope curves for world floods may be useful (Rodier and Roche 1984).

Floods up to 2000 years return periods can be estimated using regional rainfall estimates such as CRC-FORGE-based rainfall estimates (Nandakumar et al. 1997; Reed and Stewart 1989). Reed and Stewart (1989) proposed the FORGE method, which allows developing a growth curve in regional frequency analysis for a location of interest, called focal point, using recorded regional rainfall data. In the FORGE method, the annual rainfall maxima at the three rain gauges nearest to the focal point are considered first. The six highest independent standardized values are plotted using an appropriate plotting position formula. Afterwards, the six highest independent standardized rainfalls from the six stations nearest the focal point are plotted. The procedure is continued, doubling the number of stations at each step. The frequency curve is then fitted by eye balling as a line of best fit to the plotted values. The CRC-FORGE method has been applied in Australia successfully and extreme rainfall estimates from 50 to 2000 years have been derived, which can be used to obtain large to extreme design flood estimates (Nandakumar 1995; Nandakumar et al. 1997; Hill et al. 2000).

Probable maximum precipitation (PMF) is generally used to estimate PMF using a rainfall runoff model. Use of the PMP to generate the PMF has become the standard practice in many countries including the United States, China, India, and Australia (Svensson and Rakhecha 1998). The PMP can be defined as the "theoretically greatest depth of precipitation for a given duration that is physically possible over a given size storm area at a particular geographical location at a certain time of the year" (Hansen et al. 1982). Ideally, the PMP estimate should not be exceeded at a given location for a given duration and time of the year (Wang 1984). However, at many locations the measured rainfall depths have exceeded PMP estimates in the past which indicates that the PMP in reality is not associated with zero risk (Koutsoyiannis 1999).

Papalexiou and Koutsoyiannis (2006) studied the maximum precipitation depths derived by moisture maximization method at few stations in the Netherlands and concluded that probabilistic approach for the estimation of extreme precipitation value is more consistent to the natural behavior and provides better grounds for estimation. When using rainfall runoff model to estimate PMF from PMP, input values such as design losses, temporal patterns, and baseflow need to be carefully selected. Losses should be selected smaller than the minimum value observed in the largest floods in the catchment of interest. The use of zero initial loss is often recommended and the use of 1 mm h^{-1} continuing loss is usually adopted (Nathan and Weinmann 2001). In terms of selecting the design temporal patterns, more severe patterns are preferred. In estimating the uncertainties in the PMFProbable maximum flood estimates, the following sources of errors should be considered; data error in rainfall, streamflow, and losses; limited sample size in both temporal and spatial scale; model uncertainty in meteorological, hydrological and statistical models; and parameter uncertainty. The correlations of various input variables and model parameters need to be considered to avoid unrealistic estimation.

Hershfield (1961) proposed a statistical method for the estimation of PMP for USA. The World Meteorological Organization (WMO), in its various manuals and technical publications, has recommended this method for estimation of point PMP for those river basins whose daily rainfall data are available for a long period of time (World Meteorological Organization 1986).

Due to climate change, various aspects of hydrological cycle are expected to be affected including the magnitude and frequency of large to extreme floods.

There have been many studies to assess the impacts of climate changes on floods, which are generally limited up to 100 years return periods (Cunderlik and Burn 2003; Cunderlik and Ouarda 2006; Khaliq et al. 2006; El Adlouni et al. 2007; Ishak et al. 2013). Some of these nonstationary flood frequency analysis techniques are based on the assumption that due to climate change the distributional parameters are likely to change, such as the location and scale parameters becoming nonstationary. However, shape parameter does not change and hence, the use of a nonstationary probability distribution is suggested. There have been limited studies on the impact of climate change on extreme flood estimation. For example, Ousmane et al. (2012a, b) applied a nonstationary GEV distribution up to 1000 years return period to the Kemptville Creek located in Ontario, Canada, using the SWAT (Soil and Water Assessment Tool) model as rainfall-runoff model. Further research is warranted to investigate the impacts of climate change on large to extreme floods.

14.3 Methodology

The large to extreme flood regionalization (LEFR) model is based on the assumption that the standardized maximum values of the annual maximum flood series from a large number of stations in a given region can be pooled together after standardizing to allow for the across-site variation in the mean and coefficient of variation (CV) values of the recorded annual maximum flood data. The method was originally proposed by Majone and Tomirotti (2004), Majone et al. (2007) and further developed by Haddad et al. (2009, 2011a). The main advantage of the LEFR model is that it allows expected variation in CV values across the sites in the region of interest unlike the constant CV assumption in the typical index flood method (Hosking and Wallis 1997). This feature of the LEFR model enables to pool regional large and extreme flood data more effectively over many sites within a very large region.

The LEFR model assumes that the maximum observed floods $Q_{\max}$ from the annual flood series of each of the sites in a region, after standardization by the at-site mean flood and a function of the CV of the annual flood series, can be pooled similar to the station-year approach. Then, follow a single probability distribution, the standardized $Q_{\max}$ from all the sites form a homogeneous region. This is for the purpose of large to extreme flood regionalization. For example, Majone et al. (2007) developed a LEFR model, called Probabilistic Model, using flood data from 8,500 gauging stations across the world.

The LEFR model can be applied to ungauged catchments and in such case regional prediction equations for the mean and CV of the annual flood data should be developed based on observed flood data and climatic and catchment characteristics. In this application, a generalized least squares (GLS) regression was adopted to develop the prediction equations for mean and CV. The main advantages of the GLS regression over the ordinary least squares regression are that it can account for the variation in record length from site to site and the inter-station correlation. Moreover, it can differentiate between the sampling error and model error, thus offering a more realistic framework for error analysis. Further information on the GLS regression can be found in Stedinger and Tasker (1985), Madsen and Rosbjerg (1997), Haddad and Rahman (2012), and Haddad et al. (2012). The following section describes development of LEFR model for southeast Australia.

14.4 Data Preparation

To develop the LEFR model, data from 227 catchments from southeast Australia was selected. These catchments are situated in the states of Victoria and New South Wales as shown in Fig. 14.1. The streamflow data of these stations was prepared using standard procedure for filling gaps in the data, checking for outliers, rating curve error, and trends as detailed in Haddad et al. (2010). In the data preparation, initially a large number of stations were selected from the streamflow database of the New South Wales and Victoria States. A close examination revealed that many stations have unacceptable gaps in the data. There are major regulations and land use changes and hence, these stations were excluded from the study database. About 15 % stations showed trends and were excluded. Outliers were detected and censored in flood frequency analysis. Many stations exhibited a higher rating curve extrapolation error and were excluded. Finally, 227 stations were retained for this study. These are rural catchments with no major land use changes over the period of streamflow record and are not affected by major storages and dams. The catchment size of the 227 catchments ranges from 3 to 1010 km^2 with a median size of 289 km^2. The annual maximum streamflow record length ranges from 25 to 74 years with a median of 33 years. For each of the selected stations, the mean and CV values of the annual maximum flood series were obtained and at-site flood frequency analysis was carried out using FLIKE software (Kuczera 1999) adopting a log Pearson Type 3 (LP3) distribution and Bayesian parameter estimation procedure. Flood quantiles for 2, 5, 10, 20, 50, 100, and 200 years ARIs were estimated.

A total of seven climatic and catchment characteristics data were abstracted for each of the selected catchments: (i) catchment area (area, km^2); (ii) design rainfall intensity of 12–h duration and 2–year ARI (I_12_2, mm h^{-1}); (iii) design rainfall intensity of 12–h duration and 50–year ARI (I_12_50, mm h^{-1}); (iv) mean annual areal evapotranspiration (evap, mm); (v) stream density (sden, km km^{-2}); (vi) fraction of catchment forested (forest); and (vii) main stream slope (S1085, m km^{-1}).

To test the performance of the developed LEFR model, 18 catchments were selected randomly from the 227 catchments. The remaining 209 catchments were used to develop the LEFR model.

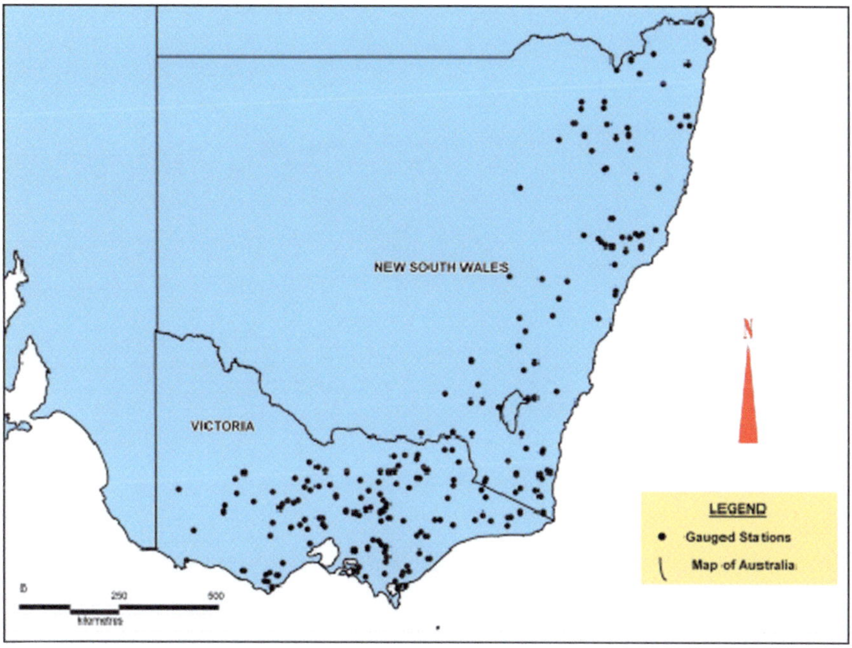

Fig. 14.1 Locations of the selected 227 catchments in the States of New South Wales and Victoria in Australia

14.5 Results

$Q_{\max}$ (referred to as Q henceforth) is the maximum data point of the annual maximum flood series data at a site, μ is the mean, s is the standard deviation and CV is the coefficient of variation of the annual maximum flood series at the site. The plot of Q/μ and CV of the selected 209 sites is shown in Fig. 14.2. The following empirical equation is used to approximate the relationship between Q/μ and CV (Eq. 14.1).

$$Q/\mu = 1 + \alpha \mathrm{CV}^{\lambda} \tag{14.1}$$

For the given dataset of 209 sites, maximum likelihood estimation gives $\alpha = 3.21$ and $\lambda = 1.42$ (with a coefficient of determination, $R^2 = 0.81$). The scatter in Fig. 14.2 can be modeled by a LEFR model of the following form (Eq. 14.2).

$$Q/\mu = 1 + f(\mathrm{ARI})\mathrm{CV}^{1.42} \tag{14.2}$$

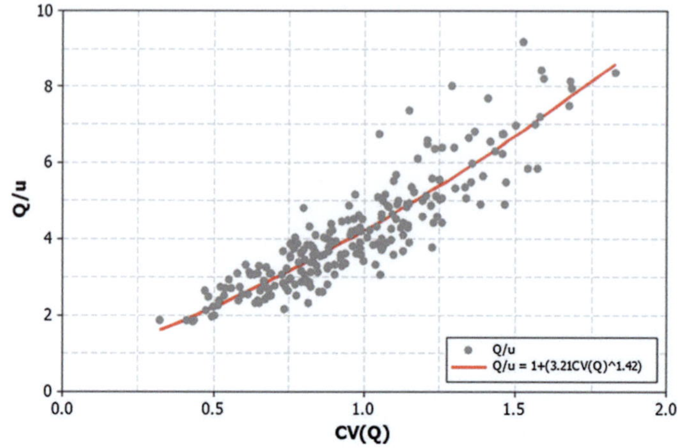

Fig. 14.2 Plot of Q/μ and CV of the selected 209 sites

where it is assumed that *f*(ARI) is a function of the average recurrence interval (ARI) of flood estimate. A standardized variable can then be defined by the following equation (Eq. 14.3).

$$Y = \frac{Q/\mu - 1}{\mathrm{CV}^{1.42}} = \frac{Q - \mu}{s\mathrm{CV}^{0.42}} \tag{14.3}$$

The standardization variable proposed by Eq. (14.3) accounts for the variation in the mean and CV. The plotting position formula, proposed by Majone and Tomirotti (2004) was applied to estimate the ARI of the $M = 209$ values of Y in the pooled dataset (Eq. 14.4).

$$\mathrm{ARI} = \frac{1}{1 - (1 - m/M)^{\frac{1}{n}}} \tag{14.4}$$

where m is the rank of the Y data, n is the average sample size, and M the number of sites, assumed to be independent in terms of maximum observed flood data. The plot of Y versus ARI is shown in Fig. 14.3. It shows that the experimental data can be interpolated by a curve whose central part can be approximated by a linear function of ln(ARI), an exponential distribution of the following form (Eq. 14.5).

$$Y = 1.1 + 0.52\ln(\mathrm{ARI}) \tag{14.5}$$

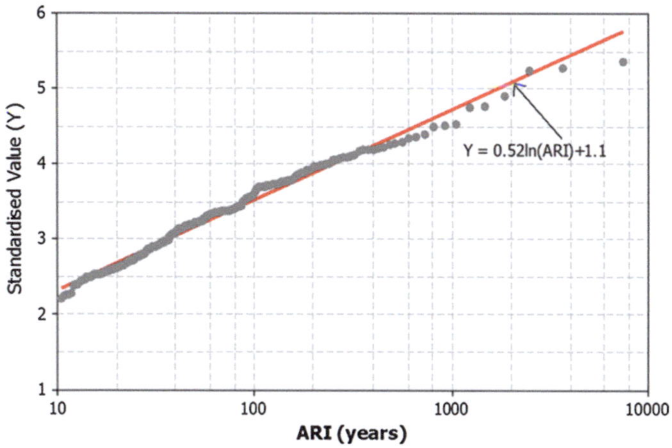

Fig. 14.3 Frequency distribution of the standardized values (Y)

Equation (14.5) can be expressed in terms of Q/μ in the following form (Eq. 14.6).

$$Q/\mu = 1 + (1.1 + 0.52\ln(\text{ARI}))\text{CV}^{1.42} \tag{14.6}$$

Equation 14.6 is the basis of the application of the LEFR model. For example, for 500, 1000 and 2000 years ARIs, the estimation equations becomes Eqs. (14.7, 14.8, and 14.9).

$$Q/\mu = 1 + (1.1 + 0.52\ln(500))\text{CV}^{1.42} \tag{14.7}$$

$$Q/\mu = 1 + (1.1 + 0.52\ln(1000))\text{CV}^{1.42} \tag{14.8}$$

$$Q/\mu = 1 + (1.1 + 0.52\ln(2000))\text{CV}^{1.42} \tag{14.9}$$

Figure 14.4 shows that the LEFR model can provide a good fit for the ARI range of 100–2000 years in CV range of 0.30–0.90. To extend the LEFR model to ARIs greater than 2000 years, data from other Australian states need to be included so that the number of stations and the resulting number of Q values in the study database increases significantly. Furthermore, the intersite dependence of the Q values needs to be accounted for in estimating the independent number of data points in the pooled database of Q. This enhancement of the LEFR model is being undertaken currently and will be published in the near future.

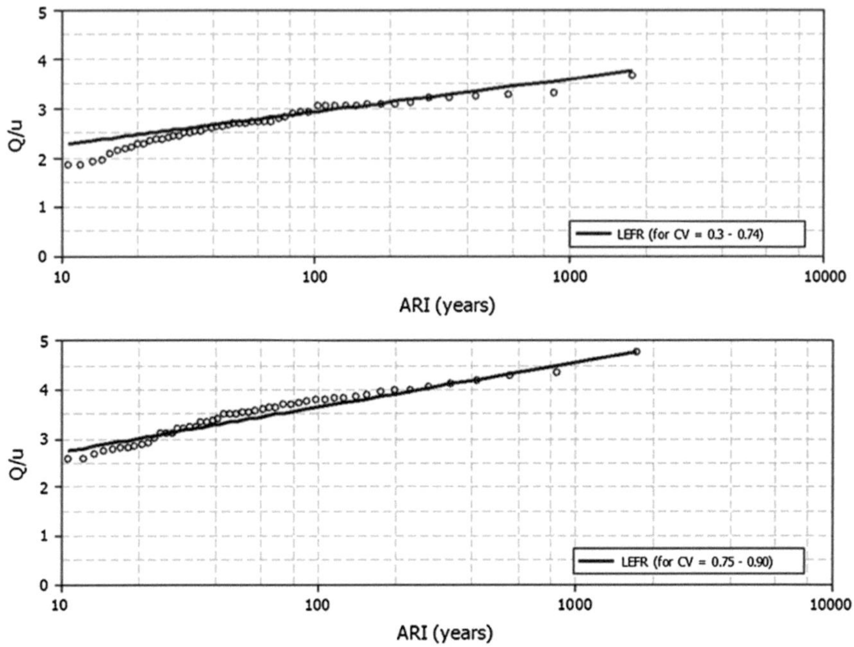

Fig. 14.4 Empirical frequency distributions of Q/μ quantiles derived from the LEFR model

To apply Eq. (14.6) for an ungauged catchment, μ and CV need to be estimated for the ungauged catchment of interest. For this case study, prediction equations were developed for μ (Eq. 14.10) and CV (Eq. 14.11) using GLS regression. Equation (14.10) has an R^2 of 0.74 and standard error of estimate of 0.31. Equation (14.11) has an R^2 of 0.64 and standard error of estimate of 0.26. In the GLS regression, the variables were selected at significance levels of at least 10 %. The standardized residual plots did not show any notable outliers and the quantile-quantile plots did not show any notable departure from the normal distribution. Hence, the regression assumptions were largely satisfied by Eqs. (14.10 and 14.11).

$$\log(\mu) = -2.99 + 1.13\log(\text{area}) + 2.00\log(I\_12\_50) + 0.35\log(\text{sden}) \quad (14.10)$$

$$\text{CV} = 1.07 + 0.63\log(I\_12\_2) - 1.26\log(\text{rain}) + 1.05\log(\text{evap}) \quad (14.11)$$

The developed LEFR model was tested for the 18 test catchments that were not used in developing the LEFR model represented by Eqs. (14.6, 14.8 and 14.9). The testing was done for ARIs up to 200 years only since the base values were estimated by at-site flood frequency analysis (Kuczera 1999) using a limited data length, 25–74 years. ARIs are expected to have a large degree of uncertainty for flood quantiles estimates by at-site flood frequency analysis using LP3 distribution beyond

Table 14.2 Summary of testing the LEFR model for 18 test catchments

ARI (years)	LEFR model (Eq. 14.6)	Median RE
50	$Q/\mu = 1 + (1.1 + 0.52\ \ln(50))\ CV^{1.42}$	21%.
100	$Q/\mu = 1 + (1.1 + 0.52\ \ln(100))\ CV^{1.42}$	30 %
200	$Q/\mu = 1 + (1.1 + 0.52\ \ln(200))\ CV^{1.42}$	35 %

200 years as applied in this study. To assess the degree of accuracy of the developed LEFR model, relative error (RE) was estimated by Eq. (14.12).

$$RE = \frac{Q_{LFR} - Q_{FFA}}{Q_{FFA}} \times 100 \quad (14.12)$$

The median RE values range from 0.21 to 0.35 for ARIs of 50–200 years (Table 14.2). This shows that as ARI increases median RE increases. This is expected as higher ARI flood estimates are associated with a higher degree of uncertainty. It should be noted here that at-site flood frequency estimate (Q_{FFA}) is subject to data error, sampling error and model error as selection of flood frequency distribution and associated parameter estimation procedure (Haddad and Rahman 2011). Hence, the RE in Eq. (14.12) provides only an approximate measure of error associated with the LEFR model. A similar degree of median RE values has been reported by other regional studies in southeast Australia (Rahman et al. 2011; Haddad et al. 2011b). It should be noted here that the intersite dependence of the selected annual maximum flood data points, Q values, need to be considered in the LEFR model as further enhancement, which is currently being investigated.

14.6 Conclusion

Large to extreme flood estimates are needed to design important hydraulic structures and for various water resources development tasks. A number of methods have been proposed for this purpose including approximate methods, rainfall runoff modeling and regionalization techniques. A large degree of uncertainty is associated with these methods since observed length of streamflow data is quite limited to develop and test these methods. This chapter presents a simple large to extreme flood regionalization (LEFR) model to estimate large to extreme floods. The method can be applied to both gauged and ungauged catchments. For the ungauged catchment application, the method is relatively easy to apply since it needs readily obtainable climate and catchment characteristics data. The method has been developed and tested initially using data from 227 catchments in New South Wales and Victoria States in Australia. The method is being enhanced by incorporating additional stations from other Australian states and by accounting for the intersite dependence of the annual maximum flood data.

Acknowledgments The work has been financially supported by the Australian Government through Geosciences Australia and Engineers Australia. The data has been obtained from Australian Rainfall and Runoff Revision Project 5 Regional Flood Methods. Authors would like to acknowledge New South Wales Department of Environment, Climate Change and Water, New South Wales, Department of Sustainability and Environment, Victoria, and Australian Bureau of Meteorology for providing data and Associate Professor Erwin Weinmann, Professor George Kuczera, Associate Professor Mark Babister, Associate Professor James Ball, Dr William Weeks, Dr Mohammad Zaman, and Dr Wilfredo Caballero for their input to the project including data preparation.

References

Aronica G, Candela A (2007) Derivation of flood frequency curves in poorly gauged Mediterranean catchments using a simple stochastic hydrological rainfall-runoff model. J Hydrol 347(1):132–142

Caballero WL, Rahman A (2014) Development of regionalized joint probability approach to flood estimation: a case study for New South Wales, Australia. Hydrological Processes, http://onlinelibrary.wiley.com/doi/10.1002/hyp.9919/pdf

Carbone D, Hanson J (2013) Floods: 10 of the deadliest in Australian history. Australian Geographic. http://www.australiangeographic.com.au/journal/the-worst-floods-in-australian-history.htm. Accessed 18 July 2013

Charalambous J, Rahman A, Carroll D (2013) Application of Monte Carlo simulation technique to design flood estimation: a case study for North Johnstone River in Queensland, Australia. Water Resour Manage 27:4099–4111

Crunch C (2012) Disaster data: a balanced perspective. Issue 27, Feb 2012, centre for research on the epidemiology and disasters, Universite Catholique de Louvain. http://www.cred.be/sites/default/files/CredCrunch27.pdf. Accessed 18 July 2013

Cunderlik JM, Burn DH (2003) Non-stationary pooled flood frequency analysis. J Hydrol 276(1–4):210–223

Cunderlik JM, Ouarda T (2006) Regional flood-duration-frequency modeling in the changing environment. J Hydrol 318(1–4):276–291

El Adlouni S, Ouarda T, Zhang X, Roy R, Bobee B (2007) Generalized maximum likelihood estimators for the non-stationary generalized extreme value model. Water Resour Res 43 W03410:13. doi:10.1029/2005WR004545

FitzGerald G, Du W, Jamal A, Clark M, Hou XY (2010) Flood fatalities in contemporary Australia (1997–2008). Emerg Med Australas 22(2):180–186

Haddad K. Aziz K, Rahman A, Weinmann PE, Ishak EH (2009) A probabilistic model for estimation of large floods in ungauged catchments: application to South-east Australia. 32nd Hydrology and water resources symp., Newcastle, 30 Nov to 3 Dec, pp 817–828

Haddad K, Rahman A, Weinmann PE, Kuczera G, Ball JE (2010) Streamflow data preparation for regional flood frequency analysis: lessons from south-east Australia. Aust J Water Resour 14 (1):17–32

Haddad K, Rahman A, Weinmann PE (2011a) Estimation of major floods: applicability of a simple probabilistic model. Aust J Water Resour 14(2):117–126

Haddad K, Rahman A, Kuczera G (2011b) Comparison of ordinary and generalized least squares regression models in regional flood frequency analysis: a case study for New South Wales. Aust J Water Resour 15(2):59–70

Haddad K, Rahman A (2011) Selection of the best fit flood frequency distribution and parameter estimation procedure—a case study for Tasmania in Australia. Stoch Env Res Risk Assess 25:415–428

Haddad K, Rahman A (2012) Regional flood frequency analysis in eastern Australia: Bayesian GLS regression-based methods within fixed region and ROI framework—quantile regression vs. Parameter regression technique. J Hydrol 430–431:142–161

Haddad K, Rahman A, Stedinger JR (2012) Regional flood frequency analysis using Bayesian generalized least squares: a comparison between quantile and parameter regression techniques. Hydrol Process 26:1008–1021

Hansen EM, Schreiner LC, Miller JF (1982) Application of probable maximum precipitation estimates. United States east of the 105th meridian. Hydrometeorological Rep. 52, National Weather Service, Silver Spring, MD, 176 pp

Hershfield DM (1961) Estimating the probable maximum precipitation. J Hydraul Div Am Soc Civ Eng 87:99–106

Hill PI, Nathan RJ, Rahman A, Lee, BC, Crowe P, Weinmann PE (2000) Estimation of extreme design rainfalls for South Australia using the CRC-FORGE method. In: Proceedings of 3rd international hydrology and water resources symposium interactive hydrology, IE Aust., Perth, Western Australia, vol 1, pp 558–563, 20–23 Nov 2000

Hosking JRM, Wallis JR (1997) Regional frequency analysis: an approach based on L-moments. Cambridge University Press, Cambridge

Ishak E, Rahman A, Westra S, Sharma A, Kuczera G (2013) Evaluating the non-stationarity of Australian annual maximum floods. J Hydrol 494:134–145

Khaliq MN, Ouarda TBMJ, Ondo JC, Gachon P, Bobee B (2006) Frequency analysis of a sequence of dependent and/or non-stationary hydro- meteorological observations: a review. J Hydrol 329(3–4):534–552

Kuczera G, Lambert MF, Heneker TM, Jennings S, Frost A, Coombes P (2006) Joint probability and design storms at the Crossroads. Aust J Water Resour 10(1):63–79

Kuczera G (1999) Comprehensive at-site flood frequency analysis using Monte Carlo Bayesian inference. Water Resour Res 35(5):1551–1557

Koutsoyiannis D (1999) A probabilistic view of Hershfield's method for estimating probable maximum precipitation. Water Resour Rcs 35:1313–1322

Madsen H, Rosbjerg D (1997) Generalized least squares and empirical Bayes estimation of regional partial duration series index-flood modeling. Water Resour Res 33(4):771–782

Majone U, Tomirotti M (2004) A trans-national regional frequency analysis of peak flood flows. L'Aqua 2:9–17

Majone U, Tomirotti M, Galimberti G (2007) A probabilistic model for the estimation of peak flood flows. Special Session 10, 32nd Congress of IAHR, Venice, Italy, 1–6 July 2007

Nandakumar N (1995) Estimation of extreme rainfalls for Victoria—application of the FORGE method. CRC for catchment hydrology working document 95/7, Monash University, Australia

Nandakumar N, Weinmann, PE, Mein, RG, Nathan RJ (1997) Estimation of extreme rainfalls for Victoria using the CRC-FORGE method (for rainfall durations 24–72 Hours). Report 97/4, Monash University

Nathan RJ, Weinmann PE, Gato S (1994) A quick method for estimation of the probable maximum flood in South East Australia. In: International hydrology and water resources symposium, water down under, November, Adelaide, Engineers Australia National Conference Publication No. 94, pp 229–234

Nathan RJ, Weinmann PE (2001) Estimation of large to extreme floods. In: Pilgrim DH (ed) Book VI, Vol. 1, Australian rainfall and runoff: a guide to flood estimation, IEAust, Canberra

Ousmane S, Ramsay A, Nistor I (2012a) Climate change impacts on extreme floods I: combining imperfect deterministic simulations and non-stationary frequency analysis. Nat Hazards 61:647–659

Ousmane S, Ramsay A, Nistor I (2012b) Climate change impacts on extreme floods II: improving flood future peaks simulation using non-stationary frequency analysis. Nat Hazards 60:715–726

Papalexiou SM, Koutsoyiannis D (2006) A probabilistic approach to the concept of probable maximum precipitation. Adv Geosci 7:51–54

PWC (2011) Economic impact of Queensland's natural disasters. Price Waterhouse Cooper (PWC), Australia
Rahman A, Weinmann PE, Hoang TMT, Laurenson EM (2002) Monte Carlo Simulation of flood frequency curves from rainfall. J Hydrol 256(3–4):196–210
Rahman A, Haddad K, Zaman M, Kuczera G, Weinmann PE (2011) Design flood estimation in ungauged catchments: a comparison between the probabilistic rational method and quantile regression technique for NSW. Aust J Water Resour 14(2):127–137
Reed DW, Stewart EJ (1989). Focus on rainfall growth estimation. In: Proceedings of 2nd national hydrology symposium, British Hydrological Society
Rodier JA, Roche M (1984) World catalogue of maximum observed floods. IAHS-AISH Publication No. 143. IAHS Press, England
Stedinger JR, Tasker GD (1985) Regional hydrologic analysis, 1. Ordinary, weighted, and generalized least squares compared. Water Resour Res 21(9):1421–1432
Svensson C, Rakhecha PR (1998) Estimation of probable maximum precipitation for dams in the Hongru River catchment. China Theor Appl Climatol 59:79–91
Wang BHM (1984) Estimation of probable maximum precipitation: case studies. J Hydraul Eng 110:1457–1472
Weinmann PE, Rahman A, Hoang TMT, Laurenson EM, Nathan RJ (2002) Monte Carlo simulation of flood frequency curves from rainfall—the way ahead. Aust J Water Resour 6 (1):71–80
World Meteorological Organization (1986) Manual for estimation for probable maximum precipitation. Operational hydrology Report No. 1 2nd edn, WMO No. 332

Chapter 15
Performance Evaluation of Synthetic Unit Hydrograph Methods in Mediterranean Climate. A Case Study at Guvenc Micro-watershed, Turkey

Tewodros Assefa Nigussie, E. Beyhan Yeğen and Assefa M. Melesse

Abstract Elements of hydrographs such as peak rate and time to peak are essential in the planning and design of hydraulic structures. Not two hydrographs are identical as not two watersheds and rainfall characteristics are identical. This calls for the development of hydrographs for every watershed. However, there are few number of gauged streams as the installation and management costs are expensive. Many models have been developed to overcome these problems, but these methods have not been widely used for they require lots of input parameters. As a result, a number of synthetic unit hydrograph (SUH) methods have been proposed as means of developing hydrographs for ungauged watersheds. However, their performance is not a much studied subject. This research was initiated to investigate the performance of Snyder's, Soil Conservation Service, Mockus, Nakayasu, Rodriguez-Valdez, and Gupta-Waymire SUH methods at Guvenc micro-watershed in Turkey. Global Mapper was used to download Digital Elevation Model (DEM), and Watershed Management System (WMS) software was used to manipulate the DEM and generate necessary data in accordance with the requirements of each SUH method. The peak discharge and the time to peak were used to compare the observed unit hydrograph (UH) with the SUHs. The results of the study showed that data from internet sources could be used as a source to generate appropriate watershed characteristics necessary to derive SUH. The comparison of the elements of SUHs showed that the SCS approach performed best in simulating the peak runoff in the study area.

T.A. Nigussie (✉) · E.B. Yeğen
Hydraulics and Water Resource Engineering Division, Faculty of Civil Engineering, Istanbul Technical University, Maslak, 34469 Istanbul, Turkey
e-mail: hiyawtewodros@gmail.com

E.B. Yeğen
e-mail: emineb@itu.edu.tr

A.M. Melesse
Department of Earth and Environment, Florida International University, Miami, FL, USA
e-mail: melessea@fiu.edu

© Springer International Publishing Switzerland 2016

A.M. Melesse and W. Abtew (eds.), *Landscape Dynamics, Soils and Hydrological Processes in Varied Climates*, Springer Geography,
DOI 10.1007/978-3-319-18787-7_15

Keywords Hydrograph · Unit hydrograph · Synthetic unit hydrograph · Instantaneous unit hydrograph · Guvenc watershed

15.1 Introduction

Water has become a scarce resource in many parts of the world as a result of population increase, expansion of agricultural, and industrial sectors. The impact of climate change and widespread water pollution have further aggravated the irregularities in the supply and demand structure of water (United Nations 1997). Due to the increasing scarcity, competition, and conflicts among uses and users of water arise. It is, therefore, necessary to make decisions about conservation and allocation of water that are compatible with social objectives such as economic efficiency, sustainability, and equity (Agudelo 2001).

Proper management of water resource requires reliable information about components of the hydrological cycle, one of the most important components being the rainfall-runoff process. The knowledge on rainfall-runoff process is of paramount importance to water resources planning and management. According to Bayazit (1988), it is important for flood caution, operation of flood-control-purposed reservoir, determination of river water potential, production of hydroelectric energy, allocation of domestic and irrigation water in drought seasons, and navigation planning in rivers. The rainfall-induced runoff process is usually given in a graphical form and shows runoff variations versus time during a given rainfall event. The elements of hydrograph such as peak flow rates and time to peak are what are very essential in the planning and design of flood-control systems and design of urban drainage systems (Granato 2012).

Not two hyetographs and hydrographs are identical as not two basins are identical resulting in various responses to rainfall. This calls for the development of hydrographs for every basin under various excess rainfall durations, which in turn needs installation of gauging stations and recording of streamflow data. However, the installation and management cost of stations is expensive making the development of hydrograph for every basin impossible (Harmel et al. 2006). There are also difficulties with the development of hydrographs resulting from the absence of precise determination of runoff from rainfall and determination of flow paths.

The unit hydrograph (UH), defined as the direct runoff from a storm that produces 1 unit of rainfall excess, has been a basic tool of rainfall-runoff computation for many decades and taken as a way of addressing some of the above-mentioned difficulties (Maidment 1993). The principal concept in the application of a UH is that each basin has one UH for a given rainfall event that does not change (in terms of its shape) unless the basin characteristics change. Given that the UH does not change in shape and represents streamflow response to 1 unit of runoff (rainfall excess) within a basin, flood hydrographs for actual storms can be simulated by multiplying the discharge ordinates from it by the rainfall excess computed from the observed rainfall record (Weaver 2003). Despite this advantage, units hydrographs developed from observed data cannot address all the difficulties in developing a storm hydrograph as it is a requirement to have gaged stations for their development.

Many statistical and conceptual models have also been developed to overcome the problems associated with the development of hydrographs, but these methods have not been widely used for they require lots of input parameters such as infiltration, evapotranspiration, soil type, soil wetness, and land-use condition. As a result, a number of synthetic unit hydrograph (SUH) methods have been proposed as means of developing hydrographs for ungauged watersheds by relating the features of a UH to the watershed characteristics. Saf (2006) stated that the most commonly used methods are the Snyder (1938), the Mockus (1957), and U.S. Soil Conservation Service (SCS) (1972) methods. According to Wilkerson (2010), SUH methods are utilized to determine runoff hydrographs for ungauged sites.

There have been very few studies on the comparative performance of various SUH methods. Bhunya et al. (2011) indicated that the traditional methods of SUH derivation such as Snyder, Taylor and Schwarz, and SCS methods do not yield satisfactory results and their application to practical engineering problems is tedious and cumbersome. Safarina et al. (2011) tried to determine the appropriate SUH development method for ungauged watersheds and concluded that the Nakayasu method is better and safe to be used in ungauged watersheds as it resulted in precise shape and hydrograph parameters. Yannopoulos et al. (2006) compared the performance of the SCS SUH, Snyder SUH, and Clark Instant UH techniques and found out that the SCS UH and Clark Instant UH techniques performed better in predicting the peak runoff and time to peak.

One can conclude, from the above review, that the performance of the methods vary from place to place and there is not a single technique that can perform best in every watershed. There has also not been much available information on comparative performance of the SUH techniques. These call for further study on the comparative study of various SUH techniques and determination of best methods for ungauged stations. This research was, therefore, initiated as an additional effort to investigate the performance of various SUH methods for developing UHs for ungauged stations. The main factor influencing the accuracy of SUH is the accuracy of determining the watershed characteristics and parameters to be used by the methods so that the SUH is close to the observed UH (Safarina et al. 2009). It was also the interest of this study to assess how secondary Digital Elevation Model (DEM) data can be used as to generate accurate characteristics required by the models.

The main objective of this study was to develop various SUH and compare their performance in simulating UH for ungauged watersheds by comparing the peak runoff and the time to peak of these models with that of the observed UH values. In addition, the reliability of data generated from online sources in characterizing the study watershed was analyzed. Snyder's, SCS, Mockus, G-W, Nakayasu, and R-V methods of SUHs were used to model the response of the study watershed and these were compared to the observed UH of Guvenc watershed, Turkey.

15.2 Materials and Methods

15.2.1 Description of the Study Area

The study was conducted in Guvenc watershed, Turkey. According to the document prepared by The Village Service General Administration of Turkey (KHGM 2000), the watershed is shared between the Guvenc and Saribeyler villages of Yenimahalle district, Ankara, Sakarya River basin, Turkey (Fig. 15.1). The outlet of the watershed is located at 40° 08′ 00″N and 32° 15′ 45″E. There is one stream gauge at 1053 m above sea level and is taken as the outlet of the study watershed. The characteristics of the study watershed determined by KHGM (2000) are summarized in Table 15.1. Watershed area (*A*), watershed perimeter (*P*), hydraulic length of the watershed (L_h), maximum basin elevation (MaxBE), minimum basin elevation (MinBE), mean basin elevation (MeanBE), mean basin slope (MBS), main stream length (MSL), total stream length (TSL), distance of centroid from outlet (DC), main stream slope (MSS) and drainage density (D_d) are presented in this table.

Based on the land-use map prepared by KHGM (2000), about 46 % of the watershed falls under grass land and about 38 % of the watershed is used as farmland. The soil on about 60 % of the watershed is shallow and susceptible to erosion.

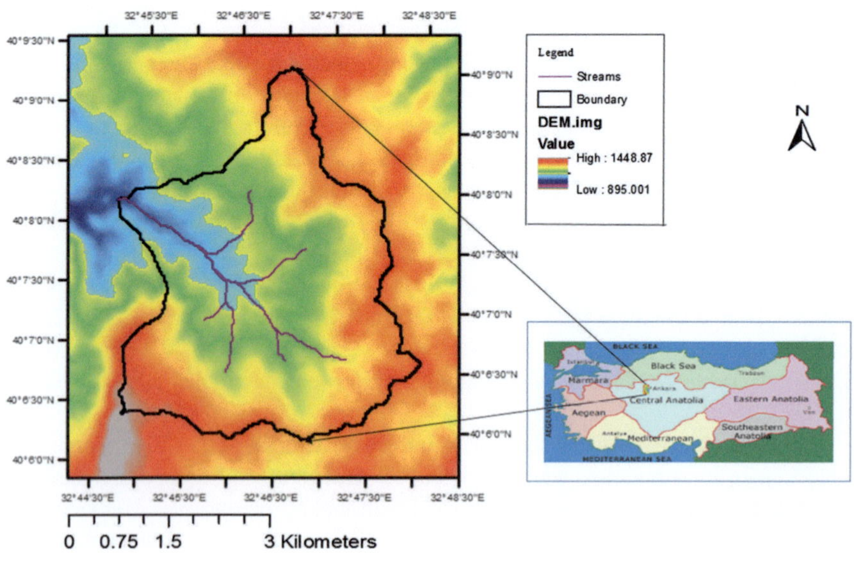

Fig. 15.1 Guvenc watershed

Table 15.1 Values of observed characteristics of the study watershed (*source* KHGM 2000)

Watershed characteristics	Observed values
Area (A), km^2	16.13
Perimeter (P), km	19
Watershed length (L), km	4.85
Watershed width (W), km	3.32
Maximum basin elevation, m	1459
Minimum basin elevation, m	1053
Mean basin elevation, m	1236
Mean basin slope, %	21
Main stream length (L_s), km	5.4
Total stream length, km	35.5
Distance of centroid, km	2.5
Main stream slope, %	4.57
Drainage density (D_d), km/km^2	2.20

15.2.2 Data Collection and Analysis

The outlet of the watershed indicated in Sect. 15.2.1 was located on Google Earth and a polygon was drawn in order to approximately determine the extent of the watershed. The polygon was then used to feed appropriate information to Global Mapper, which was, then, used to download the DEM of the watershed from the internet. The Watershed Modeling System (WMS) software was used to manipulate the downloaded DEM in order to generate appropriate characteristics of the watershed for the development of the SUH as required by the SUH methods. The details of the models used in this study are given in Sect. 15.2.3. The characteristics of the watershed developed by the WMS were compared to the characteristics of the study watershed as determined by KHGM (2000) by undertaking ground survey. The characteristics determined by KHGM (2000) are presented in Table 15.1. This was done to assess the use of secondary DEM data and software like WMS in generating information to be used for generating SUH for ungauged watersheds where data availability is limited. The SUH were developed following the appropriate steps and procedures given for each method. The peak discharge and the time to peak of the UHs were the characteristics considered for the comparison of the models to the corresponding characteristics of the observed UH.

15.2.3 Description of the Models Used

15.2.3.1 Snyder's SUH Method

The first synthetic unit hydrograph was developed by Snyder in 1938. Snyder devised two parameters: a time parameter, C_t, and a peak parameter, $C_{p,}$. These parameters are used to determine the various components of the UH.

The lag time (t_l) of the standard Snyder's SUH is given in hours as a function of C_t and other basin characteristics as follows (Ramirez 2000).

$$t_l = 0.75C_t(L_h \cdot L_c)^{0.3} \tag{15.1}$$

$$t_r = \frac{t_l}{5.5} \tag{15.2}$$

where L_h is the stream length between the outlet and the furthest point of the basin in kilometers and L_c is the distance in kilometers from the outlet to the centroid of the basin area (Bayazit 1995) and t_r is a specific effective duration of standard UH.

The desired SUH is, then, determined by taking the desired excess of rainfall duration (t_o) and the desired lag time (t_{lr}) into consideration. The desired lag time (t_{lr}) is calculated as a function of t_o as given below.

$$t_{lr} = t_l + 0.25(t_o - t_r) \tag{15.3}$$

Then the time to peak (t_p) of the desired UH is given as:

$$t_p = 0.5t_o + t_{lr} \tag{15.4}$$

The peak discharge (Q_p) in m^3/s/cm is then calculated as

$$Q_p = \frac{2.78AC_p}{t_{lr}} \tag{15.5}$$

Here, A stands for the area of the watershed and is given in km^2.

When an ungauged watershed appears to be similar to a gauged watershed, the coefficients C_t and C_p of the gauged watershed can be used in the above equations to derive the required SUH for the ungauged watershed (Ramirez 2000). This approach does not make the Snyder method easily applicable as many developing countries do not have gauged stations from where C_t and C_p could be obtained easily. However, for areas where these parameters cannot be estimated, C_t and C_p obtained from literature could be used. The value of C_p ranges between 0.56 and 0.94 and the value of C_t ranges from 1.8 to 2.2. For a mountainous region, C_t can be taken as 0.4 (Bayazit 1995). For this study, a C_p value of 0.75 (average of the given range) and C_t of 0.4 are taken as the study watershed can be taken as mountainous.

The time to base of the hydrograph (T_b) can be determined as:

$$T_b = 11.11x\frac{A}{Q_p} - 1.5W_{50} - W_{75} \tag{15.6}$$

where W_{50} is the width of the UH at half of Q_p and W_{75} is the width of the UH at 0.75 of Q_p and are given as:

$$W_{50} = \frac{2.14}{\left(\frac{Q_p}{A}\right)^{1.08}} \quad \text{and} \quad W_{75} = \frac{1.22}{\left(\frac{Q_p}{A}\right)^{1.08}} \tag{15.7}$$

Finally, the ordinates of the vertices used to develop the UH are given as (0, 0), $(t_{\rm p} - \frac{1}{3}W_{50}, 0.5Q_p)$, $(t_{\rm p} - \frac{1}{3}W_{75}, 0.75Q_p)$, $(t_{\rm p}, Q_p)$, $(t_{\rm p} + \frac{2}{3}W_{75}, 0.75Q_p)$, $(t_{\rm p} + \frac{2}{3}W_{50}, 0.5Q_p)$ and $(T_{\rm b}, 0)$.

15.2.3.2 Soil Conservation Service (SCS) Unit Hydrograph

The Soil Conservation Service method of U.S. Department of Agriculture (USDA) was derived based on the analysis of large number of recorded UHs for watersheds of various sizes (Roberson et al. 1998). In order to use this method for synthesizing UH from a certain excess rainfall, it is necessary to determine both the peak flow rate (Q_p) and time to peak ($t_{\rm p}$). The time to peak is determined using the following equation.

$$t_{\rm p} = \frac{t_o}{2} + t_l \tag{15.8}$$

where t_o is the desired excess rainfall duration (which is 1 h in this research); and t_l is the lag time from the centroid of excess rainfall to peak discharge in hours.

According to Viessman and Lewis (2002), t_l (in hours) is determined from the watershed characteristics using curve number (CN) procedure as:

$$t_l = \frac{L_{\rm h}^{0.8}\left(\frac{1000}{\rm CN} - 9\right)^{0.7}}{1900S^{0.5}} \tag{15.9}$$

where $L_{\rm h}$ is hydraulic length of watershed, i.e., length of the longest drainage path (feet); S is mean catchment slope (%); and CN is curve number determined based of characteristics of the watershed.

The peak discharge (m^3/s) for one cm of excess rainfall is then determined as:

$$Q_p = 2.08\left(\frac{A}{t_{\rm p}}\right) \tag{15.10}$$

where A is the area of the watershed (km^2) and $t_{\rm p}$ is given in hours. It should be noted that the time to base, $T_{\rm b}$, of the triangular UH is 2.67 times the time to peak or:

$$T_{\rm b} = 2.67t_{\rm p} \tag{15.11}$$

The triangular SUH with this method is then developed by taking the time to peak, the peak discharge and the time base values.

15.2.3.3 Mockus Method

This method is also another way of developing SUH for ungauged stations. The shape of the SUH in this method is accepted to be a triangular shape and it is preferred to practical applications because of this simplicity. The Kirpich's formula for calculating time of concentration given in hours in Mockus method is given as (Saf 2006):

$$T_c = 0.00032 * \frac{L_h^{0.77}}{S_h^{0.385}} \tag{15.12}$$

where L_h is hydraulic length of the watershed in meters and S_h is the harmonic slope of the channel.

S_h is determined as:

$$S_h = \left(\frac{P}{\sum_{i=1}^{P} \frac{1}{\sqrt{S_i}}} \right)^2 \tag{15.13}$$

where P is the number of segments obtained by dividing the watershed's hydraulic length defined above and S_i is the channel segment slope calculated as the ratio of elevation difference of the two end points of the segment to the length of the segment after dividing the main channel into P number of segments or:

$$S_i = \frac{h}{l} \tag{15.14}$$

where h is the elevation difference of the two end points of a segment and l is the length of the segment.

In this study, P was taken to be 10. Other parameters necessary for the development of the SUH in this method are the t_o, t_p, T_b, and Q_p. These parameters are defined and determined as follows.

$$t_o = 2\sqrt{T_c} \tag{15.15}$$

$$t_p = 0.5t_o + 0.6T_c \tag{15.16}$$

$$T_b = 2.67t_p \tag{15.17}$$

$$Q_p = \frac{0.208 * A * h_a}{t_p} \tag{15.18}$$

where t_o is the excess rainfall duration resulting in the corresponding time of concentration, T_c. t_p is time to peak, T_b is the time to base and Q_p is the peak discharge, A is the drainage are in km^2, and h_a is the depth of unit rainfall (given as 1 mm) corresponding to rainfall duration t_o.

If the value of t_o is not a whole number, the larger whole number next to the value of the calculated t_o will be taken in the calculation as long as the time of concentration is larger than 1 h. If not, t_o will be taken as T_c. In this study, however, t_o is taken as 1 h as the desire is to determine a 1 h UH.

15.2.3.4 Nakayasu Synthetic Unit Hydrograph

Nakayasu (as cited in Soemarto 1987) provided a SUH based on his research on rivers in Japan. The variables which should be obtained in order to develop UH using this technique are time to peak (t_p), peak discharge (Q_p), rising limb curve (Q_a), and decreasing limb curve (Q_d).

t_p is a function of the time of concentration (T_c) and the duration of the effective rainfall (t_o, taken as 1 h in this study, but as $0.75T_c$, otherwise), and is given by Nakayasu as:

$$t_p = 0.8t_o + T_c \tag{15.19}$$

where, if the length of the main stream <15 km,

$$T_c = 0.21L_s^{0.7} \tag{15.20}$$

or, if the length of the main stream >15 km,

$$T_c = 0.4 + 0.058L_s \tag{15.21}$$

The length of the main stream (L_s) for the study watershed was found to be 4.66 km and hence Eq. 15.20 was used.

The peak discharge, the rising limb curve and the decreasing limb curve are calculated using the following equations.

$$Q_p = \frac{AR_o}{3.6(0.3t_p + T_{0.3})} \tag{15.22}$$

$$Q_a = Q_p\left(\frac{t}{t_p}\right)^{2.4} \tag{15.23}$$

$$Q_d = Q_p * 0.3^{\left(\frac{t-t_p}{T_{0.3}}\right)} \tag{15.24}$$

where Q_p is peak discharge (m^3/s); A is area (km^2); R_o is unit rainfall (1 mm); t_p is time to peak (hour); and $T_{0.3}$ is time needed for decreasing peak discharge until 30 % of peak discharge (hour)

$T_{0.3}$ (in hours) is determined as:

$$T_{0.3} = \alpha t_g \tag{15.25}$$

where α is a constant normally taken as 2.

15.2.3.5 The R-V Geomorphological Instantaneous Unit Hydrograph (GUIH)

An extension of the UH theory is the concept of the instantaneous unit hydrograph (IUH). The IUH is the hydrograph of runoff from the instantaneous application of unit effective rain on a watershed. It is obtained when the duration of the excess rainfall is infinitesimally small (Bayazit 1995). It is a unique demonstration of a particular watershed's response to rain, independent of duration, just as the UH is its response to rain of a particular duration. Since it is not time dependent, the IUH is thus a graphical expression of the integration of all the watershed parameters of length, shape, slope condition, etc. that control such a response (Wilson 1990). IUH is the transfer function (impulse response function) of a system (Bayazit 1995) and it can be used to drive UH by routing runoff at the outlet of a watershed. The quantity and time distribution of runoff at an outlet of a watershed depend on topography of the watershed and the transmission channels (Sudharsanan et al. 2010).

GIUH, which uses geomorphological parameters, is one of the attempts to develop an IUH by relating the watershed response to the characteristics of drainage network to simulate runoff. The concept of GUIH was first introduced by Rodreguez-Iturbe and Valdes in 1979 where the peak discharge and the time to the peak were related to the morphological characteristics of the basin and a dynamic velocity based on statistical analysis. They used the geomorphological structure of a fluvial basin to interpret the runoff hydrograph as a travel time distribution, which is controlled by the hillslope and the channel network response (Mehlhorn 1998).

Gupta et al. (1980) have given a simpler and more general formulation of Rodriguez-Iturbe and Valdes's model (here called the R-V model) of drop travel times in the basin. The states of the process are defined to be $c_{i,}$ for a hillslope region draining to a stream of order i, for a stream of order i. A drop falling randomly on a catchment of order Ω can follow a finite number of paths to the outlet. The number of possible paths is determined as $2^{\Omega-1}$. For instance, for a third order basin, there are 4 possible paths given by:

Path s_1 $C_1 \rightarrow C_2 \rightarrow C_3 \rightarrow C_4$
Path s_2 $C_1 \rightarrow C_3 \rightarrow C_4$
Path s_3 $C_2 \rightarrow C_3 \rightarrow C_4$
Path s_4 $C_3 \rightarrow C_4$

If a certain stream network has a certain number of links in each order of stream, then the initial probabilities $\left(\prod_{x_i}\right)$ of these orders are determined by dividing the number of links in each order to the total number of links in the stream network.

The channel merger probability ($P_{x,y}$) from a certain order (x) of streams to another stream order (y) is taken as the ratio of the number of streams of order x draining into streams of order y to the total number of streams in order x.

The probability of following a given path, $P(S_i)$, is then given as a function of the merger probabilities as:

$$P(S_i) = \prod_{x_1} \cdot p_{x_1 x_2} \cdots p_{x_{k-1}, x_k} \tag{15.26}$$

where $\prod_{x_1}$ is the proportion of the basin area draining into the geomorphic state x_1 and p_{x_i, x_j} is the proportion of the Strahler streams in the state x_i draining into the streams in state x_j.

The cumulative density function of the travel time to the outlet, for a basin of order Ω is then where $S = \{s1, s2, \ldots\}$,

$$P(T_b < t) = \sum_{i=1}^{n} P(T_{s_i} < t) P(S_i) \tag{15.27}$$

where T_{s_i} is the travel time in path S_i, $P(S_i)$ is the probability of following this path, and $N = 1 + \Omega$ is the state of the outlet.

The travel time, T_S, is the sum of travel times in the different states which constitute path s. Its probability density function (pdf) is then the convolution of the pdfs of the individual states. For example, in the case of the third order catchment, T_S is:

$$T_{S_1} = T_{C_1} + T_{C_2} + T_{C_3} + T_{C_4} \tag{15.28}$$

where T_{C_1}, T_{C_2}, T_{C_3}, and T_{C_4}, are the times spent in states C_1, C_2, C_3, and C_4, respectively.

The corresponding pdfs are then related by:

$$f(t_{S_1}) = f(t_{C_1}) * F(t_{C_2}) * f(t_{C_3}) * f(t_{C_4}) \tag{15.29}$$

where $*$ represents the convolution operation and $f(t_{S_1})$ represents the pdf of T_{s_1}.

The GIUH is determined as a true response, which is given as:

$$h(t) = \sum_{i=1}^{N} f(T_{s_i}) P(S_i) \tag{15.30}$$

The values of the true response are drawn against a time increment of 1 unit in determining the ordinates of the GIUH. It is assumed, in this approach, that the ratio of the constant velocity, v, of the flow particles to some length, l, scale is one. This means that the time increment given as $\frac{l}{v} = 1$. Therefore, for the ordinates of 1-h GIUH, the time increment should be taken as 1 h. The ordinates of GIUH for the study watershed were, therefore, calculated by multiplying the ordinates of the true response by the area of the watershed and the unit excess rainfall and divide it by 3600 s to have the discharge (Q) in units of m^3/s. According to Sudharsanan et al. (2010), UH of any duration can be developed form the GIUH, and a 1-h UH is developed from the already generated GIUH. The ordinate of developed GIUH, first of all, is lagged by one hour. Then, the average of ordinates of GIUH and lagged GIUH gives the ordinates for 1-h UH.

15.2.3.6 G-W Geomorphological Instantaneous Unit Hydrograph (GUIH)

An attempt meant to simplify the R-V method was recommended by Gupta and Waymire in 1983.

In the G-W approach, the rainfall input injected instantaneously and uniformly at all the nodes of the stream network is routed to the basin outlet. According to Gupta and Waymire (1983), if a large number of particles are introduced instantaneously into a channel network (basin) and if the particles do not interact, then the proportion of particles arriving at the outlet of this network in time, i.e., its IUH, is equal to the pdf of some random time T for which a particle stays in the network. It is assumed that the watershed is a time variant but linear system.

In the study undertaken by Gupta et al. (1980), from the particles injected at time u, the proportion of particles arriving at the outlet at some time $t > u$ is simply given as:

$$i(u)h(t-u)\mathrm{d}u \tag{15.31}$$

where $i(u)$ is excess rainfall (cm) and $h(t)$ is the IUH (h^{-1}).

Since particles move independently of one another and since the total flow at time t consists of the contribution from all the particles that were injected between times 0 and t, runoff can be estimated by the convolution integral given below.

$$Q(t) = \int_0^t i(u)h(t-u)\mathrm{d}u \tag{15.32}$$

where $Q(t)$ is direct runoff at time t (cm/h). This equation gives the well-known linear convolution transformation between the input and the output.

It is also assumed in this approach that the links in the channel network have the same length, l, and the particles injected simultaneously at sources and junctions travel with constant velocity, v. With these assumptions, after each unit time $\left(\frac{l}{v} = 1\right)$, the particles level $i + 1$ to level i. If $i = 0$ denotes the outlet of the basin, then clearly, the particles injected at level i take exactly i time units to reach the outlet.

Gupta and Waymire (1983) also indicated that that the IUH is similar to the width function when pure translation with constant velocity is considered. The width function is a histogram showing the number of intersections ($N(x_i)$) at a certain level of time increment versus the distance (x) from the outlet and given as time increment in the units of 1 from the outlet. It is determined after the channel network is generated and is divided into a number of levels (L) by drawing horizontal lines spaced uniformly to intersect the stream network. The width function (WF) of the basin is determined as:

$$\mathrm{WF} = \frac{N(x_i)}{\sum_{i=1}^{n} N(x_i)} \tag{15.33}$$

where $N(x_i)$ is the number of intersections at the ith level of time increment.

The same procedure given in Sect. 15.2.3.5 is followed when generating a UH of any duration from the already generated GIUH.

15.3 Result and Discussion

15.3.1 Reliability of Data Generated from Online Sources in Characterizing the Study Watershed

The watershed characteristics used to describe the research watershed by KHGM (2000) were determined by using the DEM downloaded from the internet. The values of these characteristics generated by manipulating the DEM by WMS are presented in Table 15.2 under the column entitled 'Estimated Values' side by side to the observed values.

As can be seen from Table 15.2, the estimated and the observed values of A, P, MinBE and MeanBE are very close to each other. However, the MaxBE, MSL, TSL DC, MSS, and D_d values of the estimated data were found to be different from the observed values. This could be because the DEM obtained from the internet is of poor resolution. It can be inferred that the WMS was not able to identify smaller streams and source points as we move further from the outlet of the watershed to the watershed divide. These differences are not expected to have lots of impacts on the results of the study as basin mean slope, main stream length, distance of the outlet from the centroid of the watershed, and stream slope are the characteristics commonly used in many of the methods for developing SUH. Characteristics given in Table 15.2 were used to determine geomorphological characteristics of the watershed as required by each method.

Table 15.2 Values of observed and estimated characteristics of the study watershed

Watershed characteristics	Unit	Observed values	Estimated values
A	km^2	16.125	16.11
P	km	19	19
L	km	4.85	5.29
[a]L_h	km	–	5.98
MaxBE	m	1459	1409
MinBE	m	1053	1053
MeanBE	m	1236	1232
MBS	%	21	14.44
MSL	km	5.4	4.66
TSL	km	35.5	10.43
DC	km	2.5	2.8
MSS	%	4.57	3.94
D_d	km/km^2	2.20	0.65

[a]The blank cell in the table show values not determined

15.3.2 Observed Unit Hydrograph

Figure 15.2 depicts the 1 h observed unit hydrograph of the study watershed as a result of 1 mm of excess rainfall. This figure was developed using streamflow data obtained from KHGM (2000).

As can be seen from the figure, the time to peak was found to be 1.1 h and the peak flow was 4.26 $m^3/s/mm$. The base of the UH was found to be 4.5 h. This UH was used as a control to evaluate the performance of the other synthetic hydrographs developed by the the methods.

15.3.3 Snyder's Unit Hydrograph

The Snyder's UH developed using C_t value of 0.4 and C_p value of 0.75 (these values were taken from the literature as the study watershed is mountainous) is given in Fig. 15.3. The time to peak and the time to base of the desired UH were found to be 1.42 and 3.07 h, respectively. In addition, the peak runoff as a result of 1 mm excess rainfall was found to be 3.66 $m^3/s/mm$.

15.3.4 SCS Unit Hydrograph

The CN required in the SCS method for the Guvenc watershed was determined as a weighted average of CN values for various landuses and depth of soils as given by

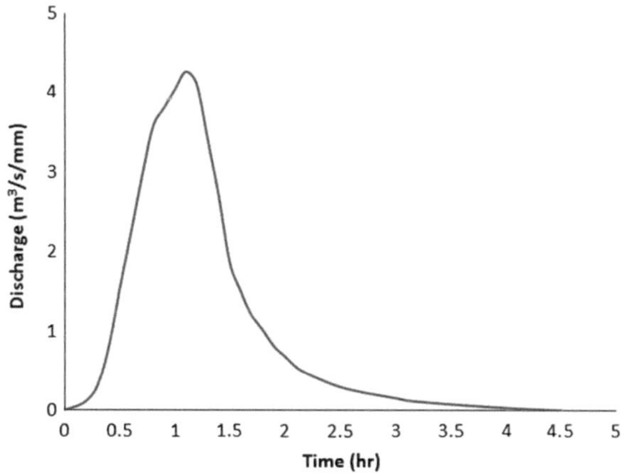

Fig. 15.2 Observed 1 h unit hydrograph of the study watershed

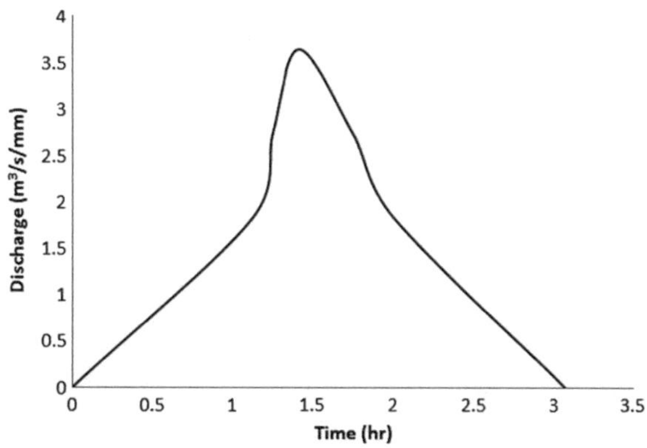

Fig. 15.3 1 h Snyder's unit hydrograph based on estimated C_t and C_p values

KHGM (2000), and was found to be 74.5. The CN was then used to determine the lag time, t_l, which was found to be 0.28 h. Based on the lag time and taking 1 h excess rainfall duration, the time to peak, the time base, and the peak discharge were found to be 0.77 h, 2.07 h and 4.31 m^3/s/mm, respectively. Figure 15.4 given below shows the SUH determined using SCS method.

15.3.5 Mockus Synthetic Unit Hydrograph

The procedures and steps given in Sect. 15.2.3.3 were followed to calculate the harmonic slope (S_h) after dividing the main stream into 10 segments with equal length, l, of 598.6 m. The elevation difference, h, between the two end points of each segment is given in Table 15.3. The S_i values in Table 15.3 were used to determine S_h and was found to be 3.82 %. The time of concentration (T_c) was then determined using S_h and it was found 0.91 h.

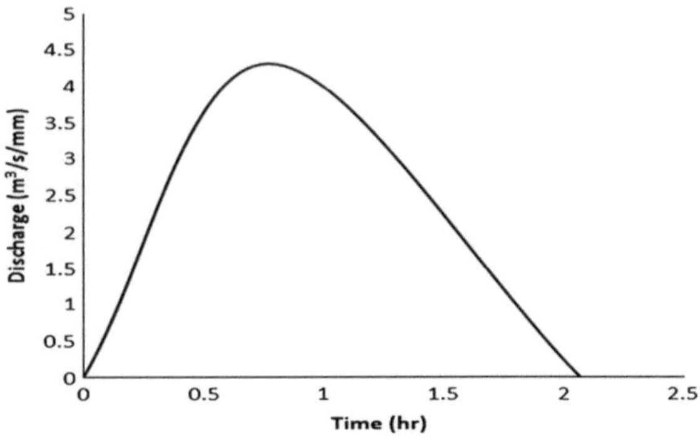

Fig. 15.4 The SUH determined using SCS method

T_c and t_o, which is taken as 1 h in this study, were used to calculate the time to peak and was found to be 1.05 h. The base time and the peak discharge, Q_p, for 1 mm excess rainfall were then calculated and found to be 2.8 h and 3.2 m^3/s/mm. The UH was drawn using Excel and is presented in Fig. 15.5.

15.3.6 Nakayasu SUH

The SUH developed using Nakayasu's method is given in Fig. 15.6. In this method, the peak discharge and the time to peak were found to be 2.7 m^3/s/mm 1.47 h, respectively.

15.3.7 Rodriguez-Valdez Geomorphological Instantaneous Unit Hydrograph

The true response of the watershed was determined from the cumulative density function of the travel time to the outlet, which in term was determined as a function

Table 15.3 S_i values for the determination of harmonic slope (S_h)

h (m)	l (m)	$S_i = \frac{h}{l}$	$\sqrt{S_i}$
10	598.6	0.017	0.129
20	598.6	0.033	0.183
17	598.6	0.028	0.169
11	598.6	0.018	0.136
19	598.6	0.032	0.178
39	598.6	0.065	0.255
37	598.6	0.062	0.249
38	598.6	0.063	0.252
58	598.6	0.097	0.311
39	598.6	0.065	0.255

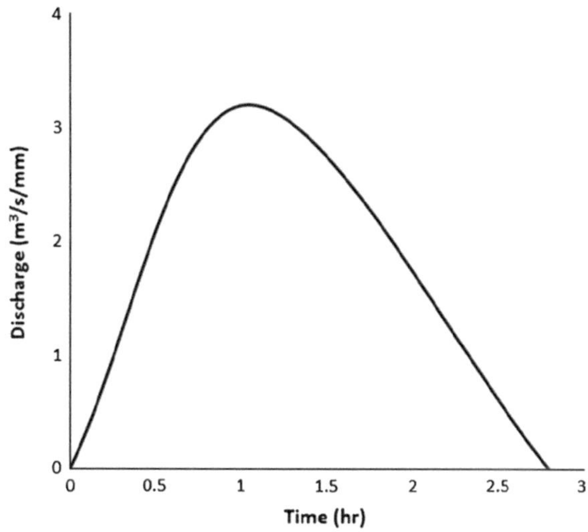

Fig. 15.5 Mockus 1 h synthetic unit hydrograph

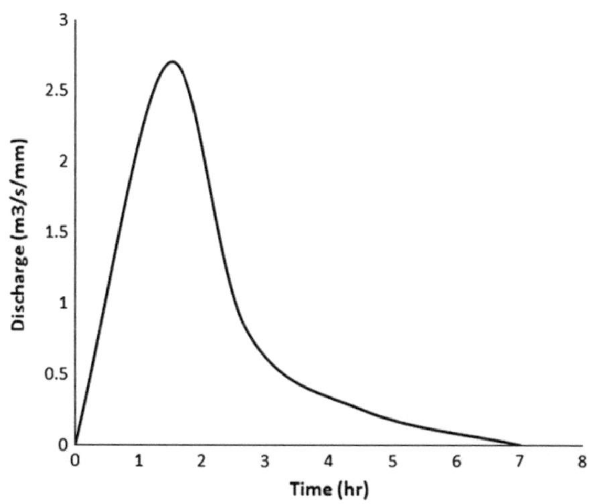

Fig. 15.6 Nakayasu synthetic unit hydrograph

of the initial probabilities, channel merger probabilities, and path probabilities. The cumulative density function of the travel time to the outlet and the true response are presented in Table 15.4.

The true response and the time increment were then used to define the ordinate of the IUH. Figure 15.7 was developed as a result.

The GIUH ordinates were then used to generate the ordinates for the UH by following the procedure given in the methodology part. The UH is depicted in Fig. 15.8.

Table 15.4 The cumulative density function and true response for the study watershed

Time increment	0	1	2	3	4	5	6	7
*f*1 * *f*2 * *f*3	0	0	0	0.3335	0.5	0.1665	0	0
*f*1 * *f*3	0	0	0.5	0.5	0	0	0	0
*f*2 * *f*3	0	0	0.3335	0.5	0.1665	0	0	0
*f*3	0	0.5	0.5	0	0	0	0	0
Response	0	0.091	0.227	0.333	0.273	0.076	0	0

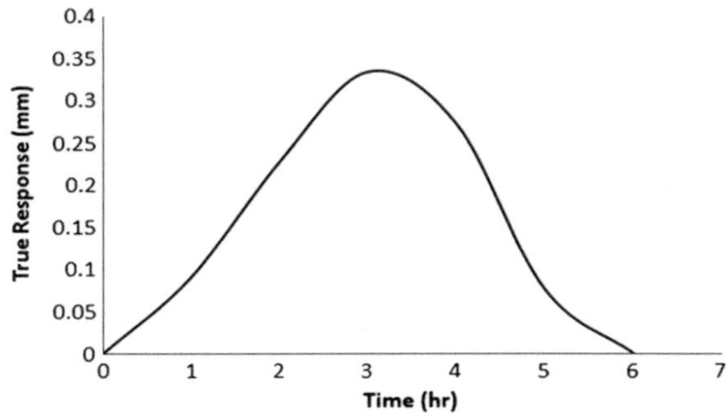

Fig. 15.7 The R-V geomorphologic instantaneous unit hydrograph

15.3.8 Gupta-Waymire Geomorphological Instantaneous Unit Hydrograph

The stream network was drawn as presented in Fig. 15.9 and six levels were used as time of increments. Then, following the procedures explained in Sect. 15.2.3.4, the number of intersections in each time increment were counted and these values are given in Table 15.5.

The width function was then calculated using Eq. 15.31 by taking the number of intersections at each level if time increment and dividing them by the sum of intersection, which was found to be 14 here. These values are presented in Table 15.5.

The coordinates of the 1 h IUH were calculated from the width function by considering 1 h time increment and 1 mm of excess rainfall for the study watershed. These coordinates were then lagged by 1 h and averaged to determine the ordinates of the 1 h geomorphologic UH as presented in the 5th column of Table 15.5. The 1 h UH is depicted in Fig. 15.10. It can be seen from Table 15.5 that the peak discharge is equal to 1.28 m^3/s/mm and the time to peak is 4 h.

15.3.9 Comparison of the Performance of Methods Used

The time to peak and the peak discharge values of the observed UH and those determined using the methods used in this study are summarizes in Table 15.6. The

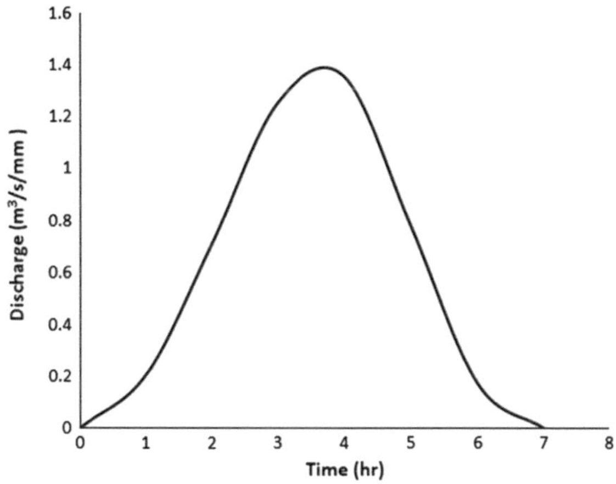

Fig. 15.8 The R-V geomorphologic unit hydrograph

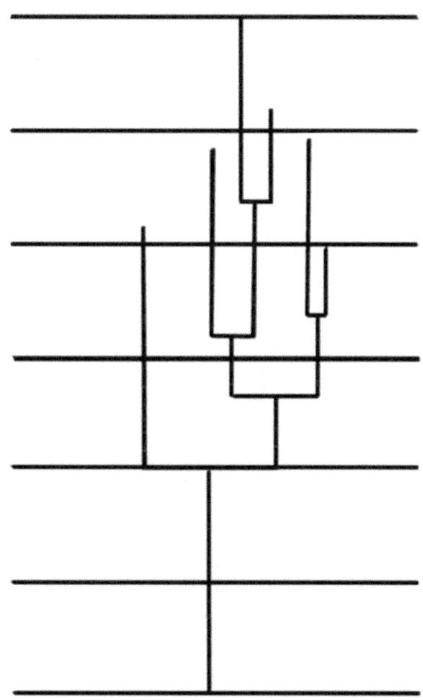

Fig. 15.9 The stream network and time increments used in calculating the width function

peak discharges presented in the table have uniform units and are results of 1 mm excess rainfall with 1 h duration. All the UHs are drawn in the same scale and are presented in Fig. 15.11 for better observation and comparison.

From the values given in Table 15.6, the SCS method is found to be the best method for estimating the peak discharge of the study watershed. All the other approaches underestimated the peak discharge, the lowest being the G-W method. Nevertheless, 87 and 76 % of the observed peak discharge were estimated by Snyder's and Mockus methods, respectively. It can be seen from Fig. 15.11 that the

Table 15.5 Width function and coordinates of the 1 h GIUH and 1 h UH for the study watershed

Time increment	Number of intersections	Width function	1 h GIUH	1 h UH
0	0	0	0	0
1	1	0.071	0.32	0.16
2	2	0.143	0.64	0.48
3	3	0.214	0.96	0.80
4	5	0.357	1.60	1.28
5	2	0.143	0.64	1.12
6	1	0.071	0.32	0.48
7	0	0	0	0.16
8	0	0	0	0

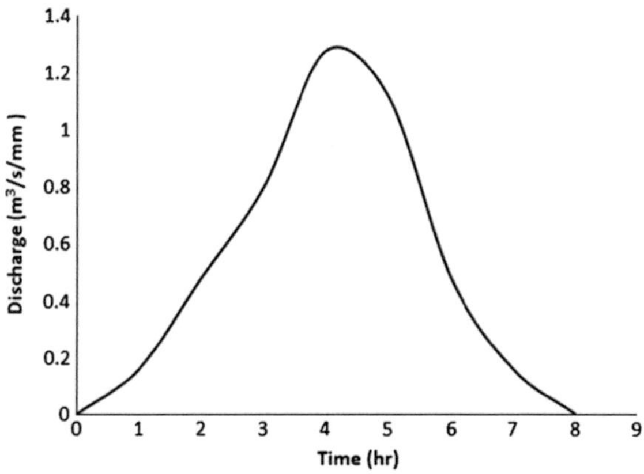

Fig. 15.10 Gupta-Waymire geomorphological unit hydrograph

Table 15.6 The time to peak and the peak discharge values of the observed and modeled UHs

Parameter	Unit hydrograph						
	Observed	Snyder's	SCS	Mockus	Nakayasu	R-V	G-W
t_p (h)	1.1	1.42	0.77	1.05	1.47	4	4
Q_p (m^3/s/mm)	4.23	3.66	4.31	3.2	2.7	1.36	1.28
Ratio to observed	1	0.87	1.02	0.76	0.64	0.32	0.30

Mockus method was the only method found to simulate the time to peak. The R-V and the G-W methods were found to be the poorest in simulating all of the elements of the UH and therefore, they may not be suitable for developing UHs for small watersheds. However, further investigation should be undertaken to reach at a final conclusion. The Nakayasu method simulated the shape of the observed UH better than the other methods.

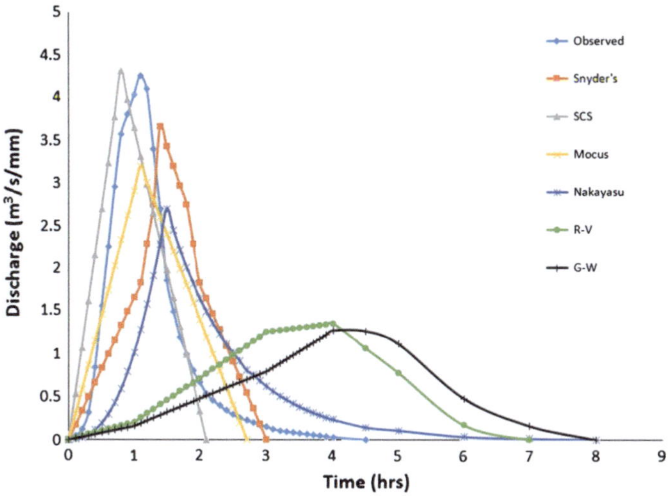

Fig. 15.11 Observed and estimated UHs for Guvenc watershed

It is noticed that there is a big difference between the calculated C_t value for the watershed by KHGM and the one obtained from literature. The C_t value determined by KHGM which is 0.166 looks significantly very small when compared to the range it was supposed to fall in (0.4–8), 0.4 being the lowest value obtained from literature for mountainous regions. This calls for an investigation of this result for the study watershed. In addition, the unsatisfactory results of the estimated values of peak discharge by the other methods could be because these models are only based on stream length parameters as opposed to the SCS methods, which is based on landuse and soil depth in addition to stream length. Moreover, all of the models used in this study were developed in areas which have different rainfall characteristics than what we have here in Turkey and the surrounding affecting values of coefficients and constants in these models and the overall performance of the models. It can also be observed from Fig. 15.11 that, unlike the others, the R-V and G-W UHs are skewed to the left and have very wide time to base.

15.4 Conclusion and Recommendation

It was observed, in this study, that watershed characterization based on secondary data is not time taking and major characteristics like main stream length, distance from centroid of the watershed, and slopes that are used by many SUH methods can be satisfactorily generated from secondary data sources and can be used for UH derivation. The results of this study show that the SCS model performed best in simulating peak discharge. The other methods did not perform well enough and, therefore, it is very difficult to apply them in practical engineering problems. The following recommendations are drawn for the future.

- The performance of the methods should be investigated under various rainfall pattern regions and geographical location so that tangible method for deriving UH for ungauged watershed could be recommended for Turkey.
- Some of the models (Snyder's, Mockus and Nakayasu) are very simple and do not require many parameters, though they did not perform satisfactorily in this study. Therefore, undertaking research in order to improve their performance by modifying them could be of great advantage.
- The shape and the time to base of the geomorphological UHs were not found to be similar to the other SUHs and therefore, the R-V and the G-W methods may not be suitable for developing UH for small watersheds.

References

Agudelo JI (2001) The economic valuation of water: principles and methods. UNESCO-IHE, Delft

Bhunya PK, Panda SN, Geol MK (2011) Synthetic unit hydrograph methods: a critical review. Open J Hydrol 5:1–8

Bayazit M (1988) Hidrolojik Modeller. I.T.Ü. Rektörlügü, Istanbul

Bayazit M (1995) Hydrology. I.T.Ü. Rektörlügü, Istanbul

Granato GE (2012) Estimating basin lagtime and hydrograph-timing indexes used to characterize stormflows for runoff-quality analysis: U.S. geological survey scientific investigations report 2012–5110. http://pubs.usgs.gov/sir/2012/5110/. Accessed 18 Oct 2013

Gupta VK, Wymire E (1983) On the formulation of analytical approach to hydrologic response and similarity at the basin scale. J Hydrol 65:95–123

Gupta VK, Wymire E, Wang CT (1980) A representation of an instantaneous unit hydrograph from geomorphology. Water Resour Res 16(5):855–862

Harmel RD, King KW, Haggard BE, Wren DG, Sheridan JM (2006) Practical guidance for discharge and water quality data collection on small watersheds. Trans ASABE 49(4):937–948

KHGM (Koy Hizmetleri Genel Mudurlugu) (2000) Hydrological information research watersheds. Ankara

Maidment DR (1993) Developing a spatially distributed hydrograph by using GIS. In: HydroGIS93: applications of GIS in hydrology and water resources. Proceedings of the Vienna conference, April 1993. IAHS Publication No. 211

Mehlhorn J, Armbruster F, Uhlenbrook S, Leibundgut C (1998) Determination of the geomorphological instantaneous unit hydrograph using tracer experiments in a headwater basin. Hydrology, water resources and ecology in headwaters. In: Proceedings of the HeadWater'98 conference held at Meran/Merano, Italy, April 1998. 1AHS Publication No. 248

Mockus V (1957) Use of storm and watershed characteristics in synthetic hydrograph analysis and application. Paper presented at American Geophysical Union, Southwest Region Meeting, Sacramento, Calif

Ramirez JA (2000) Prediction and modeling of flood hydrology and hydraulics. In: Wohl E (ed) Inland flood hazards: human, riparian and aquatic communities. Cambridge University Press, Cambridge

Roberson JA, Cassidy JJ, Chaudrhry MH (1998) Hydraulic engineering, 2nd edn. Wiley, New York

Rodriguez-Iturbe I, Valdes JB (1979) The morphologic structure of hydrologic response. Water Resour Res 15(6):1409–1420

Safarina AB, Salim HT, Hadihardaja IK, dan Syahril BK (2009) Validity analysis of synthetic unit hydrograph methods for accuracy flood discharge estimation. In: Proceeding of the 1st international conference on sustainable infrastructure and built environment in developing countries, ITB Bandung

Safarinal AB, Salim HT, Hadihardaja IK, dan Syahril BK (2011) Clusterization of synthetic unit hydrograph methods based on watershed characteristics. Int J Civ Environ Eng 11(6):76–85

Saf B (2006) Role of synthetic storms on peak flow estimation. In: Proceedings of the 2006 IASME/WSEAS international conference on water resources, hydraulics and hydrology, Chalkida, Greece, 11–13 May 2006, pp. 13–20. http://www.wseas.us/e-library/conferences/2006evia/papers/516-180.pdf. Accessed 19 Nov 2013

Snyder FF (1938) Synthetic unit-graphs. Trans Am Geophys Union 19:447–454

Soemarto CD (1987) Engineering Hydrology. Usaha Nasional, Surabaya, Indonesia

Sudharsanan R, Krishnaveni M, Karunakaran K (2010) Derivation of instantaneous unit hydrograph for a sub-basin using linear geomorphological model and geographic information systems. J Spat Hydrol 10(1):30–40

United Nations (1997) Comprehensive assessment of the freshwater resources of the world: report of the secretary-general. United Nations Department for Policy Coordination and Sustainable Development (DPCSD)

Viessman W, Lewis GL (2002) Introduction to hydrology, 5th edn. Prentice-Hall, Upper Saddle River

Weaver JC (2003) Methods for estimating peak discharges and unit hydrographs for streams in the city of Charlotte and Mecklenburg County, North Carolina. http://pubs.usgs.gov/wri/wri034108/pdf/report.pdf. Accessed 03 Nov 2013

Wilkerson JL (2010) Determination of unit hydrograph parameters for Indiana watersheds. Erosion and water pollution control. Publication No. FHWA/IN/JTRP-2010/5, SPR-3226. http://docs.lib.purdue.edu/cgi/viewcontent.cgi?article=2614&context=jtrp. Accessed on 25 Oct 2013

Wilson EM (1990) Engineering hydrology, 4th edn. Macmillan Education Ltd., Houndmills

Yannopoulos S, Katsi A, Papamichail D (2006) Comparative analysis of synthetic unit hydrograph methods using WMS rainfall—runoff process simulation. http://www.srcosmos.gr/srcosmos/showpub.aspx?aa=8329. Accessed 17 Nov 2013

Chapter 16
Flash Floods Modelling for Wadi System: Challenges and Trends

Mohamed Saber and Emad Habib

Abstract In arid regions, flash floods are the most devastating hazards especially in terms of loss of human life and infrastructures. These floods occur due to heavy rainfall for short duration, steep slope, and impervious layer as well as climate change. The combination of these factors makes it more severe and devastating. The very short warning time is the main problem of flash floods that typically leave only a few hours for civil protection services to act. Various modelling techniques and tools have been widely developed for humid area applications, however, arid and semi-arid regions have received little attention even though they suffer from lack of water resources and the threat of flash floods. Hence, there is a desperate need for developing comprehensive, innovative, and powerful physical-based models, explicitly based on the best available understanding of the physics of hydrological processes of Wadi system in such regions. Consequently, these approaches could be applied as effective and sustainable management tools for water resources and flash floods modelling. In this chapter, Wadi systems are introduced; history of flash flood and impacts are documented and hydrologic models or approaches are reviewed. Recommendations are offered for possible prediction and management of flash floods under climate change.

Keywords Flash floods · Hydrological modelling · Wadi system · Flash floods forecasting · Water resources management · Arid region hydrology

---

M. Saber (✉)
Geology Department, Faculty of Science, Assiut University, Assiut 71516, Egypt
e-mail: msaber@louisiana.edu; msaber_75@yahoo.com

M. Saber · E. Habib
Department of Civil Engineering, The University of Louisiana at Lafayette, Lafayette 70504, Louisiana, USA
e-mail: habib@louisiana.edu

© Springer International Publishing Switzerland 2016

A.M. Melesse and W. Abtew (eds.), *Landscape Dynamics, Soils and Hydrological Processes in Varied Climates*, Springer Geography,
DOI 10.1007/978-3-319-18787-7_16

16.1 Introduction

The term 'flash flood' identifies a rapid hydrological response with water levels and flow rate reaching a peak within less than 1 h to a few hours after the onset of the generating rain event (Creutin and Borga 2003; Collier 2007; Younis et al. 2008). Flash flood is a typical natural phenomenon which occurs in arid or semi-arid regions within a short duration and rapidly rising water flow level reaching maximum peak flow in a few hours. Usually, the cause is intense rainfall event or dam failure resulting in a great damage to human life and severe structural damages. One-third of the world's land surface is classified as arid or semi-arid (Şen 2008). As stated by Pilgrim et al. (1988) and according to UNESCO (1977) classification, nearly half of the countries in the world are suffering from aridity problems. Also, forty percent of Africa's population lives in arid, semi-arid, and dry sub-humid areas (WMO 2001). Flash floods are infrequent, but extremely damaging and represent a threat to human life and infrastructure, especially with the effect of regional climatic change all over the world. The hydrological conditions in these areas are extreme and highly variable, where flash floods from a single large storm can exceed the total runoff from yearly hydrographs. On the other hand, water resources are very limited, yet the availability of a sufficient supply of high-quality water is a major requirement for social, industrial, agricultural and economic development of a Wadi system. Arid regions are undergoing severe and increasing water stress due to expanding populations, increasing per capita water use, and limited water resources. Therefore, an appropriate utilization and strong management of renewable sources of water in Wadis is the only optimal choice to overcome water deficiency problems. To achieve that, understanding the characteristics of Wadi systems and various hydrological processes that affect flash flooding is important and good management support tools are desperately required for the available water resources in such areas.

Arid and semi-arid regions are challenged by three main problems as shown in Fig. 16.1. Hydrological modelling in arid and semi-arid regions is limited due to lack of monitoring network resulting in data deficiency and low quality. There is shortage of water resources. Subsurface and surface water are the main water resources for human use; but in arid regions, the available surface water is limited due to the paucity and high variability of rainfall events. Additionally, subsurface water is very important but water quality problem and aquifers depletion due to the mismanagement is challenging. Most arid regions suffer from aridity conditions most of the year except during infrequent rainfall events which culminate in flash floods that are highly devastating for human life and their properties. All these problems present practical difficulties in arid areas for water resources planning, development, and water management. Developing applicable rainfall–runoff physical-based hydrological models for flash floods modelling is a challenge.

With regard to the surface and subsurface water interactions in arid regions, Wheater et al. (1997), Telvari et al. (1998) stated that surface water and groundwater interactions depend strongly on the local characteristics of underlying alluvium and the extent of their connection to or isolation from other aquifer systems. Transmission

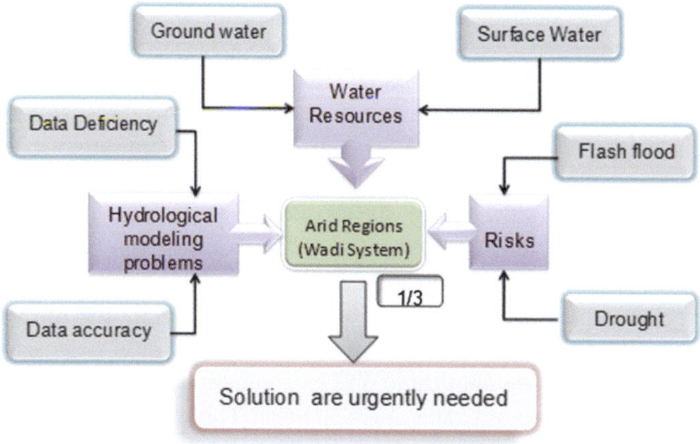

Fig. 16.1 Chart showing main problems of arid regions (Saber 2010)

losses in semiarid watersheds raise important distinctions about the spatial and temporal nature of surface water and groundwater interactions compared to humid basins. Transmission losses can be limited to brief periods during runoff events and to specific areas associated with the runoff production and downstream routing (Boughton and Stone 1985). Walters (1990), Jordan (1977) provided the evidence that the rate of loss is linearly related to the volume of surface discharge. Andersen et al. (1998) showed that losses are high when the alluvial aquifer is fully saturated, but are small once the water table drops below the surface. Sorman and Abdulrazzak (1993) provided an analysis of groundwater rise due to transmission loss for an experimental reach in Wadi Tabalah (Southwest Saudi Arabia) and they stated that about an average of 75 % of bed infiltration reaches the water table. Ephemeral streams are characterized by much higher flow variability, extended periods of zero surface flow, and the general absence of low flow except during the recession periods immediately after moderate to large high flow events (Knighton and Nanson 1997).

In arid regions, water is one of the most challenging current and future natural resources issue. The importance of water in arid regions is obviously self-evident. Water is not only the most precious natural resource in arid regions but also the most important environmental factor of the ecosystem. Generally, the majority of countries in arid and semi-arid regions depend either on groundwater (from both shallow and deep aquifers) or on desalinization for their water supply, both of which enable them to use water in amounts far exceeding the estimated renewable fresh water in the country (IPCC 1997). However, there are some important water-related problems in these countries. These include the depletion of aquifers in several areas, saline-water intrusion problems and water quality problems such as those associated with industrial, agricultural, and domestic activities. The Arabian Peninsula, as an arid to semi-arid region, suffers from low rainfall and high temperatures in addition to high humidity in the coastal areas.

Due to the severe situation of water resources paucity and flash flood threat in arid regions, more high quality research is needed, particularly to investigate hydrological processes such as spatial rainfall, infiltration, and groundwater recharge from ephemeral flows. New approaches to flood forecasting and water management are required because they represent the most important challenges of arid areas. Also, effective water management and appropriate decision support systems are needed to utilize Wadi surface flows. Modelling tools based on the availability of data and their applications are urgently needed.

With regard to the aforementioned raised issues of Wadi system in arid and semi-arid regions, the objectives of this chapter are to outline the most vital challenges in modelling flash floods at Wadi system. That would help scientists and engineers in addressing the applicable and reasonable methods and tools for flash floods forecasting, mitigation, and management. Therefore, in this chapter, the following issues regarding flash floods at Wadi system are highlighted.

1. Understanding of Wadi system characteristics and its hydrological processes including (a) topographical conditions of the catchments, (b) degree of urbanization, (b) scales or shapes of basins, (c) total amounts of rainfall and rainfall duration in upstream and mountainous areas, (d) ranges of soil permeability on the river channels, (e) variation of runoff features in Wadi basins, and (f) scarcity of data and limitation of methods for flash floods forecasting.
2. Historical flash floods and their impacts on the human lives and their properties at Wadi systems and their economical, social, and environmental risks.
3. Evaluation of existing hydrological approaches for addressing flash floods including advantages and disadvantages and introducing some case studies.
4. Current obstacles that confront researchers to develop hydrological models for flash floods forecasting and mitigation.
5. How can we fill the gap between the available and required methodologies to forecast and mitigate flash floods at Wadi systems to develop applicable methods for water resources management?
6. Discuss urgent needs for early warning systems as a component of flash flood risk management at Wadi systems.
7. Exploring the spatiotemporal variability of flash floods at Wadi systems, and the climate change effects on flash floods behaviours and frequency.
8. Present suggestions and recommendations with future trends about flash floods at Wadi systems, which could be used as key factors for flash floods modelling.

16.2 Wadi System Characteristics

Arid and semi-arid regions of the world are characterized by limited water resources, drought and flash flood threat. Wadi system characteristics and its hydrological processes are controlled by geological outcrops, geomorphologies features, soil types, and hydrological processes such as infiltration, climate, and catchments. The most obvious characteristics in the ephemeral streams in arid areas are the initial and transmission losses in addition to discontinuous occurrence of flow in both space and time. Furthermore, monitoring network and good quality data are scarce. Infrequent occurrence of flow in both space and time is important characteristics of ephemeral streams in arid region and consequently presents challenge to develop hydrological models.

According to Few et al. (2004), each flood acquires some particular and inherent characteristics of the occurrence locality such as flow velocity and depth, duration, and rate of water-level rise. Responsible factors for the short duration of the flash flood include intense rains that persist on an area for a few hours, steep slope, impermeable surfaces, and sudden release of impounded water (Georgakakos 1986). Hence, particular hydrological characteristics such as small basins, steep slopes and low infiltration capacity combined with a meteorological event contribute to flash flood formation.

Characteristics of Wadi basins are presented in Saber (2010). More than one third of the world can be described as arid region and suffer from series water shortage. Ephemeral streams in arid areas are characterized by discontinuous occurrence of flow in both space and time. In terms of geomorphology, most of the arid terrains can also be called deserts which do not have enough water and have low vegetation. Climatic conditions in arid environment are dominated by extremely high spatial and temporal rainfall variability. Initial losses and transmission losses are considered one of the important factors affecting the interaction between surface and subsurface water. There is hydrological modelling problem due to data deficiency and data accuracy. In these regions, drought and flash floods threat exit. All these characteristics and problems of the Wadi environment are serious challenges for researchers to develop hydrological approaches for water resources management and flash flood modelling in such regions.

16.2.1 Geomorphology

Geomorphologic features are important in estimation of surface runoff discharge that results after each storm rainfall. In arid regions, vegetation cover are sparse but geological features such as rock outcrops and geomorphologies are characteristics of Wadi basins and they are important factors in the hydrological processes. A significant fraction of the Earth's land surface can be described as arid. This is primarily a climatic term implying lower rainfall and distinctive vegetation adapted to the scarcity of water (Fig. 16.2). Deserts do not necessarily consist mainly of sand; some are mountainous but with usually well-exposed rocks and thinner soils. Many landforms characterize arid regions with major types of dunes characterized as (a) crescentic dune, (b) linear or longitudinal dune, (c) star dune, (d) parabolic dune and (e) dome dune.

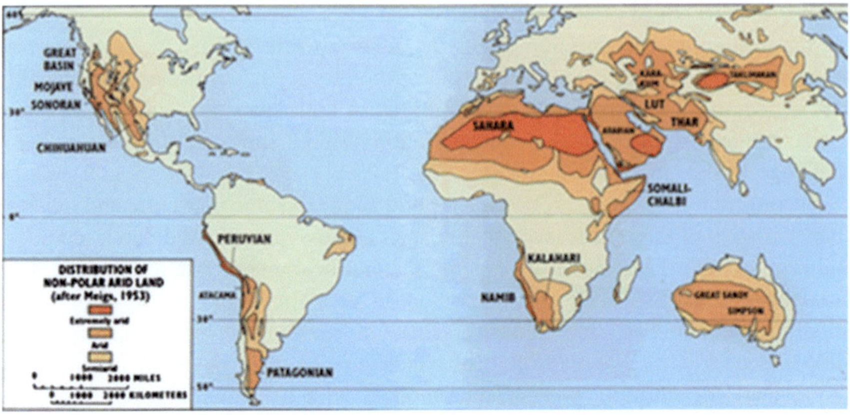

Fig. 16.2 World desert map (http://geology.com/records/sahara-desert-map.shtml)

16.3 Climatic Conditions

16.3.1 Rainfall

Rainfall is the primary hydrological input, but rainfall in arid and semi-arid areas is commonly characterized by extremely high spatial and temporal variability. Annual variability is marked where observed daily maxima can exceed annual rainfall totals (Wheater et al. 2008). Sediment yield increases during high rainfall intensity. Due to scarcity of vegetation, the formation of sediment transportation is high during rainfall. Spatial variability is directly related to local and regional topography. At high elevations, topographic rainfall occurs, and this happens especially, if surface water and nearby escarpments, cliffs, or high hills extend within 150–200 km. The rainfall initiation and intensity is augmented by windy storms prior to rainfall which initiates nucleus of condensation. At times, even though moist air is available, rainfall does not occur in the absence of winds. This is one of the main reasons for spatial and temporal variation of rainfall in arid regions (Şen 2008).

The mountains have elevations of up to 3000 m.a.s.l. and the basins have a wide range of elevation. The topographic gradient is matched by annual rainfall gradient with 30–100 mm on the Red Sea coastal plain to up to 450 mm at elevations higher than 2000 m.a.s.l. (Wheater et al. 2008).

Unlike conditions in temperate regions, the rainfall distribution in arid zones varies between summer and winter. For example, in some Arabian countries as Morocco and Egypt, rain comes during the cold winter period, while the warm summer months are almost devoid of rainfall. On the other hand, other countries like Sudan have long dry season during the winter, while the rains fall during the summer months (FAO 1989). For other arid or semi-arid areas, rainfall patterns may be very different. For example, data from arid New South Wales, Australia shows spatially extensive, low intensity rainfalls (Cordery et al. 1983) and recent research in the Sahelian zone of Africa has also indicated a predominance of widespread rainfall. This was motivated by a concern to develop improved understanding of land-surface processes for climate studies and modelling, which led to a detailed (but relatively short term) international experimental program of the HAPEX-Sahel project based in Niamey, Niger (Goutorbe et al. 1997). Although designed to study land surface/atmosphere interactions, rather than as an integrated hydrological study, it has given important information. For example, Lebel et al. (1997), Lebel and Le Barbe (1997) noted that a 100 rain gauge network was installed and reported the information on the classification of storm types, spatial and temporal variability of seasonal and event rainfall, and storm movement. Of the total seasonal rainfall, 80 % was found to fall as widespread events which covered at least 70 % of the network.

16.3.2 Temperature and Evaporation

The climatic pattern in arid zones is frequently characterized by a relatively "cool" dry season, followed by a relatively "hot" dry season and ultimately by a "moderate" rainy season. In general, there are significant diurnal temperature fluctuations within these seasons. Quite often, during the "cool" dry season, daytime temperatures peak between 35 and 45 °C and fall to 10–15 °C at night. Daytime temperatures can approach 45 °C during the "hot" dry season and drop to 15 °C during the night. During the rainy season, temperatures can range from 35 °C in the daytime to 20 °C at night. In many situations, these diurnal temperature fluctuations restrict the growth of plant species. Plant growth can take place only between certain temperature ranges. Extremely high or low temperatures can be damaging to plants. Plants might survive high temperatures, as long as they can compensate for these high temperatures by transpiration, but growth will be affected negatively. High temperatures in the surface layer of the soil result in rapid loss of soil moisture due to high levels of evaporation and transpiration (FAO 1989).

16.3.3 Atmospheric Humidity

Although rainfall and temperature are the primary factors upon which aridity is based, other factors have influence. Moisture in the air is important for the water balance in the soil. When the moisture content in the soil is higher than in the air, there is a tendency for water to evaporate. When the opposite is the case, water will condense into the soil. Humidity is generally low in arid zones. In many areas, the occurrence of dew and mist is necessary for the survival of plants. Dew is the result of condensation of water vapour from the air onto surfaces during the night when dew point temperature is reached. Dew point temperature is the temperature to which air must be cooled to reach vapour saturation to reach condensation. Mist is a suspension of microscopic water droplets in the air. Water that is collected on leaves of plants in the form of dew or mist can be imbibed through open stomata or fall onto the ground and contribute to soil moisture. The presence of dew and mist reduces evapotranspiration and help conserve soil moisture (FAO 1989). Low humidity increases the atmospheric demand for moisture and it have been found to cause partial to full stomatal closure in plants, resulting in reduced transpiration, photosynthesis, and plant productivity (Korner and Bannister 1985; Choudhury and Monteith 1986). Over longer temporal scales, humidity may influence plant evolutionary adaptations such as leaf area and cuticle thickness (Grantz 1990). These factors influence soil moisture and runoff at local scales and may influence precipitation and atmospheric circulation patterns at regional to global scales (Hansen et al. 1984).

16.4 Arid Regions

16.4.1 Aridity Index

Many aridity and semi-aridity definitions appear in the literature but none can be entirely satisfactory. The terms of "arid and semi-arid" remain somewhat imprecise and vague. Generally, aridity implies that rainfall does not support regular rainfed farming. According to Walton (1969) this definition encompasses all the seasonally hot arid and semi-arid zones classified by means of rainfall, temperature, and evaporation indices. Among the aridity indices, those based on rainfall only state that any duration without rainfall for 15 consecutive days is considered a dry period. In some regions, 21 days or more with rainfall less than one-third of what is considered normal is a dry period. There are also percentage-dependent indices such as in the case of annual rainfall less than 75 % and monthly rainfall less than 60 % of normal. Normal rainfall on any given day is typically taken to mean the average of all the observed rainfall events on a particular date. However, in an individual case, any rainfall less than 85 % is also considered a dry case (Şen 2008). Arid environments are extremely diverse in terms of their land forms, soils, fauna, flora, water balances, and human activities. Because of this diversity, no practical definition of arid environments can be derived. However, the one binding element to all arid regions is aridity. Aridity is usually expressed as a function of rainfall and temperature (FAO 1989).

Also, a drought is a departure from average or normal conditions in which shortage of water adversely impacts the functioning of ecosystems, and the resident population, whereas aridity refers to the average conditions of limited rainfall and water supplies, not to the departures there from. In general, arid zones are characterized by pastoralism and little farming but there are always exceptions (Malhotra 1984).

Aridity in climate terms relates principally to paucity of precipitation, which in turn dictates the opportunity for groundwater recharge and the sustainability of naturally occurring groundwater resources. The sporadic distributions, frequencies, intensities, and amounts of precipitation that characterize the very low precipitation areas of the world inherently constrain recharge opportunity. Coupled with ground conditions, these result in very complex and little-understood recharge processes. The simplistic boundary between arid and semi-arid reflects the onset of opportunity for direct recharge to occur as precipitation increases above 200 mm per annum (Goosen and Shayya 1999).

16.4.2 Vegetation in Arid Zone

In general, the vegetation cover is small in size, has shallow roots, and their physiological adaptation consists of their active growth. The dynamics of vegetation in regions of drifting sand have been studied primarily on the coastal dunes of temperate regions (Willis et al. 1959). The literature related to the adaptation of plants to the movement of sand in arid regions is less abundant. For instance, Bowers (1982) has indicated adaptations of species to the dune environment in the USA, especially the elongation of stems and roots. Dittmer (1959) also noted such adaptations in roots.

16.5 Arid Region Hydrology

16.5.1 Initial and Transmission Losses

During a rainfall event in arid regions, initial losses occur in the sub-basins before runoff reaches the stream networks. Transmission losses occur as water is channelled through the valley network. Initial losses are dependent on infiltration, surface soil type, land-use activities, evapotranspiration, interception, and surface depression storage. Transmission losses are important not only with respect to flow rate reduction, but also groundwater recharge of alluvial aquifers. It was suggested that two sources of transmission losses could be occurring as direct losses to the Wadi channel bed, limited by available storage, and losses through the banks during flood events as shown in Fig. 16.3.

The rate of transmission loss from a river reach is a function of the characteristics of the channel alluvium, channel geometry, wetted perimeter, flow characteristics, and depth to groundwater. In ephemeral streams, factors influencing transmission losses include antecedent moisture of the channel alluvium, duration of flow, storage capacity of the channel bed and bank, and the content and nature of sediment in the stream flow. The total effect of each of these factors on the

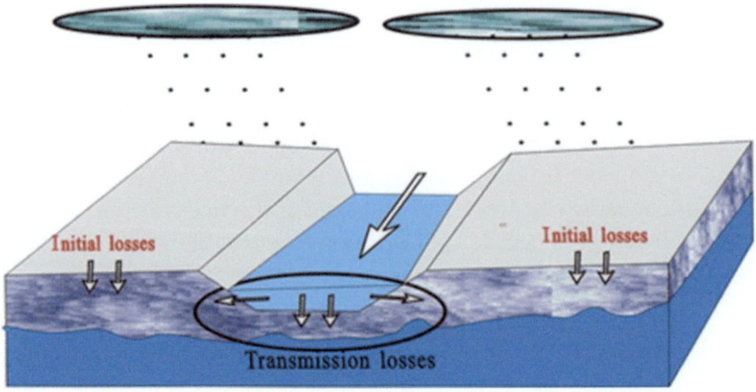

Fig. 16.3 Conceptual model showing transmission and initial losses in the Wadi system (Saber 2010)

magnitude of the transmission loss depends on the nature of the stream, river, irrigation canal, or even rill being studied (Vivarelli and Perera 2002).

16.5.2 Water Resources in Arid Regions

Water resources in arid and semi-arid regions are limited. Groundwater is the most important resource. Desalinated water and some perennial streams in some regions are the sources. However, quality and quantity of water are inadequate to cover the current needs for domestic, agricultural, industrial, and economic use in arid areas. In some Arabian countries as Saudi Arabia and Oman, they depend mainly on ground water as the main water resource. It is likely that the groundwater aquifers are affected by sea water intrusion. In the Arabian Peninsula, groundwater accounts for 40-98 % of the total freshwater resource. Of the meagre rainfall, 1–10 % becomes available as soil moisture and ends up nourishing economic crops (Rogers and Lyndon 1994; Abdel Magid et al. 1995). Thus, in a region where water is in short supply, it is imperative that a coherent water management policy be established through a system of permits, regulations, and standards to ensure supplies for sustainable development.

The assessment of water resources can be considered in four phases as summarized by Alsharhan et al. (2001); (i) the first phase is an inventory of all wells with the relevant technical data as depth, movement of groundwater, water quality and aquifer characteristics, (ii) the second phase is conservation techniques, which include control of drilling and extraction rates, the potential of water harvesting, and the construction of underground and surface water storage as dams, (iii) the third phase is based on knowledge derived from the water inventory, to evaluate cost, current and future water demand, based upon an accepted growth rate, (iv) the final

Fig. 16.4 Wadi in dry condition **a** Wadi Baih (Emirates) and **b** Eastern Desert (Egypt, March 2010)

phase is the method of increasing supply by importing water, which ranges from towing icebergs or water filled plastic pods to diverting river water long distances. The other means of importing water in the form of importing food, especially grain since a ton of grain requires a thousand tons of water to produce it.

Based on field observations at some Wadi systems in Egypt, Wadi basins suffer from drought conditions all year except during infrequent flash flood events. Figure 16.4 shows dry Wadi in the Emirates (a) and Egypt (b). They are affected by infrequent rainfall events with short durations. Flash floods occur if rainfall is severe. If the rainfall is not strong, surface water flow will be consequently low. The ephemeral streams are characterized by discontinuous flow especially at the beginning of flow as shown in Fig. 16.5a and at the cessation of the flow as depicted in Fig. 16.5b. The flash flood in the Wadi basin is a very vital natural phenomenon that are short duration with rapidly rising water flow rate and depth caused by intense rainfall or dam failure. It results in a greater danger to human life and property damages as depicted by the two events in Egypt and Saudi Arabia (Fig. 16.6a, b). It is also accompanied by landslides, mudflow, and debris flow(Saber 2010).

Fig. 16.5 Wadi channel showing **a** discontinuous flow at the starting of flash flood in the Gobi of Mongolia, 2004 and **b** at the ending of flood in Sinai, Egypt

Fig. 16.6 **a** Flash floods of January 18th, 2010 (Egypt) and **b** Wadi Aqeeq after heavy rainfall, Madina, Saudi Arabia

16.6 Historical Flash Floods and Impacts

In order to have comprehensive understanding of flash floods in Wadi system and develop modelling techniques, research in historical flash flood events is necessary. As natural phenomena, extreme floods events can be studied. Information about flash floods in the past periods can be retrieved from documentary evidence, which constitute rainfall and discharge data based on historical hydrology. Researchers are provided with an overview of the historical hydrology in Wadi system, mainly with respect to flooding and flood risk, observations and the required research priorities. Historical hydrology is introduced with various sources of data; their temporal and geographical extent in Wadi system is discussed. Future research priorities are presented.

The research development during past decades defines historical hydrology as a field situated at the interface of hydrology and environmental history, dealing mainly with documentary sources and using both hydrological and historical methodologies. It is directed towards the following three objectives:

1. Reconstruction of temporal and spatial patterns of runoff conditions as well as extreme hydrological events (floods, ice phenomena, hydrological droughts) for the period prior hydrological monitoring networks.
2. Investigation of the vulnerability of past societies and economies to extreme flash floods events.
3. Exploration oral history on social impacts of extreme hydrological events.

A crucial problem in data reconstruction is combining historical flow data derived from flood damages descriptions and flow data estimated based on water level measurements into a single comparable series. Sometimes, one has to deal with data gaps of several decades between estimated historical data and measured contemporary data.

The impact floods have on society is a main aspect of historical flood evidence. Important features are the number of people living in settlements close to the Wadi, the strategies developed for managing flash flood hazards, warning and protection systems, and measurements taken after the flash flood. Also, the identification of rainfall trends that could explain the historical evolution of flood hazard occurrence in Wadi system has to be analyzed, and the identification of the influence of urban development on the vulnerability to floods has to be carried out as well.

The main documentary data sources for historical data of flash floods are narrative written sources as visual daily weather records, governmental cities records, official economic records, newspapers, pictorial documentation, stall-keepers' and market songs, scientific papers and communications, epigraphic sources, and early instrumental records. Flash floods in arid regions are characterized by infrequent occurrence in making it difficult to predict when it will happen. It is important to show some of the past flash flood events which have hit Egypt and other Arabian countries as example for flash floods records history and their damage or impact on human life and infrastructure. From the history of flash flood events in Egypt in recent years, for example, Egypt has been affected by flooding in October 1987, December 1991, February 1992, November 1994, November 1996, March 1997, October 1997, December 1997, March 2002, and January 2010 as shown in Table 16.1.

16.7 Climate Change and Flash Floods in Arid Regions

The impact of climate change on water resources variability, especially on river flows, soil moisture, evapotranspiration, and groundwater flow has been studied by analyzing projected and downscaled climatic data and using hydrological models (Mango et al. 2011a, b; Behulu et al. 2014; Setegn et al. 2011; Setegn and Melesse 2014; Assefa et al. 2014; Melesse et al. 2009, 2011; Dessu and Melesse 2013; Grey et al. 2013).

Flash floods are likely to become more common across Wadi systems as a result of climate change. One of the most common incidences of flash flooding occurs when there is a heavy rainfall event following a period of drought. This is because the ground is very dry and cannot absorb so much water at once. It is an interest to hydrologists who focus on flash floods to assess impacts of climate change characteristics as frequency and magnitude of flash floods and impact on the natural ecosystem and society. The reason is that flash floods have become more frequent than the past decades. As example, in Egypt and Arabian countries, flash floods used to be infrequent with limited damage to property and human life losses compared to the last 10 years. For instance, flash floods happened in Egypt in 2010, 2011, and 2013 and in Saudi Arabia, almost every year. There is the need to link hydrological models with Global Circulation models (GCM) for climate prediction. Until recently, most of the hydrological models were not including climate change in flash flood applications. Integration of GIS, remote sensing, hydrological models, and climate models is needed.

Table 16.1 Flash flood history in Egypt

Detailed locations	Southern Sinai, Hurghada, and Aswan Governorate	Southern Egypt—Shalatein	Mansura, Damiette, Sinai peninsula	Aswan, Suez, and Sinai peninsula	North: Cairo region	Provinces: Aswan, Sohag, Asyut, Minya, Qena	Provinces: Asyut—Dronka (Durunka), Sohag, Qena	Sidi Abdel-Qadar	Southeastern Sinai Peninsula, Sinai
Began	18-Jan-10	9-Mar-02	10-Dec-97	17-Oct-97	2-Mar-97	13-Nov-96	2-Nov-94	5-Dec-91	17-Oct-87
Ended	19-Jan-10	11-Mar-02	14-Dec-97	23-Oct-97	04 Mar-97	25-Nov-96	08-Nov-94	06-Dec-91	19-Oct-87
Duration (days)	2	3	5	7	3	13	7	2	3
Dead	10	4	4	12	4	23	593		10
Main cause	Torrential Rain	Heavy rain	Heavy rain	Heavy rain	Heavy rain	Heavy Rain	Heavy rain	Dam/Levy, break or release	Brief torrential rain
Affected (km^2)	226,100	560	113,100	307,600	68,460	37,940	44,070	3170	12,690

http://www.dartmouth.edu/~floods/

Recently, there has been intensive global research effort in coupling atmospheric and hydrological models to improve the predictive capability of both hydrological models and General Circulation Models (Dolman et al. 2001). Bierkens and van den Hurk (2008) present an example taken from Entekhabi et al. (1999) illustrating the importance of soil moisture for simulating floods and droughts in USA. Sea surface temperature (SST) is linked to seasonal precipitation. The positive feedback between precipitation, soil moisture, and evapotranspiration may increase and prolong the magnitude and duration of floods.

To investigate the risks of flood losses, meteorological and hydrological conditions, river hydraulic characteristics, and land-use conditions must be considered and separately assessed. Each of these factors can contribute to an increased risk of flood damage, in particular, if there are negative interrelationships and synergies. Of these flood-risk factors, the meteorological conditions are considered to be of primary importance with regard to climate change. Also, vegetation and soil conditions in the catchment area, which determine water retention and evaporation processes, can be affected by climate change and thus have a feedback effect on flood development (Bronstert 2003). Flood generation and runoff is a highly nonlinear system which is affected by spatial and temporal variability of meteorology, topography, soil, vegetation, climate, groundwater conditions, and the channel drainage system. The effects of climate change on flooding and runoff processes are highly uncertain (Beven 1993).

In order to understand and assess the impact of climate change on Wadi system flash flood frequency and formation, these are the key questions:

1. Is the predicted temperature increase accompanied by an increase in precipitation as well as by an increasing number and or severity of flash floods?
2. Is there any change in the characteristics (quantity, intensity, spatiotemporal variability) and frequency of extreme precipitation events causing flash floods?

16.8 Hydrological Models

Different modelling techniques and tools have been widely used for a variety of purposes, but almost all have been primarily developed for humid area applications. However, due to scarcity of water resources and flash flood threat, developing powerful physical-based models is needed for arid and semi-arid regions.

The influence of rainfall representation on the modelling of the hydrologic response is expected to depend on complex interactions between rainfall space-time variability, the variability of the catchment soil type, landscape properties, and the spatial scale (catchment area) of the system (Obled et al. 1994; Woods and Sivapalan 1999; Bell and Moore 2000; Smith et al. 2004). The study of Michaud and Sorooshian (1994) demonstrated the sensitivity of flood peak simulation to the spatial resolution of rainfall input. River hydrograph forecasts are very dependent upon the rainfall input data. Stream flow in arid and semi-arid regions tends to be dominated by rapid responses to intense rainfall events. Such events frequently have a high degree of spatial variability coupled with poorly gauged rainfall data. This sets a fundamental limit on the capacity of the rainfall–runoff model to reproduce the observed flow (Wheater et al. 2008). Rainfall–runoff models fall into several categories as metric, conceptual, and physics-based models (Wheater et al. 1993). Metric models are typically the most simple, using observed data (rainfall and stream flow) to characterize the response of a catchment. Conceptual models impose a more complex representation of the internal processes involved in determining catchment response and can have a range of complexity depending on the structure of the model. Physics-based models involve numerical solution of the relevant equations of motion.

A set of different models is available to represent rainfall–runoff relationships, but they have limitations in the hydrologic parameters that are used to describe the rainfall–runoff process in Wadi systems (Wheater et al. 1993). Due to its significant role on Wadi basins, transmission loss in arid regions has been discussed in different literature (Jordan 1977; Lane 1994; Vivarelli and Perera 2002; Walters 1990; Andersen et al. 1998; Abdulrazzak and Sorman 1994). A few direct studies of channel transmission losses have been done in arid regions (Crerar et al. 1988; Hughes and Sami 1992), despite the fact that this process has been recognized as one of the most important components of water balance of many arid and semi-arid basins.

Transmission losses in semiarid watersheds raise important distinctions about the spatial and temporal nature of surface–groundwater interactions compared to humid basins. Because of transmission losses, the nature of surface water–groundwater interactions can be limited to brief periods during runoff events and to specific areas associated with the runoff production and downstream routing (Boughton and Stone 1985). Walters (1990), Jordan (1977) provided evidence that the rate of loss is linearly related to the volume of surface discharge, and the same finding was also confirmed by Saber et al. (2013). Andersen et al. (1998) showed that losses are high when the alluvial aquifer is fully saturated, but are small once the water table drops below the surface. Sorman and Abdulrazzak (1993) provided an analysis of groundwater rise due to transmission loss for an experimental reach in Wadi Tabalah, southwest Saudi Arabia and they stated that on the average about 75 % of bed infiltration reaches the water table.

Surface–groundwater interactions in arid and semi-arid drainages are controlled by transmission losses. In contrast to humid basins, the coupling between stream channels and underlying aquifers in semiarid regions often promotes infiltration of water through the channel bed, i.e. channel transmission losses (Boughton and Stone 1985; Stephens 1996; Goodrich et al. 1997). The relationship between Wadi flow transmission losses and groundwater recharge depend on the underlying geology. The alluvium underlying the Wadi bed is effective in minimizing evaporation loss through capillary rises. The coarse structure of alluvial deposits minimizes capillary effects. Sorey and Matlock (1969) reported that measured evaporation rates from streambed sand were lower than those reported for irrigated soils.

Table 16.2 Results of simulation of the three Wadi basins (Saber et al. 2010)

	W. Khoud	W. Ghat	W. Assiut
Area (km^2)	1874.84	649.55	7109
Slope	Steep	Steep	Gentle
Peak discharge	838 m^3/s	36 m^3/s	12.5 m^3/s
Time to peak	4–7 h	14 h	10 h
Flow duration	44 h	60 h	106 h
Runoff	88.98 %	86.22 %	92.47 %
Infiltration or transmission losses	11.02 %	13.78 %	7.53 %

A comparative study has been done between some Wadi basins in the Arabian regions of Egypt, Oman, and Saudi Arabia (Saber 2010; Saber et al. 2010). They concluded that Wadi runoff exhibits the same characteristics as the actual flash floods from the beginning of flash flood event to the end. From the distribution maps of surface runoff in the Wadi system, the discontinuous surface flow was originally depicted as one of the most import characteristics of ephemeral streams. Transmission loss is affected by the volume of surface runoff as the rate of losses is linearly related to the volume of surface discharge (Saber et al. 2010, 2013). Results of the comparative study (Table 16.2) between some Arabian Wadis, Wadi Ghat in Saudi Arabia, Wadi Assiut in Egypt, and Wadi Al-Khoud in Oman concluded that the runoff features are affected by the catchments area, slope, and rainfall evens frequency and duration (Saber 2010; Saber et al. 2010). Runoff features in the ephemeral streams are characterized by different behaviours from the runoff in the humid area based on the following results: (i) the time to peak is short (i.e. the time of flow to reach to the discharge peak); (ii) flood event time including starting and cessation is short; (iii) initial and transmission losses are considered the main source of sub-surface water recharge; and (iv) discontinuous surface flow in space and time occurrences.

The EU Project HYDRATE was established in order to better understand the hydrometeorological processes leading to flash floods. The full title of the project is "Hydrometeorological Data Resources and Technology for Effective Flash Flood Forecasting" (www.hydrate.tesaf.unipd.it). The primary objective of this project was to improve the scientific basis of flash flood forecasting by extending the understanding of past flash flood events, advancing and harmonizing a European-wide innovative flash flood observation strategy, and developing a coherent set of technologies and tools for effective early warning systems. To this end, the project acted on the organization of the existing flash flood data patrimony across Europe. Also, Japan-Egypt Hydro Network (JE-HydroNet) was established as a collaboration research between Japan and Egypt to study the water resources problems in Egypt, and one of the most important issue is developing the hydrological approaches and mitigations strategies of flash floods (Sumi et al. 2013). In arid environment, there is urgent need for a comprehensive research work that could be able to address flash floods at Wadi system question where data and capacity is lacking.

16.9 Flash Floods Modelling Challenges

It has been widely stated that the major limitation of the development of arid-zone hydrology is the lack of high-quality observations (McMahon 1979; Pilgrim et al. 1988). Flash Flood occurrences are complex since it depends on interactions between many geological and morphological characteristics of the basins including rock types, elevation, slope, sediments transport, and flood plain area. Moreover, it also depends on hydrological phenomena such as rainfall, runoff, evaporation, and surface and groundwater storage (Farquharson et al. 1992; Flerchinger and Cooly 2000; Şen 2004; Nouh 2006). In order to manage flash floods, the scientists and engineers have to apply hydrologic models coupled with climate models to predict future changes in rainfall and runoff in Wadi system. Moreover, the most important challenge to calibrate and simulate flash floods in arid regions is the deficiency in observed data. Monitoring network implementation and data collection is needed.

Long-term rainfall data availability and analysis are limited in most arid regions. This is due to (1) large inaccessible and uninhabitable areas, (2) lack of adequate network of monitoring stations, (3) non-suitability of large areas for agriculture, (4) poor data quality when data is available and (5) lack of skilled personnel for effective database management and analysis. Vulnerability to flash floods will probably increase in the coming decades due to changes in land use and the modification of the pluviometric regime associated with the evolution of climate (Parry et al. 2007; Palmer and Räisänen 2002; Rosso and Rulli 2002). As pointed out by many authors, the quality of flood prediction that is based upon hydrological simulations depends to a high degree upon the quality of the measurements and forecasts of precipitation. Also, it must be emphasized that an early warning system is indispensable for the reduction in damages associated with flash floods.

In deserts, flash floods can be particularly deadly for several reasons. First, storms in arid regions are infrequent, but they can deliver an enormous amount of rain in a very short time. Second, these rains often fall on poorly absorbent and often clay-like soil, which greatly increase the amount of runoff that rivers and other water channels have to handle. In addition, these regions may not have the infrastructure that wetter regions have to divert water from structures and roads, such as storm drains and retention basins, either because of sparse population, poverty or because residents believe that the risk of flash floods is not high enough to justify the cost. Finally, the lack of regular rain to clear water channels may cause flash floods in deserts to be headed by large amounts of debris, such as rocks, branches and logs.

Observational difficulties of flash floods at Wadi system is a barrier in hydrometeorological data collection (Viglione et al. 2010) and lack of a comprehensive archive of flash flood events hinder the development of a coherent framework for analysis of flash flood climatology, hazard, and vulnerability at Wadi system.

New research opportunities to advance hydrological models promise a better understanding of the hydrological processes that could help improve forecasting models and warning system of flash floods at Wadi system. To reach a level of understanding, it requires both database preparation for gauge networks of rainfall and runoff at Wadi system and developing physical-distributed hydrological models for representing surface runoff, losses, and evaporations functions. Consequently, it could meet the needs of flash flood warning and water management. Multidisciplinary research collaborations among hydrologists, engineers, and scientists would be central to meeting Wadi system research challenges, such as, hydrological models, GIS, and Remote Sensing techniques.

One important objective of modern flood risk management and modelling is to understand and predict the extent, location, duration, and timing of flash floods. Such forecasts must be based on complex and detailed mathematical modelling systems. To simulate flash flood, basically three different modelling components are required. The required strategy is to combine and link multidisciplinary models such as; the meteorological models, the hydrological models, and climatic models.

The challenge for the future lies in the improvement and incorporating of the above model systems to allow for early warning times based on reliable weather forecasts. Such systems would also permit the investigation of the longer term effects of climate change in the form of scenario analyses. By means of such scenarios it will be possible to analyse the behaviour of a catchment area in the event of flash flooding under selected, typical future climatic conditions in order to derive typical flash flood characteristics. These events can then be compared with present or past conditions in order to make qualitative estimates for flash floods at Wadi system under consideration.

Three main factors must be taken in consideration in order to develop the appropriate hydrological models for flash floods modelling; setting observational network of gauges at Wadi system, experimental models at the actual sites of Wadi system, and developing the mathematical models based on observational and experimental results.

16.10 Recommendations and Future Trends

In order to develop the hydrological approaches based on the physical characteristic of Wadi system for flash floods modelling, the following issues should be taken in consideration. Understanding hydrological processes of Wadi system is needed. This requires studying the geological outcrops, geomorphological features, land use, and soil types and infiltration. Watersheds analysis is essential. Experimental models are required to understand the physical properties of Wadi system. Mathematical models are needed to simulate the hydrological conditions of Wadi system. Linking meteorological, hydrological, hydrogeological, climatic models integration in an interdisciplinary approach is needed to understand and forecast flash floods in arid regions. Climatic conditions from historical data should be taken into account. Construction of observational gauges network for rainfall and discharge at Wadi system is needed. Establishment of comprehensive research projects to achieve these objectives are required. International collaboration between developed countries and developing countries is needed to achieve the goal of such project. An integrated research group must consider water demand and how to capture flash floods as a source of water. To investigate the risks of flood losses, meteorological and hydrological conditions, river hydraulic characteristics, and land-use conditions must be considered and separately assessed. Multidisciplinary research collaborations among hydrologists, engineers, and scientists would be central to meeting Wadi system research challenges, such as, hydrological models, GIS, and Remote Sensing techniques.

References

Abdel Magid HM, Abdel-Aal SI, Rabie RK, Sabrah REA (1995) Chicken manure as a biofertilizer for wheat in the sandy soils of Saudi Arabia. J Arid Environ 29:413–420

Abdulrazzak MJ, Sorman AU (1994) Transmission losses from ephemeral stream in arid region. J Irrig Drain Eng 20(3):669–675

Alsharhan AS, Rizk ZA, Nairn AEM, BAkhit DW, Alhajari SA (2001) Hydrogeology of an arid region: the Arabian Gulf and adjoining areas. Elsevier, The Netherlands, 354 pp

Andersen NJ, Wheater HS, Timmis AJH, Gaongalelwe D (1998) Sustainable development of alluvial groundwater in sand rivers of Botswana. Sustain Water Resour Under Increasing Uncertain IAHS 240:367–376

Assefa A, Melesse AM, Admasu S (2014) Climate change in upper Gilgel Abay River catchment, Blue Nile basin Ethiopia. In: Melesse AM, Abtew W, Setegn S (eds) Nile River basin: ecohydrological challenges, climate change and hydropolitics, pp 363–388

Behulu F, Setegn S, Melesse AM, Romano E, Fiori A (2014) Impact of climate change on the hydrology of upper Tiber River basin using bias corrected regional climate model. Water Resour Manage 1–17

Bell VA, Moore RJ (2000) The sensitivity of catchment runoff models to rainfall data at different spatial scales. Proc Conf Hydrol Earth Syst Sci 4(4):653–667 (Special issue on HYREX: the Hydrological Radar Experiment)

Beven KJ (1993) Prophecy, reality and uncertainty in distributed hydrological modelling. Adv Water Resour 16:41–51

Bierkens M, van den Hurk B (2008) Feedback mechanisms: precipitation and soil moisture. In: Bierkens MFP, Dolman AJ, Troch PA (eds) Climate and the hydrological cycle, 8th edn. IAHS Special publication, Mangalore

Boughton WC, Stone JJ (1985) Variation of runoff with watershed area in a semiarid location. J Arid Environ 9:13–25

Bowers JE (1982) The plant ecology of inland dunes in western North America. J Arid Environ 5:199–220

Bronstert A (2003) Floods and climate change: interactions and impacts. Risk Anal 23(3):545–557

Choudhury BJ, Monteith JL (1986) Implications of stomatal response to saturation deficit for the heat balance of vegetation. Agric For Meteorol 36:215–225

Collier C (2007) Flash flood forecasting: what are the limits of predictability? Q J R Meteorol Soc 133(622):3–23

Cordery I., Pilgrim DH., Doran DG. (1983) Some hydrological characteristics of arid western New South Wales. *In: Hydrology and Water Resources Symposium 1983*, Institution of Engineers, Australia. National Confrence Publication 83/13 287-292

Crerar S, Fry RG, Slater PM, van Langenhove G, Wheeler D (1988) An unexpected factor affecting recharge from ephemeral river flows in SWA/Namibia. In: Simmers I, Dordrecht D (eds) Estimation of natural groundwater recharge. Reidel, Dordrecht, pp 11–28

Creutin JD, Borga M (2003) Radar hydrology modifies the monitoring of flash flood hazard. Hydrol Process 17(7):1453–1456. doi:10.10002/hyp.5122
Dessu SB, Melesse AM (2013) Impact and uncertainties of climate change on the hydrology of the Mara River basin. Hydrol Process 27(20):2973–2986
Dittmer HJ (1959) A study of the root systems of certain sand dune plants in New Mexico. Ecology 40:265–273
Dolman AJ, Hall AJ, Kavvas ML, Oki T, Pomeroy JW (eds) (2001) Soil-vegetation-atmosphere transfer schemes and large-scale hydrological models, vol 270. IAHS, Mangalore, 372 pp
Entekhabi D, Asrar GR, Betts AK, Beven KJ, Bras RL, Duffy CJ, Dunne T, Koster RD, Lettenmaier DP, McLaughlin DB, Shuttleworth WJ, van Genuchten MT, Wei MY, Wood EF (1999) An agenda for land surface hydrology research and a call for the second international hydrological decade. Bull Am Met Soc 80(10):2043–2058
FAO (1989) Arid zone forestry: a guide for field technicians. Food and Agriculture Organization, Rome
Farquharson FA, Meigh JR, Sutcliffe JV (1992) Regional flood frequency analysis in arid and semi-arid areas. J Hydrol 138:487–501
Few R, Ahern M, Matthies F, Kovats S (2004) Floods, health and climate change: a strategic review. Tyndall Centre, Norwich, p 138 (Working paper 63)
Flerchinger GN, Cooly KR (2000) A ten-year water balance of a mountainous semi-arid watershed. J Hydrol 237:86–99
Georgakakos KP (1986) On the design of natural, real-time warning systems with capability for site-specific, flashflood forecasts. Bull Am Meteorol Soc 67:1233–1239
Goodrich DC, Lane LJ, Shillito RM, Miller SN, Syed KH, Woolhiser DA (1997) Linearity of basin response as a function of scale in a semiarid watershed. Water Resour Res 33(12): 2951–2965
Goosen MFA, Shayya WH (1999) Water management, purification and conservation in arid climates. In: Goosen MFA, Shayya WH (eds) Water management, vol I. Technomic Publishing Co., Lancaster, pp 1–6
Goutorbe JP, Dolman AJ, Gash JHC, Kerr YH, Lebel T, Prince SD, Stricker JNM (1997) HAPEX-SAHEL. J Hydrol 188–189 (Special issue)
Grantz DA (1990) Plant response to atmospheric humidity. Plant Cell Environ 13:667–679
Grey OP, Webber Dale G, Setegn SG, Melesse AM (2013) Application of the soil and water assessment tool (SWAT model) on a small tropical Island state (Great River Watershed, Jamaica) as a tool in integrated watershed and coastal zone management. Int J Trop Biol Conserv 62(3):293–305
Hansen J, Lacis A, Rind D, Russell G, Stone P, Fung I, Ruedy R, Lerner J (1984) Climate sensitivity: analysis of feedback mechanisms. In: Hansen JE, Takahashi T (eds) Climate processes and climate sensitivity: geophysical monographs, vol 29. AGU, Washington, D.C., pp 130–163
Hughes DA, Sami K (1992) Transmission losses to alluvium and associated moisture dynamics in a semi-arid ephemeral channel system in southern Africa. Hydrol Process 6:45–53
IPCC (1997) An introduction to simple climate models used in the IPCC second assessment report. In: Houghton JT, Filho LGM, Griggs DJ, Maskell K (eds) IPCC working group I. Available online at http://www.ippc.ch
Jordan PR (1977) Stream flow transmission losses in western Kansas. J Hydraul Div ASCE 108 (HY8):905–919
Knighton AD, Nanson GC (1997) Distinctiveness, diversity and uniqueness in arid zone river systems. In: Thomas DSG (ed) Arid zone geomorphology: process, form and change in dry lands, 2nd edn. Wiley, Chichester, pp 185–203
Korner C, Bannister UP (1985) Stomatal responses to humidity in Nothofagus menziesii. NZ J Bot 23:425–429
Lane LJ (1994) Estimating transmission losses. In: Proceedings of the special conference development and management aspects of irrigation and drainage systems: irrigation and drainage Engineering Division. ASCE, San Antonio, pp 106–113

Lebel T, Le Barbe L (1997) Rainfall monitoring during HAPEX-Sahel 2: point and areal estimation at the event and seasonal scales. J Hydrol 188–189:97–122

Lebel T, Taupin JD, D'Amato N (1997) Rainfall monitoring during HAPEX-Sahel 1: general rainfall conditions and climatology. J Hydrol 188–189:74–96

Malhotra SP (1984) Traditional agroforestry practices on arid zone of Rajasthan. In: Shamkarnarayan KA (ed) Agroforestry in arid and semi-arid zones. CAZRI, Jodhpur, pp 263–266

Mango L, Melesse AM, McClain ME, Gann D, Setegn SG (2011a). Land use and climate change impacts on the hydrology of the upper Mara River basin, Kenya: results of a modeling study to support better resource management. Hydrol Earth Syst Sci 15:2245–2258. doi:10.5194/hess-15-2245-2011 (Special issue: climate, weather and hydrology of East African Highlands)

Mango L, Melesse AM, McClain ME, Gann D, Setegn SG (2011b). Hydro-meteorology and water budget of Mara River basin, Kenya: a land use change scenarios analysis. In: Melesse A (ed) Nile River basin: hydrology, climate and water use (Chap. 2). Springer, Berlin, pp 39–68. doi:10.1007/978-94-007-0689-7_2

McMahon TA (1979) Hydrological characteristics of arid zones. In: Proceedings of symposium on the hydrology of areas of low precipitation, vol 128. IAHS, Canberra, pp 105–123

Melesse AM, Loukas AG, Senay G, Yitayew M (2009) Climate change, land-cover dynamics and ecohydrology of the Nile River basin. Hydrol Process 23(26):3651–3652 (Special issue: Nile Hydrology)

Melesse AM, Bekele S, McCornick P (2011) Hydrology of the Niles in the face of land-use and climate dynamics. In: Melesse A (ed) Nile River basin: hydrology, climate and water use. Springer, Berlin, pp vii–xvii. doi:10.1007/978-94-007-0689-7

Michaud JD, Sorooshian S (1994) Effect of rainfall-sampling errors on simulations of desert flash floods. Water Resour Res 30(10):2765–2775

Nouh M (2006) Wadi flow in the Arabian Gulf states. Hydrol Process 20:2393–2413

Obled C, Wendling J, Beven K (1994) Sensitivity of hydrological models to spatial rainfall patterns: an evaluation using observed data. J Hydrol 159:305–333

Palmer TN, Räisänen J (2002) Quantifying the risk of extreme seasonal precipitation events in a changing climate. Nature 415:512–514

Parry M, Canziani O, Palutikof J, van der Linden P, Hanson C (2007) Climate change 2007: impacts, adaptation and vulnerability. Cambridge University Press, Cambridge, 840 p

Pilgrim DH, Chapman TG, Doran DG (1988) Problems of rainfall-runoff modeling in arid and semi-arid regions. Hydrol Sci J 33(4):379–400

Rogers P, Lyndon P (1994) Water in the Arab world-perspectives and progress. Harvard University Division of Applied Sciences, Cambridge, pp 171–202

Rosso R, Rulli MC (2002) An integrated simulation method for flash flood risk assessment 2: effects of changes in land-use under a historical perspective. Hydrol Earth Syst Sci 6(3): 285–294

Saber M (2010) Hydrological approaches of Wadi system considering flash floods in arid regions. PhD. thesis, Kyoto University

Saber M, Hamagutchi T, Kojiri T, Tanaka K (2010) Hydrological modeling of distributed runoff throughout comparative study between some Arabian Wadi basins. Annu J Hydraul Eng JSCE 54:85–90

Saber M, Hamagutchi T, Kojiri T, Tanaka K, Sumi T (2013) A physically based distributed hydrological model of Wadi system to simulate flash floods in arid regions. Arab J Geosci. doi:10.1007/s12517-013-1190-0

Şen Z (2004) Hydrograph methods, arid regions, Saudi geological survey (SGS), Technical Report

Şen Z (2008) Wadi hydrology. CRC Press, New York

Setegn S, Melesse AM (2014) Climate change impact on water resources and adaptation strategies in the Blue Nile River basin. In: Melesse AM, Abtew W, Setegn S (eds) Nile River basin: ecohydrological challenges, climate change and hydropolitics, pp 389–420

Setegn S, David R, Melesse AM, Bijan D, Ragahavan S (2011) Impact of climate change on the hydro-climatology of Lake Tana basin. Ethiopia Water Resour Res 47(W04511):13. doi:10.1029/2010WR009248

Smith MB, Seo DJ, Koren VI, Reed SM, Zhang Z, Duan Q, Moreda F, Cong S (2004) The distributed model intercomparison project (DMIP): motivation and experiment design. J Hydrol 298:4–26

Sorey ML, Matlock WG (1969) Evaporation from an ephemeral streambed. J Hydraul Div Am Soc Civ Eng 95:423–438

Sorman AU, Abdulrazzak MJ (1993) Infiltration—recharge through Wadi beds in arid regions. Hydrol Sci J 38(3):173–186

Stephens DB (1996) Vadose zone hydrology. CRC Press, Boca Raton

Sumi T, Saber M, Kantoush SA (2013) Japan-Egypt hydro network: science and technology collaborative research for flash flood management. J Disaster Res 8(1):28–36

Telvari A, Cordery I, Pilgrim DH (1998) Relations between transmission losses and bed alluvium in an Australian arid zone stream. In: Wheater H, Kirby C (eds) Hydrology in a changing environment, vol II. Wiley, New York, pp 361–366

UNESCO (1977) World distribution of arid regions. Map scale: 1/25,000,000. UNESCO, Paris

Viglione A, Borga M, Balabanis P, Blöschl G (2010) Barriers to the exchange of hydrometeorological data across Europe-results from a survey and implications for data policy. J Hydrol 394:63–77. doi:10.1016/j.jhydrol.2010.03.023

Vivarelli R, Perera BJC (2002) Transmission losses in natural rivers and streams—a review. In: River symposium papers and presentations. River festival Brisbane Pty Ltd, South Brisbane

Walters MO (1990) Transmission losses in arid region. J Hydraul Eng 116(1):127–138

Walton WC (1969) Ground water resources evaluation. McGraw-Hill, New York

Wheater HS, Jakeman AJ, Beven KJ (1993) Progress and directions in rainfall-runoff modelling. In: Jakeman AJ, Beck MB, McAleer MJ (eds) Modelling change in environment systems. Wiley, Chichester, pp 101–132

Wheater HS, Woods B, Jolley TJ (1997) An integrated model of arid zone water resources: evaluation of rainfall-runoff simulation performance. In: Sustainability of water resources under increasing uncertainty, vol 24. IAHS, Mangalore

Wheater H, Sorooshian S, Sharma G (2008) Hydrological modelling in arid and semi-arid areas. Cambridge University Press, New York 195 pp

Willis AJ, Folkes BF, Yemm EW (1959) Branton burrows: the dune system and its vegetation: part II. J Ecol 47:249–288

Woods RA, Sivapalan M (1999) A synthesis of space-time variability in storm response: rainfall, runoff generation and routing. Water Resour Res 35:2469–2485

World Meteorological Organization (WMO) (2001) UNEP climate change: impacts, adaptation and vulnerability, contribution of working group ii to the third assessment report (TAR) of the intergovernmental panel on climate change (IPCC)

Younis J, Anquetin S, Thielen J (2008) The benefit of high-resolution operational weather forecasts for flash-flood warning. Hydrol Earth Syst Sci Discuss 5:345–377

Chapter 17
Seasonal Rainfall–Runoff Variability Analysis, Lake Tana Sub-Basin, Upper Blue Nile Basin, Ethiopia

Mengistu A. Jemberie, Adane A. Awass, Assefa M. Melesse, Gebiaw T. Ayele and Solomon S. Demissie

Abstract Lake Tana sub-basin of Abbay (Blue Nile) River Basin is located in the high land areas with unimodal rainy season with spatial and temporal variation of rainfall and runoff. Depending on available resources, there are many developmental plans and projects which seek the wise planning and management of water resources considering both low flows and floods. The seasonal streamflow variability analysis of the basin was performed with recorded meteorological and hydrological data. The four seasons of the year are considered for seasonality analysis. The rainfall variability is analysed using seasonality and variability measures of coefficient of variation, seasonal relative rainy days and seasonal rainfall intensity. The rainfall variability is more related with latitude and longitude. Spatial and temporal seasonal

M.A. Jemberie (✉)
Department of Civil Engineering, Adama Science and Technology University, 1888, Adama, Ethiopia
e-mail: mngst_addis@yahoo.com

A.A. Awass
Department of Hydraulics and Water Resources Engineering, Arba Minch University, Institute of Technology, Arba Minch, Ethiopia
e-mail: adaneabie@ethio.net

A.M. Melesse
Department of Earth and Environment, Florida International University, Miami, USA
e-mail: melessea@fiu.edu

G.T. Ayele
Faculty of Civil and Water Resource Engineering, Bahir Dar Institute of Technology, Bahir Dar University, 252, Bahir Dar, Ethiopia
e-mail: gebeyaw21@gmail.com

G.T. Ayele
Department of Hydraulic and Water Resources Engineering, Blue Nile Water Institute, Bahir Dar University, 252, Bahir Dar, Ethiopia

S.S. Demissie
Department of Civil and Environmental Engineering, University of California, Los Angeles, Los Angeles 90095, CA, USA
e-mail: solomon.demissie@gmail.com

© Springer International Publishing Switzerland 2016

A.M. Melesse and W. Abtew (eds.), *Landscape Dynamics, Soils and Hydrological Processes in Varied Climates*, Springer Geography,
DOI 10.1007/978-3-319-18787-7_17

rainfall variation is analysed from daily rainfall data. Seasonal runoff and streamflow variations are also analysed using HEC-HMS hydrological model to generate runoffs at required and selected points to detect spatial variation. Runoff variation for catchments with gauged stations was analysed from recorded time series streamflow data. Runoff coefficient is taken as a variability index for both generated and recorded streamflows. The runoff coefficient ranges from 0 to 1. The range is high in the dry seasons and less in the wet seasons. The average runoff coefficient value of the basin is 0.28 ranging from 0.18 to 0.36. The average seasonal runoff coefficient value from generated runoffs is 0.45, 0.3 for dry and 0.6 for wet seasons. From the results, it is shown that runoff coefficient is more dependent on antecedent soil wetness condition, land use and land covers. Catchments were categorised spatially and temporally as vulnerable, moderately vulnerable and less vulnerable to runoff based on the analysis. From hydrological data variability tests, it is clearly observed that seasonal time series data are not homogeneous, stationary and independent. Minimum flows are more stationary and homogeneous than mean and maximum flows.

Keywords Lake Tana Sub-basin · Seasonal hydrologic variability · Rainfall–runoff variability · HEC-HMS · Runoff coefficient · Blue Nile river basin

17.1 Introduction

Globally, there is growing concern that natural and anthropogenic caused climate changes are intensifying the hydrological cycle which is expected to influence river flow regimes (Hannaford and Harvey 2010). Seasonal changes in climate influence seasonal variation of meteorological and hydrological variables. Streamflow variability is highly dependent on rainfall variability.

Streamflow variability plays a key and significant role in many disciplines. In agriculture, the modern agriculture system based on irrigation is highly dependent on flow variability. Socially, the river basin is used for tourism and other socio-economic activities that are affected by flood and drought directly or indirectly related to the rainfall and streamflow variability. In engineering, flow variability in magnitude and time plays significant role for design of hydraulic and civil structures for multi purposes such as water supply, irrigation and hydropower that consider exceedance probabilities (Kwon et al. 2009).

Water resources planning and management mainly focuses on supplying a steady water supply amid hydrological variability. Such variability occurs temporally as well as spatially. The temporal variability occurs at many time scales, from hourly to daily, from daily to monthly, from monthly to seasonally and from seasonal to inter-annual and beyond. The spatial variation occurs from local to regional scales.

Engineering and management responses to flow variability depend on the time frame of the variability. Hydraulic structures such as dams typically respond to longer temporal variations. Dams are typically designed to provide water through a season or a year, storing water in high or excess flow seasons to supply in dry or low flow seasons. Whether it is natural or built reservoir, seasonal flow variation affects its storage as well as supply and demand. Storage structures mitigate the problems of drought and flood which are the consequences of seasonal hydrological variability of low flow- and high flow events. Reservoir operations in particular may benefit from a reduction in the uncertainty associated with future inflows (Kwon et al. 2009). Knowing seasonal flow variability and forecasting of hydrological variables improves water management. The seasonal variability of streamflows and lake stage (water level) are factors which play major role on the flows of the Abbay River.

Information on flood and drought seasonality is required in many practical applications in water resources engineering. Considering high flow and low flow seasons and variation is essential for developing reservoir operating rules to optimize flood prevention and water supply benefits.

Generally, Lake Tana sub-basin is characterised by enormous potential for development of irrigated agriculture, hydropower, tourism, biodiversity and recreation. For this reason, it is one of the richest sub-basins with regard to water resources. Since there is seasonal variation of hydrometeorological variables, there is fluctuation of the lake level. The lake is regulated as a reservoir with Chara Chara weir to manage seasonal flow variations for the purpose of proposed and constructed projects such as the Tana Beles Transfer that uses the regulated excess water from the lake.

The Chara Chara weir was built at the outflow of Lake Tana into the Abbay River in 1996 and expanded in 2001 for the purpose of flow regulating for downstream hydropower stations (McCartney and Shiferaw 2008). The Chara Chara weir regulates water storage in Lake Tana over a 3 m range of water level from 1784 to 1787 masl. The active storage of the lake between these levels is about 9100 Mm^3 (million m^3), which represents approximately 2.4 times the average annual outflow. The regulation for power production has modified the natural lake level regime resulting in reduced seasonal but greater inter-annual variation. The lowest level ever recorded was in June, 2003. This was a drought year in much of Ethiopia and hydropower production was constrained in many places. As a consequence of the low lake level in 2003, navigation ceased for approximately 4 months. Navigation of commercial boats ceases at lake level of 1785 masl. At this time, large areas of papyrus reed were exposed and destroyed. There was significant encroachment of agriculture on the exposed lake bed and decrease in fisheries production (Awulachew et al. 2009).

Lake Tana River basin has been a research area and many researchers have reported their findings related to climate change impact assessment, water balance, hydrological modelling and groundwater contribution assessments. The level of Lake Tana fluctuates annually and seasonally following the patterns of changes in precipitation. Estimate of the hydrological balance of the lake was reported using a mass balance approach. Water influx sources are inflows from five major rivers, subsurface inflow from the floodplain and precipitation. Outflow from the lake constitutes river discharge and evapotranspiration from the lake and analyses was on a monthly and annual basis (Wale 2008; Abreham 2009).

Prediction of net basin supply of Lake Tana was reported for different scenarios of climate change for three time windows; 2010–2039, 2040–2069 and 2070–2099.

Net basin supply is the sum of all inflow to the Lake and precipitation on the lake minus Lake Evaporation (Gebremariam 2009). The influence of topography, land use, soil and climatic condition on streamflows, soil erosion and sediment yield were examined using SWAT2005 (rainfall–runoff model). Flood forecasting approach using a quantitative precipitation forecasts and early warning was reported for Lake Tana sub-basin (Assefa et al. 2008).

In Ethiopia, although seasonal rainfall-based agriculture is the main source of economy, the need for supplemental water resource is vital. However, unless the water resource is utilized in a well-planned approach on demand and supply, its sustainability will be in danger. Lake Tana sub-basin and discharging rivers are highly dependent on seasonal rainfall. Proper planning, management and utilisation of water resources based on climate change and seasonal flow variation is essential without affecting the ecosystem (Gebremariam 2009).

Hydrology of the Nile River basin has been studied by various researchers. These studies encompass various areas including streamflow modelling, sediment dynamics, teleconnections and river flow, land use dynamics, climate change impact, groundwater flow modelling, hydrodynamics of Lake Tana, water allocation and demand analysis (Melesse et al. 2009a, b, 2011, 2014; Abtew et al. 2009a, b; Abtew and Melesse 2014a, b, c; Yitayew and Melesse 2011; Chebud and Melesse 2009a, b, 2013; Setegn et al. 2009a, b, 2010; Dessu and Melesse 2012, 2013; Dessu et al. 2014; Melesse 2011).

Water resources management has to deal with the seasonality of hydrological regimes. Generally, demand and water availability are mostly unbalanced with minimum availability of water in the winter season where demand is high. This pressure on water delivery systems increases the need to properly estimate the seasonality of hydrological processes. It is commonly assumed that seasonal as well as annual hydrological data (rainfall and streamflows) are stationary for hydrological analysis for water resources design studies although may not hold true due to non-stationary. Consequently, there is a need to evaluate the timing of hydrological regimes, its variation and check for long-term changes which could possibly violate the stationarity assumption of the annual cycle of hydrologic records (Renner and Bernhofer 2011).

It is common to observe the effects of seasonal rainfall and flow variation in most river basins. In Lake Tana sub-basin, there is significant variation of river flow discharging to the lake that adversely affects the lake level fluctuation and the downstream flow of Abbay River. Shortage and fluctuation of hydroelectric power and downstream shortage of water for different uses are the major consequences of the flow variation.

Currently, there is stiff competition for the available water resource among users at every time and place. There are many hydrological, statistical and mathematical applications using different models to solve problems related to seasonal flow variability and forecasting. It is vital to apply these techniques for decision making. The general objective of this study is to analyse the seasonal spatiotemporal rainfall–runoff variation using selected variability indices. The objectives are detecting seasonal trends in rainfall and streamflow, analysing spatial and temporal variation of rainfall–runoff using variability indices, identifying catchment hydrologic characteristics and classify catchment vulnerability to erosion and sedimentation.

17.2 Description of the Study Area

Lake Tana sub-basin which is one of the sub-basins of Abbay (Blue Nile) River is one of the most affected areas by soil erosion, sedimentation, land degradation and flooding (Setegn 2009a). The climate of Lake Tana region is 'tropical high land monsoon' with a single main rainy season from June to September. The mean annual rainfall over the catchment is 1326 mm ranging from 964 to 2000 mm with slightly more rainfall in the south and southeast than in the north of the catchment. Average annual evaporation over the lake surface is 1675 mm (Awulachew 2009).

The basin contains the largest lake in the country, Lake Tana, a natural fresh water lake which is the only lake in Ethiopia that is used for transportation with an area of 3156 km^2. The Lake Tana region is endowed with historical, cultural and natural heritages which have large tourist attractions. Consequently, the area is an important tourist destination in the country. It is estimated that close to 30,000 domestic and foreign visitors come to the area each year (Awulachew et al. 2009).

17.2.1 Location and Accessibility

Lake Tana basin is situated on the north western plateau of Ethiopia at the head waters of Abbay River, in the Amhara administrative regional state. Its geographical location extends from 10°57′N to 12°47′N latitude and from 36°53′E to 38°15′E longitude as shown in Fig. 17.1a–c. The basin area is 15,321 km^2 of which the lake covers about 20 % of the area. The basin is divided into four main sub-basins, Gilgel Abbay, Gumara, Ribb and Megech.

Under natural conditions, discharge from the lake is closely linked to rainfall and there is considerable seasonal and inter-annual variation (Kebede et al. 2006). Due to the impact of erratic and unpredictable changes in climate variables, there is a fluctuation of annual and seasonal flow. In some basins, there is decline in dry season low flows (Gebremariam 2009). As a result of high heterogeneity in habitat, the lake and surrounding riparian areas support high biodiversity and is listed in the top 250 lake regions of global importance for biodiversity (Awulachew 2009).

17.2.2 Topography

Even though the area is generally highland but elevation varies throughout the area. The altitude in the basin ranges between 1788 and 3712 m above the sea level

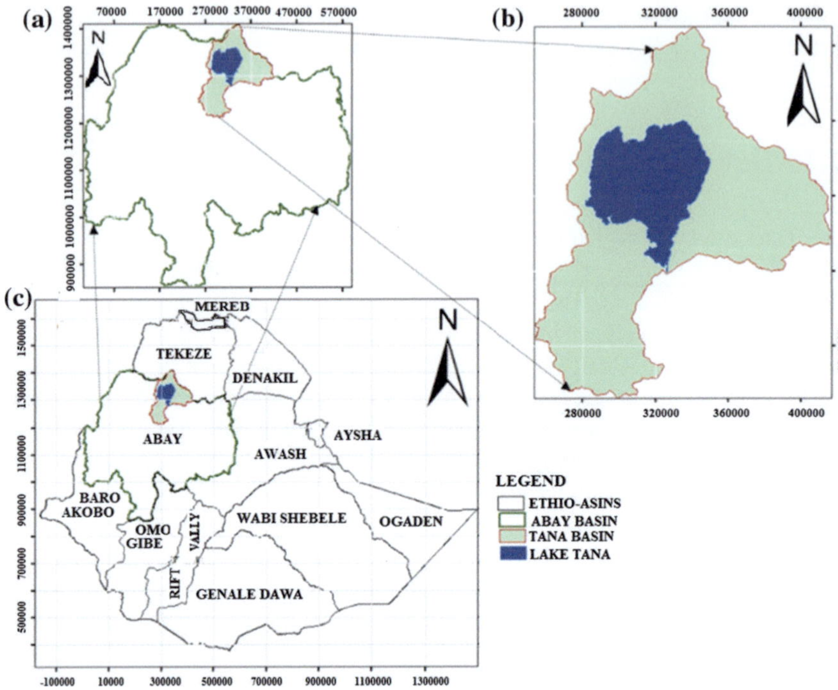

Fig. 17.1 Location of the study area **a** Abbay (Blue Nile) basin, **b** Tana sub-basin and **c** Ethiopia

(masl). It is mainly flat around the lake with a range of 1750 to 1850 masl and extending to the highlands up to 3170 masl (Fig. 17.2). Most of the basin area is flat and below elevation of 2500 masl. As the altitude variation, there is also high variation of slope ranging from 0 to 78 %. The slope is very steep at mountainous and high elevation areas and most of the basin slope is very low at low land areas near the lake. At the up streams of Gummera, Gilgel Abbay, Ribb and Megech, the slope is greater than 30 %.

17.3 Climate

17.3.1 Rainfall

Annual rainfall ranges from 964 to 2000 mm. High rainfall, 1400–2000 mm, is observed in the southern part of the lake, Gilgel Abay. Relatively lower rainfall, 1100–1400 mm, is observed in the northern part of the basin. Lowest rainfall, below 1100 mm is observed in Ribb and Megech areas (Fig. 17.3).

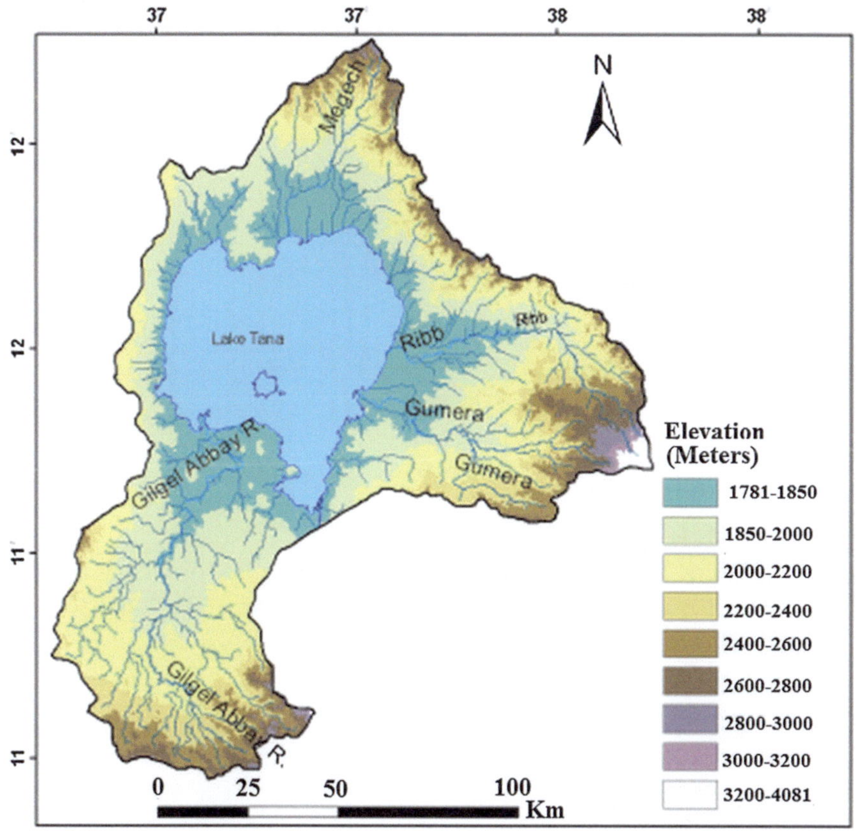

Fig. 17.2 Topography of the Lake Tana sub-basin

17.3.2 Temperature

Annual maximum and minimum temperature varies between 14 to 30 °C and 1 to 12 °C, respectively. Maximum of 25–30 °C is observed around the lake and minimum below 12 °C is observed on the highlands. Potential evapotranspiration (PET) ranges between 1160 and 1900 mm per year. Higher PET ranging from 1700 to 1900 mm is observed in the northern and eastern parts of the lake where temperature is high.

17.4 Hydrology

There are 20 hydrological gauging stations. As observed from historical flow data records, there is considerable fluctuation of flow due to rainfall and temperature variation. The mean annual runoff recorded at station Abbay near Bahir Dar is 3576.3 Mm^3. The flow is high from August to October with the peak in September.

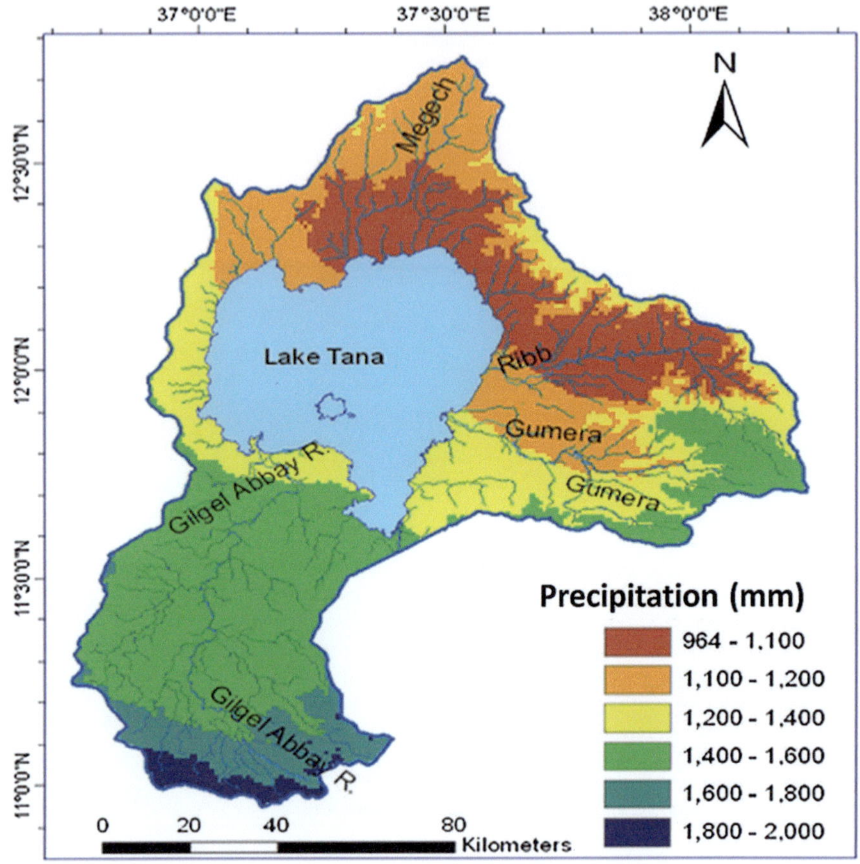

Fig. 17.3 Annual rainfall distribution in the Lake Tana sub-basin

Since there is seasonal variation of hydrometeorological variables, there is fluctuation of Lake Tana water level. A recent study on the lake's hydrology estimated the mean inflow to the lake to be 158 $m^3 s^{-1}$ (4986 $Mm^3 y^{-1}$). Moreover, the mean outflow is estimated to be 119 $m^3 s^{-1}$ (3753 $Mm^3 y^{-1}$), SMEC 2008. Under natural conditions, discharge from the lake is closely linked to rainfall and there is considerable seasonal and inter-annual variation (Kebede 2005). Table 17.1a, b depict flow statistics for Abbay basin and the four sub-basin of the Lake Tana sub-basin for the winter and summer seasons.

Naturally, the annual lake water level varies between 1785.75 and 1786.36 masl. Analyses of mean annual water levels reveals longer wet and dry cycles of approximately 6–7 years, during which mean annual water levels rise and fall, respectively. The Chara Chara weir regulates water storage in Lake Tana over a 3 m range of water levels from 1784 to 1787 masl. There is high variation of flow from season to season, year to year and also from site to site in the study area. Figure 17.4 depicts Lake Tana water level fluctuation before and after the Chara

Table 17.1 (a) Minimum, maximum, mean flows and volume for the winter season (JFM), (b) minimum, maximum, mean flows and volume for the summer season (JAS)

Gauging station	Area (km^2)	Minimum flow (in $m^3\ s^{-1}$)			Maximum flow (in $m^3\ s^{-1}$)			Mean flow (in $m^3\ s^{-1}$)			Seasonal volume ($m^3 \times 10^6$)		
		Mean	Min.	Max.	Mean	Min.	Max.	Mean	Min.	Max.	Mean	Min.	Max.
(a)													
Abbay	15320	35.2	1.1	123	114	26	259	67	19.3	165	528.23	152.16	1300.86
Gilgel Abbay	1644	2.73	0.8	11.7	7.73	2.9	30.9	4.4	1.55	17.7	34.69	12.22	139.55
Gummera	1394	0.96	0	2.75	4.23	0	14.5	2	0	8.42	15.77	0.00	66.38
Ribb	1592	0.25	0	2.55	2.28	0	10.9	0.8	0	4.7	6.31	0.00	37.05
Megech	462	0.79	0	5	1.55	0.1	8.56	1	0.02	5.61	7.88	0.16	44.23
(b)													
Abbay	15320	33.1	5.2	131	387	87	705	181	38.9	334	1427.00	306.69	2633.26
Gilgel Abbay	1644	48.9	8.5	88.1	350	242	655	163	123	203	1285.09	969.73	1600.45
Gummera	1394	14.5	0	43.4	249	172	398	102	56.9	165	804.17	448.60	1300.86
Ribb	1592	5.86	0.1	18.8	117	58	219	50	18.3	80	394.17	144.28	633.09
Megech	462	2.3	0.1	12.1	154	41	318	21	7.55	48	165.56	59.52	376.07

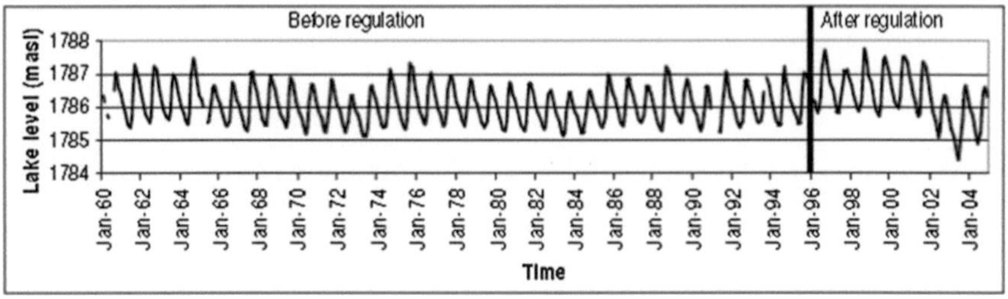

Fig. 17.4 Lake Tana level fluctuation before and after the Chara Chara weir regulation

Table 17.2 Lake Tana level fluctuation during the four seasons above reference datum of the Lake (1783.2 masl)

Seasons	Minimum level (m)			Maximum level (m)			Mean level (m)		
	Mean	Min.	Max.	Mean	Min.	Max.	Mean	Min.	Max.
Winter	2.16	1.4	2.72	2.73	2.2	3.4	2.4	1.81	3.12
Spring	1.76	0.7	2.34	2.16	1.4	2.72	1.9	1.08	2.53
Summer	1.87	0.9	2.5	3.44	2.9	4.3	2.8	2.08	3.52
Autumn	2.73	2.2	3.39	3.41	2.8	4.21	3.1	2.54	3.79

Chara weir regulation. Table 17.2 shows Lake Tana water level statistics for the four seasons of the year.

17.5 Methodology and Analysis

17.5.1 Data and Materials

For this study, spatiotemporal hydrometeorological data, daily rainfall and streamflow time series were used. Daily seasonal time series data is prepared from annual time series data. For the four seasons of the year, rainfall and flow variability analysis were done by conducting different seasonality and variability analysis measures and indices. Seasonal runoff and streamflow variations are also analysed using HEC-HMS hydrological model to generate runoffs at required and selected points to detect spatial variation.

17.5.2 Seasonal Rainfall Variability Analysis

To analyse the spatial and temporal seasonal rainfall variation, statistical and empirical methods were used. Areal rainfall was estimated using the Thiessen polygon method from recorded rainfall at 12 stations (Fig. 17.5).

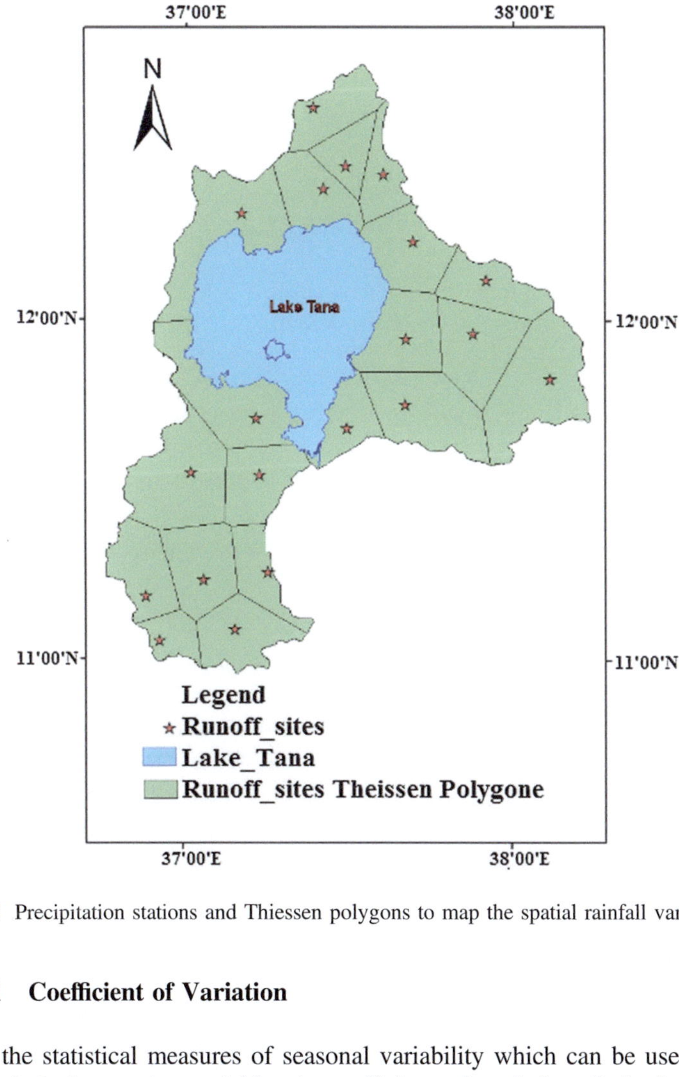

Fig. 17.5 Precipitation stations and Thiessen polygons to map the spatial rainfall variation

17.5.2.1 Coefficient of Variation

One of the statistical measures of seasonal variability which can be used for any hydrological time series variables is coefficient of variation. It is the ratio of standard deviation to the mean of time series data (Eq. 17.1).

$$\mathrm{CV} = \frac{S}{V_{\text{mean}}}, \tag{17.1}$$

where CV is seasonal coefficient of variation, S is seasonal standard deviation and V_{mean} is seasonal data mean.

17.5.2.2 Seasonal Relative Rainy Days

It is the ratio of rainy days in the season to the total number of days in the season (Eq. 17.2). It shows the variation of wet and dry days within seasons spatially as well as temporally.

$$\text{SRRD} = \frac{n * 100}{N}, \tag{17.2}$$

where SRRD is seasonal relative rainy days in %; N is total no of days in the season and n is the number of rainy days in the season.

17.5.2.3 Seasonal Rainfall Intensity

It is the ratio of cumulative seasonal rainfall to that of the number of rainy days with in a season. It indicates the spatial and temporal rainfall intensity variation of different seasons. The seasonal rainfall intensity is computed by Eq. 17.3.

$$\text{SRI} = \frac{\text{CSP}}{n}, \tag{17.3}$$

where SRI is seasonal rainfall intensity (mm d^{-1}), CSP is Cumulative Seasonal Precipitation (mm) and n is the number of rainy days with in the season (day). For the above three methods of seasonal rainfall variability measures, the values are plotted against latitude and longitude to show spatial variation and against year to show temporal variation.

17.5.2.4 Seasonal Flow Variability Analysis

Seasonal streamflow variation is analysed using Seasonal Runoff Coefficient as a measure of comparison rather than coefficient of variation. Runoff coefficient is simply the ratio of excess runoff to precipitation (Daniele 2009). To calculate the excess runoff, base flow is subtracted from recorded flow data for gauged stations. To see the variation on the whole basin, excess runoff is generated for selected points using HEC-HMS hydrological model.

17.5.2.5 Seasonal Runoff Coefficient (SRC)

Runoff coefficient is one measure or index to show the spatiotemporal runoff variability. This coefficient is used mostly to estimate runoff on data scarce sites using an empirical formula of the rational method (Eq. 17.4).

$$\text{SRC} = \frac{\sum Q_{\text{sd}}}{P_{\text{scp}} * A}, \tag{17.4}$$

where SRC is seasonal runoff coefficient, $\sum Q_{sd}$ is seasonal cumulative direct runoff coefficient, p_{scp} is seasonal cumulative precipitation and A is the catchment area.

17.5.2.6 Long-Term Data Variability

To check the long-term variability of seasonal time series data, Mann Whitney (M–W) test of homogeneity and stationarity with Wald Wolfowitz (W–W) test of independence and stationarity was conducted (Rao and Hamed 2000). From test results shown in Table 17.3, almost all of seasonal time series data for selected rivers are not homogeneous, independent and stationary.

17.6 Results and Discussion

17.6.1 Seasonal Rainfall Variability

It is common to observe spatiotemporal variation of rainfall globally, regionally and locally because of natural and anthropogenic factors. In the study area, it is observed that there is seasonal variation of rainfall which is the major cause of streamflow variation. According to the country's seasonal geographical classification, the year is classified into four seasons Winter [January, February and March (JFM)]; Spring [April, May and June (AMJ)]; Summer [July, August and September (JAS)] and Autumn [October, November and December (OND)].

The variation is analysed using variability measures of CV, SRRD and SRI along longitude and latitude for spatial variation and the period from 1997 to 2005, for temporal variation. There is significant change in coefficient of variation in all seasons based on CV values ranging from 0.85 to 7.76 with high in the winter. There is high variation in autumn season which ranges from 2.35 to 6.22 and less variation in summer ranging from 0.85 to 1.5. The rainfall coefficient of variability is relatively higher near Lake Tana. In the winter season, the variation is higher from west to east of the Lake and it is lower on the southern part of the basin. In the summer season, the value is relatively higher north of the Lake around Delgi, Gorgora and Gondar whereas at southern part of the basin near Enjibara, Dangila and Zege, variation is lower.

There is decreasing temporal trend of SRRD in autumn, spring and summer seasons but no trend in winter (Fig. 17.6a). SRI range is high in winter and low in the summer. There is no visible temporal trend of SRI variation in autumn, winter and spring but increasing trend in the summer (Fig. 17.6b).

Table 17.3 M–W test results of homogeneity and stationary at 5 % significance level

Name of river	Flow	Winter		Spring		Summer		Autumn	
	Type	α value	Comment	α value	Comment	α value	Comment	α value	Comment
	Av.	−2.35	Rejected	−3.57	Rejected	−0.42	Accepted	−1.72	Rejected
Gilgel	Min.	−2.19	Rejected	−1.48	Rejected	−0.15	Accepted	−0.26	Accepted
Abbay	Max.	−0.965	Accepted	−4.04	Rejected	−1.25	Rejected	−2.16	Rejected
	Av.	−4.46	Rejected	−4.36	Rejected	−4.46	Rejected	−2.84	Rejected
Gummera	Min.	−3.6	Rejected	−4.85	Rejected	−4.12	Rejected	−3.73	Rejected
	Max.	−3.73	Rejected	−4.49	Rejected	−2.97	Rejected	−3.24	Rejected
	Av.	−2.22	Rejected	−2.58	Rejected	−2.27	Rejected	−3.8	Rejected
Ribb	Min.	−0.73	Accepted	−0.887	Accepted	−2.06	Rejected	−1.95	Rejected
	Max.	−2.97	Rejected	−4.28	Rejected	−1.28	Rejected	−2.53	Rejected
	Av.	−4.37	Rejected	−4.34	Rejected	−3.49	Rejected	−4.06	Rejected
Megech	Min.	−4.54	Rejected	−4.18	Rejected	−4.28	Rejected	−3.23	Rejected
	Max.	−3.04	Rejected	−3.1	Rejected	−2.47	Rejected	−3.45	Rejected

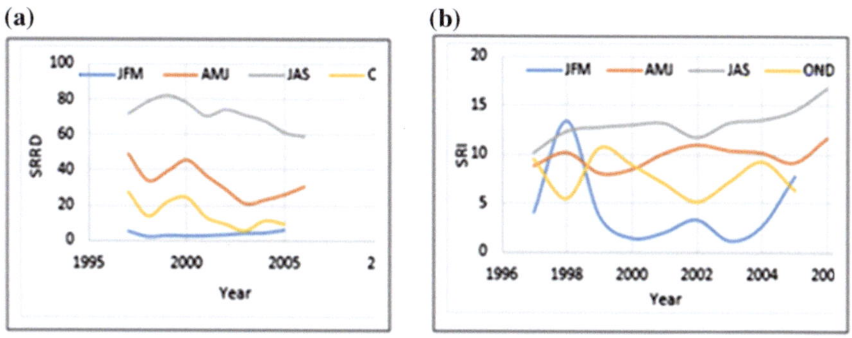

Fig. 17.6 Temporal variation of seasonal rainfall using **a** SRRD and **b** SRI indices

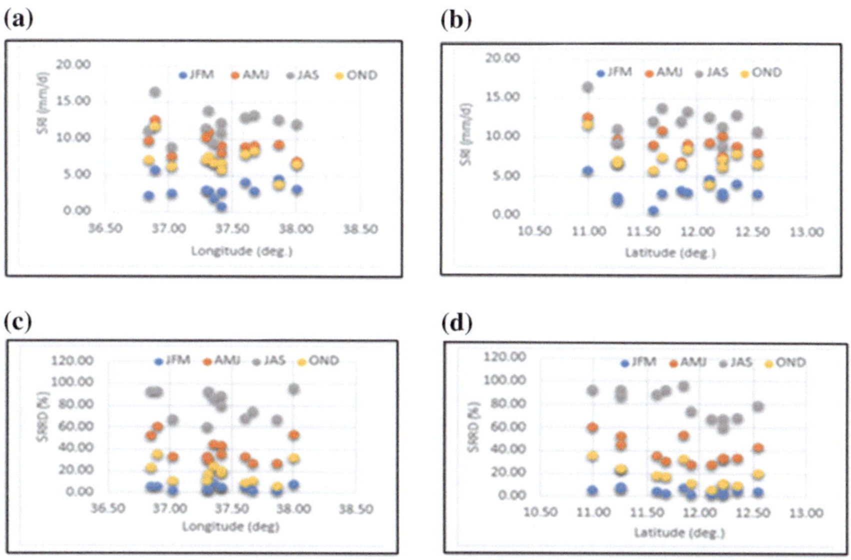

Fig. 17.7 Spatial variation of seasonal rainfall using SRI and SRRD indices

As shown in Fig. 17.7a, b, summer and spring SRI decrease with increasing longitude and latitude. Autumn and winter SRI do not have visible trend with latitude and longitude. SRRD decreases with increasing latitude and longitude for autumn, spring and summer. Winter SRRD does not have visible trend with latitude and longitude (Fig. 17.7c, d).

17.6.2 Runoff Variability Analysis

Flow simulations are done using models which are calibrated and validated using two different sets of data. Various hydrological models are used for flow simulations. The most commonly used model capable of flow predication for ungauged watershed is the Soil and Water Assessment Tool (SWAT).

The application of SWAT in predicting streamflow and sediment as well as evaluation of the impact of land use and climate change on the hydrology of watersheds has been documented by various studies (Dessu and Melesse 2012, 2013; Dessu et al. 2014; Wang et al. 2006a, b, 2008a, b, c; Wang and Melesse 2005; Behulu et al. 2014, 2013; Setegn et al. 2009a, b, 2010, 2011, 2014; Mango et al. 2011a, b; Getachew and Melesse 2012; Assefa et al. 2014; Grey et al. 2013; Mohamed et al. 2015).

The study reported in this chapter used the HEC-HMS model. For gauged catchments, the seasonal streamflow variability is easily computed from recorded flow data. For catchments outside gauging stations, excess runoff is generated using HEC-HMS model for selected points, after the model is calibrated and validated at gauged stations. The R^2 value during the calibration period was 0.76 and 0.68 during the validation period. Since the purpose of HEC-HMS model is to generate runoff, the generated runoff R^2 values are acceptable. Thus the variability is analysed using Seasonal Runoff Coefficient (SRC) as a variability measure. From the results obtained, the runoff variability pattern and trend are similar to the rainfall variability.

The spatial SRC variation along latitude and longitude is highly dependent and correlated to the SRI. All rainy days did not give excess runoff. In other words, the excess runoff is the result of infiltration excess rainfall which is the result of intensive rainfall or Antecedent Moisture Condition (AMC) of the surface. Runoff coefficient of variation and distribution likely reflects the variability of the storms analysed, mostly in terms of total precipitation, rainfall duration, rainfall intensity and antecedent wetness conditions (Ley et al. 2011). They reported runoff coefficient ranging from 0.02 to −0.69 with an average value of 0.15. From the model generated, excess runoffs, the average runoff coefficient value obtained is 0.45 ranging from 0.28 in winter and 0.64 in autumn. Figure 17.8 depict temporal variation of seasonal SRC. Distinct annual variation in SRC is clearly shown. Winter and spring SRC has increasing temporal trend.

The range of SRC variation for both gauged and ungauged sites is high in spring and autumn seasons and low in summer season. But the average value is higher in summer and autumn seasons as a result of high runoff and erosion of the summer season and high antecedent soil moisture condition of autumn season following high summer precipitation. In addition, high base flow components which are considered as runoffs have significant effect on increased SRC values. The range of SRC varies from 0 to −1 at gauged catchments. Even though the range is very high, the average runoff coefficient value for the basin is 0.28 from recorded historical flow data and it is 0.45 from generated excess runoffs using HEC-HMS model.

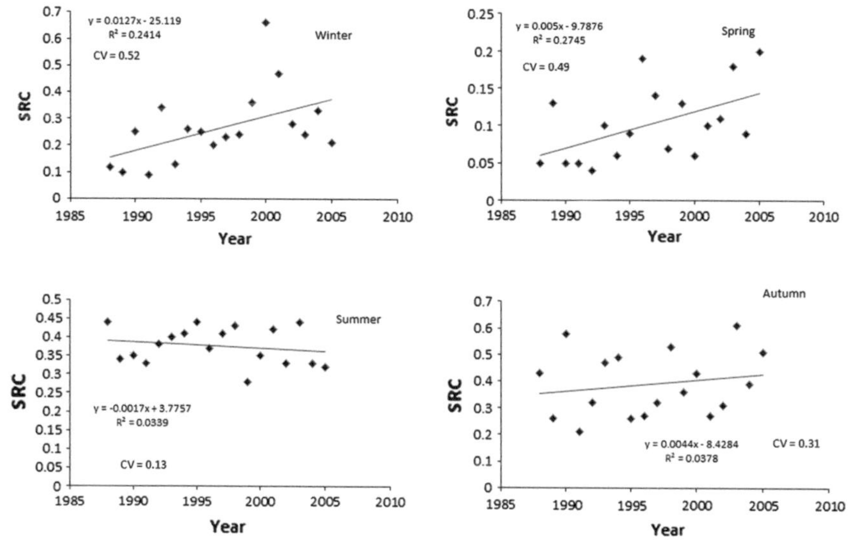

Fig. 17.8 Temporal Seasonal Runoff Coefficient Variation for winter (JFM), spring (AMJ), summer (JAS) and autumn (OND)

From both values, the average of 0.36 lies on the range of theoretical and applied values used in the rational method.

When comparing SRC values obtained from HEC-HMS results and recorded flows, the average values are over estimated and maximum values are under estimated by the model. This may be because of representing the whole catchment with limited hydrological response units (HRU) to calibrate and validate the model. The variation exceeds 1 at some places and some events which indicate that there was high runoff and erosion problem at that time at the specific place including the upstream.

The spatial SRC variation is relatively higher at altitudes from 2200 to 2400 m. Generally, the areas on southeast and east of the basin have higher runoff coefficient. The range and average runoff coefficient values are consistent with those values used for different soil groups and land uses. There is high spatial variation of runoff coefficient than temporal variation. The variation is high in winter and spring seasons. Even though there is significant spatial variation of SRC, there is temporal trend detected for winter and spring. Tables 17.4 and 17.5 summarise the obtained results.

17.6.3 Catchment Classification Based on SRC

Using SRC as a classification measure, the catchments can be seasonally classified and grouped based on their seasonal runoff coefficient values. The catchments with

Table 17.4 Spatial seasonal runoff coefficients in the sub-basin and runoff condition for the four seasons [runoff condition slight (SRC < 0.25), moderate (0.25 < RC < 0.45) and high (SRC > 0.45)]

Catchment	Seasons							
	Winter		Spring		Summer		Autumn	
	Mean SRC	Range	Mean SRC	Range	Mean SRC	Range	Mean SRC	Range
Gilgel Abbay	0.21	0.04–0.64	0.18	0.06–0.28	0.54	0.43–0.76	0.62	0.37–0.91
Gummera	0.3	0.04–0.91	0.09	0.01–0.33	0.55	0.30–0.77	0.57	0.21–0.97
Ribb	0.09	0.01–0.49	0.04	0.01–0.12	0.24	0.14–0.40	0.16	0.02–0.53
Megech	0.2	0.01–0.9	0.09	0.01–0.5	0.36	0.17–0.53	0.29	0.04–0.82
Upper Ribb	0.1	0.02–0.41	0.04	0.01–0.11	0.26	0.04–0.66	0.11	0.01–0.33
Koga	0.58	0.06–0.97	0.13	0.05–0.31	0.37	0.16–0.93	0.62	0.27–0.94
Gelda	0.44	0.09–0.93	0.21	0.01–0.94	0.72	0.31–1.15	0.77	0.18–1.12
Gemero	0.13	0.01–0.71	0.08	0.01–0.48	0.24	0.10–0.40	0.16	0.01–0.79
Amen	0.34	0.04–1.29	0.02	0.01–0.06	0.14	0.07–0.29	0.25	0.09–0.87
Garno	0.31	0.01–0.95	0.06	0.01–0.22	0.24	0.10–0.41	0.36	0.03–1.17
Angereb	0.25	0.01–0.87	0.17	0.01–0.83	0.47	0.04–1.22	0.32	0.05–0.91

Table 17.5 Temporal seasonal runoff coefficients and the runoff condition for the four seasons of the sub-basin [runoff condition slight (SRC < 0.25), moderate (0.25 < RC < 0.45) and high (SRC > 0.45)]

Year	Seasons											
	Winter			Spring			Summer			Autumn		
	Min. SRC	Mean SRC	Max. SRC	Min. SRC	Mean SRC	Max. SRC	Min. SRC	Mean SRC	Max. SRC	Min. SRC	Mean SRC	Max. SRC
1988	0	0.12	0.46	0	0.05	0.15	0.18	0.44	0.93	0.08	0.43	0.94
1989	0.01	0.1	0.51	0.01	0.13	0.83	0.1	0.34	1.01	0.05	0.26	0.53
1990	0.05	0.25	0.74	0	0.05	0.12	0.1	0.35	0.61	0.06	0.58	1.73
1991	0.03	0.09	0.23	0.01	0.05	0.17	0.14	0.33	0.64	0.01	0.21	0.58
1992	0.02	0.34	0.9	0	0.04	0.14	0.15	0.38	0.79	0.03	0.32	0.8
1993	0.01	0.13	0.6	0.01	0.1	0.24	0.21	0.4	0.76	0.09	0.47	1.04
1994	0.03	0.26	0.7	0.01	0.06	0.19	0.23	0.41	0.6	0.04	0.49	0.97
1995	0.01	0.25	1.29	0.01	0.09	0.45	0.07	0.44	1.22	0.02	0.26	0.94
1996	0.01	0.2	0.9	0.03	0.19	0.58	0.11	0.37	0.77	0.07	0.27	0.91
1997	0.04	0.23	0.76	0.02	0.14	0.54	0.16	0.41	0.75	0.03	0.32	0.96
1998	0.1	0.24	0.56	0.01	0.07	0.21	0.04	0.43	1.15	0.11	0.53	1.12
1999	0	0.36	0.85	0.02	0.13	0.48	0.12	0.28	0.62	0.05	0.36	0.91
2000	0.14	0.66	0.94	0.01	0.06	0.17	0.16	0.35	0.88	0.03	0.43	1.03
2001	0.11	0.47	0.95	0.02	0.1	0.23	0.18	0.42	1.04	0.08	0.27	0.56
2002	0.04	0.28	0.95	0.01	0.11	0.45	0.04	0.33	1.03	0.01	0.31	0.92
2003	0	0.24	0.97	0.01	0.18	0.94	0.09	0.44	1.09	0.06	0.61	0.96
2004	0.05	0.33	0.83	0.01	0.09	0.2	0.08	0.33	1.08	0.07	0.39	1.12
2005	0.1	0.21	0.63	0.02	0.2	0.5	0.06	0.32	0.69	0.01	0.51	1.04

high SRC values such as Gelda, Koga, Gilgel Abbay and Gummera are severely vulnerable for erosion and sedimentation. Low runoff coefficient values indicate that the areas have less erosion as a result of combined effects of good watershed management practices (land use and land cover change), soil type and slope. Runoff coefficient is the main driving factor for erosion and flooding. Some of the highland areas have high SRC with wide ranges at gauged catchments for all seasons even though the average values are similar to model generated runoff sites.

Previous studies have identified the erosion-prone areas on the catchment and the factors using SWAT hydrological model for different climate change scenario (Setegn et al. 2009a). According to the reported study, the main factor for erosion is slope factor which is the main driving force for the movement of surface water. Steeper and longer slopes result in high erosion rates. The second criterion is the land cover which controls the detachability and transport of soil particles and infiltration of water into the soil. The types of the soil also play a significant role for erosion depending upon their physical properties and sensitivity to erosion.

Runoff coefficient in this study is high on the areas relatively higher altitudes and slopes which supports the previous studies. Runoff coefficient is relatively high during summer seasons with smaller ranges. Depending on the results obtained, catchments can be categorized spatially and temporally as vulnerable, moderately vulnerable and less vulnerable to runoff. Catchments near south east and south parts of the lake are highly vulnerable during all seasons; whereas, highlands of some of the northern and eastern parts of the sub-basin are less vulnerable (Fig. 17.9).

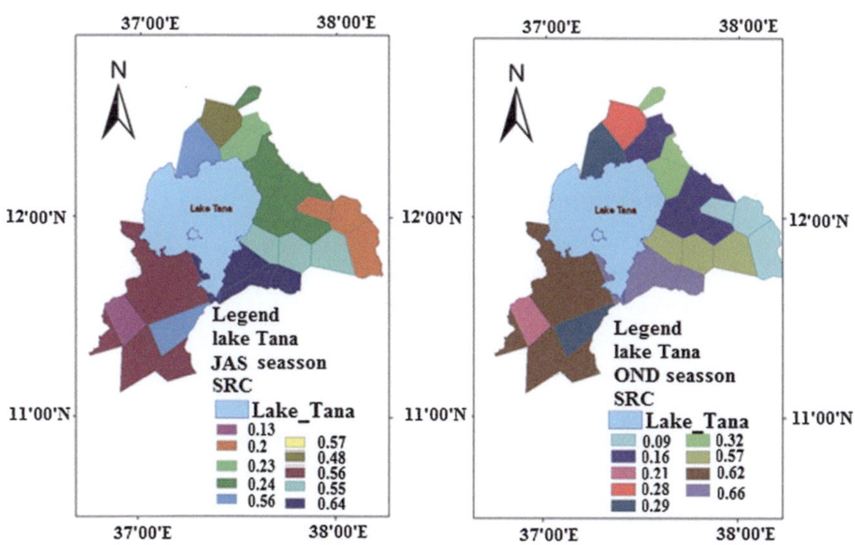

Fig. 17.9 Seasonal runoff coefficient variation and catchment classification based on runoff coefficient; summer (*left*) and autumn (*right*) seasons

17.7 Conclusion and Recommendation

17.7.1 Conclusion

Greater than 50 % of annual precipitation occurs in the summer season. Rainfall variability is dependent on altitude and there is lake effect on rainfall. The coefficient of rainfall variability is highly dependent on number of rainy days whether its intensity is high or low. There is high variation around the Lake. From the results obtained, it can be concluded that Lake Tana has slight effect on seasonal rainfall variation.

Rainfall has direct impact on runoff variability as seen from generated excess flows using hydrological lumped rainfall–runoff model, HEC-HMS. Runoff variation is more influenced by rainfall intensity than rainfall duration. The average seasonal runoff coefficient is relatively higher in autumn season. This shows that antecedent soil wetness (moisture condition) is a dominant factor for runoff coefficient. The runoff coefficient is also high in the summer and sometimes it is above the limit which shows that the rainfall magnitude, intensity and land use are significant and dominant factors (Mugabe et al. 2006). The catchment size, shape and slope are also factors for runoff coefficient variation.

From the results of seasonal rainfall variability and seasonal runoff coefficients, decision makers should take remedial measures for the areas which are exposed to high runoff and erosion. It should be considered that streamflow may not be sufficient to fulfil the water demand and it requires proper planning and management using optimization. Sectors can use the runoff coefficients which are obtained from recorded and generated flow data rather than using assumed theoretical values.

From hydrological data quality tests, it is clearly observed that seasonal time series data are not homogeneous, stationary and independent. Minimum flows are more stationary and homogeneous than mean and maximum flows.

17.7.2 Recommendation

Event runoff coefficients using storm events at relatively smaller durations can give better results. Considering future development and climate change scenario, runoff coefficient should be modified and better seasonality and variability measures used. It is recommended that researchers do seasonal regional flood frequency and catchment classification based on seasonal runoff coefficient and other measures at gridded- or semi-distributed catchment level. In data scarce areas, it is better to use seasonal runoff coefficients to estimate runoff using empirical formulae such as the rational method. Using distributed- or semi-distributed hydrological models is recommended than lumped models such as HEC-HMS for better runoff variability analysis. From the results of SRC, it is recommended to take remedial land use and soil conservation measures of upstream watersheds related with high runoff coefficient values. For better reservoir planning, operation and management of the available water resource, it is better to consider the dry and flood seasons independently. In addition, the downstream environment should be considered to receive minimum natural flows in the dry season.

References

Abreham AE (2009) Open water evaporation estimation using ground measurements and satellite remote sensing: a case study of Lake Tana, Ethiopia. MSC thesis, ITC, Enschede, The Netherlands

Abtew W, Melesse A, Desalegn T (2009a) Spatial, inter and intra-annual variability of the Blue Nile River Basin Rainfall. Hydrol Process 23(21):3075–3082

Abtew W, Melesse A, Desalegn T (2009b) El Niño southern oscillation link to the Blue Nile River Basin hydrology. Hydrol Process Spec Issue Nile Hydrol 23(26):3653–3660

Abtew W, Melesse AM (2014a) Nile River basin hydology. In: Melesse AM, Abtew W, Setegn S (eds) Nile River basin: ecohydrological challenges, climate change and hydropolitics, pp 7–22

Abtew W, Melesse AM (2014b) Climate Teleconnections and Water Management. In Nile River Basin (pp. 685-705). Springer International Publishing

Abtew W, Melesse AM (2014c) Transboundary rivers and the Nile. In: Nile River basin, Springer International Publishing, Berlin, pp 565–579

Assefa KA (2008) Flood forecasting and early warning in Lake Tana sub-basin, Upper Blue Nile, Ethiopia. WaterMill, UNESCO-IHE

Assefa A Melesse AM, Admasu S (2014). Climate change in upper Gilgel Abay River catchment, Blue Nile basin Ethiopia. In: Melesse AM, Abtew W, Setegn S (eds) Nile River basin: ecohydrological challenges, climate change and hydropolitics, pp 363–388

Awulachew SB, Tenaw M, Steenhuis T, Easton Z, Ahmed A (2009) Impact of watershed interventions on runoff and sedimentation in Gumera Watershed. CGIAR

Behulu F, Setegn S, Melesse AM, Fiori A (2013) Hydrological analysis of the Upper Tiber basin: a watershed modeling approach. Hydrol Process 27(16):2339–2351

Behulu F, Setegn S, Melesse AM, Romano E, Fiori A (2014) Impact of climate change on the hydrology of upper Tiber River basin using bias corrected regional climate model, water resources management, pp 1–17

Chebud Y, Melesse AM (2013) Stage level, volume, and time-frequency change information content of lake tana using stochastic approaches. Hydrol Process 27(10):1475–1483. doi:10.1002/hyp.9291

Chebud YA, Melesse AM (2009a) Numerical modeling of the groundwater flow system of the Gumera sub-basin in Lake Tana basin, Ethiopia. Hydrol Process Spec Issue Nile Hydrol 23 (26):3694–3704

Chebud YA, Melesse AM (2009b) Modeling Lake Stage and water balance of Lake Tana, Ethiopia. Hydrol Process 23(25):3534–3544

Daniele NM (2009) Controls on event runoff coefficients in the eastern Italian Alps. J Hydrol 375 (3–4):312–325

Dessu SB, Melesse AM, Bhat M, McClain M (2014) Assessment of water resources availability and demand in the Mara River basin. CATENA 115:104–114

Dessu SB, Melesse AM (2012) Modeling the rainfall-runoff process of the Mara River basin using SWAT. Hydrol Process 26(26):4038–4049

Dessu SB, Melesse AM (2013) Impact and uncertainties of climate change on the hydrology of the Mara River basin. Hydrol Process 27(20):2973–2986

Gebremariam H (2009) Assessment of climate change impact on the net basin supply of Lake Tana. ITC

Getachew HE, Melesse AM (2012). Impact of land use/land cover change on the hydrology of Angereb watershed, Ethiopia. Int J Water Sci 1(4):1–7. doi: 10.5772/56266
Grey OP, Webber Dale G, Setegn SG, Melesse AM (2013) Application of the soil and water assessment tool (SWAT Model) on a small tropical island state (Great River Watershed, Jamaica) as a tool in integrated watershed and coastal zone management. Int J Trop Biol Conserv 62(3):293–305
Hannaford J, Harvey CL (2010) UK seasonal river flow variability in near-natural catchments, regional outflows and long hydrometric records. In: Kirby C (ed) Role of hydrology in managing consequences of a changing global environment. British Hydrological Society Third International Symposium. British Hydrolgical Society, Newcastle, London, pp 96–102
Kebede ST (2005) Water balance of Lake Tana and its sensitivity to fluctuations in rainfall, Blue Nile basin, Ethiopia. J Hydrol 316:233–247
Kebede SY, Travi T, Alemayehu M (2006) Water balance of Lake Tana and its sensitivity to fluctuations in rainfall, Blue Nile basin, Ethiopia. J Hydrol 316:233–247
Kwon HH, Casey B, Kaiqin X, Upmanu L (2009) Seasonal and annual maximum streamflow forecasting using climate information: application to the Three Gorges Dam in Yangtze River basin, China. Hydrol Sci J 54(3):582–595
Ley R, Caspe MC, Hellebrand H, Merz R (2011) Catchment classification by runoff behaviour with self-organizing maps (SOM). Hydrol Earth Syst Sci 15:2947–2962
Mango L, Melesse AM McClain ME, Gann D, Setegn SG (2011a) Land use and climate change impacts on the hydrology of the upper Mara River Basin, Kenya: results of a modeling study to support better resource management, special issue: climate, weather and hydrology of East African Highlands. Hydrol Earth Syst Sci 15:2245–2258. doi: 10.5194/hess-15-2245-2011
Mango L, Melesse AM, McClain ME, Gann D, Setegn SG (2011b) Hydro-meteorology and water budget of Mara River basin, Kenya: a land use change scenarios analysis. In: Melesse A (ed) Nile River basin: hydrology, climate and water use, Springer Science Publisher, Chapter 2, pp 39–68, doi: 10.1007/978-94-007-0689-7_2
McCartney MP, Shiferaw A, Seleshi Y (2008) Estimating environmental flow requirements downstream of the Chara Chara weir on the Blue Nile River. In: Abtew W, Melesse AM (eds) Proceedings of the workshop on hydrology and ecology of the Nile River basin under extreme conditions, Ethiopia, Aardvark Global Publishing, June 16–19
Melesse AM (2011) Nile River basin: hydrology, climate and water use. Springer Science & Business Media, Germany
Melesse A, Abtew W, Setegn SG (2014) Nile River basin: ecohydrological challenges, climate change and hydropolitics. Springer Science & Business Media, Germany
Melesse AM, Loukas Athanasios G, Senay Gabriel, Yitayew Muluneh (2009a) Climate change, land-cover dynamics and ecohydrology of the Nile River Basin. Hydrol Process Spec Issue Nile Hydrol 23(26):3651–3652
Melesse AM, Abtew W, Desalegne T, Wang X (2009b) Low and high flow analysis and wavelet application for characterization of the Blue Nile River system. Hydrol Process 24(3):241–252
Melesse AM, Abtew W, Setegn S, Dessalegne T (2011) Hydrological variability and climate of the Upper Blue Nile River basin In: Melesse A (ed) Nile River basin: hydrology, climate and water use e, Springer Science Publisher Chapter 1, pp 3–37. doi: 10.1007/978-94-007-0689-7_1
Mohammed H, Alamirew T, Assen M, Melesse AM (2015) Modeling of sediment yield in Maybar gauged watershed using SWAT, northeast Ethiopia. CATENA 127:191–205
Mugabe FT, Hodnett MG, Senzanje A, Gonah T (2006) Spatio-temporal rainfall and runoff variability of the Runde catchment, Zimbabwe, and implications on surface water resources. Afr Water J 1:74–76
Rao AR, Hamed KH (2000) Flood frequency analysis. CRC Press, USA
Renner M, Bernhofer C (2011) Long term variability of the annual hydrological regime and sensitivity to temperature phase shifts. J Hydrol Earth syst Sci 15:1819–1833
Setegn SG, Srinivasan R, Dargahi B, Melesse AM (2009a) Spatial delineation of soil erosion prone areas: application of SWAT and MCE approaches in the Lake Tana basin, Ethiopia. Hydrol Process Spec Issue Nile Hydrol 23(26):3738–3750

Setegn SG, Srinivasan R, Melesse AM, Dargahi B (2009b) SWAT model application and prediction uncertainty analysis in the Lake Tana basin, Ethiopia. Hydrol Process 24 (3):357–367

Setegn SG, Dargahi B, Srinivasan R, Melesse AM (2010) Modelling of sediment yield from Anjeni gauged watershed, Ethiopia using SWAT. JAWRA 46(3):514–526

Setegn SG, Melesse AM, Haiduk A, Webber D, Wang X, McClain M (2014) Spatiotemporal distribution of fresh water availability in the Rio Cobre watershed, Jamaica. CATENA 120:81–90

Setegn S, Rayner D, Melesse AM, Dargahi B, Srinivasan R, (2011) Impact of climate change on the hydro-climatology of Lake Tana basin, Ethiopia. Water Resour Res 47(W04511):13. doi: 10.1029/2010WR009248

SMEC (2008) Hydrological Study of the Tana-Beles sub-basins, main report. Ministry of Water Resource, Ethiopia

Wale A (2008) Hydrological balance of Lake Tana Upper Blue Nile basin, Ethiopia. ITC, The Netherlands

Wang X, Shang S, Yang W, Melesse AM (2008a) Simulation of an agricultural watershed using an improved curve number method in SWAT. Tans Am Soc Agric Bio Eng 51(4):1323–1339

Wang X, Yang W, Melesse AM (2008b) Using hydrologic equivalent wetland concept within SWAT to estimate streamflow in watersheds with numerous wetlands. Tans Am Soc Agric Bio Eng 51(1):55–72

Wang X, Melesse AM, Yang W (2006) Influences of potential evapotranspiration estimation methods on SWAT's hydrologic simulation in a Northwestern Minnesota watershed. Trans ASAE 49(6):1755–1771

Wang X, Melesse AM (2006) Effects of STATSGO and SSURGO as inputs on SWAT model's snowmelt simulation. J Am Water Res Assoc 42(5):1217–1236

Wang X, Melesse AM (2005) Evaluations of the SWAT model's snowmelt hydrology in a Northwestern Minnesota watershed. Trans ASAE 48(4):1359–1376

Wang X, Garza J Whitney M Melesse AM, Yang W (2008c) Prediction of sediment source areas within watersheds as affected by soil data resolution. In: Findley PN (ed) Environmental modelling: new research, Ch. 7. Nova Science Publishers, Inc., Hauppauge, NY 11788, p 151–185, ISBN: 978-1-60692-034-3

Yitayew M, Melesse AM (2011) Critical water resources management issues in Nile River basin, In: Melesse A (ed) Nile River basin: hydrology, climate and water use. Springer Science Publisher, Chapter 20, pp 401–416. doi: 10.1007/978-94-007-0689-7_20

Part IV
Groundwater Flow and Aquifer Management

Chapter 18
Flood Forecasting and Stream Flow Simulation of the Upper Awash River Basin, Ethiopia Using Geospatial Stream Flow Model (GeoSFM)

Shimelis Behailu Dessu, Abdulkarim Hussein Seid, Anteneh Z. Abiy and Assefa M. Melesse

Abstract The Geospatial Stream Flow Model (GeoSFM) has been widely applied in data scarce regions for flood forecasting and stream flow simulation with remotely acquired data. GeoSFM was applied in the Upper Awash River basin (UARB) with observed input data set. GeoSFM sensitivity to observed input data quality, subbasin partition, and change in parameter were investigated. Results demonstrated that GeoSFM is sensitive to the size and number of subbasins. Among the eight model parameters, the basin loss and curve number are the most sensitive in UARB. GeoSFM evaluation using a split sample of 10 years observed daily discharge showed satisfactory performance, Nash-Sutcliff Efficiency 0.67 and 0.70, coefficient of determination, 0.60 and 0.65 for calibration and validation, respectively. The monthly average simulation captured 76 % of the observed variability over 10 years. Comparative analysis suggested increasing partitions improves performance in capturing flooding events and the single basin scenario can potentially be used for flood forecasting or resource assessment purposes. The 60 % coverage of vertisol in the basin and low quality of observed data affected model performance. Further evaluation of GeoSFM in heterogeneous soil type and land use/cover can help to identify the influence of dominant physical

S.B. Dessu (✉) · A.H. Seid
Department of Civil Engineering, Addis Ababa University, Addis Ababa, Ethiopia
e-mail: sbehailu@gmail.com

A.H. Seid
e-mail: aseid@nilebasin.org

A.H. Seid
Water Resources Management, Nile Basin Initiative Secretariat, Entebbe, Uganda

A.Z. Abiy · A.M. Melesse
Department of Earth and Environment, Florida International University, Miami, USA
e-mail: aabiy001@fiu.edu

A.M. Melesse
e-mail: melessea@fiu.edu

© Springer International Publishing Switzerland 2016

A.M. Melesse and W. Abtew (eds.), *Landscape Dynamics, Soils and Hydrological Processes in Varied Climates*, Springer Geography,
DOI 10.1007/978-3-319-18787-7_18

characteristics. In general, GeoSFM offers a competent platform for stream flow simulation and water resource assessment in data scarce regions.

Keywords GeoSFM model · FEWS-SFM model · Flood forecasting · Flow simulation · Awash river · Koka dam · Rainfall–runoff · Ethiopia

18.1 Introduction

Computer rainfall–runoff models are the go-to tools for understanding hydrological processes. These models offer fast and flexible platform to recreate historical scenarios and generate possible future ensembles. The trade-off in optimizing between capturing underlying hydrologic process and ease of application has been fading by advances in computing and data acquisition techniques. Expansion of hydrologic model domain offered variety to choose from but presented a growing challenge to identify suitable models for a particular purpose. The diversity of models can be exploited to user advantage once their strength and weaknesses are identified with respect to the intended modeling endeavor. For example, data scarcity is the major limitation to contemplate hydrological process modeling in the developing world. The growing availability of satellite-derived hydrometeorological data provides new opportunity to study watersheds of limited observed data. In response, models are being equipped with utilities to process and ingest these data.

Researchers have used different models for hydrological assessment, flow prediction, and other applications. The most commonly used model capable of predicting flows for ungauged watersheds is the Soil and Water Assessment Tool (SWAT). The application of SWAT in predicting stream flow and sediment as well as evaluation of the impact of land use and climate change on the hydrology of watersheds has been documented by various studies (Dessu and Melesse 2012, 2013a, b; Dessu et al. 2014; Wang et al. 2006a, b, 2008a, b, c; Wang and Melesse 2005; Behulu et al. 2013, 2014; Setegn et al. 2014; Mango et al. 2011a, b; Getachew and Melesse 2012; Assefa et al. 2014; Grey et al. 2013; Mohamed et al. 2015). However, SWAT has limited capability to dynamically ingest remote sensing data.

The Geospatial Stream Flow Model (GeoSFM) is developed to facilitate study of watersheds in the developing countries (Asante et al. 2008a). GeoSFM (also known as FEWS-SFM) is equipped with tools to access satellite-derived spatial and temporal input data. Despite the easy access and manipulation of remotely acquired data, verification with ground observed data remains to be critical (Dessu and Melesse 2013a; Funk et al. 2003). Neither simplicity to setup nor success histories of particular model suffices its use and implementation. Continuous testing and verification of models is essential to assist users in selection and implementation. Evaluation of GeoSFM with observed data can help to identify strength and weakness of the model, to interpret results and to improve efficiency in future applications. Exhaustive testing of a model requires sound reasoning, justifiable choices of methods, and criteria prior to practical application. Accordingly, detailed evaluation of the applicability of GeoSFM using ground observed input data set in the Upper Awash River Basin (UARB) is presented in this chapter.

GeoSFM is widely applied with satellite-derived data set in data scarce regions such as Greater Horn of Africa and Nepal. Asante et al. (2007) used the GeoSFM and TRMM (Tropical Rainfall Measuring Mission) rainfall data to develop a flood monitoring system for the Limpopo basin. They did not calibrate the model results against observed stream flow measurements, rather used the flow rates from model simulation to subjectively confirm forecasted extreme events from field sources. Mati et al. (2008) used the model to determine impact of land use/cover change in Mara River Basin. The model was also used to characterize Congo Basin stream flow using multisource satellite-derived data (Munzimi et al. 2010). Artan et al. (2007) applied GeoSFM and satellite rainfall estimates to compute runoff in the Nile River Basin and the Mekong River Basin, and reported a Nash-Sutcliff efficiency of 0.81. Shrestha (2011) used GeoSFM using observed rainfall in the Bagmati and Narayani River Basins, Nepal, and reported correlation values of 0.95 and 0.94 between observed and simulated flow rates. Shresta showed discharge estimated from satellite based rainfall follows the trend of observed values but recommended bias correction of satellite rainfall estimates. The above studies assessed hydrological variables indirectly from discharge outputs of GeoSFM based on satellite-derived spatial and temporal data inputs.

Awash River basin is currently the most developed river in Ethiopia (Fig. 18.1). The basin also represents international anthropological significance and cultural heritage (Johanson and Edey 1990; Haile-Selassie 2001; Asfaw et al. 2002). The Koka reservoir was commissioned in 1960 to store the peak summer flow from rainfall at the upstream highlands and ensure a stable discharge for power

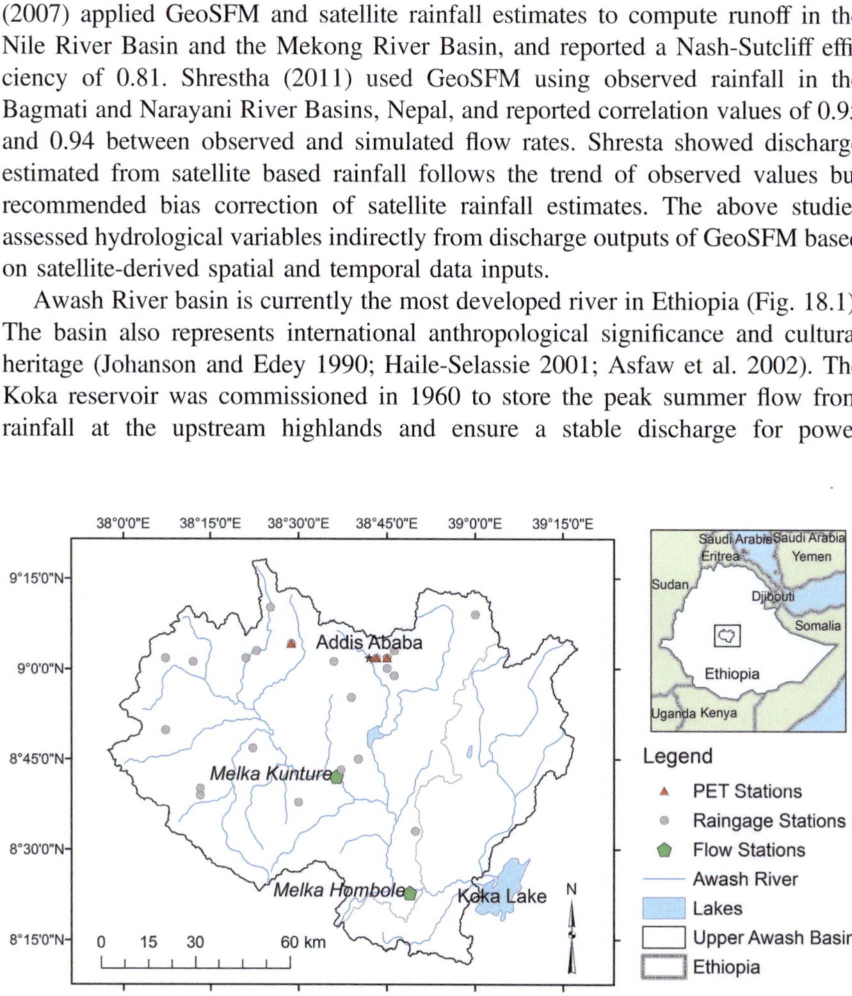

Fig. 18.1 Location and monitoring stations of the Upper Awash River Basin

generation and irrigation in the dry period. The Koka reservoir is expected to provide sufficient head and discharge for hydroelectric power generation and irrigation and ample volume for flood storage during the summer. Hydrological challenges in the Awash basin include heavy sediment inflow to the Koka reservoir, extreme inflow variability, and lack of reliable operational and management tool. The Koka reservoir has been reported to have 35 % of its storage capacity filled with sediment from the upstream highlands (Paulos 1998). This reduction in capacity poses a challenge in the management of the reservoir and flood hazard to the residents and property downstream of the dam with unpredictable release and storage patterns. Shortage of water in the dry period and devastating flood in the summer are frequent events downstream. Dry period shortage has forced rationing of power and water for the large irrigation farms downstream. In August 1996, flooding damage was estimated at $9 million and affected 75,000 residents (DPPC 1997; Edossa et al. 2010; Rogner 2000).

According to Intergovernmental Panel on Climate Change report, Eastern Africa is expected to experience more floods in the wet season through the 21st century (Christensen et al. 2007; IPCC 2007). The current hydrologic challenge in the UARB may be exacerbated by lack of capacity and infrastructure to adapt to the changing climate (Collier et al. 2008; Yanda and Mubaya 2011; Dessu and Melesse 2013b). Hydrologic models are useful to predict extreme events of low or peak flow. GeoSFM can potentially assist in timely management and operation of Koka reservoir, avoiding flooding from excess releases as well as probable water shortage. Three to seven days rainfall forecast can be used to predict the runoff response of the Awash River and manage the reservoir accordingly. Reliable hydrological estimation and forecasting of the incoming flow may improve the operation of the Koka reservoir.

Hydrologic models have been used to understand the space–time relationship of hydrological processes. The uniqueness of watersheds can be captured by the use of distributed geospatial information from soil- and land-use databases. Planning and management tools customarily use outputs from hydrological simulations to facilitate management of water resources to meet growing demand while decreasing foreseeable hazards (Dessu et al. 2014). There are numerous computer simulation models to capture rainfall–runoff process and estimate inflow discharge to a lake or reservoir.

Despite the increasing use and application of GeoSFM in data scarce regions, the performance of the model is barely scrutinized with observed input data. The purpose of this study was to assess the capability of GeoSFM to capture long-term rainfall–runoff process using observed data in the (UARB). The specific objectives were: (1) to investigate sensitivity of GeoSFM to spatial partition scale, (2) to assess sensitivity of GeoSFM parameters in simulating the rainfall runoff process of UARB, and (3) to evaluate the performance of GeoSFM to capture observed hydrological pattern. Evaluation of GeoSFM helps to identify model strength and improve its performance efficiency for future applications. This study will help to establish baseline for further application of the model in the region using observed as well as regional remote sensing data.

18.2 Description of Study Area

The water resources of Awash River are key elements of development in Ethiopia. The location and hydrological properties of the basin enabled extensive development. The Awash River Basin covers 110,000 km^2 area of the Great Rift Valley in Ethiopia (Fig. 18.1). The Awash River starts on the high plateau 3000 m above mean sea level (amsl) and flows along the East African rift valley into the Afar triangle and terminates in salty Lake Abbe over a stretch of 1200 km. The river drains major urban centers including Addis Ababa with population of more than 10.5 million (Taddese et al. 2003). The major economic activities in the basin are traditional mixed crop farming and livestock in the highlands, pastoralism in the lowlands, and commercial irrigation in the middle (Getahun 1978). Awash River Basin has been traditionally divided into four distinct physical and socioeconomic zones: Upper Basin, Upper Valley, Middle Valley, and Lower Valley (Taddese et al. 2003). The Upper Awash Basin (UARB) covers 11,500 km^2 area stretching from the headwaters to the Koka reservoir at Koka dam.

The UARB rainfall is dictated by the Inter-Tropical Convergence Zone (ITCZ). The rainfall distribution, especially in the highland areas, is bimodal with a short rainy season in March, April, and the main rains from June to September. The mean annual rainfall varies from 1600 mm at Ankober, in the highlands northeast of Addis Ababa to 160 mm at Asayita on the northern limit of the Basin. The mean annual runoff into Koka reservoir is 1660 million cubic meters (MCM). About 90 % of runoff occurs from July to October. Mean and maximum annual temperatures of at Koka dam are 20.8 and 23.8 °C, respectively.

18.3 Data and Methods

18.3.1 Description of Geospatial Stream Flow Model (GeoSFM)

GeoSFM (previously called FEWS-SFM) is a physically based, wide-area, continuous daily time-step, stream flow simulation model developed by the USGS/EROS Data Center (Artan et al. 2008). GeoSFM has been used to facilitate flood/drought early warning efforts by the Famine/Flood Early Warning System Network (FEWS-Net) (www.fews.net) and USAID. The model consists of a GIS-based module for data input, and a rainfall–runoff simulation model. The Graphic User Interface of the model runs in ArcGIS environment to prepare the necessary input data, perform simulation and display outputs (Asante et al. 2008a). Remote sensing or ground observed input data can be used to setup and run GeoSFM.

The rainfall–runoff simulation consists of three components: soil water budget, upland headwater basins routing, and river routing (Asante et al. 2008b). The runoff estimation conceptualizes the soil as composed of two main layers; an active soil layer where most of the soil–vegetation–atmosphere interactions take place and a groundwater layer. The active soil layer is further divided into an upper thin soil layer where evaporation, transpiration, and percolation take place and a lower soil layer where only transpiration and percolation occur. The runoff producing mechanisms is based on precipitation excess using the SCS curve number method (USDA-SCS 1972), rapid subsurface flow (interflow), and base flow. The surface upland routing is a physically based unit hydrograph method utilizing landscape attributes from digital elevation model (DEM) of the watershed. The interflow and baseflow components of the runoff are routed as a series of linear reservoirs. In the main river reaches, water is routed using a nonlinear formulation of the Muskingum-Cunge routing scheme (Cunge 1969; Kim and Lee 2010). GeoSFM provides the flow rate at the outlet of individual subbasins.

18.3.2 Input Data and Model Setup

GeoSFM uses spatial data to define basin physical characteristics and climate data to force simulation of rainfall–runoff process. Evaluation of GeoSFM was conducted in two steps: Input data collection and model testing. The general model setup and evaluation procedure is presented in Fig. 18.2. Required spatial inputs for model simulation are DEM of the basin, soil map, land use/cover map, and daily rainfall and potential evapotranspiration (PET) (Artan et al. 2008; Asante et al. 2008a), Fig. 18.3. Data preparation to fit GeoSFM input format was done after checking raw input data for continuity, reliability and consistency.

Topographic (scale 1:50,000 and 1:250,000) obtained from Ethiopian Mapping Agency were digitized and contours were interpolated to generate a 100 m DEM raster (Fig. 18.3a). Spot elevations and features such as rivers and streams, lakes, and roads were digitized for verification. Soil data was derived from 1:500,000 scale digitized soil unit map (Fig. 18.3b). Required soil attributes for GeoSFM were compiled from report of soil research stations in the basin (Bono and Seiler 1983), the FAO soil data base obtained from Global Soil Data Products (IGBP-DIS 2000), and guidelines presented in the GeoSFM technical manual (Asante et al. 2008b). Pellic Vertisol is the dominant soil type (>60 % of the basin). The land use/cover data was obtained from Woody Biomas Inventory project under the Ethiopian Ministry of Agriculture (Fig. 18.3c). The land use/cover attributes were reclassified to fit to the GeoSFM format. More than 75 % of the UARB is characterized by cropland with light stocks of woody plants. Land use/cover data was reclassified according to the GeoSFM format.

Observed rainfall and PET were obtained from Ethiopian Meteorological Services (Fig. 18.1). Thirty-five rain gage stations with daily rainfall record in and around the UARB were used. Daily rainfall and PET raster files were generated using inverse distance weight interpolation over the basin. Daily stream flow records derived from stage reading were obtained for the Melka Kunture and Melka Hombole stations from the Ethiopian Ministry of Water Resources Development (Fig. 18.1).

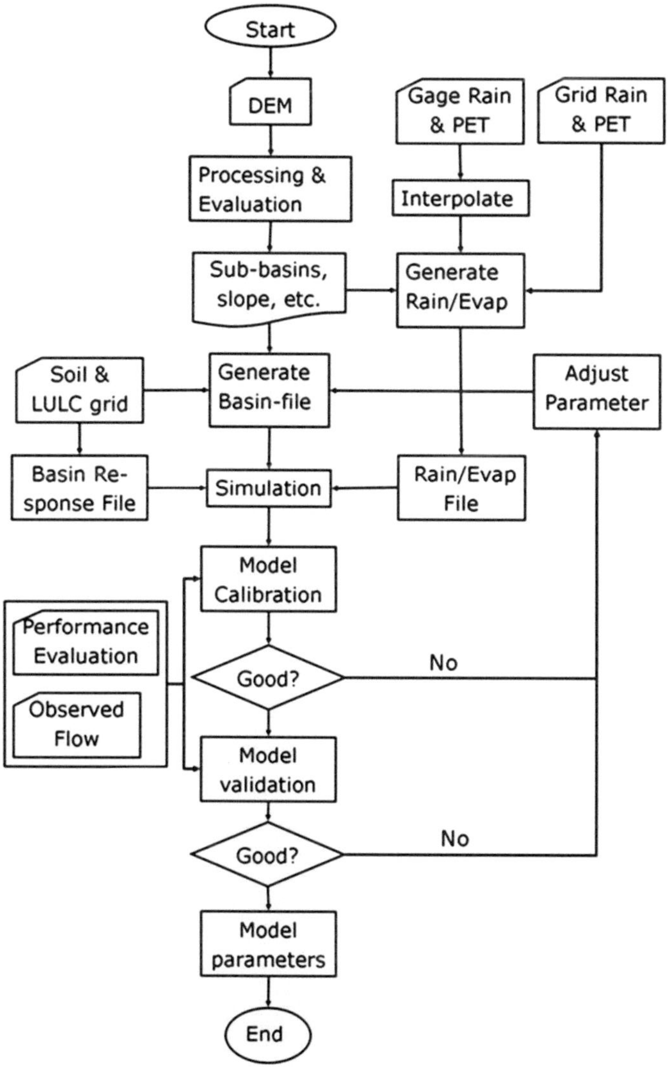

Fig. 18.2 Schematics of GeoSFM model setup

GeoSFM simulation was done for the UARB upstream of the Koka reservoir and the outlet is defined at Melka Hombole hydrologic station. Melka Hombole is at the entry of the Koka reservoir 40 km upstream of the Koka dam. To assess the models best performing module and the catchment property at different locations in the UARB, three scenarios were defined (Fig. 18.4): single basin (SB-1), 6 subbasins (SB-6), and 35 subbasins (SB-35). These three scenarios have the same basin outlet at Melka Hombole station.

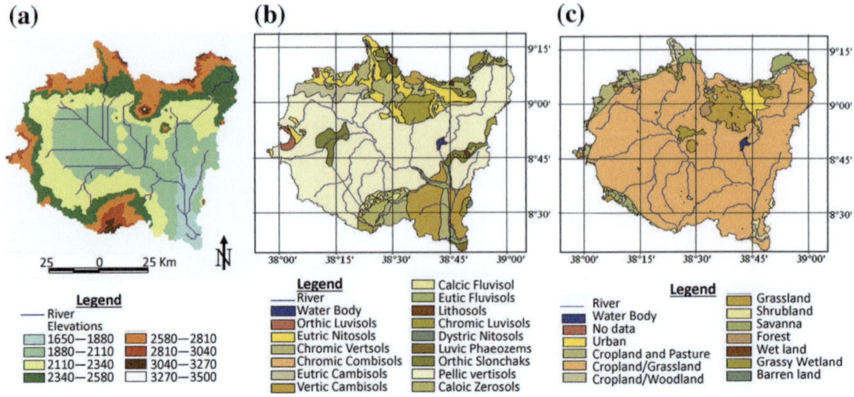

Fig. 18.3 GSM model spatial input data: **a** MRB digital elevation model (m masl), **b** soil map, **c** land use/cover map of MRB

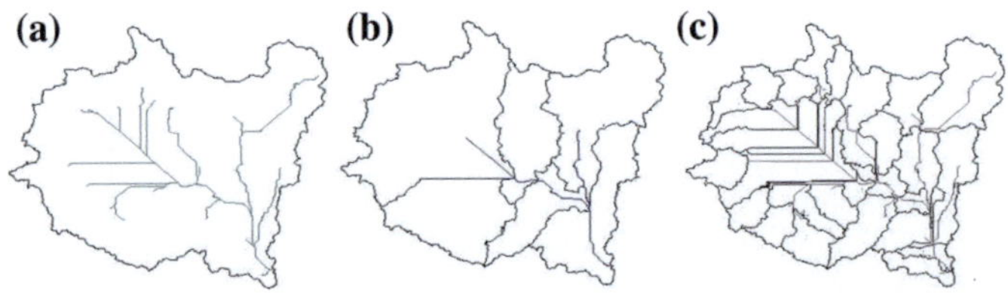

Fig. 18.4 The layout and division of the subbasins for the three scenarios. **a** Single basin (SB-1). **b** 6 subbasins (SB-6). **c** 35 subbasins (SB-35)

For scenario SB-1, GeoSFM generates the parameters for the entire catchment and the river routing component was not dominant. Scenarios SB-6 and SB-35 activate river routing module to move the inflow from the upstream subbasin(s) to the inlet of the subbasin downstream. The single basin (SB-1) scenario is used for manual parameter sensitivity analysis and for Scenario SB-6, six set of parameter were generated and these parameters were calibrated based on the physical properties of the subbasins.

GeoSFM simulation was done over 10 years (1991–2000). The first five years (1991–1995) observed flow data segment was used for calibration and the second segment (1996–2000) was used for validation of the model. Melka Hombole stations at the inlet of Koka were used in the calibration and validation of GeoSFM (Fig. 18.1). Interpolated Rainfall and PET raster data from 1990–2000 were used for simulation and daily flow records from 1991 to 2000 at Melka Hombole station were used for calibration and validation of GeoSFM. Manual calibration was performed by changing parameter values. Due to larger set of parameters (i.e., one set for each subbasin), equivalent changes were applied for all parameters during the calibration of scenario SB-6 and SB-35.

18.3.3 Model Evaluation

GeoSFM model structure was evaluated based on relative comparison of the three scenarios (SB-1, SB-6, and SB-35) and observed daily discharge hydrograph. GeoSFM sensitivity to spatial scale was done by comparing 5 years daily simulated discharge outputs before calibration. Descriptive statistics of default simulation (mean, median, maximum, and quartiles) were used to assess consistency of model output. The frequencies of daily maximum discharge among the three scenarios were compared for entire discharge range and three classes of flow (<50 m^3 s^{-1}), medium (50–100 m^3 s^{-1}) and peak (>100 m^3 s^{-1}). The percentage of minimums and maximums was considered to represent the suitability of the respective partition scenario for a particular purpose. A partition scenario with higher percentage of peak flows may produce conservative flood simulation.

SB-1 scenario was used to assess sensitivity of the eight GeoSFM model parameters: soil depth, permeability, soil water holding capacity, Manning's coefficient, curve number, top soil fraction, basin loss, and pan coefficient. The value of one parameter at a time was changed and plotted against the respective NSE (Nash-Sutcliff Efficiency). The relative rate of change of NSE was used as a measure of parameter sensitivity. A parameter displaying larger rate of change in NSE was considered more sensitive. Sensitivity analysis was done primarily on the assessment of the eight parameters used in GeoSFM. The eight parameters can be adjusted to capture the hydrologic process of a watershed (Asante et al. 2008b). Uncertainty associated with model input and model structure was evaluated based on relative performance of the three scenarios and observed daily discharge hydrograph.

The "performance," "uncertainty," and "realism" of GeoSFM simulations were assessed with respect to the observed flow data (Wagener 2003). Evaluation of GeoSFM was conducted based on statistical parameters and measures of "closeness" between the model output and the observed flow at Melka Hombole station. Mean relative error (MRE), Eq. 18.1, was used to evaluate unit independent expected deviation of the simulation from the observed flow. In flood simulation assessment, root mean square error (RMSE), Eq. 18.2, and NSE, Eq. 18.3, (Nash and Sutcliffe 1970) offer performance statistics sensitive to peak flow events. The RMSE value measures the models adequacy in matching the peak flows. The closer NSE values to one, the better the model efficiency; whereas, negative values generally indicate that the mean of the observed value is statistically better than the model result. Coefficient of determination (R^2), Eq. 18.4, evaluates the ability of GeoSFM in capturing the variability of observed flow.

$$\mathrm{MRE} = \frac{1}{n}\sum_{i}^{n}\frac{|O_i - S_i|}{O_i} \tag{18.1}$$

$$\text{RMSE} = \sqrt{\frac{1}{n}\sum_{i=1}^{n}(O_i - S_i)^2} \tag{18.2}$$

$$\text{NSE} = 1 - \frac{\sum_{i=1}^{n}(O_i - S_i)^2}{\sum_{i=1}^{n}(O_i - \bar{O})^2}; \quad \text{NSE}\ \epsilon(-\infty, 1] \tag{18.3}$$

$$R^2 = \frac{\left[\sum_{i=1}^{n}(O_i - \bar{O})(S_i - \bar{S})\right]^2}{\sum_{i=1}^{n}(O_i - \bar{O})^2\sum_{i=1}^{n}(S_i - \bar{S})^2}; \quad R^2\epsilon[0, 1], \tag{18.4}$$

where O is observed flow, S is simulated flow. Model performance was evaluated through objective functions that minimize the distance and optimize the variability between observed event and model result (Eq. 18.5). A second objective function that optimizes the NSE and R^2 to evaluate whether the simulations had optimally reproduced observed variability (White and Chaubey 2005; Dessu and Melesse 2012) of the natural hydrologic process while minimizing the overall deviation.

$$\text{ObjF} = \left\{ \begin{array}{c} \text{Optimize}\left(\sum_{j=1}^{k} R^2(O,S), \text{NSE}(O,S)\right) \\ \text{Minimize}\left(\sum_{j=1}^{k} \text{RMSE}(O,S), \text{MRE}(O,S)\right) \end{array} \right\}, \tag{18.5}$$

where k is the number of calibration sites included.

18.4 Results and Discussion

GeoSFM is developed for monitoring wide-area hydrologic events that require use of geospatial and time-series data in near-real time. The model was evaluated with respect to input data quality, spatial resolution, model parameters, and efficiency in capturing observed flow hydrograph. Three scenarios of singe (SB-1), six (SB-6), and thirty five (SB-35) subbasin partitions of UARB were setup and assessed based on their default, calibrated, and validated simulation outputs of daily and monthly average discharge. Ten years (1991–2000) of observed rainfall, evapotranspiration, and discharge data were used.

GeoSFM has been applied in Mara River Basin (13,750 km^2) (Mati et al. 2008), Limpopo River Basin (6000 km^2), (Asante et al. 2007), and Mekong River Basin (22,000 km^2), (Asante et al. 2008b). The extent of spatial discretization that the model can optimally handle is useful to exchange and utilize modeling results of these studies. The three spatial partitioning scenarios are intended to provide GeoSFM flexibility with respect to size of watershed. Over the 5-years period of simulation (1991–1996), the average daily simulation outputs for the three scenarios were more than three times the 5-years observed average daily flow at Melka Hombole, 42.7 m^3 s^{-1}. For same input data and daily simulation for 5 years, the three scenarios (SB-1, SB-6, SB-35) produced a comparatively higher daily flow frequency of (12, 42, 46 %) of the time with respect to each other. To assess the effect of discretization on different flow regimes, the three scenarios were compared for daily flow brackets of >50 and >100 $m^3 s^{-1}$. Results showed comparative daily maximum frequency (10, 36, 54 %) and (9, 30, 61 %) for flow rates greater than 50 and 100 $m^3 s^{-1}$, respectively, for scenarios SB-1, SB-6 and SB-35. Low flows are barely captured in scenario SB-1 because GeoSFM handles minimum flow as zero for peripheral and 0.1 $m^3 s^{-1}$ for interior subbasins. This model structure may be responsible for the lower frequency of relative maximum for scenario SB-1 over the simulation period. Comparatively, scenario SB-1 produced less number of peaks than scenario SB-6 that may be attributed to averaging the hydrologic variables over the entire basin. The increasing frequency of peak flows with higher flow rate brackets for scenario SB-6 may suggests that further partitioning of UARB may improve the probability of capturing flooding events. On the other hand, scenario SB-1 is invariable for change of flow range. Scenario SB-6 has a better performance in capturing high local maximum for the whole range of flow but is less variable as the flow bracket increases. In general, results of scenario comparisons suggest that performance of GeoSFM is sensitive to the scale of basin partitioning.

Compared to the observed flow hydrograph at Melka Hombole, flood peaks prevail on a more gradually varying hydrograph. Individual subbasins were investigated to locate the beginning of peak flow at the outlet of UARB. Results from scenario SB-6 suggest that the baseflow is maintained by the upper reaches of the catchment, whereas the daily peak flows are more influenced by rainfall events in the mountainous area upstream of Melka Hombole station. Hence, further division of UARB in the scenario SB-6 has improvement in the capability of GeoSFM to separately capture local variability in the basin runoff response. The third scenario offered a better spatial variability but at the cost of larger set of parameters to tune in the calibration process.

Sensitivity analysis of eight GeoSFM parameters [Soil depth (SD), permeability (k), soil water holding capacity (SWHC), Manning's coefficient (MC), curve number (CN), top soil fraction (TSF), basin loss (BL), and pan coefficient (k)] was conducted using SB-1 scenario (Fig. 18.5). Results have demonstrated that BL and CN are the most sensitive parameters followed by pan coefficient (k). The remaining five parameters did not show significant rate of change in NSE. Since only one parameter value was changed at a time, the result may not show combined sensitivity of the parameters. The sensitivity analysis can be used to refine the sensitive input data for the model as well as guiding the selection and adjustments of parameters in the calibration process. Identification of less sensitive parameters can facilitate the calibration process.

GeoSFM was calibrated and validated for the three scenarios (Fig. 18.4) over 10-years period (1991–2000). GeoSFM is developed for daily discharge simulation and short-term forecasting. Model performance evaluation was primarily based on the daily time-step simulation output. On the basis of the objective function, consistent performance efficiency was achieved for the three scenarios with average (NSE, R^2) results of (0.67, 0.70) for calibration and (0.60, 0.65) for validation of GeoSFM (Table 18.1). The RMS value obtained for the best run of scenario SB-35

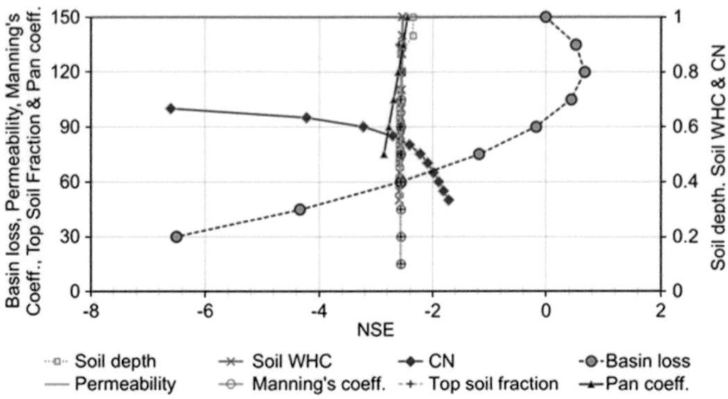

Fig. 18.5 Manual sensitivity analysis of GeoSFM parameters for the Upper Awash Basin on the basis of NSE

Table 18.1 Summary of statistical parameters after calibration and validation for the three scenarios at Melka Hombole station. Results are based on daily simulation

Statistics	Scenario, calibration (validation)		
	I	II	II
Q_{omean} ($m^3\ s^{-1}$)	42.7 [48]	42.7 [48]	46.8 [48]
Q_{smean} ($m^3\ s^{-1}$)	51.2 [60.4]	60.3 [57.6]	49 [55.6]
Mean_abs_err ($m^3\ s^{-1}$)	0.002 [0.004]	0.002 [0.003]	0.001 [0.003]
Max_abs_err ($m^3\ s^{-1}$)	397 [549]	360 [488]	365 [579]
Bias from mean ($m^3\ s^{-1}$)	−8.5 [−5.9]	−17.5 [−4.8]	−0.4 [−3.6]
RMSE ($m^3\ s^{-1}$)	44.7 [36.7]	47.8 [36.4]	22.6 [36]
NSE	0.68 [0.6]	0.63 [0.6]	0.69 [0.61]
R^2	0.69 [0.64]	0.69 [0.66]	0.71 [0.64]

is half of the other two scenarios suggesting further partitioning may help to capture the higher flow range. The maximum absolute error searches for the highest deviation of the historic record from the simulated flow. Very low bias from the mean value was observed suggesting an overall minimum variation of the mean of historical record and the simulated data.

The calibrated model overestimated the stream flow for the first rain season (April–July) while the peak flow during summer (July–September) is generally underestimated (Fig. 18.6a). The simulated flow hydrograph consistently crossed the observed flow hydrograph in mid of July and September in the ascending and recession limb. On the basis of cumulative annual flow volume and rainfall depth during the calibration period, the simulated flow mimics the rainfall pattern (Fig. 18.6b).

GeoSFM was assessed based on how good the simulated runoff repeats the historical record. For instance, peak flow of 526.2 $m^3\ s^{-1}$ was reported on 29 July 1993 but the respective rainfall value was 4.2 mm on same day and 8.8 mm for the previous day (Fig. 18.6a). Similarly, local peak flows were reported in 1992 from

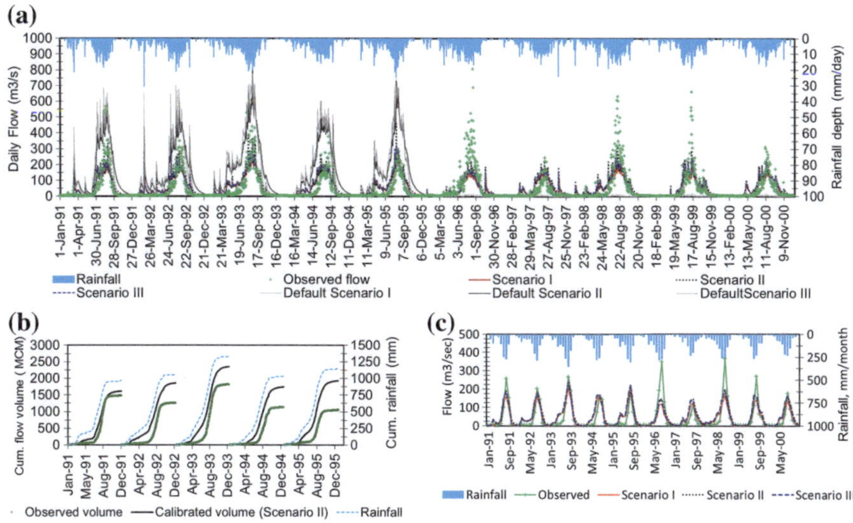

Fig. 18.6 Observed rainfall and flow at Melka Hombole versus GeoSFM simulation results. **a** Average daily rainfall, daily observed flow, calibrated and validated simulation result (1991–2000), **b** mass curve for rainfall, observed flow and calibrated flow for Scenario SB-6, and **c** monthly observed and simulated flow hydrograph

the simulated results unlike the small observed flow ratc. The peak flow is not accompanied by a higher rainfall record. The mismatch in trend or amount of the simulated and observed flow rates may be due to the stage–discharge relationship of the rating curve used to determine discharge from a stage record. According to the observer assigned at Melka Hombole station (personal communication), the staff gage was submerged during peak floods of 1996.

On the basis of model evaluation results, GeoSFM has a potential for flood simulation and forecasting. The model performance may improve with use of better spatial resolution soil and land-use input data. However, considering annual water balance, the simulated flow has overestimated the total annual flow. Results of monthly average flow analysis over the simulation period ($R^2 = 0.76$) showed that GeoSFM has a potential to be used for water resource assessment (Fig. 18.6c).

The calibrated models for each scenario were validated with 5 years (1996–2000) daily data (Table 18.1). The calibrated GeoSFM model consistently overestimated stream flow for the short rain season (April–July) and underestimated the peak flow during summer (July–September). The simulated flow hydrograph crosses the observed flow hydrograph in the mid of the September overestimating the flow in the recession limb. Moreda and Bauwens (1998) reported that UARB rainfall variability is amplified on runoff response in the summer as compared to the low flow regime (November–February). This rainfall–runoff response variability may be attributed to the dominance (>60 %) of the black cotton soil (vertisol), in the UARB. Due to the high clay content, vertisol displays a range of runoff responses to rainfall depending on the available soil moisture and season of the year (Jutzi 1988). At the onset of rainy season, vertisols swell sell until all the cracks are filled and during the heavy summer the soil maintains a thin upper saturated soil layer increasing surface runoff generating higher stream flow. This could be one of the reasons that the model overestimates the stream flow in the first rainy season and underestimate the second. Therefore, the performance of the model in such soil types may need further investigation.

Annual performance of GeoSFM was assessed to pinpoint the best and worst years of simulation of the calibration and validation process (Fig. 18.7). Annual average of the daily deviation of simulation from observed flows remained below 6 $m^3 s^{-1}$ for the calibration period but showed a wide variability in the verification (Fig. 18.7a). Results of annual evaluation of the model performance demonstrated that GeoSFM performs best for wet years (1993, 1996, 2000) as compared to relatively dry years (1995, 1997) (Fig. 18.7a) . The correlation between the observed and simulated flow values are within the range of 0.64–0.71, indicating that the model has captured the pattern of the historical record. Comparing the maximum absolute deviation of daily simulation from the observed, the minimum in 2000 is three times the catastrophic maximum flooding year of 1996 (Fig. 18.7b). The mean annual simulated flow rate was greater than the observed through the simulation period (Fig. 18.7c). In the calibration process the model performs better for the wet season; in particular, during periods of uniform daily rainfall distribution. The weak model performance for the year 1997 may be explained by the quality of observed data (Fig. 18.7d). While Melka Kunture is upstream of Melka Hombole, and it takes a maximum of 2 days for the peak to travel between the stations, the peak was not captured at Melka Hombole.

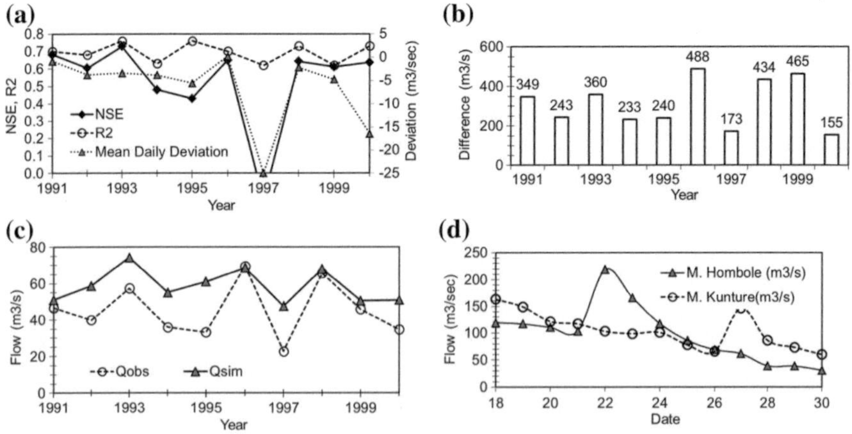

Fig. 18.7 Combined assessment of calibration and validation outputs versus observed flow at Melka Hombole station: **a** Goodness of fit, **b** maximum absolute difference, **c** mean simulation and observed flow, and **d** observed flow comparison at two stations for August 1997

18.5 Conclusion

Reliable simulation and forecasting tool may help in the management and operation of the Koka reservoir on the Awash River. GeoSM is one of such tools developed for monitoring wide-area hydrologic events that require use of geospatial and time-series data available in near-real time. This study assessed the performance of GeoSFM in stream flow and flood simulation of the UARB using ground observed data. The Awash River basin has been experiencing flooding downstream of the Koka dam. The location and hydrological properties of the basin were relatively suitable for extensive development, but catastrophic flooding has been claiming scores of human life and property damage.

Evaluation of GeoSFM was conducted on the basis of statistical measures, sensitivity to input data quality, space-time resolution of inputs and outputs, basin loss, and curve number were reported to be the most sensitive parameters. GeoSFM demonstrated satisfactory performance in daily rainfall–runoff simulation of the UARB. The model simulation also captured 76 % of the observed monthly average flow variability over 10 years. The nonlinear response of UARB to rainfall was fairly captured by the model. Summer runoff was fairly underestimated and the short rain period performance was poor. The over/underestimation may be attributed to the dominance of black cotton soil (Vertisol) in the basin, model structure, or quality of observed data. Statistical assessment of annual daily simulation indicated the potential of application of GeoSFM in data scarce regions.

The Geospatial Stream flow Model (GeoSFM) has been widely applied in flow simulation and flood forecasting with global/regional remote sensing input data in data scarce regions. Given reliable simulation performance with observed data, the application of GeoSFM for the UARB can be extended further with the increasing availability of remote sensing data. On the basis of simulation results from UARB, GeoSFM has demonstrated a fair platform for flow simulation and forecasting and hydrologic investigation in data scarce regions. Results of this study can be used as a baseline to improve the model as well as investigate its application and performance in similar hydrologic setting.

Acknowledgment This work was supported by the Ethiopian Ministry of Water Resources, Addis Ababa University, the U.S. Agency for International Development and the U.S. Geological Survey.

References

Artan G, Gadain H, Smith J, Asante K, Bandaragoda C, Verdin J (2007) Adequacy of satellite derived rainfall data for stream flow modeling. Nat Hazards 43(2):167–185

Artan GA, Asante K, Smith J, Pervez S, Entenmann D, Verdin JP, Rowland J (2008) Users manual for the geospatial stream flow model (GeoSFM). US Geological Survey

Asante KO, Artan GA, Pervez S, Ci Bandaragoda, Verdin JP (2008a) Technical manual for the geospatial stream flow model (GeoSFM). US geological survey, Reston

Asante KO, Arlan GA, Pervez S, Rowland J (2008b) A linear geospatial streamflow modeling system for data sparse environments. Int J River Basin Manage 6(3):233–241

Asante KO, Macuacua RD, Artan GA, Lietzow RW, Verdin JP (2007) Developing a flood monitoring system from remotely sensed data for the limpopo basin. IEEE Trans Geosci Remote Sens 45(6):1709–1714

Asfaw B, Gilbert WH, Beyene Y, Hart WK, Renne PR, WoldeGabriel G, Vrba ES, White TD (2002) Remains of Homo erectus from Bouri, Middle Awash. Ethiopia. Nature 416 (6878):317–320

Assefa A, Melesse AM Admasu S (2014) Climate change in Upper Gilgel Abay river catchment, blue Nile basin Ethiopia. In: Melesse AM, Abtew W, Setegn S (Eds) Nile river basin: ecohydrological challenges, climate change and hydropolitics. pp 363–388

Behulu F, Setegn S, Melesse AM, Fiori A (2013) Hydrological analysis of the Upper Tiber Basin: a watershed modeling approach. Hydrol Process 27(16):2339–2351

Behulu F, Setegn S, Melesse AM, Romano E, Fiori A (2014) Impact of climate change on the hydrology of Upper Tiber River basin using bias corrected regional climate model. Water Res Manage 1–17

Bono R, Seiler W (1983) The soil of the Suke-Hararge Research Unit (Ethiopia) classification, morphology and ecology with soil scale of 1: 5000. Soil Conservation Research Project. University of Bern, University of Switzeralnd and The United Nations University, Tokyo

Christensen JH, Hewitson B, Busuioc A, Chen A, Gao X, Held I, Jones R, Kolli RK, Kwon WT, Laprise R, Magãna Rueda V, Mearns L, Meńendez CG, Räisänen J, Rinke A, Sarr A, Whetton P (2007) Regional climate projections. In: Solomon S, Qin D, Manning M, Chen Z, Marquis M, Averyt KB, Tignor M, Miller HL (eds) Climate change: the physical science basis, contribution of working group I to the fourth assessment report of the intergovernmental panel on climate change. Cambridge University Press, Cambridge

Collier P, Conway G, Venables T (2008) Climate change and Africa. Oxford Review of Econ Policy 24(2):337–353. doi:10.1093/oxrep/grn019

Cunge JA (1969) On the subject of a flood propagation computation method (Muskingum Method). J Hydraul Res 7(2):205–230

Dessu SB, Melesse AM (2012) Modelling the rainfall–runoff process of the Mara River basin using the soil and water assessment tool. Hydrol Process 26(26):4038–4049. doi:10.1002/hyp.9205

Dessu SB, Melesse AM (2013a) Evaluation and comparison of satellite and GCMs rainfall estimates for the mara river basin, kenya/tanzania. In: Younos T, Grady CA (eds) Climate change and water resources. The handbook of environmental chemistry. Springer, Berlin. doi:10.1007/698_2013_219

Dessu SB, Melesse AM (2013b) Impact and uncertainties of climate change on the hydrology of the Mara River basin. Kenya/Tanzania. Hydrol Process 27(20):2973–2986

Dessu SB, Melesse AM, Bhat MG, McClain ME (2014) Assessment of water resources availability and demand in the Mara River Basin. CATENA 115:104–114

DPPC (1997) Food vulnerability in Ethiopia and needs for preparedness. Disaster Prevention and Prepardness Commission, Addis Ababa

Edossa D, Babel M, Das Gupta A (2010) Drought analysis in the Awash River Basin, Ethiopia. Water Res Manage 24(7):1441–1460

Funk C, Michaelsen J, Verdin J, Artan G, Husak G, Senay G, Gadain H, Magadazire T (2003) The collaborative historical African rainfall model: description and evaluation. Int J Climatol 23 (1):47–66

Getachew HE, Melesse AM (2012) Impact of land use/land cover change on the hydrology of Angereb watershed, Ethiopia. Int J Water Sci doi: 10.5772/56266, Vol. 1, 4:1-7

Getahun A (1978) Agricultural systems in Ethiopia. Agric Syst 3(4):281–293

Grey OP, Webber Dale G, Setegn SG, Melesse AM (2013) Application of the soil and water assessment tool (SWAT Model) on a small tropical Island state (great river watershed, Jamaica) as a tool in integrated watershed and coastal zone management. Int J Trop Biol Conserv 62 (3):293–305

Haile-Selassie Y (2001) Late miocene hominids from the middle Awash, Ethiopia. Nature 412 (6843):178–181

IGBP-DIS (2000) Global soil data products (compact disc). International geosphere-biosphere program, data and information services

IPCC (2007) Climate change 2007: the physical science basis. Contribution of working group i to the fourth assessment report of the intergovernmental panel on climate change. In: Solomon SD, Qin M, Manning Z, Chen M, Marquis KB, Averyt M, Tignor HL, Miller (ed) Cambridge University Press, Cambridge

Johanson D, Edey M (1990) Lucy: the beginnings of humankind. Touchstone Books

Jutzi S (1988) Deep black clay soils (Vertisols): management options for the Ethiopian highlands. Mountain research and development 153–156

Kim NW, Lee J (2010) Enhancement of the channel routing module in SWAT. Hydrol Process 24 (1):96–107

Mango L, Melesse AM, McClain ME, Gann D, Setegn SG (2011a) Land use and climate change impacts on the hydrology of the Upper Mara River Basin, Kenya: results of a modeling study to support better resource management, special issue: climate, weather and hydrology of east African highlands. Hydrol Earth Syst Sci 15:2245–2258. doi:10.5194/hess-15-2245-2011

Mango L, Melesse AM, McClain ME, Gann D, Setegn SG (2011b) Hydro-meteorology and water budget of mara River basin, Kenya: a land use change scenarios analysis, In: Melesse A (ed) Nile river basin: hydrology, climate and water use (Chap. 2). Springer Science Publisher, pp 39-68, doi: 10.1007/978-94-007-0689-7_2

Mohammed H, Alamirew T, Assen M, Melesse AM (2015) Modeling of sediment yield in maybar gauged watershed using SWAT, northeast Ethiopia. CATENA 127:191–205

Mati BM, Mutie S, Gadain H, Home P, Mtalo F (2008) Impacts of land-use/cover change on the hydrology of the Transboundary mara river, Kenya/Tanzania. Lakes Reservoirs Res Manage 13:169–177

Moreda F, Bauwens W (1998) Influences of variability of rainfall on flow regimes in central Ethiopia. IAHS publication, pp 297–306

Munzimi Y, Hansen MC, Asante KO (2010) Congo basin streamflow characterization using multi-source satellite-derived data: preliminary results. In: AGU Fall Meeting Abstracts, p 1283

Nash JE, Sutcliffe JV (1970) River flow forecasting through conceptual models part I—a discussion of principles. J Hydrol 10(3):282–290

Paulos S (1998) Water release rules for Koka Reservoire for wet seasons. Addis Ababa University, Addis Ababa

Rogner M (2000) Benefit cost consideration for inter-basin water transfers between the middle Awash River basin and the Upper Rift Valley in Ethiopia. Darmstadt University of Technology, Darmstadt

SetegnSG,MelesseAM,HaidukA,WebberD,WangX,McClainM(2014)Spatiotemporaldistributionof freshwateravailabilityintheRioCobreWatershed.JamaicaCATENA120:81–90

Shrestha MS (2011) Bias-adjustment of satellite-based rainfall estimates over the central Himalayas of Nepal for flood prediction. A Dissertation for the Degree of Doctor of Engineering Department of Civil and Earth Resources Engineering, Kyoto University, Japan

Taddese G, Sonder K, Peden D (2003) The water of the Awash River Basin: a future challenge to Ethiopia. ILRI, Addis Ababa

USDA-SCS (1972) National engineering handbook, hydrology section 4. USDA

Wagener T (2003) Evaluation of catchment models. Hydrol Process 17(16):3375–3378

Wang X, Shang S, Yang W, Melesse AM (2008a) Simulation of an agricultural watershed using an improved curve number method in SWAT. Tans Am Soc Agri Bio Engineers 51(4):1323–1339

Wang X, Yang W, Melesse AM (2008b) Using hydrologic equivalent wetland concept within SWAT to estimate streamflow in watersheds with numerous wetlands. Tans Am Soc Agri Bio Engineers 51(1):55–72

Wang X, Melesse AM, Yang W (2006) Influences of potential evapotranspiration estimation methods on SWAT's hydrologic simulation in a Northwestern Minnesota Watershed. Trans ASAE 49(6):1755–1771

Wang X, Melesse AM (2006) Effects of STATSGO and SSURGO as inputs on SWAT model's snowmelt simulation. J Am Water Res Assoc 42(5):1217–1236

Wang X, Melesse AM (2005) Evaluations of the SWAT model's snowmelt hydrology in a Northwestern Minnesota Watershed. Trans ASAE 48(4):1359–1376

Wang X, Garza J, Whitney M, Melesse AM, Yang W (2008c) Prediction of sediment source areas within watersheds as affected by soil data resolution. In: Paul N (ed) Environmental modelling: new research (Findley; ISBN: 978-1-60692-034-3), Ch. 7, p 151–185, Nova science publishers, Inc., Hauppauge, NY 11788

White KL, Chaubey I (2005) Sensitivity analysis, calibration, and validation for a multisite and multivariable SWAT Model. JAWRA J Am Water Res Assoc 41(5):1077–1089

Yanda PZ, Mubaya CP (2011) Managing a changing climate in Africa, local level vulnerabilities and adaptation experiences. Mkuki Na Nyota, Dar-Es-Salaam

Chapter 19
Regional Scale Groundwater Flow Modeling for Wakel River Basin: A Case Study of Southern Rajasthan

Himadri Biswas and Assefa M. Melesse

Abstract Rajasthan experiences varied climatic condition ranging from extreme aridity to subhumid and humid conditions. About 94 % of the total geographical area of the state falls under arid and semiarid conditions with erratic rainfall pattern. The groundwater is the main source of drinking and irrigation in the Wakel River basin but due to the overextraction of groundwater and consecutive droughts, the water table has been declining. On the basis of analyzing the hydrogeological conditions and water resources utilization of Wakel basin, a two-dimensional distributed parameters' steady and transient groundwater flow model has been developed to better understand the aquifer system. This was done by developing a suitable conceptual model and then transforming into a steady and transient-state numerical groundwater model using MODFLOW, a groundwater flow model. Using the calibrated steady-state model, three model scenarios were developed to see the effect of decreased recharge and increased pumping. The findings from this study successfully explained the overall behavior of the aquifer and its parameters associated with it for the Wakel River Basin.

Keywords Wakel River · Groundwater flow · MODFLOW · Recharge · Rajasthan

19.1 Introduction

Because of limitedsurface wateravailability and deteriorating water quality in some countries, groundwater has become the principal source of freshwater for all uses. Presently, about one-third of the world's population lives in countries that are

H. Biswas · A.M. Melesse (✉)
Department of Earth and Environment, Florida International University,
11200 SW 8th Street, Miami, FL 33199, USA
e-mail: melessea@fiu.edu

H. Biswas
e-mail: hbiswas@fiu.edu

© Springer International Publishing Switzerland 2016

A.M. Melesse and W. Abtew (eds.), *Landscape Dynamics, Soils and Hydrological Processes in Varied Climates*, Springer Geography,
DOI 10.1007/978-3-319-18787-7_19

facing water stress conditions and this fraction is expected to rise to two-third by 2025 (GWP 2000). This includes India, which has 16 % of the global population but only 3 % of the global freshwater reserves (ECIDWR 2005). The high dependence on groundwater and ever-increasing demand with increasing population for domestic, irrigation, and industrial needs has put immense pressure on this natural resource, particularly in semiarid and arid regions (Sukhija et al. 1996; de Vries and Simmers 2002). Its development and management plays a vital role in agricultural production.

Water stress is becoming acute in both urban and rural settings. Not only the quantity but also the quality of water supplied or available is being questioned. At one extreme, water is being wasted in urban areas and by industries; on the other, the rural poor lack access to safe water. According to experts, the usable water resources in several river basins will eventually be exhausted, most surface water will be polluted, and environmental deprivation will be universal. Water scarcity presents an immense threat to the lives and livelihood of communities in the Wakal River of southern Rajasthan. Subsistence farming frequently experience crop failure and livestock kills as a result of insufficient water supply for agriculture. Workers of the region have to walk several kilometers daily just to obtain water for household uses, leaving them little time to their income-generating activities. In sufficient control over groundwater pumping has resulted in an inequitable access to water as well as rapid depletion of existing supplies. Low and erratic annual rainfall in the Wakal Basin of southern Rajasthan has resulted in scarce surface and declining groundwater levels (Stiefel et al. 2007, 2008, 2009). Baseline surveys and research to determine the quantity, quality, and sustainability of water resources to increase rural economy and also to provide safe drinking water to the community is one of the basic needs of the area.

Due to increasing demand for groundwater, there has been unparalleled growth in the number of open dug wells, bore wells, diesel pumps, and electric pump sets (UNDP 2005). These factors have led to the overextraction of groundwater in some regions. Another problem associated with overdependence on groundwater is declining water quality, which is reflected in high salinity, fluoride, and nitrate concentrations in some regions of Rajasthan (ECIDWR 2005; UNDP 2005).

To overcome the groundwater scarcity problem, people in Rajasthan have traditionally been using various artificial recharge techniques such as *nadi*, *tanka*, *jhalara kund*, etc. to store water. The effectiveness of these structures is not well studied (Stiefel et al. 2007, 2008, 2009). Recently, these techniques are gaining popularity owing to increasing population, recurring droughts, and decreasing groundwater levels (Narain et al. 2005). But Rathore (2005b) has reported that these structures are built without having sound technical information on the overall context of groundwater and often key hydrological parameters are not estimated and also little or no effort is made to model hydrologic dynamics.

In response to this ever-increasing scarcity problem, a numerical groundwater flow model was developed for a small catchment in the water scarce Wakal River Basin in southern Rajasthan to better understand the climatic, geologic, and hydrologic factors affecting the aquifers in the area and predict the sustainability of the aquifers by applying different stress conditions. In addition, a conjunctive Geographic Information System (GIS) and remote sensing based approach were taken to identify the factors affecting recharge and to delineate potential recharge sites.

The specific objectives of the study are to (1) establish the effects of geology, lithology, landuse on groundwater using GIS, (2) model and simulate the potential changes in groundwater levels from different groundwater recharge and extraction rates using MODFLOW, and (3) MODFLOW evaluate the performance of the model in replicating the observed groundwater levels.

19.2 Study Area Description and Geographic Setting

19.2.1 Study Area

The Wakal River (Fig. 19.1) originates northwest of Udaipur and joins Sabarmati River near Kalakhetar village in Gujrat. This is one of the five main tributaries of the River Sabarmati—a 371 km long transboundary river that flows southward through the states of Rajasthan and Gujarat and discharges into the Gulf of Cambay in the Arabian Sea (ICID 2005). The river originates in northwest of Udaipur in the Aravalli Hills, with Mansi and Parvi Rivers as the two main tributaries (Chauhan 2007). The total catchment area of the Wakal River is 1900 km^2. Due to limited scope of the study, the entire Wakal catchment could not be investigated for groundwater modeling purpose. Rather, a small catchment located in the southern part of Wakal River basin was investigated where the groundwater demand is high for domestic and agricultural purposes (Fig. 19.1). The study area catchment lies between the latitudes of 24°11.4′ and 24°20.1′ and the longitudes of 73°19.8′ and 73°27.3′ with a total catchment area of 118 km^2.

Rainfall is the main source of water in the Wakal River basin and varies sharply from year to year. Most of the rainfall is received during the monsoon season which starts in the last week of June and lasts till mid-September (ICID 2005). Occasionally, the basin also receives some pre-monsoon rainfall in the middle of June and post-monsoon rainfall in October (Mahnot and Singh 2003). Historical rainfall data from 1965–2005 was acquired for three of the five rain gauge stations located in the Wakal River basin from the Groundwater Department, Rajasthan. These stations are located in Jhadol, Gogunda, and Kotra. Only rainfall data from Jhadol rain gauge station has been analyzed because about 52 % of the Wakal Basin lies in the Jhadol block (Chauhan 2007) including the study area catchment. It was found that Wakal River basin receives a mean annual rainfall of 645 mm. The area experienced continuous drought from 1999–2004. Meteorological drought has been defined by National Commission on Agriculture in India as a situation where the precipitation over an area decreases from normal by more than 25 % (Rathore 2005a).

Average annual temperature in Sabarmati River basin ranges from 25 to 27.5 °C (ICID 2005). For the year 2006, maximum temperature recorded in the Wakal basin was 42 °C in May while minimum temperature was 2.5 °C in January (Singh 2006).

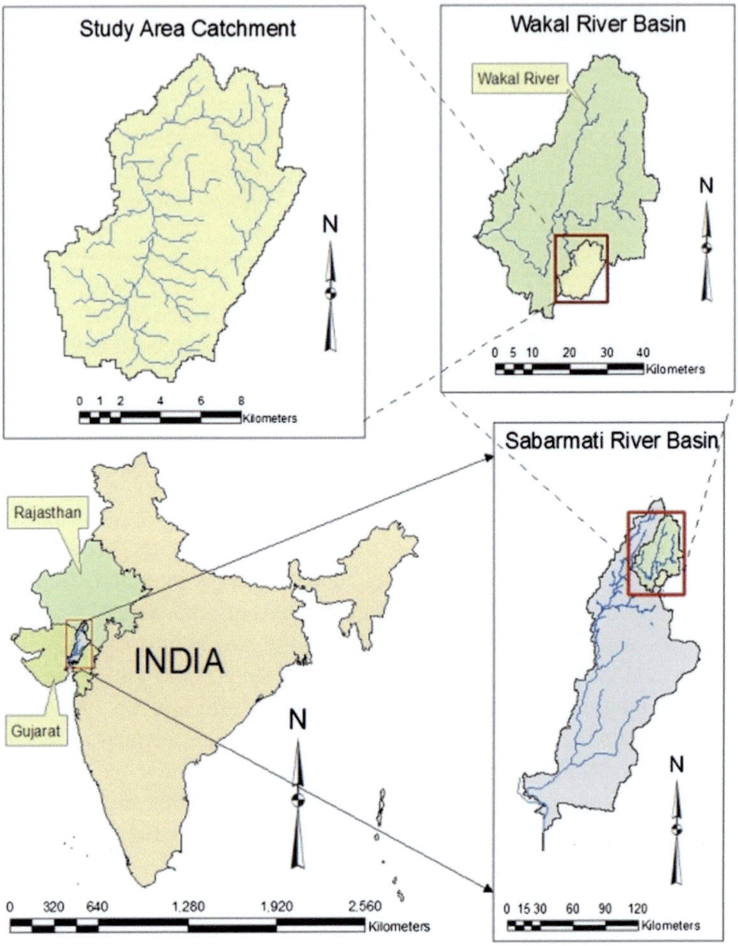

Fig. 19.1 Location of study area

Mean maximum and mean minimum relative humidity in the Wakal River basin in the year 2006 were 78 and 41 %, respectively.

The major rivers which flow through the Wakal River basin are the Wakal, Mansi, and Parvi Rivers. All the rivers and streams in the basin are typically storm channels and are, therefore, ephemeral in nature and dry up usually after the monsoon season (Kumar et al. 1999). Additionally, there are numerous ponds, lakes, and other small reservoirs scattered throughout the Wakal basin as well as in the study area. Total average annual surface water resources for the Rajasthan portion of the Sabarmati Basin which includes the Wakal River basin has been reported to be 513 million cubic meter (MCM) by Department of Irrigation, Rajasthan (ICID 2005).

19.2.2 Geomorphology

The geomorphology of the Wakal River basin and the study area catchment is largely governed by the geologic history, rock types, and structure. The Wakal basin lies in the Aravalli hill physiographic region of Rajasthan where the areas are dominantly hilly and undulating (GOR 2003). The elevation of topography in the Wakal River basin ranges from 263 m to as high as 1167 m above mean sea level (amsl) (Fig. 19.2). The study area is located in a hilly terrain with elevations ranging from 486–995 m amsl. The morphological features in the basin as well as the study area include

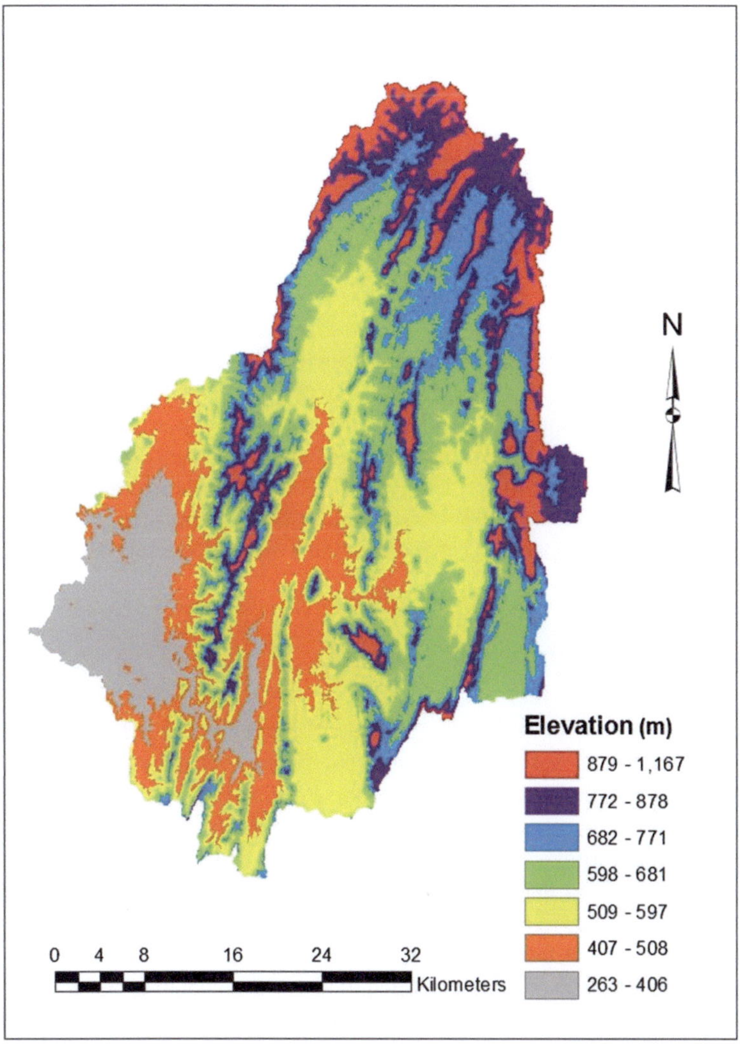

Fig. 19.2 Digital Elevation Model of Wakal sub-basin

pediments, buried pediments, and structural hills which are mostly of denudational origin with presence of valley fills of fluvial origin in some areas (GOR 2003).

19.2.3 Hydrogeology

Groundwater in the Wakal basin, including the study area catchment, exists mostly in unconfined aquifers formed by highly weathered and fractured rocks (Kumar et al. 1999) of the Delhi and Aravalli Supergroups. In addition, alluvium aquifers have also been found in some places (Chauhan 2007; Chauhan et al. 1996). However, the presence of alluvium aquifers is not known in the study area. The hard rocks essentially lack primary porosity and, therefore, movement of groundwater is mainly controlled by structural features developed within the hard rocks especially cleavages, fractures, and joints.

19.3 Methodology

19.3.1 Groundwater Modeling

Numerical groundwater models are used worldwide for various purposes such as testing hypotheses, predicting future behavior of aquifers, or organizing different hydrogeological data in a consistent conceptual framework (Anderson and Woessner 2002). A model for predictive purposes requires good field data. However, a model with inadequate data can help guide data collection activities, particularly in those areas where detailed field data are critical to the success of the model (Wang and Anderson 1982). In this research, a groundwater model has been developed that has the potential to replicate field-measured values.

19.3.2 Hydrostratigraphic Units

Geologic units of similar hydrogeologic properties are termed as hydrostratigraphic units. The study area has an unconfined aquifer system belonging to the Jharol subgroup of the Aravalli Super group of the Pre-Cambrian Eon. The aquifers are mainly composed of schistose rocks with compositional variation and quartzitic rocks of varied thickness. Bedrock lithology and subsequent structural deformation of the bedrock control the hydraulic properties of the aquifer. Drilling records of three wells were obtained from the Groundwater Department of Rajasthan in which one of the wells lies in the study area. Information for these wells is available to a depth of 50 m. However, it was found that the maximum depth of wells in the study area was 20 m, beyond which the rocks become hard and any subsequent digging requires use of explosives.

The key feature of an unconfined aquifer can be viewed as variable transmissivity with changing saturated thickness. Because the spatial variation of transmissivity in the area is not well known, it is assumed in the model that the saturated thickness of the aquifer does not change. Hence, for modeling purposes it is assumed for simplicity that the aquifer is confined and extends vertically with a uniform thickness of 20 m. The effect of this assumption is expected to be minor.

19.3.3 Boundaries

In the study area catchment, it was observed that the streams exit the catchment to the northwest. Since it is believed that groundwater flow patterns follow the surface water flow patterns, it can be inferred that groundwater also flows out of the model domain to the northwest. On the basis of this assumption, the northwestern part of the catchment was assigned a constant head boundary (Fig. 19.3). Apart from that, the entire catchment boundary was assigned a no-flow boundary assuming that the surface water divide coincides with the groundwater flow divide (Fig. 19.3).

19.3.4 Type of Model

Two-dimensional, distributed parameter, steady and transient confined aquifer models have been developed in this study. Keeping in mind the heterogeneity of the aquifer properties which play a key role in influencing groundwater flow in such hydrogeological systems, it is preferable to apply a distributed parameter model.

In a groundwater modeling study, the development of a conceptual model is followed by translation of the physical system into mathematical equations known as governing equations and boundary conditions. These mathematical equations are solved numerically by an algorithm in a computer program. This computer program is commonly referred to as computer code. Based on the model objective of this study, a relevant computer code was selected which would be able to adequately represent hydrogeological processes, hydrostratigraphy, flow and boundary conditions, etc.

The model code selected for this study is MODFLOW, a modular three-dimensional, block-centered, finite difference code developed by the United States Geological Survey (USGS) for layered aquifer systems, (Harbaugh and McDonald 1996). The groundwater flow model MODFLOW numerically approximates the solution of the partial differential equation, derived by combining an equation of continuity and Darcy's Law, for three-dimensional groundwater flow through porous media. This partial differential equation (Eq. 19.1), which is the governing equation representing the groundwater flow system in the study area based on certain boundary conditions, has to be solved to get the spatial distribution of heads. The general partial differential equation to be solved for transient conditions can be written as:

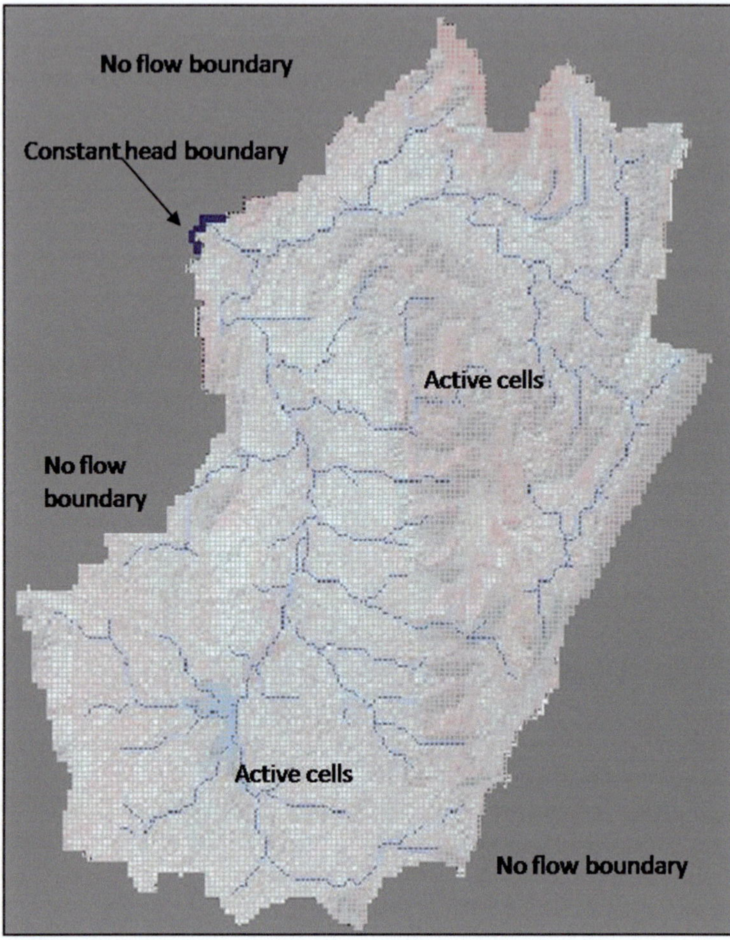

Fig. 19.3 Model domain showing horizontal extent and boundaries

$$K_x \frac{\partial^2 h}{\partial x^2} + K_y \frac{\partial^2 h}{\partial y^2} + K_z \frac{\partial^2 h}{\partial z^2} = S_s \frac{\partial h}{\partial t} - R \tag{19.1}$$

where:

Kx, *Ky*, *Kz*	Hydraulic conductivity along *x*, *y*, *z* coordinate axes
h	Potentiometric head
Ss	Specific storage
R	Volumetric flux of sources and sinks per unit volume
t	Time

For steady-state conditions the term $S_s \frac{\partial h}{\partial t}$ becomes zero and Eq. 19.1 becomes

$$K_x \frac{\partial^2 h}{\partial x^2} + K_y \frac{\partial^2 h}{\partial y^2} + K_z \frac{\partial^2 h}{\partial z^2} + R = 0 \qquad (19.2)$$

The application of Eqs. 19.1 and 19.2 are valid under the following conditions of saturated aquifer, Darcy's Law is valid and also mass is conserved.

The equations for boundary conditions can be given as:

For no-flow boundaries, the head gradient perpendicular to a boundary is set to zero and can be represented mathematically as:

$$\frac{\partial h}{\partial x} = 0 \qquad (19.3)$$

and,

$$\frac{\partial h}{\partial y} = 0 \qquad (19.4)$$

For constant head boundaries:

$$h = c, \text{ where } c \text{ is constant.} \qquad (19.5)$$

19.3.5 Rationale for Selecting MODFLOW and Model Setup

The groundwater flow model MODFLOW is well documented as well as easy to understand (McDonald and Harbaugh 1988) and is very popular and widely accepted within the modeling community. Groundwater flow in complex hydraulic conditions with various natural hydrological processes can be simulated under steady- and transient-state conditions in an irregularly shaped flow system in which the aquifers can be either confined, unconfined, or both confined and unconfined. MODFLOW is capable of simulating the effects of wells, recharge, rivers, drains, evapotranspiration, and head-dependent boundaries. Furthermore, anisotropic and spatially varying hydraulic conductivities and transmissivities can be assigned to different layers and the storage coefficient can be heterogeneous.

A small catchment located in the southern part of the Wakal River basin was selected as the domain for groundwater modeling. The horizontal extent of the model domain is shown in Fig. 19.3. The approximate areal extent of the model domain is 118 km^2. Maximum length and width of the catchment are roughly 15 and 10 km, respectively. The model domain lies within 330395–343327 Easting and 2676345–2692351 Northing. The model of the study area was constructed with one layer representing a confined aquifer. As described in the conceptual model, the

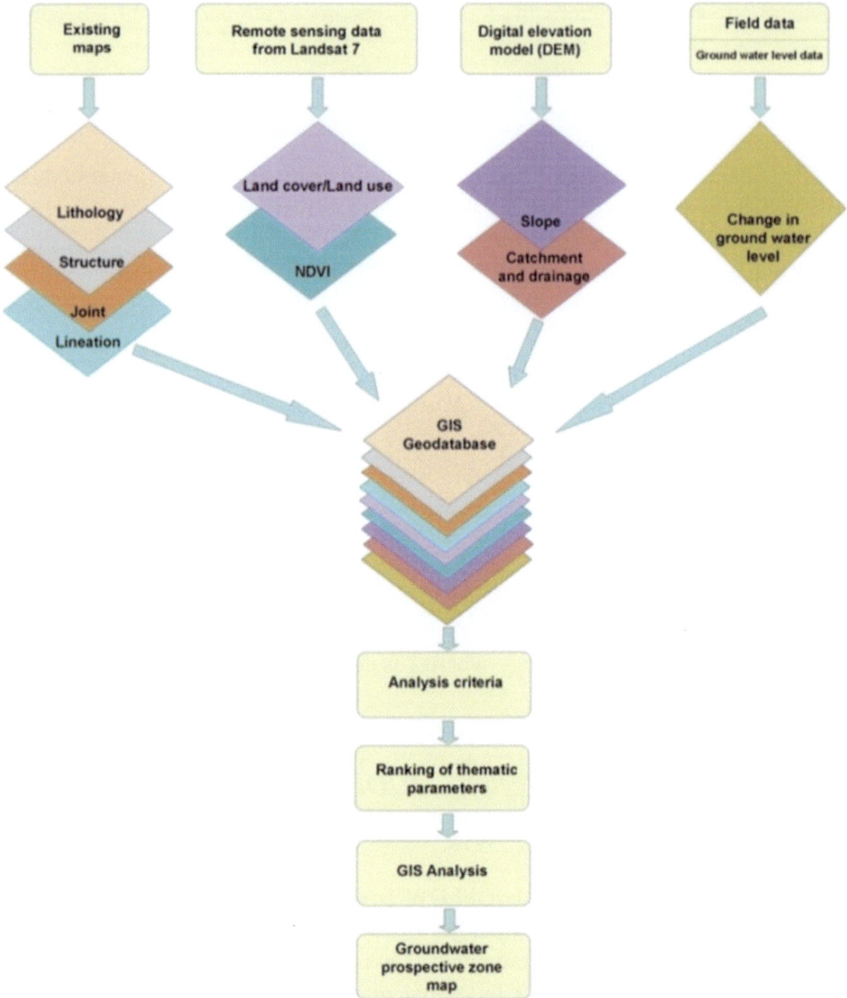

Fig. 19.4 Flowchart showing the methodology for the delineation of recharge sites

vertical extent of the model has been assumed to be of 20 m uniform thickness. The prospective groundwater recharge areas for this study were delineated using various data. Figure 19.4 shows the flowchart for delineating the recharge areas.

19.3.6 Discretization and Model Boundaries

In a numerical groundwater model, the modeling domain is replaced by a discretized domain which consists of an array of nodes and related finite difference blocks (also referred to as cells) (Anderson and Woessner 2002). The locations of these cells are represented by columns, rows, and layers. Hydraulic head is calculated at each of these cells. The model domain was discretized into 143 columns and 177 rows using 90.4 × 90.4 m grid spacing corresponding to that of the Shuttle Radar Topography Mission (SRTM) digital elevation model (DEM), which was intended to be used to assign elevations for the top and bottom of the aquifer, resulting in 25,311 grid cells. Ultimately, the elevations were irrelevant to the two-dimensional confined aquifer model. These cells were defined to be active, inactive, and constant head cells by assigning values of 1, 0, and −1, respectively, using the IBOUND array in MODFLOW. Cells outside the no-flow boundary were defined as inactive cells, cells in the constant head boundary were defined as constant head cells, and the remaining cells in the modeling domain where heads will be calculated were defined as active cells. The total number of active grid cell in the model domain is 13,616.

Based on physical features and the hydraulic conditions applicable boundary conditions were defined. No-flow boundaries were assigned along the catchment boundary except in the northwestern part where it is assumed that groundwater is flowing out of the aquifer system and is, therefore, assigned a constant head boundary below the lowest land surface elevation in the domain.

19.3.7 Initial Hydraulic Head and Hydraulic Parameters

Initial hydraulic heads are required by MODFLOW at the beginning of a flow simulation (Chiang and Kinzelbach 2003). For steady-state models, the fixed head cell values should be the actual values and the active cells can be arbitrary values. Due to a lack of field-measured data at the constant head boundary, the constant head cells were assigned a value of 488 m, which is below the lowest elevation of the land surface in the constant head boundary area.

For the transient model, however, actual values of initial hydraulic heads must be provided (Chiang and Kinzelbach 2003). Therefore, heads obtained from fully calibrated steady-state model were used as initial hydraulic heads for the transient model.

The primary hydraulic parameter required by a steady-state model is either hydraulic conductivity or transmissivity. Pumping tests were carried out at Shyampurakala and Amor in the study area and then pumping test data were analyzed by Jacob's straight line method. The transmissivity values were calculated to be 38.15 m^2/d and 15.12 m^2/d, respectively.

For the transient-state model, the calibrated values of transmissivity from the steady-state model were applied. Another important parameter required for transient-state model is the specific storage or storage coefficient values. Storage coefficient values were obtained from the literature. Storage coefficient values calculated in similar hard rock terrain range from 0.00039–0.09. Storage coefficient values were determined by the Groundwater Department of Rajasthan (GOR 1979) by pumping tests using various analysis techniques in similar hard rock terrain in the Aravalli Mountains. The transient-state model was calibrated by applying 0.00039 as a starting value for the storage coefficient.

19.3.8 Groundwater Level Data

Pre-monsoon and post-monsoon groundwater levels were collected from open dug wells in the study area during the months of late May to early August in 2006 and were used to create a groundwater level surface for the pre-monsoon and post-monsoon period using kriging. The locations of open dug wells are shown in Fig. 19.5. The wells were divided into zones 1 and 2 to reduce errors in the

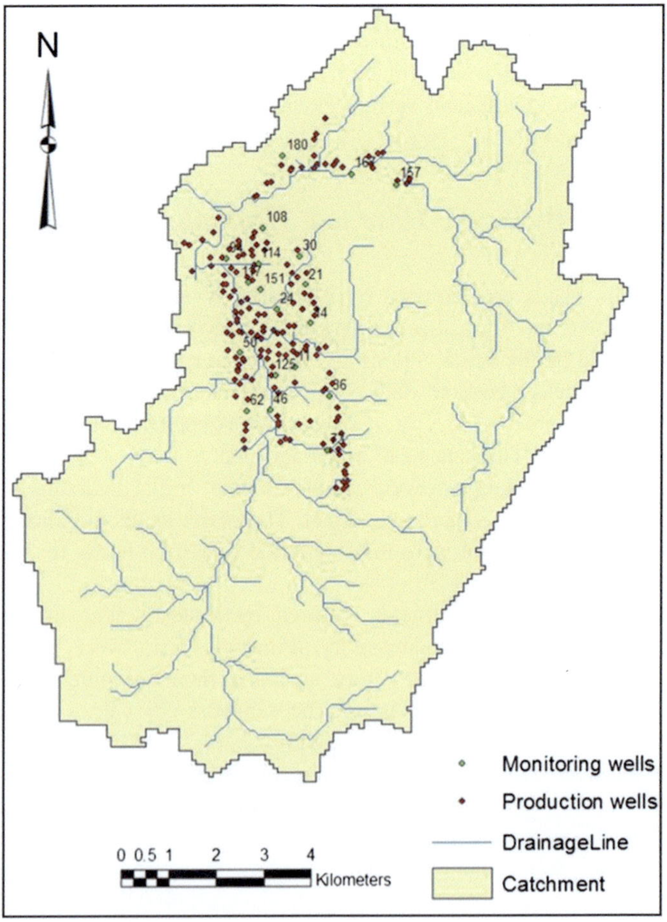

Fig. 19.5 Map showing the location of open dug wells

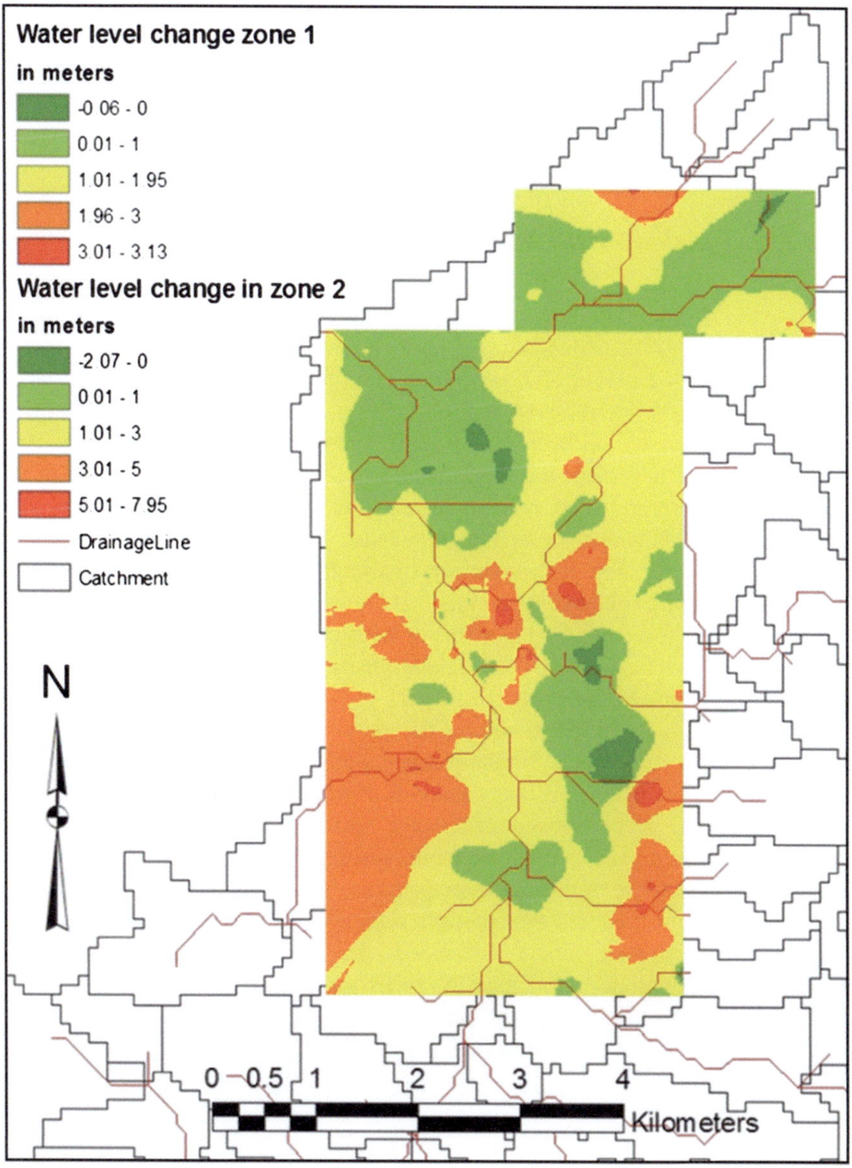

Fig. 19.6 Map showing changes in groundwater level

generation of water level surface maps. The pre-monsoon groundwater level surface was then subtracted from the post-monsoon groundwater level grid to get the change in groundwater level using the raster calculator. The change in groundwater level for the two zones is shown in Fig. 19.6.

19.4 Results and Discussion

The results of the study obtained from the data analysis followed by the steady- and transient-state calibration, model outputs, and forecasts are presented.

19.4.1 *Transient-State Model Results*

After getting similar pattern of head response over a period of 5 years, the storage coefficient values were calibrated using observed heads in 19 monitoring wells. The model was run for a period of 2 years from 2005 to 2006 and the simulated heads were compared with the observed heads. The heads used for 2005 were the same as 2006. The comparison has been shown for 8 randomly selected wells in Fig. 19.7.

The simulated and observed heads show a fairly good match in some of the wells. Wells 21, 62, and 73 show steep increases in the observed heads compared to the simulated heads which show gradual increase. Some of the wells, for example, wells 11, 14 and 62 show overprediction of simulated heads.

All these can be explained by the fact that these wells were not monitoring wells in the true sense. The wells are big diameter wells that are used by people on a daily basis. Since the wells are big, they receive plenty of water during a precipitation event directly in addition to recharge through groundwater movement. This makes the water table unusually high during the monsoon season, reaching up to the ground surface in some cases. Similarly, since these are being used by people daily, it might be possible that significant quantities of water have been withdrawn by the time groundwater level measurement were taken. All these factors may have contributed to discrepancies with the model results.

Pre- and post-monsoon regional flow patterns in the catchment are shown using water level contour map and groundwater velocity vectors in Fig. 19.8.

During transient-state model calibration, the storage coefficients had to be adjusted. This was done by creating 14 zones and adjusting values in each zone after every model run until a good match between simulated and observed heads were obtained. The final calibrated values of storage coefficient values are presented in Fig. 19.9.

19.4.2 *Model Prediction*

Management scenarios were formulated to evaluate the impacts of current pumping schedules and propose solutions to potential overdraft conditions.

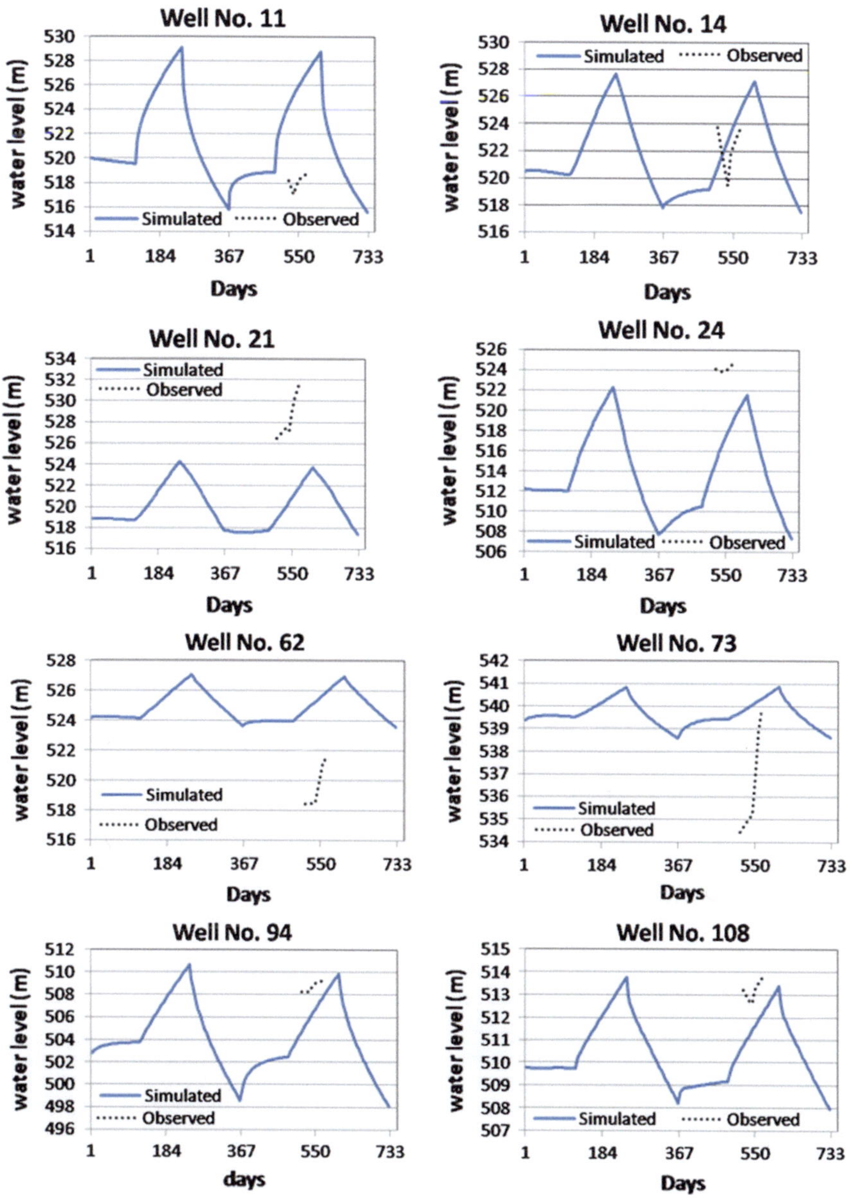

Fig. 19.7 Plot of simulated against observed heads for the year 2005–2006

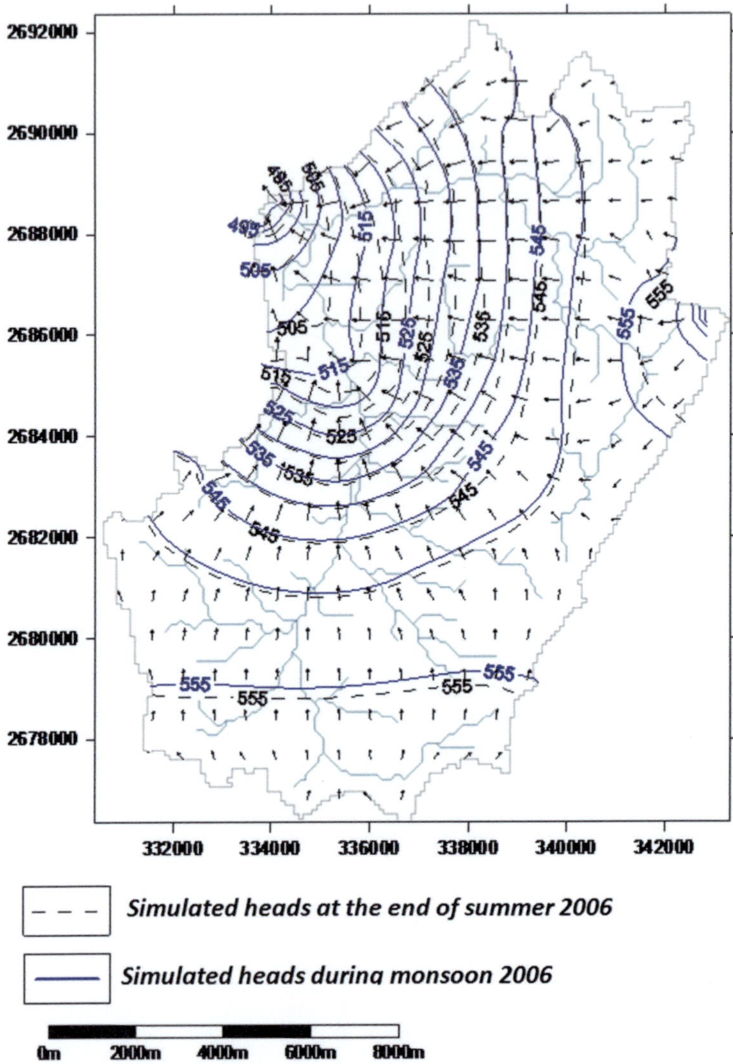

Fig. 19.8 Pre- and during-monsoon groundwater level of 2006 and groundwater velocity vectors

19.4.2.1 Model Scenario 1: 15 % Increase in Withdrawal

This scenario is predicted on recharge being unaffected by changes in withdrawal. Increases in withdrawal might allow additional recharge. The model was run to predict the regional groundwater head in the catchment until the year 2010 with an

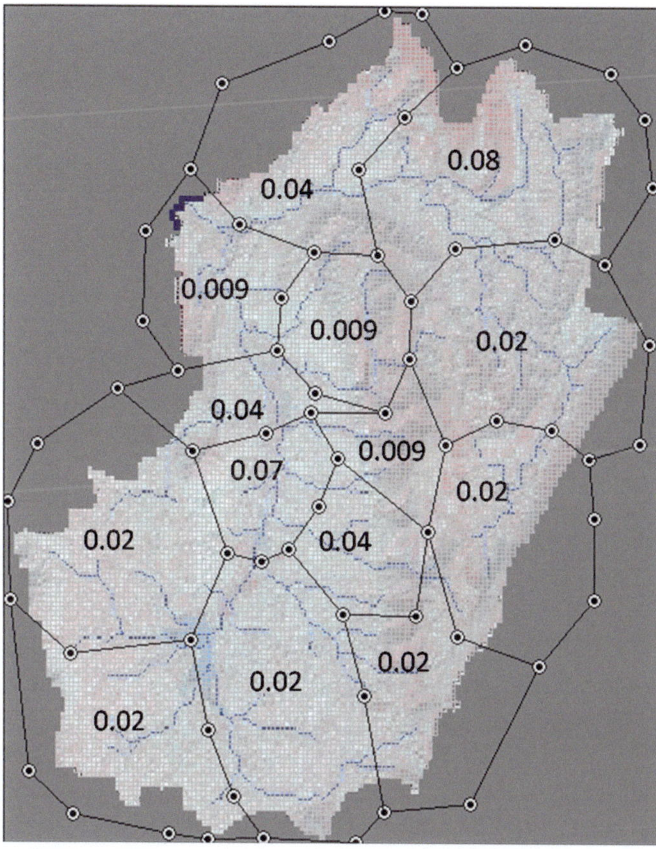

Fig. 19.9 Variation of storage coefficient in different zones

increase in abstraction by 15 % in the year 2008–2010 and normal rainfall condition. The initial and boundary conditions to calibrate transient model for 2006 has been used for this purpose. The simulated regional groundwater heads at the end of the monsoon season of 2006 and 2010 are shown in Fig. 19.10.

It can be seen that groundwater level in the eastern part of the catchment declines by the end of monsoon season in 2010. Apart from that, no significant changes in groundwater levels have been observed. The simulated heads of the monitoring wells were plotted against time to have a closer look at the response of the aquifer. Hydrographs of four of these wells are shown in Fig. 19.11. It can be seen that groundwater level in all the four monitoring wells declined in the range of 0.5 to about 4 m.

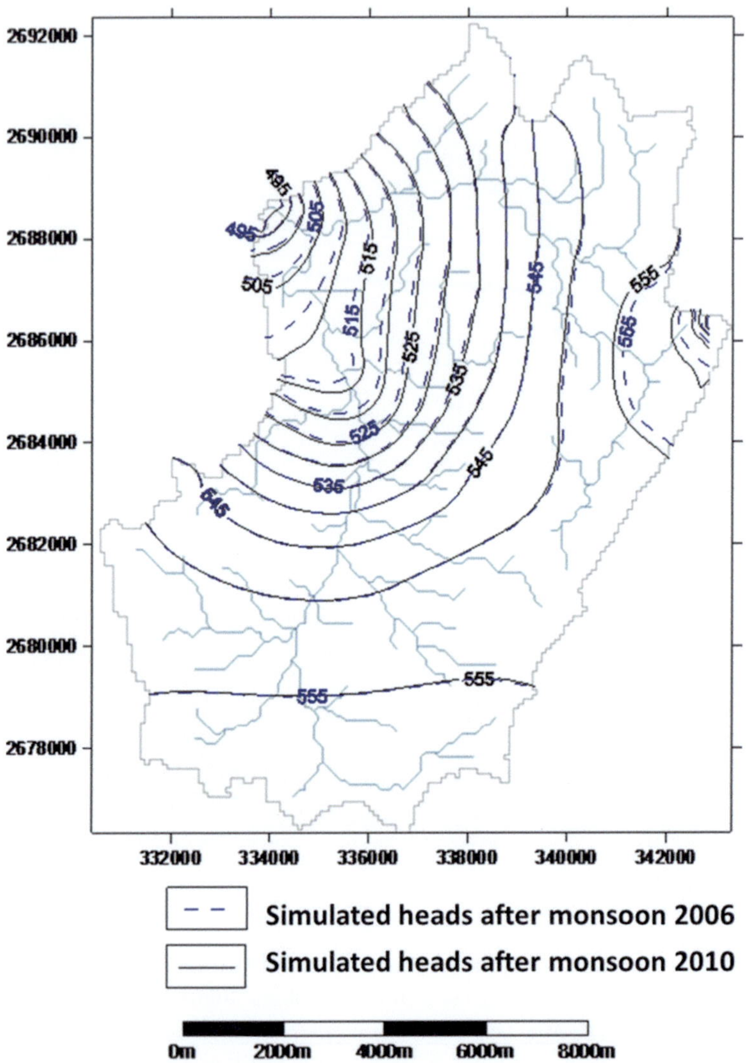

Fig. 19.10 Contour plot of simulated heads after monsoon of 2006 and 2010: Scenario 1

Although most part of the catchment will not be greatly affected by increase in withdrawal by 15 %, wells lying in the western part can have significant impact on the groundwater levels as demonstrated by the hydrograph of Well No. 151. Locally, this calls for some immediate measures to check the withdrawal situation in this area.

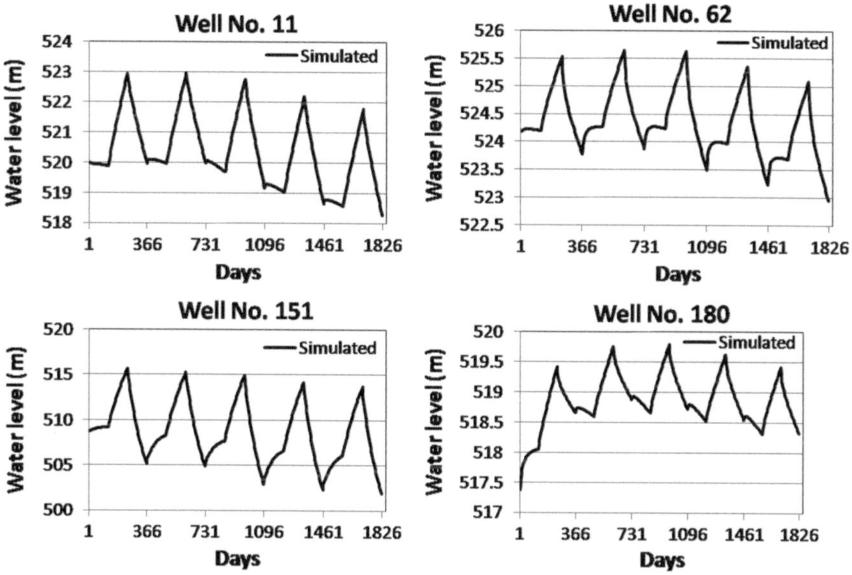

Fig. 19.11 Simulated heads of wells through December 2010 under increased pumping conditions by 15 %: Scenario 1

19.4.2.2 Model Scenario 2: Two Consecutive Drought Years in Every Four Years

This scenario is predicted on reductions in recharge proportional to reductions in rainfall brought on by drought. It is possible that reduced rainfall would not lead to recharge reductions. This is because potential recharge might be rejected when the low storage aquifer is full. But with the available data, it is not possible to know if such rejections are taking place.

Analysis of 41 years of rainfall data indicates that the study area received less than average rainfall of 645 mm in 22 years. The average rainfall for these low rainfall years was 484 mm. It was also observed that the area faced drought for 7 consecutive years from 1999–2005. In order to see the effect of drought on the groundwater levels, the model was run assuming deficit rainfall for two consecutive years in every four years until 2010. For this purpose, yearly average rainfall for deficit years was used for drought years and rainfall of 2006 was used for normal years. Abstraction was maintained at present rate. Simulated heads of four wells have been plotted against time to see the effect of drought (Fig. 19.12).

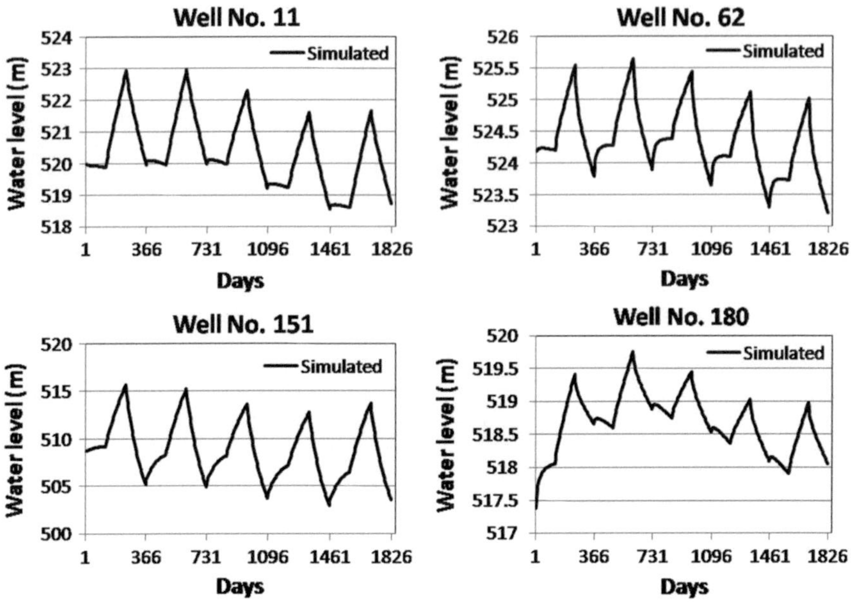

Fig. 19.12 Simulated heads of wells until 2010 under drought scenario: Scenario 2

It can be seen that groundwater levels declined in all the wells in drought years. Water levels in wells 62 and 180 declined by approximately 0.5 m. But one can clearly see that wells 11 and 151 show a decline in groundwater level by 1.5 and 3 m, respectively. In this scenario, groundwater levels recover a bit in 2010 due to normal rainfall. But if there is a prolonged drought leading to reduced recharge in the region, groundwater levels could be greatly impacted. A contour map of simulated heads after monsoon of 2006 and 2010 shows the spatial distribution of the groundwater level changes (Fig. 19.13).

19.4.2.3 Model Scenario 3: Combined Effect of Drought Years and Withdrawal Increase

This scenario is built to see the combined effect of the model scenarios 1 and 2 on the regional head distribution through 2010. The simulated regional groundwater heads at the end of the monsoon season of 2006 and 2010 are shown in Fig. 19.14.

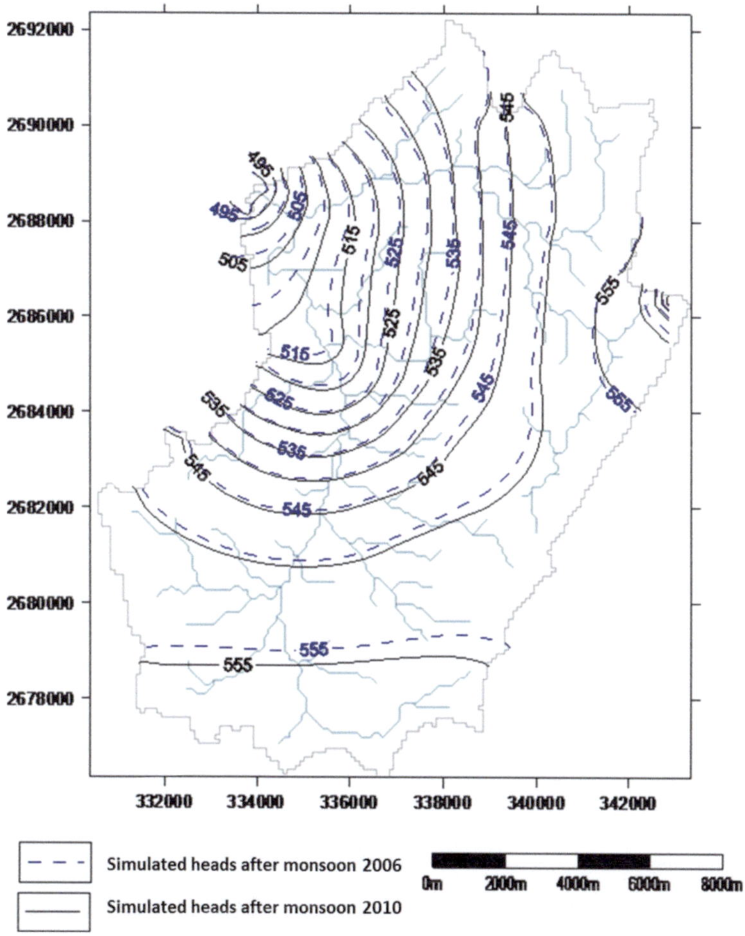

Fig. 19.13 Comparison of simulated heads after monsoon of 2006 and 2010: Scenario 2

It is clear that if the catchment faces drought years and at the same time the withdrawal is increased by 15 %, the decline in groundwater levels can be catastrophic for the entire catchment. It can be seen that the groundwater levels have declined in the entire catchment. The hydrographs of four of the monitoring wells are shown in Fig. 19.15. Well 180 shows a decline of about 1 m, but in the remaining wells groundwater declines by 2 to 9 m. Such declines can have a huge impact in the area. This problem can be further aggravated if such conditions are prolonged in the area.

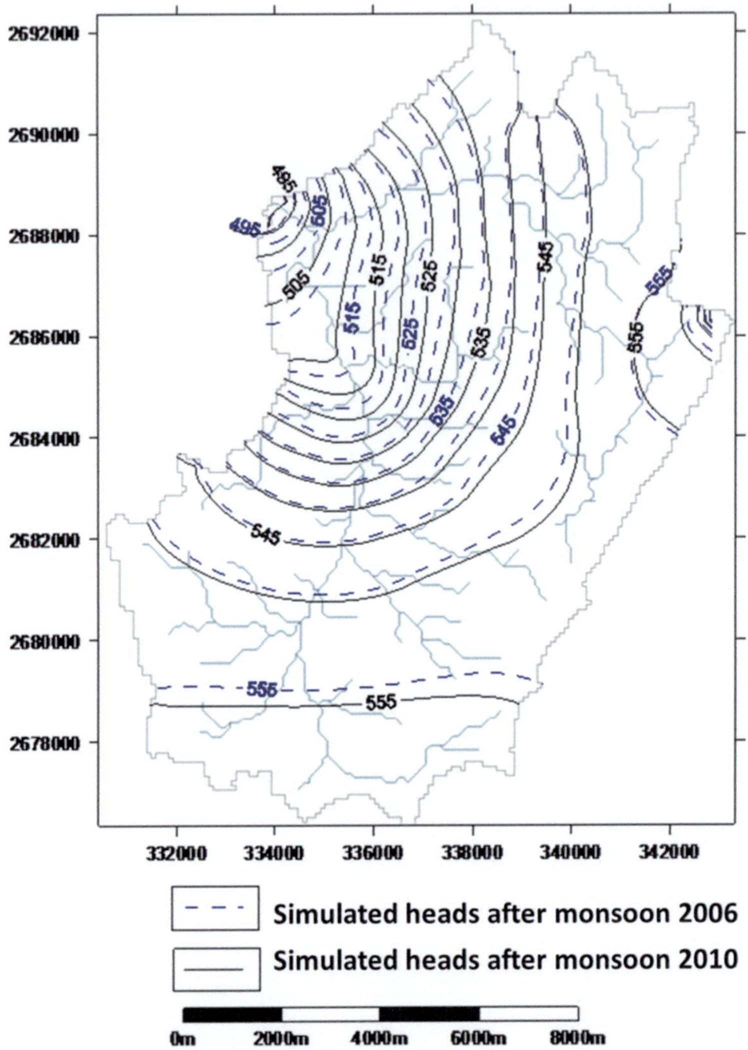

Fig. 19.14 Comparison of simulated heads after monsoon of 2006 and 2010: Scenario 3

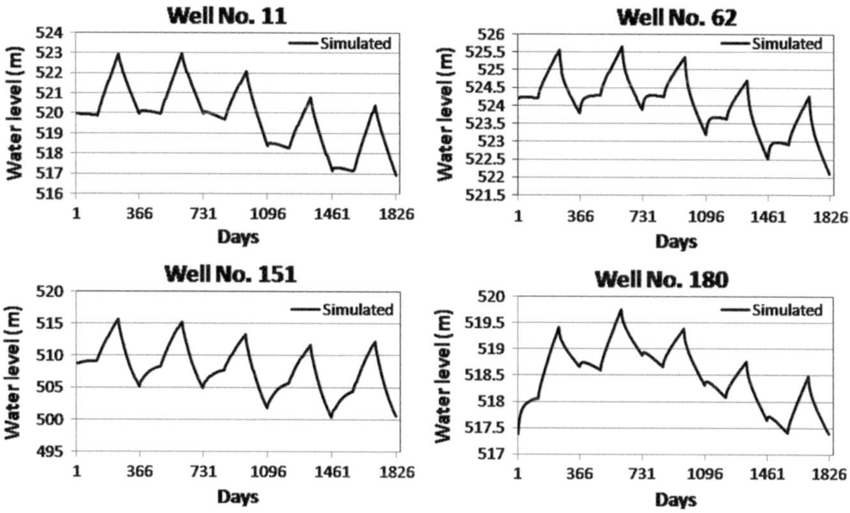

Fig. 19.15 Simulated heads of well until 2010: Scenario 3

19.5 Conclusions

Groundwater flow in the Wakel River basin was simulated using transient and steady-state numerical models. A subbasin was selected to study the groundwater flow system and the effects of groundwater development using a numerical groundwater flow model (Visual MODFLOW). Hydrological parameters obtained from pumping tests were related to each hydrostratigraphic unit to assign the distribution of parameter values within each model layer.

The transient-state model was calibrated successfully to replicate the field-measured values within acceptable error. Three different modeling scenarios were presented. The third scenario with two successive drought years in every four years and increased withdrawal by 15 % showed that the groundwater level decreased by as much as 9 m in some wells. The effect of such drought is currently uncertain however, because the runoff from the basin is unknown. Much of the precipitation falling on the basin may be rejected by the low storage capacity of the fractured rock aquifer. This groundwater modeling study in this area is one of the first detailed studies about groundwater and hydrogeology of the area. It was attempted to model the area as realistically as possible. The groundwater modeling study successfully explained the overall behavior of the aquifer and its parameters as well as processes associated with it, such as recharge. This study was also successful in developing a reasonable conceptual model which can be converted into a numerical model with a capability to reproduce field-measured data with comparatively good accuracy. This model should be refined with more detail for better accuracy. This study also demonstrated the capabilities of remote sensing and GIS for evaluation of groundwater resources and demarcation of groundwater prospect zones.

Acknowledgments Our thanks goes to World Vision India for facilitating the research in Rajasthan, India. We would also like to thank Dr. N.K. Chauhan, Mr. A.K. Bordia, and Dr. P.K. Singh for providing deep insight and knowledge as well as their significant contribution in making data and resources available to me. We are especially thankful to the villagers in the study area for their help, without which it would not have been possible to conduct this research.

We would like to acknowledge Global Water for Sustainability (GLOWS) Program for funding the research through the United States Agency for International Development (USAID).

References

Anderson MP, Woessner WW (2002) Applied groundwater modeling simulation of flow and advective transport. Academic Press, San Diego, California, p 381

Chauhan NK (2007) Hydrogeology assessment report: Wakal River Basin, India. 2007. Global Water for Sustainability Program, Florida International University, 132 p

Chauhan NK, Sharma BL, Mohemmad SA (1996) Structural geometry and strain history of the early proterozoic Aravalli rocks of Gorimari, Udaipur district, Rajasthan. J Geol Soc India 47:59–74

Chiang WH, Kinzelbach W (2003) 3D-groundwater modeling with PMWIN: A simulation system for modeling groundwater flow and pollution. Springer, Berlin, Heidelberg, New York 346p

de Vries JJ, Simmers I (2002) Groundwater recharge: an overview of processes and challenges. Hydrogeol J v10(5):17

ECIDWR (2005) Expert committee on integrated development of water resources report, 2005. Expert committee on integrated development of water resources (ECIDWR), Jaipur, pp 1–43

GOR (1979) Evaluation of hydraulic parameters of Aravalli phyllites and Erinpura granites. Government of Rajasthan, Jodhpur

GOR (2003) Ground water atlas of Rajasthan (2003), SRSAC, DST. Government of Rajasthan, Jodhpur

GWP (2000) Integrated water resources management, TAC background papers no 4. Global Water Partnership, Stockholm, p 9

Harbaugh AW, McDonald MG (1996) User's documentation for MODFLOW-96, an update to the U.S. Geological Survey modular finite-difference ground-water flow model: U.S. Geological Survey Open-File Report 96-485, 56 p

ICID (2005) Water resources assessment of Sabarmati Basin. International Commission on Irrigation and Drainage (ICID), New Delhi, India, India, p 65

Kumar MD, Chopde S, Mudrakartha S, Prakash A (1999) Chapter 5: addressing water scarcity: local strategies for water supply and conservation management in the Sabarmati Basin, Gujarat. In: Moench M, Caspari E, Dixit A (eds) Rethinking the Mosaic: investigations into local water management. Nepal Water Conservation Foundation, Kathmandu, pp 191–246

Mahnot SC, Singh PK (2003) Chapter 3: agro-climatic conditions and surface water harvesting. In: Kaul V (ed) Water Harvesting and Management. SDC/ICU, Jaipur, pp 26–34

McDonald MG, Harbaugh AW (1988) A modular three-dimensional finite-difference ground-water flow model: techniques of water-resources investigations. U.S. Geological Survey, book 6, Chap. A1

Narain P, Khan MA, Singh G (2005) Potential for water conservation and harvesting against drought in Rajasthan, India. Working paper 104 (drought series: paper 7). International Water Management Institute (IWMI), Colombo, pp 1–25

Rathore MS (2005a) State level analysis of drought policies and impacts in Rajasthan, India. Working paper 93. (Drought series: paper 6). International Water Management Institute, Colombo, pp 1–29

Rathore MS (2005b) Groundwater exploration and augmentation efforts in Rajasthan. Institute of Development Studies, Jaipur, pp 1–33

Singh PK (2006) Climatology data of Udaipur. Department of soil and water conservation engineering. Maharana Pratap University of Agricultural and Technology (MPUAT), Udaipur (unpublished)

Stiefel J, Melesse AM, McClain M, Price R (2007) Rainwater harvesting in Rajasthan, India: recharge estimation using tracers, 13th international rainwater catchment systems conference and 5th international water sensitive urban design conference, 21–23 Aug 2007. Sydney, 8 p (CD proceeding)

Stiefel JM, Melesse AM, McClain ME, Price RM, Chauhan NK (2008) The impact of artificial recharge from rainwater harvesting structures on the groundwater of nearby wells in rural Rajasthan. International groundwater conference on groundwater dynamics and global change, 11–14 Mar 2008, Jaipur, India

Stiefel J, Melesse A, McClain M, Price RM, Anderson AP, Chauhan NK (2009) Effects of rainwater harvesting induced artificial recharge on the groundwater of wells in Rajasthan, India. Hydrogeol J 17(8):2061–2073

Sukhija BS, Nagabhushanam P, Reddy DV (1996) Groundwater recharge in semi-arid regions of India: an overview of results obtained using tracers. Hydrogeol J v4(3):50–71

UNDP (2005) Looking to the future. groundwater management in Rajasthan: issues, perspectives & policy. A National Consultation Organised by Jal Bhagirathi Foundation and UNDP, Jaipur, 16 p

Wang HF, Anderson MP (1982) Introduction to groundwater modeling: finite difference and finite element methods. W.H. Freeman, San Francisco, p 256

Chapter 20
Water Resources Assessment and Geographic Information System (GIS)-Based Stormwater Runoff Estimates for Artificial Recharge of Freshwater Aquifers in New Providence, Bahamas

M. Genevieve Diamond and Assefa M. Melesse

Abstract The Bahamas is a small island nation that is dealing with the problem of freshwater shortage. All of the country's freshwater is contained in shallow lens aquifers that are recharged solely by rainfall. The country has been struggling to meet the water demands by employing a combination of overpumping of aquifers, transport of water by barge between islands, and desalination of sea water. In recent decades, new development on New Providence, where the capital city of Nassau is located, has created a large area of impervious surfaces and thereby a substantial amount of runoff with the result that several of the aquifers are not being recharged. A geodatabase was assembled to assess and estimate the quantity of runoff from these impervious surfaces and potential recharge locations were identified using a Geographic Information System (GIS). This study showed that runoff from impervious surfaces in New Providence represents a large freshwater resource that could potentially be used to recharge the lens aquifers on New Providence.

Keywords Artificial recharge · New Providence · Bahamas · Freshwater · Aquifers · GIS · Freshwater scarcity

Abbreviations

ARC	Antecedent runoff condition
DEM	Digital elevation model
ESRI	Environmental Systems Research Institute
FAO	Food and Agriculture Organization
FIS	Fractional impervious surface

M.G. Diamond (✉) · A.M. Melesse
Department of Earth and Environment, Florida International University, Miami, FL, USA
e-mail: diamondg@fiu.edu

A.M. Melesse
e-mail: melessea@fiu.edu

© Springer International Publishing Switzerland 2016

A.M. Melesse and W. Abtew (eds.), *Landscape Dynamics, Soils and Hydrological Processes in Varied Climates*, Springer Geography,
DOI 10.1007/978-3-319-18787-7_20

GIS	Geographic information system
NDVI	Normalized difference vegetation index
NP	New Providence
NRCS	Natural Resources Conservation Service
OAS	Organization of American States
SRTM	Shuttle Radar Topography Mission
SSI	Stormwater Solutions Incorporated
TNC	The Nature Conservancy
TR-55	Technical Report 55
USDA	United States Department of Agriculture
USGS	United States Geological Survey

20.1 Introduction

20.1.1 Overview

The availability of freshwater for domestic, agricultural, and industrial uses is one of the most serious issues facing much of the world today. The Millennium Ecosystem Assessment found that more than one billion people live in areas that do not have sustainable freshwater supplies, and are meeting their needs through overdraft or engineered transfers of water (Millennium Ecosystem Assessment 2005). Freshwater is a vital resource, since life cannot be sustained without it. The trend in overuse of existing freshwater reserves is alarming and countries need to address it in their long-range planning.

The supply of freshwater is distributed unevenly across the planet, and is governed by factors such as geomorphology, geology, climate, and location. The use of freshwater varies among countries in the developed and developing world. While we tend to view freshwater as freely available and renewable, in fact the time required for liquid water to move among the various storage areas of the planet, the oceans, rivers, aquifers, lakes, and atmosphere, can exceed a human lifetime. It is therefore vital that this resource be carefully managed.

While clean freshwater resources on the planet become more scarce, it is being recognized generally that nations must develop integrated water resource management plans to ensure that there will be sufficient freshwater for use by all peoples today and for future generations (Rosegrant et al. 2002). These plans must encompass all phases of water management and must acknowledge and accommodate the various sectors of water use: social (human), economic, and environmental (Rosegrant et al. 2002).

Potable freshwater is in high demand across the globe and the nature of its uneven distribution presents unique problems in its management (Wescoat and White 2003). It is often difficult to maintain a balance between the needs of farmers for water to irrigate crops and the needs of a river to maintain a healthy riparian population (Millennium Ecosystem Assessment 2005). It is difficult to overestimate the economic value of water as it is a vital component in almost every undertaking of industry, agriculture, and recreation. In fact, as freshwater becomes scarcer, it becomes more of an economic good in itself as nations are forced to find alternate means of obtaining freshwater supplies.

20.1.2 Artificial Recharge of Freshwater Aquifers

It has been estimated that as much as 97 % of all liquid freshwater on the planet is contained in natural subsurface storage areas known as groundwater aquifers (Shiklomanov 1993). The use of groundwater for all purposes has been steadily increasing over the last century as a result of a number of causes, mainly overuse of surface water, especially for irrigation of crops, which has led to the decline of this resource (Brown 2006). Along with this depletion nations have seen the increase of impervious surfaces in the form of buildings, roads, parking lots, and other built areas. As many of the world's largest groundwater aquifers become depleted, nations are seeking ways to supplement their supplies of freshwater. Artificial recharge of these aquifers has emerged as an effective means to accomplish this. Artificial recharge refers to any type of system designed for the purpose of directing water from the surface; several of these systems are in use in different countries. These can generally be described as direct surface, direct subsurface, and indirect methods (Asano 1985).

Direct surface methods include surface spreading by flooding or diverting streams, by constructing spreading basins for rainfall–runoff, and by using methods such as ridge and furrow irrigation for recharge of irrigation waters. Direct surface recharge can be accomplished by constructing dams and weirs to reroute rivers and streams to areas of recharge such as those used in the Netherlands as described by Peters (1998), where water from the Rhine and Meuse Rivers is redirected to basin recharge sites in coastal sand dunes. These basins have the added advantage of acting as a means for helping to clean the water of silt and impurities as it seeps through the sand and gravel before it enters the saturated zone.

Direct subsurface methods include injecting water into the aquifer by means of injection wells.

Indirect methods of artificial recharge include the practice of induced recharge in which water is induced to enter the aquifer from a stream bed by pumping the water out at a location of a suitable distance from the well thus "inducing" water to enter the aquifer from above to replace the water removed by the pump. Other indirect methods include aquifer modifications such as bore blasting, and groundwater conservation structures such as dams and sealing of fractures in the aquifer (Bouwer 2002).

20.1.3 Technologies in Use in the Caribbean Region

The OAS Sourcebook (1997) lists a number of nations in the Latin America and Caribbean region that employ various methods of artificial recharge for the purpose of storing surface water for future use as well as for harvesting rainwater and reuse of reclaimed wastewater. These include infiltration basins, canals, and drainage wells for surface runoff. In the San Juan River basin in Argentina, artificial recharge was successfully accomplished using a combination of infiltration basins and canals to recharge the groundwater aquifer in the Valley of Tulum. In Barbados the limestone aquifer is recharged by way of drainage wells called suckwells. These wells are dug into the rock until an adequate fissure is reached through which the water is “sucked” into the subsurface. The suckwells are fed by runoff from road surfaces. The Barbados aquifer is also recharged by effluent from septic tank soakaways. Jamaica conserves excess runoff from road surfaces by treating it for settling of suspended solids in areas upgradient of sinkholes through which the water is directed to the karstic limestone aquifers. Jamaica recorded total recharge of four million cubic meters over 18 months during which time some wells showed marked decrease in salinity levels (OAS 1997).

20.1.4 Use of GIS to Select Sites for Artificial Recharge Facilities

Selecting suitable sites for recharge is vital to the success of the operation. The technology of GIS has been used successfully in selecting sites for effective artificial recharge. Using various digital layers, such as land use and land cover, soil type and depth, geology, and others it is possible to pinpoint locations where recharge will be successful. Jothiprakash et al. (2003) used GIS in Tamilnadu, India to delineate potential zones for artificial recharge when it was found that existing recharge ponds were not effective due to the presence of large quantities of clay in the areas, which seriously reduces the percolation capacity of the soil. Using map layers such as permeability, soil depth, and geology, the authors were able to pinpoint more suitable areas for recharge sites. Biswas (2008) and Biswas et al. (2007) demonstrated that GIS was an effective tool for siting of recharge areas in the Wakal River basin, India, and Ghayoumian et al. (2007) made use of GIS methods to determine the best recharge areas in a coastal aquifer in southern Iran with layers depicting land use, geomorphology, slope, infiltration rates, depth to groundwater, and soil permeability. In this latter study, the purpose of artificial recharge was to improve the quality of groundwater in the coastal aquifer, which was vulnerable to saltwater intrusion. The effectiveness of rainwater harvesting structures in the Wakel River basin was also studied using tracers (Stiefel et al. 2007, 2008, 2009).

20.2 Water Issues in the Bahamas

Small island developing states such as the Bahamas will be affected by global issues like climate change: more frequent droughts and seawater intrusion caused by hurricanes. For these nations, it is imperative to find solutions to freshwater shortages that they will, no doubt, face in the future. The United Nations has determined that the Bahamas and much of the Caribbean region are facing water stress or scarcity in the near future (UNEP 2008)

20.2.1 Freshwater in New Providence

In the Bahama Island of New Providence all natural freshwater stored as groundwater comes from rain but much of the island has been converted to impervious surface by development and road building. Altered land use has resulted in increased flooding in residential and business areas and a huge loss of water that is being wasted as government directs its resources and attention to removing it from the island as quickly as possible, which is being done by directing it to the ocean via drains, or by deep injection below the surface.

Plans for securing freshwater in the future mainly focus on desalination by reverse osmosis (RO); government intends to produce by RO most of the freshwater needed (Water and Sewerage, 2014). As fuel costs rise, water will become more expensive since the Bahamas has no natural energy resources available and depends on imports. The technology of RO is generally suitable for areas that are rich in cheap fuel (Einav et al. 2002) and it is not a practical solution for New Providence.

Much of the freshwater use on New Providence is for tourism, which is the main source of income for the island as well as the nation. Should this industry weaken or fail, water use will drop, but it will be difficult for government to continue subsidizing production by RO or other artificial means as a result of reduced income. If all aquifers were healthy and operational, it would represent a large resource for the residents. However, at this time, several of the aquifers have become saline and will require years of remediation to restore them to usefulness. This is where a coordinated and carefully implemented artificial recharge plan can be most beneficial, so that all aquifers can be restored to usefulness and health by the long-term addition of freshwater from rain. If tourism continues in its present role as the largest component of the economy, the present rate of freshwater use will continue or increase, with concurrent increase in both financial and environmental costs.

20.3 Study Area

The Bahama Islands (Fig. 20.1) are composed of carbonates precipitated from the ocean, and of sediments carried by wind and water and deposited over time (Sealey 1995). As the ocean levels rose and fell during and between glaciations, the surfaces were in turn exposed and eroded by wind and water, and submerged and acted upon by the same elements creating a type of karst landscape (Carew and Mylroie 1997). One of the main landforms in the Bahamas is the eolian ridges that formed on the ocean sides reaching up to 60 m in some areas. The highest point in the islands is Mount Alvernia on Cat Island at 63 m. The other commonly occurring landform is that of lowland areas which occur between these ridges. Most of the surface is made of Pleistocene limestone on the interiors of the islands, while in coastal regions limestone of Holocene age is often found. The Pleistocene rock is covered with a red calcrete or terra rossa paleosol, unless it has been removed by erosion (Carew and Mylroie 1997). Within the Pleistocene limestone freshwater aquifers have been formed by rain that seeped down through the porous surface and settled on the saltwater. Freshwater aquifers also form in the Holocene sands in the southeastern islands and in the coastal areas of the northern islands. The size of the aquifer is limited by the size of the island. Holocene sand aquifers form in strands and beach sands. Extraction is more difficult from these aquifers, but there is potential for the retention of large amounts of freshwater in them (Whitaker and Smart 2004). The Lucayan limestone aquifer is predominant in the islands of the Bahamas. Extraction from these aquifers is generally simple, via drilled wells, and in some low areas, by shallow, hand dug wells. Those that are near to coastlines can discharge directly into "bights" or "creeks", and thence into the ocean. Since most of the land area of

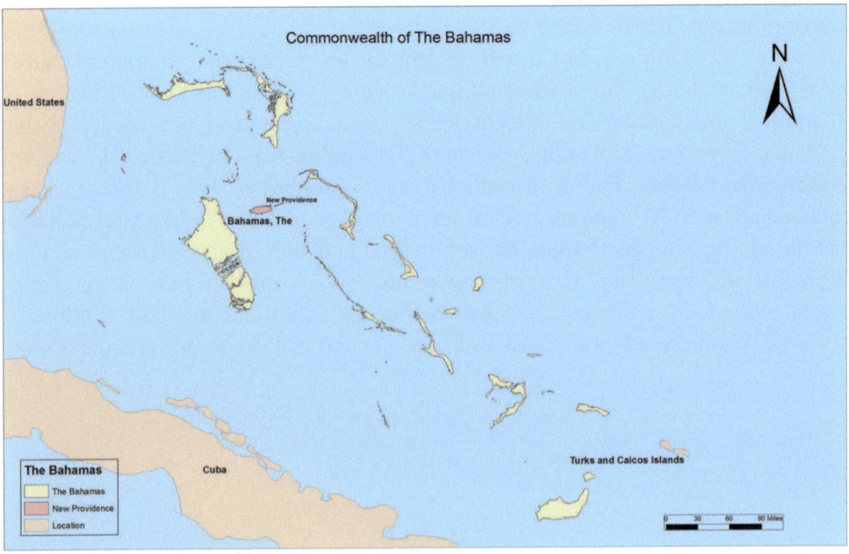

Fig. 20.1 Location map of New Providence (NP), Bahamas

the Bahamas is near sea level, the vadose zone in most areas is only about 1 m thick. The porosity of these surfaces means that falling rain quickly infiltrates the soil producing little runoff. In both types of aquifers, evapotranspiration can occur directly from the aquifer due to the shallowness of the vadose zone. Often tree roots penetrate the aquifer, drawing water directly from within. Also, because of their shallowness, the aquifers are extremely vulnerable to pollution from the surface (Roebuck et al. 2004). The Bahamas is considered to be a water poor country with 61.92 m^3 a year available per capita (FAO 2003).

The island of New Providence, where this study is centered, is one of the northernmost of the Bahama Islands; it sits in an east–west direction on the edge of the Great Bahama Bank opposite to Andros, across the deep extension of the Atlantic Ocean known as the Tongue of the Ocean. It is one of the smallest islands in the Bahamas chain but it is home to more than two-thirds of the residents of the country. New Providence has a total area of 200 km^2, or 51,200 acres (Cant and Weech 1980). A prominent eolian ridge runs along the north side of the island and a lesser ridge along the south side, with low-lying areas between those ridges. Figure 20.2 illustrates the distribution of the ridges. There are a number of lakes on New Providence, the largest of which is Lake Killarney, a shallow, sometimes brackish lake near the western end of the island. Much of the island has been developed with densest population centered in the capital city of Nassau, where many impervious surfaces have been created in the form of roads, buildings, parking lots, and shopping centers. In the natural state, this island surface is highly permeable and water quickly sinks into the calcareous and sandy soils so that no natural perennial streams have formed.

The island receives a generous amount of rainfall which varies over the year as well as from year to year. There is a definite wet and dry seasonal variation with most rainfall occurring during the summer months from May to October. Rain events are generally short and intense with sunshine returning shortly after the end of the rain. Annual rainfall totals vary from 100 cm to more than 170 cm. The rainfall data for the island of New Providence was obtained from the Bahamas Meteorological Office

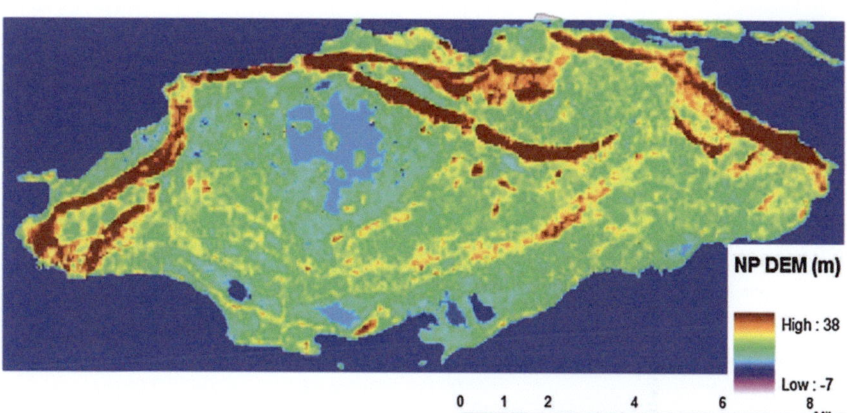

Fig. 20.2 Elevation map of New Providence, Bahamas

covering the period from 1971 to 2000 (Bahamas Department of Meteorology 2008). Average rainfall during this period was 140 cm. The measurements were taken at Nassau International Airport, which was the only location from which complete figures were available, and therefore the only rainfall data used in this study.

20.3.1 Water Resources

The hydrogeology of New Providence is similar to the other islands, that is, Pleistocene limestone throughout most of the island with Holocene sands in coastal regions having Ghyben–Hertzberg type lens aquifers. There are nine aquifers which have a depth of greater than 20 ft thickness located on New Providence. They are distributed throughout the island with a total acreage of 17,503 acres. The total freshwater storage capacity of the New Providence aquifers has been estimated at 120.45 million cubic meters (MCM). The largest of these, the Western New Providence aquifer, is in the western section of the island and has a total storage capacity of 88.82 MCM. The Prospect/Grants Town aquifer is the second largest with 8.73 MCM of freshwater storage capacity, followed by the Blair and Pinewood Gardens aquifer with 7.45 MCM, Blue Hills Ridge with 6.23, South Beach with 4.41, and South Lake Killarney with 3.49 MCM. There are three small aquifers, Golden Gates, Cow Pen Road, and East of Sea Breeze with a total of 1.33 MCM combined (Cant and Weech 1980). If all aquifers were producing, the maximum available water is estimated at 36,441 m^3 per day (Cant and Weech 1980) (Table 20.1).

Water usage on New Providence is mainly for domestic purposes since there is little industry other than tourism. However, this is a high demand sector which raises water usage above that needed for the resident population. The tourism industry has grown rapidly over the last several decades and the water demand has risen concurrently. While the resident population of the country doubled from just over 150,000 persons to about 300,000 during the period from 1975 to 2000 (Dept. of Statistics 2000), water usage increased more than five times in the same period (Water and Sewerage 2014).

Table 20.1 Capacity of aquifers on New Providence

Aquifer	Approximate volume (m^3)	Max lens thickness (cm)	Average lens thickness (cm)
Blair and Pinewood Gardens	7,450,000	9.14	6.35
East of Sea Breeze	222,000	5.84	5.08
South Beach	4,406,000	6.10	5.08
Golden Gates	616,000	5.59	5.08
Blue Hills Ridge	6,229,000	9.40	6.35
Prospect to Grants Town	8,727,000	7.62	6.35
Cow Pen Road	493,000	6.35	5.08
South Lake Killarney	3,489,000	6.35	5.08
Western New Providence	88,816,000	12.70	7.62

Source Cant and Weech (1980)

The water usage of about 11 million imperial gallons per day for New Providence is supplied from a variety of sources, with less than one-fourth from groundwater. All groundwater pumping is from the Windsor wellfield in the western section of the island (Middleton 2008).

20.3.2 Other Water-Related Issues

The creation of impervious surfaces has created an additional problem—that of flooding of neighborhoods in low-lying areas. The area receiving the most serious and frequent flooding events is a housing development in the southeastern area of the island called Pinewood Gardens (SSI 2009). This area is most problematic as a result of its location between the high ridge on the north and a second ridge on the south, as well as the low level of most of the land. Other areas that experience flooding on a regular basis are West Bay Street, which is often inundated by flood water draining from the Chippingham area located on the high ridge running along the north central part of the island, the area along West Bay Street called Rocky Point, where streets are often closed for days at a time as a result of standing floodwater, and Stuart Cove at the southwestern end of the island, which also experiences frequent road closings due to stormwater flooding (SSI 2009) Historically, the method of dealing with these areas of flooding has been to drill drainage wells to try and remove the water from the flooded streets, but repeated maintenance problems have made these attempted solutions ineffective.

20.4 Methodology

The method used in this analysis to calculate runoff is the curve number method that was developed by the United States Department of Agriculture, USDA Soil Conservation Service (later the Natural Resources Conservation Service or NRCS) and first published as Technical Report 55 (TR-55) in January 1975 under the title Urban Hydrology for Small Watersheds (USDA, 1986). The curve number method utilizes curve numbers which were determined on the basis of a number of factors including hydrologic soil group (HSG), cover type and treatment, hydrologic condition, and antecedent runoff condition (ARC). Sufficient data were not available to determine all of these factors, e.g., ARC, thus curve numbers were chosen from the lookup table of TR-55 for urban areas on the basis of the data in hand. The Bahamas Public Works Department provided geographic maps of the New Providence roads, plots, and a set of aerial photographs for the island that were made in the year 2000. These layers were used to determine the area of impervious surfaces and to create a GIS layer to be used in the calculation.

20.4.1 GIS and Runoff Calculation

The runoff was calculated using the equation:

$$Q = \frac{(P - I_a)2}{(P - I_a) + S} \tag{20.1}$$

where

Q runoff (mm)
P rainfall (mm)
S potential maximum retention after runoff begins (mm) and
I_a initial abstraction (mm)

Initial abstraction (I_a) includes all water that is lost before runoff begins, and includes water that is retained on the surface in depressions and water that is held by vegetation. It also includes evaporation and infiltration. Studies of many small watersheds have shown that I_a can be approximated by

$$I_a = 0.2S \tag{20.2}$$

In order to remove I_a as an independent parameter, Eq. 20.2 was substituted into Eq. 20.1 yielding the following:

$$Q = \frac{(P - 0.2S)2}{(P + 0.8S)} \tag{20.3}$$

S is related to the soil and land cover of the watershed through the curve number (CN), which has a range of 0–100. S is determined using the equation

$$S = \frac{25{,}400}{\text{CN}} - 254 \tag{20.4}$$

A weighted curve number for a watershed can be calculated using the equation

$$\text{CN(weighted)} = \frac{\sum \text{CN} * \text{Area}}{\sum \text{Area}} \tag{20.5}$$

20.4.2 Suitable Recharge Areas

In any management plan for artificial recharge of aquifers, it is necessary to choose locations based on a number of factors. One of the most important of these factors is the slope of the land as this determines the direction in which the water will flow and where it will accumulate. Slope can be calculated in GIS using a digital elevation model (DEM). Other factors, such as infiltration rate, land use/land cover, and geology are important factors as well as location of aquifers and depth to water table. On the small island of New Providence there is little variation in most of these factors other than slope and land use. The ridges that run along the north and south sides of the island effectively direct runoff either to the sea or to the low-lying areas between them. The built areas such as roads and other impervious structures determine the quantity of runoff that will be directed.

For this study, slope was determined using a 90 m Shuttle Radar Topography Mission (SRTM) raster layer. Since the island has no rivers or perennial streams, a roads network was used as a stream network in the runoff analysis. Additional data in the form of an Excel spreadsheet of drainage wells were provided by the Public Works Department and were converted to GIS layers for the purpose of illustrating the areas of the island that are subject to flooding.

20.4.3 Dataset

Data were gathered from a number of sources and in a variety of formats. For the analysis of runoff, it was necessary to understand the surface of the land and the fate of rain that falls upon it. To form a comprehensive picture of the island, geographic layers were needed that represent land cover, elevation, and locations of freshwater aquifers. To analyze runoff it was necessary to determine areas of impervious surfaces by means of zoning and use, and paved road. These were obtained from the Bahamian Government 2004 ARC Info dataset with aerial photographs, two foot contours, and roadway coverages supplied by the Bahamas Ministry of Works and Transportation.

The GIS layers of land cover and freshwater lens for the entire Bahamas were obtained from The Nature Conservancy (TNC) where they had been created by digitizing topographic maps supplied by the Department of Lands and Surveys from 1968 to 1975 at 1:10,000 scale, along with political boundary layers. Excel spreadsheets with drainage wells and catch basins to the sea were provided by the Ministry of Works and Transport.

20.4.4 Spatial Data Preparation

The first step in preparing the GIS data was to extract the data for New Providence from all the layers using ArcMap extraction tools to obtain political boundaries, land cover, freshwater lens, and DEM.

In order to calculate the amount of runoff, impervious surfaces had to be identified as follows. The areas designated as “human altered lands” were extracted from the land cover layer and used as a mask to select roads and plots. Initial

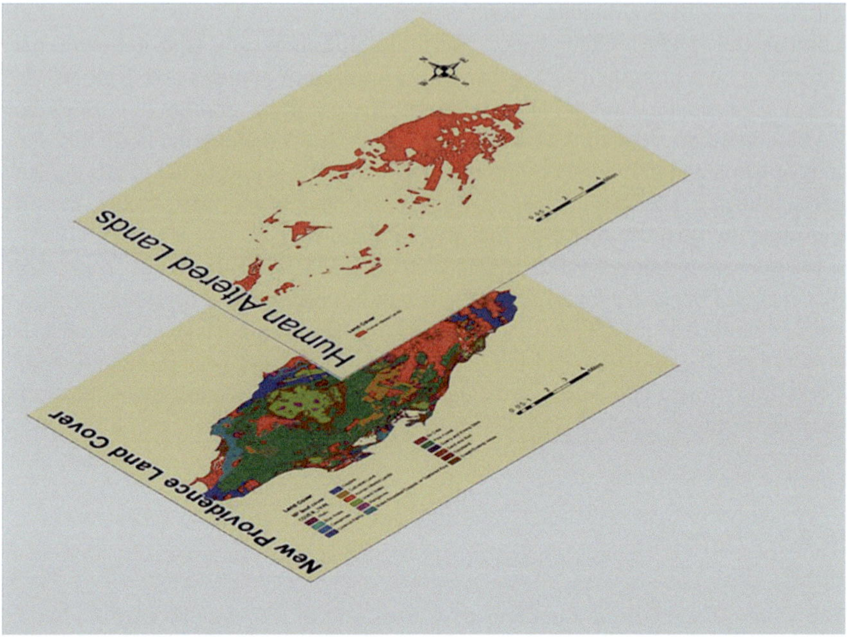

Fig. 20.3 Human altered lands extraction

selection of plots and roads was made: roads were selected that fell within areas designated as human altered lands on the land cover layer; plots were selected in the same manner as shown in Figs. 20.3 and 20.4. The roads layer as obtained was a polyline layer, which did not represent the area of the roadways, so a 5 m buffer was applied to both sides of the selected roads. The figure was chosen as an average since the width of the roads was not available and in the heavily developed downtown areas, most of the easements have been paved over, becoming part of the roadways, so these areas were included in the buffer.

After these initial selections were made, the next step was to verify land use by determining which of the selected plots actually had structures on them. Aerial images were overlaid on selected plots and inspected for structures as shown in Fig. 20.5. By this method a total of 45,705 plots were found with structures. These were then classified according to zoning/use and area of plot, as well as size of structure. Finally, curve numbers were added to the table of roads and plots following the TR-55 lookup table.

20.4.5 Curve Numbers

Plots were assigned curve numbers based on zoning and land use and verified using aerial photographs. Designation of plots zoning/land use was made using the lookup table as follows:

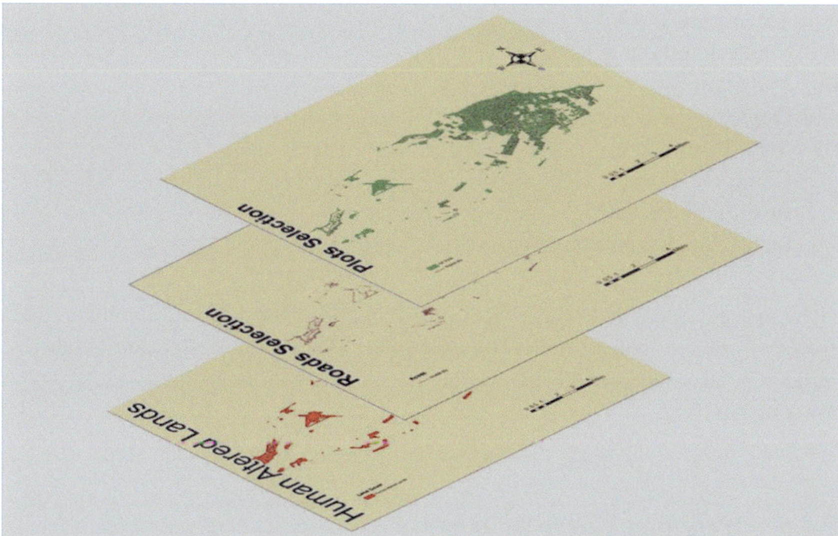

Fig. 20.4 Roads and plots extraction

Fig. 20.5 Selection of plots using aerial image overlay

T Townhouses
SF Single family
MF Multifamily
D Duplex
MU Mixed use
C Commercial
IP Institutional public
IG Institutional semi-public

The process described in the TR-55 flowchart was used to select suitable curve numbers from the lookup table. The land cover types were designated either residential or nonresidential and then further classified according to individual type as shown in Fig. 20.6.

20.4.6 Weighted Curve Number

In some areas, it was necessary to calculate weighted curve numbers because of lack of data for individual surfaces. Weighted curve numbers were used for the roads network since two types of roadways exist on NP—roads with gutters and curbs and roads with swales. The roads in the older and more heavily populated parts of the city of Nassau are of the type having curbs and gutters, whereas roads in more recently developed areas tend to have more swale areas. Therefore, a weighted curve number was calculated for the combined roadways using the formula described above. A weighted curve number was also calculated for duplex properties based on the area of the lots. Worksheet 2 from the TR-55 manual was used in this process. Once curve numbers had been determined for each land cover type, the map table was edited in ArcMap to add curve numbers, precipitation, S value, runoff depth, and runoff volume in order to complete the calculations.

20.4.7 Selection of Recharge Locations

A slope percentage layer was generated in ArcMap from the DEM layer using the Spatial Analyst tool. The selection of suitable recharge locations was based on three factors: elevation, slope, and proximity of aquifers. The raster calculator was used to generate a layer that met the specification [Elevation] ≥ 7 and [slope_perc] < 5. Seven meters of elevation was chosen in order to allow for the proximity of the aquifers to sea level. Slope percent of less than 5 % was chosen as this has been shown to be suitable for recharge surfaces (Ghayoumian et al. 2007) (Fig. 20.7).

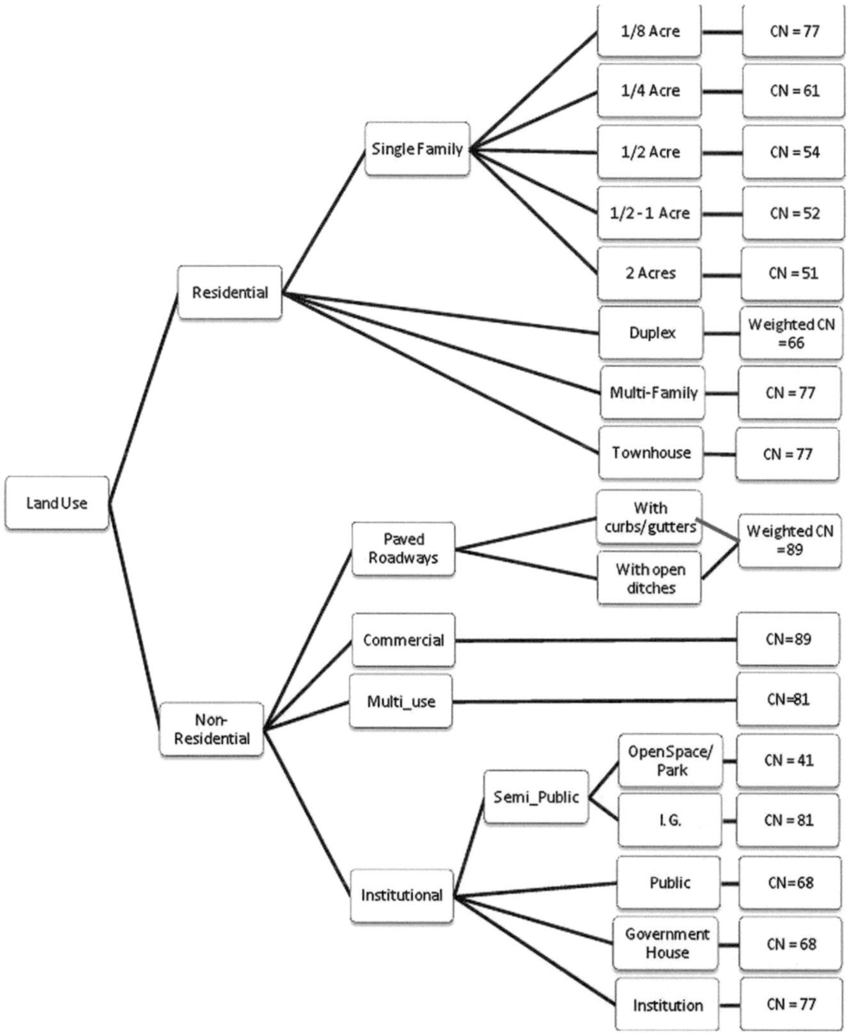

Fig. 20.6 Curve number tree

20.5 Results and Discussions

This analysis shows that a substantial amount of runoff is created by the impervious surfaces on New Providence with highest depths in the downtown area of Nassau.

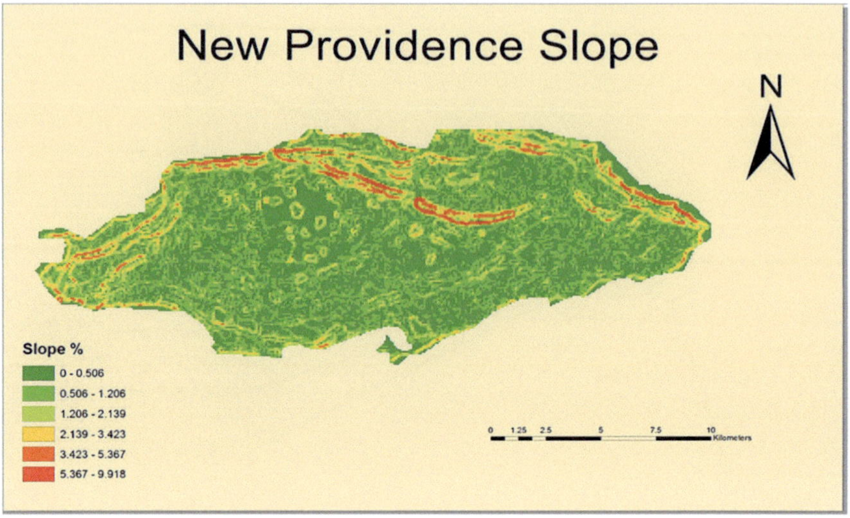

Fig. 20.7 Calculation of slope

20.5.1 Runoff Estimation

Runoff depth and volume were calculated in ArcGIS using the curve number method as outlined above. The runoff depth map is presented in Fig. 20.8.

Runoff depth on impervious surfaces is shown to be greatest in the most intensely developed areas of the City of Nassau. In these areas, almost all the surface has been paved or built over and most of the rainfall is converted to runoff and here the runoff depth ranges between 1.037 and 1.36 m. Since many of these areas are also at higher elevations, they are the source of flood water in the lower elevations between the ridges.

Volume of runoff as calculated on impervious surfaces totaled 70.0 MCM on a yearly basis. Not all of this runoff will be available for artificial recharge as some of it will be lost to the ocean in direct runoff and some will remain on the surface to be evaporated. Approximately, 25 percent of rainfall eventually reaches the groundwater aquifer (Sealey 1995) under normal conditions where vegetation is present, therefore it can be conservatively estimated that 17.50 MCM could be added to the freshwater supply on New Providence each year if this resource was carefully managed. Since this runoff is generated on the impervious surfaces that are clustered in and around the city of Nassau, those aquifers in the eastern area will benefit from harvesting of this water. Furthermore, since the surfaces that are generating this runoff are mostly devoid of vegetation, it can be surmised that this conservative estimate is much lower than what may be captured by efficient management.

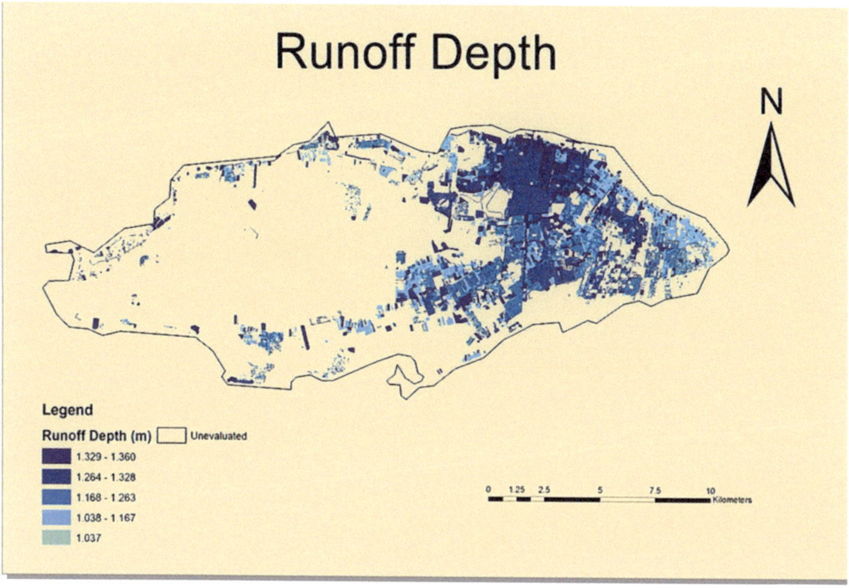

Fig. 20.8 Runoff depth calculation

20.5.2 Suitable Recharge Locations

Determining suitable locations for artificial recharge of aquifers will be influenced by several characteristics of the watershed (Ghayoumian et al. 2007). Depth to aquifer is one parameter that must be considered. In New Providence, depth to groundwater is generally very shallow and can even be too shallow to be suitable for recharge since excessive ponding of water would occur and evaporation would claim much of it. It would be more desirable to locate recharge areas in places where there is enough depth so that the natural process of filtering of impurities from the water as it passes through the subsurface would be maximized. Areas suitable for artificial recharge were selected on the basis of elevation of 7 m or greater and slope less than five percent and overlaying an aquifer. The results are shown in Fig. 20.9. The eastern section of New Providence has been most intensely developed with a mixture of residential, commercial, and industrial uses. Zoning regulations have been lax or nonexistent and the result is that much of the land has been completely covered with impervious surfaces (SSI 2009). Much of the western section of the island is held by the Bahamas Government and is forested with pine. This is a protected area that overlays the largest freshwater aquifer, which is normally recharged, therefore little opportunity or need for artificial recharge exists there. The areas in the eastern section (indicated in brown) (Fig. 20.9), are where artificial recharge would be most beneficial as the aquifers there are the ones that have been subjected to over pumping and pollution from the surface. Here is where

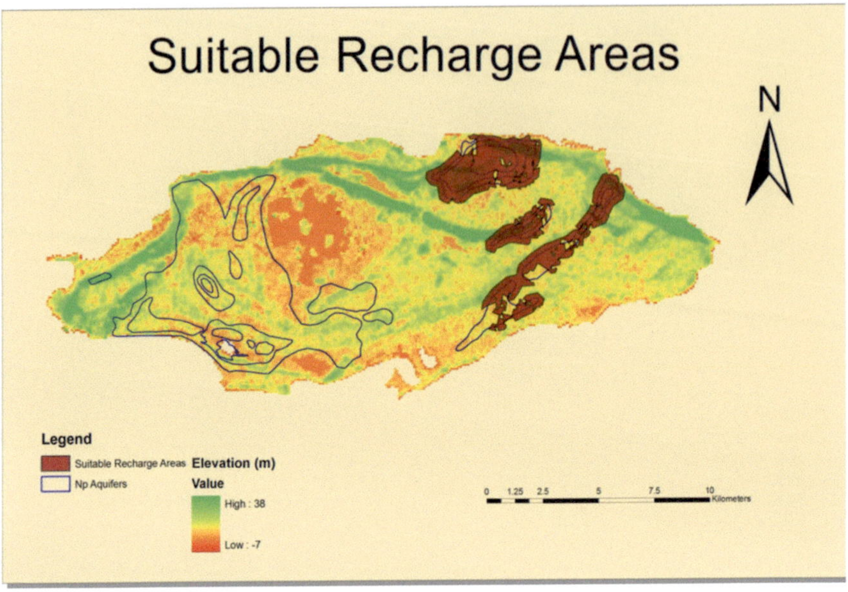

Fig. 20.9 Areas suitable for artificial recharge of groundwater aquifers

the need for artificial recharge exists as well as the opportunity for relief from flooding because of runoff from the impervious surfaces. However, as a result of the intense development, any plan for artificial recharge will be met with a number of challenges.

The first of these challenges will be cost, both to secure enough land to construct spreading basins as well as to prepare the surface, grade roadways, and construct channels to direct the runoff to those basins. One option would be to use the aquifer storage and retrieval method. Since a large number of drainage wells already exist (Fig. 20.10), it may be possible to convert some of these to recharge and add filters to help purify the water. There is little doubt that many of these wells are contributing to pollution of the aquifers (SSI 2009) and this method presents the opportunity to reverse that.

20.6 Conclusions

Because of the rapid and generally unregulated development on New Providence, and especially in the capital city of Nassau, much of the surface has been transformed from being highly transmissive to highly impervious to the movement of water to the aquifers. This transformation has led to the creation of large quantities of surface runoff from roads, parking lots, and buildings and has drastically altered the natural water cycle of this island. This study has shown that the runoff from

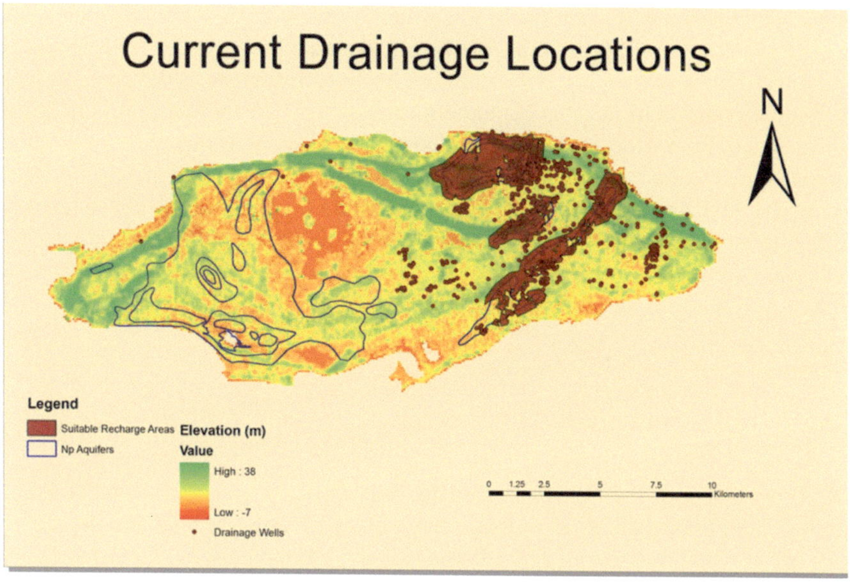

Fig. 20.10 Location of current drainage in areas prone to flooding

these impervious surfaces is substantial and, if captured in suitable locations, could be used to recharge the aquifers that are currently unusable and therefore return them to usefulness. However, because the rainfall is unevenly distributed over the year, during the summer months the runoff depth will be higher and more water will be available. The challenge will be to find ways to capture this water and store it in spreading basins and wetlands for recharge to the aquifers.

It will require intervention by the government agencies involved addressing this and there are obstacles that will have to be overcome in order to accomplish it. The intensity of development in the city of Nassau presents a major challenge as land for recharge is scarce. It will require the understanding and cooperation of the population along with drastic action of government to acquire land for recharge. It will require an investment of capital to reroute flows of water from the channels formed by roadways such as Blue Hill Road to areas designated for recharge. It will also require expertise in engineering to design and implement a system for this rerouting. However, because the quantity of water potentially available for recharge is vast and many benefits will accrue to the population from the increased natural freshwater, and to the government in savings of revenue, the effort and expense to carry out a plan to capture runoff for recharge of aquifers are well worth it.

Presently, the water resources of New Providence are overseen by two agencies, each charged with separate responsibilities. Water and Sewerage is concerned with delivering potable water to users by any means necessary. Therefore, much of the water used on the island is produced by desalination. As the cost of fuel continues to rise, production by desalination will become more expensive. Further, all nations may soon be faced with shortage of fossil fuels as the world's supply dwindles, and many nations are in the process of developing alternate fuel supplies. The Bahamas, as a small island developing state, does not have the resources to develop alternate fuel supplies and will have to continue to import fossil fuels for as long as it can afford to do so. In the meantime, the cost of producing freshwater by artificial means will continue to rise. The second agency involved with management of water, the Ministry of Works and Transportation, will continue its efforts to cope with flooding that will worsen as development continues and the residents will continue to bear the burden. There are a number of steps that may be taken to restore the damaged aquifers and alleviate flooding.

20.6.1 Recommendations

The Government of the Bahamas must develop a water management plan that will involve these two agencies as well as environmental health agencies and the public. This runoff water must be recognized as a valuable resource for all the residents and ways to capture and preserve it must be sought. A plan to capture the runoff will entail the procurement of land for the creation of spreading basins as well as a careful study of the flow patterns created by roadways and other impervious surfaces so that channels to direct that flow can be constructed. In the heavily developed areas, this will be especially challenging as these areas are so densely populated, but it is precisely these densely populated areas that create most of the runoff.

The lack of zoning regulations has been a major contributing factor in the creation of the vast areas of impervious surfaces and this must be addressed. Any future development projects must be required to provide pervious areas to prevent the creation of more runoff and greater loss of water. Existing landholders should be given incentives for the creation of structures to capture rainfall. These incentives could be in the form of tax relief, rebates, or even direct payment.

Existing drainage wells should be inspected and suitability for their use as recharge wells should be determined. Those deemed suitable should be refitted with filters in order to deliver runoff to the aquifers and to prevent further pollution. No doubt many of the wells currently being used are contributing to pollution and a reversal of this by the addition of filters would aid in the recovery of those aquifers.

Even if artificial recharge of aquifers were accomplished, there would still be excess runoff water that would need to be managed. The excess water could be captured using other methods of rainwater harvest such as rooftop collection. Presently, Government does not encourage the public to harvest rainwater although this would represent a large savings in the amount of freshwater that WSC is providing to its customers. Water harvested by rooftop collection need not be used for drinking but could be a substitute for household use such as flushing toilets, washing clothes and automobiles, and watering lawns. A program should be started to give incentives to property owners to install rainwater collection systems on their buildings.

Finally, government must involve and educate the public to build awareness of the need for conservation and the benefits of artificial recharge. Only with the full understanding and cooperation of the population any management plan will succeed. Constructed wetlands and recharge areas need not be only utilitarian, but can add esthetic value to a neighborhood as well.

20.6.2 Future Research

In order to successfully manage the runoff created by impervious surfaces on New Providence, further research is needed. A project must be implemented to map the floodplains created by the transformation of the surface to an impervious rather than transmissive type and of the channels created by roadways. The flow of water has been drastically altered by these surfaces and an understanding of the dynamics of runoff will be an aid in planning for future development as well as for corrective measures that may be needed for existing developments that experience severe flooding during heavy rainfall events. This will also be useful to demonstrate the problem and influence the attitude of the population. In addition, a study of the types and sources of pollutants that are being carried by the runoff from these surfaces should be done in order to raise public awareness of the necessity of taking action.

One of the limitations to this analysis was the age of the data. The land cover layer was digitized using maps originally drawn in the 1970s and the zoning map is a decade old. It will be necessary to reclassify the land use and land cover of New Providence in order to obtain more accurate results. Although resources were not available to accomplish the ground truthing needed for this, an analysis using remote sensing was undertaken.

It has been shown that remote sensing methods can be used successfully to map impervious urban areas (Melesse et al. 2007; Braun and Herol 2003). Using a Landsat image obtained from the Global Land Survey of the website, glovis.usgs.gov, an unsupervised classification was done using eight classes. Three types of impervious areas were identified and runoff was calculated on the total of these areas.

A second assessment was made using the technique to derive fractional impervious surface (FIS) from the scaled normalized vegetation difference index (NDVIs) as described by Melesse and Wang (2007):

$$\mathrm{FIS} = 1 - \left[\frac{\mathrm{NDVI}_i - \mathrm{NDVI}_{\mathrm{low}}}{\mathrm{NDVI}_{\mathrm{high}} - \mathrm{NDVI}_{\mathrm{low}}}\right]^2 \tag{20.6}$$

Table 20.2 Runoff results in GIS, remote sensing—classified and FIS

Runoff volume	
Method	Volume (m^3)
GIS	53,781,935
Remote sensing—classified	84,986,700
Remote sensing—FIS	116,456,800

(NDVIs range between 0 and 1)where $NDVI_i$ is the NDVI value for pixel I, $NDVI_{low}$ and $NDVI_{high}$ are values for bare soil and dense vegetation, respectively (Melesse et al. 2007). FIS shows the degree of imperviousness at a pixel level; FIS value of 1 shows 100 percent imperviousness for that pixel. By continuing the calculation of runoff with the weighted curve number method the results were obtained as shown in Table 20.2. Although this assessment is rudimentary, it illustrates the potential that exists for further development of this work and the possibility that the runoff resource is much greater than determined by this study.

This study has shown that there is a valuable resource on the island of New Providence in the form of freshwater from rainfall that is currently being wasted, which, with planning and careful management, can be captured and made available for the population now and in the future.

Acknowledgments The grant for this project was provided by USDA Cooperative State Research, Education, and Extension Service (Grant Number USDA—CSREES 2006-51160-03409) through the Agroecology Program at Florida International University. Thanks to Bahamas Department of Meteorology for providing information on rainfall data, to Bahamas Ministry of Works and Transportation for GIS and drainage data The Nature Conservancy for additional GIS data. Thanks to Daniel Gann at Florida International University GIS Center for his willing assistance and invaluable advice.

References

Asano T (1985) Artificial recharge of groundwater. Stoneham, MA: Butterworth

Bahamas Department of Meteorology (2008) Bahamas rainfall statistics. Nassau

Biswas H (2008) Numerical groundwater flow modeling in the Wakal River Basin, India. Thesis. Florida International University

Biswas H, Melesse AM, McClain M (2007) Remote sensing and-GIS based approach for delineation of groundwater prospect zones in a semi-arid area in Rajasthan, India, Annual AWRA meeting, Albuquerque, NM

Bouwer H (2002) Artificial recharge of groundwater: hydrogeology and engineering. Hydrogeol J 10:121–42. Web http://www.springerlink.com/content/u0r3xtxnnfqm4p6y/. 1 Aug 2009

Braun M, Herol M (2003) Mapping imperviousness using NDVI and linear spectral unmixing of ASTER data in the Cologne-Bonn region (Germany). Proceedings of the SPIE 10th international symposium on remote sensing. Barcelona, Spain, 8–12 Sept 2003

Brown LR (2006) Plan B 2.0 rescuing a planet under stress and a civilization in trouble. W. W. Norton & Company, New York

Cant RV, Weech PS (1980) Water resources evaluation of the Bahamas. Ministry of Works and Utilities, Tech. Nassau, Bahamas

Carew JL, Mylroie JE (1997) Geology of the Bahamas. In: Vacher HL, Quinn TM (eds) Geology and hydrogeology of carbonate Islands. Elsevier, St. Louis, 91–139
Department of Statistics Census (2000) The commonwealth of the Bahamas. www.statistics.bahamas.gov.bs. Accessed 13 Feb 2011
Einav R, Harussi K, Perry D (2002) The footprint of the desalination processes on the environment. Desalination 152:141–154
FAO (2009) Summary table: renewable water resources in the world by country. Aquastat. Food and Agriculture Organization of the United Nations, 13 May 2003. Web. http://www.fao.org/nr/water/aquastat/water_res/waterres_tab.htm. 15 Feb 2009
Ghayoumian J, Saravi MM, Feiznia S, Nouri B, Malekian A (2007) Application of GIS techniques to determine areas most suitable for artificial groundwater recharge in a coastal aquifer in southern Iran. J Asian Earth Sci 30:364–374
Jothiprakash V, Marimuthu G, Muralidharan R, Senthilkumar N (2003) Delineation of potential zones for artificial recharge using GIS. J Indian Soc Remote Sens 31(1):37–47
Melesse A, Wang X (2007) Impervious surface area dynamics and storm runoff response (book chapter). Remote sensing of impervious surfaces, CRC Press/Taylor & Francis, Dallas, pp 369–384
Melesse AM, Weng Q, Thenkabail PS, Senay GB (2007) Remote sensing sensors and applications in environmental resources mapping and modeling. Sensors 7:3209–3241
Middleton R (2008) Bahamas water and sewerage production 2004–2008. Microsoft excel file
Millennium Ecosystem Assessment (2005) Ecosystems and human well-being: synthesis. Island Press, Washington, DC
OAS (1997) Source book of alternative technologies for freshwater augmentation in Latin America and the Caribbean. OAS, Washington
Peters JH (1998) Artificial recharge groundwater. Taylor & Francis, Dallas
Roebuck L, Ortiz T, Pochatila J (2004) Water resources assessment of the Bahamas. Rep. US Army Corps of Engineers, 22 Nov 2004. Web http://www.sam.usace.army.mil/en/wra/Bahamas/BAHAMASWRA.pdf. 6 Feb 2011
Rosegrant MW, Cai X, Cline SA (2002) World water and food to 2025 dealing with scarcity. Detroit: Intl Food Policy Research Inst. http://www.ifpri.org/sites/default/files/publications/water2025.pdf. International Food Policy Research Institute, 24 Sept 2003. Web http://www.ifpri.org. 30 Aug 2009
Sealey NE (1995) Bahamian landscapes an introduction to the geography of the Bahamas. Grand Rapids: Media
Shiklomanov I (1993) World fresh water resources. In: Gleick PH (ed) Water in crisis: a guide to the world's fresh water resources, USGS Website. Oxford University Press, New York
Stiefel J, Melesse AM, McClain M, Price R (2007) Rainwater harvesting in Rajasthan, India: recharge estimation using tracers. 13th international rainwater catchment systems conference and 5th international water sensitive urban design conference. Sydney, Australia, 8 p (CD proceeding), 21–23 Aug 2007
Stiefel JM, Melesse AM, McClain ME, Price RM, Chauhan NK (2008) The impact of artificial recharge from rainwater harvesting structures on the groundwater of nearby wells in rural Rajasthan. International groundwater conference on groundwater dynamics and global Change, Jaipur, India, 11–14 Mar 2008
Stiefel J, Melesse A, McClain M, Price RM, Anderson AP, Chauhan NK (2009) Effects of rainwater harvesting induced artificial recharge on the groundwater of wells in Rajasthan, India. Hydrogeol Journal 17(8):2061–2073
Stormwater Solutions, Inc. (SSI) (2009) New Providence Island Stormwater Masterplan—Final Report. Unpublished
UNEP (2008) Vital water graphics—an overview of the state of the world's fresh and marine waters, 2nd edn. UNEP, Nairobi, Kenya. ISBN 92-807-2236-0
U.S. Department of Agriculture (USDA) (1986) Soil conservation service, engineering division. CPESC, Inc. Urban Hydrology for Small Watersheds, Washington, DC. http://www.cpesc.org/reference/tr55.pdf

Water and Sewerage (2014) Web http://www.wsc.com.bs/History.aspx. 20 Apr 2014
Wescoat JL Jr, White GF (2003) Water for life water management and environmental policy (Cambridge studies in environmental policy), Cambridge University Press, New York
Whitaker FF, Smart PL (2004) Hydrogeology of the Bahamian Archipelago. In: Vacher HL, QuinnGeology TM (eds) Hydrogeology of Carbonate Islands. Elsevier, St. Louis, pp 183–216

Chapter 21
Groundwater Vulnerability Analysis of the Tana Sub-basin: An Application of DRASTIC Index Method

Anteneh Z. Abiy, Assefa M. Melesse, Yewendwesen Mengistu Behabtu and Birlew Abebe

Abstract In the Blue Nile River basin and particularly in the Tana sub-basin, Ethiopia, groundwater development and use for various purposes is increasing. This calls for understanding of how the groundwater system is functioning including its vulnerability to contamination. In this chapter, the vulnerability of the groundwater resources in the Tana sub-basin, upper Blue Nile River basin is evaluated with the application of the DRASTIC index method. Geographic Information System (GIS) application is implemented to prepare the different layers of the DRASTIC model and to calculate the DRASTIC index. Accordingly, the high DRASTIC index values indicate the sites highly susceptible to pollution in the presence of contaminant sources. The DRASTIC index in the Tana sub-basin is in the range of 66–120. The highest DRASTIC index is indicated in the southern part of the Tana sub-basin. The southern part, where the higher DRASTIC index is recorded, is known with highly porous vesicular basalt, lower soil thickness, and near-surface water table. The mountainous regions and the floodplain in the eastern part of the sub-basin exhibit lower vulnerability. The thick clay layers, which cover the floodplains of the sub-basin, have a significant role in purifying the water before it reaches the groundwater system.

Keywords Groundwater · Vulnerability · Pollution · GIS · DRASTIC index · Tana sub-basin · Upper Blue Nile River basin

---

A.Z. Abiy (✉) · A.M. Melesse
Department of Earth and Environment, Florida International University, Miami, FL 33199, USA
e-mail: aabiy001@fiu.edu

A.M. Melesse
e-mail: melessea@fiu.edu

Y.M. Behabtu
Abbay Basin Authority, Bahir Dar, Ethiopia

A.Z. Abiy · B. Abebe
Tana Sub-Basin Office, Bahir Dar, Ethiopia

© Springer International Publishing Switzerland 2016

A.M. Melesse and W. Abtew (eds.), *Landscape Dynamics, Soils and Hydrological Processes in Varied Climates*, Springer Geography,
DOI 10.1007/978-3-319-18787-7_21

21.1 Introduction

21.1.1 The Concept of Aquifer Vulnerability

Groundwatervulnerabilityrepresenting the spatial variability of the susceptibility of groundwater contamination, irrespective of the presence of pollutant loading, indicates how the natural media and landscape could affect or control the susceptibility of the groundwater system from pollution (Chilton 2006; Michael et al. 2005; Palmer et al. 1995; Vrba and Zoporozec 1994). The National Research Council (1993) has explained that the groundwater vulnerability is a concept, which is not directly measurable, but a probability about the future contamination possibility inferred from various and possibly interchangeable and/or evitable measurements.

It is derived from the concept that the hydrologic materials and the physical environments such as the soil, land cover, geology and topography facilitate a natural filtration, irrespective of the nature of the contaminant, and hence, such analysis results to the pollution attributed to the ***intrinsic vulnerability.*** Such vulnerability analysis does not consider any sources of contaminant and the specific nature of contaminants, but rather it focuses on the inherent geologic, hydrologic, and hydrogeological features of the natural environment. Hence, it represents the degree of ***susceptibility***, or the ***aquifer sensitivity***, or the ***natural vulnerability*** of the aquifer material. Other considerations of the groundwater vulnerability analysis incorporate the nature and spatial distribution of a specific contaminant, where the result refers the ***integrated vulnerability*** of the aquifer material to the specified contaminant, and hence, this result is the ***specific vulnerability*** of the aquifer to the contaminant under consideration. In both cases, the analysis considers vertical transport of the contaminants along with percolating water (Gogu and Dassargues 2000; Gogu et. al. 2003; Vrba and Zoporozec 1994; The National Research Council 1993; Stigter et al. 2006).

Considering the heterogeneity and anisotropy of the hydrologic materials and the physical environment, the natural protection to groundwater from pollution varies from place to place. Thus, in a basin study, it refers to the fact that some land area in a basin is more vulnerable to pollution than others. Thus, the vulnerability analysis considers, in the presence of contaminants, the groundwater in some land area is likely to be polluted than another location within the river basin. Accordingly, the First Law of Groundwater Vulnerability refers that "***All groundwater is vulnerable to contamination***": Vulnerability is a relative concept, therefore. Hence, it is necessary to consider the effect on groundwater quality over longer time spans and greater distances (The National Research Council 1993). At the same time, in any case of groundwater vulnerability analysis, "***uncertainty is inherent***"—the Second Law of Groundwater Vulnerability. Overall, the goal of a vulnerability study is to define the spatial distribution, extent, and degree of severity of the groundwater contamination susceptibility, while the natural complications would lead to uncertainty to the analysis, but the vulnerability to contamination is likely to happen.

Aquifer vulnerability maps are beneficial to planners, developers, and governments who can utilize them for development planning, identifying potentially highly sensitive areas, and prioritizing and designing monitoring programs (Berkhoff 2007; Vrba and Zoporozec 1994; Aller et al. 1987). The National Research Council (1993) has identified that the groundwater vulnerability analysis results are effectively used in different applications. The first application is in ***policy analysis and development***. It is also known as a strong input and guidance to decision supportive system to ***program management*** to allocate resources. The third use of groundwater vulnerability analyses is its application to inform ***land use decisions***. In addition, its use as ***education and awareness creation*** tool in regions hydrologic resources is also highlighted with emphasis.

Vulnerability mapping does not replace site-specific investigations, but can act as a guide to determine the intensity of site investigation needed and to emphasize highly vulnerable areas for management practices. The advantages of vulnerability studies in river basin studies are to:

- make local assessment and identify areas susceptible to contamination,
- incorporate design criteria, among others, for groundwater monitoring network in the sub-basin,
- identify degree of groundwater contamination possibility and spatial variability, in the presence of contaminant sources,
- define approaches in dealing with existing groundwater pollutant sources, and project the fate of groundwater development plan in the sub-basin, and
- make environmentally sound decisions regarding land use and groundwater protection interventions.

The objective of the study reported in this chapter is to identify the spatial variability of the vulnerability of the groundwater resources because of the physical environment. It is intended to provide an indicative and supportive tool to decision making in relation to development activities and the sustainability of good quality of the groundwater resources in the sub-basin. While groundwater data are often scanty in the area, such information is addendum to indicate hot spots for planning and implementation of groundwater monitoring network program in the area.

21.1.2 Methods of Groundwater Vulnerability Analysis

There are numerous approaches of groundwater vulnerability analysis techniques, approaches, and concepts. Depending on the assumptions and nature of analysis, they can be categorized into three groups of groundwater vulnerability analysis methods: the physical process-based methods, statistical methods, and the overlay–index methods (The National Research Council 1993; Worrall and Besien 2005; Wang and Yang 2008). Other researchers, such as Arthur et al. (2007), consider empirical and hybrid models and hence five categories of methods. Considering that the first three are the main ones to all other methods, a brief description for the three methods is included herewith.

I. **Physical process-based methods**: These are modeling approaches using the physical process of contaminant loading and transport through various parts of the physical environment. It targets to estimate the contaminant transport and distribution in space and time based on the simulation of the physical processes that take place in the groundwater system. Hence, it tries to address how the physical environment protects the groundwater system and how the contaminants behave within the groundwater flow path. For this, it requires an advanced knowledge on conceptualization and relation of the conceptual models to mathematical manipulations. In addition, intensive database of various kinds is required to get good results of these simulation approaches. Under the conditions of sufficient database, sufficient expertise knowledge, and experience, these approaches are advantageous for reliable modeling results than other approaches (Gogu and Dassargues 2000; Barbash and Resek 1996; Thapinta and Hudak 2003; The National Research Council 1993).
II. **Statistical method**: These methods attempt to define the contaminant concentration distribution or probabilities of contamination. It accounts for the correlation between spatial variables, such as aquifer properties and sources of contamination and contaminant occurrences known by measured data and monitoring (The National Research Council 1993; Babiker et al. 2005). This method requires quite voluminous data, and it requires the area under consideration to have some level of contamination.
III. **Overlay–Index method**: This method considers the relative importance of the physical environment to control the spatial distribution of vulnerability of the groundwater in an area. It combines the different physical attributes of the area under consideration. While the type of physical environment attributes to be deployed in the analysis varies with the nature and/or assumption of the models, these methods allow a subjective judgment of the relative weight of the attributes to the users (The National Research Council 1993; Thapinta and Hudak 2003). However, the index methods are most widely applied approaches to groundwater vulnerability analysis. Since it is known to be suitable for regional studies, it requires easily accessible/available data, such as topography, soil, land use/land cover, geology, and depth to groundwater table. As compared to the process-based methods, these methods require less data and have less complexity and they are easy to understand and apply for evaluation and application. The most widely applied techniques of overlay–index method, among the many, are as follows:

   - GOD (Foster 1987a, b),
   - DRASTIC (Aller et al. 1987),
   - SEEPAGE (Moore and John 1990),
   - AVI (Van Stempvoort et al. 1993),
   - SINTACS (Civita 1994),
   - ISIS (Civita and De Regibus 1995),

- EPIK (Doerfliger and Zwahlen 1997),
- The German method (von Hoyer and Söfner 1998),
- IRISH (Daly and Drew 1999),

For this specific study, the application of the DRASTIC model is considered.

21.2 Description of the DRASTIC Model

21.2.1 General

Developed under the US Environmental Protection Agency (US-EPA) in association with the National Water Well Association (Aller et al. 1987), the DRASTIC model is the widely accepted and used groundwater vulnerability analysis method (Worrall and Kolpin 2004; Babiker et al. 2005). The DRASTIC model results are used for groundwater pollution controlling and regulation in different scales (Rupert 2001). The DRASTIC model has been used to produce maps in many parts of the USA, Israel, Nicaragua, Portugal, South Africa, and South Korea. Among the many, models developed and used by different studies (Hamza et al. 2007; Piscopo 2001; Fritch et al. 2000; Secunda et al. 1998) have indicated the fact that the DRASTIC model is effective and simple for representation of the different aquifer setups. These studies demonstrate sound reasons to apply the method in basin-level groundwater pollution control, monitoring, and management planning. Its capability to allow users for spatial data management, analysis, and visualization makes DRASTIC a good alternative, among others, to groundwater vulnerability analysis. In addition, the DRASTIC model can be easily transferred to other catchments considering the presence of sufficient data and basic knowledge of the area under consideration for analysis.

However, the DRASTIC has limitations in that it does not effectively incorporate pollutant sources and/or the effect of future pollutant sources to pollution. Further, it does not allow evaluation of general groundwater vulnerable regions (Barbash and Resek 1996). In some specific geologic setups, such as karastic aquifers and sinkholes, the DRASTIC model does not have systems to account for the vulnerability (Arthur et al. 2007).

21.2.2 Features of DRASTIC

DRASTIC is an acronym for seven physical factors that determine the possibility of groundwater contamination, in the presence of distributed source contaminants, namely:

1. **D** = **D**epth to water
2. **R** = net **R**echarge
3. **A** = **A**quifer media
4. **S** = **S**oil media
5. **T** = **T**opography (slope)
6. **I** = **I**mpact of the vadose zone media
7. **C** = hydraulic **C**onductivity of the aquifer

It calculates an index value based on the sum of the product of the relative of importance (weight (w)) by the rate of (r) of the seven factors. The weight of the factors is a relative to each other in order of importance from 1 to 5, whereas the rate enables the ranking of the ranges found in each DRASTIC feature in the range of 1 to 10. Accordingly, the DRASTIC index (DR_i), or the degree of susceptibility of the aquifers to pollution or the vulnerability, also called the pollution potential of the aquifers in and a watershed, irrespective of the presence of contaminants is determined by the DRASTIC index.

$$\mathrm{DR}_i = D_w D_r + R_w R_r + A_w A_r + S_w S_r + T_w T_r + I_w I_r + C_w C_r \tag{21.1}$$

where

DR_i = DRASTIC index
w = weight of respective DRASTIC factor
r = rate of respective DRASTIC factor

The relative weight refers to indicate the nature of the DRASTIC factor to channel surface-distributed contaminant inclusion and transport within the groundwater system. For example, in the presence of surface contaminants, near-surface groundwater table decreases the creep length where contaminants should travel before it joins into the groundwater system. In this case, the groundwater table is suitable to furnish the susceptibility of groundwater pollution; thus, it receives the highest relative weight. Likewise, degraded land with steep slope but low permeability favors surface runoff where contaminants' chance to join the groundwater system is less likely. In this case, the relative weight of net recharge at that specific area of the watershed should be lower.

Overall, values assigned to the weights within a watershed are spatially different and it is dependent on the relative importance to trigger groundwater contamination. If the condition of a factor facilitates suitable media of groundwater contamination, the factor at the specific location of the watershed will receive larger weight. However, same DRASTIC factor at a different location is in such a situation to prevent infiltration, increase creep length of increase travel time before the contaminants join to the groundwater system that factor is less important to cause pollution and the relative weight at the specific location will be lower.

Presence of a given DRASTIC factor is important to enhance filtration of the contaminants and/or able to decrease the potential to join the groundwater system. Hence, those factors that prevent contaminants from joining into the groundwater

Table 21.1 Relative weight of DRASTIC factors/features

Features	Weight
Depth to water	5
Net **r**echarge	4
Aquifer media	3
Soil media	2
Topography (slope)	1
Impact of the vadose zone media	5
Hydraulic conductivity of the aquifer	Not used in this study

receive lower DRASTIC weight. Based on the scientific study results, the relative weight of the DRASTIC factor is defined by Aller et al. (1987) (Table 21.1).

In cases where knowledge on the different parameters of a specific area of interest is sufficient, the effective weights for each parameter can be different. The calculation results in a five-class range of DRASTIC index. The highest the index value is, the highest the possibility of the specific site within the watershed is susceptible to pollution in the presence of contaminant sources.

21.3 Description of the Tana Sub-basin

21.3.1 Location of the Sub-basin

The Tana sub-basin is located in the northwest highland of Ethiopia, forming the source region of the Blue Nile River (Fig. 21.1). It is one of the sixteen sub-basins of the Blue Nile River basin in Ethiopia. The sub-basin has a total area of 15,200 km^2, and the Lake Tana covers 3156 km^2. Description of the Blue Nile River basin and reports of various hydrologic studies in the basin are available in various documents (Chebud and Melesse 2009a, b, 2013; Setegn et al. 2009a, b, 2010; Melesse et al. 2009a, b, 2014; Abtew et al. 2009a, b; Abtew and Melesse 2014a, b, c; Melesse 2011).

21.3.2 Climate

The Tana sub-basin has the annual average areal maximum and minimum temperatures that vary from 22 to 29.5 °C and 8.5 to 16 °C. Rainfall in the Tana sub-basin and its surrounding region is characterized by a unimodal pattern that falls within a relatively similar time span of three months of June to August. The sub-basin exhibits variability in the area distribution of rainfall, ranging from 865 to 2300 mm (Fig. 21.2).

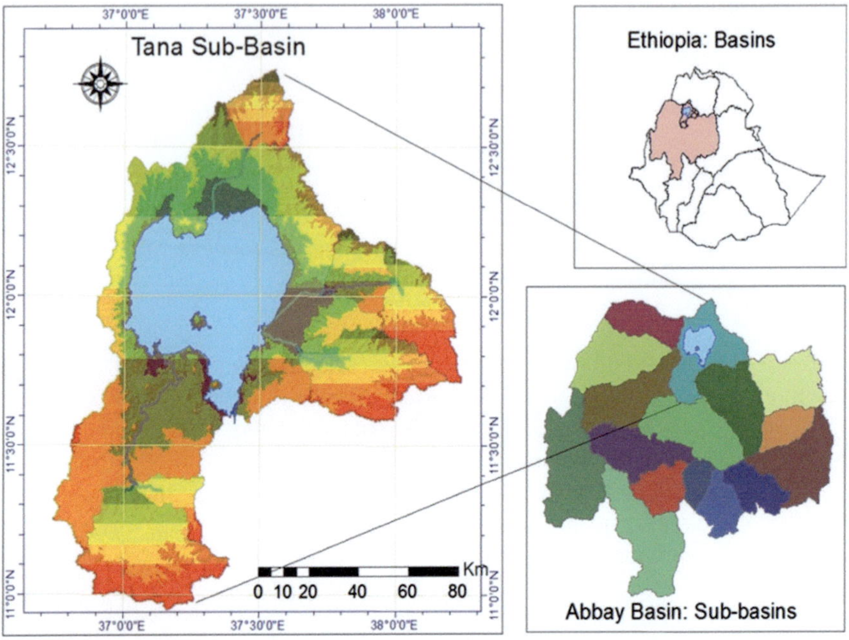

Fig. 21.1 Location map of the Tana sub-basin

The areal distribution of the rainfall represents the highest annual precipitation in the area surrounding southern section of the sub-basin, reaching up to 2300 mm/year. Likewise, the sub-basin's lower rainfall is below 900 mm/year for areas around the northwest of the sub-basin.

21.3.3 *Physiography and Drainage*

The Tana basin and northwestern Ethiopian Plateau is represented by a subsided basin and plateau margin that generally covers contrasted physiographic areas. Terrain diversification throughout the area is the reflection of the geotectonic evolution and overall tectonic processes that are responsible for the formation of the Tana basin. Furthermore, the present-day contrasting topography prevalent to the project area is the result of recurring and prolonged geotectonics and geomorphologic processes, such as faulting, uplift, subsidence, weather, and erosion.

Land class division for groundwater studied is mainly based on the combined differences of altitude, climate, vegetation and geological processes, and hence, in accordance with this, the project area can broadly be divided into four major physiographic divisions:

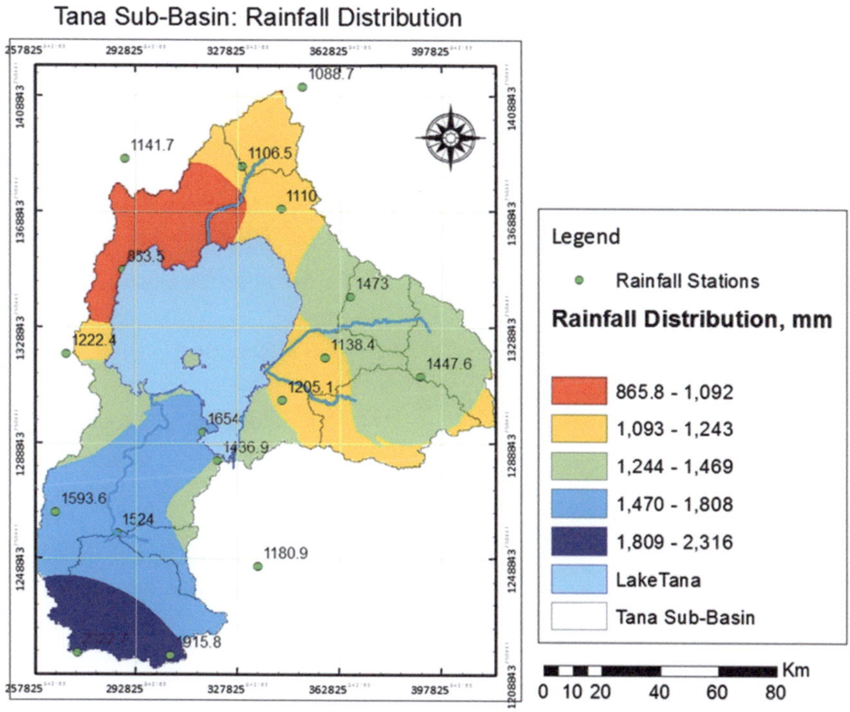

Fig. 21.2 Spatial distribution of precipitation in the Tana sub-basin

- Mountainous terrain with steep slope,
- Mountainous terrain with moderately dissected valleys,
- Rolling to hilly terrain, and
- Flat- to low-lying terrain.

Mountainous terrain with steep slope: This physiographic region corresponds to the major watersheds beginning zones in all sides of the sub-basin. The most known and hydrologically significant area of this physiographic class lies on the Choke Mountain belt at Adama area in the south, Guna Mountain in the west, and the Semen Mountain in the north of the Lake Tana area. The steep slope in this area is the result of geologic formation history which is associated with shield volcanism and that has favored the formation of steep slope. However, this physiographic region has gained the current landform due to erosion and land reformation. Figure 21.3 shows the digital elevation model (DEM) of the study area.

Mountainous Terrain with Moderately Dissected Valleys: Expressed by an outstanding geologic feature developed as a hogback, the dominant part of the sub-basin is defined by mountainous area with dissected valley. The uncollapsed tectonically uplifted basin margin, which is the result of the shield volcanism, with a different basalt geology stands out against a collapsed basin, which creates

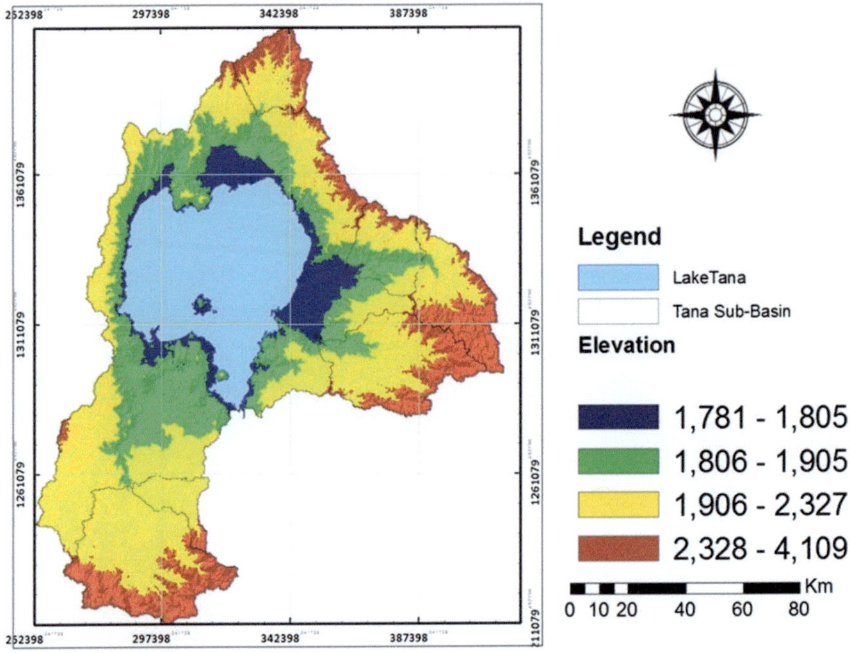

Fig. 21.3 Digital elevation model (DEM) of the study area (m amsl)

contrasting topography. Pik et al. (2003) have indicated that prolonged erosion, making the area a site of the source for the sediment transported to the Sudanese plain, has given rise to the steep mountainous terrain into moderately dissected valleys, where land reformation process has developed numerous local watersheds. It is important to note that there are complex geologic and tectonic processes, whose imprints are still preserved and are modified by subsequent prolonged weather and erosion processes. The uplifted blocks are a remarkable scenic panorama, creating strong topographic contrast of mountains. The presence of different flows on the mountain slope section testifies the chronological order (time sequence) of the rock formations.

Rolling to Hilly Terrain: A rolling to hilly terrain is the characteristic of the area underlain by predominantly scoria/scoriaceous basalt and trachytes and aphanitic basalt rock geology. It is the result of volcano-tectonic effect. Development of cinder cones due to scoria extrusions and uplifted blocks represent rolling to hilly landscape, such as in the west-central and south part, northeastern of the area.

Similar type of terrain has been developed due to intense fracturing at the early phases of basin subsidence, to the east of Lake Tana. Prolonged weather and erosion processes have modified the landform into smooth topped hills with rolling sides throughout those areas. In addition, the relative gentle nature of the slope in this physiographic class makes it a hydrologic pass through the hillside hydrologic continuum from the several dissected valley of the steep mountainous physiographic class.

Flat- to low-lying terrain: Flat- to low-lying terrain is typical of tectonic controlled extensional areas that extend along the perimeter of the lake area. The most flat areas, known as floodplains, include Fogra floodplain, Dembia floodplain, Gilgel Abbay floodplain, and low-lying small plain areas in the sub-basin. Lake Tana 1800 m above sea level is 75 km long and 60 km wide. Its 3600-km^2 surface is dotted with more than 30 islands. This region is also characterized by sag area within the floodplain and delineates the low-lying physiographic setup of the sub-basin. Studies indicate that the lake body is a deep hallow that has passed millennia's as deposition pool for the sediment transported from the surrounding zone. Hence, it accounts for flat-lying region of the physiographic class.

21.3.4 Drainage Pattern

The Lake Tana sub-basin is a north to south elongated, cylindrically curved watershed with eastward prolongation of a land mass forming the Lake Tana water body an apparent centripetal point (Fig. 21.4). The lake water body, also, forms outlet point for the drainage of the sub-basin. With this, the sub-basin has four major drainage systems that are accounted as tributary rivers of the lake water body.

The longest stream channel, also containing the largest areal coverage of the sub-basin, is the Gilgel Abbay River. The Lake Tana sub-basin is the headwater of the Blue Nile River basin, and under the consideration of the geometric setup of the Gilgel Abbay River catchment, it is known as the source of the Blue Nile River. The northern and northwest parts of the basin are drained largely by four rivers, namely Dirma, Megech, Gumara, and Arno-Garno. The eastern part is drained by Ribb, Gumara, and Gelda River channels. The western part of the sub-basin contains several streams, most of them joining the lake water body directly.

The drainage pattern in the Tana sub-basin is dominantly characterized by grand dendrite pattern. However, at some specific locations, in closer look at the junction of the tributary rivers with the mainstream channel, drainage patterns with typical box and angular pattern are dominant. All the major river channels, especially in the upper reach of the river system, have such box and angular drainage patterns.

Associated with the concurrence of box and angular drainage patterns, with prolonged geomorphologic process, which includes tectonic reorientation, erosion, and deposition, in the upper reach of the stream channels, the landform is a rolling terrain forming numerous micro-watersheds. The saturation zone of these micro-watersheds acts as the source of the big rivers, in localities forming a perennial stream and others as annular ephemeral springs.

As the lake forms, a centripetal point to the drainage pattern of the river system in the sub-basin, the terrestrial body along the periphery of the water body has a gentle slope with characteristic nature of drainage pattern in soft geologic formations. All the upper reach drainage channels are chunked forming the four major

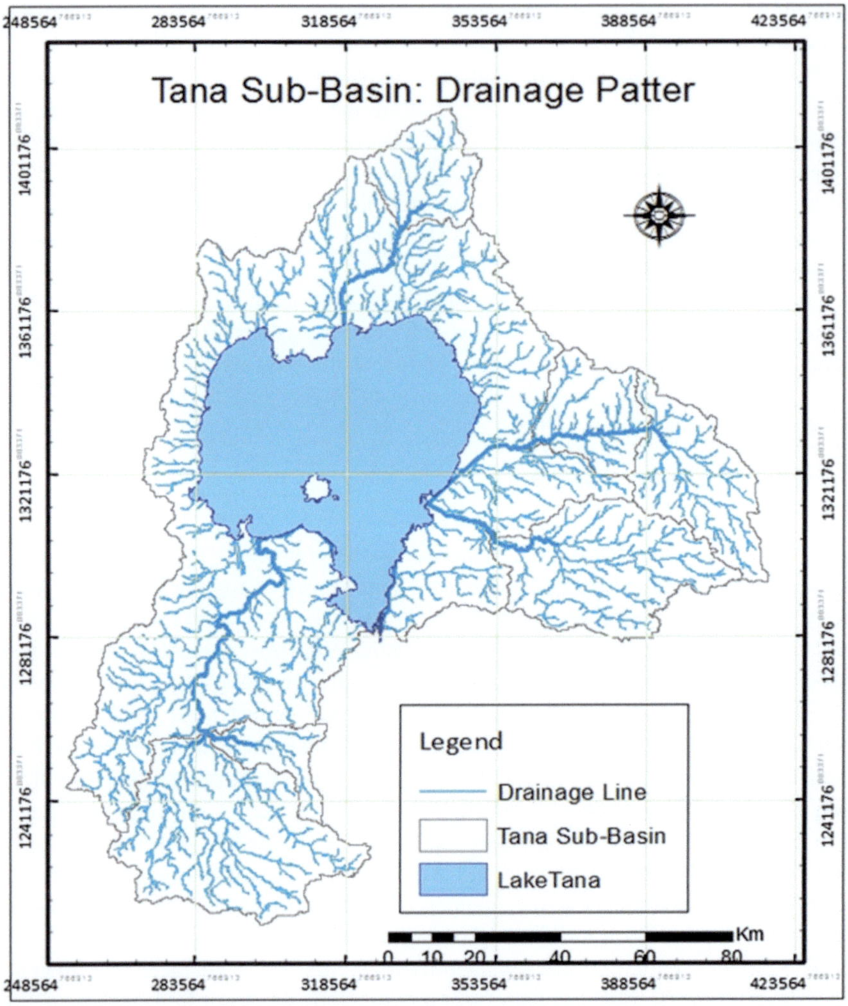

Fig. 21.4 Drainage pattern of the Tana sub-basin

rivers, while some more short streams join directly to the lake water body. These major rivers flow with meandering drainage system along the flat plains of the lower reach of the catchment. Besides the meandering nature of the stream channel, there exists extensive stream channel outflow leading to the formation floodplain. The integrated impact of the flood laid deposition; stream channel deposition, the meandering nature of the stream channels at the lower reach is highly dynamic. The junction of the lower end of streams, as it joins into the lake water body shifts repeatedly, and erratically. In the floodplain, subsurface drainage systems with formation of pipe flow are common. Formation of potholes is prominently observed.

21.3.5 Geology and Groundwater Occurrence

The dominant part of the sub-basin covers two major geological formations, the Termaber basalts and basalts related to volcanic center. The Termaber basalt covers the northern and western parts and southern end of the sub-basin (Fig. 21.5). The lowland along the periphery of the Lake Tana water body is covered by thick Quaternary alluvial deposit, lacustrine deposits, and colluvium. Basalts related to volcanic centers predominantly contain scoriaceous basalt and cover the southern part of the Tana sub-basin. Despite the age of formation, Amba Aiba basalts, Lateriteon Amba Alaji rhyolite, and Ashangi basalts are sparsely distributed in the sub-basin.

Following the lithology and structural implications, there are three major aquifer systems: the Tertiary Volcanics (mostly including the Ashangi, Aiba and Termaber basalts), the Quaternary Basalt aquifer and the Quaternary alluvial deposits (SMEC 2007).

The alluvial aquifer, located at the downstream of the different catchments, receive groundwater recharged from lateral groundwater inflows from the volcanic aquifers of the upper catchments, and direct percolation from the flood. The volcanic aquifer of Quaternary vesicular basalt and Tertiary scoriaceous basalt are recharged from rainfall and lateral flow from upstream areas. The recharged water

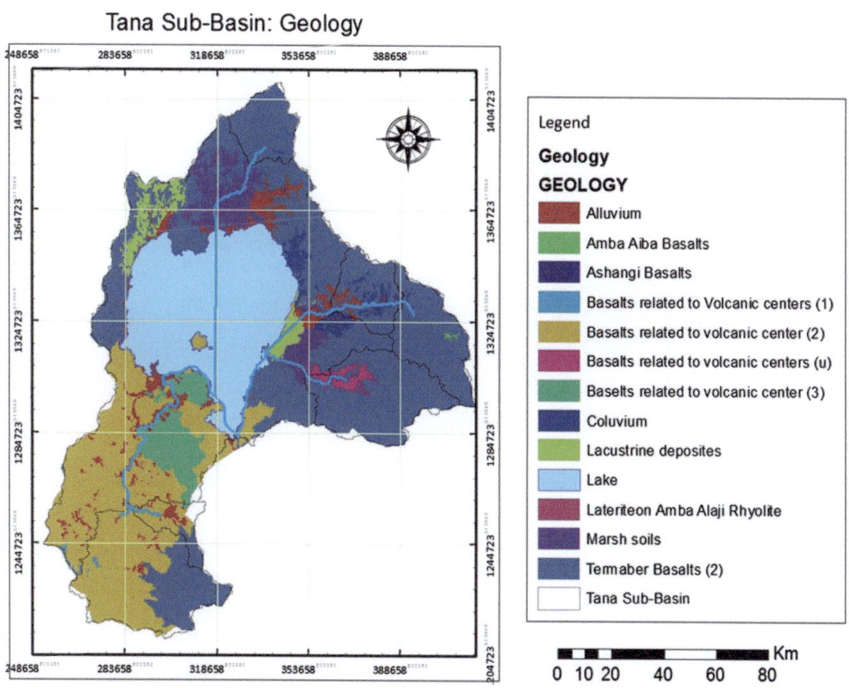

Fig. 21.5 Geological map of the Tana sub-basin (*Source* BCEOM and Associates 1998)

in these aquifers discharges to springs, wetlands, and streams and directly into the southern part of Lake Tana.

According to the hydrogeological study by Abbay River Basin Integrated Development Master Plan Project (BCEOM and Associates 1998), the Tertiary basalts and recent lava flows of the southern part of the sub-basin are grouped as extensive aquifer with fracture permeability, showing the highest groundwater potential.

21.3.6 Soil

The type and description of the soils in the Tana sub-basin is originally developed by BCEOM and Associates (1998), while further description and hydrologic grouping is elaborated by the study in SMEC (2007). Figure 21.6 shows the soil map of the basin. Based on this, the soils in the Tana sub-basin are indicated herewith.

Alisols: Alisols are soils that have a higher clay content in the subsoil than in the topsoil as a result of pedogenetic processes (especially clay migration) leading to an argic subsoil horizon. Alisols have a low base saturation at certain depths and high

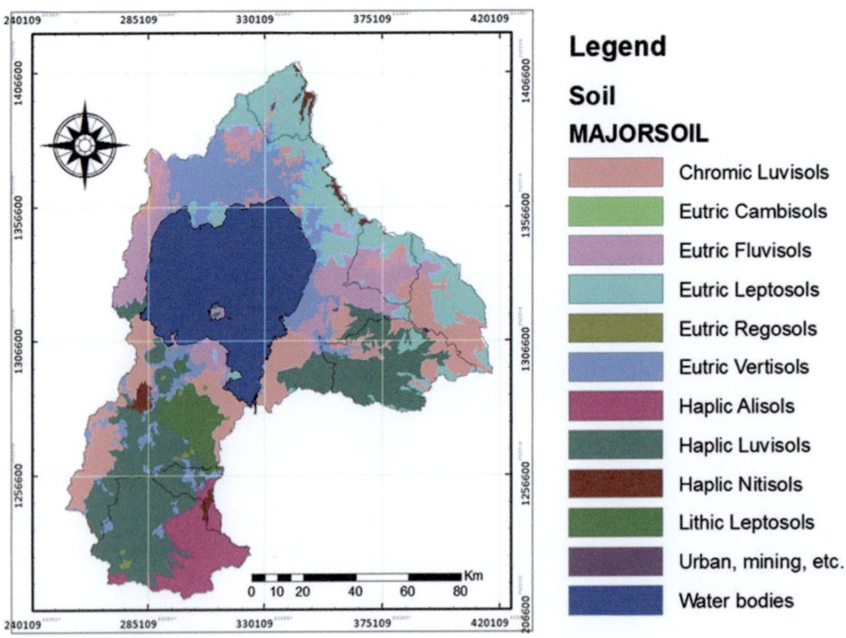

Fig. 21.6 Spatial distribution of the soils in the Tana sub-basin (*Source* BCEOM and Associates 1998)

activity. They are most common in hilly or undulating topography and in humid tropical, humid subtropical and monsoon climates.

Cambisols: Cambisols combine soils with at least an incipient subsurface soil formation. Transformation of parent material is evident from structure formation and mostly brownish discoloration, increasing clay percentage, and/or carbonate removal.

Fluvisols: Fluvisols accommodate genetically young, azonal soils in alluvial deposits. The name "fluvisols" may be misleading in the sense that these soils are not confined only to river sediments (Latin fluvius, river); they also occur in lacustrine and marine deposits and environmentally found in alluvial plains, river fans, valleys, and tidal marshes on all continents and in all climate zones; many fluvisols under natural conditions are flooded periodically.

Leptosols: Leptosols are very shallow soils over continuous rock and soils that are extremely gravelly and/or stony. Leptosols are particularly common in mountainous regions. Leptosols are found in all climate zones (mostly found in hot or cold dry regions), in particular in highly eroded areas.

Luvisols: Luvisols are soils that have higher clay content in the subsoil than in the topsoil. Luvisols have high-activity clays throughout the argic horizon and a high base saturation at certain depths. These soils are most common in flat or gently sloping land in cool temperate regions and in warm regions with distinct dry and wet seasons.

Nitisols: Nitisols are deep, well-drained, red, tropical soils with diffuse horizon boundaries and a subsurface horizon with more than 30 % clay and moderate-to-strong angular blocky structure elements. Weather is relatively advanced, but nitisols are far more productive than most other red, tropical soils. Nitisols are predominantly found in level to hilly land under tropical rain forest or savannah vegetation.

Regosols: Regosols are extensive in eroded lands, particularly in arid and semiarid areas and in mountainous terrain. They are found in all climate zones without permafrost and at all elevations.

Vertisols: Vertisols are churning, heavy clay soils with a high proportion of swelling clays. These soils form deep wide cracks from the surface downward when they dry out, which happens in most years. They are found in depressions and level to undulating areas, mainly in tropical, subtropical, and semiarid to sub-humid and humid climates with an alternation of distinct wet and dry seasons. The dominant vegetation is savannah, natural grassland, and/or woodland.

21.3.7 Source of Contaminants to the Groundwater in the Tana Sub-basin

The groundwater in the Tana sub-basin can be contaminated from fertilizers, pesticides application from the intensive cropland covering more than 75 % of the sub-basins land mass. Other sources of contamination can be from geologic materials due to prolonged water logging in the floodplains of the area. Pollutant from septic systems, open landfills, and industrial west disposals can be the other sources of contamination, mainly from in urban centers.

21.4 DRASTIC Model for the Tana Sub-basin

21.4.1 Data Layer Preparation

The data preparation of DRASTIC layers is required to define the rate of the layers reflecting their respective relative significance. This rate is variable to the specific conditions of the thematic layers, and hence, the values are differed with spatial variability of the nature of the entities in the thematic layers. Based on this, the layers applied in this model are described below:

D: Depth to water table: A depth to water table map of the groundwater resources in the Tana sub-basin is developed based on the static water table record from 49 groundwater tapping wells. The surface map, prepared by interpolation of the static water table, is reclassified into ranges correlated with the values of the rate and weight, as indicated in Table 21.2.

The product of the rate and index, as shown in Fig. 21.7, is the index value from the depth to water table.

R: Net Recharge: Hydrological information, such as spatial distribution of groundwater recharge in the Tana sub-basin is quite limited. Optional methods to determine the spatial distribution of the rate of importance of net recharge in the model are applied by different studies, such as Piscopo (2001) and Ta'any et al. (2013). According to this study, the possibility of the area to gain recharge due to combined effect of source of recharge (rainfall) and media to facilitate recharge (soil permeability and slope) is considered to substitute net recharge, where the DRASTIC rate is calculated based on the following equation:

$$\begin{aligned}\text{Recharge rating} &= \text{Slope factor} + \text{Rainfall factor}\\ &\quad + \text{Soil permeability factor}\end{aligned} \tag{21.2}$$

Considering this approach, the net recharge rate for the DRASTIC model for the Tana sub-basin is calculated (Table 21.3; Fig. 21.8).

Table 21.2 Assigned values of depth to water table range and ratings

Factors	Range	Rating
Depth to water table (m)	0–5	10
	5–15	9
	15–30	7

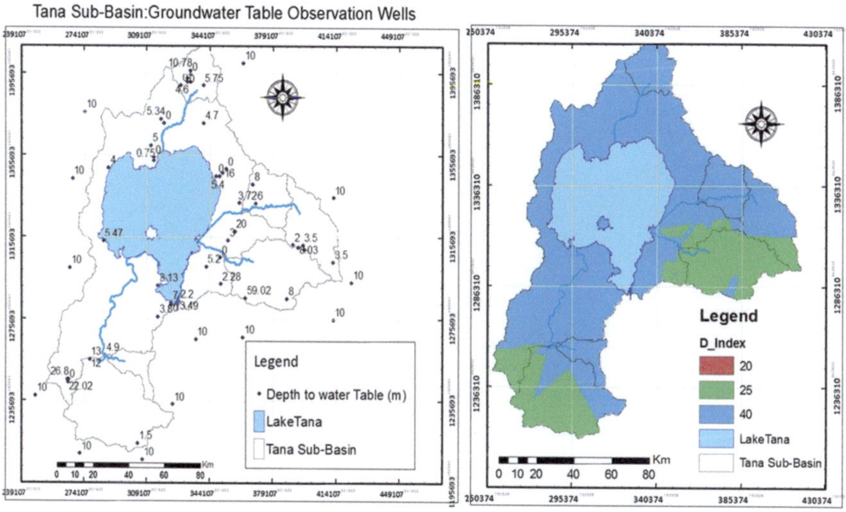

Fig. 21.7 Depth to water table: spatial distribution (*left*) and index value (*right*)

Table 21.3 Assigned values of net recharge (mm) range and ratings

Factors	Range	Rating	Typical rating
Net recharge (mm)	170–350	5	
	350–400	7	
	400–600	8	
	600+	10	

A: Aquifer Media: The DRASTIC index of the aquifer media is calculated based on the geological information of the sub-basin (Table 21.4; Fig. 21.9).

S: Soil Media: The soil map of the Tana sub-basin (Table 21.5; Fig. 21.10) is used to calculate the spatial distribution of the DRASTIC index values attributed to the soil media.

T: Topography (***slope***): The digital elevation model of the area is used to calculate the percent rise and define the associated DRASTIC index. After calculating the rate based on the slope map, the index map, indicated below, is prepared as a product of the rate with the weight (Table 21.6; Fig. 21.11).

I: Impact of the vadose zone media: After a closed evaluation of the soil and geology of the sub-basin, the impact of the vadose zone media in groundwater contamination is defined. This has been defined based on the soil map layer. Accordingly, the index value for the impact of the vadose zone media is indicated in Table 21.7 and Fig. 21.12.

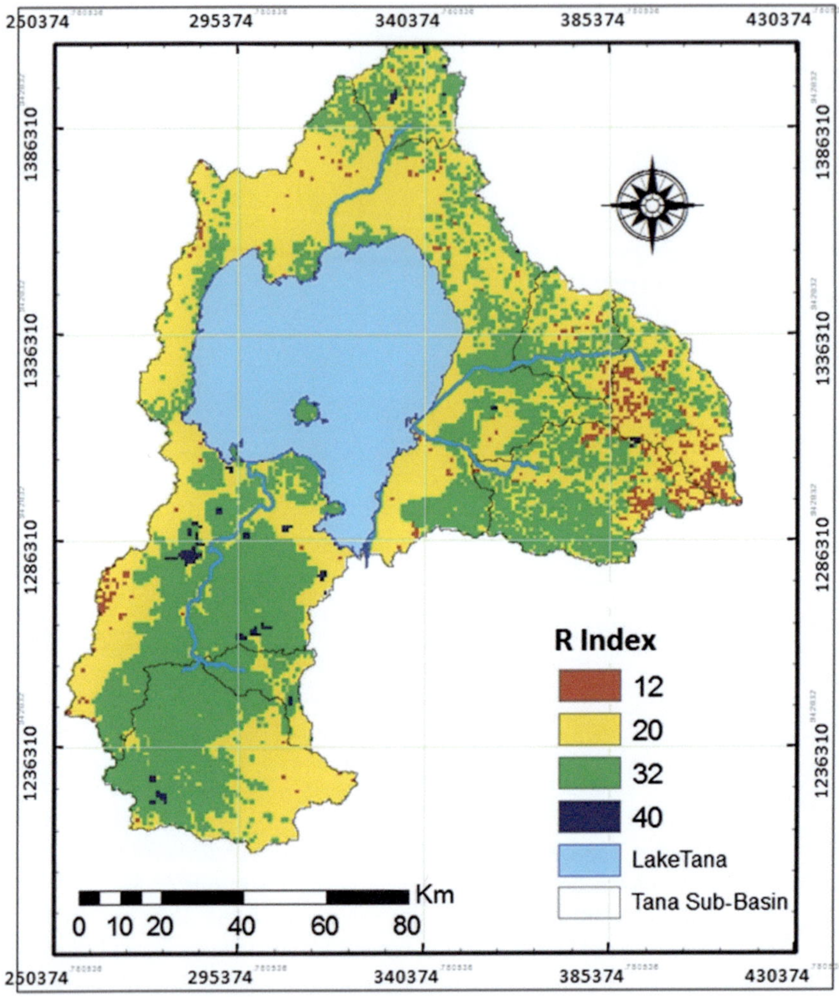

Fig. 21.8 Net recharge index value of the Tana sub-basin

21.5 Results and Discussion

The DRASTICaquifer vulnerability map (Fig. 21.13) indicates that the groundwater resource vulnerability to contamination in the Tana sub-basin is in the range of 66–117.

Low DRASTIC index values, which are in the range of 66–76, are recorded in topographic high areas of the sub-basin. This accounts for 36 % of the area coverage of the sub-basin covering around 5263.9 km^2. Based on the groundwater resource potential knowledge of the sub-basin, this part of the area is characterized

Table 21.4 Values of aquifer media range and ratings

Factors	Range	Rating	Typical rating
Aquifer media	Massive shale	1–3	2
	Metamorphic/igneous	2–5	3
	Weathered metamorphic/igneous	3–5	4
	Glacial till	4–6	5
	Bedded sandstone, limestone, and shale sequence	5–9	6
	Massive sandstone	4–9	6
	Massive limestone	4–9	6
	Sand and gravel	4–9	8
	Basalt	2–10	9
	Karst limestone	9–10	10

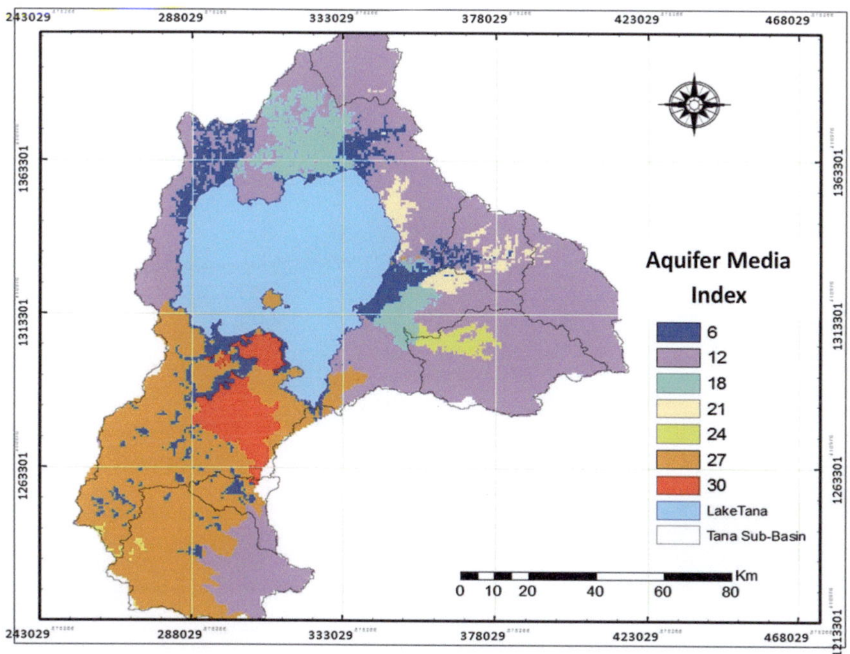

Fig. 21.9 Geological map and aquifer media index value

by shallow groundwater system with limited potential to large-scale development requirements. These factors of DRASTIC index calculation, depth to groundwater, aquifer media, and soil cover consider the low vulnerability of the groundwater resource in this part of the sub-basin.

Two vulnerability index classes, as indicated in Table 21.8, refer to the vulnerability index ranging from 77 to 92. The area defined by this DRASTIC index is

Table 21.5 Assigned values of soil media range and ratings

Factors	Range	Rating
Soil media	Thin or absent	10
	Gravel	10
	Sand	9
	Peat	8
	Shrinking and/or aggregated clay	7
	Sandy loam	6
	Loam	5
	Silty loam	4
	Clay loam	3
	Muck	2
	Non-shrinking and non-aggregated clay	1

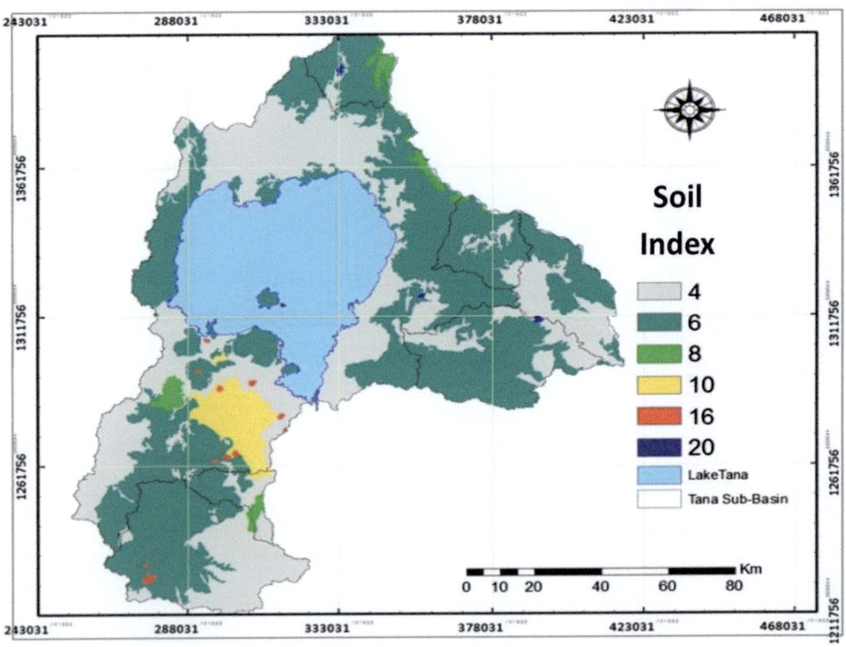

Fig. 21.10 Soil map and index value of the soils in the Tana sub-basin

located in the middle slope areas of the eastern and northern parts of the Tana sub-basin. Some localities of this area, such as the northern part have very shallow groundwater table, however the geologic material and thick flood layer deposit have a strong control on the decline in the index value. The groundwater table in some part of the area defined by this DRASTIC index is near surface due to the confide nature of the aquifers. Hence, it indicates that the index value is valid.

Table 21.6 Assigned values of topography/slope (%) range and ratings

Factors	Range	Rating	Typical rating
Topography/slope (%)	0–2.8	10	
	2.8–8.5	9	
	8.5–14.8	8	
	14.8–21.9	7	
	21.9–29.7	6	
	29.7–38.2	5	
	38.2–48.8	4	
	48.8–96.2	3	
	96.2+	1	

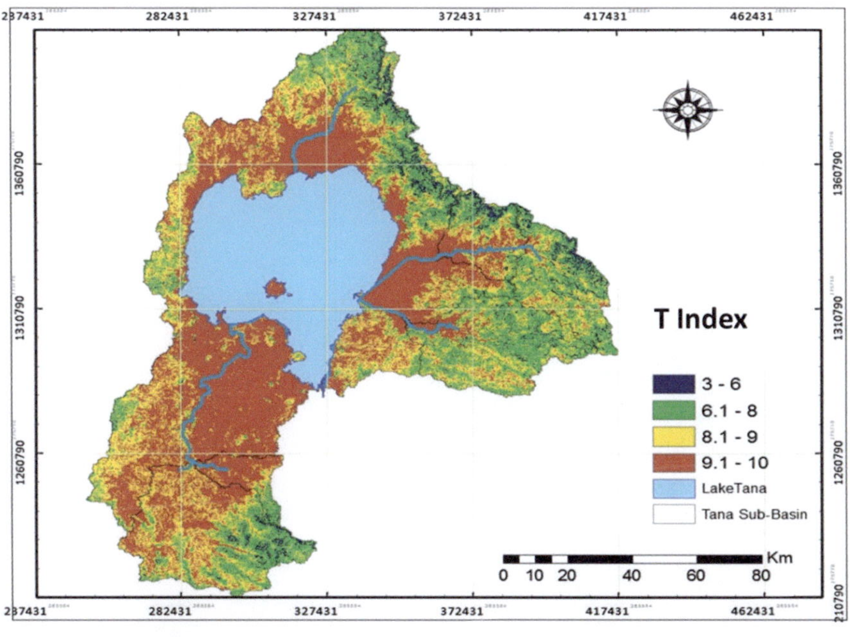

Fig. 21.11 Topography index value of the Tana sub-basin

Sites of the highest DRASTIC index value in the range of 93–117 lie mainly on the southern part of the sub-basin. This part of the study area has very near-surface water table, highly porous geologic material, and thin soil layer. In relative to other parts of the study area, this part of the sub-basin exhibits gentile to flat-lying topography, where the impact of the slope is in a way to favor groundwater pollution, in the presence of pollutants. Based on this, the southern Tana area is the most vulnerable part of the sub-basin. It is known that this part of the sub-basin is a huge groundwater reserve.

Table 21.7 Assigned values of impact of the vadose zone media range and ratings

Factors	Range	Rating	Typical rating
Impact of the vadose zone media	Confining layer	1	1
	Silty/clay	2–6	3
	Shale	2–5	3
	Limestone	2–7	6
	Sandstone	4–8	6
	Bedded limestone, sandstone, shale	4–8	6
	Sand and gravel with significant silt and clay	4–8	6
	Metamorphic/igneous	2–8	4
	Sand and gravel	6–9	8
	Basalt	2–10	9
	Karst limestone	8–10	10

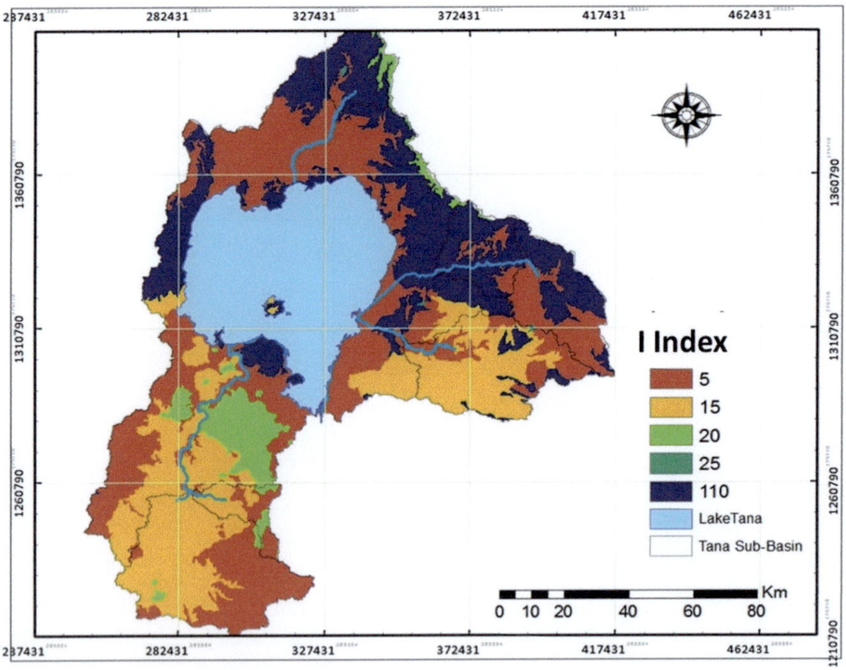

Fig. 21.12 Impact of vadose zone media index value of the Tana sub-basin

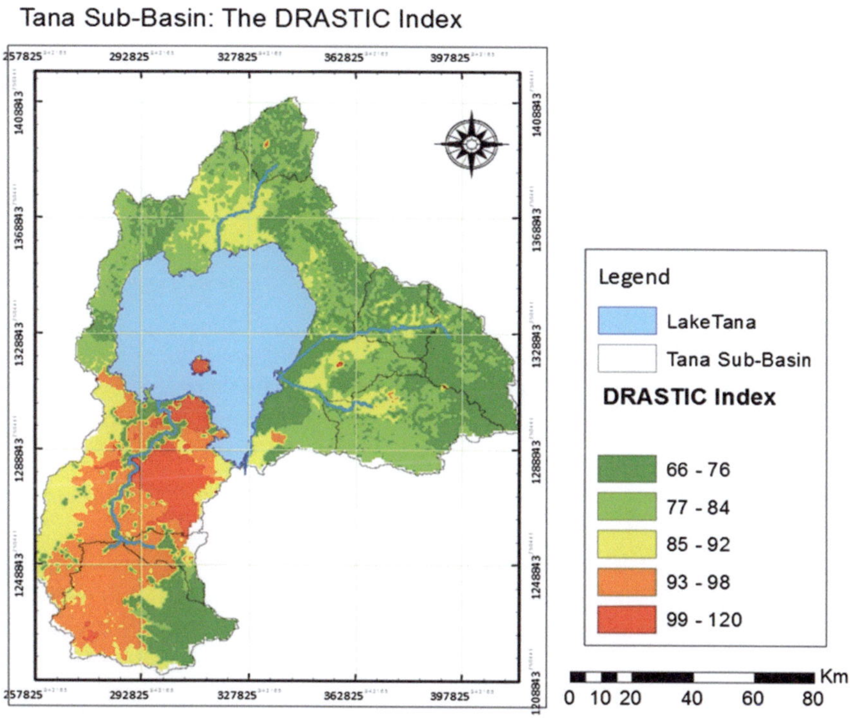

Fig. 21.13 Spatial distribution of the DRASTIC index in the Tana sub-basin

Table 21.8 Summary of the DRASTIC index values and area coverage

S. no	DRASTIC index range	Total count of cells	Proportion	Area (km^2)
1	66–76	5435	0.36	5263.83
2	77–84	4447	0.29	4306.94
3	85–92	3939	0.26	3814.94
4	93–98	573	0.04	554.95
5	99–120	992	0.06	863.91
Total		15,286	1.00	14,804.57

The study reveals that the dominant part of the groundwater vulnerability in the Tana sub-basin is defined by DRASTIC index value of 96 (Fig. 21.14). This corresponds to the total area of 3306.561 km^2.

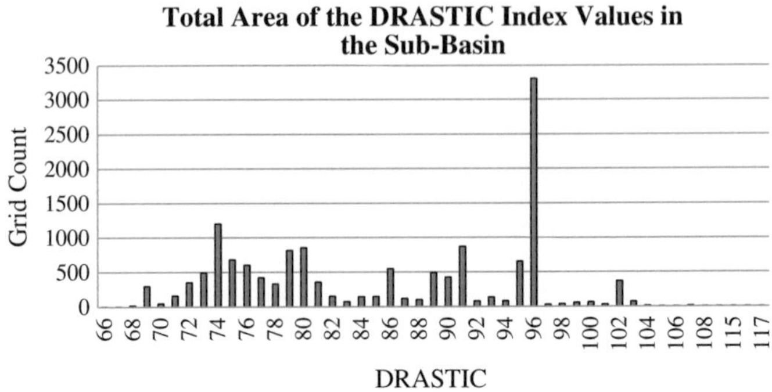

Fig. 21.14 Count of values for DRASTIC index of the Tana sub-basin

21.6 Conclusion and Recommendation

The vulnerability index and its spatial distribution indicate that the zones with high groundwater potential of the sub-basin have the highest susceptibility to pollution. This is controlled by the highest nature of permeability of the fractured vesicular basalt aquifers with thin soil cover. A good indicator of this is the recurrence of increasing turbidity of the groundwater wells and springs shortly after high-intensity rainfall event.

Aquifers in urban areas, such as Gondar, are designated in the highest DRASTIC index. Although the groundwater table is relatively deep, and the geology and soil cover favors to maintain low vulnerability index, the land use have the highest contribution in enhancing the susceptibility of the groundwater to pollution.

Based on this study, we have noticed that the geology, soil cover, and land uses have significant influence on the vulnerability of the overall groundwater resource in the Tana sub-basin. Other DRASTIC parameters are implied by these three governing parameters.

Therefore, any developmental intervention in these high groundwater reserve zones of the Tana sub-basin, such as to the south of the Lake Tana will have a significant influence in maintaining a good quality of groundwater. This area provides the water supply for Bahir Dar and the surrounding area, and agricultural intensifications and application of chemicals can gradually deteriorate the groundwater quality.

Acknowledgments This report is an improved and customized version of the first author's research and development work in the Tana Beles Integrated Water Resources Management Project of the Abbay Basin Authority, Ethiopia. This is a World Bank-funded project, and we are grateful to the World Bank for the financial support. We extend our greatest acknowledgement to the Ministry of Water and Energy and the Abbay Basin Authority for providing data in order to carry out in this study.

References

Abtew W, Melesse AM (2014a) Nile River basin hydrology. In: Melesse AM, Abtew W, Setegn S (eds) Nile River basin: ecohydrological challenges, climate change and hydropolitics, pp 7–22

Abtew W, Melesse AM (2014b) Climate teleconnections and water management. In Nile River basin, pp 685–705. Springer International Publishing, Berlin

Abtew W, Melesse AM (2014c) Transboundary rivers and the Nile. In: Nile River basin, pp 565–579. Springer International Publishing

Abtew W, Melesse AM, Desalegn T (2009a) Spatial, inter and in tra-annual variability of the Blue Nile river basin rainfall. Hydrol Process 23(21):3075–3082

Abtew W, Melesse AM, Desalegn T (2009b) El Niño southern oscillation link to the Blue Nile river basin hydrology. Hydrol Process Spec Issue: Nile Hydrol 23(26):3653–3660

Aller LT, Bennett HJR, Lehr R, Petty J, Hackett G (1987) DRASTIC: a standardized system for evaluating groundwater pollution potential using hydrogeological settings. EPA 600/2–EP87/036. National Water Wells Association, Ada, Oklahoma

Arthur JD, Wood HAR, Baker AE, Cichon JR, Raines GL (2007) Development and implementation of a bayesian-based aquifer vulnerability assessment in Florida. Nat Resour Res 16(2)

Babiker IS, Mohamed MAA, Hiyama T, Kato K (2005) A GIS-based DRASTIC model for assessing aquifer vulnerability in Kakamigahara heights, Gifu prefecture, Central Japan. Sci Total Environ 345:127–140

Barbash JE, Resek EA (1996) Pesticides in groundwater: distribution, trends, and governing factors. Ann Arbor Press

BCEOM and Associates (1998) Abbay basin master plan

Berkhoff K (2007) Groundwater vulnerability assessment to assist the measurement planning of the water framework directive—a practical approach with stakeholders. Hydrol Earth Syst Sci Discuss 4:1133–1151

Chebud YA, Melesse AM (2009a) Numerical modeling of the groundwater flow system of the Gumera sub-basin in Lake Tana basin, Ethiopia. Hydrol Process Spec Issue: Nile Hydrol 23 (26):3694–3704

Chebud YA, Melesse AM (2009b) Modeling lake stage and water balance of Lake Tana, Ethiopia. Hydrol Process 23(25):3534–3544

Chebud Y, Melesse AM (2013) Stage level, volume, and time—frequency change information content of Lake Tana using stochastic approaches. Hydrol Process 27(10):1475–1483. doi:10.1002/hyp.9291

Chilton J (2006) Assessment of aquifer pollution vulnerability and susceptibility to the impacts of abstraction. In: Schmoll O, Howard G, Chilton J, Chorus I (eds) World Health Organization. Protecting groundwater for health: managing the quality of drinking-water sources. IWA Publishing, London

Civita M (1994) Le carte della vulnerabilità degli acquiferi all'inquinamento. Teoria e practica [Aquifer vulnerability map to pollution. Theory and application]. Pitagora, Bologna:13

Civita M, De Regibus C (1995) Sperimentazione di alcune metodologie per la valutazione della vulnerabilità degli acquiferi [Experimentation of a methodology for mapping the value of the aquifer vulnerability]. Pitagora, Bologna, Q Geol Appl 3:63–71

Daly D, Drew D (1999) Irish methodologies for karst aquifer protection. In: Beck B (ed) Hydrogeology and engineering geology of sinkholes and karst. Balkema, Rotterdam, pp 267–272

Doerfliger N, Jeannin PY, Zwahlen F (1999) Water vulnerability assessment in karst environments: a new method of defining protection areas using a multi-attribute approach and GIS tools (EPIK method). Environ Geol 39(2):165–176

Foster SSD (1987a) Fundamental concepts in aquifer vulnerability, pollution risk and protection strategy. In: van Duijvenbooden W, van Waegeningh HG (eds) Vulnerability of soil and groundwater to pollutants. Proc Inf TNO Comm Hydrol Res (38):69–86

Foster SSD (1987b) Fundamental concepts in aquifer vulnerability, pollution risk and protection strategy. In Gogu RC, Dassargues A (2000) Current trends and future challenges in groundwater vulnerability assessment using overlay and index methods. Environ Geol 39 (6):549–559

Fritch TG, McKnight CL, Yelderman JC Jr, Arnold JG (2000) An aquifer vulnerability assessment of the paluxy aquifer, central Texas, USA, using GIS and a modified DRASTIC approach. Environ Manage 25:337–345

Gogu RC (2000) Advances in groundwater protection strategy using vulnerability mapping and hydrogeological GIS databases. PhD Thesis, Faculty of Applied Sciences, University of Liège, Belgium, pp 153

Gogu RC, Dassargues A (2000) Current trends and future challenges in groundwater vulnerability assessment using overlay and index methods. Environ Geol 39(6):549–559

Gogu RC, Hallet V, Dassargues A (2003) Comparison of aquifer vulnerability assessment techniques. Application to the Néblon river basin (Belgium). Environ Geol 44:881–892

Melesse AM (2011) Nile river basin: hydrology, climate and water use. Springer Science & Business Media

Melesse AM, Loukas AG, Senay G, Yitayew M (2009a) Climate change, land-cover dynamics and ecohydrology of the Nile river basin. Hydrol Process Spec Issue: Nile Hydrol 23(26):3651–3652

Melesse AM, Abtew W, Desalegne T, Wang X (2009b) Low and high flow analysis and wavelet application for characterization of the Blue Nile river system. Hydrol Process 24(3):241–252

Melesse A, Abtew W, Setegn SG (2014) Nile river basin: ecohydrological challenges, climate change and hydro politics. Springer Science & Business Media

Michael, JF., Thomas, ER., Michael, GR., & Dennis, RH. (2005). Assessing groundwater vulnerability to contamination: Providing scientifically defensible information for decision makers. US geological survey circular 1224, US Department of the Interior, US Geological Survey. USGS Publishing Network

Moore and John S (1990) SEEPAGE: A system for early evaluation of the pollution potential of agricultural groundwater environments, USDA. SCS, Northeast Technical Center, Geology Technical Note 5

Palmer RC, Holman IP, Robins NS, Lewis MA (1995) Guide to Groundwater Vulnerability Mapping in England and Wales, National Rivers Authority

Pik R, Marty B, Carignan J, Lavé J (2003) Stability of the Upper Nile drainage network (Ethiopia) deduced from (U–Th)/He thermochronometry: implications for uplift and erosion of the Afar plume dome. Earth Planet Sci Lett 215:73–88

Piscopo G (2001) Groundwater vulnerability map, explanatory notes—Castelreagh catchment. Department of Land and Water Conservation, New South Wales, Australia

Rupert MG (2001) Calibration of the DRASTIC groundwater vulnerability mapping method. Groundwater 39(4):625–630

Secunda S, Collin M, Melloul AJ (1998) Groundwater vulnerability assessment using a composite model combining DRASTIC with extensive land use in Israel's Sharon region. J Environ Manage 54:39–57

Setegn SG, Srinivasan R, Dargahi B, Melesse AM (2009a) Spatial delineation of soil erosion prone areas: application of SWAT and MCE approaches in the Lake Tana basin, Ethiopia. Hydrol Process Spec Issue: Nile Hydrol 23(26):3738–3750

Setegn SG, Srinivasan R, Melesse AM, Dargahi B (2009b) SWAT model application and prediction uncertainty analysis in the Lake Tana basin, Ethiopia. Hydrol Process 24(3):357–367

Setegn SG, Bijan Dargahi B, Srinivasan R, Melesse AM (2010) Modelling of sediment yield from Anjeni gauged watershed, Ethiopia Using SWAT. JAWRA 46(3):514–526

SMEC (2007) Hydrological study of the Tana-Beles Sub Basins, Technical report (Ministry of Water Resources, Addis Ababa

Stigter TY, Ribeiro L, CarvalhoDill AMM (2006) Evaluation of an intrinsic and a specific vulnerability assessment method in comparison with groundwater salinisation and nitrate contamination levels in two agricultural regions in the south of Portugal. Hydrogeol J 14:79–99

Ta'any RA, Alaween MA, Al-Kuisi MM, Al-Manaseer NM (2013) GIS based model of groundwater vulnerability and contamination Risk of Wadi Kufrinja catchment area, Jordan. World Appl Sci J 24(5):570–581

Thapinta A, Hudak PF (2003) Use of geographic information systems for assessing groundwater pollution potential by pesticides in Central Thailand. Environ Int 29:87–93

The National Research Council (1993) Groundwater vulnerability assessment: predicting relative contamination potential under conditions of uncertainty. ISBN 0-309-58508-2

Van Stempvoort D, Ewert L, Wassenaar L (1993) Aquifer Vulnerability Index: A GIS Compatible Method for Groundwater Vulnerability Mapping. Canadian Water Resources Journal 18(1)

von Hoyer M, Söfner B (1998) Groundwater vulnerability mapping in carbonate (karst) areas of Germany. Fed Inst Geosci Nat Resources, Hannover, Archiv no. 117854, pp 38

Vrba J, Zoporozec A (1994) Guidebook on mapping groundwater vulnerability. IAH Int Contrib Hydrogeol 16:131 (Hannover7 Heise)

Wang L, Yang YS (2008) An approach for catchment-scale groundwater nitrate risk assessment from diffuse agricultural sources: a case study in the Upper Bann, Northern Ireland. Hydrol Process 22:4274–4286

Worrall F, Besien T (2005) The vulnerability of groundwater to pesticide contamination estimated directly from observations of presence or absence in wells. J Hydrol 303:92–107

Worrall F, Kolpin DW (2004) Aquifer vulnerability to pesticide pollution—combining soil, land-use and aquifer properties with molecular descriptors. J Hydrol 293:191–204

Chapter 22
Groundwater Recharge and Contribution to the Tana Sub-basin, Upper Blue Nile Basin, Ethiopia

Anteneh Z. Abiy, Solomon S. Demissie, Charlotte MacAlister, Shimelis B. Dessu and Assefa M. Melesse

Abstract The Tana sub-basin exhibits one of the huge groundwater reserve zones in Ethiopia. While there is effective three months of precipitation, Lake Tana receives year-round recharge from the four major rivers in the sub-basin: Gilgel Abay, Gumara, Ribb, and Megech. Hence, the groundwater contribution to the Lake Tana is quite important to the aquifer-dependent ecosystem and environmental flow requirement in the area. However, the contribution of the groundwater resource to the Lake Tana water body is not well studied, and useful information for development planning and management is not available. The purpose of this study was to develop information that can better inform decision making on groundwater abstraction and/or lake water uses in the Tana sub-basin considering the volume of the groundwater contribution to the Lake Tana water body. Accordingly, the study combined stream flow simulation and baseflow separation. The stream flows at the junction of the lake water body are generated by the application of Soil and Water

A.Z. Abiy
Abbay Basin Authority, Bahir Dar, Ethiopia

S.S. Demissie
Department of Civil & Environmental Engineering,
University of California, Los Angeles, Los Angeles 90095, CA, USA
e-mail: solomon.demissie@gmail.com

C. MacAlister
International Research Development Center, Ottawa, ON, Canada
e-mail: cmacalister@idrc.ca

S.B. Dessu
Department of Civil Engineering, Addis Ababa University, Addis Ababa, Ethiopia
e-mail: sbehailu@gmail.com

A.Z. Abiy (✉) · A.M. Melesse
Departments of Earth and Environment, Florida International University,
11200 SW 8th Street, Miami, FL 33199, USA
e-mail: aabiy001@fiu.edu

A.M. Melesse
e-mail: melessea@fiu.edu

© Springer International Publishing Switzerland 2016

A.M. Melesse and W. Abtew (eds.), *Landscape Dynamics, Soils and Hydrological Processes in Varied Climates*, Springer Geography,
DOI 10.1007/978-3-319-18787-7_22

Assessment Tool (SWAT). The groundwater contribution of the streams is defined by baseflow separation using a reclusive digital filter method. The groundwater contribution to the Lake Tana water is through lateral flow from the vadose zone and return flow to shallow aquifers. The contribution of the four major rivers' catchment area; Gilgel Abay, Gumera, Ribb, and Megech Rivers, is 718, 414, 451, and 350 mm/year, respectively. This accounts up to 60 % of the total annual in flow to the Lake Tana water body through stream flow.

Keywords Lake Tana · Baseflow · Groundwater · Recharge · SWAT · T-plot · Reclusive digital filter · Blue Nile River

22.1 Introduction

The Tana sub-basin exhibits one of the huge groundwater reserve zones in the country (Tenalem et al. 2008). The groundwater contributes a wide array of socioeconomic importance including a source of water supply to the people in the sub-basin, maintains a sustainable aquifer-dependent ecosystem in the wetlands of the sub-basin and recharge to the Lake Tana water body. Groundwater uses in the sub-basin are rising at an increasing rate. Likewise, the dry season stream flow, which is the discharge of groundwater as baseflow, is declining. Hence, knowledge that would reasonably estimate the overall groundwater contribution to the stream flow and Lake Tana water body is important to develop a systematic water resource planning and management.

The knowledge of recharge–discharge relations of aquifers in the Tana sub-basin is of primary importance in defining the impact of water resources development in the stream flow and overall recharge to the Lake Tana water body. Most importantly, it is useful in defining the dry season flows for environmental flow requirements analysis and other water allocation scenarios. The annual recharge–discharge analysis would be important for devising exploitation and management strategies of the aquifers by linking water management and recharge facilities. This includes, but not limited to integrated planning and implementation of water supply projects, groundwater pump irrigation projects, and watershed management practices.

The Lake Tana receives water from direct precipitation and year-round recharge from the following four major rivers: Gilgel Abay, Gumara, Ribb, and Megech. The lake water body has an area of 3,000 km^2, and the total sub-basin is 15,000 km^2. Studies indicate that the sub-surface interaction of the Lake Tana water body to the surrounding aquifers is imminent (Kebede et al. 2011; Zemedagegnehu et al. 2007; SMEC 2008). Therefore, the portion of the groundwater contribution to the Lake Tana water is through baseflow.

Hydrology of the Nile River basin has been studied by various researchers. These studies encompass various areas including stream flow modeling, sediment dynamics, teleconnections and river flow, land use dynamics, climate change impact, groundwater flow modeling, hydrodynamics of Lake Tana, water allocation, and demand analysis (Melesse et al. 2009a, b, 2011, 2014; Melesse 2011; Abtew et al. 2009a, b; Abtew and Melesse 2014a, b, c; Yitayew and Melesse 2011; Chebud and Melesse 2009a, b, 2013; Setegn et al. 2009a, b, 2010; Dessu and Melesse 2012, 2013; Dessu et al. 2014).

A comprehensive review of the studies on the hydrology of the Tana sub-basin is reported by consultants under the Ministry of Water Resources of Ethiopia. This review (Sogreah and Geomatrix 2013) indicates that the groundwater potential in the sub-basin is a significant part of the water balance. However, most studies consider the upper small gauged part of the sub-basin in their water balance calculation. Others that calculate the water balance for the entire sub-basin focus on the surface water part of the sub-basin water balance. Other recent studies on the hydrology of the lake and groundwater systems show Lake Tana stage variability and fluctuations, and the contribution of the groundwater (Chebud and Melesse 2009a, b, 2013).

However, water resources management and sustainability of water resources development require information derived from time series patterns that would indicate alternative directions of development in space and time. In this study, we have considered hydrological modeling approach to calculate the groundwater contribution to the Lake Tana water body using the concept of top model described in Beven and Kirkby (1979) and Schneiderman et al. (2007).

The goal of the study was to develop information that can better inform decision making on groundwater abstraction and/or lake water uses in the Tana sub-basin. Accordingly, the specific objectives of the study reported in this chapter were to (1) model and simulate river flows at the mouth of the rivers, (2) estimate groundwater contribution through baseflow separation, and (3) analyze baseflow trends over time.

22.2 Study Area Description

The Tana sub-basin, located in the northwestern highland of Ethiopia, is the source of the Blue Nile River, and it is one of the sixteen sub-basins of the Abay River Basin.[1] Uni-modal rainfall patterns, ranging from 865 mm/year to 2300 mm/year, characterize the sub-basin[1]. The annual average areal maximum, and minimum temperature varies from 22 to 29.5 °C and 8.5 to 16 °C, respectively.

The topography of Lake Tana varies and the elevation ranges from 1780 to 4000 m. a.s.l. Four major physiography classes control the drainage pattern of the sub-basin[1].

[1]For details, readers are referred to read Chap. 21.

The soil and land uses of the Blue Nile basin, developed by the Ministry of Water Resources Ethiopia (BCEOM 1999), were used to characterize the Lake Tana sub-basin[1]. Detailed study area description is shown in Chap. 21 of this book.

22.3 Methods

The study reported in this chapter attempted the following. First, using Soil and Water Assessment Tool (SWAT) (Arnold et al. 1998), river flows are calibrated and validated for four major river basins contributing to the Lake Tana using gauged data. Second, based on the calibrated model, river flows at the mouth of the rivers (entry point to the lake) are estimated. Third, the baseflow component of the river hydrographs was separated from the direct flow, and groundwater contribution to the lake is estimated. Finally, trends of the baseflow over the period of time are analyzed. Figure 22.1 shows the flowchart of the methodology followed.

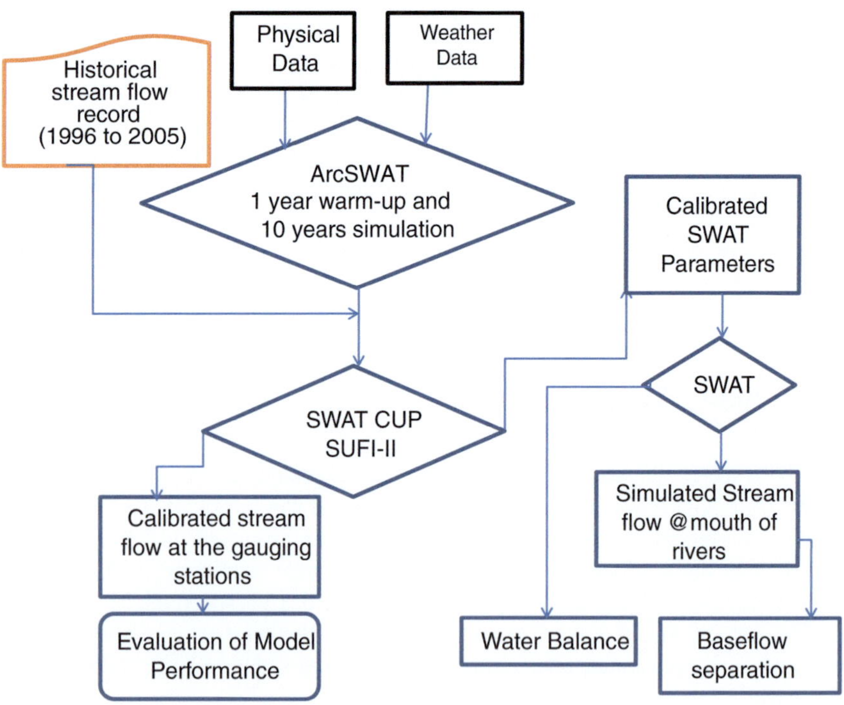

Fig. 22.1 Methodology flowchart

22.3.1 Soil and Water Assessment Tool (SWAT)

The SWAT (Arnold et al. 1998) considers multiple inputs of the physical environment and the hydro-climatic parameters. The physical data required for hydrological modeling with SWAT include digital elevation model (DEM), soil and land use/land cover data. Global Shuttle Radar Topography Mission (SRTM) DEM data of 90-m resolution were used to characterize the topography of the river basins. Soil and land cover information of the area are extracted from the Abay River Basin Development Master Plan (BCEOM 1999). The hydrometeorological data used for flow estimation include daily precipitation and maximum and minimum temperature collected from the gauging stations in the area.

The application of SWAT in predicting stream flow and sediment as well as evaluation of the impact of land use and climate change on the hydrology of watersheds has been documented by various studies (Dessu and Melesse 2012, 2013; Dessu et al. 2014; Wang et al. 2006, 2008a, b, c; Wang and Melesse 2005, 2006; Behulu et al. 2013, 2014; Setegn et al. 2009a, b, 2010, 2011, 2014; Mango et al. 2011a, b; Getachew and Melesse 2012; Assefa et al. 2014; Grey et al. 2013; Mohammed et al. 2015).

SWAT Modeling Design: The four major tributary rivers that drain into the Lake Tana are Gilgel Abay, Gumara, Ribb, and Megech (Fig. 22.2). As indicated in the figure, stream flows are generated for each river at the outflow points of the lake (1, 2, 3 and 4). The SWAT[2] model is run for 11 years; one-year warm-up period, and 10-year simulation period from 1995 to 2005.

Based on the gauged flow data, SWAT was calibrated at selected stations shown in green dot in Fig. 22.2. In Fig. 22.2, numbers indicate outlet points of the rivers, and the shaded area is the model boundary. Accordingly, outlet number 1 is outlet of the Gilgel Abay River, 2 for Gumara, 3 for Ribb, and 4 is for Megech River basins. The drainage area of the four rivers accounts for 8389 km^2 terrestrial land of the Lake Tana sub-basin (Table 22.1).

Each of the four rivers' model outputs is calibrated in SWAT-CUP statistical tool with the application of sequential uncertainty fitting procedure (SUFI-2) (SWAT-CUP 2009). In every round of one thousand iterations, the SUFI-2 provides the most fitting range of the sensitive parameters. It also helps to knock out non-sensitive parameters. Table 22.2 shows the calibration parameters used based on sensitivity analysis.

Once the model is calibrated in the SWAT CUP environment, the calibrated model package is transposed to the SWAT module to rerun the SWAT based on the calibrated parameters to simulate flows for ungauged sections of the basins.

---

[2]The detailed description of the SWAT modeling process is not incorporated.

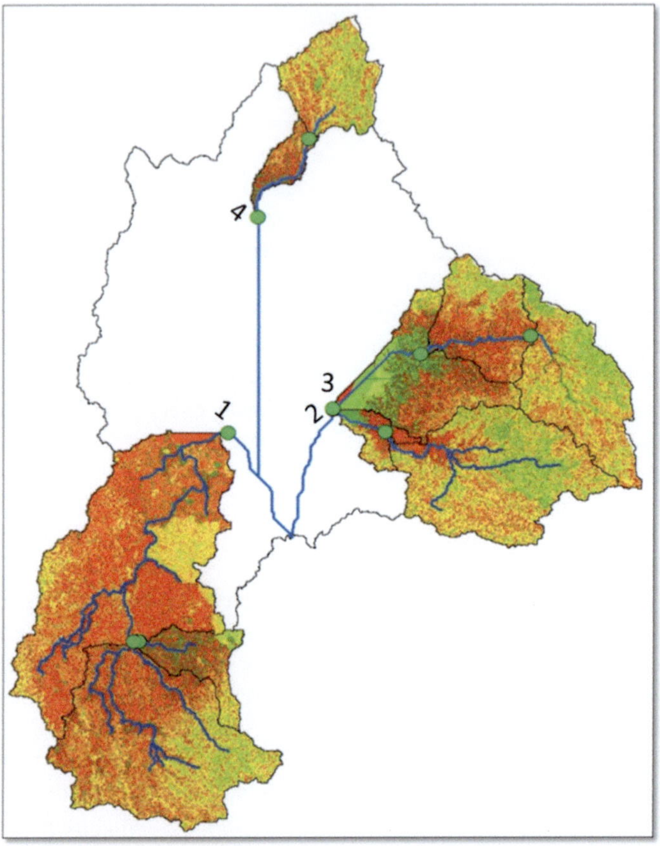

Fig. 22.2 Drainage boundary of the four major rivers in the Tana sub-basin

Table 22.1 Total modeled area of the major rivers

Major River basin	Area (km^2)
Gilgel Abay	4178
Gumera	1418
Ribb	2132
Megech	661
Total model area	**8389**

22.3.2 Groundwater Recharge Analysis

Baseflow is the genetic component of stream flow originating primarily from groundwater, springs, and seeps or other persistent, slowly varying sources. Baseflow is distinguished from surface and/or shallow subsurface runoff, which is generally assumed to be the direct response to a given precipitation event. During

Table 22.2 Calibrated parameters used for SUFI-II

Variable	Definition	[a]Method of calibration	[b]Range/Percent
GW_DELAY	Groundwater delay (days)	Replace	1-180 days
ALPHA_BF	Baseflow alpha factor (days)	Replace	1-180 days
SURLAG	Surface runoff lag time (days)	Replace	1-180 days
GWQMN	Threshold depth of water in the shallow aquifer [m]	Replace	1-200 mm
LAT_TTIME	Lateral flow travel time (days)	Replace	1–180 days
ESCO	Soil evaporation compensation factor	Replace	0.2–0.99
EPCO	Plant uptake compensation factor	Replace	0.2–0.99
CN2	Initial SCS CN II value	Replace	40–85
Depth	Soil layer depths (mm)	Percent	50–150 %
BD	Bulk density moist (g/cc)	Percent	50–150 %
AWC	Average available water (mm/mm)	Percent	50–150 %
KSAT	Saturated conductivity (mm/h)	Percent	50–150 %
RCHRG_DP	Deep aquifer percolation fraction	Replace	0–1.0
REVAPMN	Depth of water in the aquifer for revap [mm]	Replace	0–500 mm
GW_REVAP	Groundwater "revap" coefficient	Replace	0–0.2

[a]"Replace" indicates values were replaced within an initial range published in the literature
[b]"Percent" indicates values were determined by adjusting the base initialization default variables by a certain percentage

the dry season, unmanaged stream flow may be composed entirely of baseflow and thus consists primarily of groundwater discharge. Over all, in an area, the total annual baseflow is considered the total recharge to the groundwater during rainy season of the same hydrological year.

The recharge of groundwater can be estimated by different methods, which includes the following:

- **Catchment Water Balance**: The annual precipitation less annual runoff and annual actual evapotranspiration give change in storage of water underground. Most part of the change in storage is considered as annual recharge. The application of this method requires obtaining actual evapotranspiration data.
- **Baseflow separation/Hydrograph analysis**: This method applies the separation of the dry season flow from mean annual hydrograph of a river at the given hydrometric station. The baseflow, which is normally considered as recharge, is the area below the line/curve that separates baseflow of the mean annual hydrograph.
- **Recharge area separation approach**: This involves delineation of recharge and discharge areas in each groundwater basin and estimation of recharge from precipitation in the area.
- **Groundwater Flow/Darcy Approach**: The product of transmissivity (m^2/year) of the aquifer, the gradient of groundwater table, and the estimated flow channel

width (m) for the aquifer gives the annual recharge of groundwater. The knowledge of the spatial distribution, homogeneity/heterogeneity of the aquifer, and the slope of the groundwater table is critically important in the application of this method.

- **Groundwater table fluctuation**: This method estimates the annual recharge potential from the mean annual water table fluctuations and the aquifer parameter called storativity/storage coefficient, which is specific for each aquifer. According to this method, the product of mean annual water table fluctuation and aquifer storativity is equal to the mean annual recharge. Application of this method requires regular monitoring of groundwater, knowledge of the storativity of the aquifer, and a long-term recording of the water table depth in the aquifer.

In this study, catchment water balance and baseflow separation/hydrograph analysis are implemented successively. The steam flow is generated by the water balance approach, after simulating the four rivers' hydrology using SWAT. SWAT is used to simulate hydrology of the four major rivers of the sub-basin by locating the outlet at the junction of the Lake Tana. With the first approach, the annual water budget of every major river at the junction of the Lake Tana, including recharge to shallow aquifers, deep recharge, and evaporation from the groundwater resource, is evaluated. With the second approach, the annual distribution of the volume of the contribution of the shallow groundwater aquifers to the four major streams of the sub-basin and to the Lake Tana is evaluated.

22.3.3 The Baseflow Separation Method

The aim of baseflow separation is to distinguish two stream flow components: baseflow (groundwater discharging into the stream) and quick flow (surface runoff and interflow). This process provides considerable smoothing. Hence, in the frequency spectrum of a hydrograph, long waves will be more likely to be associated with baseflow while the high-frequency variability of the stream flow will primarily be caused by direct runoff. It should therefore be possible to identify the baseflow by low-pass filtering of the hydrograph. In this study, to estimate the groundwater contribution to the major rivers in the Tana sub-basin, a baseflow separation by recursive digital filtering technique (Eckhardt 2005) is considered. This technique partitions the hydrographs into two components: direct runoff and baseflow, as defined by the following mass balance relation:

$$y_k = f_k + b_k \tag{22.1}$$

where y = total stream flow, f = direct runoff, b = baseflow, and k = time step number.

The shape of a hydrograph is determined by factors such as topography, climate, seasonal variations, bedrock geology, land use, surface water storage such as lakes, and any artificial controls on stream flow. The recursive digital filter applies the baseflow separation process using statistical filtering of the low flows, which are mainly baseflows. The low flows are assumed totally generated from the shallow aquifer groundwater. Accordingly, the recursive filters are also called "infinite impulse response filters," which conduct a geometric progression of a single disturbance that echoes around the feedback loop.

For the purpose of evaluation of the contribution of the baseflow component from stream flow, Eq. (22.2) as described by Nathan and McMahon (1990) is used as follows:

$$f_{k=\propto} f_{k-1} + \frac{(1+\propto)}{2}(y_k - y_{k-1}) \tag{22.2}$$

where f_k is the filtered quick response at the kth sampling instant, y_k is the original stream flow, and α is the filter parameter. The filtered baseflow is thus defined as $y_k - f_k$.

22.4 Result and Discussion

22.4.1 Stream Flows

There is widespread agreement that good correspondence of measured and simulated stream flow at the catchment outlet is not a sufficient criterion for the validity of a physically-based hydrologic model, but that additional knowledge concerning catchment internal processes is needed (Beven and Kirkby 1979). Yet, for many catchments, such information is not available and the model's performance can only be assessed by comparing simulated and measured stream flow. The stream flows at different locations were generated and calibrated with a reasonable level of accuracy measures (Table 22.3). Figure 22.3 shows the modeling result of the flows for the four river basins. Accordingly, the model indicates reasonable agreement that would be fairly enough to generate information for regional planning. The Nash-Sutcliffe

Table 22.3 Efficiency of estimation of the flows of the SWAT model

Major river catchment	Gauging station	R^2	NSE
Gilgel Abay	Gilgel Abay @Pikolo Bridge	0.70	0.67
	Koga @Merawi	0.40	0.30
Gumera	Gumera @ Bridge	0.61	0.58
Ribb	Upper Ribb	0.38	0.37
	Lower Ribb @ Bridge	0.69	0.62
Megech	Megech @Azezo	0.38	0.37

Efficiency (NSE) (Nash and Sutcliffe 1970) is used to evaluate the performance of the model in replicating the observed data for the four contributing river basins.

The NSE is a normalized statistic that determines the relative magnitude of the residual variance ("noise") compared to the measured data variance ("information") (Nash and Sutcliffe 1970). NSE indicates how well the plot of observed versus simulated data fits the 1:1 line. Nash-Sutcliffe efficiencies range from $-\infty$ to 1. Essentially, the closer to 1, the more accurate the model is. NSE of 1 corresponds to a perfect match of modeled to the observed data. NSE of zero indicates that the model predictions are as accurate as the mean of the observed data, and values of $-\infty < \text{NSE} < 0$ indicate that the observed mean is better predictor than the model.

It is shown that the NSE values for flow simulated at Gilgel Abay @Bikolo Bridge, Gumera @ Bridge, and Lower Ribb @ Bridge are higher compared to the others indicating good model performance. Figure 22.3 shows predicted and observed flows for the four rivers at gauging stations and at the mouth of the rivers (lake entry point).

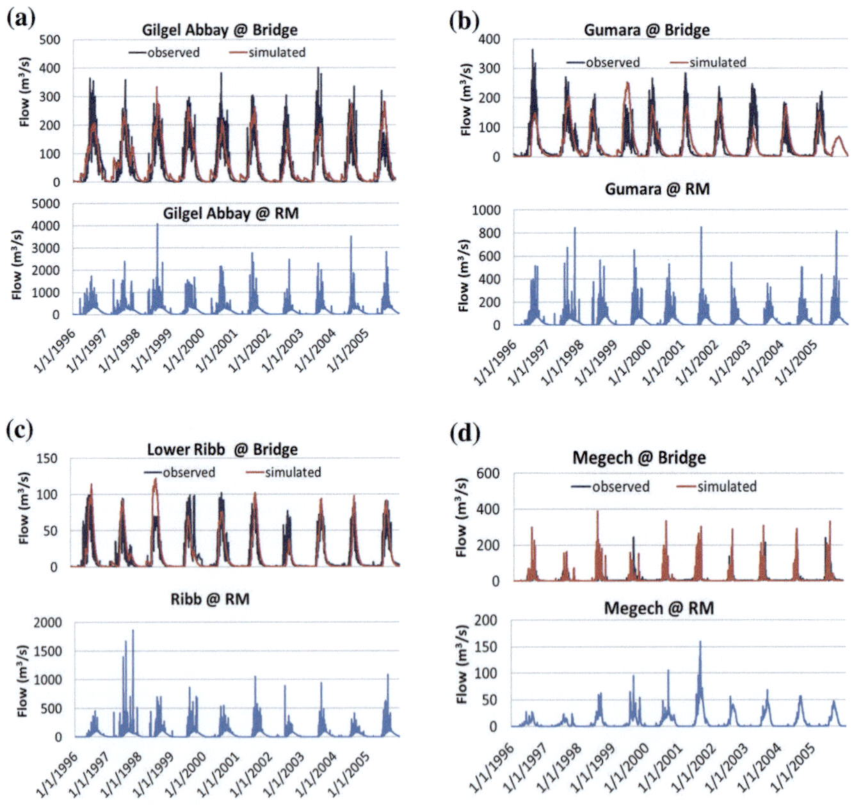

Fig. 22.3 Simulated flow based on calibrated SWAT at flow gauging stations and river mouths (RM): **a** Gilgel Abay, **b** Gumara, **c** Ribb, and **d** Megech Rivers

Flow estimation at the bridge of every river represents the gauging stations as represented by green dots in the upstream of every river catchment in Fig. 22.3. Flow simulation at the "river mouth" (RM) represents the flow estimated at the junction of the lake, as referred by numbers 1, 2, 3, and 4 for rivers of Gilgel Abay, Gumara, Ribb, and Megech, respectively, in Fig. 22.2.

22.4.1.1 A Recession Analysis/Baseflow Separation

The results of the application of the model for the four major rivers in the Tana sub-basin by using the flow data generated by the SWAT inverse model are presented in Figs. 22.4, 22.5, 22.6, and 22.7. As described by Nathan and McMahon (1990), the most acceptable baseflow separation conducted by several datasets was in the range of 0.9–0.95. For better evaluation of the baseflow separation, it is important that the low flows of the SWAT dataset are equal to the baseflow separation results of the TIMPLOT. Hence, several trials are applied to define the value of the filter parameter.

Two interrelated curves are generated for every river flow at the junction of Lake Tana as shown in Figs. 22.4, 22.5, 22.6, and 22.7. The first curve is a detailed view of the curves' overlap as the digital filter separates the low flow from the high flow. The importance of this model is to observe how the curves of the baseflow and total flow at the low flow and return flow conditions fit, so that the digital filter can explicit the total groundwater contribution during the high flow period. The assumption in this case is that the low flows and return flow that generates the recession curves are fully the contributions of the groundwater recharge.

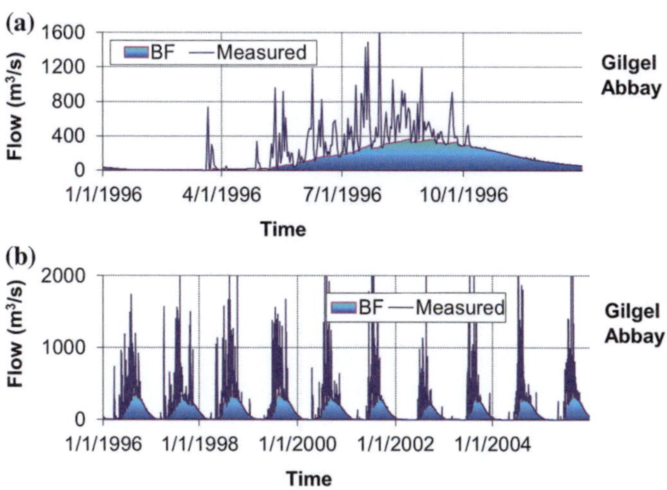

Fig. 22.4 Baseflow separation for Gilgel Abay River basin: **a** curve fitting for the recession and low flows and **b** ten years, 1996–2005, daily flow at the mouth of the river

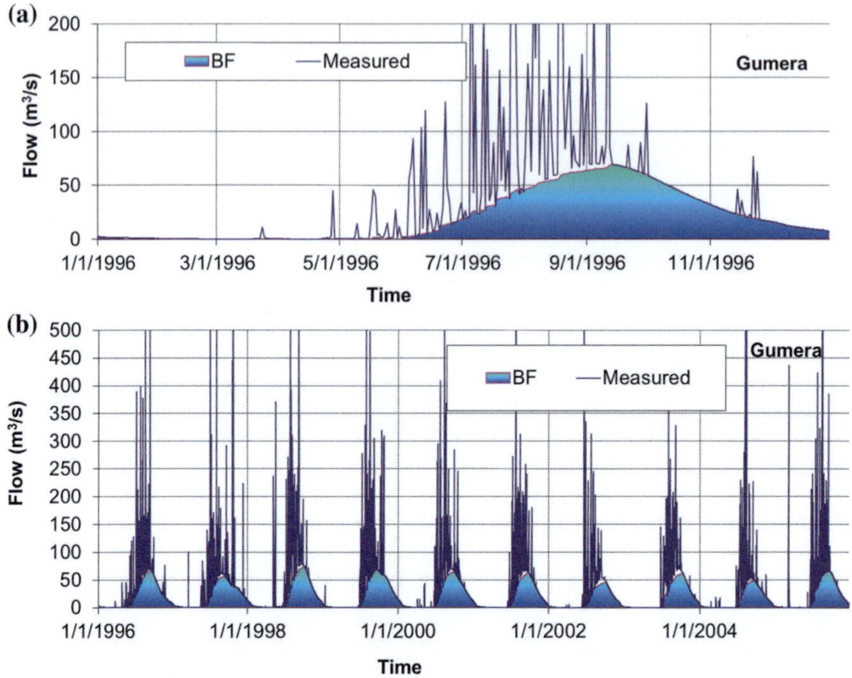

Fig. 22.5 Baseflow separation for Gumera River basin: **a** curve fitting for the recession and low flows and **b** ten years, 1996–2005, daily flow at the mouth of the river

22.4.2 Baseflow Trends of the Rivers

As indicated in Fig. 22.8, the Gilgel Abay River has the largest share of baseflow contribution to the Lake Tana. Also, the baseflow comparison curve entails that the Gilgel Abay flow persists longer for an extended period after the end of the rainy seasons.

Figure 22.9 indicates that three of the rivers show a declining trend of baseflow contribution to the Lake Tana over the recorded period. Megech River shows a slight increase in the baseflow contribution. The Angereb River reservoir is upstream of the Megech outflow point to Lake Tana, and this might have some contribution to the groundwater contribution to the lake.

The slope of the baseflow considered as the rate of change of baseflow decline or increase in the rivers entails that the Gilgel Abay River has the largest rate of decline of baseflow contribution. Considering Gilgel Abay is the largest contributor and the fastest running to dry, the groundwater contribution in the Lake Tana is declining considerably.

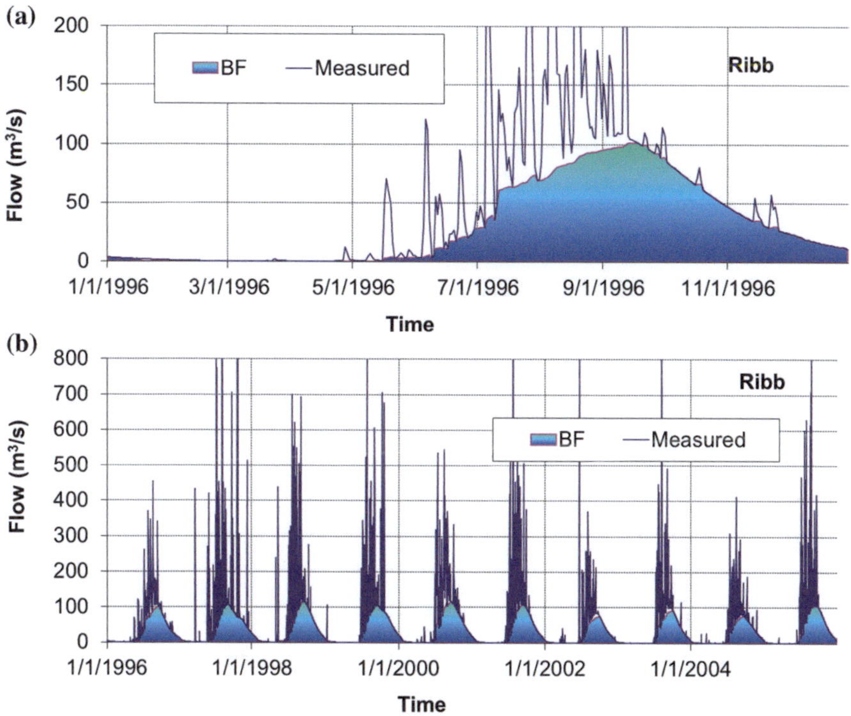

Fig. 22.6 Baseflow separation for Ribb River basin: **a** curve fitting for the recession and low flows and **b** ten years, 1996–2005, daily flow at the mouth of the river

22.4.3 Water Balance and Groundwater Contribution

The catchment water balance components help to evaluate annual recharge to deep groundwater aquifers, evaporation loss from the recharge water, and baseflow and interflow from shallow aquifers. This water balance, incorporating surface water parameters and deep groundwater recharge, is developed from the summary of the ten-year average water balance results from the calibrated SWAT model (Table 22.4).

The groundwater contribution is the sum of the lateral flow (Latf) from the vadose zone, and return flow (Retf) contribution from the shallow aquifers zone. According to this, the total annual inflow into the lake, from the major rivers catchment area, is 1951.36 mm. This is generated from the total catchment area of the four major rivers of 8389 km^2. Out of the total of 1951.36 mm, 717.94 mm is generated from the Gilgel Abay River catchment. Despite its smallest contribution of the groundwater to the lake, Megech has the highest share of the lateral flow contribution. It is remarkable that lateral flow is a shallow aquifers contribution and can be enriched by different practices of proper soil and water conservation exercises.

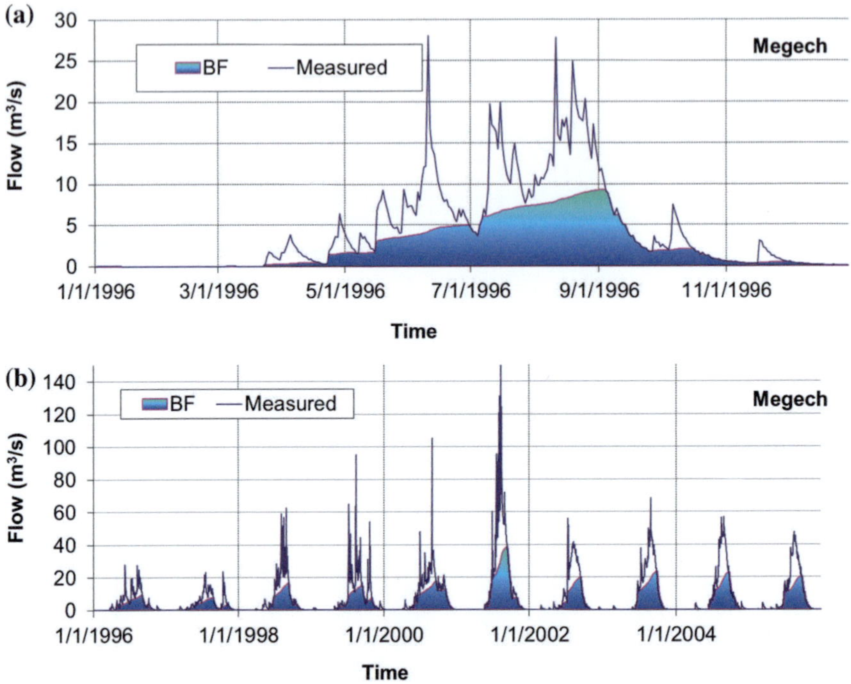

Fig. 22.7 Baseflow separation for Megech River basin: **a** curve fitting for the recession and low flows and **b** ten years, 1996–2005, daily flow at the mouth of the river

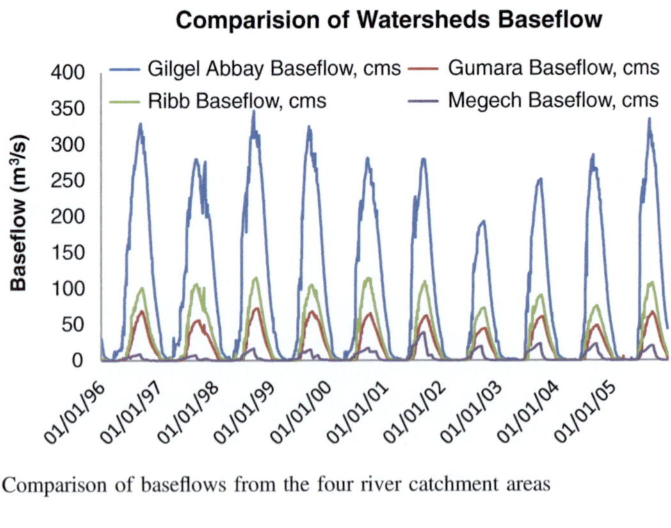

Fig. 22.8 Comparison of baseflows from the four river catchment areas

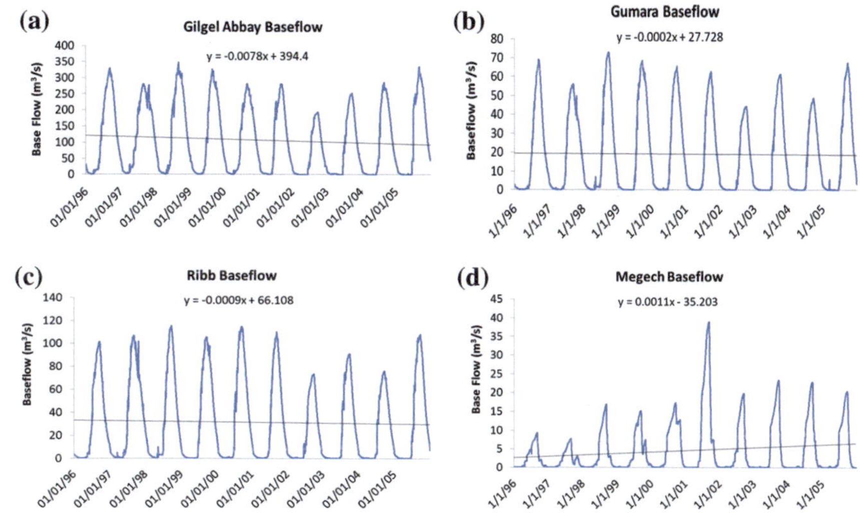

Fig. 22.9 Temporal trend of baseflows: **a** Gilgel Abay, **b** Gumara, **c** Ribb, and **d** Megech Rivers

Table 22.4 Annual water balance of the major river basin at the junction (mouth of the river or lake entry point) of the Lake Tana

Major River Catchment	Annual water balance components (mm)								
	Pcp	Runoff	Latf	Perc	Retf	Rdep	Revap	ET	Total groundwater contribution
Gilgel Abay	2140.6	798.42	16.08	753.04	701.86	37.65	13.21	566.9	717.94
Gumera	1423.7	398.93	27.68	437.64	404.00	21.88	11.76	557.1	431.68
Ribb	1407.1	392.66	31.84	453.25	419.45	22.66	11.14	531.2	451.29
Megech	1250.8	21.35	229.6	407.16	120.46	15.22	75.71	593.71	350.06

Pcp precipitation, *Latf* lateral flow, *Perc* percolation, *Retf* return flow, *Rdep* deep recharge, *Revap* evaporation from shallow surface, and *ET* evapotranspiration

It was found that the deep groundwater recharge seems smaller compared to the shallow aquifer contribution. A study by Ministry of Water and Energy on the detailed groundwater investigation in Lake Tana and Beles River sub-basins indicates that the geological structures in all of the four river catchments are intricate and deep groundwater migration out of the Tana sub-basin is suspected. This is observed in the southern part at the boundary of Tana and Beles sub-basins. The large section of the Lake Tana sub-basin lacks detailed geological information for a better understanding of the deep groundwater circulation in the sub-basin.

22.5 Conclusion and Recommendations

The SWAT model was used to estimate flows at the mouths of four major rivers and hence their contribution of the hydrology of Lake Tana. The baseflow component of this contribution was partitioned, and the groundwater flow recharge was estimated. It was found that the groundwater contribution from the total annual inflow into the lake is up to 60 %. Also, it is indicated that the groundwater contribution from the Gilgel Abay River basin is the highest of all the four rivers. Other rivers have relatively less but considerable volume of groundwater contribution to the lake water body.

The baseflow contribution trends of the rivers in the Tana sub-basin indicated a decreasing pattern over time. The highest groundwater contributor, Gilgel Abay River, also showed the highest rate of decline of baseflow. This implied not only a decline in the groundwater reserve zone in the area, but also a gradual limitation of the annual recharge to the Lake Tana. This also shows the aquifer-dependent ecosystem of the area is highly threatened by the declining pattern of the streams' low flow condition.

Land management practice with focus on restoration of upland recharge is necessary for reducing the negative consequences of seasonal water shortages in the upstream catchments of the Tana sub-basin. As long as sustainability of the water resources is a major issue of concern in the area, we recommend that the watershed management practices in the region should reinforce the realism of sustainable recharging mechanisms. Also, we also recommend the importance of acknowledging groundwater resources monitoring in the area. Monitoring and legal enforcement of participation of all concerned stockholders are important to a sound groundwater information management system, so that the monitoring can be effusively utilized to devise timely solutions and management directions.

Acknowledgments This hydrological model is developed by the use of data collected from the International Water Management Institute Addis Ababa. The World Bank funds the Tana Beles Integrated Water Resources Development Project. The results of the SWAT model are consulted with Professor Tammo S Steenhuis, Drs Zach Easton, and Daniel Fucka.

References

Abtew W, Melesse AM (2014a) Nile River Basin hydology. In: Melesse AM, Abtew W, Setegn S (eds) Nile River Basin: ecohydrological challenges, climate change and hydropolitics, pp 7–22

Abtew W, Melesse AM (2014b) Climate teleconnections and water management. In: Nile River Basin, Springer International Publishing, Berlin, pp. 685–705

Abtew W, Melesse AM (2014c) Transboundary rivers and the Nile. In: Nile River Basin, Springer International Publishing, Berlin, pp. 565–579

Abtew W, Melesse AM, Desalegn T (2009a) Spatial, inter and intra-annual variability of the Blue Nile River Basin rainfall. Hydrol Process 23(21):3075–3082

Abtew W, Melesse AM, Desalegn T (2009b) El Niño southern oscillation link to the Blue Nile River Basin hydrology. Hydrol Process Spec Iss: Nile Hydrol 23(26):3653–3660

Arnold JG, Srinivason R, Muttiah RR, Williams JR (1998) Large area hydrologic modeling and assessment part I: model development. J Am Water Resour Assoc 34(1):73–89

Assefa A, Melesse AM, Admasu S (2014) Climate change in Upper Gilgel Abay River catchment, Blue Nile Basin Ethiopia. In: Melesse AM, Abtew W, Setegn S (eds) Nile River Basin: ecohydrological challenges, climate change and hydropolitics, pp 363–388

BCEOM (1999) Abay River Basin integrated master plan, main report. Ministry of Water Resources, Addis Ababa

Behulu F, Setegn S, Melesse AM, Fiori A (2013) Hydrological analysis of the Upper Tiber Basin: a watershed modeling approach. Hydrol Process 27(16):2339–2351

Behulu F, Setegn S, Melesse AM, Romano E, Fiori A (2014) Impact of climate change on the hydrology of Upper Tiber River Basin using bias corrected regional climate model. Water Res Manage 1–17

Beven KJ, Kirkby MJ (1979) A physically-based, variable contributing area model of basin hydrology. Hydrol Sci Bull 24:43–69

Chebud YA, Melesse AM (2009a) Numerical modeling of the groundwater flow system of the Gumera Sub-Basin in Lake Tana Basin, Ethiopia. Hydrol Process Spec Iss: Nile Hydrol 23 (26):3694–3704

Chebud YA, Melesse AM (2009b) Modeling lake stage and water balance of Lake Tana, Ethiopia. Hydrol Process 23(25):3534–3544

Chebud Y, Melesse AM (2013) Stage level, volume, and time-frequency change information content of Lake Tana using stochastic approaches. Hydrol Process 27(10):1475–1483. doi:10.1002/hyp.9291

Dessu SB, Melesse AM (2012) Modeling the rainfall-runoff process of the Mara River Basin using SWAT. Hydrol Process 26(26):4038–4049

Dessu SB, Melesse AM (2013) Impact and uncertainties of climate change on the hydrology of the Mara River Basin. Hydrol Process 27(20):2973–2986

Dessu SB, Melesse AM, Bhat M, McClain M (2014) Assessment of water resources availability and demand in the Mara River Basin. CATENA 115:104–114

Eckhardt K (2005) How to construct recursive digital filters for baseflow separation. Hydrol Process 19:507–515

Getachew HE, Melesse AM (2012) Impact of Land use/land cover change on the hydrology of Angereb Watershed, Ethiopia. Int J Water Sci 1(4):1–7. doi:10.5772/56266

Grey OP, Webber DG, Setegn SG, Melesse AM (2013) Application of the soil and water assessment tool (SWAT Model) on a small tropical island state (Great River Watershed, Jamaica) as a tool in Integrated Watershed and Coastal Zone Management. Int J Trop Biol Conserv 62(3):293–305

Kebede S, Admasu G, Travi Y (2011) Estimating ungauged catchment flows from Lake Tana floodplains (Ethiopia): isotope hydrological approach. Isotop Environ Health Stud 47:1–16

Mango L, Melesse AM, McClain ME, Gann D, Setegn SG (2011a) Land use and climate change impacts on the hydrology of the upper Mara River Basin, Kenya: results of a modeling study to support better resource management, special issue: climate, weather and hydrology of East African Highlands. Hydrol Earth Syst Sci 15:2245–2258. doi:10.5194/hess-15- 2245-2011

Mango L, Melesse AM, McClain ME, Gann D, Setegn SG (2011b) Hydro-meteorology and water budget of Mara River basin, Kenya: a land use change scenarios analysis. In: Melesse A (ed) Nile River Basin: hydrology, climate and water use. Springer, Berlin, pp 39–68. doi:10.1007/978-94-007-0689-7_2

Melesse AM (2011) Nile River Basin: hydrology, climate and water use. Springer Science & Business Media, Berlin

Melesse AM, Loukas AG, Senay G, Yitayew M (2009a) Climate change, land-cover dynamics and ecohydrology of the Nile River Basin. Hydrol Process Spec Iss: Nile Hydrol 23(26):3651–3652

Melesse AM, Abtew W, Desalegne T, Wang X (2009b) Low and high flow analysis and wavelet application for characterization of the Blue Nile River system. Hydrol Process 24(3):241–252

Melesse AM, Abtew W, Setegn S, Dessalegne T (2011) Hydrological variability and climate of the Upper Blue Nile River Basin In: Melesse AM (ed) Nile River Basin: hydrology, climate and water use. Springer, Berlin, pp 3–37. doi:10.1007/978-94-007-0689-7_1

Melesse A, Abtew W, Setegn SG (2014) Nile River Basin: ecohydrological challenges, climate change and hydropolitics. Springer Science & Business Media, Berlin

Mohammed H, Alamirew T, Assen M, Melesse AM (2015) Modeling of sediment yield in Maybar gauged watershed using SWAT, northeast Ethiopia. CATENA 127:191–205

Nash JE, Sutcliffe JV (1970) River flow forecasting through conceptual models part I—a discussion of principles. J Hydrol 10(3):282–290

Nathan RJ, McMahon TA (1990) Evaluation of automated techniques of baseflow and recession analyses. Water Resour Res 26(7):1465–1473

Schneiderman EM, Steenhuis TS, Thongs DJ, Easton ZM, Zion MS, Mendoza GF,Walter MT, Neal AL (2007) Incorporating variable source area hydrology into the curve number based generalized watershed loading function model. Hydrol Process 21:3420–3430

Setegn SG, Srinivasan R, Dargahi B, Melesse AM (2009a) Spatial delineation of soil erosion prone areas: application of SWAT and MCE approaches in the Lake Tana Basin, Ethiopia. Hydrol Process Spec Iss: Nile Hydrol 23(26):3738–3750

Setegn SG, Srinivasan R, Melesse AM, Dargahi B (2009b) SWAT model application and prediction uncertainty analysis in the Lake Tana Basin, Ethiopia. Hydrol Process 24(3):357–367

Setegn SG, Dargahi B, Srinivasan R, Melesse AM (2010) Modelling of sediment yield from Anjeni Gauged Watershed, Ethiopia using SWAT. JAWRA 46(3):514–526

Setegn S, David R, Melesse AM, Bijan D, Ragahavan S (2011) Impact of climate change on the hydro-climatology of Lake Tana basin, Ethiopia. Water Resour Res 47, W04511. doi:10.1029/2010WR009248

Setegn SG, Melesse AM, Haiduk A, Webber D, Wang X, McClain M (2014) Spatiotemporal distribution of fresh water availability in the Rio Cobre Watershed, Jamaica. CATENA 120:81–90

SMEC (2008) Hydrological study of the Tana-Beles Sub Basins. Technical report, Ministry of Water Resources, Addis Ababa

Sogreah and Geomatrix (2013) Ministry of Water and Energy Consulting Services for Detailed Groundwater Investigations & Monitoring In Tana and Beles Sub-Basins Final Stage 1 Report —Volume Ii—Part 5: Hydrological Survey

SWAT-CUP (2009) SWAT Calibration and uncertainty programs—a user manual

Tenalem A, Molla D, Stefan W (2008) Hydrogeological framework and occurrence of groundwater in the Ethiopian aquifers. J Afr Earth Sc 52(2008):97–113

Wang X, Melesse AM (2005) Evaluations of the SWAT model's snowmelt hydrology in a Northwestern Minnesota Watershed. Trans ASAE 48(4):1359–1376

Wang X, Melesse AM (2006) Effects of STATSGO and SSURGO as Inputs on SWAT Model's Snowmelt Simulation. J Am Water Resour Assoc 42(5):1217–1236

Wang X, Melesse AM, Yang W (2006) Influences of potential evapotranspiration estimation methods on SWAT's hydrologic simulation in a Northwestern Minnesota Watershed. Trans ASAE. 49(6):1755–1771

Wang X, Shang S, Yang W, Melesse AM (2008a) Simulation of an agricultural watershed using an improved curve number method in SWAT. Tans Am Soc Agri Biol Eng 51(4):1323–1339

Wang X, Yang W, Melesse AM (2008b) Using hydrologic equivalent wetland concept within SWAT to estimate streamflow in watersheds with numerous wetlands, Tans Am Soc Agri Biol Eng 51(1):55–72

Wang X, Garza J. Whitney M. Melesse AM, Yang W (2008c) Prediction of sediment source areas within watersheds as affected by soil data resolution. In: Findley PN (ed) Environmental

modelling: new research. Nova Science Publishers, Inc., Hauppauge., pp 151–185. ISBN: 978-1-60692-034-3)
Yitayew M, Melesse AM (2011) Critical water resources management issues in Nile River Basin, In: Melesse AM (ed) Nile River Basin: hydrology, climate and water use. Springer, Berlin, pp 401–416. doi:10.1007/978-94-007-0689-7_20
Zemedagegnehu E, Sileshi Y, Tuinhof A (2007) Groundwater resources in Lake Tana Sub-Basin and adjacent areas. World Bank Rapid Assessment

Part V
Sediment Dynamics and Soil Management

Chapter 23
Sediment Production in Ravines in the Lower Le Sueur River Watershed, Minnesota

Luam A. Azmera, Fernando R. Miralles-Wilhelm and Assefa M. Melesse

Abstract Ravine erosion is an important occurrence of soil erosion processes. Although its sediment yield is significant, ravine erosion is not usually accounted in routines for predicting soil loss. This study focuses on quantifying explicitly the sediment budget of deeply incised ravines in the lower Le Sueur River watershed, in southern Minnesota. High-rate-gully-erosion equations along with the universal soil loss equation (USLE) were implemented in a numerical modeling approach that is based on a time integration of the sediment balance equations. The model estimates the rates of ravine width and depth change and the amount of sediment periodically flushing from the ravines. Using ArcGIS and ArcHydro tools, topographic characteristics of the study ravines were derived from the LiDAR-based digital elevation model (DEM) of the area. A 30-year survey and sediment data from similar research was used as a case study in an attempt to justify the theoretical framework of the model and characterize model parameters. The model output for the case study was found to compare favorably with the estimates found in previously reported studies. Components of the sediment budget in the study ravines were simulated with the model and results suggest that the ravine walls are the major sediment source in the ravines. A sensitivity analysis revealed that the erodibility coefficients of the gully bed and wall, the local slope angle, and the Manning's coefficient are the key parameters controlling the rate of sediment production. Recommendations to guide further monitoring efforts in the watershed and increased detail modeling approaches are highlighted as a result of this modeling effort.

L.A. Azmera (✉)
Department of Biological and Environmental Engineering, College of Agriculture and Life Sciences Cornell University, Ithaca, NY, USA
e-mail: Laa89@cornell.edu

F.R. Miralles-Wilhelm · A.M. Melesse
Department of Earth and Environment, Florida International University, Miami, FL, USA
e-mail: Miralles@fiu.edu

A.M. Melesse
e-mail: melessea@fiu.edu

© Springer International Publishing Switzerland 2016

A.M. Melesse and W. Abtew (eds.), *Landscape Dynamics, Soils and Hydrological Processes in Varied Climates*, Springer Geography,
DOI 10.1007/978-3-319-18787-7_23

Keywords Ravine erosion · Le Sueur river · Sediment transport, USLE · LiDAR · GIS

23.1 Introduction

Soil erosion has long been considered to accelerate over the development of human activity in history. Soil erosion caused by water has been one of the prominent processes causing reduced soil quality and reduced water-holding capacity of agricultural areas leading to decreased agricultural yield. Soil erosion also has off-site impact on larger spatial scales than the agricultural fields. Sediment yield from agricultural watersheds causes sedimentation in watercourses and reservoirs, damaging engineering structures, and rapidly increase water turbidity. Sediments are also carriers of pollutants such as nutrients, pathogens, and toxic substances. Increased sediment loading to watercourses and reservoirs had resulted in poor water quality in the USA. According to the US Environmental Protection Agency (EPA)'s most recent list, there are about 40,000 impaired water bodies in USA. Sediment and nutrients together are the major concern for approximately 13,000 of these water bodies. Soil conservation planning and development of effective sediment control strategies hence are the main constraints in catchment management planning.

A number of studies used laboratory, field scales, and modeling studies to understand soil erosion and sediment dynamics in various regions (Defersha and Melesse 2012a, b; Defersha et al. 2011, 2012; Maalim and Melesse 2013; Maalim et al. 2013; Setegn et al. 2010; Melesse et al. 2011; Msagahaa et al. 2014; Wang et al. 2008; Mekonnen and Melesse 2011; Setegn et al. 2009; Mohammed et al. 2015). These studies found that the effect of land management is critical in reducing soil erosion and sediment flux. Critical to the development of such management systems is the identification of the potentially significant sources and quantifying the sediment yield from each source. The Le Sueur River watershed in Minnesota is one example where the hydrology of the system had been artificially altered for agricultural purposes and its geomorphic characteristics continue to be naturally modified. Large amounts of sediments are being produced in this watershed and loading to the Le Sueur River, increasing the turbidity of the watercourse.

The Le Sueur River is located in the south-central part of the state, (Fig. 23.1). Its watershed is one of the twelve major watersheds of the Minnesota River Basin. According to reports of Minnesota Pollution Control Agency (MPCA), significant stretches of the Le Sueur and Minnesota Rivers are claimed to be turbid under the Clean Water Act. Recent sediment gauging efforts indicate that the Le Sueur River is the primary sediment contributor (24–30 %) of the Minnesota River (MPCA et al. 2007). The Minnesota River is also one of the major tributaries of the upper Mississippi River (Fig. 23.1). The Minnesota River contributes 85–90 % of suspended sediment to Lake Pepin (Kelley and Nater 2000). Lake Pepin is a natural

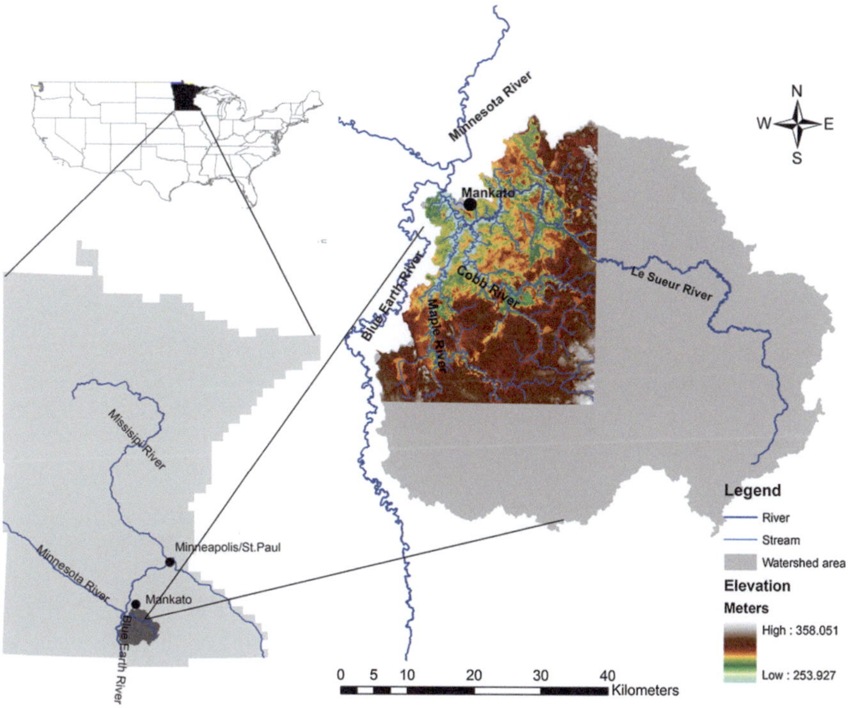

Fig. 23.1 Geographic location of the Le Sueur River watershed

impoundment 80 km downstream of the metropolitan area of Minneapolis–St. Paul along the Mississippi River. Lake Pepin is an important recreational and commercial resource for the region. However, its impaired water quality has recently become a serious concern for pollution control agencies such as MPCA. Analyses of sediment cores in Lake Pepin indicate that sediment loads in the lake have increased tenfold since the onset of European settlement in the mid-1800s (Engstrom et al. 1997).

In the Le Sueur and Minnesota Rivers, turbidity levels are high and call for management actions. Understanding the location and magnitude of sediment sources is essential for guiding management decision to reduce sediment loading and improve water quality. As a first step to quantifying the sediment budget of the river watershed, an effort has been carried out to define the potential sediment source locations. A study by Gran et al. (2009) shows that the major sediment sources to the Le Sueur River are upland-derived sediment, high bluffs, terraces, and large permanent gullies–ravines. Figure 23.2 shows sediment sources on digital elevation model (DEM) of lower Le Sueur valley.

The next step in understanding the sediment dynamics of the Le Sueur River would be to establish an integrated sediment budget of the river watershed. Once the sediment sources are identified and the sediment budget components are quantified, best management practices can be introduced to reduce the sediment

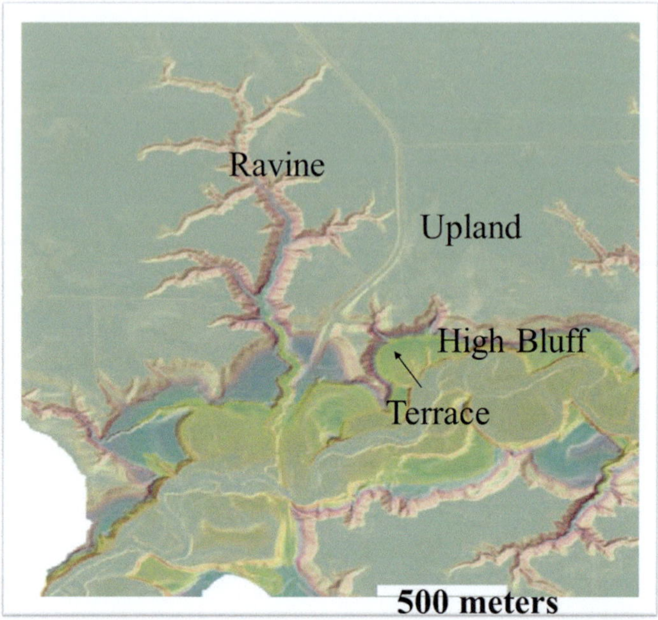

Fig. 23.2 Sediment sources in lower Le Sueur valley LiDAR. *Source* Gran et al. (2009)

loading. This, however, needs a refined estimate of sediment produced from each source. The erosion processes in the main three identified sources need to be addressed individually so that the integrated sediment budget would give a better understanding of the sediment dynamics in the watershed.

Ravine erosion is an important soil erosion process in the Le Sueur River valley, because deeply incised ravines are prevalent features in this part of Minnesota. Although their sediment contribution is believed to be significant, scarce quantitative research on the area of ravine erosion has been done so far in this area. Ravine erosion is not usually accounted in routines schemes for predicting soil loss from watersheds. The development stages, rate of growth, and the quantification of the sediment volume produced from ravines is also an important topic in erosion research; however, limited number of erosion and sediment transport models attempt to explicitly describe these processes. This might be because many previous studies dealing with soil erosion by water had concentrated on rill and interrill erosion. For this reason, it seemed relevant to undertake a new study of ravine erosion with special attention to quantifying the sediment budget of ravines.

The Soil Science Society of America defines permanent gullies as channels often too deep to easily ameliorate with ordinary farm tillage equipment, typically ranging from 0.5 to as much as 25–30 m depth (Soil Science Society of America 2001). Ravines in the lower Le Sueur valley have similar geomorphic features as permanent gullies. But the ravines in this area seem to have a deeper and wider valley, they connect relatively larger areas in the landscape, and they have a vegetation cover of dense trees. There is also considerable erosion activity within the ravine valleys which make them a potential major sediment producing source in the Le Sueur River watershed. A more detailed description of the study ravines is given in the following sections of this report.

Despite their minor morphological differences, the general theory of sediment dynamics of wide and deep permanent gullies is believed to apply to ravines. Therefore, based on the review of previous studies of gully erosion, this study provides an insight to the sediment dynamics.

Thus, the objective of this study is to (1) identify the main sediment sources inside ravines to help constrain the contribution of different sources, (2) estimate rates of sediment production inside ravine, in an effort to quantify the contribution of ravine sediment load to the Le Sueur River, and (3) calculate the rate of ravine growth in terms of its rate of change in width and depth.

23.2 Gully Erosion

Gully development and the rate of erosion in gullies is a well-documented topic in erosion research. Most previous studies dealing with soil erosion by water had concentrated on rill and interill erosion. Recent studies, however, have given attention to gully erosion and consider gullies as another possible substantial source of sediments. According to the review by Poesen et al. (2003) and the data collected in 56 different catchments located in different parts of the world, soil loss by gully erosion accounted from 10 to 94 % of total sediment yield caused by water. Moreover, in a review of the fingerprinting method of identifying the origin of sediments within catchments to determine the relative contribution of potential sources, the contribution of gully erosion accounted 80 % in Australia, 90–98 % in New South Wales, 60–70 % in Chinese Loess Plateau, and 70 % in an Ethiopian highland (Valentin et al. 2005).

In addition to being a substantial sediment source, gullies also aggravate water erosion by increasing the connectivity in the landscape hence promoting redistribution of sediment within the catchment and effective delivery of sediment from uplands to lowlands and water courses (Poesen et al. 2003). The eroded sediment volume also increases with the density of active gullies within a catchment. For example, to study the impact of the presence of concentrated active gullies on the specific sediment yield of a catchment, Poesen et al. (2003) used reservoir sedimentation data of 22 selected Spanish catchments and survey of gullies within a 5 km radius of the reservoir or river channels draining to the reservoir. The study found that for catchments with no gullies, the mean specific yield was 0.74 ton ha^{-1} $year^{-1}$ and for those with numerous gullies, it was 9.61 ton ha^{-1} $year^{-1}$, and catchments with some gullies had mean specific yield of 2.97 ton ha^{-1} $year^{-1}$.

23.2.1 *Gully Types*

The two main types of gullies are ephemeral gully and permanent or classical gullies. As defined by the Soil Science Society of America, ephemeral gullies are small channels eroded by concentrated overland flow that can be easily filled by normal tillage only to reform again in the same location by additional runoff events. Permanent gullies on the other hand are permanent features in the landscape and are often defined for agricultural land in terms of channels too deep to easily ameliorate with ordinary farm tillage equipment (Soil Science Society of America 2001). Poesen et al. (2003) summarize some criteria used to distinguish the rills from gullies, such as "the square foot criterion" Poesen (1993): a gully has a minimum cross-sectional area of 929 cm^2; Brice (1966): a minimum gully depth and width criteria of 0.3 and 0.6 m; and Imeson and Kwaad (1980): a minimum gully depth criterion of 0.5 meters. However, there is no specific clear-cut definition to the upper limit of gullies.

23.2.1.1 Gully Development

Gully development is a threshold phenomenon which is controlled by temporal changes in flow hydraulics, rainfall, soil type, land use, topography, climate, and weather (Poesen et al. 2003; Valentin et al. 2005). Sidorchuk (1999) explains that the main causes of gully initiation in a landscape are anthropogenic factors: changes in land use such as clearing of native forests, tilling of fallow lands, and associated change of hydrological conditions in their rainfall- runoff system. According to Sidorchuk (1999), gully development has two stages: the initial stage where the gully development rate is very rapid and the last stage where the gully size is near stable and reaches its maximum value. In the initial stage, gully morphological characteristics are not stable, the hydraulic erosion is predominant and rapid mass movement occurs on the gully sides and gully bottom. In a second stage, the rate of gully development decreases, and the gully is assumed to be in its final morphological equilibrium. At this stage, sediment transport and sedimentation are the main erosion processes in the gully, its width increases due to lateral erosion and slow mass movement transforms the gully sides. The last stage occupies the largest part of a gully's life time, whereas the initial stage accounts only 5 % of gully's life time. Major morphological characteristics of the gully, however, are formed during the initial stage (Sidorchuk 1999).

23.2.2 *Controlling Factors*

The magnitude of soil loss from gullies highly depends on the size of the study area considered. For study areas ranging between 1 and 10 ha or more, gully erosion becomes important and its contribution to sediment yield might be more than that of rill and interill erosion (Poesen et al. 2003). Furthermore, environmental factors such as topography, soil type, land use, gully type, climate, and weather also control gully erosion. The magnitude of gully erosion in a study area also fluctuates depending on the time scale of the study period. The variation may be attributed to the changes in land use and other environmental factors during the study period (Poesen et al. 2003).

23.2.3 Techniques of Measuring Gully Erosion

Several attempts have been used in the past to estimate and monitor gully erosion. In earlier studies such as by Crouch (1990), gully erosion rate measurements involved the use of erosion pins and ground surveys, where the study area would be surveyed in defined intervals of time. However, this method was found to be time-consuming and difficult to apply to the study of large areas with high gully density over a long period of time. In more recent studies of short-term monitoring of gully-headcut or gully-wall retreat, Vandekerckhove et al. (2001a, b) regularly measured the distance between the edge of the gully head or gully wall and benchmark pins. In an effort to study long-term growth of valley-bottom gully, Thomas et al. (2004) used annual surveys of the gully perimeter over 30 years and produced a three-dimensional surface for each topographic survey to estimate the increase in gully volume.

Photogrammetric techniques were also used in other gully erosion researches. In the studies by Betts and DeRose (1999) and Martinez-Casanovas (2003) for example, DEMs were constructed from sequential aerial photographs for measuring and monitoring the volume of sediment lost by gully erosion in a geomorphologically unstable environment. For medium-term scale, an analysis of high-altitude aerial photographs in combination with ground measurements was also used to quantify the volumetric gully-head retreat rates for permanent gullies in Spain (Vandekerckhove et al. 2003).

Martínez-Casasnovas et al. (1998) also applied geographic information system (GIS) techniques to analyze DEM of 25 m resolution derived from multi-date aerial photographs to assess erosion rates in the gully system of northeast Spain. The gully system in this study is characterized by vertical sidewalls and is 11–60 m deep and 75–350 m wide. The study found out that the linear retreat rate of gully walls and maximum rate of channel incision was in the order of meters per year, while the average sediment production rate was in the order of tones ha^{-1} $year^{-1}$.

In studies carried out in small catchments in Qiaogou, China, Wu and Cheng (2005) used a high-accuracy Global Positioning System (GPS) to measure the morphological parameters of gullies to investigate the short-term erosion rates of hill slope gullies, slope–area relationships, and thresholds of hill slope gully initiation.

A method based on dendrochronology was also developed as an alternative to the traditional methods mentioned above. This method uses trees impacted by gully erosion revealing information on the history of the erosion process by datable deviations of their normal growth pattern (Vandekerckhove et al. 2001a). However, the authors report that methodological problems limit the application of the dendrochronological estimation of gully erosion.

23.2.4 Gully Erosion and Sediment Transport Models

The historical development of research in gully erosion modeling is reviewed by Bull and Kirkby (1997). In the review, an attempt has been made to trace the development of gully erosion models, from the first stochastic models in the 1970s to the more recent approaches of process-based representations of the system for understanding the theory behind gully initiation in the 1980s (Bull and Kirkby 1997; Kirkby and Bull 2000; Merritt et al. 2003). Another overview regarding a number of existing erosion and sediment transport catchment-scale models deserving specific mention include the review by Merritt et al. (2003). The literature comprehensively reviews a range of models that have been used to simulate aspects of erosion, sediment generation, and sediment transport through a landscape at a catchment scale. The models reviewed range significantly in the erosion process they represent, the manner in which these processes are described and the temporal and spatial scales of application for which they were developed (Merritt et al. 2003). Furthermore, the review points out that, if alternate erosion sources contribute significantly to the generation of sediment (e.g., permanent gullies), then such processes need to be represented explicitly in the selected model. However, most of the catchment-scale erosion models do not account for gully erosion as a process explicitly. Table 23.1 provides a summary of the reviewed catchment-scale erosion models and the processes they explicitly represent. It can be noted that only four out of the seventeen reviewed models incorporate routines to account for gully erosion.

In the review by Merrit et al. (2003), four models, the AGricultural Non-Point Source Pollution model (AGNPS), the Chemicals, Runoff and Erosion from Agricultural Management Systems (CREAMS), the Hydrological Simulation Program–FORTRAN (HSPF), and the Sediment River Network model (SEDNET), were examined to assess whether the gully erosion routine explicitly represents the gully sediment generation, sediment transport in gullies, gully growth rates, and the sediment loading at the gully outlet. Although these four models are capable of estimating gully sediment yield, sediment generation, and transport in the gully, the capabilities to simulate gully growth rates and change in gully dimensions over time are not included in these models. SEDNET in particular was developed as a tool to help identify the major sources of sediment to a stream network, location of sediment entrainment, and the dominant erosion process contributing sediment to the network at the catchment scale. (Merritt et al. 2003). However, its applicability to

Table 23.1 Process represented in the models reviewed after Merritt et al. (2003)

Model	Rainfall-runoff	Land surface sediment			Gully	In-stream sediment			Sediment associated water quality	
		G	T	D		G	T	D	Land	In-stream
AGNPS	Yes	Yes	No	No[a]	Yes	Yes	Yes	Yes	Yes	Yes
ANSWERS	Yes	Yes	Yes	Yes	No	No	No	No	No	No
CREAMS	Yes	Yes	Yes	Yes	Yes	No	No	No	Yes	No
EMSS	Yes	No[b]	No	No	No	Yes	Yes	Yes	No	No
GUEST	Yes	Yes	Yes	Yes	No	No	No	No	No	No
HSPF	Yes	Yes	Yes	Yes	Yes	Yes	Yes	Yes	Yes	Yes
IHACRES-WQ	Yes	No	No	No	No	Yes	Yes	Yes	Yes	Yes
IQQM	Yes	No	No	No	No	No	No	No	No	No
LASCAM	Yes	Yes	No	No	No	Yes	Yes	Yes	Yes	Yes
LISEM	Yes	Yes	No	No	No	Yes	Yes	Yes	No	No
MIKE-11	Yes	Yes	Yes	Yes	No	Yes	Yes	Yes	Yes	Yes
PERFECT	Yes	Yes	No	No	No	No	No	No	Yes	No
SEDNET	Yes	Yes	No	No[a]	Yes	Yes	Yes	Yes	Yes	Yes
SWRRB	Yes	No	No	No	No	Yes	Yes	Yes	Yes	Yes
TOPOG	Yes	Yes	Yes	Yes	No	No	No	No	No	No
USLE	No	Yes	No	No	No	No	No	No	No	No
WEPP	Yes	Yes	Yes	Yes	No	Yes	Yes	Yes	No	No

G, sediment generation: *T*, sediment transport; *D*, deposition
[a]Requires a sediment delivery ratio (SDR) to compute sediment yield from gross erosion
[b]Uses prescribed loads for a land use type

this particular study is substantially limited by the extensive data requirements, namely a grid of mean annual rainfall, soil erodibility, a grid of gully density, and a description of the mean characteristics for each link.

Woodward (1999) describes the Ephemeral Gully erosion Model (EGEM) which is a modification of the Agricultural Research Service Ephemeral Gully Erosion Estimate (EGEE) to meet the Natural Resources Conservation Service (NRCS) needs. The EGEM has two major components: the hydrology component which uses the NRCS curve number, drainage area, watershed flow length, average watershed slope, and 24-h rainfall and standard NRCS temporal rainfall distributions to estimate peak discharge rates and runoff volumes. A combination of empirical relationships and physical process equations is used as the erosion component to compute the width and depth of the ephemeral gully. However, this model was built on the assumption that ephemeral gullies typically erode to the tillage depth, limited to 18 inch or less and further work is needed to involve the capability to simulate erosion in branching gully systems which limits its application to large permanent gullies such as the ravines in the lower Le Sueur River.

Sidorchuck (1999) attempted to model gully erosion based on a thorough description of the physics involved in the process. This study introduced the concept that the gully undergoes through two stages as it changes its morphology. The two stages are then modeled in two types of gully erosion models: the dynamic model which predicts the rapid changes of gully morphology at the initial stage of gully development, and the static model which calculates the morphometric parameters of a more stable gully (Sidorchuk 1999). Dynamic gully erosion model uses the mass conservation and deformation equations to characterize the factors that control the rate of gully incision (water flow velocity, depth, turbulence, temperature, soil texture and mechanics, and vegetation cover). As the gully stabilizes, sediment transport and sedimentation are the main processes at the gully bottom. The static model represents the change in the longitudinal profile of the ravine. The sediment flux in the gully is defined by the equation of mass conservation and the change in gully bottom, the sediment budget is estimated by the equation of deformation. Both model stages are two-dimensional (space-time); they attempt to represent the change in gully in time and distance. The models would well represent the sediment budget in the study ravines in the lower Le Sueur River, if sediment and water data of the ravines were measured along the ravine length in a set of time intervals. However, the data available for this study have been measured at the ravine head and outlet only; hence, the applicability of Sidorchuk's gully erosion model is limited at the time of this study.

Another approach to modeling gully erosion is the high-rate gully erosion equations described in the study by Torri and Borselli (2003). This study presents an approach to gully erosion based on a mass balance equation derived for a dynamically developing gully system. This model formulation attempts to estimate the sediment budget of a gully using sediment generation rates of the gully wall, gully bed, and sediment being deposited in the gully. The derived equations link gully widening rate to gully deepening rate during peak discharge. The model assumes a one-dimensional flow along the centerline of the stream channel. It also assumes a prismatic channel and does not take into account the cross-stream variations due to variable channel cross-sectional features. However, the model has the advantage of compatibility with currently available data in the Le Sueur River gullies and can provide useful insight into the relative contribution of different components of the sediment budget in gullies with limited measured data.

23.3 Research Methodology

23.3.1 Study Area Description

The Le Sueur River is located in the south-central part of Minnesota. Its watershed is one of the twelve major watersheds of the Minnesota River Basin with a total area of approximately 2880 km^2. The Le Sueur River flows northward to its confluence with the Blue Earth River. About 5 km north, the Blue Earth River joins the Minnesota River at the city of Mankato, MN, and flows northward to its confluence with the Mississippi River at the Twin Cities—Minneapolis/St. Paul. The drainage network of the Le Sueur River watershed is defined by the main channel of the river and its major tributaries: the Mapple River and the Big Cobb River and smaller streams. There also exist an extensive network of artificial drainage ditches and tile drainage installed to aid water infiltration in the agricultural fields. According to the MPCA, the Le Sueur River is the primary contributor of suspended sediments to the Minnesota River (23–30 %) (Minnesota Pollution Control Agency et al. 2007).

A major part of the Le Sueur River watershed area has low gradient to flat uplands. The study of Gran et al. (2009) shows that the lower reaches of the river and its major tributaries are currently incising, and the knick points are migrating upstream causing a high relief to the incised portion of the watershed. High bluffs border many of the outer bends along the main stream of the Le Sueur River. Deeply incised ravines are also prevalent especially toward the lower reaches of the river.

Agriculture is the dominant land use within the watershed of the Le Sueur River (87 %) (Minnesota Pollution Control Agency et al. 2007). Corn and soy bean are the most common crops grown. However, the poorly drained soils in the agricultural fields would make it impossible to grow crops without artificial drainage mechanism. Farmers commonly use subsurface drainage tiles to minimize runoff and increase infiltration in the agricultural fields. Installation of the subsurface drainage network and surface ditches in the landscape consequently has completely altered the hydrology of the watershed. It rapidly increased the vertical hydraulic conductivity of the agricultural fields to create optimum soil moisture for the crops. It also increased the horizontal hydraulic conductivity so that water would flow easily and more rapidly to ditches, ravines, or the river.

Although the artificial drainage network has enhanced the crop production in the area, approximately 89 % of the wetlands were lost through drainage. The rapid movement of water through the watershed also increased the pollutant and sediment transport and loading to the water channels. Concentrated flow from several drainage tiles is directed into the ravines. The concentrated flow from the tiles may not carry significant sediment load to the ravines when compared to surface runoff, but it possibly could affect the sediment dynamics and production inside the ravines.

Ravines in the lower Le Sueur valley act as runoff and sediment pathways linking the uplands (agricultural fields) and the river valley bottom. The lower reaches of the Le Sueur River are currently incising, leading to migration of the knick point upstream, (Gran et al. 2009). In response to knick point migration, most of the ravines in the lower Le Sueur are changing their morphological characteristics (length, depth, width, area, and volume) and are periodically flushing significant sediments. Hence, through and below the major knick zones, ravines are believed to be not only a link but also an important sediment source.

23.3.2 Data Collection

The first task of this research consisted of field data collection. The collected data were mapped using GIS layers for further analysis. Following this, a numerical model to study the sediment transport was developed. These activities are described in more detail in the following sections of this chapter.

Field Topographic Observations: The main objective of the field work in summer 2008 was to observe erosion activities inside the two study ravines along the lower reaches of the Le Sueur and collect data to help establish a more accurate estimate of the ravine sediment production. Two ravines were selected as study sites because of the availability of installed field instrumentation. The study sites are located about 6 km south of the city of Mankato, MN, and 2 miles west of State Route-22 (SR-22). The headcut of one of the study ravines is located south of the County Route-90 (195th ST, CR-90) and continues southward parallel to SR-22 to join the Le Sueur River. The second ravine starts few meters west of County Route-8 (Monks Ave, CR-8), continues southwards, and joins the Le Sueur River. In this report, the ravine on CR-90 is denoted as CR-90 and the second ravine as CR-8. The location of the study sites is depicted in Fig. 23.3.

In 2007, the field research team at the National Center for Earth-Surface Dynamics (NCED), St. Anthony Falls Lab at the University of Minnesota, had started to monitor the water discharge and water quality of the two ravines. Two ISCO autosamplers were installed at the outlet of the two ravines and one at a culvert a few meters downstream of the headcut of the ravine on CR-90. The ISCO autosamplers recorded the water-level every 15 min and took water samples during

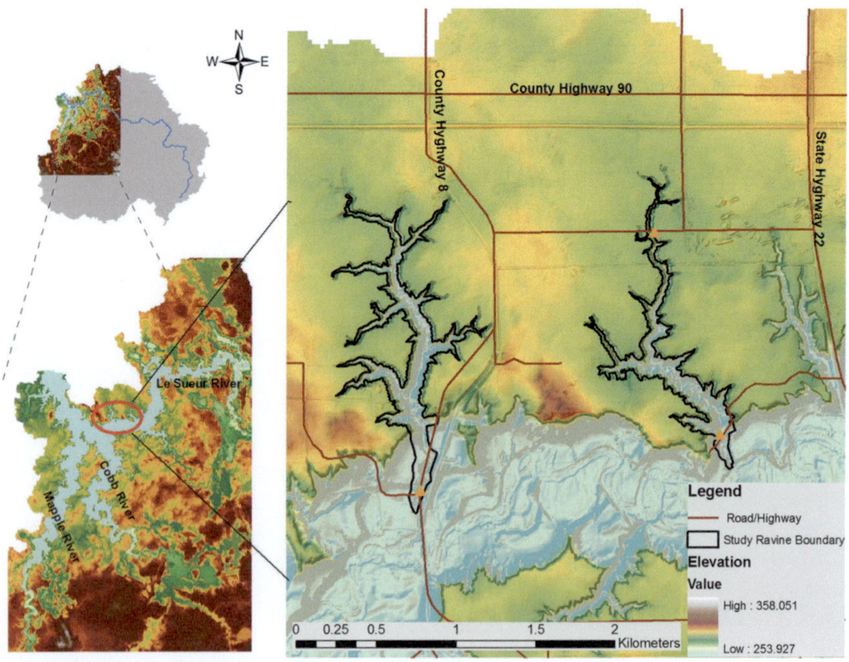

Fig. 23.3 Geographic location of study ravines in the lower Le Sueur watershed

storm events. The Water Resource Center of Minnesota State University monitors both gauges. Water quality data for storm events between the months of April and August 2008 was obtained from this office. A comparison of the suspended sediment concentration measurements taken during these storm events showed that a significant amount of sediment is being flushed out of these ravines. For the ravine on CR-90 for example, Fig. 23.4 shows a plot of the total suspended solids (TSS). Measurements taken during the 30 storm events show that TSS concentrations at the mouth of the ravine are one of the magnitudes higher than the TSS measurements taken near the ravine head.

To help map the major sources of sediment inside the ravines, a GPS was used to record the longitude and latitude location of bluffs and terraces. Soil samples from some of major bluffs and major terraces were also collected to determine the grain size distribution. A comparative estimate of the geometry such as the surface area and slope of major bluffs, surface area, and depth of major terraces, was also recorded. There are also bluffs located inside tributaries of the ravines. The end of the tributary where it meets the mainstream can be as small and narrow as 30 cm and lead to a 12-m-high bluff at the head. The water sources for the tributaries are either from tile drains or concentrated overland flow. The velocity and water depth of the stream in the ravines vary. At some locations, the water depth was very shallow, but at others, it reached up to knee high.

Along the mainstream and tributaries of ravine CR-8, 17 major bluffs were located. The largest of all has a surface area about 190 square meters, with sandy deposits and some vegetation cover. Ravine CR-90 has relatively larger bluffs; the location and the geometry of 19 of these major bluffs were recorded. The largest bluff in this ravine has a height of 12 m and surface area of 240 m^2. Most bluffs in both ravines have a very steep surface slope and were actively eroding. Figure 23.5 shows the ravines and the corresponding upstream drainage areas.

The collected data such as the GPS coordinates for bluff and terrace locations, pictures, estimated bluff heights and widths, terrace height and top area, general stratigraphy as well as other remarks were composed in a tabular format in

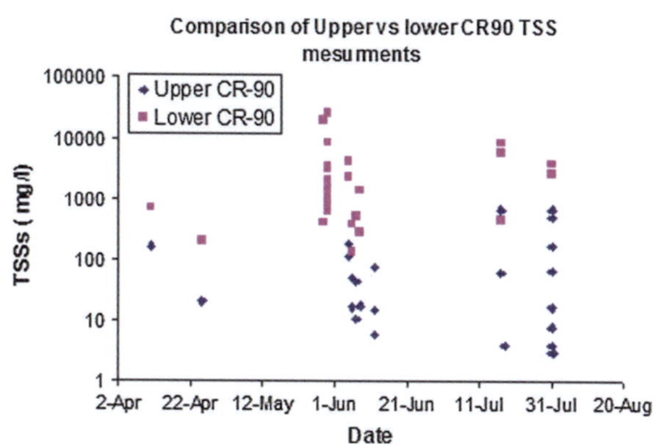

Fig. 23.4 Comparison of TSS measurements in 2008 of the upper and lower gauges on CR-90

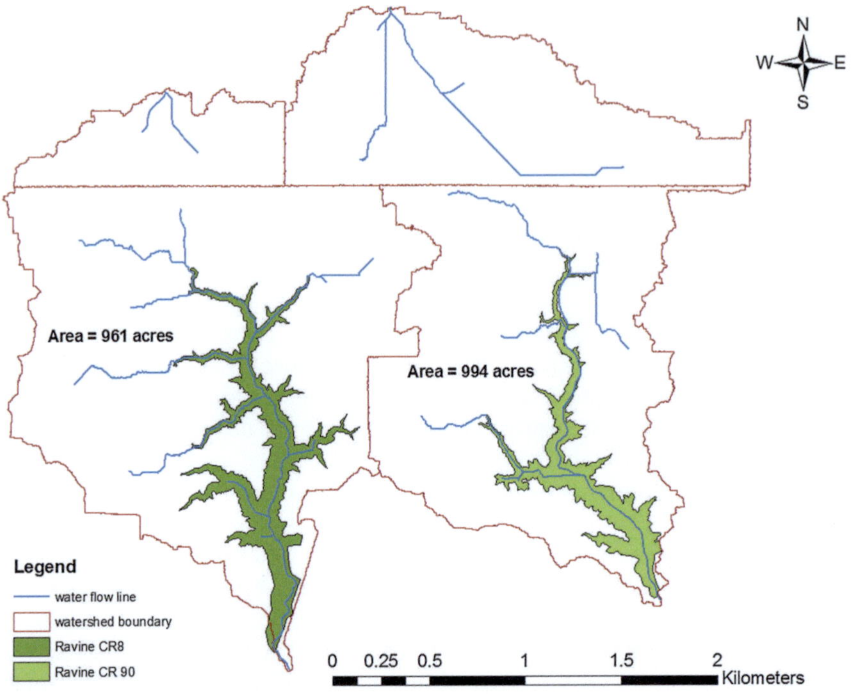

Fig. 23.5 Map of study ravines and their watersheds

spreadsheets (Fig. 23.6). Using the coordinates and GIS, a map of major bluff and terrace locations along the two study ravines was produced.

Drainage Tiles: Observations of different tile drains include metal, concrete, and plastic with slits and collector drains. The collector drains have a diameter of 27–36 inches. They have smaller tile drains connected to them and the plastic drains with slits are able to collect water from the sides as it seeps through the ground. Other tile drains range from as small as 5 in. to as big as 14 in. Some tile drains caused deep valleys in the walls of the two ravines. This incision on the slopes could also increase the sediment production in the ravines. The ravine walls were also incised to about 1.5 m at some locations. The incision destabilizes the ravine walls, and walls fail by slumping on the fluvial channel. The location of control structures such as culverts and bridges along the ravines was recorded in GPS.

Ravine Geometry Estimations: An accurate estimate of the ravine geometry is the first step to calculating the sediment production and loading to the Le Sueur River. The aspects of morphology of these study ravines in the lower Le Sueur valley are similar to that of permanent gullies. Compared to the geometry of gullies, the ravines in the study area have deeper and wider valley, steep banks, eroding bluffs, large terraces, and vegetation cover of dense trees. These ravines also connect relatively large areas in the landscape. The total channel length of the ravine along the CR-90, for example, is approximately 3700 m, measuring from the

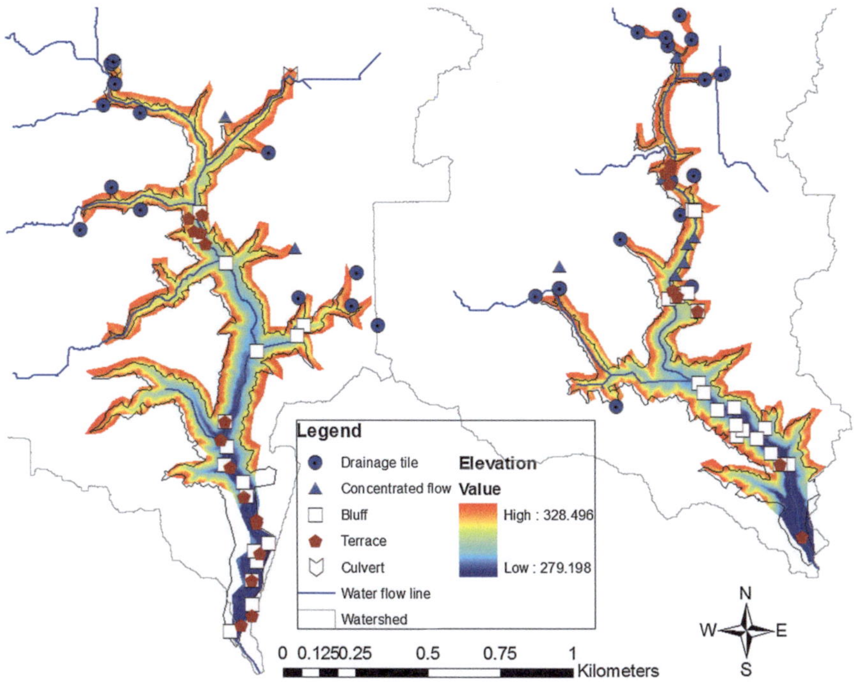

Fig. 23.6 Map of bluffs, terraces, and entry points of concentrated flow. The concentrated flow comes from overland flow from the agricultural fields and the tile drainage into the ravine

headcut to its mouth along the ravine valley including the stream length of its major tributaries. This ravine drains an agricultural area of about 990 acres. The topography surrounding the study ravines has a very gentle slope of 0–2 %. A DEM at 3 meter resolution of the Le Sueur River watershed was obtained from the Blue Earth County. Using GIS tools of spatial analyst and ArcHydro, topographic characteristics of the two ravines were derived from the LiDAR. The area of the watershed, channel length, ravine surface area, and average slope was directly calculated in ArcGIS. The mean width of the ravines was calculated by dividing the ravine planimetric area by the channel length. Similarly, the mean depth was estimated by dividing the ravine volume by its planimetric area. Table 23.2 summarizes the results from the calculations performed in ArcGIS.

23.3.3 Numerical Model Development

In an effort to quantify the sediment budget of deeply incised ravines in the lower Le Sueur River watershed, gully erosion equations developed by Torri and Borselli (2003) and the USLE model were arranged into a numerical model.

Table 23.2 Summary of ravine morphometric parameters

Description[a]	Ravine CR-90	Ravine CR-8
Drainage area (acres)	994	961
Channel length (m)	3760	4900
Ravine planimetric area (acres)	56	80
Volume (million cubic meters)	9.26	5.79
Mean gully top width (m)	61	66
Mean gully depth (m)	41	18
Mean gully bank slope (%)	133	55
Average longitudinal slope (%)	32	39.5
Average upland Slope (%)	1.97	2.12
Vegetation cover	Dense-tree	Dense-tree
ISCO autosampler	At the ravine head and outlet	At the ravine outlet[a]

[a]Estimations are the results of calculations in GIS

In this model formulation, the sediment budget of the ravines was calculated as the difference between the storage of sediment and the sum of sediments loads derived from the agricultural fields, ravine sidewalls, terraces, and ravine bed. Using the available sediment and water flow data for the two study ravines, the DEM of the area, along with reasonable assumptions of some parameters, the model was run to estimate the sediment budget in the study ravines of the Le Sueur River.

23.3.3.1 Theoretical Framework of the Model

Given the limited data available, this study aims to use as few parameters as possible, yet preserve the physical gully erosion process description. The equations developed by Torri and Borselli (2003) for high-rate gully erosion use a few parameters and hence were adopted in the numerical model presented in this study.

The main channel of the ravine/gully is assumed to be prismatic with a rectangular cross section. Figure 23.7 shows the sketch of the gully cross section, the sediment sources, and storages. The channel has a length L, a depth D, and width W. The amount of sediment leaving the channel in a given time interval is ΔQ_{sto}.

The sediment budget is calculated as the difference between the sediment storage and sediment fluxes from the agricultural fields, ravine sidewalls, and ravine bed. It is given by Eq. (23.1).

$$\Delta Q_{\mathrm{STO}} = Q_{\mathrm{IN}} - Q_{\mathrm{Out}} + Q_{\mathrm{L}} + Q_{\mathrm{S}} + Q_{\mathrm{B}} - Q_{\mathrm{SED}} \qquad (23.1)$$

where

Q_{IN} Sediment from upstream entering the gully (kg day^{-1})
Q_{OUT} Sediment budget, sediment leaving the gully (kg day^{-1})
Q_{L} Lateral Flux, sediment from uplands entering the gully (kg day^{-1})
Q_{S} Sediment flux from gully banks (kg day^{-1})

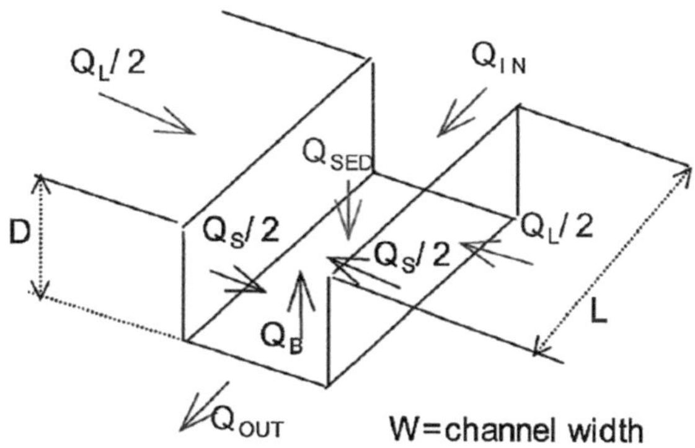

Fig. 23.7 Sketch of gully cross section with sediment sources.(Torri and Borselli 2003)

Q_{B}	Upward flux, sediment from gully bed (kg day^{-1})
Q_{SED}	Downward flux, sediment settling in the gully (kg day^{-1})
Q_{STO}	Sediment storage in the flow (kg day^{-1})

For a small time interval, the variation of sediment momentaneously suspended in water passing through a small segment of channel is then given as follows:

$$\frac{\partial q_{\mathrm{STO}}}{\partial t} = q_{\mathrm{IN-OUT}} + q_{\mathrm{L}} + q_{\mathrm{S}} + q_{\mathrm{B}} - q_{\mathrm{SED}} \tag{23.2}$$

The basic sediment rate equations for sediment loads from gully sides, gully bed, and settling sediment are given by Torri and Borselli (2003) as follows:

$$q_S = 2k_S(z)(e_f p - p_{\mathrm{cr}})D \tag{23.3}$$

$$q_B = k_B(z)(p - p_{\mathrm{cr}})W \tag{23.4}$$

$$q_{\mathrm{SED}} = \frac{q_{\mathrm{o,STO}}}{W} u_{\mathrm{SED}} + \rho D \frac{\partial W}{\partial t} \tag{23.5}$$

where

W	Mean gully width (m)
D	Mean gully depth (m)
ρ	Soil bulk density (kg m^{-3})
k_S	Coefficient of soil erodibility of gully walls (day^2 m^{-2})
k_B	Coefficient of soil erodibility of gully bed (day^2 m^{-2})
u_{SED}	Sedimentation velocity in a turbulent flow (m s^{-1})
p	Flow aggressiveness (kg day^{-2})
p_{cr}	Critical flow aggressiveness (kg day^{-2})
e_{f}	Efficiency coefficient, the ratio between the force exerted by flow on gully banks and force exerted on the gully bed (dimensionless)
q_S	Rate of sediment load from gully banks (kg day^{-1})

q_B Rate of upward flux, sediment from gully bed (kg day^{-1})
q_{SED} Rate of downward flux, sediment settling in the gully (kg day^{-1})
q_L Rate of lateral flux (kg day^{-1})
$q_{IN\text{--}Out}$ The difference between rate of sediment entering from upstream and leaving the gully (kg day^{-1})
$q_{o,STO}$ Sediment momentaneously suspended in water passing through a small segment of channel (kg day^{-1})

Equations for the other remaining sediment rates of the sediment budget are not explicitly given on the referenced literature. However, the rate of sediment from upstream entering the gully and leaving the gully can be calculated using the measured total suspended sediment at the ravine head and mouth. The measured TSS in mass/volume was multiplied by the measured water discharge in volume/time to obtain suspended sediment discharge in mass/time. To estimate the lateral sediment flux coming from the uplands, the USLE and the RUSLE models were used. The estimates, however, were essentially equivalent; hence, the USLE model was adopted.

23.3.3.2 The Universal Soil Loss Equation—USLE

The universal soil loss equation is a widely used regression model for predicting soil erosion. It is an empirical model used to predict soil loss due to sheet and rill erosion. The equation was developed from over 10,000 plot-years of runoff and soil loss data, collected on experimental plots of agricultural land in 23 states by the US Department of Agriculture (Simons and Senturk 1992). Measurements of precipitation, runoff, and soil loss associated with 42 stations were continuously collected for a period of 5–30 years or more. Field plots of 72.6 ft long on a 9 % uniform slope in bare fallow soil and tilled were arbitrarily selected to serve as a reference for evaluation. The model is based on the field data collected from these field plots and simulated rainfall data (Simons and Senturk 1992). The empirical equation of the USLE is given as follows:

$$A = RKLSCP \tag{23.6}$$

where
A Soil loss in tones per unit area per year
R Rainfall and runoff erosivity index for a geographic location
K Soil erodibility factor
LS Slope steepness and length (topographic) factor
C Cropping and management factor
P Erosion control practices such as contouring or terracing

The computed soil loss A has a time period of R and soil loss dimensions of K. It has units of tones per unit area per year. A more detailed descriptions of the USLE equation and its terms can be found in Smith and Wischmeier (1957) and Wischmeier and Smith (1978). The LS, C, and P are all dimensionless. Values of each of the factors were estimated using field data and Agricultural Handbook No. 537 procedures and tables.

The R factor depends on the frequency distributions of annual, seasonal, or annual–maximum storms. It is predicted on a probability basis. In the Agriculture Handbook by Wischmeier and Smith (1978), an isoerodent map for average annual values of the rainfall erosion index is given. From the figure in the reference, R factor of 135 was used for the study area.

The K factor, which is the soil erodibility factor, was found to be a function of percent of silt, percent of coarse sand, soil structure, permeability of soil, and percent of organic matter. The soil erodibility nomograph in Wischmeier and Smith (1978) is used to determine K factor for top soils or subsoil horizons.

The LS – topographic factor was defined as the ratio of soil loss from any slope and length to soil loss from a 72.6 ft plot length at a nine percent slope, with all other conditions the same (Simons and Senturk 1992). The slope length is the distance from the point of overland flow origin to the point where either slope decreases to the extent that deposition begins or runoff water enters a well-defined channel (Smith and Wischmeier 1957). The slope-effect chart was used to determine the LS value for this study, yielding a LS value of 0.32 in this study.

The cropping-management factor C is defined as the ratio of soil loss from land cropped under specific conditions to corresponding loss from tilled, continuously fallow ground. (Smith and Wischmeier 1957). The factor depends on the type of vegetation cover, crop season, and management techniques. Its value ranges between 0 and 1.0 approximately. Based on the values used in similar studies, a C value of 0.28 was adopted for the two study ravines.

The USLE predicts the gross soil loss from sheet and rill erosion per a unit area. To calculate the sediment yield per unit area, the USLE predictions must include the factors of delivery ratio and the watershed area. The sediment yield is given by the following equation:

$$Y = \frac{E(DR)}{Ws} \quad (23.7)$$

where

Y Sediment yield in tones per unit area (tones per acre)
E Gross soil erosion in tones
DR Delivery ratio
Ws Area of the watershed in acres

The delivery ratio DR is the ratio of sediment delivered at a downstream point in the watershed to erosion from the area above that point. This ratio considers deposition in watershed and by definition is less than unity. The value of the delivery ratio can be approximated by estimating the amount of soil loss A that will be deposited within a watershed depending on the nature of the land surface (Smith and Wischmeier 1957).

23.3.4 Governing Equations of Gully Erosion Rate

23.3.4.1 Gully Width and Depth

To see how the gully width and gully-bottom change in relation to the sediment budget, it is important to calculate the rate of change in width and depth during peak flows. The equations developed by Torri and Borselli (2003) link gully widening to gully deepening rates. The rate of change of gully width and depth during peak flows is expressed, respectively, as follows:

$$\frac{\partial W}{\partial t} = \frac{q_S}{\rho D} \tag{23.8}$$

where
ρ is the soil bulk density,
D is the gully depth, and
q_s is the sediment contribution from the sidewalls per unit of channel length per unit of time

The gully depth changes due to the balance between deposition on and detachment from the gully bed. It is assumed that lateral sediment inputs are considered negligible with respect to the amount from the gully sidewall and bed during peak discharge—at the most important phase of concentrated erosion (Torri and Borselli 2003). The rate of gully depth change over time is then given by the following equation:

$$\frac{\partial D}{\partial t} = \frac{1}{\rho} \frac{q_B - q_{SED}}{W} \tag{23.9}$$

Torri and Borselli (2003) presented an approach to gully erosion based on the general equation derived from theoretical consideration. The derived equations link gully widening rate to gully deepening rate.

For a fast and turbulent peak flow, it is assumed that the sedimentation velocity is nearly zero. Equations for Q_{SED} and $\frac{\partial D}{\partial t}$ are then modified for turbulent flow and the rate of depth change is given by the following equation:

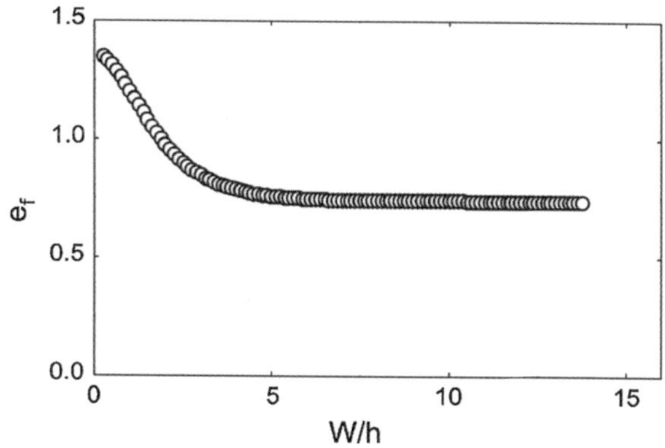

Fig. 23.8 Efficiency coefficient as a function of the ration between channel width and water flow depth from Chow (1973). Adopted from (Torri and Borselli 2003)

$$\frac{\partial D}{\partial W} = \frac{k_b}{2k_s e_f} - \frac{D}{W} \tag{23.10}$$

The k_s and k_b coefficients indicate soil erodibility and e_f is an efficiency parameter defined as the ratio between the force exerted by the flow on the walls and force exerted on the bed. Calculating the exact value of an efficiency coefficient is important but complicated. Torri and Borselli (2003) suggest that the efficiency coefficient as a function of the ratio between channel width and water flow depth can be read from a graph of e_f and *W/h* plot given by Chow (1973) (Fig. 23.8).

Torri and Borselli (2003) used the Laplace polynomial expansion technique to get one solution to the above equation and is given as follows:

$$D - D_0 = \frac{k_B}{4k_s e_f}(W - W_0) \tag{23.11}$$

where D_0 and W_0 are the depth and width reached before the flow starts digging the soil layer characterized by the erodibility k_b and k_s.

It is important to note that the above equation holds during peak flow. For later erosion developments, the relationship between width and depth needs to be modified. Furthermore, the flow aggressiveness p, which is any measure of flow erositivity, is given using two of the most commonly used estimators, the stream power and flow shear stress. Torri and Borselli (2003) provide the equations for flow aggressiveness using the equations for gully bed and walls as follows:

Assuming p is unit stream power,

$$p(t) = \frac{p_A g Q \sin\gamma}{W(t)} \tag{23.12}$$

where p_A is water density, g is acceleration due to gravity, and γ is the local slope angle, assuming p is a unit stream power.

The rate of change of gully width is then modified to be

$$\frac{\partial W}{\partial t} = \frac{2k_S(e_f\rho_A gQ\sin\gamma - Wp_{cr})}{\rho W} = \frac{C_{sp}Q\sin\gamma}{W} - \frac{2k_s p_{cr}}{\rho} \tag{23.13}$$

where C_{sp} is a composite parameter, it has a dimension of $[L^{-1}]$, and it is given as follows:

$$C_{sp} = \frac{2k_S e_f \rho_A g}{\rho} \tag{23.14}$$

Assuming the term $2k_s p_{cr}/\rho$ is small with respect to the first addendum, the above equation simplifies into

$$W = \sqrt{C_{sp}\sin\gamma \int_{\Delta t_{eff}} Q\,dt + W_0^2} \tag{23.15}$$

where W_0 is the channel width before peak discharge, and Δt_{eff} is the time interval during which the flow is erosive and close to peak discharge.

These sediment load equations were solved by an iterative scheme for each time step of the available flow data.

23.3.4.2 Input Requirements

All parameters required by the model to simulate the sediment budget change and soil loss from the ravines are summarized in Table 23.3.

Table 23.3 Model input parameters

Notation	Description	Units
D	Initial depth of gully	m
W	Initial width of gully	m
Q	Water flow rate	$m^3\ s^{-1}$
ρ	Soil bulk density	$kg\ m^{-3}$
γ	Local slope angle	Percent
n	Manning's roughness coefficient	–
Q_{IN}	Sediment from upstream entering the gully	$kg\ day^{-1}$
Q_{OUT}	Sediment leaving the gully.	$kg\ day^{-1}$
Q_L	Sediment from uplands entering the gully.	$kg\ day^{-1}$
C_{sp}	Composite parameter	m^{-1}
k_S	Coefficient of soil erodibility of gully walls	$day^2 m^{-2}$
k_B	Coefficient of soil erodibility of gully bed	$day^2 m^{-2}$
e_f	Efficiency coefficient—the ratio between the force exerted by flow on gully banks and force exerted on the gully bed	–

Model Assumptions and Limitations

- The model is limited to the processes of incision and widening only.
- Lengthwise growth of the gully system is assumed to be negligible within single runoff event. Hence, gully-headcut retreat rate is not computed.
- Cross-stream variations induced by cross-sectional geometrical features such as constrictions or expansions or obstructions by woody debris or rocks are neglected.
- The channel is assumed to be prismatic.
- Further study is needed to involve the capability to simulate erosion in branching gully systems.

23.3.5 Characterization of Coefficient of Soil Erodibility

The coefficients of soil erodibility of both gully wall and gully bed are normally determined in the field. However, due to the lack of existing field data, an alternative approach was followed in which numerical estimation was done using the Meyer-Peter and Muller (1948) formulation for bed load transport. The Meyer-Peter and Muller estimates of the sediment load from the gully bed were compared with the estimates given by Torri and Borselli's equation of Q_B, and a k_B value was calculated. According to (Meyer-Peter and Müller 1948) formulation valid for sediment diameters between 0.23 and 28.6 mm, the bed load sediment discharge Q_S is given as follows:

$$Q_S = 8\frac{1}{\rho^{1/2}}\frac{1}{(\gamma_s - \gamma)}(\tau_b - \tau), \quad \text{when } \tau_b \geq \tau_c$$
$$Q_S = 0, \quad \text{when } \tau_b \leq \tau_c \tag{23.16}$$

τ_b is the bed shear stress computed as follows:

$$\tau_b = \frac{n^2\gamma}{h^{1/3}}\left(U^2 - V^2\right) \tag{23.17}$$

and τ_c is the critical shear stress given as follows:

$$\tau_c = 0.047(\gamma_s - \gamma)D_m \tag{23.18}$$

Figure 23.9 shows the estimation of the coefficient k_b, by close matching the estimates of gully bed sediment load between the unit stream power formulation, Eq. (23.12) with the estimate using the Meyer-Peter and Muller Eq. (23.17).

To study how well the Q_s estimates using the two approach matches, a best-fit regression line is plotted in Fig. 23.10. The slope of the regression line indicates the relative relationship between the two Q_s estimates. This resulted in a "best match"

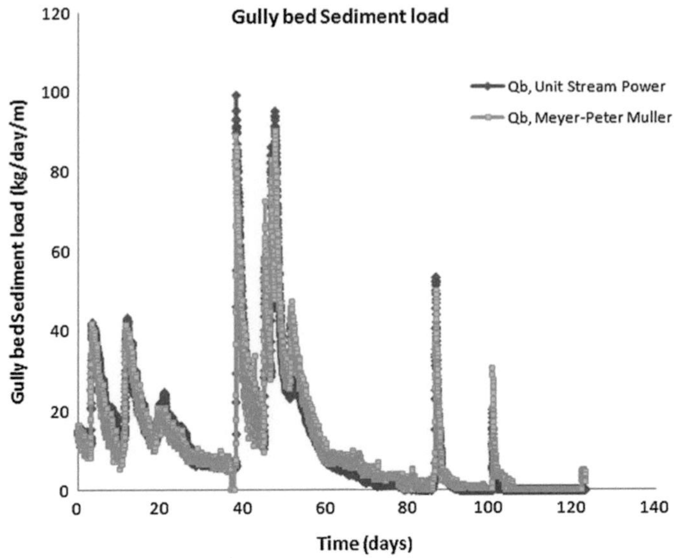

Fig. 23.9 Ravine bed sediment load calculation. Using Torri and Borcelli's unit stream power and Meyer-Peters and Muller equations

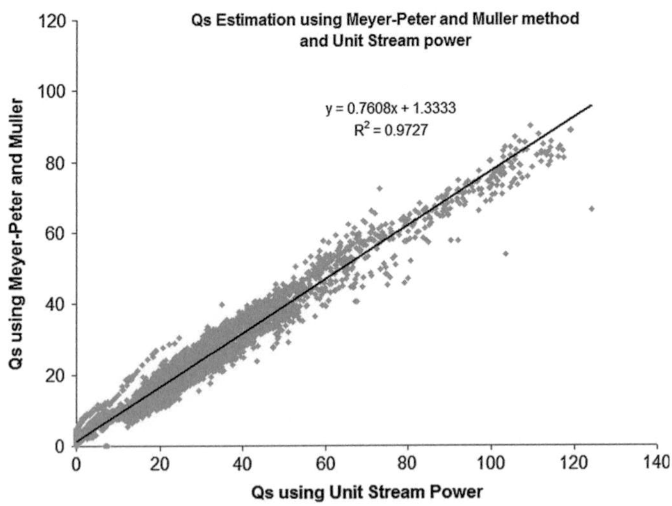

Fig. 23.10 Comparison of the fits of the two equations for estimating sediment load from gully bed

that yields a coefficient of soil erodibility of $k_B = 1.97 \times 10^{-16}$ day^2 m^{-2}. Moreover, to study the relationship between cross section and gully width, Torri and Borselli (2003) introduce a ratio *R* which is defined as the ratio of k_s to k_B. If the ratio *R* is characterized by a sufficiently large standard deviation, the relation between gully cross section and width can be described by linear equations. R value of 1 ± 0.35 was used in the reference; a ratio of 1.35 was adopted in this study; hence, $k_s = 1.35k_B$.

Model efficiency was calculated following Nash and Sutcliffe (1970)'s approach. The Nash–Sutcliffe efficiency (NSE) is an indicator of model performance. NSE is a normalized statistic that determines the relative magnitude of the residual variance ("noise") compared to the measured data variance ("information") (Nash and Sutcliffe 1970). NSE indicates how well the plot of observed versus simulated data fits the 1:1 line (Moriasi et al. 2007). NSE is computed as shown in the equation below:

$$\mathrm{NSE} = 1 - \left[\frac{\sum_i^n \left(Y_i^{\mathrm{obs}} - Y_i^{\mathrm{sim}}\right)^2}{\sum_i^n \left(Y_i^{\mathrm{obs}} - Y_i^{\mathrm{mean}}\right)^2}\right] \tag{23.19}$$

where

$Y_{\mathrm{i}}^{\mathrm{obs}}$ is the *i*th observation for the constituent being evaluated

$Y_{\mathrm{i}}^{\mathrm{sim}}$ is the *i*th simulated value for the constituent being evaluated

Y_i^{mean} is the mean of observed data for the constituent being evaluated

n is the total number of observations

NSE values range between $-\infty$ and 1.0, where efficiency of 1.0 indicates a perfect match of the simulated value to the observed data and efficiency values of ≤0.0 indicate that the mean observed value is a better predictor than the simulated value, hence unacceptable performance (Moriasi et al. 2007).

23.4 Results and Discussion

23.4.1 Test Case Study

The theoretical framework of the model was verified using a dataset from the literature. The dataset used in the research by Thomas et al. (2004) was used to build case study simulations in the model. The study was carried out on valley-bottom gully in western Iowa to estimate the annual growth rate of a permanent gully over 30 years. The gully was mapped and surveyed several times during the 30 year time. Daily discharges of water and suspended sediment through the gully were also recorded nearly continuously for over 36 years. The flow and suspended sediment data of the study site in Iowa were obtained directly from the authors of this reference. Using the monthly average flow and TSS data, the gully erosion model was then run for a time period of 1964–2000.

Application of the model yielded estimates for total soil loss from the valley-bottom gully, change in gully width, depth, and bank slope that agreed reasonably well with the estimations from the referenced literature. Comparison of model outputs with the estimates given in the referenced literature is shown in the plots presented in Fig. 23.11. The modeled values were plotted against the data obtained from the literature. The mean gully width estimates were on an average within 99 % of those reported in the literature, with a difference ranging between −2 and 5 %. The corresponding R^2 value is 0.89. The mean gully depth estimates have a difference of −0.4 % on average and ranging between −17 and 13 %. Differences were larger for the gully bank slope: −5 % on an average and raging between −5 and 15 %.

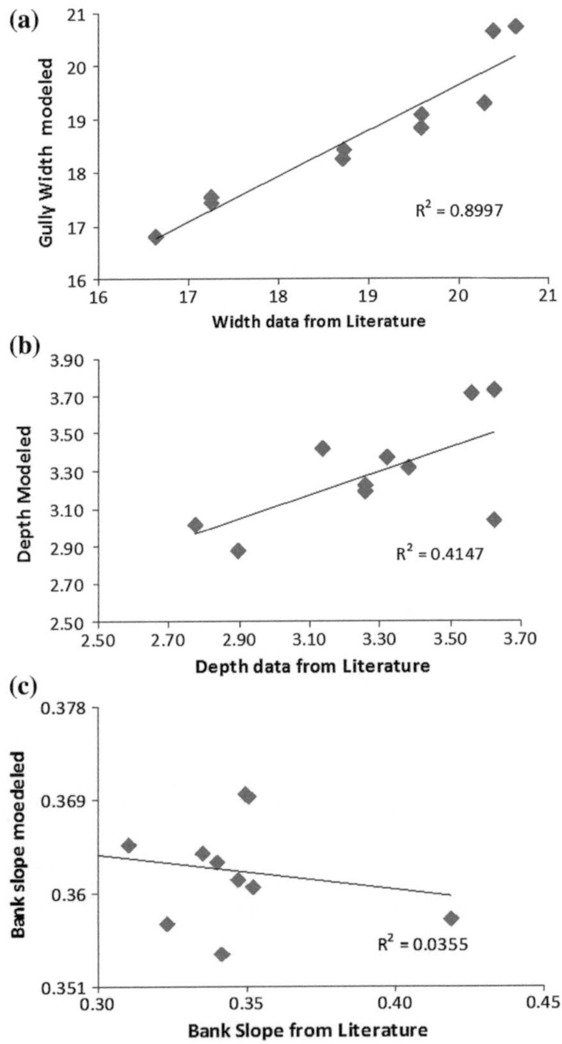

Fig. 23.11 Comparison of modeled and observed **a** gully width, **b** gully depth, and **c** bank slope

For gully width, depth, and bank slope simulations, the Nash–Sutcliffe efficiency (NSE) was calculated as an indicator of performance. Agreement between model estimations of gully width and estimates in the referenced literature corresponds to model efficiency of 0.94. But model efficiency was low for depth and bank slope simulations. (0.30 for gully depth and −0.05 for bank slope). Figures 23.12, 23.13, and 23.14 show the results of the model application (Model) and comparison with field results (Data) reported in Thomas et al. (2004).

Moreover, model estimates for the total sediment yield of the gully were approximately equivalent to the values reported in the literature. Thomas et al. (2004) estimate that an average of 3.2×10^5 kg of sediment removed from the gully annually. Model estimates were 3.21×10^5 kg of sediment per year.

The simulated mean monthly growth rate and the mean monthly runoff have a power relation which is in agreement with the rates calculated in the referenced literature. However, the relation in the model simulation is slightly different than the once calculated in the literature (Fig. 23.15).

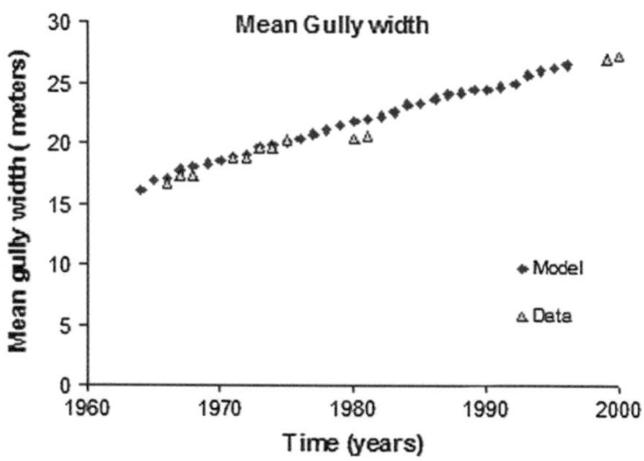

Fig. 23.12 Mean gully width growth over time

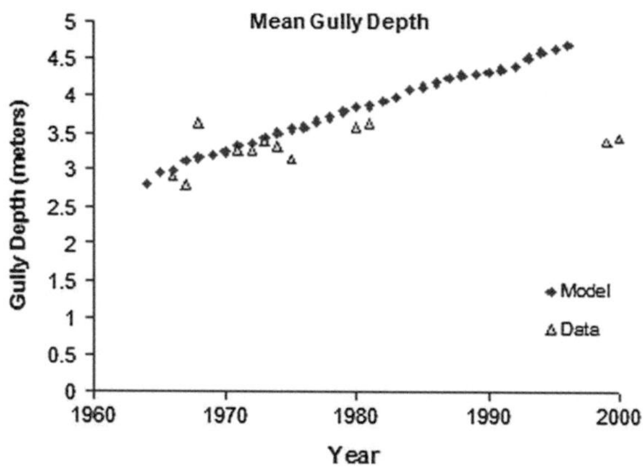

Fig. 23.13 Mean gully depth over time

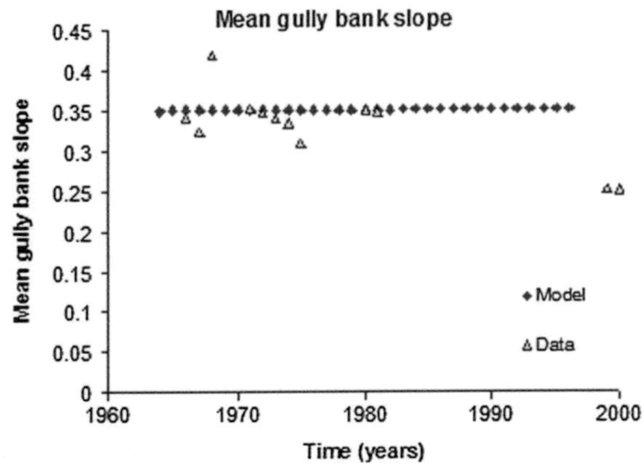

Fig. 23.14 Mean gully bank slope over time

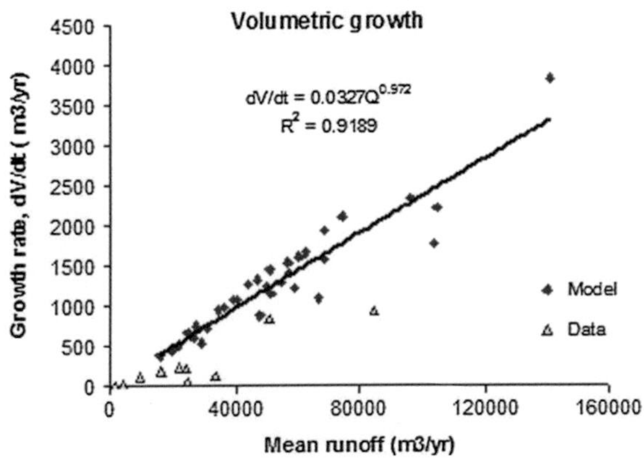

Fig. 23.15 Volumetric growth of gully versus water discharge

23.4.2 Model Estimations for Study Ravines

After the successful run for the test case in Thomas et al. (2004), the model used the available sediment and flow data of the two study ravines in the Le Sueur watershed to simulate the growth rate, width, and depth change of the CR-8 and CR-90 ravines. The results are presented in the plots on Figs. 23.16 and 23.17.

23.4.2.1 Ravine CR-90

The simulation of the ravine width and depth change of Ravine CR-90 shows that the storm events during April to October 2008 had triggered a very small increase in both gully width and depth. Though in small magnitude, the ravine increased in width rather than depth. The rate of change of gully depth is mostly negative for CR-90, implying there was deposition or temporary sediment storage within the ravine.

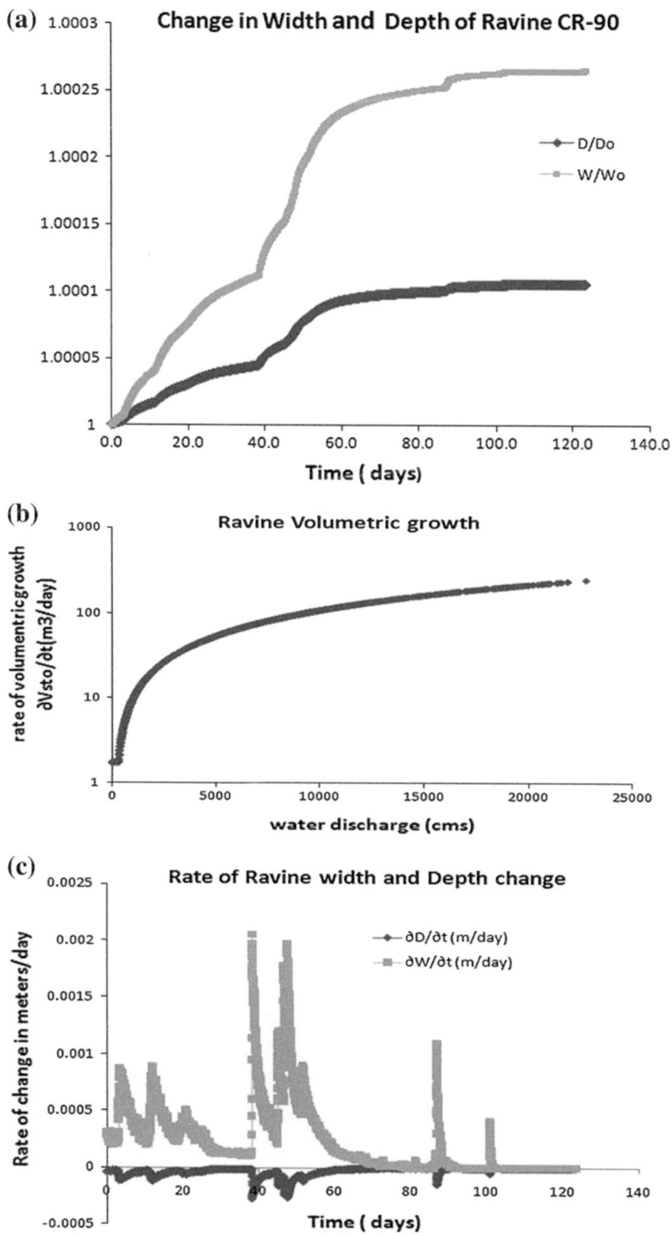

Fig. 23.16 Model simulations for CR-90: **a** change in width and depth of ravine CR-90, **b** estimation of volumetric growth of ravine CR-90, and **c** rate of ravine width and depth change of ravine CR-90

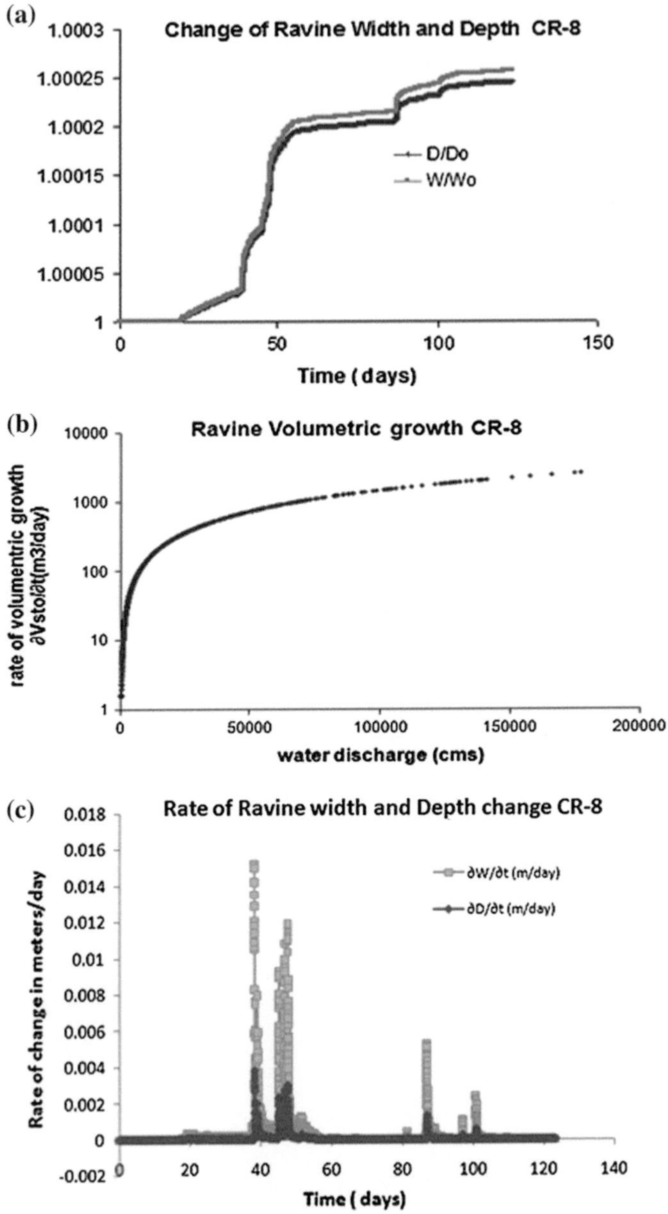

Fig. 23.17 Model simulations for ravine CR-8: **a** change of width and depth of ravine CR-8, **b** volumetric growth of ravine CR-8, and **c** rate of change of ravine width and depth of ravine CR-8

23.4.2.2 Ravine CR-8

The change of mean width and depth of Ravine CR-8 was different than CR-90 that the ravine seems to be increasing in depth and width in approximately the same magnitude. The rate of depth change was positive for this ravine, which shows the ravine was incising. But in both cases, the magnitudes of change were small for the measured storm events of summer 2008.

Table 23.4 summarizes the model estimates of sediment contribution of each sediment sources and the total sediment loading of the study ravines during the period of April to October 2008. The sediment yield of ravines was calculated using the ravine area and the total sediment loading.

23.4.3 Sensitivity Analysis

In order to assess the relative importance of each variable, a sensitivity analysis was conducted to study the effect of a change in the input on the model output. The sensitivity of the model to the various parameters was evaluated by increasing and decreasing the input parameter values by 50 %. The relative changes in total sediment yield associated with these perturbations are then listed in Table 23.5 in a decreasing order of their sensitivity. To quantify the degree of sensitivity of each

Table 23.4 Summary of sediment loading and sediment contribution of sources inside ravines

Description	Ravine CR-90	Ravine CR-8
Drainage area (acres)	994	961
Ravine area (acres)	56	80
Sediment budget (kg)		
Sediment contribution from ravine walls	162,438 53.4 %	77,478 32.8 %
Sediment contribution from ravine bed	133,811 44.0 %	152,963 64.7 %
Sediment contribution from upstream	0 0 %	0 0 %
Sediment contribution from uplands	8198 2.7 %	5916 2.5 %
Sediment deposited in ravine	162,438 53.4 %	77,478 32.8 %
Total sediment loading to Le Sueur (kg)	141,997	158,871
Ravine yield (kg/ha)	6213	4883
Percentage of upland-driven sediment	6 %	4 %
Percentage of Ravine driven sediment	94 %	96 %[a]

[a]The sediment load estimates are for the period April to October 2008, where measurements were taken

Table 23.5 Sensitivity of parameters

Notation	Description	Units	Initial value	Sensitivity coefficient		Change (%) in soil loss after change of	
				$S_{50\ \%}$	$S_{-50\ \%}$	50 %	−50 %
n	Manning's roughness coefficient	–	0.04	2.52	2.06	204	−82
k_B	Coefficient of soil erodibility of gully bed		1.97E−16	0.95	0.93	47	−47
p_{cr}	Critical flow aggressiveness		1.05E+14	−0.10	−0.06	−4	4
k_s	Coefficient of soil erodibility of gully walls		2.66E−16	0.00	0.00	0	0
e_f	Efficiency coefficient	–	0.7	0.00	0.00	0	0
Ravine dimensions							
γ	Local slope angle	°	0.31	0.91	0.91	44	−47
L	Stream length	m	3760	−0.05	−0.08	−2	6
D	Initial depth of gully	m	40	0.00	0.00	0	0
W	Initial width of gully	m	60	0.00	0.00	0	0
ρ	Soil bulk density.	kg day^{1}	1510	–	–	–	–
	Gully bank slope	%	1.33	0.00	0.00	0	0
Upland-driven erosion							
Q_L	Sediment from uplands entering the gully	tones year^{-1}	951	0.07	0.04	3	−3
TSS_{IN}	Sediment from upstream entering the gully	mg l^{-1}	142.355	0.00	0.00	0	0
TSS_{OUT}	Sediment leaving the gully	mg l^{-1}	31.053	0.00	0.00	0	0

parameter, an expression for sensitivity coefficient S used by McCuen and Snyder (1986) was selected for its simplicity and was applied in this study. The sensitivity coefficient S is the ratio of the relative output change to the relative input change. The following equation describes the sensitivity ratio.

$$S = \left(\frac{O_2 - O_1}{O_{\overline{12}}}\right)\left(\frac{I_2 - I_1}{I_{\overline{12}}}\right)^{-1} \tag{23.20}$$

where

O_2 Output of model using input parameter of maximum value of I_2
O_1 Output of model using input parameter of minimum value of I_1
O_{12} Average of the outputs O_1 and O_2
I_{12} Average of the outputs I_1 and I_2

Baffaut et al. (1997) explain that the sensitivity index obtained using the above method is independent of the magnitude of the input and the output; hence, it can be used to compare the sensitivity of the model to different variables, but it does not account for interaction between variables. This study, however, is limited to the broad assumption that the parameters are independent of each other. The analysis performed using Eq. (23.19) was intended to provide an estimate of the sensitivity of the simulation results to the model parameters.

Two sensitivity coefficients using both the minimum and maximum values of the input parameters were evaluated for each input parameter, and S values are also listed in two sensitivity coefficients using both the minimum and maximum values of the input parameters were evaluated for each input parameter, and S values are also listed in Table 23.5 The minimum input value is assumed to be 50 % of the defined value and a maximum value as 50 % more than the defined value.

Sediment derived from the gully sides was responsible for about 55 % of the total sediment yield in this gully. Hence, parameters related to these sources have higher sensitivity to the model output. One of the most important parameters is the erodibility coefficient of the gully bed. However, this parameter is also sensitive to the manning's coefficient n as it was estimated using the Meyer-Peter and Muller equations for bed load transport. Table 23.6 shows parameterization of the erosion model and sensitivity analysis results.

23.5 Conclusions and Recommendations

23.5.1 Conclusions

This study has presented a numerical model for quantifying soil loss from two gauged study ravines in the lower Le Sueur River, Minnesota. Despite the simplification, it was possible to compare the model estimations with data from

Table 23.6 Parameterization of the erosion model and sensitivity analysis of its parameters

Notation	Description	Units	Initial value	Sensitivity coefficient		Change(%) in soil loss after change of	
				$S_{50\ \%}$	$S_{-50\ \%}$	50 %	−50 %
n	Manning's roughness coefficient	–	0.04	2.52	2.06	204	-82
k_B	Coefficient of soil erodibility of gully bed		1.97E−16	0.95	0.93	47	−47
p_{cr}	Critical flow aggressiveness		1.05E+14	−0.10	-0.06	−4	4
k_S	Coefficient of soil erodibility of gully walls		2.66E−16	0.00	0.00	0	0
e_f	Efficiency coefficient	–	0.7	0.00	0.00	0	0
Ravine dimensions							
γ	Local slope angle	°	0.31	0.91	0.91	44	−47
L	Stream length	m	3760	−0.05	−0.08	−2	6
D	Initial depth of gully	m	40	0.00	0.00	0	0
W	Initial width of gully	m	60	0.00	0.00	0	0
ρ	Soil bulk density	kg day^{1}	1510	–	–	–	–
	Gully bank slope	%	1.33	0.00	0.00	0	0
Upland-driven erosion							
Q_L	Sediment from uplands entering the gully.	tones year^{-1}	951	0.07	0.04	3	−3
TSS_{IN}	Sediment from upstream entering the gully	mg l^{-1}	142.355	0.00	0.00	0	0
TSS_{OUT}	Sediment leaving the gully	mg l^{-1}	31.053	0.00	0.00	0	0

literature. An investigation of a case study in a gully system in Iowa (Thomas et al. 2004) using the numerical model developed in this study yielded estimates for total soil loss and rates of change of gully morphology that agreed reasonably well with the estimations from the referenced literature.

The results presented in the previous sections lead to answer to the research questions and objectives. The modeling results suggest that about 94–96 % of the sediment loading to the Le Sueur from the two study ravines originates inside the ravines. The ravines act as a link connecting the uplands (of which agriculture is the main activity) and the main river stream. However, the model estimates of the contribution of the sediment derived from the uplands and routed through the ravines are minimal when compared to the sediment produced from the sources inside the ravines.

In this study, it was also possible to identify the sediment sources and quantify the sediment budget of the study ravines. Field observations showed that there are a number of eroding bluffs and terraces inside ravines which are the major sediment sources in the ravine. This was in agreement with the model estimations of the sediment budget. The sediment budget was calculated as the difference between the sediment storages and sediment fluxes from the agricultural fields, ravine sidewalls, and ravine bed. In ravine CR-90, the major sediment source was the ravine walls, contributing 54 % of the total sediment yield and 44 % was derived from the ravine bed. For ravine CR-8, 65 % of the total sediment yield was derived from the ravine bed and ravine bed erosion contributed 33 %.

TSS data from the two monitoring gauges show extremely short-lived, very high sediment loads to the Le Sueur River. Direct discharge of water to ravine increases erosion activity inside ravine. Plunge pools are noticed at the outlet of the tile drainages and points of entry of overland flow. Moreover, the concentrated flow created deep incisions along the walls of the ravine causing amass wasting of the steep ravine valley walls.

Furthermore, it was noted that the ravines are widening at a higher rate rather than incising. There were no data available to compare the rates of change of the ravine width and depth. However, the model was able to give an idea on how the width and depth have changed during the storm events of April to October 2008.

23.5.2 Recommendations

This research is expected to serve as the gateway for investigations into the ravine sediment contribution to the total sediment budget of the Le Sueur River watershed. This study focused on modeling only two of the gauged ravines in the lower reaches of the river, and fundamentally aimed at quantifying the sediment budget using the available data from the gauges. However, there are more than 90 ravines in the river watershed, and gauging these ravines is prohibitively expensive. Hence, the results and conclusions drawn from this research will be extrapolated to the ravines throughout the river watershed so that more accurate ravine sediment loading to the river can be accounted in the sediment budget of the Le Sueur River watershed.

The model is limited to the processes of incision and widening only. However, lengthwise growth of the gully system within single runoff event should be considered to better understand the ravine change over time. Another assumption made in this study was that the ravine channel was considered to be prismatic and with uniform cross section. However, the cross-stream variation induced by geometrical features such as constrictions or expansions or obstructions by debris or rocks should also be considered to better represent the erosion problem in the system. The use of cross-sectional data of the ravines would help verify the model estimates. Moreover, the model needs to involve the capability to simulate erosion in branching gully systems. Furthermore, the present data available for the gauged ravines is for the last one year only. Using data of longer period time, in the future the model can be calibrated and validated for the study area. Testing of the model in a more complex environment also presents a future challenge.

Acknowledgments We would like to acknowledge the National Center for Earth-Surface Dynamics (NCED) for funding the field research and St. Anthony Falls Lab at the University of Minnesota for providing the work space.

References

Baffaut C, Nearing MA, Ascough JC II, Liu B (1997) The WEPP watershed model: II. Sensitivity analysis and discretization on small watersheds. Trans ASAE 4(4):935–943

Betts HD, DeRose RC (1999) Digital elevation models as a tool for monitoring and measuring gully erosion. Int J Appl Earth Obs Geoinf 1(2):91–101

Brice JB (1966) Erosion and deposition in the loess-mantled Great plains, Medecine creek drainage basin, Nebraska. U.S. Geological Survey Professional Paper 352H, 235–339

Bull LJ, Kirkby MJ (1997) Gully processes and modeling. Prog Phys Geogr 21(3):354–374

Chow VT (1973) Open-channel hydraulics. McGraw-Hill, Singapore 680

Crouch RJ (1990) Erosion processes and rates for gullies in granitic soils Bathurst, New-South-Wales, Australia. Earth Surf Process and Landf 15(2):169–173

Defersha M, Melesse AM (2012a) Effect of rainfall intensity, slope and antecedent moisture content on sediment concentration and sediment enrichment ratio. CATENA 90(2012):47–52

Defersha M, Melesse AM (2012b) Field-scale investigation of the effect of land use on sediment yield and surface runoff using runoff plot data and models in the Mara River basin, Kenya. CATENA 89:54–64. doi:10.1016/jCATENA.2011.07.010

Defersha MB, Quraishi S, Melesse AM (2011) Interrill erosion, runoff and sediment size distribution as affected by slope steepness and antecedent moisture content. Hydrol Earth Syst Sci Dis 7(4):6447–6489

Defersha M, Melesse AM, McClain M (2012) Watershed scale application of WEPP and EROSION 3D models for assessment of potential sediment source areas and runoff flux in the Mara River Basin, Kenya. CATENA 95:63–72

Engstrom DR, Almendinger JE, Wolin JA (1997) Historical changes in sediment and phosphorus loading to the Upper Mississippi River. In: Final research report prepared for the Metropolitan Council Environmental Services, St. Croix Watershed Research Station, Science Museum of Minnesota, Marine on St. Croix, MN

Gran KB, Belmont P, Day SS, Jennings C, Johnson A, Parker G, Perg L, Wilcock PR (2009) Geomorphic evolution of the Le Sueur River, Minnesota, USA, and implications of current sediment loading. In: James LA, Rathburn SL, and Whittecar GR (eds) Management and Restoration of Fluvial Systems with Broad Historical Changes and Human Impacts: Geological Society of America Special Paper 451. doi: 10.1130/2008.2451(08).

Imeson AC, Kwaad FJPM (1980) Gully types and gully prediction, KNAG Geografisch Tijdschrift XIV(5):430–441

Kelley DW, Nater EA (2000) Historical sediment flux from three watersheds into Lake Pepin, Minnesota, USA. J Environ Qual 29(2):561–568

Kirkby MJ, Bull LJ (2000) Some factors controlling gully growth in fine-grained sediments: a model applied in southeast Spain. CATENA 40(2):127–146

Maalim FK, Melesse AM, Belmont P, Gran K (2013) Modeling the impact of land use changes on runoff and sediment yield in the Le Sueur Watershed, Minnesota using GeoWEPP. CATENA 107:35–45

Maalim FK, Melesse AM (2013) Modeling the impacts of subsurface drainage systems on Runoff and Sediment Yield in the Le Sueur Watershed, Minnesota. Hydrol Sci J 58(3):1–17

McCuen RH, Snyder WM (1986) Hydrologic modeling statistical methods and application. Prentice Hall

Martínez-Casasnovas JA (1998) Soil–landscape–erosion. Gully erosion in the Alt Penedes-Anoia (Catalonia, Spain). A spatial information technology approach: spatial databases, GIS and remote sensing. PhD thesis, University of Lleida, Lleida, Spain

Martinez-Casanovas JA, Anton-Fernandez C, Ramos MC (2003) Sediment production in large gullies of the Mediterranean area (NE Spain) from high-resolution digital elevation models and geographical information systems analysis. Earth Surf Process Landf 28(5):443–456

Mekonnen M, Melesse A (2011) Soil erosion mapping and hotspot area identification using GIS and remote sensing in northwest Ethiopian highlands, near Lake Tana, In: Melesse A (ed) Nile River Basin: hydrology, climate and water use. Springer, Berlin, pp 207–224. doi:10.1007/978-94-000689-7_10

Melesse AM, Ahmad S, McClain M, Wang X, Lim H (2011) Sediment load prediction in large rivers: ANN approach. Agric Water Manage 98:855–886

Merritt WS, Letcher RA, Jakeman AJ (2003) A review of erosion and sediment transport models. Environ Model Softw 18(8–9):761–799

Meyer-Peter E, Müller R (1948) Formulas for bed-load transport. In: Proceedings of the 2nd meeting of the International Association for Hydraulic Structures Research, Delft, Netherlands, pp 39–64

Minnesota Pollution Control Agency (MPCA), Department of Agriculture Minnesota, State University Minnesota, Water Resources Center Manko, and Environmental Services Metropolitan Council (2007) Summary of surface water quality monitoring 2000–2005, State of The Minnesota River, p 20

Mohammed H, Assen M, Alamirew T, Melesse AM (2015) Modeling of sediment yield in Maybar gauged watershed using swat, northeast Ethiopia. CATENA 127:191–205

Moriasi DN, Arnold JG, Liew MWV, Bingner RL, Harmel RD, Veith TL (2007) Model evaluation guidelines for systematic quantification of accuracy in watershed simulations. Trans ASABE 50(3):885–900

Msagahaa J, Ndomba PM, Melesse AM (2014) Modeling sediment dynamics: effect of land use, topography and land management, In: Melesse AM, Abtew W, Setegn S (eds) Nile River Basin: ecohydrological challenges, climate change and hydropolitics, pp 165–192

Nash JE, Sutcliffe JV (1970) River flow forecasting through conceptual models part I—a discussion of principles. J Hydrol 10(3):282–290

Poesen J (1993) Gully typology and gully control measures in the European loess belt. In: Wickerek S (ed) Farm land erosion in temperate plains environment and hills. Elsevier Science Publishers, Amsterdam, pp 221–239

Poesen J, Nachtergaele J, Verstraeten G, Valentin C (2003) Gully erosion and environmental change: importance and research needs. CATENA 50(2–4):91–133

Setegn SG, Srinivasan R, Dargahil B, Melesse AM (2009) Spatial delineation of soil erosion prone areas: application of SWAT and MCE approaches in the Lake Tana Basin, Ethiopia. Hydrol Process Spec Iss: Nile Hydrol 23(26):3738–3750

Setegn SG, Bijan Dargahi B, Srinivasan R, Melesse AM (2010) Modelling of sediment yield from Anjeni Gauged Watershed, Ethiopia using SWAT. JAWRA 46(3):514–526

Sidorchuk A (1999) Dynamic and static models of gully erosion. CATENA 37(3–4):401–414

Simons DB, Senturk F (1992) The universal soil loss equation. Sediment Transport technology water and sediment dynamics. Water Resources Publications, Littleton, pp 496–502

Smith DD, Wischmeier WH (1957) Factors affecting sheet and rill erosion. Trans Am Geophys Union 38:889–896

Soil Science Society of America (2001) Glossary of soil science terms, Madison, WI

Thomas JT, Iverson NR, Burkart MR, Kramer LA (2004) Long-term growth of a valley-bottom gully, western Iowa. Earth Surface Process Landforms 29(8):995–1009

Torri D, Borselli L (2003) Equation for high-rate gully erosion. CATENA 50(2–4):449–467

Valentin C, Poesen J, Yong L (2005) Gully erosion: impacts, factors and control. CATENA 63(2–3):132–153

Vandekerckhove L, Muys B, Poesen J, Weerdt BD, Coppe´ N (2001a) A method for dendrochronological assessment of medium-term gully erosion rates. CATENA 45(2):123–161

Vandekerckhove L, Poesen J, Govers G (2003) Medium-term gully headcut retreat rates in Southeast Spain determined from aerial photographs and ground measurements. CATENA 50 (2–4):329–352

Vandekerckhove L, Poesen J, Wijdenes DO, Gyssels G (2001b) Short-term bank gully retreat rates in Mediterranean environments. CATENA 44(2):133–161

Wang X, Garza J, Whitney M, Melesse AM, Yang W (2008) Prediction of sediment source areas within watersheds as affected by soil data resolution, In: Findley PN (ed) Environmental modelling: new research. Nova Science Publishers, Inc., Hauppauge, pp 151–185. ISBN: 978-1-60692-034-3

Wischmeier WH, Smith DD (1978) Predicting rainfall erosion losses—a guide to conservation planning. In: Agriculture handbook No 532, U.S Department of Agriculture

Woodward DE (1999) Method to predict cropland ephemeral gully erosion. CATENA 37(3–4):393–399

Wu Y, Cheng H (2005) Monitoring of gully erosion on the Loess Plateau of China using a global positioning system. CATENA 63(2–3):154–166

Chapter 24
Effect of Filter Press Mud on Compaction and Consistency of Aquert and Fluvent Soils in Ethiopia

Abiy Fantaye, Abebe Fanta and Assefa M. Melesse

Abstract Soil compaction on sugar cane fields has been found to reduce yield and productivity. Influence of filter press mud (FPM), a residue obtained by filtration of the mud in cane juice clarification process, on compaction and consistency of Fluvent and Aquert soils was investigated on experimental plots. For compaction study, standard proctor test was employed and consistency parameters [liquid limit (LL) by drop cone method and plastic limit (PL)] were determined. Infield penetration resistances and basic infiltration were measured. The maximum dry bulk densities (MDBD) of the soils after standard proctor compaction were 1.42 and 1.30 g cm^{-3} at 28 and 29 % critical moisture content (CMC), respectively. For Fluvent, the CMC was at PL and on Aquerts at 66 % of the PL. For both soils, total porosity, degree of saturation and air-filled porosity were significantly improved by FPM application only at the CMC. Moreover, FPM increased cone index (CI) of the soil. On all experimental fields, the CI was below the critical value, but CI increment was observed around 20–25 cm depths after tillage. Basic infiltrations of the soils were 6.5 for Fluvent and 4.4 mm/h for Aquert, but effect of FPM was statistically non-significant.

Keywords Filter press mud · Maximum dry bulk densities · Critical moisture content · Consistency

---

A. Fantaye
RTS P.O. Box 15, Wonji, Ethiopia

A. Fanta
Haramaya University, P.O. Box 138, Dire Dawa, Ethiopia
e-mail: fantaalemitu@gmail.com

A.M. Melesse (✉)
Department of Earth and Environment, Florida International University, Miami, FL 33199, USA
e-mail: melessea@fiu.edu

© Springer International Publishing Switzerland 2016

A.M. Melesse and W. Abtew (eds.), *Landscape Dynamics, Soils and Hydrological Processes in Varied Climates*, Springer Geography,
DOI 10.1007/978-3-319-18787-7_24

24.1 Introduction

Filter press mud (FPM) is a residue obtained by filtration of the mud in cane juice clarification process as by-product (Blackburn 1991; Fauconnier 1993). It consists of several organic compounds and plant nutrients among which P nutrient is the principal constituent (Blackburn 1991).

Thus, FPM is also considered as fertilizer especially when the soil is deficient in P or topsoil is removed (Blackburn 1991). It can be also source of N (Fauconnier 1993), improves soil organic matter for cane production (Blackburn 1991). Moreover, literature shows that the response of cane to applied FPM is, beyond its nutrient supply, it improves the soil physical properties (Blackburn 1991). However, in all the Ethiopian sugar estates, at present, it is dumped in pits without any use at commercial level.

Before the 1970s, decomposed and dry FPM was applied to cane fields at a rate of 15 t ha^{-1} (ARS 1981). However, the use of FPM was abandoned after 1970s, thinking it will aggravate development of ground water table (Aschalew 1981; ARS 1981), moreover, lack of concrete information regarding the rate to be applied, when and where to apply the fresh and/or dried FPM and the lack of appropriate equipment for spreading the bulky amount of FPM (Tariku 2001) hindered the resumption.

At Wonji-Shoa, the major soil physicochemical constraints to promote production are heavy clay soil and soil compaction (Tariku 2001), seepage from reservoirs and high water table on some places and occurrence of iron chlorosis (Mukerji 2000). Compaction reduces mainly the percentage of macro-pores and partly of mesopores, decreases the pore continuity and increases horizontal orientation of soil pores (Frede 1987; Pagliai 1987). This may reduce soil available water (Kay 1990) and limit root growth (Dexter 1987). On sugar cane field, compaction reduces aeration of roots, root growth and decreases absorption of P (Rao 1990).

Reports show yield loss in sugar cane due to compaction. In Colombia, a minimum of 10 % cane yield reduction was observed due to inter-row compaction and a maximum of 42 % when there was direct damage to the stools, similar results were also reported in South Africa.

Soil compaction problems become a growing concern in the agricultural industry as the level of mechanization increases. The sugar plantations of Ethiopia are one of the most mechanized farms in Ethiopia (Rahmeto 2000). To minimize soil compaction, most of the field operations at Wonji-Shoa, like moulding, fertilizer application and harvesting, are done in dry soil condition and the soils are subsoiled during seedbed preparation to improve the bulk density to the desired depth (ADPM 1993). The effect of tillage on soil moisture availability and storage as well as on grain yield has been studied by Berhe et al. (2013a, b).

However, moisture range for safe operation was not established, only drying of fields was considered as solution. This approach suffers attaining the desired output due to the shallow ground water and unexpected rain in the harvesting season.

Compaction on topsoil can be alleviated when the field is ploughed out. Deep tillage or subsoiling is required to break hard pans at depth. Studies indicate that subsoiling is recommended for the Fluvisol, but it is not recommended for Vertisols, as it is likely to result in more compaction (Ahmad 1996).

Organic matter reduces soil compaction by increasing stability of soil. In addition, the greater amount of molecules of moisture retained around particles of soils with high organic matter content helps such soils to withstand compaction (Paul 1974). The maximum bulk density of soil is significantly negatively correlated with the organic carbon and the silt content (Argon et al. 2000).

Organic materials incorporated to improve soil physical properties and reduce soil compaction include FPM (Paul 1974), wheat straw (Gue'rif 1979), peatmoss (Ohu et al. 1985), green manure (Macrae and Mehuys 1985), corn residue (Gupta et al. 1987), farmyard manure and slightly or highly humified peats (Zhang et al. 1997) and sugar cane residue (cane leaf) (Barzegar et al. 2000).

Soil compaction is often measured indirectly. The usual procedure is to determine the change in soil physical parameter or a set of parameters in response to a compaction force. It includes measuring the change in porosity, density, pore size distribution, conductivity to air and water, soil moisture retention, stress–strain, cone index (CI) and visual appearance of soil fabric (Freitag 1971). Prominently soil bulk density, porosity and CI are used to measure soil compaction. The complexity of the phenomenon of soil compaction requires using a combination of soil mechanical characteristics to assess soil trafficability, workability and plant growth restrictions as related to compaction. Soil bulk density is the most popular parameter to assess the degree of soil compaction that can be related to porosity and void ratio that have more meaning as related to soil behaviour.

Porosity is the most meaningful parameter to describe soil compaction status because; it gives a measure of the soil volume available to plant root and the capacity of the soil to store water. The total pore and void ratio can be calculated from the bulk density and they have the advantage of being dimensionless and dependent of particle density (Soane et al. 1981).

Soane et al. (1981) used the cone penetrometer for assessing soil compaction under wheels. Cone resistance effects due to compaction may be much larger than bulk density (Voorhees et al. 1978). Moreover, cone penetrometer allows faster and easier readings than bulk density measurements (USDA 2003).

In this study, the application of FPM before land preparation and incorporated by the tillage operation and incorporated after land preparation and left for one year was investigated with the objectives of determining the influence of FPM on soil maximum dry bulk density (MDBD) and critical moisture content (CMC), and its relation with the soil consistence on the Aquert and Fluvent soils of Wonji-Shoa sugar estate, Ethiopia.

24.2 Materials and Methods

24.2.1 Site Description

24.2.1.1 Geographic Location

The study was conducted in the central part of the East Africa rift valley system at 39° 10′–39° 20′ longitude and 8° 30′–8° 35′ latitudes at an altitude of 1540 m above sea level (masl) at Wonji-Shoa in Ethiopia (Fig. 24.1).

Wonji-Shoa is the first sugar estate in Ethiopia established in 1954, which is located in the central part of east African rift valley near Awash River, comprising Wonji and Shoa sugar factories with crushing capacity of 1400 and 1600 ton cane per day, respectively. The two factories produce 64,812 t of sugar on average annually. In addition, 20,921 t of molasses (90° brix), 177,283 t bagasses and 21,713 t of FPM are produced annually as by-product.[1]

24.2.1.2 Climate

Mean annual rainfall at Wonji-Shoa is 830 mm with mean air temperature of 20.8 °C. The climate is subtropical monsoon (Cwal) according to Koppen, or semiarid (AD'ad') in Thornthwaite classification system.

24.2.1.3 Soils

On the basis of soil water holding capacity and CMCs, the sugar estates classify the soils into five major soil management units (Kuipers 1961). Soil management units are designated using capital letters and numbers as subscript designate as A_1, A_2, BA_2, $B_{1.4}$ and C_1 soil management units. Moreover, the sugar estates group these five soil management units into two; A_1, A_2 and BA_2 soil management units which are clay soils with high moisture holding capacity and high CMC having serious water logging and drainage problem unless proper management is practiced. Thus, the fields are cultivated with sugar cane after fallow and *Crotalaria junceae* is grown in fallow period as green manure. The second group includes $B_{1.4}$ and C_1 soil management units that are coarser in texture and has a natural drainage with very little or no drainage problem; therefore, there is no fallow period, and cultivation in cane after cane production schedule.

[1]*Summarized from factory conference 2003 ten years average.*

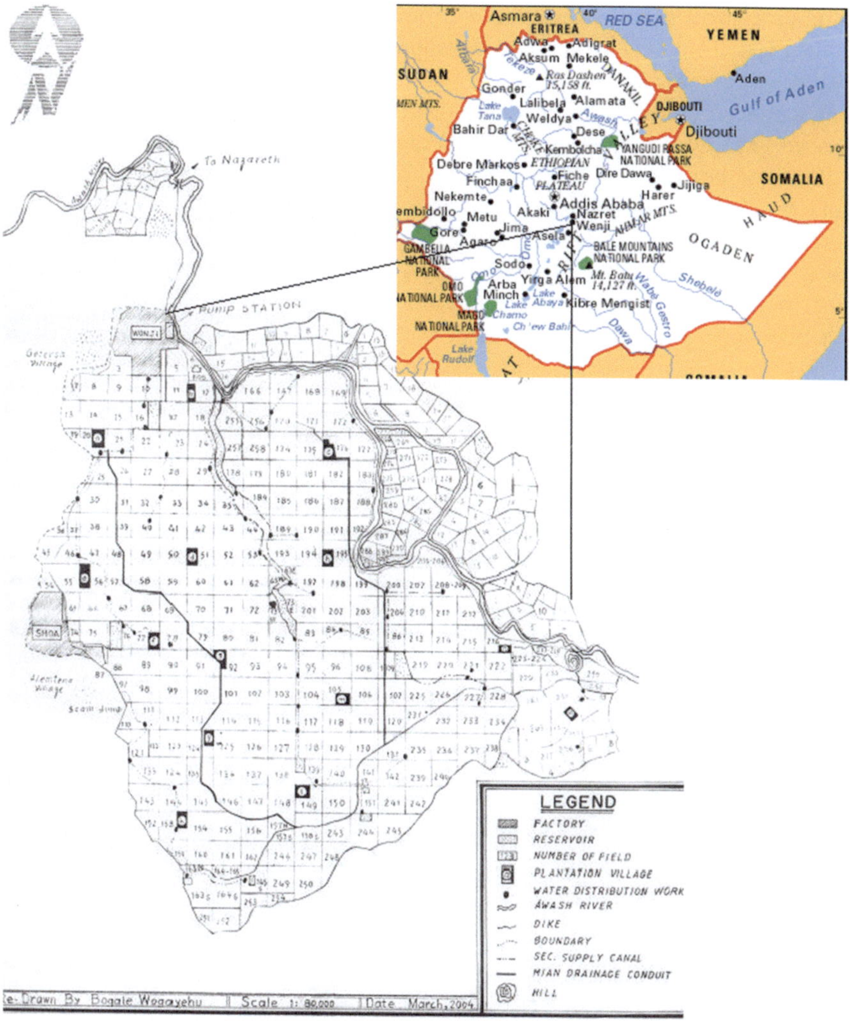

Fig. 24.1 Map of the study area

24.2.2 Experimental Procedures

The experiment was conducted on two selected fields on clay soils (Aquert) and coarse texture soils (Fluvent) by applying FPM. Additional samples were collected from similar experiment fields that were treated with FPM one year ago. The treatments were surface incorporation of FPM by the conventional land preparation method at a rate of 60, 120 t ha^{-1} and untreated control. These rates were selected based on the previous experiment results (Paul 1974; ARS 1979; Barzegar et al. 2000).

From composite soil samples taken from the experiment fields at the beginning of the experiment, soil texture was determined by hydrometer method (Boyoucous 1927), specific gravity of soil by pycnometer method (Liu and Evett 1984) from which particle density was calculated. The organic carbon and N contents of the applied FPM was analysed by digestion and dry combustion methods.

Standard proctor compaction tests (Liu and Evett 1984) and from the Atterburg limit the liquid limit (LL) by drop cone method and the plastic limit (PL) as described by Liu and Evett (1984) were made in laboratory following 2*2*3 factorial (2 soil types, 2 time of application and 3 levels of FPM) randomized complete block design (RCBD) with three replications. The analysis was carried out in soil test laboratory of Transport and Construction Design Sh. Co. (TCDSCo) at Addis Ababa.

In the field, the penetration resistance of the soil and the basic water infiltration rate were measured using 30° cone penetrometer and double-ring infiltrometer, respectively. The penetration resistance was measured at known moisture content. The CI is calculated as the penetration resistance in kg per unit base area of the cone in cm^2 ($kg\ cm^{-2}$) which is equivalent to bar.

Total porosity of soil (*f*) at MDBD was calculated using the formula:

$$f = 1 - \frac{P_b}{P_s} \tag{24.1}$$

where

P_b bulk density

P_s particle density

Degree of saturation (*S*) at CMC was calculated as follows:

$$S = \frac{WP_b/P_w}{f} \tag{24.2}$$

where

W gravimetric moisture content

P_b bulk density

P_w density of water

Air-filled porosity (f_a) at MDBD and CMC was calculated as follows:

$$f_a = f(1 - S) \tag{24.3}$$

Similar studies made elsewhere were reviewed to select appropriate maximum and critical bulk density of these soils for cane cultivation that can represent the situation of Wonji-Shoa.

24.2.3 Statistical Analysis

The data were analysed using MSTATC statistical package (MSTAT 1988). For penetration resistance, covariance analysis was employed taking moisture content of soil as covariate. Then, least significant differences (LSD) were calculated to compare pairs of means at 5 and/or 1 % level of probabilities.

24.3 Results and Discussion

24.3.1 Initial Soil Analysis

From profile, pits opened and soil analysed from the coarse texture and clay soils qualify in Fluvent and Aquert suborder following soil taxonomy (Soil Survey Staff 1996). On both soils, the ground water table was shallow. In the Fluvents, the water level fluctuated from 45 to 70 cm, while on Aquert soil, it was between 30 and 50 cm depth from the surface. But on Fluvents, there was sand subsoil layer below the sandy loam top soil; thus, it drains quickly. The physicochemical properties of the Ap horizon are presented in (Table 24.1). The FPM applied about 32 % OC, and the physicochemical composition of FPM is presented in Table 24.2.

24.3.2 Soil Bulk Density–Moisture Curve from Standard Proctor Test

24.3.2.1 Maximum Dry Bulk Density

Analysis of variance indicated that soil type, method and time of FPM application and its interaction with soil have significant effect ($P < 0.05$) on MDBD (Table 24.3). MDBD of the two soil types have inherent differences. As shown in Fig. 24.2, the MDBD for Fluvent was 1.42 g cm^{-3} became 1.39 and 1.42 g cm^{-3} at 60 and

Table 24.1 Soil physicochemical properties at the experimental site

Parameter	A_2 (Aquert)	C_1 (Fluvent)
Sand (%)	16	38
Silt (%)	30	22
Clay (%)	54	40
Textural class	Clay	Clay loam
Particle density (g/cm^3)	2.25	2.31
Organic C (%)	1.68	1.07
pH (1:2.5 H_2O)	7.6	7.6
EC (dS/m) (1:2.5 H_2O)	0.265	0.122

Table 24.2 Chemical composition of FPM

	pH	EC (dS/m)	N (%)	P (%)	K (%)	OC (%)	C/N
FPM	6.1	2.85	1.20 ± 0.05	0.73 ± 0.06	0.45 ± 0.08	32.81 ± 3.00	27

Table 24.3 Effect of FPM soil type and time of FPM application on MDBD and CMC

Parameter		MDBD	CMC
Main effect	FPM	ns	ns
	Soil type	**	**
	Time	**	**
Interaction	FPM*Soil	ns	ns
	FPM*Time	**	**
	Soil*Time	**	**
	FPM*Soil*Time	ns	ns

NB **means statistically significant at $P < 0.01$; *ns* means statistically non-significant

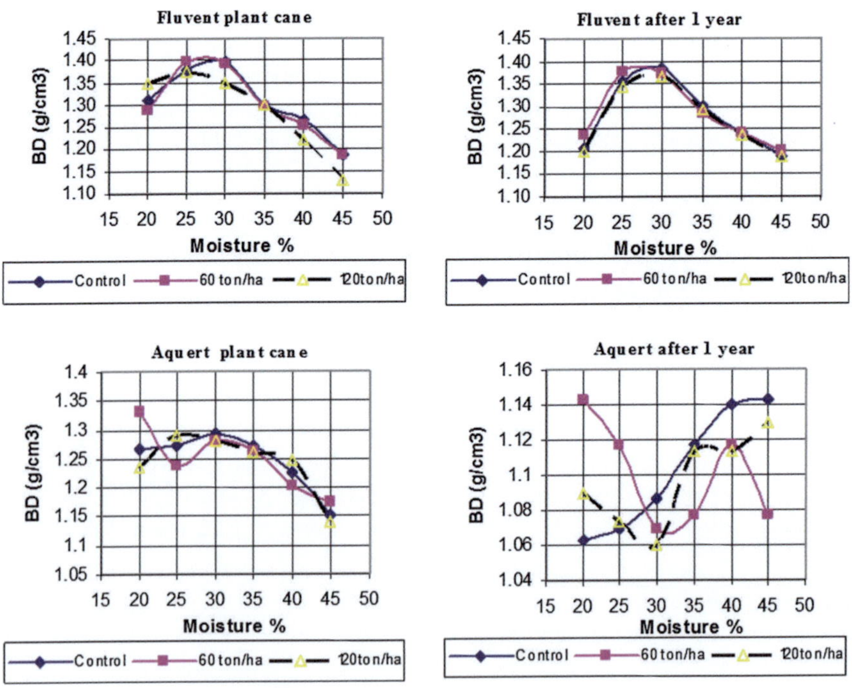

Fig. 24.2 Standard proctor compaction curve of Fluvent and Aquert soils

120 t ha^{-1} FPM applications, respectively. For Aquert, the MDBD was 1.303 g cm^{-3}, and this became 1.128 and 1.297 g cm^{-3} for 60 and 120 t ha^{-1} FPM applications, respectively. However, the bulk density obtained at the rate of 60 and 120 t ha^{-1}. FPM application rates were statistically non-significant. Aquerts have dome-shaped

moisture–density curve, while Aquert have M-shaped. Probably, this could be explained by the degree of saturation at CMC.

One year after FPM application, statistically different ($P < 0.05$) MDBD reduction is obtained only for Aquerts, which is 1.153 and 1.143 g cm^{-3} for 60 and 120 t ha^{-1} FPM, respectively, w hile for Fluvent, it is 1.417 and 1.387 g cm^{-3} for 60 and 120 t ha^{-1} FPM application, respectively (Table 24.4). Application of FPM reduced MDBD of soil even when it was applied on green manure fields of Aquert. The two rates of application investigated had effect on MDBD after a year. Therefore, this study has shown the long-term effect of FPM.

There was no information regarding the maximum and critical bulk density of the soils for cane cultivation in Wonji/Shoa. Therefore, similar studies made elsewhere were reviewed to select the appropriate value that can represent the situation of Wonji/Shoa. The literature shows that the critical bulk density values for sugar cane production vary greatly from soil to soil. Previous studies made on sugar cane production indicate critical bulk density range of 1.4 g cm^{-3} (Rao 1990) up to 1.8 and 1.9 g cm^{-3} (Blackburn 1991). However, according to Hartemink (1998), the threshold values were reported to be about 1.3 g cm^{-3} for the Fluvisols and 1.2 g cm^{-3} for Vertisols in Papua New Guinea for the top soils and were slightly higher for subsoil. The result obtained from both soils of Wonji/Shoa is just above the threshold values obtained by Hartemink (1998). The threshold values obtained by Rao (1990) may be applicable only for C_1 soil cycles.

So, due to the variability of the threshold values and the maximum bulk density among different soil types, the value of bulk density measured in the field did not show the magnitude of soil compaction unless it is associated with MDBD and critical soil bulk density. Therefore, as shown in Fig. 24.2, bulk density has to be reported as a proportion of the MDBD and the critical soil bulk density of the soil has to be stated when it is used as an indicator of soil compaction level. Thus, the threshold values of Hartemink (1998) (1.3 g cm^{-3} for the Fluvisols and 1.2 g cm^{-3} for Vertisols) were used for calculating the ratio for the soils of Wonji-Shoa.

24.3.2.2 Critical Moisture Content

The CMC was found to be 28.0 % for Fluvent, and this changed to 27.6 % at both rates of FPM application. On Aquert, the CMC was 29.0 %, which became 32.33 and 28.66 % at 60 and 120 t ha^{-1} FPM applications, respectively (Fig. 24.3). The

Table 24.4 Interaction effect of FPM, soil type and time on MDBD

FPM (t ha^{-1})	Aquert (A_2 SMG)		Fluvent (C_1 SMG)	
	After application	After one year	After application	After one year
0	1.303^b	1.153^c	1.420^a	1.403^a
60	1.287^b	1.133^c	1.390^a	1.407^a
120	1.297^b	1.143^c	1.417^a	1.387^a

LSD value = 0.0535. $^{a, b, c}$: values with similar letter are non-significant at $P < 0.05$

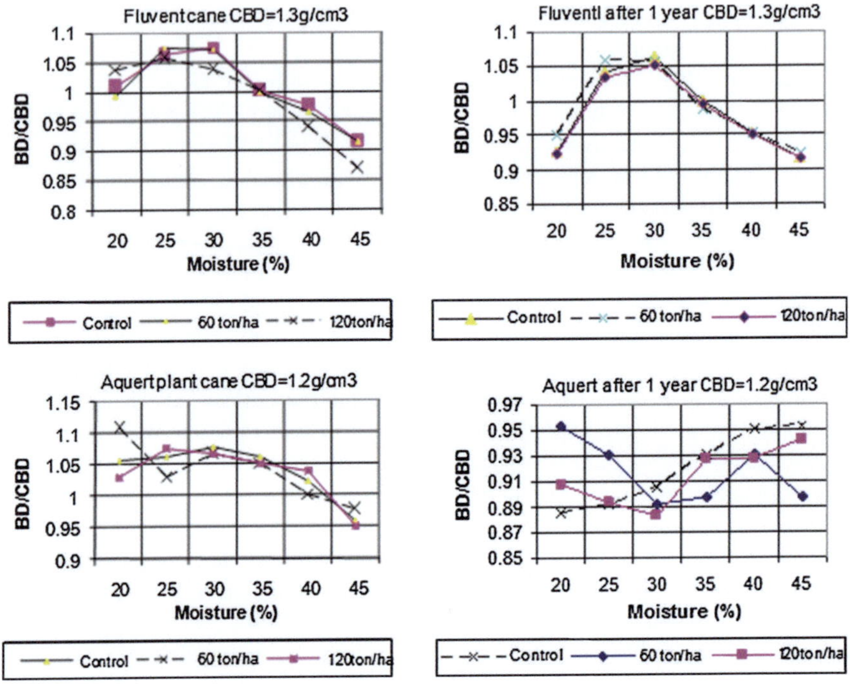

Fig. 24.3 Bulk density as a proportion of critical bulk density

difference was statistically significant at $P < 0.01$. The main effect of year after application and its interaction with soil was only significant for Aquert ($P < 0.01$), which was found to be 33 and 42 % at 60 and 120 t ha^{-1} FPM, respectively. The main effect of FPM treatments and the interaction effect of soil, FPM and year were statistically non-significant.

24.3.3 Soil Consistence

Fluvent soils have PL at 28.95 % and LL at 49.63 % moisture content and the plasticity index (PI) was 20.55. Application of FPM did not change significantly the LL, PL and PI of the soil. For Aquert soils, the PL was at 46.38 % and the LL was at 78.00 % moisture content and the PI was 35.31. Application of FPM increased the LL significantly to 81.61 and 80.4 % and the PL to 48.7 % at 60 and 120 t ha^{-1} application rates, respectively. Therefore, the PI was reduced to 32.8 and 31.7 at the respective FPM rates.

In Fluvent, the CMC was at the PL (CMC = 28 %, PL = 28 %), but for Aquert, it was at 62 % of the PL (CMC = 29 %, PL = 46 %). The type and percentage of clay showed similar effect as that of compaction. As studied by Barzegar et al. (2000), when the compaction load is increased starting from lower than the proctor test load to standard Proctor test load, as the load increases, the CMC was reduced to 0.8 PL. Similarly, this study has shown as the expansiveness nature of the soil and per cent of clay increase the CMC is reduced below the PL. This shows sensitivity of Aquerts to soil compaction than Fluvents. Moreover, incorporating FPM on Aquerts reduced compaction caused by heavy load at water content lower than the PL. Similar result was also obtained by Barzegar et al. (2000) by incorporating sugar cane residue at a rate of 60 t ha^{-1}. The PI of Fluvent is lower (20) than Aquert (35) showing that Aquerts are more plastic than Fluvents. Fluvents and Aquerts treated with FPM a year ago manifested increased plasticity.

24.3.4 Total Porosity at MDBD

After compacting the soils to maximum attainable bulk density at 25, 30, 35, 40 and 45 % moisture content, Fluvents had porosity of 40.1, 40.3, 43.8, 46.1 and 49.5 %, while for Aquerts, the values were 43.8, 43.0, 43.9, 45.6 and 48.8 % at respective moisture per cents (Fig. 24.4). Total porosity being the function of both particle and

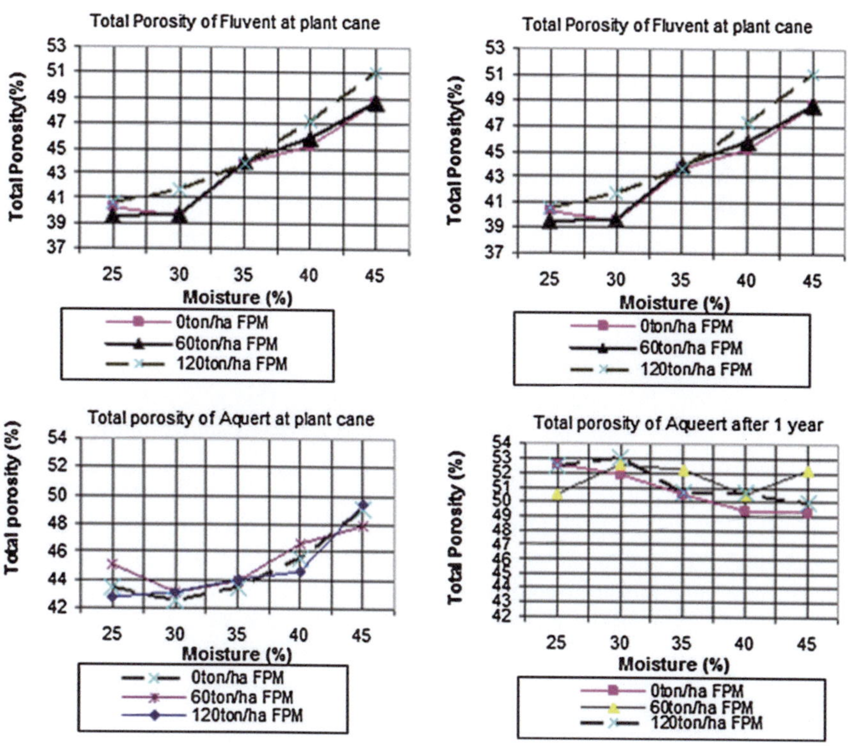

Fig. 24.4 Total porosity after standard proctor compaction

bulk density as shown in Eq. 24.1, the higher porosity expected from Fluvent is compensated by higher bulk density obtained under standard proctor compaction; as a result, the effect of particle density is reduced. Application of FPM did not significantly change porosity at all the moisture contents. However, application of FPM affected porosity only at CMC due to its effect on reducing MDBD.

FPM incorporated before a year on Fluvents resulted in 42.7 % total porosity, and on similar field of Aquerts, the total porosity was 51.2 %. According to Rao (1990) for sugar cane, the optimum soil porosity is 50 %. On both soils, porosity after standard proctor compaction was lower than recommended by Rao (1990) except for Aquerts treated before a year. This shows that on Aquerts application of FPM has cumulative effect over time in improving the total porosity even after compaction.

24.3.5 Degree of Saturation at CMC

Degree of saturation after standard proctor compaction of Fluvent was about 100 %, and application of FPM at 60 and 120 t ha^{-1} did not bring any significant difference. In fields treated with FPM a year ago, saturation was around 100 % at all levels of FPM application. For Aquerts, the degree of saturation was 96.1 %, and similar to Fluvents, FPM treatment at 60 and 120 t ha^{-1} did not change the degree of saturation significantly. However, on fields treated before a year, the values were 77.7, 74.1 and 75.7 % at 0, 60 and 120 t ha^{-1}, respectively (Table 24.5).

The moisture content, at which the soils became saturated, after standard proctor compaction, was 30 % for Fluvent and 45 % for Aquert. These values were just at the PLs of the soils (Tables 24.5 and 24.6). This shows that before the PL is reached cohesiveness and internal friction of the soils contribute in resisting soil compaction. After the PL when the soil particles just start to slide over one another and the cohesiveness is reduced, the soil moisture reinforces the soil fabrics from further increment of bulk density. This explains why the proctor compaction curve is dome shaped which the maximum bulk density is obtained at the junction of reduced cohesiveness with the minimum amount of moisture. This is clearly obtained on Fluvent soils and its CMC is just at the PL (PL = 29 %) as shown in the standard proctor test curve.

In Aquerts, however, there were two maximum bulk densities observed especially when the soils had higher organic matter content. Therefore, the combined effect of clay type and the amount of organic matter content has changed the soils' cohesiveness and moisture content interaction. Results of the experiment have shown that the first maximum bulk density for these soils was at 0.62 PL (28 % moisture) and the second maximum density was obtained just at the PL (PL = 46 %). The first CMC could be even lower than the value obtained in this experiment since the graph showed an increasing trend at the beginning of compaction test. It is recommended to undertake detailed study considering lower moisture ranges.

Table 24.5 Degree of saturation of Fluvent (C_1 soil management unit) at different gravimetric moisture contents and bulk density

Gravimetric moisture content (%)	Bulk density (g/cm^3)													
	0.90	0.95	1.00	1.05	1.10	1.15	1.20	1.25	1.30	1.35	1.40	1.45	1.50	1.55
25	0.37	0.40	0.44	0.48	0.52	0.57	0.62	0.68	0.74	0.81	0.89	0.97	1.00	1.00
26	0.38	0.42	0.46	0.50	0.55	0.60	0.65	0.71	0.77	0.84	0.92	1.00	1.00	1.00
27	0.40	0.44	0.48	0.52	0.57	0.62	0.67	0.74	0.80	0.88	0.96	1.00	1.00	1.00
28	0.41	0.45	0.49	0.54	0.59	0.64	0.70	0.76	0.83	0.91	0.99	1.00	1.00	1.00
[a]29	0.43	0.47	0.51	0.56	0.61	0.66	0.72	0.79	0.86	0.94	1.00	1.00	1.00	1.00
30	0.44	0.48	0.53	0.58	0.63	0.69	0.75	0.82	0.89	0.97	1.00	1.00	1.00	1.00
31	0.46	0.50	0.55	0.60	0.65	0.71	0.77	0.84	0.92	1.00	1.00	1.00	1.00	1.00
32	0.47	0.52	0.56	0.62	0.67	0.73	0.80	0.87	0.95	1.00	1.00	1.00	1.00	1.00
33	0.49	0.53	0.58	0.64	0.69	0.76	0.82	0.90	0.98	1.00	1.00	1.00	1.00	1.00
34	0.50	0.55	0.60	0.65	0.71	0.78	0.85	0.93	1.00	1.00	1.00	1.00	1.00	1.00
35	0.52	0.56	0.62	0.67	0.73	0.80	0.87	0.95	1.00	1.00	1.00	1.00	1.00	1.00
36	0.53	0.58	0.63	0.69	0.76	0.82	0.90	0.98	1.00	1.00	1.00	1.00	1.00	1.00
37	0.55	0.60	0.65	0.71	0.78	0.85	0.92	1.00	1.00	1.00	1.00	1.00	1.00	1.00
38	0.56	0.61	0.67	0.73	0.80	0.87	0.95	1.00	1.00	1.00	1.00	1.00	1.00	1.00
39	0.57	0.63	0.69	0.75	0.82	0.89	0.97	1.00	1.00	1.00	1.00	1.00	1.00	1.00
40	0.59	0.65	0.71	0.77	0.84	0.92	1.00	1.00	1.00	1.00	1.00	1.00	1.00	1.00
41	0.60	0.66	0.72	0.79	0.86	0.94	1.00	1.00	1.00	1.00	1.00	1.00	1.00	1.00
42	0.62	0.68	0.74	0.81	0.88	0.96	1.00	1.00	1.00	1.00	1.00	1.00	1.00	1.00
43	0.63	0.69	0.76	0.83	0.90	0.98	1.00	1.00	1.00	1.00	1.00	1.00	1.00	1.00
44	0.65	0.71	0.78	0.85	0.92	1.00	1.00	1.00	1.00	1.00	1.00	1.00	1.00	1.00
45	0.66	0.73	0.79	0.87	0.94	1.00	1.00	1.00	1.00	1.00	1.00	1.00	1.00	1.00
46	0.68	0.74	0.81	0.89	0.97	1.00	1.00	1.00	1.00	1.00	1.00	1.00	1.00	1.00
47	0.69	0.76	0.83	0.90	0.99	1.00	1.00	1.00	1.00	1.00	1.00	1.00	1.00	1.00

(continued)

Table 24.5 (continued)

Gravimetric moisture content (%)	Bulk density (g/cm^3)													
	0.90	0.95	1.00	1.05	1.10	1.15	1.20	1.25	1.30	1.35	1.40	1.45	1.50	1.55
48	0.71	0.77	0.85	0.92	1.00	1.00	1.00	1.00	1.00	1.00	1.00	1.00	1.00	1.00
49	0.72	0.79	0.86	0.94	1.00	1.00	1.00	1.00	1.00	1.00	1.00	1.00	1.00	1.00
50	0.74	0.81	0.88	0.96	1.00	1.00	1.00	1.00	1.00	1.00	1.00	1.00	1.00	1.00
Total porosity	0.61	0.59	0.57	0.55	0.52	0.50	0.48	0.46	0.44	0.42	0.39	0.37	0.35	0.33
Void ratio	1.57	1.43	1.31	1.20	1.10	1.01	0.93	0.85	0.78	0.71	0.65	0.59	0.54	0.49

[a]Plastic limit

Table 24.6 Degree of saturation of Aquert (A_2 soil management unit) at different gravimetric moisture contents and bulk density

Gravimetric moisture content (%)	Bulk density (g/cm^3)													
	0.90	0.95	1.00	1.05	1.10	1.15	1.20	1.25	1.30	1.35	1.40	1.45	1.50	1.55
25	0.38	0.41	0.45	0.49	0.54	0.59	0.64	0.70	0.77	0.84	0.92	1.00	1.00	1.00
26	0.39	0.43	0.47	0.51	0.56	0.61	0.67	0.73	0.80	0.88	0.96	1.00	1.00	1.00
27	0.41	0.44	0.49	0.53	0.58	0.63	0.69	0.76	0.83	0.91	1.00	1.00	1.00	1.00
28	0.42	0.46	0.50	0.55	0.60	0.66	0.72	0.79	0.86	0.94	1.00	1.00	1.00	1.00
29	0.43	0.48	0.52	0.57	0.62	0.68	0.74	0.81	0.89	0.98	1.00	1.00	1.00	1.00
30	0.45	0.49	0.54	0.59	0.65	0.70	0.77	0.84	0.92	1.00	1.00	1.00	1.00	1.00
31	0.46	0.51	0.56	0.61	0.67	0.73	0.80	0.87	0.95	1.00	1.00	1.00	1.00	1.00
32	0.48	0.53	0.58	0.63	0.69	0.75	0.82	0.90	0.98	1.00	1.00	1.00	1.00	1.00
33	0.49	0.54	0.59	0.65	0.71	0.78	0.85	0.93	1.00	1.00	1.00	1.00	1.00	1.00
34	0.51	0.56	0.61	0.67	0.73	0.80	0.87	0.95	1.00	1.00	1.00	1.00	1.00	1.00
35	0.52	0.58	0.63	0.69	0.75	0.82	0.90	0.98	1.00	1.00	1.00	1.00	1.00	1.00
36	0.54	0.59	0.65	0.71	0.77	0.85	0.92	1.00	1.00	1.00	1.00	1.00	1.00	1.00
37	0.55	0.61	0.67	0.73	0.80	0.87	0.95	1.00	1.00	1.00	1.00	1.00	1.00	1.00
38	0.57	0.62	0.68	0.75	0.82	0.89	0.98	1.00	1.00	1.00	1.00	1.00	1.00	1.00
39	0.58	0.64	0.70	0.77	0.84	0.92	1.00	1.00	1.00	1.00	1.00	1.00	1.00	1.00
40	0.60	0.66	0.72	0.79	0.86	0.94	1.00	1.00	1.00	1.00	1.00	1.00	1.00	1.00
41	0.61	0.67	0.74	0.81	0.88	0.96	1.00	1.00	1.00	1.00	1.00	1.00	1.00	1.00
42	0.63	0.69	0.76	0.83	0.90	0.99	1.00	1.00	1.00	1.00	1.00	1.00	1.00	1.00
43	0.64	0.71	0.77	0.85	0.92	1.00	1.00	1.00	1.00	1.00	1.00	1.00	1.00	1.00
44	0.66	0.72	0.79	0.87	0.95	1.00	1.00	1.00	1.00	1.00	1.00	1.00	1.00	1.00
45	0.67	0.74	0.81	0.89	0.97	1.00	1.00	1.00	1.00	1.00	1.00	1.00	1.00	1.00
[a]46	0.69	0.76	0.83	0.90	0.99	1.00	1.00	1.00	1.00	1.00	1.00	1.00	1.00	1.00
47	0.70	0.77	0.85	0.92	1.00	1.00	1.00	1.00	1.00	1.00	1.00	1.00	1.00	1.00

(continued)

Table 24.6 (continued)

Gravimetric moisture content (%)	Bulk density (g/cm^3)													
	0.90	0.95	1.00	1.05	1.10	1.15	1.20	1.25	1.30	1.35	1.40	1.45	1.50	1.55
48	0.72	0.79	0.86	0.94	1.00	1.00	1.00	1.00	1.00	1.00	1.00	1.00	1.00	1.00
49	0.73	0.81	0.88	0.96	1.00	1.00	1.00	1.00	1.00	1.00	1.00	1.00	1.00	1.00
50	0.75	0.82	0.90	0.98	1.00	1.00	1.00	1.00	1.00	1.00	1.00	1.00	1.00	1.00
Total porosity	0.60	0.58	0.56	0.53	0.51	0.49	0.47	0.45	0.42	0.40	0.38	0.36	0.34	0.31
Void ratio	1.50	1.37	1.25	1.15	1.05	0.96	0.88	0.80	0.73	0.67	0.61	0.55	0.50	0.45

[a]Plastic limit

The resistance of Aquerts for compaction at lower degree of saturation shows that the soils can withstand compaction by being cohesive when the particles are close enough to develop the cohesive force equivalent to the applied one; in addition, there were some considerable amount of pores that survived under applied forces by standard proctor compaction. This shows that, there could be a possibility of further compaction if these soils are subjected to higher compaction pressure. For such soils, reduction of the axle load may lead to control of soil compaction. Therefore, it is recommended to test the response of the soils by varying compaction pressure too. In Fluvent, the soils are saturated at the end of every standard proctor test trial; thus, the water in the pores was responsible for resisting further reduction in volume.

24.3.6 Air-Filled Porosity at MDBD and CMC

Air-filled porosity in Fluvents was 1.4 % and that of Aquerts was 3.2 % after standard proctor compaction. Application of FPM did not change the air-filled porosity on both soils; however, the Aquerts treated before a year had 12 % air-filled porosity and application of 60 and 120 t ha^{-1} FPM did not change it significantly.

Since the air-filled porosity varies with moisture content and total porosity (Tables 24.7 and 24.8) to compare the two soil types, it is better to consider the air-filled porosity of the soils at similar physical state. The air-filled porosity of these soil cycles at PL, when they have bulk density of 1.1 g cm^{-3}, Fluvent had 20 % air-filled porosity (PL = 29 % moisture) and Aquerts only 6 % (PL = 46 % moisture content) (Tables 24.7 and 24.8). This shows that Aquerts have less air-filled porosity than Fluvents due to the higher water holding capacity when they are at the same consistency; therefore, aeration could be a limiting factor in Aquerts when they are compacted.

Aeration is the major limiting factor in compacted soil, especially when the soil has higher moisture. In both soils, the ground water is very near to the surface. Particularly, on Aquerts the water level does not recede greater than 40 cm from the surface. Therefore, this situation will aggravate the problem of plant root suffocation and problems related with salinity.

24.3.7 Physical Properties of Soil Measured from the Experimental Fields

For the two soil types, the basic infiltration and the time elapsed to attain the basic infiltration were statistically non-significant. The value of basic infiltration measured in Fluvents was 6.489 mm h^{-1} and attained after 132 min. On Aquerts, the basic infiltration was 4.356 mm h^{-1} and attained after 127 min.

Table 24.7 Air-filled porosity of Fluvent (C_1 soil management unit)

Gravimetric moisture content (%)	Bulk density (g/cm^3)													
	0.9	0.95	1	1.05	1.1	1.15	1.2	1.25	1.3	1.35	1.4	1.45	1.5	1.55
25	0.39	0.35	0.32	0.28	0.25	0.21	0.18	0.15	0.11	0.08	0.04	0.01	0.00	0.00
26	0.38	0.34	0.31	0.27	0.24	0.20	0.17	0.13	0.10	0.06	0.03	0.00	0.00	0.00
27	0.37	0.33	0.30	0.26	0.23	0.19	0.16	0.12	0.09	0.05	0.02	0.00	0.00	0.00
28	0.36	0.32	0.29	0.25	0.22	0.18	0.14	0.11	0.07	0.04	0.00	0.00	0.00	0.00
[a]29	0.35	0.31	0.28	0.24	0.21	0.17	0.13	0.10	0.06	0.02	0.00	0.00	0.00	0.00
30	0.34	0.30	0.27	0.23	0.19	0.16	0.12	0.08	0.05	0.01	0.00	0.00	0.00	0.00
31	0.33	0.29	0.26	0.22	0.18	0.15	0.11	0.07	0.03	0.00	0.00	0.00	0.00	0.00
32	0.32	0.28	0.25	0.21	0.17	0.13	0.10	0.06	0.02	0.00	0.00	0.00	0.00	0.00
33	0.31	0.28	0.24	0.20	0.16	0.12	0.08	0.05	0.01	0.00	0.00	0.00	0.00	0.00
34	0.30	0.27	0.23	0.19	0.15	0.11	0.07	0.03	0.00	0.00	0.00	0.00	0.00	0.00
35	0.30	0.26	0.22	0.18	0.14	0.10	0.06	0.02	0.00	0.00	0.00	0.00	0.00	0.00
36	0.29	0.25	0.21	0.17	0.13	0.09	0.05	0.01	0.00	0.00	0.00	0.00	0.00	0.00
37	0.28	0.24	0.20	0.16	0.12	0.08	0.04	0.00	0.00	0.00	0.00	0.00	0.00	0.00
38	0.27	0.23	0.19	0.15	0.11	0.07	0.02	0.00	0.00	0.00	0.00	0.00	0.00	0.00
39	0.26	0.22	0.18	0.14	0.10	0.05	0.01	0.00	0.00	0.00	0.00	0.00	0.00	0.00
40	0.25	0.21	0.17	0.13	0.08	0.04	0.00	0.00	0.00	0.00	0.00	0.00	0.00	0.00
41	0.24	0.20	0.16	0.12	0.07	0.03	0.00	0.00	0.00	0.00	0.00	0.00	0.00	0.00
42	0.23	0.19	0.15	0.10	0.06	0.02	0.00	0.00	0.00	0.00	0.00	0.00	0.00	0.00
43	0.22	0.18	0.14	0.09	0.05	0.01	0.00	0.00	0.00	0.00	0.00	0.00	0.00	0.00
44	0.21	0.17	0.13	0.08	0.04	0.00	0.00	0.00	0.00	0.00	0.00	0.00	0.00	0.00
45	0.21	0.16	0.12	0.07	0.03	0.00	0.00	0.00	0.00	0.00	0.00	0.00	0.00	0.00
46	0.20	0.15	0.11	0.06	0.02	0.00	0.00	0.00	0.00	0.00	0.00	0.00	0.00	0.00
47	0.19	0.14	0.10	0.05	0.01	0.00	0.00	0.00	0.00	0.00	0.00	0.00	0.00	0.00

(continued)

Table 24.7 (continued)

Gravimetric moisture content (%)	Bulk density (g/cm³)													
	0.9	0.95	1	1.05	1.1	1.15	1.2	1.25	1.3	1.35	1.4	1.45	1.5	1.55
48	0.18	0.13	0.09	0.04	0.00	0.00	0.00	0.00	0.00	0.00	0.00	0.00	0.00	0.00
49	0.17	0.12	0.08	0.03	0.00	0.00	0.00	0.00	0.00	0.00	0.00	0.00	0.00	0.00
50	0.16	0.11	0.07	0.02	0.00	0.00	0.00	0.00	0.00	0.00	0.00	0.00	0.00	0.00

[a]Plastic limit

Table 24.8 Air-filled porosity of Aquert (A_2 soil management unit)

Gravimetric moisture content (%)	Bulk density (g/cm^3)													
	0.90	0.95	1.00	1.05	1.10	1.15	1.20	1.25	1.30	1.35	1.40	1.45	1.50	1.55
25	0.38	0.34	0.31	0.27	0.24	0.20	0.17	0.13	0.10	0.06	0.03	0.00	0.00	0.00
26	0.37	0.33	0.30	0.26	0.23	0.19	0.16	0.12	0.09	0.05	0.01	0.00	0.00	0.00
27	0.36	0.32	0.29	0.25	0.22	0.18	0.14	0.11	0.07	0.04	0.00	0.00	0.00	0.00
28	0.35	0.31	0.28	0.24	0.20	0.17	0.13	0.10	0.06	0.02	0.00	0.00	0.00	0.00
29	0.34	0.30	0.27	0.23	0.19	0.16	0.12	0.08	0.05	0.01	0.00	0.00	0.00	0.00
30	0.33	0.29	0.26	0.22	0.18	0.14	0.11	0.07	0.03	0.00	0.00	0.00	0.00	0.00
31	0.32	0.28	0.25	0.21	0.17	0.13	0.10	0.06	0.02	0.00	0.00	0.00	0.00	0.00
32	0.31	0.27	0.24	0.20	0.16	0.12	0.08	0.05	0.01	0.00	0.00	0.00	0.00	0.00
33	0.30	0.27	0.23	0.19	0.15	0.11	0.07	0.03	0.00	0.00	0.00	0.00	0.00	0.00
34	0.29	0.26	0.22	0.18	0.14	0.10	0.06	0.02	0.00	0.00	0.00	0.00	0.00	0.00
35	0.29	0.25	0.21	0.17	0.13	0.09	0.05	0.01	0.00	0.00	0.00	0.00	0.00	0.00
36	0.28	0.24	0.20	0.16	0.12	0.08	0.04	0.00	0.00	0.00	0.00	0.00	0.00	0.00
37	0.27	0.23	0.19	0.15	0.11	0.06	0.02	0.00	0.00	0.00	0.00	0.00	0.00	0.00
38	0.26	0.22	0.18	0.14	0.09	0.05	0.01	0.00	0.00	0.00	0.00	0.00	0.00	0.00
39	0.25	0.21	0.17	0.12	0.08	0.04	0.00	0.00	0.00	0.00	0.00	0.00	0.00	0.00
40	0.24	0.20	0.16	0.11	0.07	0.03	0.00	0.00	0.00	0.00	0.00	0.00	0.00	0.00
41	0.23	0.19	0.15	0.10	0.06	0.02	0.00	0.00	0.00	0.00	0.00	0.00	0.00	0.00
42	0.22	0.18	0.14	0.09	0.05	0.01	0.00	0.00	0.00	0.00	0.00	0.00	0.00	0.00
43	0.21	0.17	0.13	0.08	0.04	0.00	0.00	0.00	0.00	0.00	0.00	0.00	0.00	0.00
44	0.20	0.16	0.12	0.07	0.03	0.00	0.00	0.00	0.00	0.00	0.00	0.00	0.00	0.00
45	0.20	0.15	0.11	0.06	0.02	0.00	0.00	0.00	0.00	0.00	0.00	0.00	0.00	0.00
[a]46	0.19	0.14	0.10	0.05	0.01	0.00	0.00	0.00	0.00	0.00	0.00	0.00	0.00	0.00
47	0.18	0.13	0.09	0.04	0.00	0.00	0.00	0.00	0.00	0.00	0.00	0.00	0.00	0.00

(continued)

Table 24.8 (continued)

Gravimetric moisture content (%)	Bulk density (g/cm^3)													
	0.90	0.95	1.00	1.05	1.10	1.15	1.20	1.25	1.30	1.35	1.40	1.45	1.50	1.55
48	0.17	0.12	0.08	0.03	0.00	0.00	0.00	0.00	0.00	0.00	0.00	0.00	0.00	0.00
49	0.16	0.11	0.07	0.02	0.00	0.00	0.00	0.00	0.00	0.00	0.00	0.00	0.00	0.00
50	0.15	0.10	0.06	0.01	0.00	0.00	0.00	0.00	0.00	0.00	0.00	0.00	0.00	0.00

[a]Plastic limit

24.3.7.1 Penetration Resistance

Penetration resistance of the soil was measured to the depth of 45 cm on both soil types after the cane was planted. As per the recommendation of ASAE (1983), mean of six measurements made using 30° cone angle was used. Higher CI value is obtained at 20–25 cm depth on both soil types.

The average CI value for the depth for 0–45 cm and adjusted for moisture content on Fluvents was 1.296 and 1.515 kg cm^{-2} on Fluvent and Aquert soils, respectively; however, the difference was statistically non-significant. But application of FPM has significantly increased the CI of both soil types. On Fluvent soils, the CI became 1.1515 and 2.092 kg cm^{-2}, while on Aquerts, it became 1.746 and 2.273 kg cm^{-2} at 60 and 120 t ha^{-1}, respectively (Fig. 24.5).

Usually, roots cannot penetrate soil compacted to 20 kg cm^{-2} or more; in some soils, there are potential penetration problems at 9.66 kg cm^{-2} (USDA 2003). CI value of 2 kg cm^{-2} was suggested by Torres and Villegas (1993), Dick (1987) as normal value for soil cropped with sugar cane. The CI of the soils of Wonji/Shoa has the peak value obtained from 20 to 25 cm depth of the soil as shown in Table 24.9 that might be caused by tillage operation. However, the CI value does not exceed the critical vale of 9.66 kg cm^{-2} suggested by USDA soil quality institute (USDA 2003).

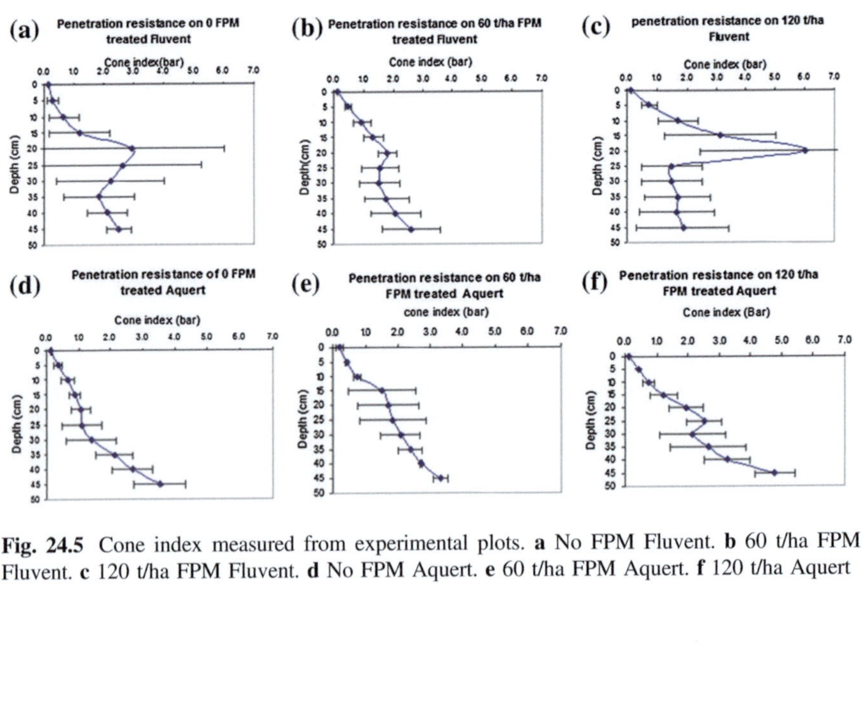

Fig. 24.5 Cone index measured from experimental plots. **a** No FPM Fluvent. **b** 60 t/ha FPM Fluvent. **c** 120 t/ha FPM Fluvent. **d** No FPM Aquert. **e** 60 t/ha FPM Aquert. **f** 120 t/ha Aquert

Table 24.9 Effect of FPM on soil penetration resistance

Parameter		Cone index (bar)
FPM (t ha^{-1})	O	1.405
	6O	1.630
	12O	2.182
	Sig.	**
	LSD	0.606
Depth (cm)	0	0.1721
	5	0.4540
	10	0.9029
	15	1.546
	20	2.573
	25	1.948
	30	1.903
	35	2.181
	40	2.511
	45	3.202
	Sig.	**
	LSD	0.332
Interaction	FPM*Depth	ns

Sig. = Significance
** statistically significant at $P < 0.01$

24.4 Conclusion and Recommendation

FPM applied a year ago resulted in reduced MDBD on both soil types. Doubling the application rate of FPM from 60 to 120 t ha^{-1} did not lead further reduction of MDBD or change CMC.

Due to high variability of MDBD among soil types, measured the bulk density only is not sufficient to be an indicator of soil compaction for all soil types. Rather soil compaction level should be presented as a proportion of the measured bulk density to the soils' bulk density/MDBD and the threshold values for the specific crop under cultivation need to be stated.

Tillage and cultivation operations can reduce soil compaction only if they are accomplished at moisture below the CMC. The current standard practice of the estate that employs traffic and field operation after drying the soil functions only when the drying process reduced the soil moisture below the CMC. Drying the field when the initial moisture content is above the critical moisture may aggravate soil compaction by bringing the moisture status to the CMC. Thus, it is a recommended norm for field operation from drying the soil to avoiding the CMC at the time of field operation.

Fluvent soil has dome-shaped moisture–density curve under standard proctor test, but Aquerts have M-shaped curve. Increased load at moisture lower than CMC can cause severe compaction on Aquerts than Fluvents. Thus, reducing the compaction load is recommended especially on Aquerts. FPM application has significantly increased the penetration resistance of the soils. Moreover, increased CI was observed at 20–25 cm depth on the experimental fields. Since FPM has significant effect on the penetration resistance of the soils, to avoid soil layer having higher penetration resistance, the soil moisture at these depths has to be monitored during tillage operation especially on FPM-treated fields. Moreover, to manage soils having compaction problem properly, soil classification should be made to great group level (soil taxonomy) since the presence or absence of hard layer and moisture regime is the major classification criteria vertisol and entisol.

References

Agricultural Division Plantation Manual (ADPM) (1993) Wonji-Shoa sugar plantation operation manual. Wonji, Ethiopia

Agricultural Research and Services (ARS) (1979) Annual report of 1978/79. Wonji, Ethiopia

Agricultural research and services (ARS) (1981) Annual report of 1980/81. Wonji, Ethiopia

Ahmad N (1996) Management of vertisols in rainfed conditions. In: Ahmad N, Mermut A (eds) Vertisols and technologies for their management. Elsevier, Amsterdam, pp 91–123

American Society of Agricultural Engineering (ASAE) (1983) Agricultural engineers yearbook of standard. St. Joseph, MI

Argon A, Garcia MG, Filgueira RR, Pachepsky YA (2000) Maximum compactibility of Argentina soils from the proctor test: the relationship with organic carbon and water content. Soil Tillage Res 56:197–204

Aschalew A (1981) The effect of filter press mud on water holding capacity of the five major soil cycles and its use as source of fertilizer on specific soil cycles. Ethiopian Sugar Corporation, Agricultural Research Services, Wonji, Ethiopia (unpublished)

Barzegar AR, Asoodar MA, Ansari M (2000) Effectiveness of sugarcane residue incorporation at different water contents and the proctor compaction load in reducing soil compactibility. Soil Tillage Res 57:167–172

Berhe F, Melesse AM, Fanta A, Alamirew T (2013a) Characterization of the effect of tillage practices on furrow irrigation hydraulics and soil water storage. Catena 110:161–175. doi:10.1016/j

Berhe F, Fanta A, Alamirew T, Melesse AM (2013b) The effect of tillage practices on and water use efficiency and grain yield. Catena 100:128–138

Blackburn F (1991) Sugarcane: tropical agriculture series. Longman Singapore Publishers Private Ltd., Singapore, pp 95–129

Boyoucous GJ (1927) The hydrometer as a new method for the mechanical analysis of soil. Soil Sci 23:343–353

Dexter AR (1987) Mechanics of root growth. Plant Soil 97:401–406

Dick RG (1987) Cane transport developers aim for optimum high flotation high capacity vehicle. BSES Bull 20:10–15

Fauconnier R (1993) Sugarcane: the tropical agriculturalist (trans: MacCimmun ED, Tindall CTA). Macmillan, Hong Kong

Frede GH (1987) The importance of pore volume and pore geometry to soil aeration. In: Monnier G, Goss MJ (eds) Compaction and regeneration. Balkema, Rotterdam, pp 25–30

Freitag DR (1971) Methods of measuring soil compaction. In: Compaction of agricultural soils ASAE monograph, Chart 3, pp 47–103

Gue'rif J (1979) Mechanical properties of straw: the effect on soil. In: Grossbrd E (ed) straw decay and its effect on disposal and utilization. Wiley, Chichester, pp 75–81

Gupta SC, Schneider EC, Larson WE, Hadas A (1987) Influence of corn residue on compaction and compaction behaviour of soils. Soil Sci Soc Am J 51:207–212

Hartemink AE (1998) Soil chemical and physical properties as indicators of sustainable land management under sugarcane in Papaua New Guinea. Geoderma 85:280–306

Kay BD (1990) Rates of change of soil structure under different cropping systems. Adv Soil Sci 12:1–52

Kuipers H (1961) Water content at pF 2 as a characteristic in soil-cultivation research in the Netherlands. Neth J Agric Sci 9(1):27–35

Liu C, Evett JB (1984) Soil properties testing, measurement and evaluation. Printice-Hall, New Jersy

MacRae RJ, Mehuys GR (1985) The effect of green manuring on the physical properties of temperate-area soils. Adv Soil Sci 3:71–94

MSTAT (1988) MSTAT-C user's Manual. Michigan State University, East Lansing

Mukerji (2000) Rehabilitation optimization and expansion of agriculture and factory. In: Interim report: sugar rehabilitation and expansion, vol I. Addis Ababa (unpublished)

Ohu JO, Raghavan GSV, Mckyes E (1985) Peatmoss effect on the physical and hydraulic characteristics of compacted soils. Trans ASAE 28(5):420–424

Pagliai M (1987) Micromorphometric and micromorphological investigation on the effect of compaction by pressures and deformations resulting from tillage and wheel traffic. In: Monnier G, Goss MJ (eds) Compaction and regeneration. Balkema, Rotterdam, pp 31–38

Paul CL (1974) Effect of filter press mud application on the availability of macro and micronutrients. In: Proceedings of the 15th international congress society sugarcane technology, vol 2, pp 568–575

Rahmeto A (2000) Review of research on sugarcane mechanization in the Ethiopian sugar estates. In: Ambachew D, Girma A (eds) Review of sugarcane research in Ethiopia. Research and Training Service, Delhi, pp 205–216

Rao PN (1990) Recent advances in sugarcane, 1st edn. KCP, India

Soane BD, Blackwell PS, Dikson JW, Painter DJ (1981) Compaction by agricultural vehicles: a review I: soil and wheel characteristics. Soil Tillage Res 207–37

Soil Survey Staff (1996) Keys to soil taxonomy, 6th edn. NRCS-USDA, Washington, D.C.

Tariku G (2001) Recurrent problems of the plantation department of Wonji/Shoa sugar estate, Wonji/Shoa sugar factory agricultural division. Wonji, Ethiopia (unpublished)

Torres JS, Villegas F (1993) Differentiation of soil comapaction and cane stool damage. Sugar Cane 1:7–11

USDA (2003) Soil compaction detection, prevention, and alleviation. In: Agronomy technical note no 17 natural recourse conservation service. Soil Quality Institute, Bhopal. http://www.soils.usda.gov/sqi

Voorhees WB, Senst CG, Nelson WW (1978) Compaction and soil structure modification by wheel traffic in the northern Corn Belt. Soil Sci Soc Am J 344–349

Zhang H, Hartge KH, Ringe H (1997) Effectiveness of organic matter incorporation in reducing soil compactibility. Soil Sci Soc Am J 61:239–245

Chapter 25
Effect of Filter Press Mud Application on Nutrient Availability in Aquert and Fluvent Soils of Wonji/Shoa Sugarcane Plantation of Ethiopia

Abiy Fantaye, Abebe Fanta and Assefa M. Melesse

Abstract Effect of filter press mud (FPM) application on nutrient availability in Aquert and Fluvent soils at Wonji-Shoa Sugarcane Plantation, located at central part of the East African Rift Valley, was studied during the 2003/04 growing season. The rate of FPM application was 0, 60 and 120 ton/h. Field monitoring was made for ninety days. As per the management practice of the Sugar Estate, the Aquert soil was green manured with *Crotalarea junceae* prior to the treatment application. Soil samples were collected and analyzed every fortnight for organic carbon (OC) and available N. Available P, exchangeable bases, Fe, and Mn content of the soil were monitored at the beginning and end of the experiment. Incorporation of FPM significantly increased the OC content in both soils. Nonetheless, the accumulation of OC was more in Aquert than Fluvent soils, due to the initial higher OC obtained from Crotalatea j., higher clay content and moist condition of the Aquert soil. The application of FPM promoted the available N though the availability of N ion, however, in both soils, the content declined during the first 45 days after application. Gradually, significant increase in the available N was observed (starting from 45th day of application up to the 90th day). This indicates that FPM having C/N ratio of 27 resulted in priming effect. The trend analysis made indicated that proportional increase in available P can be obtained with increasing rate of FPM application, though available K was found to be lower in FPM treated soils. The application of FPM buffered (reduced) the soil pH to 7.6. The highest rate of FPM application (120 ton/ha) slightly improved the soil electrical conductivity (EC). Furthermore, the study indicated that the application of FPM had the ability to reduce the soils exchangeable Na content. Thus, FPM has

A. Fantaye (✉)
Research and Training Section, P.O. Box 15, Wonji, Ethiopia

A. Fanta
Haramaya University, P.O. Box 138, Dire Dawa, Ethiopia
e-mail: fantaalemitu@gmail.com

A.M. Melesse
Department of Earth and Environment, Florida International University, Miami, FL, USA
e-mail: melessea@fiu.edu

© Springer International Publishing Switzerland 2016

A.M. Melesse and W. Abtew (eds.), *Landscape Dynamics, Soils and Hydrological Processes in Varied Climates*, Springer Geography,
DOI 10.1007/978-3-319-18787-7_25

the potential to amend sodicity and abnormal pH of soils. However, it was noted that the application of FPM had reduced Fe/Mn ratio indicating its possible ability to slow down or reduce Fe uptake by plants. Therefore, prior checking for available K and Fe has to be made to ensure unnecessary consequences on sugarcane plants.

Keywords Filter press mud · Filter cake · Aquert · Fluvent · Plant cane · Ratoon · Planting season · Priming effect

25.1 Introduction

Commercial sugar production in Ethiopia commenced in 1954 from sugarcane (*Saccharum officinarum* L.) by the Netherlands Company called Handles-Vereening Amsterdam (HVA) at Wonji plain in the central part of East African Rift Valley. In 1962 expanding the cane fields the Company established Shoa Sugar Factory in the vicinity forming the Wonji/Shoa Sugar Estate. The Wonji and Shoa Factories have crushing capacity of 1400 and 1600 tons cane per day, respectively. The mean sugarcane yield is estimated to be 174 ton/ha with factory sugar yield of 11.80–12.00 % (Blackburn 1991). On average, the two factories produce 64,812 tons of sugar annually and as a by product the factories produce 21,713 tons of filter press mud (FPM) annually.[1]

Despite the increasing sugar demand and the need for increasing productivity, the estate face production constraints like; low soil organic matter management practice resulting in high usage of chemical fertilizers, groundwater rise and poor drainage condition that can lead to the development of salt affected soil (Gebeyehu 2001) and low cane juice phosphate content (Elias and Ambachew 2002).

Sugarcane is a long duration exhaustive crop. It has been found that 85 tons of crops remove 122.24 and 142 kg NPK ha^{-1}, respectively from soil (Bokhtiar et al. 2001a, b). One of the main constraints for its good yield is its high nutritional requirements along with increased cost of fertilizers (Gholve et al. 2001). As a result, sugarcane growers, all over the world, have become aware of the importance of maintaining soil quality through the use of organic amendments. Soil quality is defined largely by soil function or use, and represents a composite impact of physical, chemical and biological properties on an ecosystem (Doran et al. 1996). These authors defined soil health as the continued capacity of soil to function as a vital living system, within ecosystem and land-use boundaries, to sustain biological productivity. Soil amendments used by growers include sugarcane by-products such as trash, filter cake, bagasse and fly ash, all by-products of the sugarcane industry. Other sources of organic soil amendments include poultry and cattle manure, pine bark chips and green manuring (Antwerpen et al. 2003).

[1]Ten years mean summarised from data presented at the annual factory conference 2003.

The utilization of industrial wastes, like press mud, on agricultural fields, is gaining popularity. Press mud is a good source of organic matter, NPK, and micronutrients. As a result, press mud has established its importance in improving fertility, productivity and other physical, biological and chemical properties of agricultural soils (Jamil et al. 2008). Press mud can serve as a good source of organic manure (Bokhtiar et al. 2001a, b) an alternate source of crop nutrients and soil ameliorates (Razzaq 2001). Filter cake increases cation exchange capacity for thirty months after its application (Rodella et al. 1990) and its residual effect remains in the soil after four years of application (Viator et al. 2002).

FPM is a byproduct generated during juice clarification process. It contains several organic compounds and plant nutrients, among which phosphors is the principal one. It is also good source of organic carbon (OC) and nitrogen. At Wonji/Shoa the fields are fertilized only with N in the form of inorganic fertilizer urea. If FPM is applied it can supply both N and P in addition to OC. In economic terms, if we consider the 21,713 tons of FPM as a source of N (assuming 1.3 % N in FPM), it can supply N equivalent to 613.6 tons of urea or 1447.5 tons of DAP. Moreover, literature show that the response of cane to applied FPM is beyond its nutrient supply because of its effect on soil physical properties (Prasad 1976; Blackburn 1991).

In most of the sugar estates, around the world, FPM is a useful fertilizer (Blackburn 1991). It is used to improve soil in flooded conditions in India (Sing et al. 1995), for seed set production in Cuba (Cepero-Garcia et al. 1989), as fertilizer in combination with Vinasse in Argentina (Scandaliaris et al. 1990) and with rice husk in India (Channabasavanna and Setty 1995) and as organic source of plant nutrient in Lousiana (Hallmark et al. 1998). However, in Wonji/Shoa Sugar Estate, for years, it is dumped in pits without any use at commercial level due to lack of concrete information with regard to the rate to be applied, when and where to apply and the condition of FPM, fresh or processed wet or dry (Gebeyehu 2001). Since the aim of fertilization program is to supply nutrient to the plant understanding availability of nutrient is very important.

The effect of tillage on soil moisture availability and storage as well as on grain yield has been studied by Berhe et al. (2013a, b). Numerous researchers have reported on the beneficial effects of trash, filtercake, bagasse, poultry manure, cattle manure, pine bark chips and green manures on the biological, chemical and physical properties of soils. Information on the effect of these soil improvers on the biological properties of soils in Wonji/Showa Sugar Estate is either limited or non-existent. Furthermore, spiralling prices coupled with short availability of fertilizers and depletion of available nutrients and organic matter due to continuous cane cropping with inorganic fertilizers (Kumar and Verma 2002) necessitates the integrated use of organic and mineral fertilizer resources. Thus, the objective of this study was to evaluate the effect of FPM application on the nutrient availability of Fluvent and Aquert soils of Wonji/Shoa Sugar Estate.

25.2 Materials and Methods

25.2.1 Site Description

The study was conducted at Wonji/Shoa, Ethiopia, located in the central part of the East Africa Rift Valley (See Fig. 24.1 in Chap. 24). Geographically, the site is located at 39° 10′–39° 20′ longitude and 8° 30′–8° 35′, latitudes and has an altitude of 1540 m above sea level (masl). Wonji/Shoa has semiarid climate with annual rainfall of 830 mm and mean air temperature of 20.8 °C; its mean relative humidity is 55 %. The Estate currently cultivates more than 7000 ha of sugarcane.

The Sugar Estates classify the soils into five major soil cycles on the basis of soil water holding capacity and critical moisture contents. Capital letters and numbers as subscript designate the soils as A_1, A_2, BA_2, $B_{1.4}$ and C_1 soil cycles. This classification approach was adopted from the Netherlands finder research that indicated water content at pF_2 is a characteristic in soil-cultivation (Kuipers 1961).

Moreover, the sugar estates group the five soil cycles into two; A_1, A_2 and BA_2 soil cycles which are fine textured clay soils with high moisture holding capacity and high critical moisture content having serious water logging and drainage problem; requiring proper management practiced. Thus, the fields are cultivated cane after fallow and *Crotalarea junceae* is grown during the fallow period as green manure. The second group comprises $B_{1.4}$ and C_1 soil cycles that are coarser in texture and have a natural drainage with very little or no drainage problem; hence there is no fallow period, and cultivation for cane production is continuous. When these two groups classified according to Soil Taxonomy, the fine textured clay soils are classified in the suborder Aquert and the cores texture soils as Fluvent.

25.2.2 Experimental Design

Two fields, from Aquert and Fluvent soils were selected for the experiment. Then the experimental plots were arranged in split-split plot layout with three replications. The soil cycles were the main plots, FPM applied at three different levels (0, 60 and 120 ton/ha) were the sub plots and time of nutrient release monitoring, every fortnight after the application of FPM, was taken as sub-subplot. Each plot had 6 furrows, and samples were collected from the central 4 rows. The spacing between the sub plots was 1.5 m and 2.9 m between replications.

The treatments were assigned randomly to the plots using random numbers. The treatments were applied following the plantation operation schedule just after *C. junceae.* cutting for Aquert soils and after sub-soiling for Fluvent.

25.2.3 Soil Analysis

To secure baseline information, composite samples from 0 to 30 cm depth were taken and analyzed for particle size distribution, texture, adopting hydrometer method (Boyoucos 1927); pH and EC were measured at 1:2.5 soil to water ratio using JENWAY 3345 ion meter and JENWAY 4310 conductivity meter, respectively. Organic C was determined by the method of Walkley and Black (1934). Available N (NO_3^- and NH_4^+) was determined following the analytical procedure outlined by Tekalign et al. (1991). For available P determination, extraction was made following the procedure of Olsen et al. (1954) and the reading was taken at 720 nm wave length on CECIL CE1001 series 1000 spectrophotometer. Available K was extracted with Morgan's solution and the amount was determined using flame photometer (JENWAY PFP 7). From the exchangeable bases extracted, determination of Ca and Mg was made by employing EDTA titration method. Na was determined from the extract using JENWAY PFP 7 flame photometer. CEC was determined by ammonium acetate method as described by Sahlemedhin and Taye (2000), and then exchangeable sodium percentage (ESP) was calculated. The analyses were carried out in the central laboratory of Research and Training Services (RTS) at Wonji Sugar Estate. Available Fe and Mn were determined using atomic absorption spectrophotometer at 248.3 and 279.5 nm wavelengths respectively using slit width of 0.2 nm. The extraction was made in RTS central laboratory and the reading was made in the laboratory of Geological Survey of Ethiopia (GSE) in Addis Ababa.

After applying the treatments, soil samples were collected and analyzed for organic C and available N every fortnight throughout the study period, 90 days, to monitor the organic matter mineralization and N release. At the end of the experiment, P, Ca, Mg, K, Na and micronutrients Fe and Mn were determined in addition to OC and available N following similar analytical procedures at the beginning of the experiment.

Before FPM application, N, P and K contents of FPM and that of the experimental soils were analyzed by digestion and OC content determined by dry combustion method.

25.2.4 Statistical Analysis

The data was analyzed using MSTAT statistical package (MSTAT 1988). Then least significant differences (LSD) were calculated to compare pairs of means at 5 and/or 1 % levels of probabilities.

25.3 Results and Discussion

25.3.1 Carbon Mineralization and Availability of Nitrogen

The Aquert and Fluvent soils had OC content of 1.68 and 1.07 %, respectively as shown in Table 25.1 before the application of FPM. The high OC content of the Aquert soil was due to the cumulative effect of *crotalaria* L. incorporation. After application of FPM which had 32.81 % OC (Table 25.2), the OC content went to 2.31 and 3.04 % for the Aquert soil and 1.25 and 1.39 % for Fluvent soil at application rate 60 and 120 ton/ha of FPM, respectively. In the Aquert soil, OC accumulation was note; this is in line with the findings of Lynch and Cotnoir (1956) that soils having higher clay content and higher exchangeable capacities (CEC), tend to accumulate OC in them whenever organic matter is applied.

Similar to the finding of Tiwari et al. (1998), the application of FPM on the soils of Wonji/Shoa improved the available nitrogen level. As shown in Fig. 25.1, the highest effect was on Fluvent. Moreover, when FPM application rate was doubled the NO_3 content of the soil increased on Fluvent, but on *crotalaria* L. incorporated fields, Aquert soil, the amount of both NO_3-N and NH_4-N in the soil didn't change significantly at all levels of FPM application.

There were variations in the OC contents during the study period (Table 25.3 and Fig. 25.2). At 30th day after the application of FPM, the soil OC content was reduced on both soil types due to the fast decomposition. At 45th day after the application of FPM, the available nitrogen (NO_3^- and NH_4^+) was reduced to a

Table 25.1 Physicochemical properties of aquert and fluvent soils of Wonji/Shoa sugar estate

Parameter	Aquert (A_2)	Fluvent (C_1)
Sand %	16	38
Silt %	30	22
Clay %	54	40
Textural class	Clay	Clay loam
Particle density (g/cm^3)	2.25	2.31
Organic C (%)	1.68	1.07
pH (1:2.5 H_2O)	7.6	7.6
EC (dS/m) (1:2.5 H_2O)	0.265	0.122
CEC (cmol$_c$ kg^{-1})	40.67	34.22
Exchangeable Ca (cmol$_c$ kg^{-1})	31.20	26.34
Exchangeable Mg (cmol$_c$ kg^{-1})	5.30	1.95
Exchangeable Na (cmol$_c$ kg^{-1})	0.92	0.44
Exchangeable K (cmol$_c$ kg^{-1})	1.61	0.97
Available N (NO_3^- + NH_4^+) (ppm)	$41.09^{(19.61 + 21.48)}$	$28.02^{(14.01 + 14.01)}$
Available P (ppm)	6.04	2.46
Available K (ppm)	208.5	137.00
Exchangeable sodium percent (ESP)	2.26	1.28

Table 25.2 Chemical composition of FPM

	pH	EC (dS/m)	N (%)	P (%)	K (%)	OC (%)	C/N
FPM	6.1	2.85	1.20 ± 0.05	0.73 ± 0.06	0.45 ± 0.08	32.81 ± 3.00	27

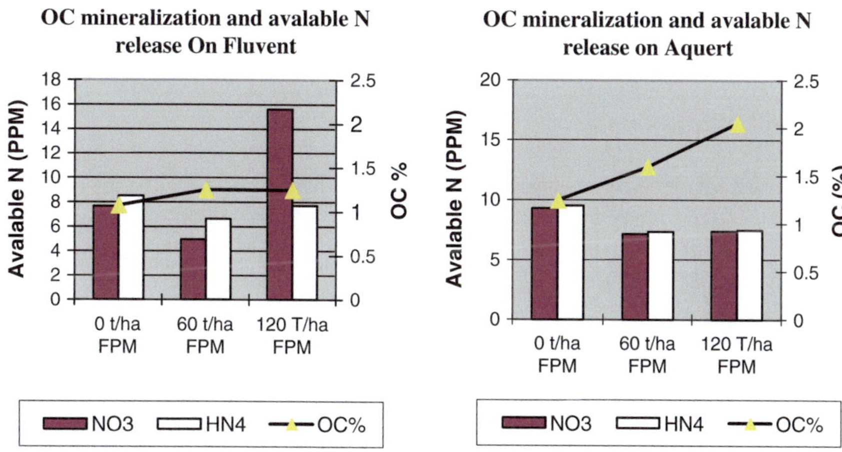

Fig. 25.1 Effect of FPM amount on available N and organic carbon

Table 25.3 Interaction effect of soil type and number of days after FPM application on available N

Days after application	NO_3^- (ppm)		HN_4^+ (ppm)	
	Aquert	Fluvent	Aquert	Fluvent
15	14.32	12.2	15.25	14.94
30	7.08	6.22	6.69	7.31
45	4.04	2.49	4.36	2.18
60	7.31	8.40	7.93	9.02
75	7.00	4.36	6.38	4.59
90	11.98	7.00	10.58	6.61
LSD (5 %)	2.212		2.590	

minimum level. FPM incorporation had resulted in a priming effect. After the 45th day to the 90th day, the available nitrogen content of both soils started to increase progressively.

The findings of Thonnissen et al. (2000) indicated decline in the content of NO_3^- nitrogen after 35–55 days of the application of FPM; in this study too, after 30 days from application of FPM, soil available N declined. Earlier than this period, NH_4^+

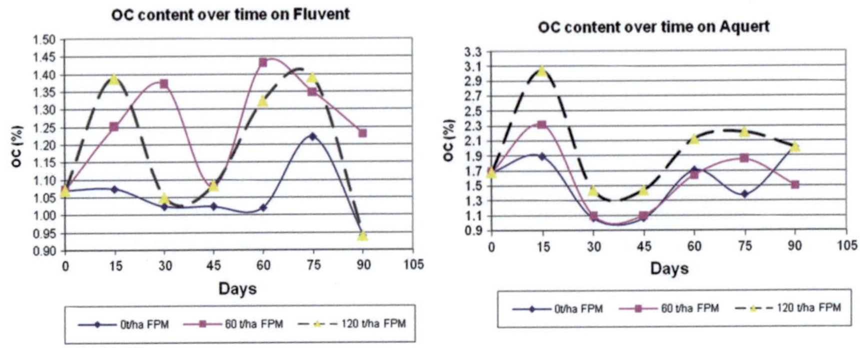

Fig. 25.2 Effect of FPM on soil organic carbon content

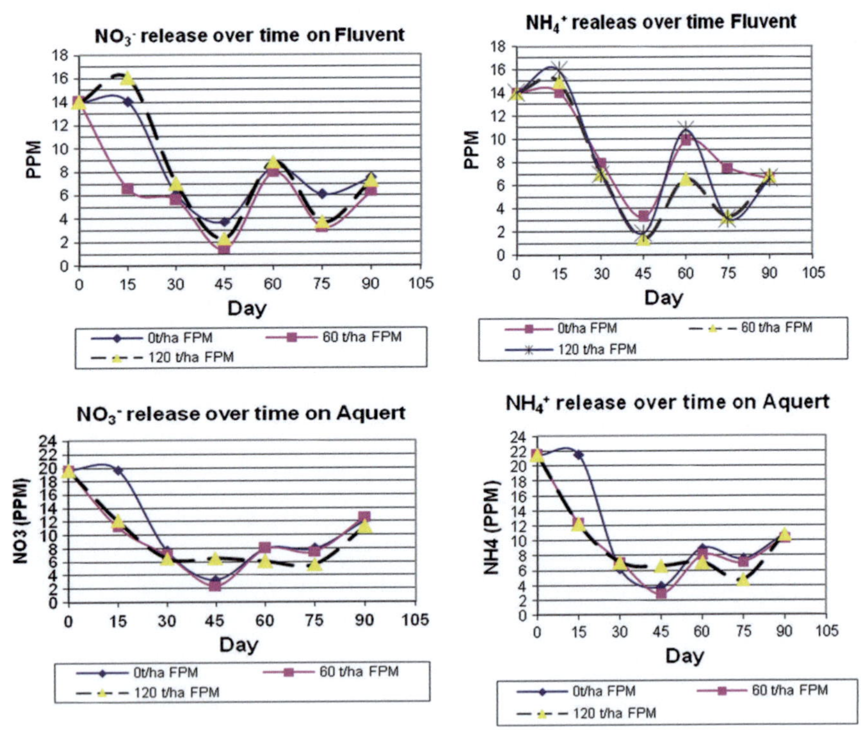

Fig. 25.3 Trend of NO_3^- and NH_4^+ N

nitrogen was dominant form of available N (which is the first product of organic matter decomposition), and after 60 days NO_3^- started to dominate the form of available N over NH_4^+ nitrogen (Fig. 25.3). This shows that there is fast decomposition of OC during this period; as a result, the OC content of the soil reduced in subsequent days.

25.3.2 Soil PH and Electrical Conductivity

The pH (1:2.5, soil:H_2O) of Aquert soil was 7.96 and that of the Fluvent soil was 7.56; the difference is statistically significant ($P > 0.01$). Application of FPM changed the pH level to 7.77 and 7.66 on Aquert while on Fluvent it became 7.58 and 7.61 at the application rate of 60 and 120 ton/ha of FPM, respectively (Fig. 25.4). Application of FPM has reduced the pH in Aquert soil. But in Fluvent soil, even if the difference was not significant, pH had an increasing trend (Table 25.4). Effect of FPM in increasing low pH soils was also confirmed by Naidu and Syers (1992) on Oxisol soils of Fiji in which the pH was increased from 5.0 to 6.0 when FPM applying.

This study showed that FPM had buffed the pH of the soils at Wonji/Shoa to 7.6 especially at highest FPM application rate of 120 ton/ha. FPM application brought the pH closer to neutrality. Thus FPM can be used to amend abnormal soil pH.

The electrical conductivity (EC) of soils was below the critical value of 4 dS/m. The Fluvent soils had EC of 0.193 dS/m. Application of FPM at the rates of 60 and 120 ton/ha increased the EC to 0.236 and 0.272 dS/m, respectively. The Aquert soils, which had 0.339 dS/m EC, exhibited 0.382 and 0.502 dS/m at application rates of 60 and 120 ton/ha FPM, respectively. Extremely low EC is not good for the stability of the soil structure, thus increasing the soil EC up to 1 dS/m is vital to make soil Na tolerance (Fig. 25.5).

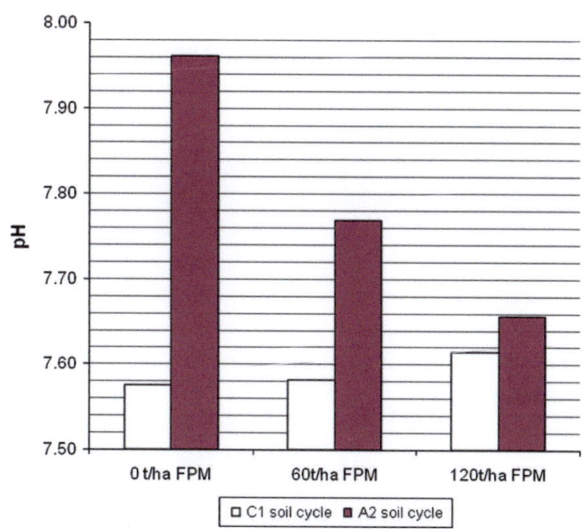

Fig. 25.4 FPM buffering soil pH to 7.6

Table 25.4 Interaction effect of FPM and soil type on soil pH

FPM (ton/ha)	Soil type	
	Fluvent	Aquert
0	7.575[c]	7.961[a]
60	7.581[c]	7.769[b]
120	7.614[c]	7.657[bc]

LSD (0.05) = 0.1409; [a, b, c]: pH values with similar letter are non significant at $P \leq 0.05$

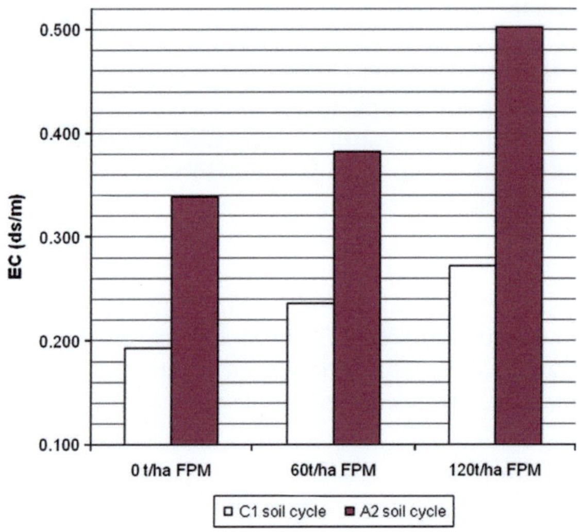

Fig. 25.5 Effect of FPM on electrical conductivity of soil

25.3.3 *Effect of FPM on Available P*

In the untreated Fluvent soils, the available P content is much lower than that of the Aquert soils. This was due to the high clay content of Aquert soils and effect of *C. junceae, green manure that* had contributed to the increased level of available P. Available P content analyzed from composite samples obtained from the experimental fields indicated that there was a considerable increment in the amount of available P with increasing rate of FPM application. Ten times increment (from 2.5 to 25 ppm) in Fluvents soils and more than three times increment (from 10 to more than 30 ppm) in Aquerts soils were obtained when the two soils were treated with 120 ton/ha of FPM (Fig. 25.6). These levels of Olsen P, which are brought above 25 ppm through the application of FPM was the highest levels for sugarcane production (Blackburn 1991; Hartemink 1998). Such an increment in P in the Aquert soils was due to the improved pH of the soil in addition to the amount of P released from the applied FPM.

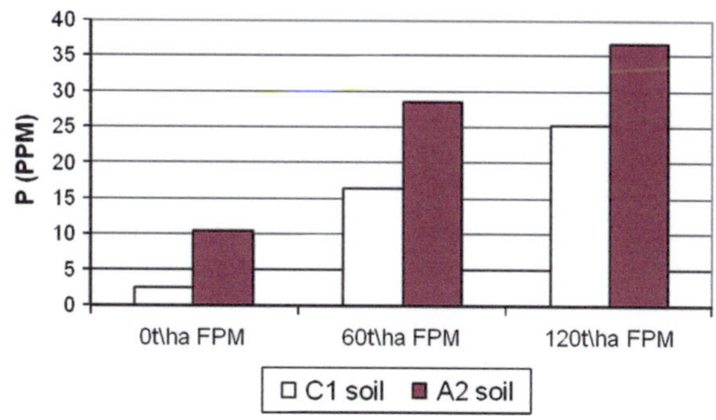

Fig. 25.6 Effect of FPM on available P content of the soils

25.3.4 Effect of FPM on Exchangeable Cations

Calcium is the dominant exchangeable cation on both soils and larger amount was found in Aquert soils when FPM was applied at the rate of 120 ton/ha. Magnesium is the second dominant basic cation found in both soils at all rate of FPM. Higher amount is found in Aquert soils than Fluvent soils. One of the reasons for the slight increase in EC on Aquert soil with increasing FPM application could be due to the increase in amount of calcium and magnesium salts.

Even if the amount of sodium in both soils was within acceptable range (ESP < 15 %), the amount in Aquert soils was more than twice than that of Fluvent soil. However, application of FPM had reduced the exchangeable Na content of both soils (Fig. 25.7). Moreover, attention has to be given to the Aquert since these soils have shallow water table with poor drainage condition and high clay content that can aggravate salinity and sodicity development.

The K content of Fluvent soils was 1.23 meq/100 g and it became 1.36 meq/100 g and 1.29 meq/100 g at the application of 60 and 120 ton/ha of

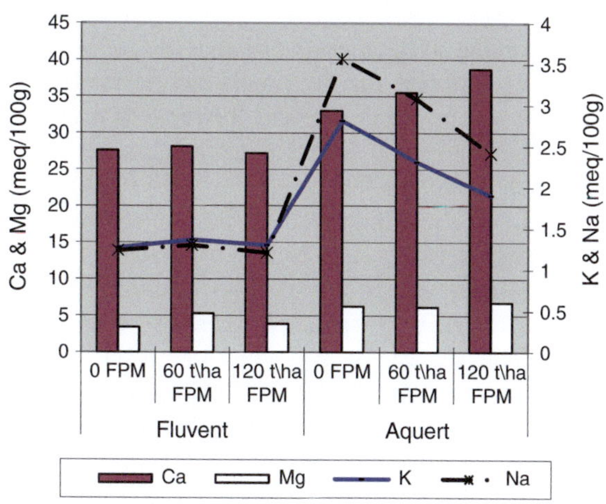

Fig. 25.7 Exchangeable bases content after FPM application

FPM, respectively, while in Aquert soils the level of K was 2.81 meq/100 g and it became 2.31 meq/100 g and 1.89 meq/100 g at the application rates of 60 and 120 ton/ha of FPM, respectively. The amount of exchangeable K of soils at Wonji/Shoa as determined by Gebeyehu (1994) was between 0.45 meq/100 g and 2.03 meq/100 g in vertisol with mean value of 1.42 and in Fluvesol it was between 0.49 meq/100 g and 1.45 meq/100 g with mean value of 0.79 meq/100 g. In both soils the amount obtained was above the range obtained by Gebeyehu (1994). There were different critical values of exchangeable K recommended for sugarcane production; 0.26 meq/100 g by Innes (1959), and 0.17 meq/100 g by Leverington et al. (1962). The amounts determine in both soils were well above the critical values.

Even if the amount of available K in the soils was above the critical value required for the growth of sugarcane, application of FPM reduced the available K content of the soils. This is due to Ca, Na and K interaction in ion exchange. According to Bond (2004), in the presence of Ca^{++}, Na^{+} and K^{+}; and Ca^{++} exchange occurs at constant value, near zero K^{+} concentration. When Na^{+} is in exchange with Ca^{++}; K^{+} content will be reduced. Thus due to the exchange process the K content of the Aquert soils (2.8 meq/100 g) is below the Na content (3.5 meq/100 g) and with increasing rate of FPM application, the amount of K and Na had been reduced by 36 % (to 1.8 meq/100 g) and 31 % (to 2.4 meq/100 g), respectively.

25.3.5 Effect of FPM on Iron (Fe) and Manganese (Mn) Content of the Soil

Increasing the rate of FPM application from 60 to 120 ton/ha on Fluvent soils increased the Mn content and reduced Fe level. But on Aquert soils, 60 ton/ha of FPM increased the Fe content and at 120 ton/ha of FPM, the Fe content declined, while at all levels of FPM, the Mn content didn't change in on Aquert soils (Fig. 25.8).

According to the reports of Pandeya (1993) and Pandeya and Singh (1997) the stability constants for the complexes formed between iron species existing in ambient soil environment, and fulvic acids was found to be the highest for FPM when it is compared with poultry manure, sewage silage and farm yard manure, which show the highest availability of Fe from FPM. However, the result of this study didn't show improvement in the Fe content.

Sugarcane plants require larger amount of Fe than Mn. In 100 tons of cane 2–10 kg of Fe in the form of Fe_2O_3 and 0.2–1.0 kg of Mn is removed (Fauconnier 1993). Comparing the two soils there was more Fe than Mn in Fluvent soils, but in Aquert soil, the reverse was true. As a result, the Fe to Mn ratio of the Fluvent is lower than the Aquert. In general, application of FPM at higher rate slightly reduced Fe but no effect on Mn content of the soil so the Fe:Mn ratio is reduced.

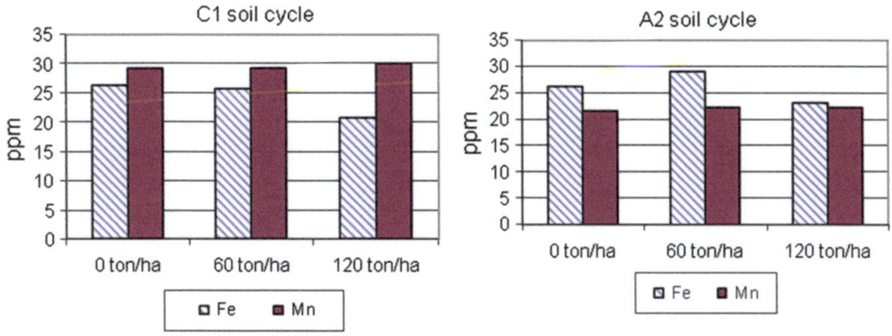

Fig. 25.8 Fe and Mn contents of the soil 90 days after FPM application

Fig. 25.9 Ratio of Fe to Mn in C_1 and A_2 soil cycles

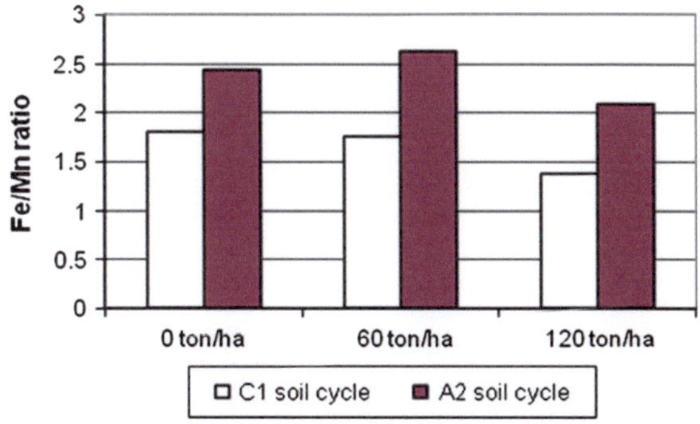

Incorporation of *Crotalaria* on Aquerts soil might contributed to the soil Fe content, however, to give concrete recommendation it has to be studied further.

Mn has an antagonistic effect on Fe uptake, therefore not only the amount of Fe in the soil that determines the availability of Fe to the plant but the ratio of Fe to Mn. Fe chlorosis is one of the problem at Wonji/Shoa Sugar Estate. As shown in Fig. 25.9, FPM application at higher rate of 120 ton/ha reduces the Fe to Mn ratio that will have possible reduction of Fe availability to sugarcane.

25.4 Conclusion

FPM application can improve the available P content of the soil markedly and supplies N. Moreover, it buffers soil pH. FPM has supplied large amount of Ca and Mg to the base saturation of the soils and reduced the amount of Na from the exchangeable base pool of the soil. Therefore this study has indicated FPM has good potential in supplying P, N and OC to the soil and can amend abnormal soil pH and sodicity. However, at higher rates it has reduced the exchangeable K and

Fe/Mn ratio of the soil. Therefore to ensure balanced nutrient supply of the soil prior to application of FPM availability of these elements has to be checked. Since organic matter addition has long term effect, such investigation has to continue on permanent experimental plots to see the effect on ratoon cane.

References

Antwerpen RV, Haynes RJ, Meyer JH, Hlanze D (2003) Assessing organic amendments used by sugarcane growers for improving soil chemical and biological properties. Proc S Afr Sugar Technol Assoc 2003:77

Berhe F, Melesse AM, Fanta A, Alamirew T (2013a) Characterization of the effect of tillage practices on furrow irrigation hydraulics and soil water storage. Catena 110:161–175

Berhe F, Fanta A, Alamirew T, Melesse AM (2013b) The effect of tillage practices on and water use efficiency and grain yield. Catena 100:128–138

Blackburn F (1991) Sugarcane. Tropical agriculture series. Longman, London

Bokhtiar SM, Paul GC, Rashid MA, Rahman ABM (2001a) Effect of pressmud and oganic nitrogen on soil fertility and yield of sugarcane grown in high Ganges river flood plain soils of Bangladesh. Indian Sugar L1:235–240

Bokhtiar SM, Paul GC, Rashid MA, Rahman ABM (2001b) Effect of pressmud and organic nitrogen on soil fertility and yield of sugarcane grown in high Ganges river flood plain soils of Bangladesh. Indian Sugar L1:235–240

Bond WJ (2004) Competitive exchange of K^+, Na^+ and Ca^+ during transport through soil. Aust J Soil Res 35(4):739–754

Boyoucos GJ (1927) The hydrometer as a new method for the mechanical analysis of soil. Soil Sci 23:343–353

Cepero-Garcia S, Davila-Iriarte A, Cairo-Cairo P (1989) Potential for using filter cake as a fertilizer in seed cane banks in dark plastic soils on the northern coast of Villa Clara. Centro-Agricola 16(4):78–86

Channabasavanna AS, Setty RA (1995) Effect of pressmud and rice husk on growth and yield of main and ratoon crop of sugarcane (*Saccharum officinarum*). Indian J Agron 40(4):670–672

Doran JW, Sarrantonio M, Liebig MA (1996) Soil health and sustainability. In: Sparks DL (ed) Advances in agronomy, vol 56. Academic press, New York

Fauconnier R (1993) Sugarcane. In: MacCimmun (trans, ed) The tropical agriculturalist. Macmillan, Hong Kong (Tindall, CTA)

Gebeyehu T (1994) The status of potassium in soils and its uptake by sugar cane varieties at Wonji/Shoa Sugar Estate. M.Sc. thesis, Alemaya University, Ethiopia

Gebeyehu T (2001) Recurrent problems of the plantation department of Wonji/Shoa sugar estate. Wonji/Shoa Sugar Factory Agricultural Division, Wonji, Ethiopia (unpublished)

Gholve SG, Kumbhar SG, Bhoite DS (2001) Recycling of various conventional and nonconventional organic sources in adsali sugarcane (*Saccharum officinarum* L.) planted with different patterns. Indian Sugar L1(1):23–27

Hallmark WB, Brown LP, Hawkins GL, Judice J (1998) Effect of municipal, fish and sugar mill wastes on sugarcane yields. La Agric 41(1):9–10

Hartemink AE (1998) Soil chemical and physical properties as indicators of sustainable land management under sugarcane in Papaua New Guinea. Geoderma 85:280–306

Innes RF (1959) The potash manuring of sugarcane. Proceedings of the 10th Congress of the International Society of Sugarcane Technologists, pp 441–450

Jamil M, Qasim M, Zia MS (2008) Ultilization of pressmud as organic amendment to improve physico-chemical characteristics of calcareous soils under two legume crops. J Chem Soc Pak 30(4):577–583

Kuipers H (1961) Water content at pF2 as a characteristic in soil- cultivation research in the Netherlands. Neth J Agric Sci 9(1 Feb)
Kumar V, Verma KS (2002) Influence of use of organic manures in combination with inorganic fertilizers on sugarcane and soil fertility. Indian Sugar L11(3):177–181
Leverington KC, Sedl JM, Burge JR (1962) Some problems in predicting potassium requirement of Sugarcane. Proceedings of the International society of Sugarcane Technologists, 11, 123–130
Lynch DL, Cotnoir LJ (1956) The influence of clay minerals on the breakdown of certain organic substrates. Soil Boil Biochem 20:367–370
MSTAT (1988) MSTAT-C users manual, Michigan State University
Naidu R, Syers JK (1992) Influence of sugarcane millmud, lime and phosphorous, on soil chemical properties and the growth of leucaena leucocephala in an oxisols from Fiji. Bioresour Technol 41(1):65–70
Olsen SR, Cole V, Watenabe FS, Dean LA (1954) Estimation of available phosphorus in soils by extraction with sodium bicarbonate. USDA Cir No. 939
Pandeya SB (1993) Ligand competition method for determining stability constants of fluvic acid iron complexes. Geoderma 58:219–231
Pandeya SB, Singh AK (1997) Discontinuous spectrocolorimeteric titration method for determining stability constants of fulvic acid–iron complexes. Aust J Soil Res 35:1279–1290
Pracad M (1976) Response of sugarcane to filter pressmud and N, P, and K fertilizers. I. Effect on plant composition and soil chemical properties. Agric J 68:539–542
Razzaq A (2001) Assessing sugarcane filtercake as crop nutrients and soil health ameliorant. Pak Sugar J 21(3):15–18
Rodella AA, Silva LCFDA, Filho JO (1990) Effect of filter cake application on sugarcane yields. Turrialba 40:323–326
Sahlemedhin S and Bekele T (eds) (2000) Procedures for soil and plant analysis, technical paper No. 74, National Soil Research Centre Ethiopian Agricultural Research Organization, Addis Ababa, Ethiopia
Scandaliaris J, Dantur CN, Perez-Zamora F (1990) Influence of date of fertilization with filter press cake and vinasse on the response of ratoon sugarcane. Revista Industrial Y-Agrricola de Tucuman 67(1):69–77
Tadesse T, Haque I, Aduayi EA (1991) Soil, plant, water fertilizer, animal manure and compost analysis. Working document No. B13. Soil Science and Plant Nutrition Section, ILCA, Addis Ababa, Ethiopia
Tesfaye E, Ambachew D (2002) Survey of available phosphorous status in soils of Wonji/Shoa and Metehara sugarcane plantations (unpublished)
Thonnissen C, Midmore DJ, Ladha JK, Olk DC, Schmidhalter U (2000) Legume decomposition and nitrogen release when applied as green manures to tropical vegetable production systems. Agron J 92:253–260
Tiwari RJ, Bangar KS, Nema GK, Sharma RK (1998) Long term effect of pressmud and nitrogeneous fertilizers on sugarcane and sugar yield on a typic chromustert. J Indian Soc Soil Sci 46(2):243–245
Viator RP, Kovar JL, Hallmark WB (2002) Gypsum and compost effects on sugarcane root growth, yield, and plant nutrients. Agron J 94:1332–1336
Walkley A, Black IA (1934) An examination of the Degtjareff method for determining organic carbon in soils: Effect of variations in digestion conditions and of inorganic soil constituents. Soil Sci. 63:251–263

Chapter 26
Spatial Runoff Estimation and Mapping of Potential Water Harvesting Sites: A GIS and Remote Sensing Perspective, Northwest Ethiopia

Mulatie Mekonnen, Assefa M. Melesse and Saskia D. Keesstra

Abstract Freshwater resources scarcity is becoming a limiting factor for development and sustenance in most parts of Ethiopia. The Debre Mewi watershed, in northwest Ethiopia, is one of such areas where the need for supplemental water supply through rainwater harvesting is essential. Suitable water harvesting sites were identified through overlay analysis considering both social and technical parameters, such as land use/land cover, slope gradient, soil texture, flow accumulation and stakeholders' priority. This was performed with the integration of GIS and remote sensing applications. Knowledge of runoff resulting from rainfall is most important for designing any water harvesting structure. Direct field-level measurement of runoff is always good, but it is time consuming, labour intensive and expensive. In conditions where direct measurement of runoff could not be possible, remote sensing technology and GIS combined with runoff models are proven to be effective. In this study, the remotely sensed satellite data (Quickbird2) provided spatial information on land use/land cover. Precipitation was obtained from the nearest meteorological station, and soil data were acquired form laboratory analysis. The GIS tools were used to store, manipulate and estimate runoff depth, surface storage and runoff volume, applying Soil Conservation Service (SCS) Curve Number (CN) formula. The direct runoff volume estimated using SCS-CN model is 146,697 m^3 for the month of August, at Debre Mewi watershed,

M. Mekonnen (✉)
Bureau of Agriculture, Natural Resources Conservation and Management Department, Amhara Region, Bahir Dar, Ethiopia
e-mail: mulatiemekonneng@gmail.com

A.M. Melesse
Department of Earth and Environment, ECS 339 Florida International University, 11200 SW 8th Street, Miami, USA
e-mail: melessea@fiu.edu

S.D. Keesstra
Soil Physics and Land Management Group, Wageningen University, P.O. Box 47, 6700 AA Wageningen, The Netherlands
e-mail: saskia.keesstra@wur.nl

© Springer International Publishing Switzerland 2016

A.M. Melesse and W. Abtew (eds.), *Landscape Dynamics, Soils and Hydrological Processes in Varied Climates*, Springer Geography,
DOI 10.1007/978-3-319-18787-7_26

which covers about 508 ha. The result was compared with measured values, and closer relationship was found. This indicates that there is enough runoff water to be harvested for different uses. Remote sensing was found to be a very important tool in providing input parameters. GIS was also found to be a very important tool in mapping and integrating the different variables, in the process of runoff estimation and suitable water harvesting sites selection.

Keywords Water harvesting · Runoff estimation · GIS · Remote sensing · Amhara region · Blue Nile basin · Ethiopia

26.1 Introduction

Water is essential to humans, animals and plants. It is used in many different ways for food production, drinking, domestic and industrial activities. It is also a part of the larger ecosystem on which biodiversity depends. Water harvesting has been employed for thousands of years to irrigate and restore productivity to the land, provide drinking water to both human and animals, minimize risk in drought prone areas, increase groundwater recharge and reduce storm water discharges. Water supply is considered to be one of the most critical problems facing humanity today (Hinrichson 2003; World Water Assessment Programme 2003). About 300 million people in Africa, a third of the continents population, are living under water scarcity (RELMA 2005).

Ethiopia is known as the "Water Tower of East Africa" referring to the high plateaus and mountain ranges of the Amhara Region, where one of Africa's most important rivers, Blue Nile (Abay in Ethiopia), has its source in Sekela, Gishe Abay. Today, the Blue Nile is a thick brown liquid, carrying with it millions of tons of Ethiopian soil into other African countries where it piles up sediments several hundred metres thick. As the capacity of the country to harvest and use the excess water is very poor, most of the water flows out through transboundary rivers to neighbouring countries. Consequently, the country suffers from water scarcity triggered hazards, such as more frequent crop failure, food insecurity, drought and famine (Teshome 2003). This is the effect of decades of deforestation and over-exploitation of natural resources caused by a rapidly growing rural population and poor land use practices. It is also a major indication of a lack of effective management of the most precious resource that Ethiopia has: water.

Deforestation for agriculture combined with the resultant erosion is the main cause for lack of water in Ethiopia in general and in Amhara region in particular. This is because as soil is eroded away from an area, it is less able to absorb water when rain falls. Fresh water adequacy in rural areas of the Amhara National Regional State is 41 % and over 59 % lack adequate water source (BOFED 2007). Now more than ever, water harvesting is needed to ensure water needs in these water scarce areas.

Decline in vegetation cover, inappropriate land use, overgrazing and other practices for the last several years played a great role for the dry up of rivers and streams at Debre Mewi Watershed where the study is carried out. The only source of water for human as well as domestic animals consumption is either rain or groundwater. As a result, hand-dug wells are serving as a source of drinking water for watershed communities. But, lack of water for livestock consumption is one of the serious problems in the area (SWHISA 2007). Because of this, people of the watershed are travelling a long distance with their cattle to find water.

Communities in the watershed identified their problems and prioritized as soil fertility reduction, soil erosion, shortage of water for livestock consumption and high price of agricultural inputs (SWHISA 2007). Based on this Participatory Rural Appraisal (PRA) survey and problem prioritization, Adet Woreda Office of Agriculture (AWOA) and Sustainable Water Harvesting and Institutional Strengthening in Amhara (SWHISA) project planned to construct communal farm ponds to harvest water for watershed communities.

A good knowledge or estimation of the expected amount of runoff in a given area is important in planning water harvesting schemes. Reliable prediction of surface runoff from a catchment area is a prerequisite for implementing water harvesting. Remote sensing and GIS, in combination with appropriate runoff models, provide ideal tools for the estimation of direct runoff volume in a catchment (Miloradov and Marjanovic 1991; Demayo and Steel 1996).

Hence, this study was conducted to estimate surface runoff using Soil Conservation Service Curve Number (SCS-CN) model and identify suitable runoff water harvesting sites with the integration of GIS techniques at Debre Mewi Watershed of the upperBlue Nile River basin with specific objectives: (1) to estimate runoff amount in the watershed using SCS-CN model combined with GIS tools and information derived from high-resolution satellite imagery as an input for planning water harvesting and (2) to identify suitable runoff water harvesting sites using remotely sensed data together with field investigations and GIS applications.

26.2 Water Resource in Ethiopia

The geographical location of Ethiopia and its endowment with favourable climate provides a relatively higher amount of rainfall in the country. Much of the water, however, flows across the borders through transboundary rivers to the neighbouring countries. There arc 12 major river/drainage basins in Ethiopia where seven are transboundary. The total annual runoff from these basins is estimated at about 111 billion m^3 (Ministry of Water Resources 2001). According to Kassahun (2007), the nation's annual surface runoff is 122 billion m^3 and groundwater potential is 2.6 billion m^3 with the average rainfall of 1090 mm. The major rivers carry water and sediments and drain mainly to the arid regions of neighbouring countries. There are also eleven major lakes with a total area of 750,000 ha (Ministry of Water Resources 2001).

The study area is located in the upper Blue Nile River basin which has ample amount of annual rainfall. Hydrology of the Nile River basin has been studied by various researchers. These studies encompass various areas including stream flow modelling, water harvesting technology evaluations, sediment dynamics, teleconnections and river flow, land use dynamics, climate change impact, groundwater flow modelling, hydrodynamics of Lake Tana, water allocation and demand analysis (Melesse et al. 2009a, b, 2011; Abtew et al. 2009a, b; Abtew and Melesse 2014; Yitayew and Melesse 2011; Chebud and Melesse 2009a, b, 2013; Setegn et al. 2009a, b, 2010).

26.3 Rain Water Harvesting and Site Selection

Water harvesting is the collection and storage of water either directly in the formof precipitation and runoff, or indirectly in the form of groundwater, surface spring or river (Pacey and Cullis 1991). The water is later used for domestic, agricultural or industrial purposes. Water harvesting systems have existed for thousands of years in many parts of the world and are a main source of water for many communities. The main benefits obtained from water harvesting systems are to secure water supply for domestic use, to contribute to water and soil conservation, and to reduce erosion. More precisely, water harvesting reduces the dependence on groundwater supply for water use, can reduce flooding in certain areas (capture and storage of runoff) and improve household economic situation on the long term.

There are many water harvesting techniques; the choice of which to use is dependent upon the water source, catchment area, storage and use. There are a number of different classifications of water harvesting techniques (Reij et al. 1988). Techniques based on water sources include runoff, direct rainwater, snow, surface spring, river and groundwater. Techniques based on catchment area include small catchments (roofs of all kinds and field micro-catchments) and large catchments (on the ground within the field). Techniques based on storage include above-ground tanks of all kinds, cisterns, ponds, reservoirs and wells. Finally, techniques based on water use include domestic use (drinking water, household purposes and garden irrigation) and agricultural use (irrigation and animal consumption).

Water harvesting structures of various types are known by different names in the country such as farm ponds, small earthen dams, irrigation tanks and percolation tanks. The most common catchment runoff water harvesting structures are of two types: (a) embankment type ponds for hilly and rugged terrain, and (b) excavated or dugout type farm ponds for flat topography.

Earth fill dams can be constructed at suitable sites to harvest and store surface runoff for irrigation, floods moderation, sediment control, etc. It is an earthen embankment constructed across a water course with adequate spillway for disposal of excess storm flow. A dugout pond is formed by excavating a pit and forming an embankment around the pit by excavated earth. Dugout ponds could be either fed by surface runoff or groundwater aquifer. Surface water ponds are most common farm ponds.

There are many important aspects to consider when choosing a site for the implementation of a water harvesting scheme. Before selecting water harvesting pond sites, first, it is important to make sure that the area meets certain physical characteristics. These characteristics include topography, drainage area, soil texture and land use. A good pond site contains (l) soil with sufficient clay to hold water and reduce seepage, (2) level topography that provides for economical construction, and (3) an adequate runoff water supply (Frasier and Myers 1983). Moreover, when searching for an appropriate location, one should keep in mind that the area should satisfy local people's needs. It should also be located at the lowest point in the contributing catchment area and near the point of use in order to minimize piping and canal requirements.

26.3.1 Soil Texture

The suitability of a pond construction site is mostly dependent on the soil type. The site should have a soil type and composition that holds water economically reducing seepage problem. Clay or silty clay soils are excellent for ponds; sandy clays are usually satisfactory and coarse-textured sand, and sand–gravel mixtures do not hold water well and are unsuitable for ponds (FAO 1991). Limestone or shale areas are also not suitable because of possible fractures, which create leakages.

26.3.2 Topography

Slope of the land is important in site selection and implementation of all ground-based runoff water harvesting systems, especially farm ponds. Water harvesting is not recommended for areas where slopes are greater than 5 % due to uneven distribution of runoff and large quantities of earth work, which is not economical (FAO 1991). Kuiper (1999) also found that the area where the water harvesting system is intended to be located can be characterized by gentle slopes having less than 5% to minimize soil erosion and earth work for economical implementation.

26.3.3 Stakeholders' Priority

A water harvesting site is acceptable if it fits into the people's interest and applicable if implemented where the problem exists. In addition to this, field visits and the local people's knowledge about the area is an important factor in identifying suitable water harvesting sites together with technical skills. FAO (1991) reported that stakeholders' priority is an important factor in selecting water harvesting areas.

26.3.4 Land Use/Land Cover

In selecting water harvesting pond sites, the land use/land cover of the area is an important parameter. Settlement and forest areas, crop lands and areas of rock outcrop are not suitable to construct communal pond and collect runoff water for livestock consumption. Communal grazing lands, which reduce ownership and accessibility problems, are preferable.

26.4 Runoff Estimation

Surface runoff is rainwater that flows directly into a stream or it is that portion of rainfall which appears as flowing water in the drainage network of a watershed during and following a rainfall event. According to Morgan (1995), surface runoff refers to the portion of rainwater that is not lost to interception, infiltration, evapotranspiration or surface storage and flow over the surface of the land to a stream channel. A good knowledge or estimation of the expected amount of runoff in a given area is important in planning rain water harvesting schemes.

Surface runoff depends on a great number of factors, such as rainfall characteristics, watershed characteristics, soil physical characteristics (depth, texture, structure), land cover/land use and soil moisture conditions prior to rainfall events (USDA 1985; FAO 1991). Surface runoff estimation using plot-scale experiments provide useful information on the effect of land cover, slope and field management on runoff potentials (Defersha and Melesse 2011; Defersha et al. 2012).

26.4.1 Land Use/Land Cover and Surface Runoff

Land cover is one important parameter that affects surface runoff. When rain falls, the first drops of water are intercepted by the leaves and stems of the vegetation. This is usually referred to as interception storage, which leads to evaporation loss from the canopy. The root systems of vegetation and organic matter in the soil increase soil porosity, thus allowing more water to infiltrate. In general, an area densely covered with vegetation yields less runoff than bare ground. An increase in the vegetation density results in a corresponding increase in interception storage, retention and infiltration rates which consequently decrease the volume of runoff.

26.4.2 Hydrologic Soil Group

The nature of a soil where the rain falls is a very important determinant of how much runoff can be expected. Soil texture plays a great role in determining infiltration rate and water holding capacity of a soil. The runoff potential and infiltration characteristics of soils are important to determine the hydrologic soil groups, which is vital to find Curve Number (CN) values. The USDA-SCS (1986) has classified soil into four hydrologic groups: A, B, C and D based on the information of soil texture and defines them as follows. Hydrologic soil group (HSG) A are soils having low runoff potential and high infiltration rates, even when thoroughly wetted (sand, loamy sand or sandy loam). HSG B are soils having moderate infiltration rates when thoroughly wetted (silt loam or loam). HSG C are soils having slow infiltration rates when thoroughly wetted (sandy clay loam). HSG D are soils having high runoff potential, very slow infiltration rates when thoroughly wetted and consisting chiefly of clay soils (clay loam, silty clay loam, silty clay, clay).

26.4.3 Rainfall

Rainfall is a major source of runoff, and it is a major factor that influences the volume of runoff. Depending on time and budget availability, it is often suggested to set up an on-site micro-meteorological station to collect rainfall data. This allows for better runoff estimation. On the other hand, if monitoring site setup is not possible, alternative sources of rainfall data from nearby meteorological station should be considered. The limited data availability is a common challenge in developing countries since there are sparse meteorology stations.

26.4.4 Catchment Area

A catchment area is an extent of land where water from rainfall drains downhill into a body of water or an outlet. It acts as a funnel, collecting all the water within the area and channelling it into a water way. The size (area) of a drainage basin or a watershed is one important variable affecting runoff. If a catchment is very large, its surface runoff yield per unit area is low, and if a catchment is small in size, its surface runoff yield is higher (Morgan 2005). Determining the size of the drainage area that contributes to runoff is a basic step in runoff analysis. It can be determined either from field survey using GPS or at the office using topographic maps, aerial photographs and satellite images.

26.5 Materials and Methods

26.5.1 The Study Area

Debre Mewi Watershed is about 508 ha of land located in Adet Woreda, Western Gojam Zone, Amhara National Regional State. Geographically, it is located between 11° 20′ 10″N and 11° 21′ 50″N latitude, and 37° 24′ 35″E and 37° 25′ 55″ E longitude. It is about 40 km from the capital city of Amhara Region (Fig. 26.1).

The nearest meteorological station at Adet town is 10 km from the study area. Based on 20-year rainfall data analysis of this station, the area is characterized by unimodal pattern with peak in July and August and receives on average 1100 mm of precipitation annually. About 80–90 % of the rainfall falls in the main rainy season which starts in June/July and extends to August/September. The average minimum and maximum temperature of the area ranges from 8.7 to 25.4 °C, respectively.

Laboratory analysis of soil samples indicates that the area has six soil types according to FAO classification system with Eutric Fluvisols and Nitosols on higher elevation and lower slope lands and Pellic Vertisols on lower and flat plains. Eutric Fluvisols are found along river banks, whereas Eutric Vertic Cambisols and Eutric

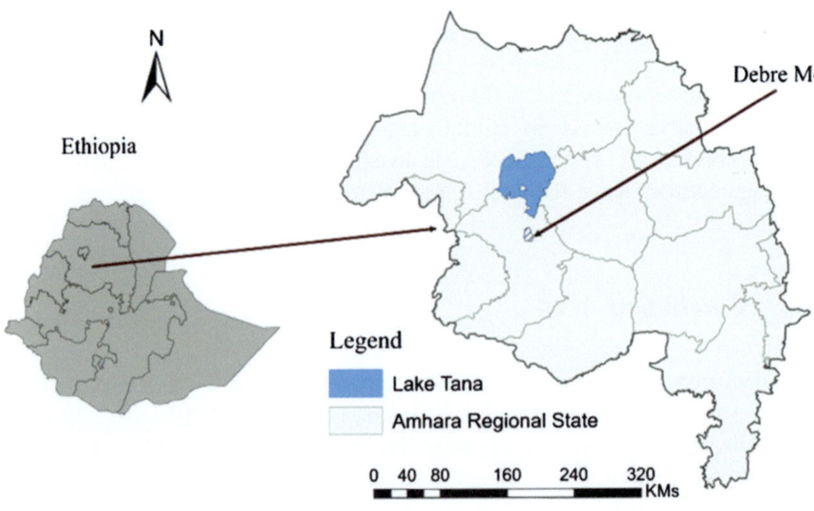

Fig. 26.1 Location map of the study area

Aquic Vertisols are dominant on averagely sloped areas. Hilly and steep lands are dominated by Eutric Cambisols. Soils in the area contain high percentage of clay. Texture is classified as clay, clay loam and heavy clay.

In the study area, five major land use/land cover types have been identified using both US Quickbird2 satellite image and field observation. The land use/land cover types comprise 70.3 % crop land (agricultural area), 6.4 % bush land, 19.6 % grazing land, 2.5 % eucalyptus plantation and 1.2 % built-up area.

Since slope is the most important terrain characteristic and plays a vital role in runoff process, it is very important to have an understanding of its spatial distribution in the study area. Therefore, the slope map of the study area was derived from DEM, which is prepared by digitizing 20-m interval contour lines from 1:50,000 scale topographic map of Adet Woreda. The watershed is characterized by a slope ranging from 1 to 57 % and elevation ranging from 2194 to 2360 m.

The rural economy of the people in the watershed is based on agricultural production. Only rain fed crops such as barley, tef, maize and wheat can grow during the rainy season. In some parts of the watershed, grass pea (Guaya) grows by residual moisture. The other source of income is livestock. Livestock population is very high, which includes goat, sheep and donkey. There is grazing pressure in the watershed.

26.5.2 Methodology

During the fieldwork, the study area, watershed and stakeholders' priority water harvesting sites were delineated using GPS with the aid of 1:50,000 topographic map and 1:4000 US Quickbird2 image map. Informal discussions with farmers working on their plots and formal discussions with watershed community representatives were done to collect information and select stakeholders' priority water harvesting sites in the watershed.

Data collected during the field work was processed together with remotely sensed satellite image data. First, required parameter maps of the model were prepared and the model was run. Second, runoff amount was quantified and suitable runoff water harvesting sites were selected. Finally, model validation result, discussion, conclusion and recommendations are given.

The rapid rural appraisal technique of the topographic transect walk method was employed for the assessment of the natural resource base of the watershed. In order to obtain as much detail information as possible, the transect walk was applied four times in two directions, east to west and south to north. During the transect walk, recording land use/land cover types, slope gradient, slope length, soil characteristics and drainage patterns was done over a distance of about 1.73 km in east–west and 3.26 km in north–south directions. It also provided a good opportunity for informal discussions with farmers working on their plots.

To assess and compute runoff amount, the US Department of Agriculture SCS-CN model, which requires rainfall and a watershed coefficient as input, was used (USDA-SCS 1972). The watershed coefficient, known as the runoff CN, represents the runoff potential of the land cover–soil complex. It is determined from the combination of land use and soil runoff potential (hydrologic soil group). CN values range between 0 and 100. A CN value of 100 indicates that all rainfall is transformed into runoff, while zero value indicates no direct runoff is generated.

The SCS-CN model is commonly used because it is simple to use, yields reasonable estimates of runoff, easy to parameterize and is a suitable method for ungauged watersheds (USDA 1985). Melesse et al. (2003) found that the SCS-CN model can estimate total runoff volume with an average efficiency of 98 %. The required input to estimate the amount of runoff in a watershed is data on soils, land use/land cover, rainfall and catchment area (USDA 1985; FAO 1991). After estimating the surface runoff amount using SCS-CN model, suitable water harvesting sites for communal farm ponds were selected. According to Frasier and Myers (1983) and FAO (1991), the most important parameters to be considered to select suitable water harvesting sites were stakeholders' priority, land use/land cover, soil texture and slope. Field-level collected runoff data for validation and soil texture data for soil texture mapping were obtained from Adet Agricultural Research Center, and rainfall was obtained from Ethiopian Meteorological Agency. Ground truth data were also collected at field level. Figure 26.2 shows the methodological flow chart to identify suitable water harvesting site.

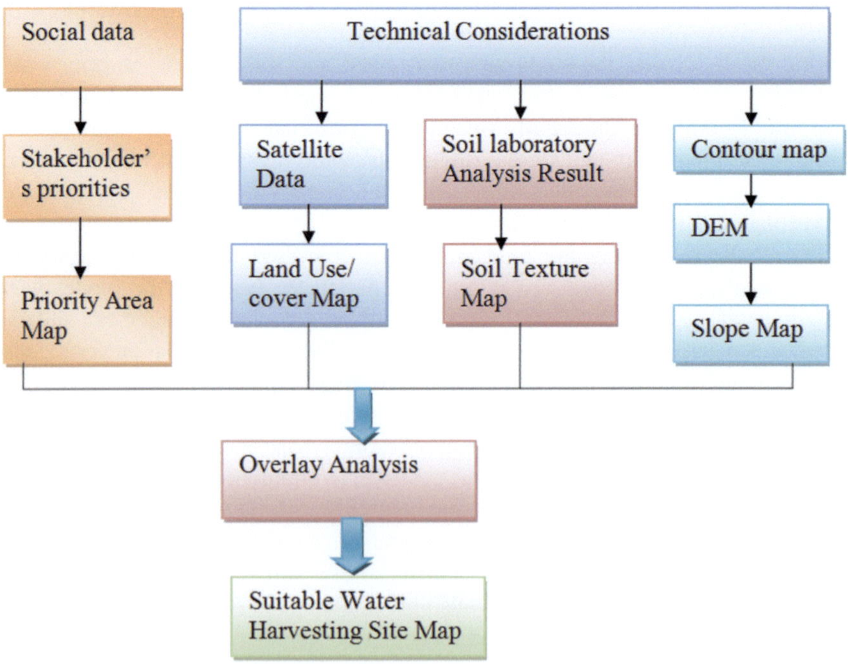

Fig. 26.2 Methodological flow chart to identify suitable runoff water harvesting

26.6 Results and Discussions

26.6.1 The Curve Number Model Input Analysis

To run the CN model and estimate runoff amount atDebre Mewi Watershed, the important input parameters such as hydrologic soil group, land use/land cover, CN, antecedent moisture conditions (AMC) and rainfall were derived as follows based on USDA-SCS (1972) and USDA (1985).

26.6.1.1 Hydrologic Soil Group (HSG)

Soil texture map of the watershed was prepared using field survey data and laboratory analysis. However, the information needed to determine a CN is the HSG, which can be determined based on soil texture: HSG A for sand, loamy sand or sandy loam; HSG B for silt loam or loam; HSG C for sandy clay loam; and HSG D for clay loam, silty clay loam, silty clay or clay texture soils (USDA-SCS 1986). Hence, the HSG was determined based on soil texture obtained from soil sample laboratory analysis and it falls under HSG D, because texture of the soil in the study area is clay, clay loam and heavy clay.

26.6.1.2 Land Use/Land Cover and Curve Number

Land use/land cover is another important parameter that affects surface runoff. An increase in the vegetation density results in a corresponding increase in interception losses, retention and infiltration rates which consequently decrease the volume of runoff. Hence, land use/land cover of the study area was prepared by digitizing the satellite image US Quickbird2. CN represents the runoff potential of land covers and soils of the watershed. After identifying HSG and land cover types of the watershed, the corresponding CN values were assigned from CN table values (Table 26.1) based on USDA-SCS (1986).

Table 26.1 CN values for the different land use/land cover types

Land use/land cover descriptions	CN values for HSG D
Crop land (cultivated agricultural land, wheat, barley…)	91
Grazing land (poor condition, ground cover <50 %, continuous grazing)	89
Bush land	83
Eucalyptus plantation (good cover)	85
Built-up area (farmsteads and rural residence)	93

Source USDA-SCS (1986)

26.6.1.3 Antecedent Moisture Condition (AMC)

Antecedent moisture condition is used as an index of wetness in a particular area. It can be determined using cumulative last five consecutive day's rainfall (Guta and Panigraphy 2008). If rainfall is less than 35 mm, it is AMC I, which has lowest runoff potential and dry soil condition. If the cumulative rainfall is between 35 and 52.5 mm, it is AMC II. AMC III is characterized by highest runoff potential and wet soil condition with a rainfall of greater than 52.5 mm. In the study, the last five days' rainfall was found to be 38.5 mm. Therefore, AMC II was selected and adjustment of CN was done accordingly. Finally, one month rainfall data (August, 2008) was obtained from Adet meteorological station, which is 320 mm, and the runoff depth (Q) and the volume of runoff for the watershed were calculated (Table 26.2).

26.6.2 Runoff Volume

The total estimated runoff amount of the watershed for the month of August was about 146,697 m^3, and the measured runoff amount was 244,704 m^3 for the same month (Table 26.2). It is determined by integrating hydrologic soil group, land use/land cover, CN, rainfall and antecedent moisture condition. Crop or agricultural land contributed the highest runoff volume followed by grazing land, whereas built-up area contributed the lowest (Fig. 26.3).

To validate the predicted runoff volume obtained by the model, the ratio between predicted value and measured value is used. According to Morgan (2005), the acceptable ratio value ranges between 0.5 and 2. In this study, the ratio of predicted to measured runoff data was found to be 0.59, which is within the acceptable range of 0.5–2. Even if the estimated runoff is within the range of the acceptable ratio, it was less than the measured value. The reduction of the estimated amount could be because of difference in the characteristics of the small measurement plot (180 m^2)

Table 26.2 Model estimated runoff for the month of August, 2008 in the study area

Land use/cover	Area (ha)	Av. runoff (Q) in mm	Total discharge for the month (m^3)
Crop land	357.5	291.7	104,283
Grazing land	99.87	285.2	284,578
Bush land	32.46	265.3	8583
Eucalyptus plantation	13	271.9	3528
Built-up area	6.2	298.2	1846
Total		Estimated	146,697
		Measured	244,704
		Ratio	0.59

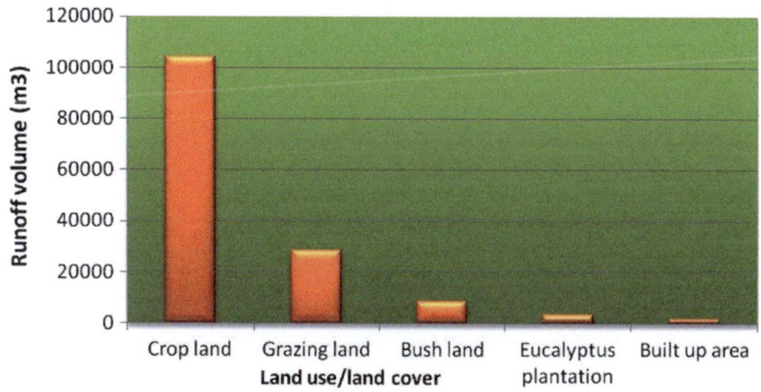

Fig. 26.3 Runoff volume (m^3) of the study area by land use/land cover category

and the total watershed (5,080,000 m^2). But, it can be concluded that the predicted runoff can represent the actual runoff in the area.

26.7 Water Harvesting Site Selection

26.7.1 Water Harvesting Site Selection Input Analysis

Derivations of the major factors contributing to the selection of suitable water harvesting sites were done as documented in Frasier and Myers (1983) and FAO (1991) as follows.

26.7.1.1 Slope

Slope gradient is an important variable in selecting runoff water harvesting sites. Hence, the slope map of the study area, which is derived from digital elevation model, was classified into two classes, as slope less than 5 % and slope greater than or equal to 5 % (Fig. 26.4).

26.7.1.2 Stakeholders' Priority

The watershed community representatives selected water harvesting sites based on their own interest and criteria. The criteria the stakeholders used include proximity to their animal stocking, communal land ownership, possibility of diverting runoff from natural drainage ways (alternative water sources) and availability of runoff water. The area selected by the watershed communities is delineated using hand-held GPS and mapped as shown in Fig. 26.5.

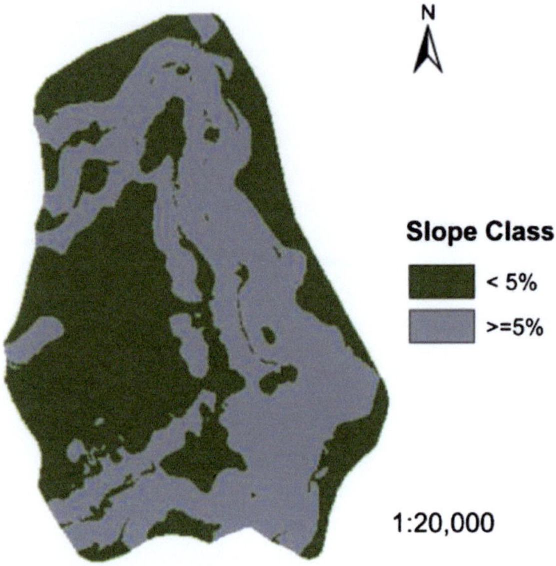

Fig. 26.4 Slope class map for suitable water harvesting site selection

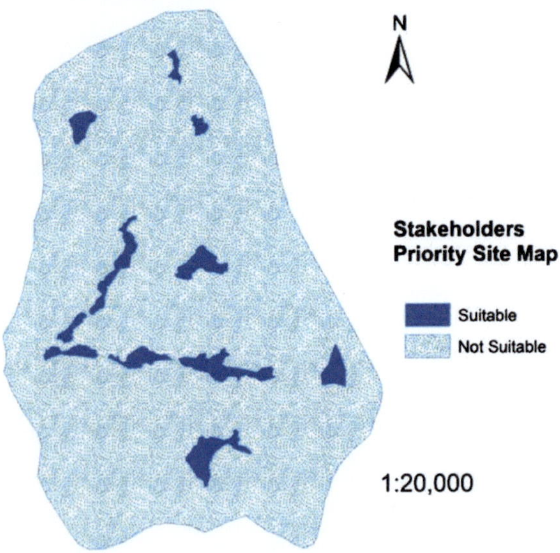

Fig. 26.5 Stakeholders' priority water harvesting site map in the study area

26.7.1.3 Soil Texture

The suitability of a pond site depends on the ability of the soil to hold water. Soils having clay or silty clay textures are excellent for ponds; sandy clays are satisfactory. Coarse-textured sand and sand–gravel mixtures do not hold water well and are unsuitable for ponds (FAO 1991). Based on laboratory analysis of soil samples,

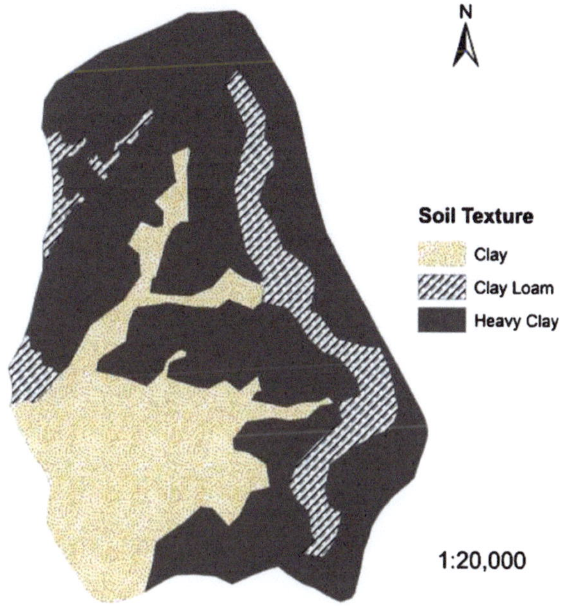

Fig. 26.6 Soil texture map of the study area

texture of the soil in the study area is clay, heavy clay and clay loam throughout the horizons. Hence, soil texture map of the study area is mapped as shown in Fig. 26.6.

26.7.1.4 Land Use/Land Cover Selection

Communal grazing lands are the best land use types to construct communal water harvesting ponds and collect runoff water for livestock consumption because of their proximity to animals stocking and communal land ownership. Land use/land cover map of the study area was digitized from US QuickBird2 satellite image. From the digitized land use/land cover map, communal grazing lands were identified and mapped as indicated in Fig. 26.7.

26.7.1.5 Flow Accumulation

Flow accumulation is the amount of rain that fell on the surface, upslope from each cell. It can determine how much rain has fallen within a given catchment. Cells with high flow accumulation are areas of concentrated flow and may be used to identify areas of high runoff amount (Jenson and Domingue 1988). It is important in ensuring the outlet of a catchment that receives a substantial amount of flow from upland cells which helps to identify potential sites for water harvesting. Hence, flow accumulation is derived from digital elevation model (Fig. 26.8) and considered as a factor. After generating the required input factors as indicated above, all vector

Fig. 26.7 Grazing land map of the study area

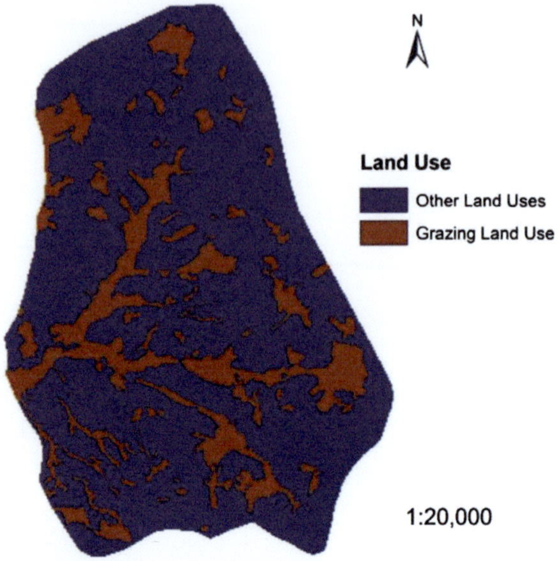

Fig. 26.8 Flow accumulation map of the study area

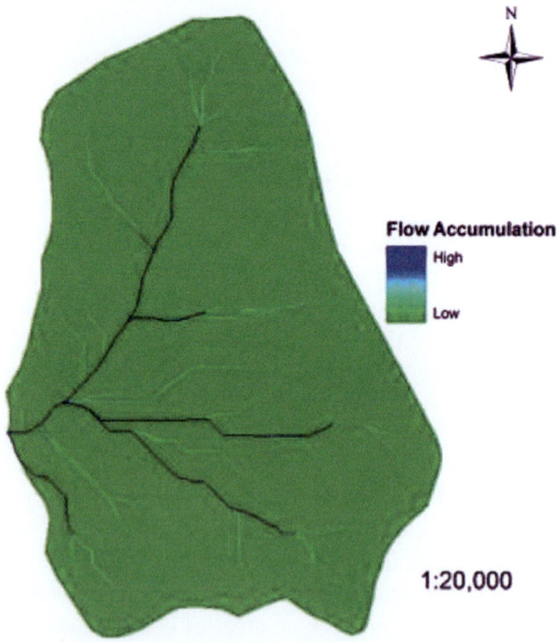

maps were changed to raster maps and intersected in ArcGIS raster calculator to find suitable areas for runoff water harvesting, which satisfy both controlling parameters.

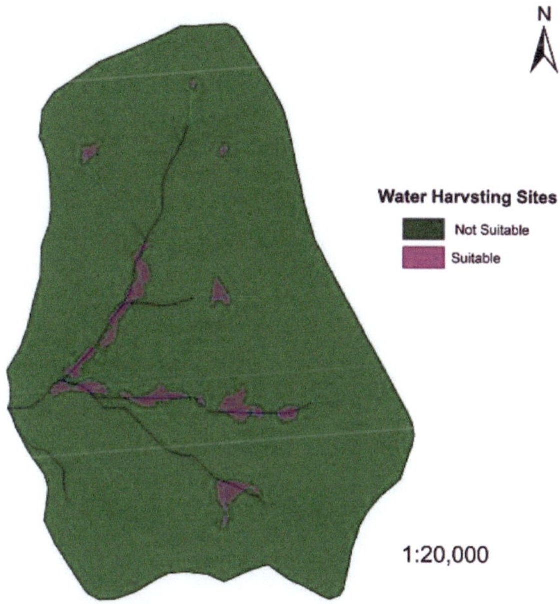

Fig. 26.9 Suitable water harvesting site map of the study area

26.7.2 Suitable Water Harvesting Areas

Suitable water harvesting areas for communal pond construction were identified using GIS by intersecting the required parameters such as stakeholders' priority site map, soil texture map, flow accumulation, slope map and land use map of the study area (Fig. 26.9). Some of the areas were not included in the final suitable area map even if they were selected by the local people due to technical considerations.

In order to verify the availability of enough runoff draining catchment area and runoff amount at the selected areas, flow accumulation was used. As indicated in Fig. 26.9, flow accumulation map of the study area is intersected with other parameters and the selected suitable runoff water harvesting areas were found where there is high flow accumulation.

26.8 Conclusions

Based upon the above results, it is possible to draw the following conclusions. Estimated runoff using SCS-CN model is comparable with the runoff amount measured by the conventional method. Therefore, it is possible to predict potential surface water to be harvested for the development purposes. Suitable areas for runoff water harvesting have been identified and mapped considering both social and technical aspects. From the results of this study, it is clear that the SCS-CN model is a powerful model for the qualitative as well as quantitative assessment of runoff amount in a watershed with reasonable accuracy.

DEM, which is derived from digitizing contours of 1:50,000 topographic sheets, is found to be a very good data source to develop a better slope gradient. GIS is a very important tool in selecting suitable runoff water harvesting sites and estimating a spatially distributed runoff by mapping and integrating different variables.

Acknowledgments We would like to thank the Amhara National Regional State—Bureau of Agriculture (ANRS-BOA) and SWHISA for the financial support. We extend our thanks to Dr. Birru Yitaferu for his constructive and useful assistance during the study and also to Ethiopia Abesha, Nardos Mulatie, Kaleab Mulatie, Mahilet Mulatie, Ayen Mulu, Tigist Getachew and Birara Chekol for their genuine support in creating conducive work environment. We deeply appreciate and respect the small scale farmers of Debre Mewi Watershed for giving us all the assistance we required and who are struggling all their lives for a better future.

References

Abtew W, Melesse AM (2014) Nile River basin hydology. In: Melesse AM, Abtew W, Setegn S (eds) Nile River basin: ecohydrological challenges, climate change and hydropolitics, pp 7–22

Abtew W, Melesse A, Desalegn T (2009a) Spatial, inter and intra-annual variability of the Blue Nile River basin rainfall. Hydrol Process 23(21):3075–3082

Abtew W, Melesse A, Desalegn T (2009b) El Niño southern oscillation link to the Blue Nile River basin hydrology. Hydrol Process Spec Issue: Nile Hydrol 23(26):3653–3660

BOFED (2007) Amhara national regional state bureau of finance and economic development. Annual statistical bulletin. Bahir Dar, Ethiopia

Chebud YA, Melesse AM (2009a) Numerical modeling of the groundwater flow system of the Gumera Sub-Basin in Lake Tana basin, Ethiopia. Hydrol Process Spec Issue: Nile Hydrol 23 (26):3694–3704

Chebud YA, Melesse AM (2009b) Modeling lake stage and water balance of Lake Tana, Ethiopia. Hydrol Process 23(25):3534–3544

Chebud Y, Melesse AM (2013) Stage level, volume, and time-frequency change information content of Lake Tana using stochastic approaches. Hydrol Process 27(10):1475–1483. doi:10.1002/hyp.9291

Defersha MB, Melesse AM (2011) Field-scale investigation of the effect of land use on sediment yield and surface runoff using runoff plot data and models in the Mara River basin, Kenya. CATENA 89:54–64

Defersha MB, Melesse AM, McClain M (2012) Watershed scale application of WEPP and EROSION 3D models for assessment of potential sediment source areas and runoff flux in the Mara River Basin, Kenya. CATENA 95:63–72

Demayo A, Steel A (1996) Data handling and presentation. In: Chapman D (ed) Water quality assessment. A guide to the use of biota, sediments and water in environmental monitoring, 2nd edn. United States environmental programme. United States Environmental, Scientific and Cultural organization, World Health Organization, London

FAO (1991) A manual for the design and construction of water harvesting schemes for plant production. Water Harvesting (AHL/MISC/17/91), Rome

Frasier GW, Myers LE (1983) Handbook of water harvesting. United States Department of Agriculture (USDA), Agriculture Handbook No. 600

Guta PK, Panigraphy S (2008) Predicting the spatio—temporal variation of runoff generation in India using remotely sensed input and Soil Conservation Service Curve Number model. Indian Research Organization, Ahmadabad 380 015, India

Hinrichson D (2003) A human thirst. World Watch 16(1):12–18

Jenson SK, Domingue JO (1988) Extracting topographic structure from digital elevation data for geographic information system analysis. Photogram Eng Remote Sens 54(11):1593–1600

Kassahun D (2007) Rainwater harvesting in Ethiopia: capturing the realities, exploring opportunities. FSS research report No. 1. Forum for Social studies, Addis Ababa, Ethiopia

Kuiper JR (1999) The role of rain fed farm ponds in sustaining agriculture and soil conservation in the dry high valley region of Cochabamba, Bolivia. Design considerations and post impoundment analysis. M.Sc. Thesis. University of North Texas, Denton, USA

Melesse AM, Graham WD, Jordan JD (2003) Spatially distributed watershed mapping and modeling. GIS based storm runoff response and hydrograph analysis, Part 2. J Spat Hydrol 3 (2):1–28

Melesse AM, Loukas Athanasios G, Senay Gabriel, Yitayew Muluneh (2009a) Climate change, land-cover dynamics and ecohydrology of the Nile River basin. Hydrol Process Spec Issue: Nile Hydrol 23(26):3651–3652

Melesse AM, Abtew W, Desalegne T, Wang X (2009b) Low and high flow analysis and wavelet application for characterization of the Blue Nile River system. Hydrol Process 24(3):241–252

Melesse AM, Abtew W, Setegn S, Dessalegne T (2011) Hydrological variability and climate of the Upper Blue Nile River basin. In: Melesse A (ed) Nile River basin: hydrology, climate and water use, Chap 1. Springer Science Publisher, pp 3–37. doi:10.1007/978-94-007-0689-7_1

Millorddov M, Marjanovic P (1991) Geographic information systems in environmentally sound river basin development. In 3rd Rhine—Danube workshop, proceedings, 7–8 Oct. Technische universteit Delft, Delft, the Netherlands

Ministry of Water Resources (2001) Initial national communication of Ethiopia to the United Nations framework convention on climate change (UNFCCC). A report submitted to the conference of parties of the UNFCCC under the GEF support climate change enabling activities project of Ethiopia, Addis Ababa, Ethiopia

Morgan (1995) Soil erosion and conservation. Addison-Wesley, Longman, 198 pp

Morgan RPC (2005) Soil erosion and conservation, 3rd edn. National Soil Resource Institute, Cranfield University

Pacey A, Cullis A (1991) Rainwater harvesting: the collection of rainfall and runoff in rural areas. Intermediate Technology Publications, London, UK

Reij C, Mulder P, Begemann L (1988) Water harvesting for plant production. World Bank, Washington DC, USA

RELMA (2005) Potential for rainwater harvesting in Africa, vol 1. In: ICRAF and UNEP

Setegn SG, Srinivasan R, Dargahil B, Melesse AM (2009a) Spatial delineation of soil erosion prone areas: application of SWAT and MCE approaches in the Lake Tana basin, Ethiopia. Hydrol Process Spec Issue: Nile Hydrol 23(26):3738–3750

Setegn SG, Srinivasan R, Melesse AM, Dargahil B (2009b) SWAT model application and prediction uncertainty analysis in the Lake Tana basin, Ethiopia. Hydrol Process 24(3):357–367

Setegn SG, Bijan Dargahi B, Srinivasan R, Melesse AM (2010) Modelling of sediment yield from Anjeni Gauged watershed, Ethiopia using SWAT. JAWRA 46(3):514–526

SWHISA (2007) Participatory rural appraisal report of Debre Mewi watershed. IRG–09, Bahir Dar, Ethiopia (unpublished)

Teshome W (2003) Irrigation practices, state interpenetration and farmers life-worlds in Drought-prone Tigray. PhD Dissertation, Wageningen, the Netherlands

USDA (1985) Soil conservation service: National Engineering Handbook, Sect 4—Hydrology, Washington, DC

USDA-SCS (1972) National engineering handbook, Sect 4—Hydrology. USDA-SCS, Washington, DC, USA

USDA-SCS (1986) Urban hydrology for small watersheds. Technical release 55. http://www.info.usda.gov/CED/ftp/CED/tr55.pdf

World Water Assessment Programme (2003) Water for people, water for life. United Nations Educational, Scientific and Cultural Organizations (UNESCO), and Berghan Books, New York, USA

Yitayew M, Melesse AM (2011) Critical water resources management issues in Nile River basin. In: Melesse AM (ed) Nile River basin: hydrology, climate and water use, Chap 20. Springer Science Publisher, pp 401–416. doi:10.1007/978-94-007-0689-7_20

Part VI
Climate Change Impact on Sediment and Water Dynamics

Chapter 27
Climate Change Impact on the Hydrology of Weyb River Watershed, Bale Mountainous Area, Ethiopia

Alemayehu A. Shawul, Tena Alamirew, Assefa M. Melesse and Sumedha Chakma

Abstract Identifying local impact of climate change at a watershed level and quantitative estimates of hydrological effects of climate change is crucial for solving potential water resource management problems. The aim of this study was to downscale the global circulation model (GCM) to Weyb watershed for analyzing the impact of climate change on hydrological variability. The SDSM was used to downscale the GCM output and it has accurately replicated the observed series both for A2a and B2a scenarios. SWAT model was calibrated and validated to simulate streamflow. For each future time horizon, the change in mean annual maximum and minimum temperature has indicated a slight increment from the base period both for A2a and B2a scenarios. The projected precipitation revealed the average seasonal precipitation could increase in dry season and decrease in wet season both for A2a and B2a scenarios for the 2020s, 2050s, and 2080s time period. The change in annual streamflow with exception to 2080s shows an increase by 0.83 % and 0.64 % in 2020s and 2050s, respectively, under A2a scenario, and by 5.02 % in 2080s under B2a scenario. Subsurface flow parameters were found to be more sensitive to the streamflow of the watershed. The annual streamflow might be reduced by 1.5 % in 2080s under A2a scenario and could also be reduced by 1.14 %

A.A. Shawul (✉)
Department of Natural Resource Management, Madawalabu University, P.O.Box 247, Bale Robe, Ethiopia
e-mail: aabate50@gmail.com

T. Alamirew
Water and Land Resources Center (WLRC), Addis Abeba, Ethiopia
e-mail: tena.a@wlrc-eth.org

A.M. Melesse
Department of Earth and Environment, Florida International University, Modesto A. Maidique Campus, Miami, FL 33199, USA
e-mail: melessea@fiu.edu

A.A. Shawul · S. Chakma
Civil Engineering Department, IIT Delhi, Hauz Khaz, New Delhi 110016, India

© Springer International Publishing Switzerland 2016

A.M. Melesse and W. Abtew (eds.), *Landscape Dynamics, Soils and Hydrological Processes in Varied Climates*, Springer Geography,
DOI 10.1007/978-3-319-18787-7_27

and 0.99 % in 2020s and 2050s, respectively, under B2a scenario. The result has revealed an increase of streamflow on dry season and reduction on the wet and intermediate seasons which has similar pattern with the rainfall. The change in the amount and distribution of rainfall and level of temperature would affect agricultural productivity and water utilizations in the region.

Keywords Climate change · Hydrological modeling · SWAT · SDSM · Bale Mountains · Genale-Dawa River Basin · Ethiopia

27.1 Introduction

Water is the lifeblood of the planet, and the state of this resource affects all natural, social and economic systems. Water serves as the fundamental link between the climate system, the human society, and the environment. Climate change is severely impacting the hydrological cycle and consequently, water management. This will in turn have significant effects on human development and security (IPCC 2007). The scientific assessment by the Intergovernmental Panel on Climate Change (IPCC) concludes that since the late nineteenth century, anthropogenic-induced emissions of gases such as carbon dioxide (CO_2) that trap heat in the atmosphere in the manner of a greenhouse have contributed to an increase in global mean surface air temperature of about 0.3–0.6 °C over the past century and by about 0.2–0.3 °C over the last 40 years (Nicholls et al. 1996). Without further action to reduce greenhouse gases (GHG) emission, the global average surface temperature is projected to be likely increased further by 1.8–4.0 °C in this century.

In Africa, the average annual temperature has been rising steadily. During the twentieth century, the continent saw increase in temperatures of about 0.5 °C. Countries in the Nile Basin have an increase of about 0.2–0.3 °C per decade during the second half of the century. Future warming is likely to be greatest over the interior of semiarid margins of the Sahara and central southern Africa (Eriksen et al. 2008). Confidence in projecting changes in the direction and magnitude of climate extremes depends on many factors including the type of extreme, the region and the season, the amount and quality of observational data, the level of understanding of the underlying processes, and the reliability of their simulation in models (IPCC 2012). Although climate change is expected to have adverse impacts on socio-economic development globally, the degree of the impact will vary across nations. A systematic study, analyzing the impact of climate-induced scenarios on water resources availability and adaptation strategies on the sub-basin as the basic unit of assessment, is still missing (Faramarzi et al. 2013). Developing countries in general and least developed countries like Ethiopia in particular are more vulnerable to the adverse impacts of climate variability and change. This is due to their low adaptive capacity, its economic and geographic settings, and high sensitivity of socioeconomic systems to climate variability and change (NMSA 2007). The recent flooding occurrences and the frequent drought in Ethiopia can be sited as concrete evidences for these impacts.

Mountainous watersheds are the origin for many of the largest rivers in the world and represent major sources of water availability for many countries (Sanjay et al. 2010). They represent not only the local water resources but also considerably influence the runoff regime of the downstream rivers. Farm Africa-SOS Sahel Ethiopia (2007) described that among the prior ecological services of the Bale Mountain National Park (BMNP), afro-alpine ecosystem is one of its hydrological systems. Moreover, BMNP is a source of over 40 streams on which more than 10 million people are dependent. The importance of the hydrological services that the area provides to southeastern Ethiopia and parts of Somalia and Kenya has gradually been recognized and its conservation is now a primary purpose of the park. Climate change has already become a global issue and a concern for all caring for the future. There are some studies on climate change impact on water resources of river basins in Ethiopia (Lijalem 2006; Abdo et al. 2009; Melesse et al. 2009; Setegn et al. 2011; Yirefu 2012). Despite the fact that the impact of different climate change scenarios is projected at a global scale, the exact type and magnitude of the impact at a small watershed scale remain untouched in most parts of the world. Therefore, identifying localized impact of climate change at a watershed level and quantitative estimates of hydrological effects of climate change is crucial. This also gives an opportunity to define the degree of vulnerability of local water resources to climate change and plan appropriate adaptation measures that must be taken ahead of time.

The Weyb River is a multi-scope river upon which different water-based projects depend on its flow. The projects include many existing and proposed irrigation schemes, tourism, and fish farming at the different parts of the river. Hence, estimation of monthly, seasonal, and long-term runoff yield and future climate change impact on the hydrology of the watershed may help to identify the best and sustainable land use and management options. Therefore, the output of this study may be utilized to plan and implement effective land and water resources development and management in Bale Mountainous area. Against these backgrounds, the general objective of this study is therefore to downscale the global circulation model (GCM) output in particular temperature and precipitation to watershed level for analyzing the impact of climate change on streamflow variability of the Weyb River. The specific objectives of the study are to develop temporal climate change scenarios (precipitation and temperature), for the Weyb River watershed, for the future until 2099. Also, quantify the climate change impact on streamflow of the Weyb River for three time horizons until 2099.

27.2 Description of Study Area

Weyb watershed is found in southeastern part of Ethiopia in Genale-Dawa basin, and it is located between 6° 30′–7° 30′N latitudes and 39° 30′–41° 02′E longitudes as shown in Fig. 27.1. It covers a total drainage area of 4123.8 km^2. The Weyb

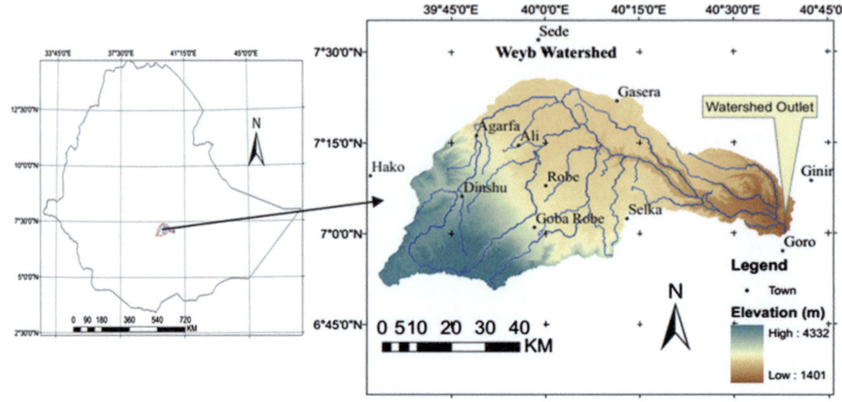

Fig. 27.1 GCM grid box for Ethiopia (*left*), location map, river networks, and elevation (m.a.s.l) of the Weyb watershed (*right*)

River originates from the northern flanks of the Bale Mountains and first flows generally northeastward then flows to east and southeastward for the remainder of its course. Finally, it joins with Genale and Dawa rivers near Ethiopia–Somalia border strengthening its journey to the Indian Ocean. It originates from an elevation of 4343 m above mean sea level, in the Bale Mountains extreme points locally called Sanetti to an elevation of 1530 m at the watershed outlet. The upper most part of the watershed is covered with the afro-alpine ecosystem which is known to be the largest such area in Ethiopia.

27.3 General Circulation Model (GCM) and Climate Scenario

Output from HadCM3 (Hadley Centre Coupled Model, version 3) GCM model was used for this impact study. The model results are available for the A2 and B2 scenarios, where A2 is referred as the medium-high emission scenario and B2 as the medium-low emission scenario. It is freely downloaded from the Canadian Institute for Climate Studies (CICS) Web page. This model is selected for two reasons. First, the model is widely applied in many climate change impact studies, and second, it provides large-scale daily predictor variables which can be used for statistical downscaling model (SDSM). The relative performance of GCMs depends on the size of the region as small regions at sub-grid scale are less likely to be well described than large regions at continental scale. It also depends on location as the level of agreement between GCM outputs varies a lot from region to region, and on variables such as regional precipitation which has high spatially variation and is more difficult to model than regional temperature (Carter et al. 2001).

Figure 27.1 shows a GCM with 2.5° latitude 3.75° longitude grid size from which the grid box for the study area is selected. The watershed is located in African window of GCM grid box between 5° 00′N and 7° 30′N latitude, and 37° 30′E and 41° 15′E longitude.

27.3.1 SDSM Model Setup

27.3.1.1 Screening the Downscaling Predictor Variables

The choice of appropriate downscaling predictor variables was undertaken through the screen variable option of SDSM using correlation analysis, partial correlation analysis, and scatter plot. Partial correlation analysis was done for selected predictors to see the level of correlation with each other. This statistics identify the amount of explanatory power of the predictor to explain the predictand. Finally, the scatter plot was carried out in order to identify the nature of the association (linear, nonlinear, etc.), whether or not data transformation(s) may be needed, and the importance of outliers.

27.3.1.2 Calibration, Validation, and Scenario Generation

The model was calibrated for precipitation and maximum and minimum temperature (predictand variables) along with a set of predictor variables, and computed the parameters of multiple regression equations with an optimization algorithm. Daily data were used for model calibration for data representing the current climate condition of the period 1985–1995. Areal values of predictand variables from each meteorological station in the watershed were used for calibrating the predictor variables. For monthly models, different model parameters were used for each month (Wilby and Dawson 2007). Validation was done based on seven years simulation for years 1996–2001. Validation of the model was performed using the results of weather generator and independent observed data that were used for calibration. Finally, scenarios were generated for precipitation, and maximum and minimum temperature predictand variables, both for base and future period, using HadCM3 GCM model output for the two emission scenarios (A2 and B2 SRES emission scenarios). Each time series data of the observed climate variable was linked to the regression model weights to generate the synthetic time series data into a series of ensembles. The average of those 20 ensembles was used. The future scenarios were developed by dividing the future time series into three periods of 30 years; 2020s (2011–2040), 2050s (2041–2070), and 2080s (2071–2099).

27.4 Hydrological Modeling with SWAT

The semi-distributed physically based model SWAT was used for simulation of streamflow of Weyb watershed to meet the goals of this study. The SWAT model was found to be applicable and efficient for simulation of streamflow in Bale Mountainous area of Shaya River watershed (Shawul et al. 2013). In the land phase of hydrological cycle, SWAT simulates the hydrological cycle based on the water balance Eq. (27.1).

$$\mathrm{SW_t} = \mathrm{SW_0} + \sum_{i=1}^{t} \left(R_{\mathrm{day}} - Q_{\mathrm{surf}} - E_{\mathrm{a}} - W_{\mathrm{seep}} - Q_{\mathrm{gw}}\right) \tag{27.1}$$

where $\mathrm{SW_t}$ is the final soil water content (mm), $\mathrm{SW_0}$ is the initial soil water content on day i (mm), t is the time (days), R_{day} is the amount of precipitation on day i (mm), Q_{surf} is the amount of surface runoff on day i (mm), E_{a} is the amount of evapotranspiration on day i (mm), W_{seep} is the amount of water entering the vadose zone from the soil profile on day i (mm), and Q_{gw} is the amount of return flow on day i (mm).

To estimate surface runoff, two methods are available. These are the SCS curve number procedure USDA Soil Conservation Service (USDA 1972) and the Green and Ampt infiltration method as cited in Neitsch et al. (2005). In this study, the SCS curve number method was used to estimate the surface runoff. The Penman-Monteith method (Monteith 1965) was used for the estimation of potential evapotranspiration (PET). The SCS curve number is described by Eq. (27.2).

$$Q_{\mathrm{surf}} = \frac{\left(R_{\mathrm{day}} - 0.2S\right)^2}{\left(R_{\mathrm{day}} + 0.8S\right)} \tag{27.2}$$

where Q_{surf} is the accumulated runoff or rainfall excess (mm), R_{day} is the rainfall depth for the day (mm), and S is the retention parameter (mm). More detailed descriptions of the different model hydrological components are listed in SWAT user's manual (Neitsch et al. 2005).

The application of SWAT in predicting stream flow and sediment as well as evaluation of the impact of land use and climate change on the hydrology of watersheds has been documented by various studies (Dessu and Melesse 2012, 2013; Dessu et al. 2014; Wang et al. 2006, 2008a, b, c; Wang and Melesse 2006, 2005; Behulu et al. 2013, 2014; Setegn et al. 2009a, b, 2010, 2011, 2014; Mango et al. 2011a, b; Getachew and Melesse 2012; Assefa et al. 2014; Grey et al. 2013; Mohamed et al. 2015).

27.4.1 SWAT Model Inputs

27.4.1.1 Digital Elevation Model

Topography was defined by a digital elevation model (DEM) which describes the elevation of any point in a given area at a specific spatial resolution as a digital file. Furthermore, sub-basin parameters such as slope, slope length, and defining of the stream network with its characteristics such as channel slope, length, and width was derived from the DEM. DEM with a resolution of 30 m was used, which was obtained from ASTER GDEM official Web site.

27.4.1.2 Soil Properties

Basic physico-chemical properties of major soil types in the watershed were mainly obtained from the following sources; Genale-Dawa River Basin integrated resources development master plan soil database and digital soil map from Ethiopian Ministry of Water and Energy (MoWE). Soil and Terrain Database for northeastern Africa CD-ROM (FAO 1998) was also sourced. In addition to these sources, some soil properties were calculated based on available soil parameters and through direct field observations.

27.4.1.3 Land Use/Land Cover Map

Land use or land cover (LULC) is one of the important spatial input data for the SWAT model that affects surface water runoff, evapotranspiration, soil erosion, and other hydrological process in a given watershed. The LULC map and datasets were obtained from MoWE, Genale-Dawa River Basin master plan which was produced between 2004 and 2007 and from field investigations.

27.4.1.4 Meteorological Data

Meteorological data were collected from National Meteorology Service Agency (NMSA) for stations at Bale Robe, Goba, Dinsho, Agarfa, Sinana, Adaba, Homa, Ali, Gassera, Goro, Ginir, and Hunte. Relation between meteorological stations was evaluated through cross-correlation and double mass curve analysis. Furthermore, homogeneity test of time series data was evaluated using Rainbow and XLSTAT. Meteorological data from the 11 stations and period of records are presented in Table 27.1.

Table 27.1 Inventory of meteorological data records

S. no.	Stations name	Period of record	Latitude (°N)	Longitude (°E)	Altitude (m)
1.	Robe	1984–2011	7.13	40	2464
2.	Dinsho	1981–2007	7.1	39.78	3072
3.	Goba	1998–2007	7.02	40	2545
4.	Agarfa	1988–1997	7.26	39.81	2360
5.	Sinana	1982–2008	7.07	40.31	2364
6.	Ginir	1980–2012	7.13	40.71	1929
7.	Adaba	1980–2010	7	39.4	2415
8.	Goro	2001–2010	7	40.47	1806
9.	Gassera	1980–2010	7.36	40.18	2337
10.	Homaa	1988–2010	7.14	39.94	2505
11.	Hunte	1980–2011	7.09	39.42	2413

27.4.1.5 Hydrological Data

Daily river discharge data of the Weyb River watershed and some of its tributaries (Shaya, Tegona, and Denka) were obtained from the Hydrology Department of the MoWE. Daily river discharges of the gauging stations were utilized for performing sensitivity analysis, calibration, and validation of the SWAT model. Average annual observed flow series of the station passed the homogeneity test for range and maximum of cumulative deviation.

An automated baseflow separation and recession analysis technique (Arnold and Allen 1999) was employed to separate the baseflow and surface runoff using the total daily streamflow records.

27.4.2 SWAT Model Setup

27.4.2.1 Watershed Delineation

The watershed delineation process includes five major steps, DEM setup, stream definition, outlet and inlet definition, watershed outlets selection and definition, and calculation of sub-basin parameters. For the stream definition, the threshold-based stream definition option was used to define the minimum size of the sub-basin.

27.4.2.2 Hydrological Response Units

The land area in a sub-basin may be divided into Hydrological Response Units (HRU). SWAT uses the concept of HRU. A HRU is the total area in the sub-basin with a particular land use, management, and soil. The HRU distribution in this study was determined by assigning multiple HRU to each sub-basin.

27.4.2.3 Sensitivity Analysis

Sensitivity analysis was made using a built-in SWAT sensitivity analysis tool that uses the Latin Hypercube One-factor-At-a-Time (LH-OAT). The inputs were the observed daily flow data, the simulated flow data, and the sensitive parameter in relation to flow with the absolute lower and upper bound and default type of change to be applied (method application) was used. LH-OAT combines the OAT design and LH sampling by taking the Latin Hypercube samples as initial points for OAT design (Van Griensven 2005).

27.4.2.4 Model Calibration, Validation, and Performance Evaluation

Calibration is the process whereby model parameters are adjusted to make the model output match with observed data. Manual and automatic calibration method was applied. First, the parameters were manually calibrated until the model simulation results are accep

table as per the model performance measures. Next, the final adjusted parameter values of manual calibration were used as the initial values for the autocalibration processes. The general procedure of calibrating the SWAT model was performed as reported in Santhi et al. (2001). The model was run for six years, January 1, 1986, through December 31, 1991; however, the first year of the recording period was used for stabilization of model runs.

Validation is comparison of the model outputs with an independent dataset without making further adjustments to the parameters used for calibration. Validation was performed for a period of four years (January 1, 1992, to December 31, 1995) on monthly basis. Three statistical model performance evaluation measures, namely regression coefficient (R^2), percent difference (D), and Nash and Sutcliffe simulation efficiency (E_{NS}) (Nash and Sutcliff 1970) were used both for calibration and validation of SWAT model.

27.5 Analysis of Climate Change Impacts on Streamflow

The impact of climate change on water resources variability, especially on river flows, soil moisture, evapotranspiration, and groundwater flow, has been studied by analyzing projected and downscaled climatic data and using hydrological models (Mango et al. 2011a, b; Behulu et al. 2014; Setegn et al. 2011, 2014; Assefa et al. 2014; Melesse et al. 2009, 2011; Dessu and Melesse 2012; Grey et al. 2013).

The climate model output from HadCM3, which was downscaled to the Weyb watershed level, was used to simulate streamflow. Simulation of streamflow corresponding to future climate change was done using the SWAT model. The downscaled climate scenario (which consists of maximum and minimum temperature, and precipitation) with an estimated potential evapotranspiration was used as an input to the model. The analysis of streamflow was carried out for three time horizons in the future. These periods were the 2020s, (2011–2040); the 2050s, (2041–2070); and the 2080s, (2071–2099).

27.5.1 Mann-Kendall Trend Test

The Mann-Kendall test is based on the calculation of Kendall's tau measure of association between two samples, which is itself based on the ranks of the samples. This method is a non-parametric rank-based procedure that has been commonly used to assess if there is a trend in a time series hydro-meteorological data (Hamed 2008; Karpouzos et al. 2010). The Mann-Kendall test was applied to test trends in future climatic variables. Mann-Kendall test is expressed by Eq. (27.3).

Let $X_1, X_2, \ldots, X_n$ represent n data points where X represents the data point at time j. The Mann-Kendall statistic (S) is given by Eq. (27.3):

$$S = \sum_{k=1}^{n-1} \sum_{j=k+1}^{n} \operatorname{sign}(X_j - X_k)$$

$$\textbf{For } \operatorname{sign}(X_j - X_k) = \begin{cases} 1 & \text{if } (X_j - X_k) > 0 \\ 0 & \text{if } (X_j - X_k) = 0 \\ -1 & \text{if } (X_j - X_k) < 0 \end{cases} \tag{27.3}$$

A positive value of S is an indicator of an increasing trend, and a negative value indicates a decreasing trend.

27.6 Climate Change Scenarios and Climate Projection

27.6.1 Minimum Temperature

A2a emission scenario maintained higher variability in minimum temperature pattern throughout the century than the B2a emission scenario. As shown in Table 27.2, the minimum temperature could decrease throughout the century in dry and wet seasons and could increase in intermediate seasons. Historical average minimum temperature indicated that the lowest record was obtained in the dry season (November, December, January, and February) and it could continue to be cooler in the future time periods. The minimum temperature could be cooler in the intermediate season as have been indicated by HadCM3A2a and HadCM3B2a

Table 27.2 Absolute change in mean seasonal minimum temperature in (°C) at different time horizons from base period

Scenario	Seasons in the 2020s			Seasons in the 2050s			Seasons in the 2080s		
	Dry	Intermediate	Wet	Dry	Intermediate	Wet	Dry	Intermediate	Wet
A2a Scenario	−0.34	1.14	−0.3	−0.33	1.2	−0.51	−0.17	1.22	−0.89
B2a Scenario	−0.39	1.12	−0.3	−0.39	1.11	−0.43	−0.22	1.22	−0.67

scenarios. Locally, the dry season is known as Bega, the intermediate season, Belg, and the wet season, Kiremt. The seasonal classifications were made based on the local rainfall patterns, Dry season (Nov, Dec, Jan and Feb), Intermediate season (Mar, Apr, May and Jun) and Wet season (Jul, Aug, Sept and Oct), therefore, it is applicable only in south-eastern part of Ethiopia.

Relatively, a larger absolute monthly change from the baseline minimum temperature was found in the month of April for both A2a and B2a emission. As shown in Figs. 27.2 and 27.3, the change is observed for all the three time horizons (2020s, 2050s, and 2080s). For the months of February, March, April, and May, the model has indicated an increase in minimum temperature and decrease for the months of July, August, October, November, and December. In both extreme conditions (increase or decrease), the change in minimum temperature is higher in the 2080s. Largest decrease in average minimum temperature was observed in the month of October about −2.01 °C for 2080s at the end of the century. For each time horizon, the change in mean annual minimum temperature has a slight increase, 0.16 %, 0.12 %, and 0.048 °C for 2020s, 2050s, and 2080s, respectively, from the base

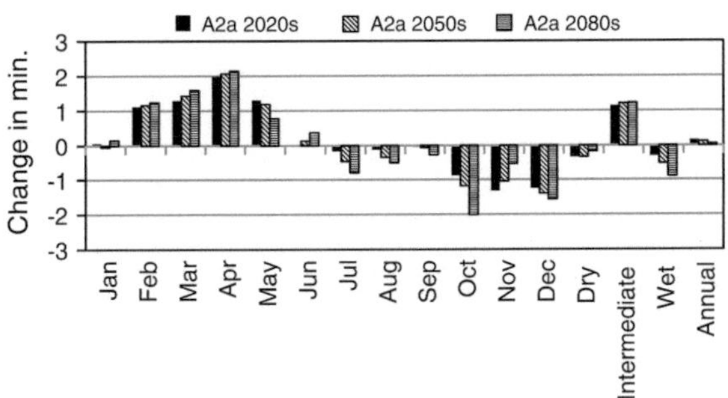

Fig. 27.2 Future changes in average monthly and seasonal minimum temperature (°C) from the base period for A2a scenario

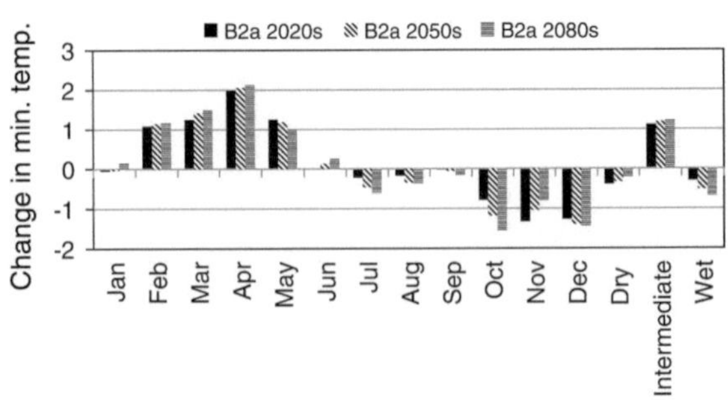

Fig. 27.3 Future changes in average monthly and seasonal minimum temperature (°C) from the base period for B2a scenario

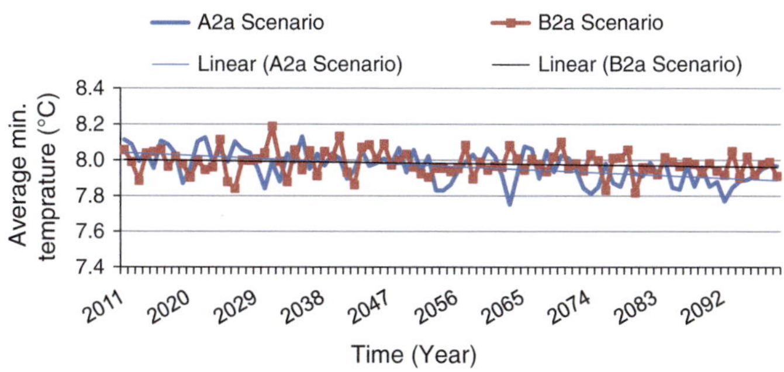

Fig. 27.4 Future pattern of average annual minimum temperature (°C)

period level for A2a scenario and 0.14 °C, 0.13 °C, and 0.10 °C for 2020s, 2050s, and 2080s, respectively, from the base period level for B2a scenario.

Generally, the change in average monthly minimum temperature may range between −1.32 °C in November and +1.98 °C in April for the coming 2020s (2011–2040); between −1.4 °C in December and +2.06 °C in April for 2050s (2041–2070), and −1.56 °C in December and +2.14 °C in April for 2080s (2071–2099) for the A2a scenario. For B2a scenario, the change in average monthly minimum temperature may range between −1.32 °C in November and +1.98 °C in April for 2020s (2011–2040); between −1.28 °C in December and +2.01 °C in April for 2050s (2041–2070); and between −1.55 in October and +2.12 °C in April for 2080s (2071–2099). The projected changes in minimum temperature values for A2a and B2a future scenarios show both increase and decrease are extreme in 2080s at the end of twenty-first century. Overall, as shown in Fig. 27.4, the average minimum temperature has a decreasing trend in the future time periods until the end of century.

27.6.2 Maximum Temperature

The projected mean seasonal maximum temperature shows an increasing trend in dry and wet seasons but decrease in the intermediate season in future time horizons for both A2a and B2a scenarios. The relative changes of maximum temperature from the base period for both scenarios for each season are presented in Table 27.3. The projected mean annual maximum temperature in the 2020s indicated that the change will be by +0.03 and +0.02 °C for A2a and B2a scenarios, respectively, which is negligible. The changes in the 2050s could be +0.09 and +0.07 °C for A2a and B2a scenarios, respectively, and in the 2080s, the increment could be +0.21 and +0.16 °C for A2a and B2a scenarios, respectively. The change in mean annual maximum temperature is negligible because increase and decrease of maximum temperature occur in different months. For instance in the months of May, June,

Table 27.3 Absolute change in mean seasonal maximum temperature (°C) at different time horizons from base period

Scenario	Seasons in the 2020s			Seasons in the 2050s			Seasons in the 2080s		
	Dry	Intermediate	Wet	Dry	Intermediate	Wet	Dry	Intermediate	Wet
A2a Scenario	0.09	−0.066	0.07	0.37	−0.25	0.17	0.73	0.51	0.4
B2a Scenario	0.06	−0.044	0.05	0.27	−0.21	0.14	0.54	−0.35	0.29

September, and November, decrease in maximum temperature is observed and an increase in the months of January, February, April, July, August, and December. The projected mean monthly maximum temperature has larger magnitude of increment in the month of April 2080s (+1.92 °C) and (+1.53 °C) for A2a and B2a scenarios, respectively. On the other hand, a larger decrease in mean monthly maximum temperature might occur on May 2080s (−2.55 °C) and (−1.96 °C) for A2a and B2a scenarios, respectively as shown in (Figs. 27.5 and 27.6).

The variability of maximum temperature is higher for A2a scenario than B2a scenario and for both scenarios a linear trend in slight increase is indicated until the end of twenty-first century (Fig. 27.7). Comparatively, A2a which is the high emission scenario shows higher maximum temperature trend by the end of the next century than the B2a scenario. Generally, the future scenarios have shown slightly increasing trend on maximum temperature and decreasing trend on minimum temperature.

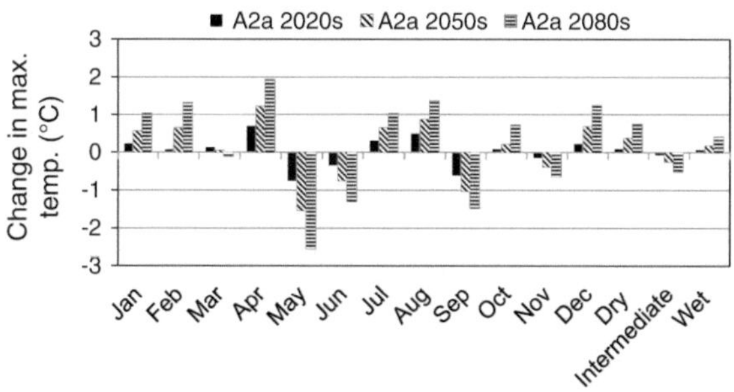

Fig. 27.5 Change in average monthly and seasonal maximum temperature (°C) from the base period for A2a scenario

Fig. 27.6 Change in average monthly and seasonal maximum temperature (°C) from the base period for B2a scenario

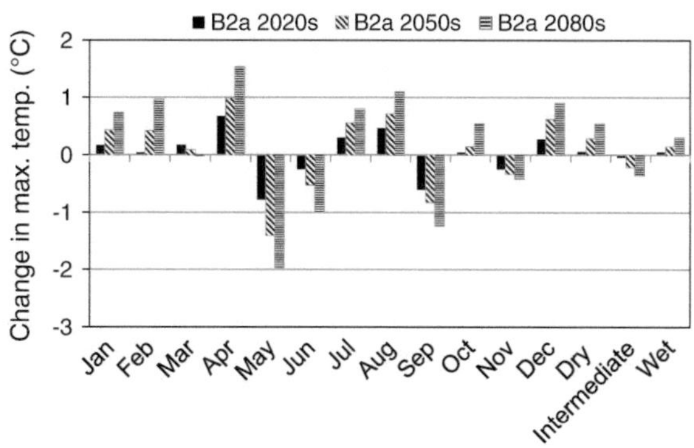

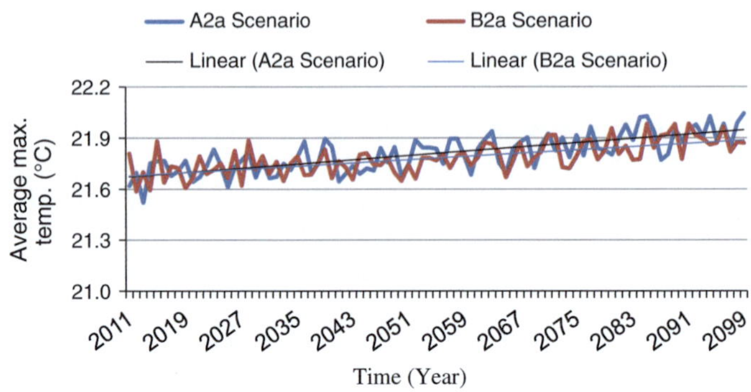

Fig. 27.7 Future trend of average annual maximum temperature (°C)

27.6.3 Precipitation

As described in the IPCC Third Assessment Report (McCarthy et al. 2001), the projected future changes in mean seasonal rainfall in Africa are less well defined. The diversity of African climates, the high rainfall variability, and a very sparse observational network make the predictions of future climate change difficult at the sub-regional and local scales. The generated precipitation scenarios have indicated an increase of precipitation in the dry season and decrease in the wet season (the main rainy season) for both A2a and B2a scenarios with overall increase in mean annual precipitation for both A2a and B2a scenarios (Table 27.4).

In all future time horizons, for the months of November and December, the projected rainfall amounts could increase with large magnitude compared with the baseline period, for both A2a and B2a scenarios. But for the rainiest months of the year, August and September, the model shows a decrease in each future time horizon for both A2a and B2a scenarios. For the months of January, February, March, and April, the projected rainfall amount increased from the baseline period except for January and February 2080s which showed a decrease for both A2a and B2a scenarios. On the other hand, for the months of May, June, July, and October, the projected rainfall amount in all future time horizons exhibits a negligible change from the baseline period for both A2a and B2a scenarios. Figure 27.8 shows pattern of future mean annual precipitation with a range of 909.01 mm to 1163.47 mm for years 2040 and 2061, respectively, for A2a scenario, and 954.85 mm to 1136.57 mm for years 2042 and 2090, respectively, for B2a scenario. It shows substantial variability of total average annual precipitation from year to year

Table 27.4 Percent difference in mean annual precipitation at different time horizons from the base period in (%)

Scenario	2020s	2050s	2080s
A2a scenario	6.56	7.75	5.96
B2a scenario	7.24	7.26	9.28

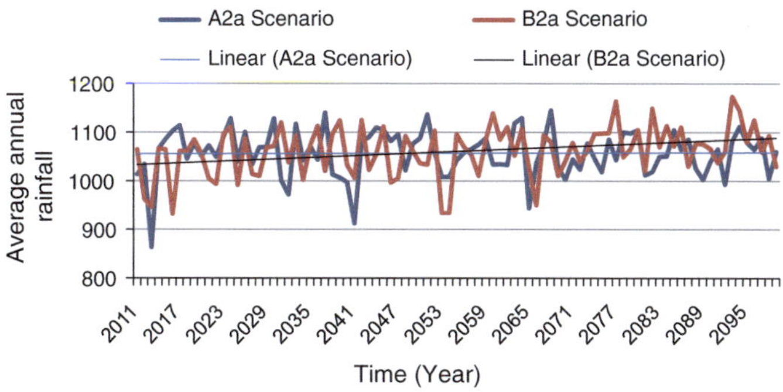

Fig. 27.8 Future pattern of annual precipitation

throughout the century. Shongwe et al. (2009) reported a general increase in rainfall in the tropics, which is also the case in this study. In East Africa, Faramarzi et al. (2013) reported a generally wetter climate and a large spatial variation from a decrease of 25–50 % to an increase of 25–50 %. Generally, comparatively high increase of rainfall in the dry season could have positive impact on pastoral region of the study area and it could affect the highland areas negatively due to this season being the only main crop harvesting period.

27.6.4 Mann-Kendall Trend Test of Future Climate (2011–2099)

Based on the standardized test statistics, it is possible to conclude that Mann-Kendall test has revealed a statistically significant trend for both future precipitation and temperature at the 5 % significant level. Precipitation for B2a scenario and maximum temperature for both A2a and B2a scenarios have shown an increasing trend for the future until year 2099. Precipitation for A2a scenario and minimum temperature for B2a scenario have shown decreasing trend but the trends were not significant at 5 % level as shown in Table 27.5.

Table 27.5 Mann-Kendall trend test for future annual precipitation and temperature A2a and B2a scenarios

	Kendall's tau	*p*-value	Alpha	Sen's slope	Trend
Precipitation A2a	−0.044	0.542	0.05	−0.111	Not significant
Precipitation B2a	0.196	0.007	0.05	0.604	Significantly increasing
Temp max A2a	0.52	<0.0001	0.05	0.003	Significantly increasing
Temp max B2a	0.479	<0.0001	0.05	0.003	Significantly increasing
Temp min A2a	−0.349	<0.0001	0.05	−0.002	Significantly decreasing
Temp min B2a	−0.113	0.119	0.05	−4.36E−04	Not significant

27.7 Simulation of the Hydrology of the Watershed

27.7.1 Watershed and HRU Delineations

From a minimum user-defined threshold area, 27 sub-basins were delineated in the Weyb watershed area of 4123.8 km^2. Each sub-basin boundary marks the end of a reach, the end point of accumulation for all flow from upstream, which then is fed into a downstream sub-basin and reach (Fig. 27.9). Once the main reach and the longest paths/tributaries are formed, the model uses other physical parameters (soil, land use, and land slope) to define HRUs.

From the assumed threshold values for HRU delineation of 10 %, 20 %, and 20 (%) for land use, soil, and slope, respectively, 276 HRUs were identified in 27 sub-basins. As shown in Table 27.6, the larger part of the watershed is covered by two soil types Eutric Vertisol and Haplic Luvisol, 21.13 and 20.27 %, respectively, of the middle to lower end of the watershed. On the other hand, Calcic Vertisol and Calcaric Cambisol cover relatively smaller portion of the watershed 2.89 and 2.03 %, respectively.

The dominant land use type within the study watershed is Agricultural Land-Close-grown (intensively cultivated cereal crops such as wheat and barley) about 35.68 % of the total watershed area. Residential-Med to Low Density (towns) and Residential-Low Density (farm villages) cover about 0.03 and 0.66 % of the total watershed area Table 27.7.

Depending on the maximum and standard deviation of land slope in the watershed, three slope classes were considered by dividing land slope classes as:

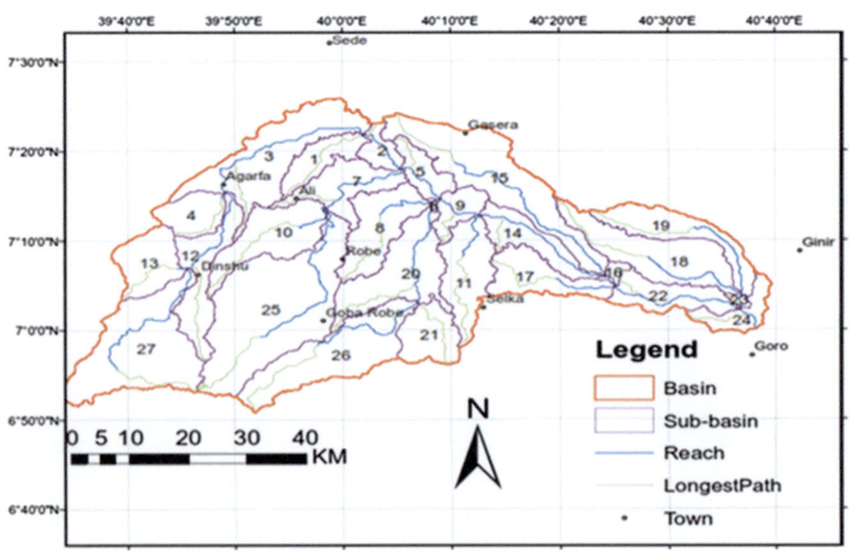

Fig. 27.9 Sub-basin delineation in the Weyb watershed by SWAT model

Table 27.6 Major soil of Weyb watershed and areal coverage

No.	Soil unit name	Area (ha)	% of watershed
1.	Eutric Vertisol	87,148.6	21.13
2.	Haplic Luvisol	83,596.4	20.27
3.	Vertic Luvisol	42,021.7	10.19
4.	Chromic Cambisol	44,315.9	10.75
5.	Regosol	27,133.9	6.58
6.	Eutric Cambisol	20,091.3	4.87
7.	Calcaric Cambisol	11,904.6	2.89
8.	Dystric Cambisol	61,595.5	14.94
9.	Calcic Vertisol	8365.5	2.03
10.	Leptosol	26,205	6.35

Table 27.7 Major land use of Weyb watershed and areal coverage

Land use	SWAT codes	Area (ha)	% of watershed
Agricultural land-close-grown	AGRC	147,152.6	35.6
Agricultural land-generic	AGRL	58,713.2	14.2
Range-grasses	RNGE	42,079.5	10.2
Pasture	PAST	66,232.7	16.1
Forest-mixed	FRST	29,632	7.2
Forest-evergreen	FRSE	883.6	0.2
Residential-med/low density	URLD	117.9	0.03
Residential-low density	URML	2723	0.66
Range-brush	RNGB	68,450.5	16.6
Water	WATR	9.6	0.001

class1 (0–10 %), class2 (10–25 %), and class3 (25–9999 %). Maximum value of the slope in SWAT database has a default value of 9999 %. From land slope classifications, about 64.81 % of the watershed area has slope 0–10 %, 23.1 % of the area has 10–30 %, and the rest 12.09 % has 25–78 % slope.

27.7.2 *Sensitivity Analysis*

The SWAT model considered the twenty-seven flow parameters for the analysis of sensitivity from which twenty-one of them were found to be relatively sensitive. Among the sensitive flow parameters, groundwater flow parameters were the most sensitive. Baseflow alpha factor (Alpha_Bf) [days], Threshold water depth in the shallow aquifer for flow (Gwqmn) [mm], Soil evaporation compensation factor (Esco), Initial curve number (II) value (Cn2), Soil depth (Sol_Z) [mm], Threshold water depth in the shallow aquifer for "revap" (Revapmin) [mm], Available water capacity (Sol_Awc) [mm water/mm soil] and Groundwater "revap" coefficient (Gw_Revap) were found to be the most sensitive hydrological parameters for the simulation of streamflow in the Weyb watershed. A brief description of each these hydrological parameters are listed in the SWAT model user's manual (Neitsch et al. 2005).

27.7.3 SWAT Model Calibration and Validation

27.7.3.1 Model Calibration

The automatic baseflow separation technique based on the daily streamflow data measured at the outlet of the watershed indicated that about 61 % of the total water yield is contributed from subsurface water source.

Model parameters were first calibrated manually followed by automatic calibration using ParaSol (Parameter Solutions), an autocalibration tool which is embedded in SWAT 2009. The calibration processes considered 8 flow parameters (Table 27.8) and values were varied iteratively within the allowable ranges until satisfactory agreement between measured and simulated streamflow was obtained. The autocalibration processes significantly improved model efficiency. The result from different statistical method of model performance evaluation met the criteria with R^2, E_{NS}, and D values of 0.81, 0.75, and 23, respectively.

An intensive hydrological calibration resulted in good SWAT predictive efficiency at the monthly time step when compared to measured flow data. The hydrograph of observed and simulated flow indicated that the SWAT model is capable of simulating the hydrology of Weyb mountainous watershed. Though rigorous calibration was undertaken for flow, there is still slight overestimation during the dry season flows of observed series both at calibration and validation periods

Table 27.8 Calibrated flow parameter values and variation methods

Flow parameters	Lower and upper bounds	Calibrated values	Variation methods
Alpha_Bf; baseflow alpha factor [days]	0.0–1.0	0.96	Replacement
Cn2; initial curve number (II) value	±25.0	−10.9	Multiply
Esco; soil evaporation compensation factor	0.0–1.0	0.16	Replacement
Gw_Delay; groundwater delay [days]	±10.0	5.47	Addition
Gwqmn; threshold water depth in the shallow aquifer for flow [mm]	±1000.0	854.02	Addition
Revapmn; threshold water depth in the shallow aquifer for "revap" [mm]	±100.0	92.71	Addition
Sol_Awc; available water capacity [mm water/mm soil]	±25.0	15.43	Multiply
Sol_Z; soil depth [mm]	±25.0	6.17	Multiply

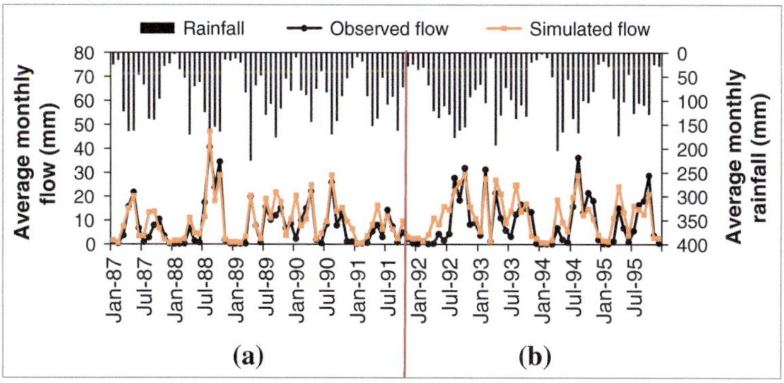

Fig. 27.10 Hydrograph of observed and simulated monthly flow for the **a** calibration period 1987–1991 and **b** validation period 1992–1995

(Fig. 27.10). The under prediction of flow during peak events and over prediction during low flows by the SWAT model has been reported in many studies (Thripati et al. 2003; Gassman et al. 2007; Sathian and Symala 2009).

27.7.3.2 Model Validation

It was found that the SWAT model has strong predictive capability with R^2, E_{NS}, and D values of 0.65, 0.59, and 20, respectively. Statistical model efficiency criteria fulfilled the requirement of $R^2 > 0.6$ and $E_{NS} > 0.5$ which is recommended by SWAT developer (Santhi et al. 2001). This indicates that the model parameters represent the processes occurring in the Weyb watershed as good as possible given the quality of available data. It can be used to predict watershed response for various outputs. The model validation results for monthly flow (Fig. 27.10) indicate generally a good fit between measured and simulated output and a slight overestimation of the low flows and underestimation of the peak flows were observed at the validation period. Since the model performed as well in the validation period, as for the calibration period, the set of optimized parameters listed in Table 27.8 during calibration can be taken as the representative set of parameters for the Weyb watershed.

27.8 Impact of Climate Change on the Hydrology

The ultimate objective of downscaling is to generate an estimate of meteorological variables corresponding to a given scenario of future climate. These meteorological variables are used as basis for hydrological impact assessment. Simulation of 1985–2005 period was used as a base period against which the change in future climate is evaluated. Daily precipitation and minimum and maximum temperature in SWAT are adjusted on monthly basis using results from the GCM for the future three periods of 30 years: 2020s (2011–2040), 2050s (2041–2070), and 2080s (2071–2099). Historical or base period was then rerun with the adjusted climate inputs. Other climate variables as wind speed, solar radiation, and relative humidity were assumed to be constant throughout the future simulation periods. Simulated monthly streamflow for the future time horizons exhibited larger variation in some months relative to the baseline period but generally it shows a double peak mode due to the bimodal type of rainfall in the area. With respect to individual months, absolute difference or change in depth of streamflow (mm) between the baseline and future time periods for A2a and B2a scenarios indicates decrease for January, February, March, April, and August months and increase for May, June, November, and December in all time horizons (2020s, 2050s, and 2080s). The variation in monthly streamflow was negligible in July, September, and October. The mean seasonal streamflow could increase in the dry season (November, December, January, and February) by 3.73 mm, 4.62 mm, and 3.28 mm. Intermediate season (March, April, May, and June) mean flow is expected to decrease by −4.05, −1.01, and −1.66 mm. Wet season (July, August, September, and October) is also expected to decrease by 0.87, −2.41, and −4.41 mm in the 2020s, 2050s, and 2080s, respectively, under A2a scenario as shown in Table 27.9. Similarly, the

Table 27.9 Average monthly, seasonal, and annual difference in streamflow between the baseline and future scenarios in percent (%)

Month	A2a Scenario			B2a Scenario		
	2020 (%)	2050 (%)	2080 (%)	2020 (%)	2050 (%)	2080 (%)
Jan	−35.1	−29.18	−34.04	−27.17	−35.2	−25.05
Feb	−16.2	−13.88	−40.62	−9.77	−19.54	−23.65
Mar	−61.65	−61.2	−66.25	−61.2	−64.13	−60.32
Apr	−29.47	−24.53	−23.32	−31.9	−30.02	−22.26
May	11.67	18.33	17.22	9.65	13.59	18.48
Jun	85.2	89.51	87.72	81.32	83.11	92.34
Jul	2.22	2.17	2.82	−0.4	0.55	2.82
Aug	−11.98	−16.31	−16.42	−16.01	−14.74	−10.41
Sep	7.67	0.76	−3.74	5.59	3.88	10.28
Oct	8.87	6.02	0.97	8.43	6.54	10.32
Nov	52.24	53.95	53.53	50.32	52.03	63.78
Dec	41.09	42.27	45.21	40.94	46.98	64.8
Dry season	16.03	15.66	11.12	15.39	13.46	24
Intermediate season	−6.12	−1.53	−2.51	−8.15	−6.49	−0.05
Wet season	0.96	−2.66	−4.87	−1.4	−1.69	2.53
Annual	0.83	0.64	−1.5	−1.14	−0.99	5.02

mean seasonal streamflow could increase in dry season by 4.54, 3.97, and 7.08 mm and decrease in intermediate season by −5.39, −4.29, and −0.03 mm. In the wet season, decrease is expected by −1.27, −1.53, and 2.29 mm in the 2020s, 2050s, and 2080s, respectively, under B2a scenario.

As shown in Table 27.9, the percent difference for the months of January, February, March, April, and August in simulated streamflow shows significant decrease for all the three time horizons under A2a and B2a scenarios. The month of June shows the largest increase of streamflow, more than 81 %, for all time horizons under both A2a and B2a scenarios. Future streamflow may also increase largely for the months of November and December. The maximum reduction in streamflow could occur for the month of March, where the mean monthly flow could be reduced by 61.6, 61.2, and 66.2 % in the 2020s, 2050s, and 2080s, respectively, for A2a scenario. Similarly, streamflow could be reduced for B2a scenario by 61.2, 64.1, and 60.3 % in the 2020s, 2050s, and 2080s, respectively. Comparatively, the reduction of runoff could be larger in the intermediate season, in the region within which runoff contribution is medium. On the other hand, the flow in the dry season "Bega" could largely increase in all future time horizons for A2a and B2a scenarios. In the wet season "Kiremt" the main rainy season in the region, the change in streamflow is predicted to be moderate. Likewise, the change in annual streamflow with exception of the 2080s shows an increase by 0.83 and 0.64 % in the 2020s and 2050s, respectively, under A2a scenario, and by 5.02 % in the 2080s under B2a scenario. Faramarzi et al. (2013) have reported that East Africa is projected to have increase in discharge between 10 and 20 %. This study in the contrary predicts annual streamflow reduction by 1.5 % in the 2080s under A2a scenario and could also be reduced by 1.14 and 0.99 % in the 2020s and 2050s, respectively, under B2a scenario.

It has been discussed in Faramarzi et al. (2013), Setegn et al. (2011) that streamflow is sensitive to rainfall variability and this is in agreement with our study increase in rainfall results streamflow to increase. Variability of streamflow which will result from climate change in the Bale highlands not only affects the livelihood of people in the Bale zone but also all inhabitants at the far downstream areas in Somalia whose life depends on the flow of Weyb River. Drought has been one of the most frequent climate-related phenomena occurring across large portions of the African continent, often with devastating consequences on agricultural production and food security (Rojas et al. 2011). Particularly the downstream parts of the study area in southeastern Ethiopia and Somalia experience frequent drought. As projected in this study, the reduction of streamflow in some months in the intermediate and wet seasons may cause water scarcity in the region, and larger increment of streamflow from the baseline scenario may cause flooding to flood plain zones. It is crucial to consider and implement integrated water resources management to meet the alarmingly increasing water demand due to high population growth rate.

27.9 Conclusions

The downscaled climatic parameters (precipitation, maximum temperature, and minimum temperature) were used as an input to the SWAT hydrological model in order to simulate streamflow. The result of climate projection revealed that the SDSM model has good ability to replicate the historical maximum and minimum temperature, and precipitation. Mann-Kendal trend test shows linear decreasing trend for minimum and increasing trend for maximum temperature through the end of this century. The projected mean monthly maximum temperature has larger magnitude of increment in the 2080s time horizon. The Mann-Kendal trend test of downscaled precipitation shows significant increasing trend for B2a scenario; however, the trend was decreasing for A2a scenario. An increase of rainfall comparatively is higher in the dry season which could have positive impact on pastoral region of the study area and but will negatively affect harvesting from agricultural fields. Seasonally, unlike maximum and minimum temperature, precipitation does not manifest a systematic increase and decrease in all future time horizons for both A2a and B2a scenarios. The main rainy season (wet) and intermediate rainy season indicate a decreasing trend; on the other hand, the dry season showed an increasing trend. The mean seasonal streamflow might increase in dry season and decrease in intermediate and wet seasons. The streamflow was found to be sensitive to the rainfall variability as a result the increase in rainfall at the dry season has resulted to an increase of streamflow at the same season and it has revealed reduction of streamflow on the wet and intermediate seasons. Generally, there will be high seasonal and monthly variation of streamflow than on the annual basis. The change in the amount and distribution of rainfall and temperature would affect agricultural productivity and water utilizations in the region. It is recommended to apply different GCMs and emission scenarios so as to make comparison between different models as well as to explore a wide range of climate change scenarios that would result in different hydrological impacts.

Acknowledgements This research was undertaken by the financial contribution from IDRC (International Development Research Center) under ICTWCC project through University of Nairobi. Therefore, we highly appreciate the ICTWCC project. We would also like to thank Madawalabu University for the facilities.

References

Abdo KS, Fiseha BM, Rientjes THM, Gieske ASM, Haile AT (2009) Assessment of climate change impacts on the hydrology of Gilgel Abay catchment in Lake Tana basin, Ethiopia. Hydrol Process 23:3661–3669

Arnold JG, Allen PM (1999) Automated methods for estimating baseflow and ground water recharge from streamflow records. J Am Water Resour Assoc 35(2):411–424

Assefa T, Anwar A, Melesse AM, Admasu S (2014) Climate change in Upper Gilgel Abay River catchment, Blue Nile Basin Ethiopia. In: Melesse AM, Abtew W, Setegn S (eds) Nile River Basin: ecohydrological challenges, climate change and hydropolitics, pp 363–388

Behulu F, Setegn S, Melesse AM, Fiori A (2013) Hydrological analysis of the Upper Tiber Basin: a watershed modeling approach. Hydrol Process 27(16):2339–2351
Behulu F, Setegn S, Melesse AM, Romano E, Fiori A (2014) Impact of climate change on the hydrology of Upper Tiber River Basin Using bias corrected regional climate model. Water Resour Manage 1–17
Carter TR, La Rovere EL, Jones RN, Leemans R, Mearns LO, Nakicenovic N, Pittock AB, Semenov SM, Skea J (2001) Developing and applying scenarios. In: McCarthy JJ, Canziani OF, Leary NA, Dokken DJ, White KS (eds) Climate change 2001: impacts, adaptation, and vulnerability. Contribution of Working Group II to the Third Assessment Report of the Intergovernmental Panel on Climate Change. Cambridge University Press, Cambridge, pp 145–190
Dessu SB, Melesse AM (2012) Modeling the rainfall-runoff process of the Mara River Basin using SWAT. Hydrol Process 26(26):4038–4049
Dessu SB, Melesse AM (2013) Impact and uncertainties of climate change on the hydrology of the Mara River Basin. Hydrol Process 27(20):2973–2986
Dessu SB, Melesse AM, Bhat M, McClain M (2014) Assessment of water resources availability and demand in the Mara River Basin. CATENA 115:104–114
Eriksen S, O'Brien K, Rosentrater L (2008) Climate change in Eastern and Southern Africa impacts, vulnerability and adaptation. University of Oslo, GECHS Report 2008:2
FAO/UNESCO-ISWC (1998) The world reference base for soil resources. Rome
Faramarzi M, Abbaspour KC, Vaghefi SA, Farzaneh MR, Zehnder AB, Srinivasan R, Yang H (2013) Modeling impacts of climate change on freshwater availability in Africa. J Hydrol 480:85–101
Farm Africa-SOS Sahel Ethiopia (2007) Participatory natural resource management programme and Oromia Bureau of Agriculture and Rural Development. Six Months Report, 45p
Gassman WP, Reyes MR, Green CH, Arnold JG (2007) The soil and water assessment tool: historical development, applications, and future research direction. Trans ASABE 50(4): 1211–1250
Getachew HE, Melesse AM (2012) Impact of land use/land cover change on the hydrology of Angereb watershed, Ethiopia. Int J Water Sci 1, 4:1–7. doi:10.5772/56266
Grey OP, Webber Dale G, Setegn SG, Melesse AM (2013) Application of the soil and water assessment tool (SWAT model) on a small tropical island state (Great River Watershed, Jamaica) as a tool in integrated watershed and coastal zone management. Int J Trop Biol Conserv 62(3):293–305
Hamed KH (2008) Trend detection in hydrologic data: the Mann-Kendall trend test under the scaling hypothesis. J Hydrol 349:350–363
IPCC (2007) Climate change 2007: the physical science basis. Contribution of Working Group I to the Fourth Assessment Report of the Intergovernmental Panel on Climate Change. In: Solomon S, Qin D, Manning M, Chen Z, Marquis M, Averyt KB, Tignor M, Weybr HL (eds). Cambridge University Press, Cambridge, and New York, 996p
IPCC (2012) Managing the risks of extreme events and disasters to advance climate change adaptation. A Special Report of Working Groups I and II of the Intergovernmental Panel on Climate Change. In: Field CB, Barros V, Stocker TF, Qin D, Dokken DJ, Ebi KL, Mastrandrea MD, Mach KJ, Plattner GK, Allen, Tignor M, Midgley PM (eds). Cambridge University Press, Cambridge, UK and New York, 582p
Karpouzos DK, Kavalieratou S, Babajimopoulos C (2010) Trend analysis of precipitation data in Pieria Region (Greece). Euro Water 30:31–40
Lijalem Z (2006) Climate change impact on Lake Ziway watershed water avialability, Ethiopia. MSc thesis. University of Applied Sciences Cologne, Germany, 123p
Mango L, Melesse AM, McClain ME, Gann D, Setegn SG (2011a) Land use and climate change impacts on the hydrology of the upper Mara River Basin, Kenya: results of a modeling study to support better resource management. Hydrol Earth Syst Sci 15:2245–2258, doi:10.5194/hess-15-2245-2011 (Special issue: Climate, weather and hydrology of East African Highlands)

Mango L, Melesse AM, McClain ME, Gann D, Setegn SG (2011b) Hydro-meteorology and water budget of Mara River basin, Kenya: a land use change scenarios analysis (Chap. 2). In: Melesse AM (ed) Nile River Basin: hydrology, climate and water use. Springer Science Publisher, Heidelberg, pp 39–68. doi:10.1007/978-94-007-0689-7_2

McCarthy JJ, Canziani OF, Leary NA, Dokken DJ, White KS (2001) Climate change 2001: impacts, adaptation, and vulnerability. Contribution of Working Group II to the Third Assessment Report of the Intergovernmental Panel on Climate Change

Melesse AM, Loukas AG, Senay G, Yitayew M (2009) Climate change, land-cover dynamics and eco-hydrology of the Nile River Basin. Hydrol Process 23(26):3651–3652

Melesse A, Bekele S, McCornick P (2011) Hydrology of the Niles in the face of land-use and climate dynamics. In: Melesse AM (ed) Nile River Basin: hydrology, climate and water use. Springer Science Publisher, Heidelberg, vii-xvii. doi:10.1007/978-94-007-0689-7

Mohammed H, Alamirew T, Assen M, Melesse AM (2015) Modeling of sediment yield in Maybar gauged watershed using SWAT, northeast Ethiopia. CATENA 127:191–205

Monteith JL (1965) Evaporation and the environment. In: XIXth symposium of the Society of Experimental Biologists Swansea on the state and movement of water in living organisms. Cambridge University Press, Cambridge

Nash JE, Sutcliff JV (1970) River flow forecasting through conceptual models, part I—a discussion of principles. J Hydrol 10:282–290

Neitsch SI, Arnold JG, Kinrv JR, Williams JR (2005) Soil and water assessment tool, theoretical documentation, version 2005. USDA Agricultural Research Service Texas A and M Black land Research Center, Temple

Nicholls N, Gruza GV, Jouzel J, Karl TR, Ogallo LA, Parker DE (1996) Observed climate variability and change. In: Houghton, JT, Meira Filho LG, Callander BA, Harris N, Kattenberg A, Maskell K (eds) Climate change 1995: the science of climate change. Contribution of Working Group I to the Second Assessment Report of the Intergovernmental Panel on Climate Change. Cambridge University Press, Cambridge

NMSA (2007) Climate change national adaptation programme of action (NAPA) of Ethiopia. National Metrological Agency, Addis Ababa

Rojas O, Vrieling A, Rembold F (2011) Assessing drought probability for agricultural areas in Africa coarse resolution remote sensing imagery. Remote Sens Environ 115:343–352

Sanjay K, Jaivir T, Vishal S (2010) Simulation of runoff and sediment yield for a Himalayan watershed using SWAT model. J Water Resour Prot 2(3):267–281

Santhi C, Arnold JG, Williams JR, Dugas WA, Srinivasan R, Hauck LM (2001) Validation of the SWAT model on a large river basin with point and nonpoint sources. J Am Water Resour Assoc 37(5):1169–1188

Sathian KK, Symala P (2009) Application of GIS integrated SWAT model for basin level water balance. Indian J Soil Conserv 37(2):100–105

Setegn SG, Srinivasan R, Dargahi B, Melesse AM (2009a) Spatial delineation of soil erosion prone areas: application of SWAT and MCE approaches in the Lake Tana Basin, Ethiopia. Hydrol Process 23(26):3738–3750

Setegn SG, Srinivasan R, Melesse AM, Dargahi B (2009b) SWAT model application and prediction uncertainty analysis in the Lake Tana Basin, Ethiopia. Hydrol Process 24(3): 357–367

Setegn SG, Dargahi B, Srinivasan R, Melesse AM (2010) Modelling of sediment yield from Anjeni gauged watershed, Ethiopia using SWAT. J Am Water Resour Assoc 46(3):514–526

Setegn SG, Rayner D, Melesse AM, Dargahi B, Srinivasan R (2011) Impact of climate change on the hydroclimatology of Lake Tana basin, Ethiopia. Water Resour Res 47:W04511

Setegn SG, Melesse AM, Haiduk A, Webber D, Wang X, McClain M (2014) Spatiotemporal distribution of fresh water availability in the Rio Cobre Watershed, Jamaica. CATENA 120:81–90

Shawul A, Alamirew T, Dinka MO (2013) Calibration and validation of SWAT model and estimation of water balance components of Shaya mountainous watershed, Southeastern Ethiopia. Hydrol Earth Syst Sci Discuss 10:13955–13978. doi:10.5194/hessd-10-13955-2013

Shongwe ME, van Oldenborgh GJ, van den Hurk BJJM, de Boer B, Coelho CAS, van Aalst MK (2009) Projected changes in mean and extreme precipitation in Africa under global warming. Part I: Southern Africa. JClim 22:3819–3837

Tripathi MP, Panda RK, Raghuwanshi NS (2003) Identification and prioritization of critical sub-watersheds for soil conservation management using the SWAT model. Bio-syst Eng 85 (3):365–379

USDA Soil Conservation Service (SCS) (1972) National engineering handbook (Section 4: Chaps. 4–10). In: Hydrology

Van Griensven A (2005) Sensitivity, auto-calibration, uncertainty and model evaluation in SWAT 2005. UNESCO-IHE, 48p

Wang X, Melesse AM (2005) Evaluations of the SWAT model's Snowmelt hydrology in a Northwestern Minnesota watershed. Trans ASAE 48(4):1359–1376

Wang X, Melesse AM (2006) Effects of STATSGO and SSURGO as inputs on SWAT model's snowmelt simulation. J Am Water Resour Assoc 42(5):1217–1236

Wang X, Melesse AM, Yang W (2006) Influences of Potential evapotranspiration estimation methods on SWAT's hydrologic simulation in a Northwestern Minnesota watershed. Trans ASAE 49(6):1755–1771

Wang X, Shang S, Yang W, Melesse AM (2008a) Simulation of an agricultural watershed using an improved curve number method in SWAT. Tans ASABE 51(4):1323–1339

Wang X, Yang W, Melesse AM (2008b) Using hydrologic equivalent wetland concept within SWAT to estimate streamflow in watersheds with numerous wetlands. Trans ASABE 51 (1):55–72

Wang X, Garza J, Whitney M, Melesse AM, Yang W (2008c) Prediction of sediment source areas within watersheds as affected by soil data resolution (Chap. 7). In: Findley PN (ed) Environmental modelling: new research. Nova Science Publishers, Inc., Hauppauge, pp 151–185. ISBN:978-1-60692-034-3

Wilby RL, Dawson CW (2007) Statistical downscaling model (SDSM 4.2)—a decision support tool for the assessment of regional climate change impacts, version 4.2, user's manual, 67p

Yirefu B (2012) Development of future climate change scenarios and its impact on stream flow of Mille River in Awash River basin, Ethiopia. MSc thesis, Haramaya University, Ethiopia, 98p

Chapter 28
Climate Change Impact on Sediment Yield in the Upper Gilgel Abay Catchment, Blue Nile Basin, Ethiopia

Anwar A. Adem, Seifu A. Tilahun, Essayas K. Ayana, Abeyou W. Worqlul, Tewodros T. Assefa, Shimelis B. Dessu and Assefa M. Melesse

Abstract According to Intergovernmental Panel on Climate Change (IPCC) future projections, precipitation and temperature will increase over eastern Africa in the coming century. This chapter presents basin-level impact of climate change on sediment yield in Upper Gilgel Abay catchment, Blue Nile Basin, Ethiopia, by downscaling HadCM3 global climate model using Statistical Downscaling Model (SDSM). IPCC-recommended baseline period (1961–1990) was used for baseline scenario analysis. Future scenario analysis was performed for the 2020s, 2050s, and 2080s. Globally, HadCM3 model is widely applied for climate change studies and it consists of A2 (medium high emission) and B2 (medium low emission) scenarios. Impact assessment on sediment yield was done by Soil and Water Assessment Tool (SWAT) hydrological model. SWAT model performance in simulating daily

A.A. Adem (✉)
Tana Sub-Basin Office, Blue Nile Authority, Bahir Dar, Ethiopia
e-mail: unnxuss@gmail.com

S.A. Tilahun · E.K. Ayana · A.W. Worqlul · T.T. Assefa
School of Civil and Water Resources Engineering, Bahir Dar University, Bahir Dar, Ethiopia
e-mail: seifuad@yahoo.co.uk

E.K. Ayana
e-mail: ekk45@cornell.edu

A.W. Worqlul
e-mail: Abeyou_wale@yahoo.com

T.T. Assefa
e-mail: ttaffese@gmail.com

E.K. Ayana
Geospacial Data and Technology Center, Bahir Dar University, Bahir Dar, Ethiopia

S.B. Dessu
Department of Civil Engineering, Addis Ababa University, Addis Ababa, Ethiopia
e-mail: sbehilu@gmail.com

A.M. Melesse
Department of Earth and Environment, Florida International University, Miami, USA
e-mail: melessea@fiu.edu

© Springer International Publishing Switzerland 2016

A.M. Melesse and W. Abtew (eds.), *Landscape Dynamics, Soils and Hydrological Processes in Varied Climates*, Springer Geography,
DOI 10.1007/978-3-319-18787-7_28

sediment yield for the study area was satisfactory with Nash–Sutcliffe Efficiency (NSE) of 0.58 and 0.51 for calibration and validation periods, respectively. Mean annual changes of precipitation and temperature (maximum and minimum) were applied to quantify these impacts. The result of downscaled precipitation and temperature reveals a systematic increase in all future time periods for both A2 and B2 scenarios. These increases in climate variables are expected to result in increase in mean annual sediment yield of 11.3, 16.3, and 21.3 % for A2 scenario and by 11.0, 14.3, and 11.3 % for B2 scenario for the 2020s, 2050s, and 2080s, respectively. This increase in sediment yield is double the increase in stream flow due to climate change for all time periods. Future work need to consider also impact of land use change on the catchment for future sustainable development plan.

Keywords IPCC · Climate change · HadCM3 · SWAT · Upper Gilgel Abay · A2 and B2 scenarios · SWAT · Blue Nile basin

28.1 Introduction

Climate change impacts such as increased drought, desertification, increased severe weather, and increased coastal flooding could have serious impacts on food supply, especially in developing countries (Houghton 2002). It has significant effects on soil erosion rates for a variety of reasons (Nearing et al. 2004). The erosive power of rainfall and wind is primary (Nearing 2001; Boardman et al. 1990). The second factor is indirect dominant pathway of influence on erosion rates from changes in plant biomass as a result of climate change.

The reasoning behind the mechanisms by which climate affects biomass and by which runoff and erosion are affected by biomass changes is complex (Pruski and Nearing 2002a). Faster accumulation of the necessary growing degree-days for crop maturity refers to warmer temperature which can increase biomass production rates (O'Neal et al. 2005). This is due to anthropogenic increases in atmospheric carbon dioxide concentration which translate to an increase in soil surface canopy cover and more importantly biological ground cover (Nearing et al. 2004). On the other hand, increases in soil and air temperature and moisture will likely cause faster rates of residue decomposition due to increase in microbial activity. An increase in biomass production may also be due to increased precipitation, and with higher temperatures, evaporation rates may increase. Increase in rainfall would tend to lead to higher soil moisture levels, which in turn may influence infiltration and runoff amounts and rates (Pruski and Nearing 2002b).

Erosion and sediment transport processes are sensitive to changes in climate and land cover and to a wide range of human activities because of their close links to land cover, land use, and the hydrology of a river basin (Walling 2009). These include forest cutting and land clearing, expansion of agriculture, land use changes, mineral extraction, urbanization and infrastructural development, sand mining, dam and reservoir construction, and soil conservation and sediment control programs

(Walling 2006). Recent concern about the impact of global change on the earth system has emphasized the impact of climate change resulting from increased emission of greenhouses gases and associated global warming (Walling 2009). The concern is not shared in the Great Horn of Africa like Ethiopia. In the second half of this century, increased rainfall is predicted for the Horn of Africa (Parry 2007). Soil loss from sheet and rill erosion will and cause accelerated erosion and deposition in Lake Tana. Consequently, research on how climate affects erosion and impacts the sediment yield of the catchment is important.

Across a large river basin, both the magnitude of changes and direction in erosion and sediment yield could vary. In this context, modeling the future sediment yield using projected changes of local-scale climate variables and a physically based distributed modeling is necessary to plan properly for future climate conditions (Walling 2009). In spite of the implication and significance of Lake Tana and its basin for the continued dependence of the people in the basin, only little is done in this case (Gebremariame 2009). Blue Nile River Basin is one of the most sensitive basins to climate change and water resources variability (Kim and Kaluarachchi 2009). Nevertheless, water availability effects of climate change on water resource analysis, management, and policy formulation in Lake Tana Basin have not been addressed adequately (Setegn et al. 2011). Similarly, climate change impact studies on sediment yield are also rare in the Upper Blue Nile basin except the study which was done by Mekonnen and Tadele (2012). This may be due to data scarcity in the region.

Hydrology of the Nile River Basin has been studied by various researchers. These studies encompass various areas including stream flow modeling, sediment dynamics, teleconnections and river flow, land use dynamics, climate change impact, groundwater flow modeling, hydrodynamics of Lake Tana, water allocation, and demand analysis (Melesse et al. 2009a, b, 2011b; Abtew et al. 2009a, b; Abtew and Melesse 2014a, b, c; Melesse 2011; Dessu and Melesse 2012, 2013; Dessu et al. 2014; Yitayew and Melesse 2011; Chebud and Melesse 2009a, b, 2013; Setegn et al. 2009a, b, 2010).

A number of studies used laboratory, field scales, and modeling studies to understand soil erosion and sediment dynamics in various regions (Defersha and Melesse 2011, 2012, Defersha et al. 2011, 2012; Maalim and Melesse 2013; Maalim et al. 2013; Setegn et al. 2010; Melesse et al. 2011a, 2014; Wang et al. 2008c; Mekonnen and Melesse 2011; Msagahaa et al. 2014; Setegn et al. 2009a; Mohamed et al. 2015).

Mekonnen and Tadele (2012) analyzed the combined effects of land use and climate change on soil erosion and stream flow in the Upper Gilgel Abay River catchment using SWAT hydrological model. The result showed that the combined effect of trend-based land use and climate change (2011–2025) will increase the average annual sediment load by 29 % and significantly affects soil erosion compared with other scenario. GCM of European Centre Hamburg Model (ECHAM5, A1B emission scenario) was applied.

The main objective of this chapter is to quantify the impact of climate change on sediment yield in the catchment of Upper Gilgel Abay using SWAT hydrological model. Water erosion is a major part of land degradation in Upper Gilgel Abay catchment that affects the physical and chemical properties of soils and resulting in on-site nutrient loss and off-site sedimentation of water ways (Mekonnen and Tadele 2012). This objective was approached by downscaling selected global climate model (GCM) to reproduce changes of local-scale climate variables. The findings in this research could help in examining the applicability of GCMs under various geographical and physiographic settings.

The application of SWAT in predicting stream flow and sediment as well as evaluation of the impact of land use and climate change on the hydrology of watersheds has been documented by various studies (Dessu and Melesse 2012, 2013; Dessu et al. 2014; Wang et al. 2006, 2008a, b, c; Wang and Melesse 2005, 2006; Behulu et al. 2013, 2014; Setegn et al. 2009a, b, 2010, 2011, 2014; Mango et al. 2011a, b; Getachew and Melesse 2012; Assefa et al. 2014; Grey et al. 2013; Mohamed et al. 2015).

28.2 Study Area

Gilgel Abay River is the largest tributary in Lake Tana sub-basin covering an area of about 1654 km^2. It originates from the highland spring of Gish-Abay town in northeastern Ethiopia. It contributes 44.2 % of the flows into Lake Tana (Wale 2008).

Geographically, Upper Gilgel Abay catchment is the northern part of the Upper Blue Nile basin which is the southern part of Lake Tana sub-basin at latitudes and longitudes between 10° 56′ 53″ to 11° 21′ 58″N and 36° 49′ 29″ to 37° 23′ 34″E, respectively (Fig. 28.1).

The elevation of Upper Gilgel Abay catchment ranges from 1890 to 3514 m above mean sea level (amsl). The highest elevation of the catchment is located at

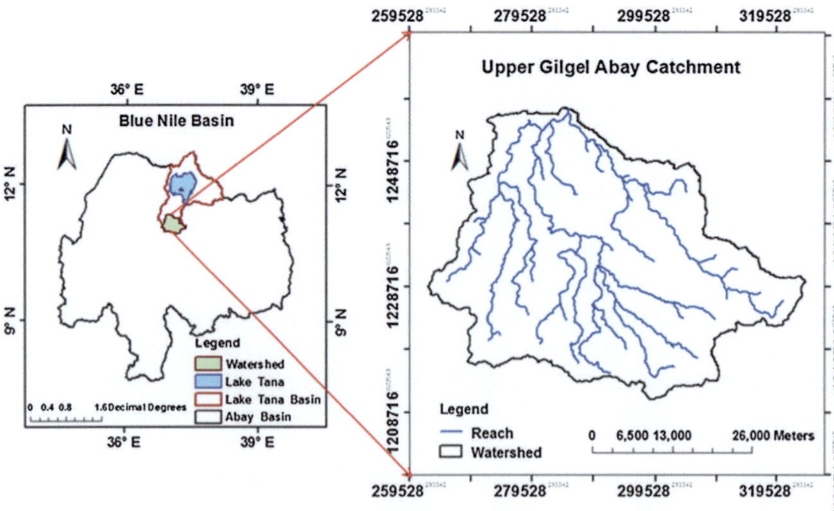

Fig. 28.1 Location of Upper Gilgel Abay catchment

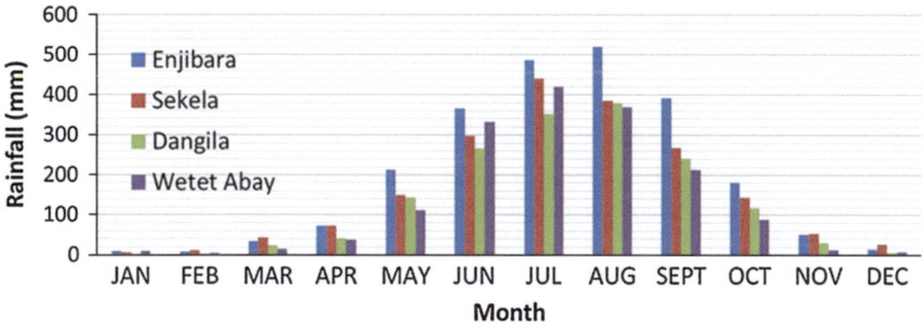

Fig. 28.2 Mean monthly rainfall distribution of Upper Gilgel Abay catchment (1994–2008)

the southeastern tip. Nearly two-third of the catchment has an elevation ranging from 1890 to 2427 m amsl extending from the center to the northern tip, the outlet of the river.

Mean monthly rainfall (1994–2008) distribution at Injibara, Sekela, Dangila, and Wetet Abay rainfall stations indicates that the study area has one peak in a year. Therefore, Upper Gilgel Abay catchment lies in mono-modal climate class according to Ethiopian climate classification with respect to rainfall regimes (Tadege 2001). The main rainfall season is from June to September and accounts for 70–90 % of the annual rainfall (Abdo et al. 2009).

Mean annual areal rainfall varies from 1624 to 2349 mm. Most of the study area has mean annual rainfall between 1842 and 1986 mm (Fig. 28.2).

The temperature of the study area is dependent on elevation where temperature decreases with increase in elevation. For instance, mean monthly maximum temperature of Wetet Abay at elevation of 1915 m amsl varies from 24.3 to 31.3 °C and Gundil at elevation of 2574 m amsl varies from 18.3 to 24.9 °C (Fig. 28.3). Generally, daily variation between maximum and minimum temperatures is high as compared to the seasonal variation of temperature in the study area.

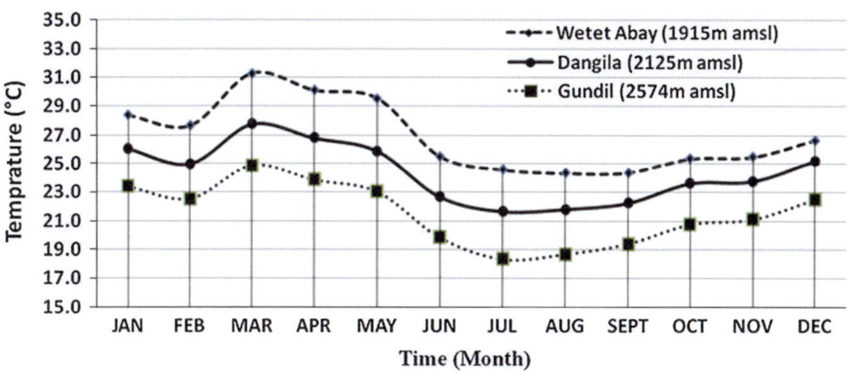

Fig. 28.3 Mean monthly temperature variation at three stations with elevation differences (1994–2008)

28.3 Data Source

28.3.1 Meteorological Data

Meteorological variable such as rainfall and maximum and minimum temperatures is required as a predictand to downscale the global climate GCM data to local climate variables. Ten meteorological stations with 15 years of daily data were selected for this study (Fig. 28.4). For deriving statistical relationships between the

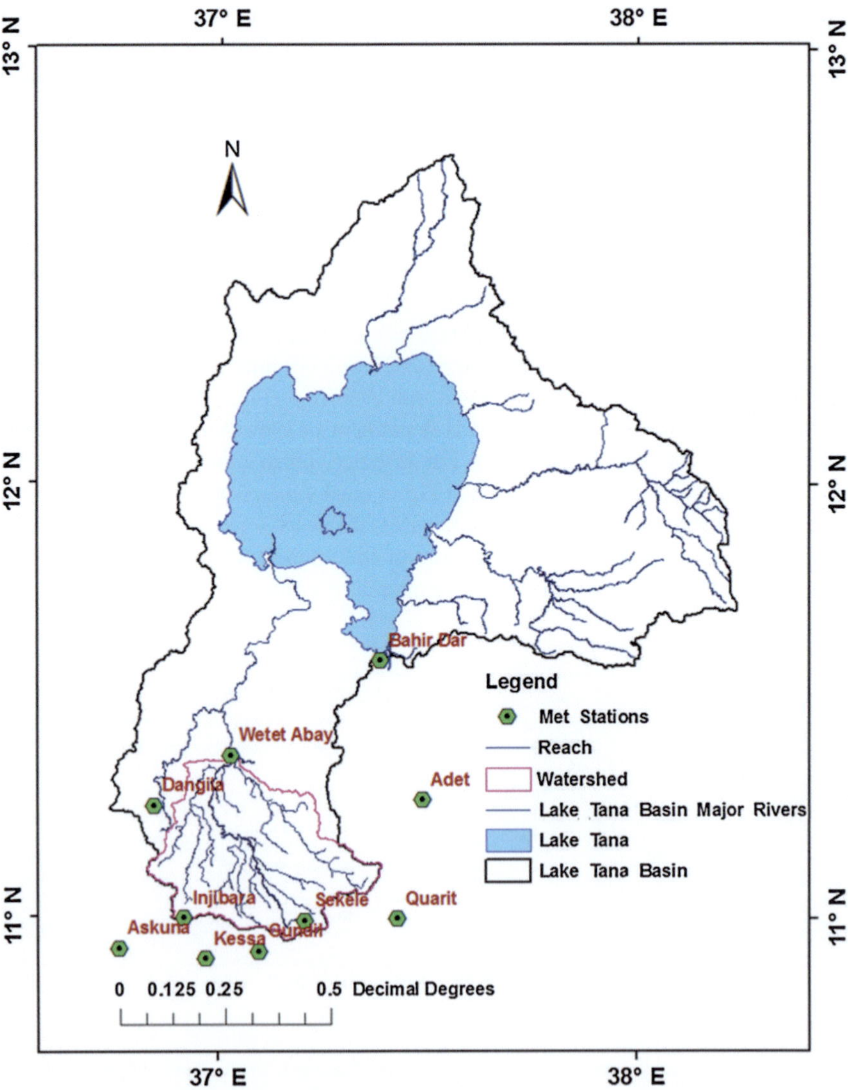

Fig. 28.4 Meteorological station locations

predictand and predictor only, Bahir Dar station data are used as this is the only station that satisfies the baseline historical data period recommended by Inter Governmental Panel on Climate Change (IPCC), 1961–1990. Meteorological variables from all the ten stations were used for stream flow and sediment yield modeling in SWAT.

28.3.2 GCM Data

GCM data are required to project and quantify the relative change of climate variables between current and future time periods. One of the global circulation models, HadCM3, was used for this study because the model is widely applied in many climate change studies and the model provides daily predictor variables which can be used for the Statistical Downscaling Model (SDSM). The predictor variables are supplied on a grid box by grid box basis as large as 2.5° latitude by 3.75° longitude. On entering the location of the study area, the correct grid box was calculated and a zip file was downloaded (http://www.cics.uvic.ca/scenarios/sdsm/select.cgi).

To apply SDSM to GCM data, both observed predictand and GCM data should ideally be available on the same grid spacing. Individual predictor and predictand files (one variable to each file, time series data only) are denoted by the extension *. DAT (Wilby et al. 2002). The predictor represents large-scale atmospheric variables, whereas the predictand represents local surface variables such as temperature and precipitation.

28.3.3 Hydrological Data

Stream flow was used to calibrate and validate the SWAT model simulation. Daily stream flow data were obtained from Ministry of Water and Energy (MoWE) for Gilgel Abay River near Merawi town. The sediment data from 1983 to 2007 were sparse, one or two data in each year. In this study, this data were used to develop discharge–sediment rating curve.

Sediment rating curves display the rate of sediment transport as a function of stream flow. The rate of sediment transport is given in terms of sediment discharge or alternatively as a flux-averaged concentration. Sediment concentration is converted to load to make the rating curve relation in ton $days^{-1}$ with flow in cumecs (m^3 s^{-1}). Using the sediment load data, the rating curve was developed by fitting power function for the relation between suspended sediment and stream flow (Fig. 28.5). The rating for daily sediment load is expressed by Eq. 28.1 with R^2 of 0.98.

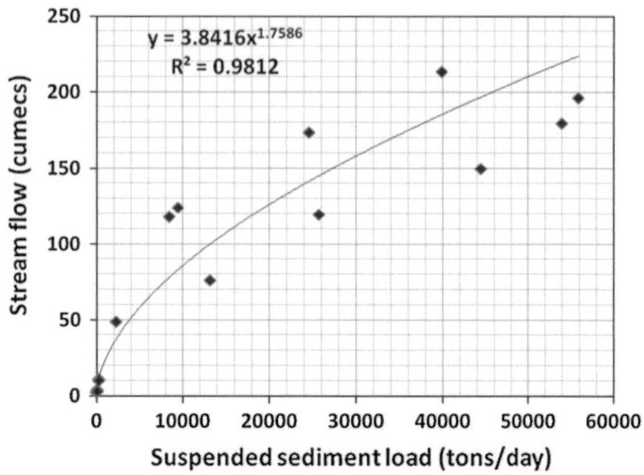

Fig. 28.5 Discharge–Sediment rating curve for Upper Gilgel Abay catchment

$$Q_S = 3.841Q^{1.758} \tag{28.1}$$

where Q_S is suspended sediment in ton days^{-1} and Q is discharge in cumecs.

28.3.4 Spatial Data

Digital elevation model (DEM), land use/land cover, and soil are the three spatial data inputs required by SWAT model. DEM describes the elevation of any point in a given area at a specific spatial resolution as a digital file. DEM is required by SWAT to delineate the watershed into a number of sub-watersheds or sub-basins and to analyze the drainage pattern of the watershed as slope, stream length, and width. The DEM was obtained from the NASA Shuttle Radar Topographic Mission (SRTM) with a resolution of 90 m by 90 m (http://srtm.csi.cgiar.org/SELECTION/inputCoord.asp).

Land use/land cover and soil data were acquired from MoWE Metadata section, data used for Abay River master plan study by Le Bureau, central d'Etudes Pourles Equipments d'outre-Mer (BCEOM) in 1996–1998. The land use/land cover and soil map scale used for the master plan study were 1:250,000 (Gebrekristos 2008).

As per BCEOM, there are three land uses, namely agriculture (AGRL), agro-pastoral (AGRC), and Agro-Sylvicultural (FRSD) in the catchment (Fig. 28.6a). Eutric Regosols, Eutric Vertisols, Haplic Alisols, Haplic Luvisols, and Haplic Nitisols are the major soils in the study area (Fig. 28.6b). Information regarding soil physical and chemical properties was acquired from FAO-HWSD Version 1.1 (FAO-Harmonized World Soil Database). Using these physical and chemical properties, soil database parameters were developed using pedotransfer functions.

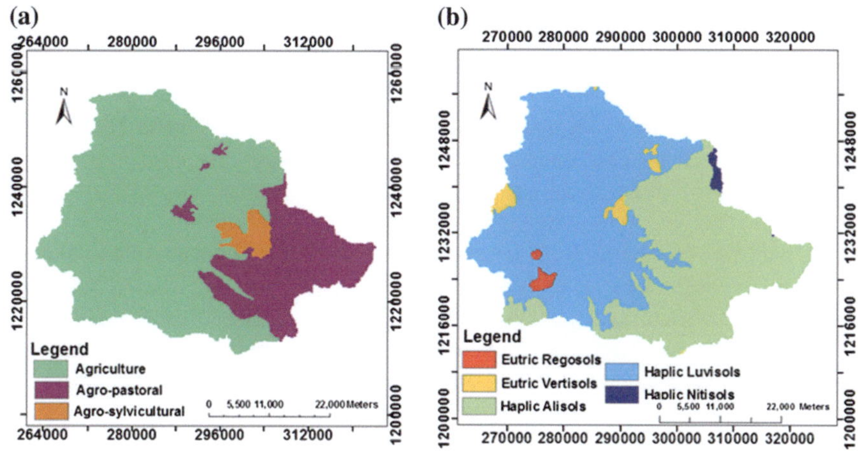

Fig. 28.6 Spatial inputs for SWAT **a** land use and **b** soils

28.4 Methodology

The IPCC Special Report on Emissions Scenarios (SRES), the HadCM3 global climate model, and the downscaling technique applied in HadCM3 are the major part of this impact assessment. Sediment modeling using SWAT model is the second part of the methodology with final objective of climate change impact analysis.

28.4.1 Climatological Baseline

In order to assess the implications of future climate changes on the environment, society, and the economy, it is first necessary to have information about current or recent conditions as a reference point or baseline (Carter et al. 1999; McCarthy et al. 2001). Baseline information is important for (1) characterizing the prevailing conditions under which an exposure unit functions and to which it must adapt; (2) describing average conditions, spatial and temporal variability and anomalous events, some of which can cause significant impacts; (3) calibrating and testing impact models across the current range of variability; (4) identifying possible ongoing trends or cycles; and (5) specifying the reference situation with which to compare future changes (Carter et al. 1999).

According to the above importance of baseline information, this study considered the suggested IPCC climatological baseline period (1961–1990). For the study area, only Bahir Dar meteorological station fulfills IPCC baseline period requirement because the observed data cover the range from 1961 to 1990. Therefore, these data were used as a predictand for downscaling.

28.4.2 SRES

IPCC described that climate scenarios are plausible representations of the future that are consistent with assumptions about future emissions of greenhouse gases and other pollutants. Our understanding of the effect of increased atmospheric concentrations of these gases on global climate has certain assumptions. These assumptions include future trends in energy demand, emissions of greenhouse gases, land use change as well as assumptions about the behavior of the climate system over long timescales. The Intergovernmental Panel on Climate Change–Task Group on Data and Scenario Support for Impact and Climate Assessment (IPCC–TGICA) classified climatic scenarios into three main types based on how they are constructed (Carter 2007). These are synthetic scenarios, also known as incremental scenarios, analogue scenarios, and climate model-based scenarios.

The IPCC SRES in replacing the old IPCC scenarios (IS92) identify 40 different scenarios following four families of storylines (Santos et al. 2008). Each storyline represents a distinctly different direction for future developments such as demographic, socioeconomic, technological, and environmental developments. The four qualitative storylines yield four sets of scenarios called families (A1, A2, B1, and B2). The main characteristics of the two SRES storylines and scenario families which are considered in HadCM3 global climate model are described as follows (Nakicenovic et al. 2000).

The A2 storyline and scenario family describe a very heterogeneous world. The underlying theme is self-reliance and preservation of local identities. Fertility patterns across regions converge very slowly, which results in continuously increasing global population. Economic development is primarily regionally oriented, and per capita economic growth and technological changes are more fragmented and slower than in other storylines. The B2 storyline and scenario family describe a world in which the emphasis is on local solutions to economic, social, and environmental sustainability. It is a world with continuously increasing global population at a rate lower than A2, intermediate levels of economic development, and less rapid and more diverse technological change than in the B1 and A1 storylines. While this scenario is also oriented toward environmental protection and social equity, it focuses on local and regional levels.

28.4.3 Statistical Downscaling Techniques

General circulation models (GCMs) indicate that rising concentrations of greenhouse gases will have significant implications on the climate at global and regional scales (Wilby and Dawson 2007). Due to their coarse spatial resolution and inability to resolve important sub-grid scale features such as clouds and topography, GCMs are restricted in their usefulness for local impact studies by their coarse spatial resolution. GCMs depict the climate using a three-dimensional grid over the globe, typically having a horizontal resolution of between 250 and 600 km, 10–20 vertical layers in the atmosphere, and sometimes as many as 30 layers in the oceans (Nakicenovic et al. 2000). Their resolution is thus quite coarse relative to the scale of exposure units in most impact assessments. Several methods have been adopted for developing regional GCM-based scenarios at the sub-grid scale, a procedure known as "regionalization" or "downscaling." Two different approaches to downscaling are possible which are dynamical and statistical (Hewitson and Crane 1996).

The typical application in dynamic (nested model) downscaling is to drive a regional dynamic model at mesoscale or finer resolutions with the synoptic-scale and larger scale information from a GCM (Giorgi and Mearns 1991; Jenkins and Barron 1997). Detailed information at spatial scales down to 10–20 km, and at temporal scales of hours or less, may be achieved in such applications (Hewitson and Crane 1996). Such models are computationally demanding and are not easily accessible, but in the long term, this technique is likely to be the best approach that needs to be encouraged, whereas statistical (empirical) downscaling is computationally efficient in comparison with dynamical downscaling. It is a practical approach for addressing current needs in the climate change research community especially in many of the countries liable to be most sensitive to climate change impacts (Hewitson and Crane 1996). Many of the processes, which control local climate as topography, vegetation, and hydrology, are not included in coarse resolution GCMs. The development of statistical relationships between the local scale and large scale may include some of these processes implicitly. Under the broad empirical/statistical downscaling techniques, the following three major techniques have been developed. These are weather classification or typing schemes, transfer function or regression models, and stochastic weather generator methods.

Regression models are conceptually simple means of representing linear or nonlinear relationships between local climate variables (predictands) and the large-scale atmospheric forcing predictors, Wilby et al. (2004). Commonly applied methods include canonical correlation analysis (CCA) (von Storch et al. 1993), artificial neural networks (ANN) which are skin to nonlinear regression (Crane and Hewitson 1998), and multiple regression (Murphy 1999).

For this particular study, a type of regression model, SDSM, was used. SDSM is widely applied in many regions of the world over a range of different climatic conditions. It permits the spatial downscaling of daily predictor–predictand relationships using multiple linear regression techniques. The predictor variables provide daily information concerning the large-scale state of the atmosphere, while the *predictand* describes conditions at the site scale (CCIS 2008).

SDSM is a decision support tool that facilitates the assessment of regional impacts of global warming by allowing the process of spatial scale reduction of data provided by large-scale GCMs (Wilby et al. 2002). It is best described as a hybrid of the stochastic weather generator and regression-based methods. This is because large-scale circulation patterns and atmospheric moisture variables are used to linearly condition local-scale weather generator parameters as precipitation occurrence and intensity (Wilby et al. 2002).

Users are allowed to simulate through combinations of regressions and weather generators sequences of daily climatic data for present and future periods by extracting statistical parameters from observed data series (Gagnon et al. 2005). The stochastic component of SDSM permits the generation of 100 simulations. The SDSM software reduces the task of statistically downscaling daily weather series into seven discrete steps (Wilby and Dawson 2007). These are (1) quality control and data transformation; (2) screening of predictor variables; (3) model calibration; (4) weather generation using observed predictors; (5) statistical analyses; (6) graphing model output; and (7) scenario generation using climate model predictors.

HadCM3 is downscaled using SDSM with IPCC climatological baseline of 1961–1990. Observed historical climate variables from 1961 to 1980 were used for calibration and from 1981 to 1990 for validation.

28.5 SWAT Sediment Modeling Approach

SWAT is a river basin or watershed scale model developed by Dr. Jeff Arnold for the United Sates Department of Agriculture (USDA) Agricultural Research Service (ARS). It is physically based, conceptual, and computationally efficient model that operates on a daily time step at basin scale (Arnold et al. 2000). It is designed to predict the impact of land management practices on water, sediment, and agricultural chemical yields in large complex watersheds with varying soils, land use, and management conditions over long periods of time (Neitsch et al. 2005). Benefits of SWAT model include application in watersheds with no monitoring data as stream flow. Also, the relative impact of changes in management practices, climate, and vegetation on water quality or other variables of interest can be quantified (Neitsch et al. 2005).

SWAT uses a two-level disaggregation scheme. A preliminary sub-basin identification is carried out based on topographic criteria, followed by further discretization using land use and soil-type considerations. Areas with the same soil type and land use form a hydrologic response unit (HRU), a basic computational unit assumed to be homogeneous in hydrologic response to land cover change. SWAT model has been applied for various catchment areas ranging from 0.015 km^2 (Mapfumo et al. 2003) to as large as 491,700 km^2 (Arnold et al. 2000).

Precipitation is the hydrology input in the SWAT watershed system. Flows and water quality parameters are routed in the model on the basis of HRU to each sub-basin and subsequently to the watershed outlet. In this study, SWAT 2009 model integrated with Arc GIS techniques was used to simulate sediment yield of the study area.

SWAT uses two phases for the simulation of sediment load transport, land phase, and routing phase. Land phase simulation calculates the amount of sediment, water, pesticide, and nutrient loading into the main channel in each basin. SWAT calculates the surface erosion and sediment yield caused by rainfall and runoff within each HRU with the Modified Universal Soil Loss Equation (MUSLE), William (1975). MUSLE is a modified version of the Universal Soil Loss Equation (USLE) developed by Wischmeier and Smith (1978). While the USLE uses rainfall as an indicator of erosive power of energy, MUSLE uses the amount of runoff to simulate erosion and sediment yield. The modified universal soil loss equation is determined by Eq. 28.2 (Williams 1995).

$$\text{sed} = 11.8 \cdot (Q_{\text{surf}} \cdot q_{\text{peak}} \cdot \text{area}_{\text{hru}})^{0.56} \cdot K_{\text{USLE}} \cdot C_{\text{USLE}} \cdot P_{\text{USLE}} \cdot LS_{\text{USLE}} \cdot \text{CFRG} \tag{28.2}$$

where sed is sediment yield on a given day (metric tons), Q_{surf} is the surface runoff volume (mm ha^{-1}), q_{peak} is the peak runoff rate (m^3 s^{-1}), area_{hru} is the area of the HRU (ha), K_{USLE} is the USLE soil erodibility factor (0.013), C_{USLE} is the USLE cover and management factor, P_{USLE} is the USLE support practice factor, LS_{USLE} is the USLE topographic factor, and CFRG is the coarse fragment factor. The detailed descriptions of the different model components can be found in Neitsch et al. (2001).

The routing phase of SWAT defines the movement of water, nutrients, sediment, and pesticides through the channel network of the watershed into the outlet. Variable storage routing or Muskingum River routing can be used as routing technique for flow and sediment loading. For this study, the variable storage routing method was used. The sediment-routing model that simulates sediment transport in the channel network consists of two components operating simultaneously, deposition and degradation (Arnold et al. 1995). The amount of deposition and degradation is based on the maximum concentration of sediment in the reach and the concentration of sediment in the reach at the beginning of the time step. The final amount of sediment in the reach is determined by Eq. 28.3.

$$\text{sed}_{\text{ch}} = \text{sed}_{\text{ch},i} - \text{sed}_{\text{dep}} + \text{sed}_{\text{deg}} \tag{28.3}$$

where sed_{ch} is the amount of suspended sediment in the reach (metric tons days^{-1}), $\text{sed}_{\text{ch},i}$ is the amount of suspended sediment in the reach at the beginning of the time period (metric tons days^{-1}), sed_{dep} is the amount of sediment deposited in the reach segment (metric tons days^{-1}), and sed_{deg} is the amount of sediment re-entrained in the reach segment (metric tons days^{-1}). The amount of sediment transported out of the reach is calculated by Eq. 28.4.

$$\text{sed}_{\text{out}} = \text{sed}_{\text{ch}} \times \frac{V_{\text{out}}}{V_{\text{ch}}} \tag{28.4}$$

where sed_{out} is the amount of sediment transported out of the reach (metric tons days^{-1}), sed_{ch} is the amount of suspended sediment in the reach (metric tons days^{-1}), V_{out} is the volume of outflow during the time step (m^3), and V_{ch} is the volume of water in the reach segment (m^3). The detailed descriptions of the different model components can be found in Neitsch et al. (2001).

28.5.1 SWAT Model Setup

28.5.1.1 Watershed Delineation

Watershed delineator tool in ArcSWAT allows the user to delineate the watershed and sub-basins using digital elevation model (DEM). Flow direction and accumulation are the concept behind the definition of the stream network of the DEM in SWAT. The monitoring point was added manually, and the numbers of sub-basin were adjusted accordingly. Finally, the catchment area was delineated as 1654 km^2 for Upper Gilgel Abay River catchment and 19 sub-basins are formed for the whole catchment.

28.5.1.2 HRU Analysis

HRU analysis helps to load land use map and soil map and also incorporate classification of HRU into different slope classes. The land use map as well as soil map overlapped 100 % with the delineated watershed, and 74 HRUs were formed. Spatial inputs of slope, soil, and land use were used to define the catchment. Ninety meter resolution SRTM DEM topographic data were used for slope classification. Extracted topographic map with the catchment boundary shows an elevation range from 1890 to 3514 m amsl.

28.5.1.3 Weather Data Definition

Available meteorological records for precipitation, minimum and maximum temperatures, relative humidity and wind speed (1994–2008), and location of meteorological stations were prepared based on ArcSWAT 2009 input format and integrated with the model using weather data input wizards. Bahir Dar meteorological station data were defined and used as weather generator for this study.

28.5.2 Sensitivity Analysis

Sensitivity analysis is a simple technique for assessing the effect of uncertainty on the system performance. It is also a measure of the effect of change of one parameter on another. Sensitivity analysis was undertaken by using a built-in tool in SWAT-CUP that uses the global sensitivity design method. SWAT-CUP uses t-test to rank the most sensitive parameter that corresponds to greater change in output response. To improve simulation results and thus understand the behavior of the hydrologic system in Upper Gilgel Abay River catchment, sensitivity analysis was conducted using the six sediment parameters.

28.5.3 Calibration and Validation

Model calibration involves modification of input parameters and comparison of predicted output with observed values until a defined objective function is achieved (James and Burges 1982). Parameters identified in sensitivity analysis that influence significantly the simulation result were used to calibrate the model. For this study, the measured stream flow and sediment yield data obtained from discharge–sediment rating curve of Gilgel Abay River at Merawi were used to calibrate the model by SWAT-CUP using Sequential Uncertainty Fitting version-2 (SUFI-2) algorithm (Abbaspour 2012) from 1996 to 2004 by skipping two years data as a warm-up period. The rest of the data were used for validation.

SUFI-2 algorithm was selected for calibration for this study from two other algorisms (ParaSol and GLUE) for its better performance in fitting parameters for Gilgel Abay, Gumara, Ribb, and Megech catchments (Setegn et al. 2009b). In SUFI-2, parameter uncertainty accounts for all sources of uncertainties such as uncertainty in driving variables as rainfall, conceptual model, parameters, and measured data.

28.6 Hydrological Models in Climate ChangeImpact Studies

Operational hydrologists use hydrological models to seek solutions for water resource problems, such as flood protection, frequency and duration of extreme hydrological events (floods and droughts), irrigation, hydraulic structure design, hydropower evaluation, and water supply design, under different climatic conditions at one location or at different locations (Xu 1999).

The advantages of hydrological models in climate change impact studies include (1) models tested for different climatic and physiographic conditions and structured for use at various spatial scales and dominant process representations are readily available; (2) GCM-derived climate perturbations at different levels of downscaling can be used as model input; (3) a variety of responses to climate change scenarios can be modeled, including transient ones; and (4) the models can convert climate change output to relevant water resources variables related to, for example, reservoir operation, groundwater recharge, or irrigation demand (Schulze 1997).

The general procedure for estimating the impacts of hypothetical climate change on hydrological behavior is used to (1) determine the parameters of a hydrological model in the study catchment using current climatic inputs and observed river flows for model validation; (2) perturb the historical time series of climatic data according to some climate change scenarios; (3) simulate the hydrological characteristics of the catchment under the perturbed climate using a calibrated hydrological model; and (4) compare the model simulations of the current and possible future hydrological characteristics (Xu 1999).

Climate simulations from the GCMs could be used directly to drive hydrologic models which in turn could be used to evaluate the hydrologic and water resource effects of climate change. Accordingly, methods of simple alteration of the present conditions are widely used by hydrologists. In the simple alteration method, the generation of climate scenarios consists of two steps (Xu 1999). First, estimate average annual changes in precipitation and temperature using either GCM results or historical measurements of change, or personal estimates (typically, $\Delta T = +1$, +2 and +4 °C and $\Delta P = 0, \pm 10, \pm 20$ %). Second, adjust the historical temperature series by adding ΔT, and for precipitation, by multiplying the values by $(1 + \Delta P/100)$.

28.7 Results and Discussion

28.7.1 Selected Predictor Variables

Table 28.1 shows months with best predictand and predictor correlations for precipitation and minimum and maximum temperatures. Predictand and predictor have good correlations showing the predictor has best performance to downscale the global climate variables to local-scale climate variable compared to others. For instance, relative humidity at 500 hPa and near-surface relative humidity have good performance to downscale precipitation rather than other predictors. Also, relative humidity at 500 hPa was very good predictor for the months of January, February, April, August, November, and December, whereas precipitation for the months of June, September, and October was efficiently downscaled with near-surface relative humidity.

28.7.2 Baseline Climate Analysis

Baseline scenario analysis was performed for Bahir Dar station for a 30-year period from 1961 to 1990. HadCM3 was downscaled for daily baseline period for two emission scenarios (A2 and B2), and some of the statistical properties of the downscaled data were compared with daily observed data. One of the criteria commonly used in evaluating the performance of any useful downscaling is whether the historical or observed condition can be replicated or not.

Table 28.1 Months with best predictand and predictor correlations for precipitation and minimum and maximum temperatures

Predictand	Predictor	Description	Month correlated
Precipitation	ncepr500	Relative humidity at 500 hPa	Jan, Feb, Apr, Aug, Nov, Dec
	nceprhum	Near-surface relative humidity	Jun, Sep, Oct
Maximum temperature	ncepp5_u	500 hPa zonal velocity	Feb, Apr
	ncepp500	500 hPa geopotential height	Jan, Oct, Nov, Dec
	nceptemp	Mean temperature at 2 m	Mar, May, Jun, Jul, Aug, Sep
Minimum temperature	ncepp500	500 hPa geopotential height	Feb, Mar, Apr, May, Jun, Jul, Aug
	ncepshum	Surface specific humidity	Sep, Oct, Nov, Dec
	nceptemp	Mean temperature at 2 m	Jan

The downscaled baseline daily temperatures show good agreement with observed data. However, due to the conditional nature of daily precipitation, downscaled values have less concurrence with observed daily data. In conditional models, there is an intermediate process between regional forcing and local weather. Local precipitation amounts depend on consecutive occurrence of wet and dry sequences which in turn depend on regional-scale predictors such as humidity and atmospheric pressure (Wilby et al. 2004). Additionally, complicated nature of precipitation processes and its distribution in space and time is the other reason for low concurrence. Climate model simulation of precipitation has improved over time but is still problematic (Bader et al. 2008). It has a larger degree of uncertainty than temperature (Thorpe 2005). This is because rainfall is highly variable in space and so the relatively coarse spatial resolution of the current generation of climate models is not adequate to fully capture that variability. Coefficient of determination (R^2) for daily observed versus simulated (downscaled) clearly shows the difference between unconditional and conditional models for both calibration and validation (Table 28.2).

28.7.3 Future Climate Projection

Future climate was projected using the HadCM3 (A2 and B2) global circulation model. The projection generates 20 ensembles of daily temperature and 100 ensembles of daily precipitation variables. These ensembles are averaged out in order to represent the characteristics of all those ensembles. The analysis was done for three 30 years of data ranges based on recommendation of the World

Table 28.2 Downscaled daily precipitation and maximum and minimum temperatures simulated and observed relations

	R^2	
	Calibration (1961–1980)	Validation (1981–1990)
Precipitation	0.24	0.25
Maximum temperature	0.66	0.64
Minimum temperature	0.57	0.56

Meteorological Organization (WMO), the 2020s (2011–2040), 2050s (2041–2070), and 2080s (2071–2099). The generated scenarios were evaluated individually for each baseline predictand as follows.

28.7.3.1 Precipitation

Rainfall is expected to experience a mean annul increase of 2.21, 2.23, and 1.89 % for A2 scenario for the 2020s, 2050, and 2080, respectively. Mean annual increase for B2 scenario is 2.06, 1.85, and 0.36 % for the 2020s, 2050s, and 2080s, respectively. These values show that increase in rainfall is not uniform; instead, it differs from time period to time period.

A rainfall projection for Bega season, October to January, shows increase for the two emission scenarios except 2080s for B2 scenario, whereas, Belg season, February to May, projection shows a decreasing mean monthly rainfall for the first two months, February and March, and increasing for the last two months of the season, April and May, for A2 scenario. In the case of B2 scenario for the Belg season, rainfall will increase in May for 2020s. The two months, June and July, of the Kiremt season (June to September) rainfall decrease for 2080s as is for April and May. Except increasing trend for September, Kiremt season more or less has nearly constant rainfall (Fig. 28.7a, b).

The projected mean monthly rainfall for this study has a similar pattern to that of the work of Abdo et al. (2009) and (deBoer 2007) using HadCM3 and ECHAM5/MPI-OM global climate models in the same catchment. These works were done on Gilgel Abay catchment and northern Ethiopian highlands. Both of the studies agreed on mean monthly rainfall decrease in May, June, and July and increase in September, October, and November compared to the baseline period. Similarly in IPCC third assessment report of McCarthy et al. (2001), rainfall is predicted to increase in December through February and decrease in June through August in parts of East Africa under intermediate warming scenarios. This IPCC report supports conclusions of this study on increasing mean monthly rainfall from December to February and some decrease from June to August for both A2 and B2 scenarios.

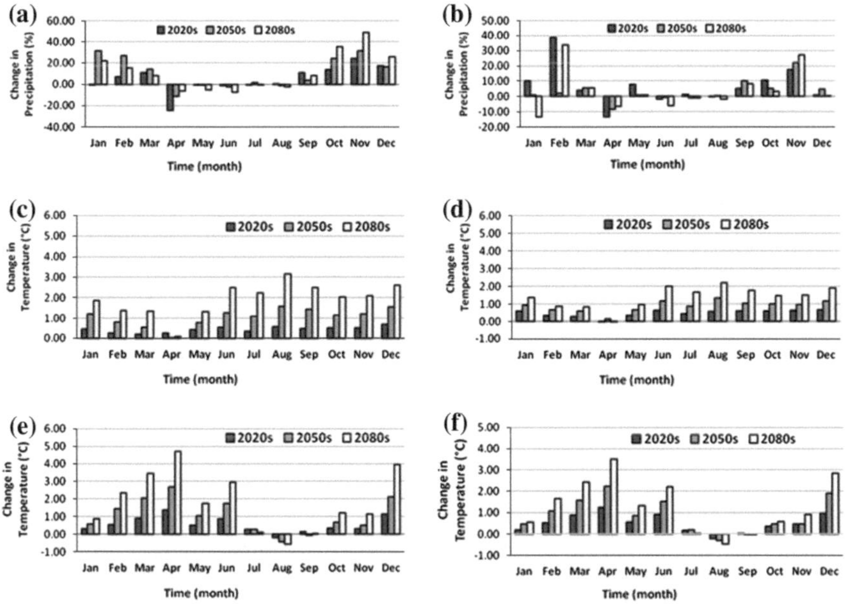

Fig. 28.7 Future changes in mean monthly precipitation for **a** A2 and **b** B2 scenarios; maximum temperature for **c** A2 and **d** B2 scenarios; minimum temperature for **e** A2 and **f** B2 scenarios from the baseline period

28.7.3.2 Maximum Temperature

The projected mean annual maximum temperature shows increasing trend for all time periods by 0.43, 1.05, and 1.92 °C for A2 scenario for the 2020s, 2050s, and 2080, respectively. B2 scenario also shows an increase of mean annual maximum temperature of 0.47, 0.87, and 1.38 °C for the 2020s, 2050s, and 2080, respectively. As compared to B2 scenario, A2 scenario has higher increasing trend. The increase includes all months mostly for all time periods except for April (Fig. 28.7c, d).

28.7.3.3 Minimum Temperature

The projected minimum temperature shows an increasing trend in all time periods. Both the A2 and B2 emission scenarios generate similar changes for future minimum temperature. For A2 scenario, mean annual minimum temperature increases 0.55, 1.06, and 1.83 °C and for B2 scenario 0.50, 0.87, and 1.29 °C for the 2020s, 2050s, and 2080, respectively. Mean monthly variation of minimum temperature is higher than maximum temperature. For both A2 and B2 emission scenarios, minimum temperature is expected to increase from October to June. Not much change in July and September and decreasing trend in August is expected (Fig. 28.7e, f).

Table 28.3 Ranges of projected changes of precipitation and maximum and minimum temperatures from the present condition in wet season

Scenarios	Time period	Precipitation (%)	Maximum temperature (°C)	Minimum temperature (°C)
A2	2020s	−1.5 to 10.8	0.34 to 0.57	−0.2 to 0.87
	2050s	−2.3 to 3.9	1.1 to 1.59	−0.41 to 2.74
	2080s	−7.1 to 8.0	2.23 to 3.15	−0.54 to 2.94
B2	2020s	−1.8 to 5.0	0.43 to 0.63	−0.23 to 0.9
	2050s	−0.8 to 10.1	0.86 to 1.34	−0.31 to 1.53
	2080s	−5.9 to 8.1	1.67 to 2.21	−0.47 to 2.21

Generally, the projected minimum and maximum temperatures in all time periods is within the ranges projected by IPCC which reported average temperature rises of 1.4–5.8 °C toward the end of this century. Maximum and minimum temperatures changes relate to IPCC emission scenario story lines. Increase for A2 scenario is greater than B2 scenario because A2 scenario represents a medium high scenario which produces more carbon dioxide concentration than the B2 scenario which represents a medium low scenario. Ranges of projected changes of precipitation and maximum and minimum temperatures from the present condition in Kiremt (wet season) are shown in Table 28.3.

Impact assessments on stream flow and sediment yield were performed with the Soil and Water Assessment Tool (SWAT) hydrological model. Results of mean annual changes of precipitation and temperature (maximum and minimum) obtained from this section were applied to quantify these impacts.

28.7.4 Hydrological Model Calibration and Validation

Sensitivity analysis (discussed in methodology part) was carried out for 6 sediment parameters using SWAT-CUP global sensitivity analysis. The most sensitive parameters were linear parameter for calculating the maximum amount of sediment (SPCON), USLE equation support particle factor (USLE_P), and exponent parameter for calculating sediment re-entrained channel sediment routing (SPEXP) (Table 28.4).

Upper Gilgel Abay SWAT model was therefore calibrated and validated for stream flow and sediment yield using the sensitive parameters of discharge (Chap. 29 of this book) and sediment. Including two years as a warmup period, periods 1994–2004 and 2005–2008 were used for both stream flow and sediment yield

Table 28.4 Calibrated fitted values of sediment-sensitive parameters

Rank	Parameter	Fitted value
1	SPCON.bsn	0.0037
2	USLE_P.mgt	0.7095
3	SPEXP.bsn	1.9135

calibration and validation, respectively. According to Zhang et al. (2007), to minimize the effect of the user's estimates of initial state variables such as soil and water content and surface residue, the warmup period allows the model to cycle multiple times. Upper Gilgel Abay stream flow and sediment yield gauged near Merawi town were used to calibrate the hydrological model from January 1, 1996, to December 31, 2004, using SWAT-CUP SUFI-2 algorithm as described in the methodology part. Therefore, the calibration was performed with the nine sensitive parameters of stream flow (Chap. 29 of this book) and three parameters of sediment yield. After several iterations, the three fitted parameter values were gained (Table 28.4).

SWAT-CUP in parallel gives the best simulation of stream flow and sediment yield using the above fitted parameter values. Using calibration fitted parameter values, simulation was done for the validation period and its performance checked with the observed data. Performance of the best simulation of sediment yield result which used these fitted parameter values for daily calibration and validation is shown in Table 28.5 and Fig. 28.8. Figure 28.9a, b depict daily simulated and observed sediment yield with stream flow for calibrations and validation periods.

Santhi et al. (2001) stated that efficiency values greater than or equal to 0.50 are considered adequate for SWAT model application. Setegn et al. (2009b) also stated on the model performance that can be judged as satisfactory if R^2 is greater than 0.6 and Nash–Sutcliffe Efficiency (NSE) is greater than 0.5. Hence, it is observed that SWAT exhibited strong performance in representing the hydrological conditions of the catchment.

Table 28.5 Calibrated model simulation performance

Criteria	Calibration (1996–2004)	Validation (2005–2008)
Coefficient of determination (R^2)	0.59	0.56
Nash–Sutcliffe efficiency (NSE)	0.58	0.51
RMSE-observations standard deviation ratio (RSR)	0.65	0.70

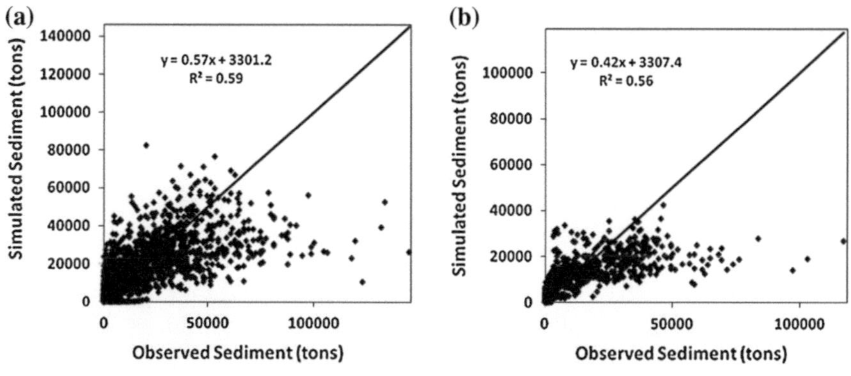

Fig. 28.8 Scatter plot of simulated versus observed daily sediment load **a** calibration and **b** validation

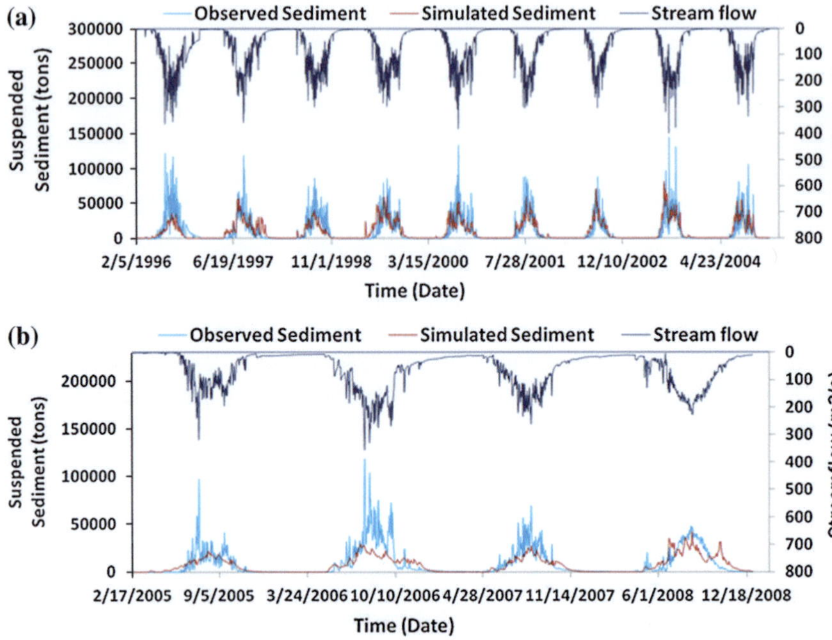

Fig. 28.9 Observed and simulated daily suspended sediment in comparison with observed stream flow for Upper Gilgel Abay catchment **a** calibration and **b** validation periods

Results of daily sediment yield simulation are fair except in cases of extreme sediment loads as shown in Fig. 28.8 for both calibration and validation periods. Considering the difficulty of developing good sediment rating curves, the model outputs are appreciable. The modeling results indicate that the Upper Gilgel Abay SWAT model simulates stream flow and sediment flow with reasonable accuracy.

Mean annual sediment transport load of observed and simulated values are 20.8 and 19.0 metric tons ha^{-1}, respectively (1996–2008). Although close, the model estimates are lower than the observed sediment load. This may be due to limited observed data. The model underestimation on mean monthly data is shown in Fig. 28.10. It is observed in high flow months such as July, August, and September,

Fig. 28.10 Mean monthly observed and simulated sediment load (1996–2008)

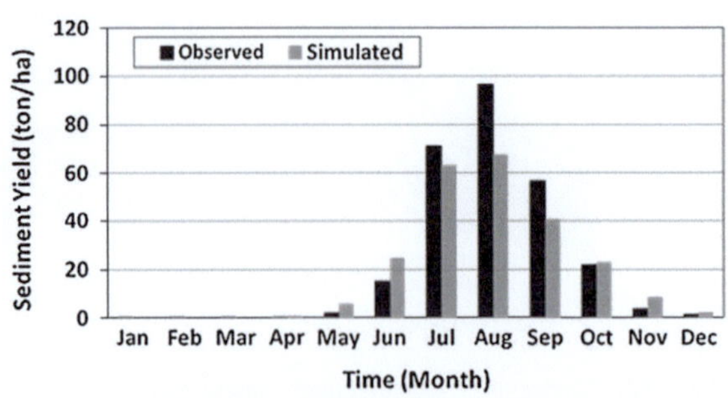

and the model underestimates and overestimates in October and November which are low flow months.

28.7.5 *Climate Change Impact on Sediment Yield*

Mean annual percentage change of sediment yield has a similar pattern to that of stream flow. Mean annual sediment yield is expected to increase by 11.3, 16.3, and 21.3 % in the 2020s, 2050s, and 2080, respectively for A2 scenario. It is also expected to increase by 11.0, 14.3, and 11.3 % in the 2020s, 2050s, and 2080s, respectively, for B2 scenario (Table 28.6). Overall, the mean annual sediment yield increase follows the trend of the mean annual stream flow increase. Sediment yield increases more than linearly with an increase in stream flow (Naik and Jay 2011). Therefore, the impact of climate change on sediment yield is greater than on stream flow. The study area receives most of its rainfall in the wet season of June through September. This is the season which may be impacted highly with climate change.

Monthly sediment change is higher from July through October for both A2 and B2 scenarios (Fig. 28.11). The highest sediment yield change is observed on July which is the month that mean monthly sediment peaks are observed for both scenarios. For this month, sediment yield increases by 4.2, 6.4, and 11.4 % for the 2020s, 2050s, and 2080s time horizons, respectively. For A2 scenario, increases are 4.0, 5.3, and 4.9 % for the 2020s, 2050s, and 2080s, respectively. Similar to stream flow, percentage change of sediment on low flow season months is higher than that of high flow season, but the mass changes as compared to the high flow season are very low.

The result of this study on impact of climate change on sediment yield is comparable to that of Mekonnen and Tadele (2012) who analyzed the combined effects of land use and climate change on soil erosion and stream flow in the Upper Gilgel Abay River catchment using SWAT hydrological model. The climate model they used was GCM of European Centre Hamburg Model (ECHAM5, A1B emission scenario). The result showed that the combined effect of trend-based land use and climate change scenario is projected to increase the average annual sediment load by 29 % for 2011–2025.

Table 28.6 Ranges of projected changes of sediment yield from the present condition in wet season

Scenarios	Time period	Change ranges of sediment yield (%)
A2	2020s	−0.7 to 28.6
	2050s	6.4 to 30.6
	2080s	11.4 to 35.9
B2	2020s	−1.1 to 28.2
	2050s	5.3 to 27.2
	2080s	4.9 to 22.8

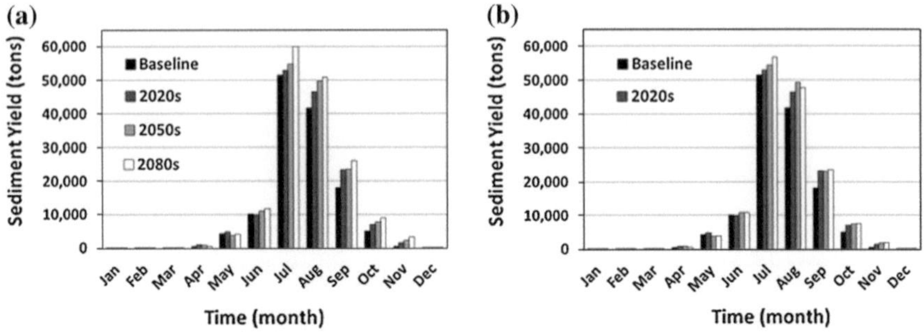

Fig. 28.11 Mean monthly sediment for **a** A2 and **b** B2 scenarios

28.7.6 Uncertainties on Studying Climate Change and Its Impact

Uncertainties in projected changes in a hydrological system arise from internal variability of the climate; uncertainty in future greenhouse gas and aerosol emissions; the translation of these emissions into climate change by global climate models; hydrological model uncertainty; uncertainty in sufficiency of field data at all scales; and uncertainty of downscaling techniques (Bates et al. 2008; Xu et al. 2005).

In general, the sources of uncertainties of climate scenarios are multiple. The climate system itself is too complex to be represented in a numerical model and contains a number of assumptions and parameterizations that each climate modeling center approaches differently. Uncertainties in climate scenarios and GCM outputs may be much larger (Abdo et al. 2009).

The uncertainties related to SDSM model in this work could be that relationship between the predictor and predictand is achieved by only considering the data statistical condition. A major drawback is that the model does not take into consideration the physical nature of the catchments. Also, high-quality data are required for model calibration, and the model is highly sensitive to the choice of predictor variables and empirical transfer schemes (Bekele 2009).

Parameter uncertainties and structural deficiencies in the hydrological models that are used for impact assessments are the other sources of uncertainties. This is why different hydrological models may give different stream flow results for a given input (Abdo et al. 2009).

28.8 Conclusion

Awareness and understanding of the impacts of climate change is critical globally, continentally, nationally as well as at specific catchment scale. GCMs, which estimate the impacts of anthropogenic emissions on climate, play a significant role in creating awareness and understanding of climate change. GCMs are restricted in their usefulness for local climate change and its impact studies by their coarse spatial resolution. This is why one needs to downscale GCMs specifically to the study area. This research which studied the Upper Gilgel Abay River catchment contributes to the understanding of climate change impact on stream flow and sediment yield in the future. Results of climate change on Upper Gilgel Abay River catchment using HadCM3 A2 and B2 emission scenarios downscaled by SDSM model showed that temperature and rainfall are expected to increase in all time periods.

Impact assessment of climate change on sediment yield was quantified by modeling sediment yield from 1994 to 2008. The sediment yield simulation for calibration (1996–2004) and validation (2005–2008) was satisfactory as discussed in the result and discussion section. Therefore, the model performed well for the study area.

Results of climate change impact assessment on sediment yield clearly showed that sediment loss will increase in all time periods. This change also is expected to be larger than change in stream flow. By 2100, one can expect increase of sediment yield by 21.3 and 11.3 % for A2 and B2 scenarios of HadCM3. These results of sediment yield change show that relatively small mean annual increase of rainfall will produce high stream flow and sediment yield. The change will be far greater when increasing land degradation is considered.

Acknowledgments We would like to acknowledge the Blue Nile Water Institute (BNWI) and Tana Sub-Basin Organization (TaSBO) for their financial support. We are also thankful to the Ministry of Water and Energy (MoWE) and the National Meteorological Agency (NMA) Bahir Dar Branch Directorate for their help in providing necessary data for the study.

References

Abbaspour KC (2012) SWAT-CUP 2012: SWAT calibration and uncertainty programs—a user manual. http://www.neprashtechnology.ca/Downloads/SwatCup/Manual/Usermanual_Swat_Cup_2012.pdf. Accessed 21 Dec 2014

Abdo KS, Fiseha BM, Rientjes THM, Gieske ASM, Haile AT (2009) Assessment of climate change impacts on the hydrology of Gilgel Abay catchment in Lake Tana basin, Ethiopia. Hydrol Process 23:3661–3669

Abtew W, Melesse AM (2014a) Nile River basin hydology. In: Melesse AM, Abtew W, Setegn S (eds) Nile River basin: ecohydrological challenges, climate change and hydropolitics, pp 7–22

Abtew W, Melesse AM (2014b) Climate teleconnections and water management. In: Nile River basin. Springer International Publishing, pp 685–705

Abtew W, Melesse AM (2014c) Transboundary rivers and the Nile. In: Nile River basin. Springer International Publishing, pp 565–579

Abtew W, Melesse A, Desalegn T (2009a) Spatial, inter and intra-annual variability of the Blue Nile River basin rainfall. Hydrol Process 23(21):3075–3082

Abtew W, Melesse A, Desalegn T (2009b) El Niño southern oscillation link to the Blue Nile River basin hydrology. Hydrol Process Spec Issue: Nile Hydrol 23(26):3653–3660

Arnold JG, Williams JR, Maidment DR (1995) Continuous-time water and sediment-routing model for large basins. J Hydraul Eng 121:171–183

Arnold JG, Muttiah RS, Srinivasan R, Allen PM (2000) Regional estimation of base flow and groundwater recharge in the Upper Mississippi River basin. J Hydrol 227:21–40

Assefa A, Melesse AM, Admasu S (2014) Climate change in Upper Gilgel Abay River catchment, Blue Nile basin Ethiopia. In: Melesse AM, Abtew W, Setegn S (eds) Nile River basin: ecohydrological challenges, climate change and hydropolitics, pp 363–388

Bader D, Covey C, Gutowski W, Held I, Kunkel K, Miller R, Tokmakian R, Zhang M (2008) Climate models: an assessment of strengths and limitations. US Department of Energy Publications, p 8

Bates B, Kundzewicz ZW, Wu S (2008) Climate change and water. Intergovernmental Panel on Climate Change (IPCC)

Behulu F, Setegn S, Melesse AM, Fiori A (2013) Hydrological analysis of the Upper Tiber basin: a watershed modeling approach. Hydrol Process 27(16):2339–2351

Behulu F, Setegn S, Melesse AM, Romano E, Fiori A (2014) Impact of climate change on the hydrology of Upper Tiber River basin using bias corrected regional climate model. Water Resour Manage 1–17

Bekele HM (2009) Evaluation of climate change impact on Upper Blue Nile Basin reservoirs (case study on Gilgel Abay reservoir, Ethiopia). A thesis submitted in partial fulfillment of the requirements for the degree of Masters of Science in hydraulics and hydropower engineering of Arba-Minch University

Boardman J, Evans R, Favis-Mortlock DT, Harris TM (1990) Climate change and soil erosion on agricultural land in England and Wales. Land Degrad Dev 2:95–106

Carter (2007) Guidelines on the use of scenario data for climate impact and adaptation assessment v2. http://www.ipcc-data.org/guidelines/TGICA_guidance_sdciaa_v2_final.pdf. Accessed 21 Dec 2014

Carter TR, Hulme M, Lal M (1999) Guidelines on the use of scenario data for climate impact and adaptation assessment v1. http://unfccc.int/resource/cd_roms/na1/v_and_a/Resoursce_materials/Climate/ScenarioData.pdf. Accessed 21 Dec 2014

CCIS (2008) Frequently asked questions. SDSM background. http://www.cics.uvic.ca/scenarios/index.cgi?More_Info-SDSM_Background. Accessed 21 Dec 2014

Chebud YA, Melesse AM (2009a) Numerical modeling of the groundwater flow system of the Gumera sub-basin in Lake Tana basin, Ethiopia. Hydrol Process Spec Issue: Nile Hydrol 23 (26):3694–3704

Chebud YA, Melesse AM (2009b) Modeling lake stage and water balance of Lake Tana, Ethiopia. Hydrol Process 23(25):3534–3544

Chebud Y, Melesse AM (2013) Stage level, volume, and time-frequency change information content of Lake Tana using stochastic approaches. Hydrol Process 27(10):1475–1483. doi:10.1002/hyp.9291

Crane RG, Hewitson (1998) Doubled CO_2 precipitation changes for the Susquehanna basin: down—scaling from the genesis general circulation model. Int J Climatol 18:65–76

Xu C, Wid'en E, Halldin S (2005) Modelling hydrological consequences of climate change—progress and challenges. Adv Atmos Sci 22:789–797

Deboer B (2007) The impact of climate change on rainfall extremes over Northeast Africa. KNMI, Royal Netherlands Meteorological Institute, De Bilt, Netherlands, p 5

Defersha MB, Melesse AM (2011) Field-scale investigation of the effect of land use on sediment yield and surface runoff using runoff plot data and models in the Mara River basin, Kenya. CATENA 89:54–64

Defersha MB, Melesse AM (2012) Effect of rainfall intensity, slope and antecedent moisture content on sediment concentration and sediment enrichment ratio. CATENA 90:47–52

Defersha MB, Quraishi S, Melesse AM (2011) Interrill erosion, runoff and sediment size distribution as affected by slope steepness and antecedent moisture content. Hydrol Proc 7 (4):6447–6489

Defersha MB, Melesse AM, McClain M (2012) Watershed scale application of WEPP and EROSION 3D models for assessment of potential sediment source areas and runoff flux in the Mara River basin, Kenya. CATENA 95:63–72

Dessu SB, Melesse AM (2012) Modeling the rainfall-runoff process of the Mara River basin using SWAT. Hydrol Process 26(26):4038–4049
Dessu SB, Melesse AM (2013) Impact and uncertainties of climate change on the hydrology of the Mara River basin. Hydrol Process 27(20):2973–2986
Dessu SB, Melesse AM, Bhat M, McClain M (2014) Assessment of water resources availability and demand in the Mara River basin. CATENA 115:104–114
Gagnon SB, Singh B, Rousselle J, Roy L (2005) An application of the statistical downscaling model (SDSM) to simulate climatic data for streamflow modelling in Québec. Can Water Resour J 30:297–314
Gebrekirstos ST (2008) Watershed modeling of Lake Tana basin using SWAT. MSc thesis, ArbaMinch University, Ethiopia
Gebremariame ZH (2009) Assessment of climate change impact on the net basin supply of Lake Tana water balance. ITC MSc thesis, Enschede, The Netherlands
Getachew HE, Melesse AM (2012) Impact of land use/land cover change on the hydrology of Angereb watershed, Ethiopia. Int J Water Sci 1, 4:1–7. doi:10.5772/56266
Giorgi F, Mearns LO (1991) Approaches to the simulation of regional climate change: a review. Rev Geophys 29:191–216
Grey OP, Webber Dale G, Setegn SG, Melesse AM (2013) Application of the soil and water assessment tool (SWAT Model) on a small tropical island state (Great River watershed, Jamaica) as a tool in integrated watershed and coastal zone management. Int J Trop Biol Conserv 62(3):293–305
Hewitson BC, Crane RG (1996) Climate downscaling: techniques and application. Clim Res 7:85–95
Houghton DD (2002) Introduction to climate change: lecture notes for meteorologists. Secretariat of the World Meteorological Organization, Geneva, Switzerland
James LD, Burges SJ (1982) Selection, calibration, and testing of hydrologic models. In: Hydrologic modelling of small watersheds, pp 437–472
Jenkins GS, Barron EJ (1997) Global climate model and coupled regional climate model simulations over the eastern United States: GENESIS and RegCM2 simulations. Global Planet Change 15:3–32
Kim U, Kaluarachchi JJ (2009) Climate change impacts on water resources in the Upper Blue Nile River basin, Ethiopia1. JAWRA J Am Water Resour Assoc 45:1361–1378
Maalim FK, Melesse AM (2013) Modeling the impacts of subsurface drainage systems on runoff and sediment yield in the Le Sueur watershed, Minnesota. Hydrol Sci J 58(3):1–17
Maalim FK, Melesse AM, Belmont P, Gran K (2013) Modeling the impact of land use changes on runoff and sediment yield in the Le Sueur watershed, Minnesota using GeoWEPP. CATENA 107:35–45
Mango L, Melesse AM, McClain ME, Gann D, Setegn SG (2011a). Land use and climate change impacts on the hydrology of the upper Mara River Basin, Kenya: results of a modeling study to support better resource management. Special issue: climate, weather and hydrology of East African Highlands. Hydrol Earth Syst Sci 15:2245–2258. doi:10.5194/hess-15-2245-2011
Mango L, Melesse AM, McClain ME, Gann D, Setegn SG (2011b) Hydro-meteorology and water budget of Mara River basin, Kenya: a land use change scenarios analysis. In: Melesse A (ed) Nile river basin: hydrology, climate and water use, Chap 2. Springer Science Publisher, pp 39–68. doi:10.1007/978-94-007-0689-7_2
Mapfumo E, Chanasyk DS, Baron VS (2003) Patterns and simulation of soil water under different grazing management systems in central Alberta. Can J Soil Sci 83:601–614
Mccarthy JJ, Canziani OF, Leary NA, Dokken DJ, White KS (2001) Climate change 2001: impacts, adaptation, and vulnerability: contribution of Working Group II to the third assessment report of the Intergovernmental Panel on Climate Change. Cambridge University Press, Cambridge
Mekonnen M, Melesse A (2011) Soil erosion mapping and hotspot area identification using GIS and remote sensing in northwest Ethiopian highlands, near Lake Tana. In: Melesse A (ed) Nile

River basin: hydrology, climate and water use, Chap 10. Springer Science Publisher, pp 207–224. doi:10.1007/978-94-007-0689-7_10
Mekonnen K, Tadele K (2012) Analyzing the impact of land use and climate changes on soil erosion and stream flow in the upper Gilgel Abbay catchment, Ethiopia. Ohrid, Republic of Macedonia, pp 1–13
Melesse AM (2011) Nile River basin: hydrology, climate and water use. Springer Science & Business Media
Melesse AM, Loukas AG, Senay G, Yitayew M (2009a) Cli mate change, land-cover dynamics and ecohydrology of the Nile River basin. Hydrol Process 23(26):3651–3652
Melesse A, Abtew W, Desalegne T, Wang X (2009b) Low and high flow analysis and wavelet application for characterization of the Blue Nile River system. Hydrol Process 24(3):241–252
Melesse AM, Ahmad S, McClain M, Wang X, Lim H (2011a) Sediment load prediction in large rivers: ANN approach. Agric Water Manage 98:855–886
Melesse A, Abtew W, Setegn S, Dessalegne T (2011b) Hydrological variability and climate of the Upper Blue Nile River basin. In: Melesse A (ed) Nile River basin: hydrology, climate and water use, Chap 1. Springer Science Publisher, pp 3–37. doi:10.1007/978-94-007-0689-7_1
Melesse A, Abtew W, Setegn SG (2014) Nile River basin: ecohydrological challenges, climate change and hydropolitics. Springer Science & Business Media
Mohammed H, Alamirew T, Assen M, Melesse AM (2015) Modeling of sediment yield in Maybar gauged watershed using SWAT, northeast Ethiopia. CATENA 127:191–205
Msagahaa JJ, Ndomba PM, Melesse AM (2014) Modeling sediment dynamics: effect of land use, topography and land management. In: Melesse AM, Abtew W, Setegn S (eds) Nile River basin: ecohydrological challenges, climate change and hydropolitics, pp 165–192
Murphy J (1999) An evaluation of statistical and dynamical techniques for downscaling local climate. J Clim 12:2256–2284
Naik PK, Jay DA (2011) Distinguishing human and climate influences on the Columbia river: changes in mean flow and sediment transport. J Hydrol 404:259–277
Nakicenovic N, Alcamo J, Davis G, De Vries B, Fenhann J, Gaffin S, Gregory K, Grubler A, Jung TY, Kram T (2000) Special report on emissions scenarios: a special report of working group III of the Intergovernmental Panel on Climate Change. Pacific Northwest National Laboratory, Richland, WA (US), Environmental Molecular Sciences Laboratory (US)
Nearing MA (2001) Potential changes in rainfall erosivity in the US with climate change during the 21st century. J Soil Water Conserv 56:229–232
Nearing MA, Pruski FF, O'Neal MR (2004) Expected climate change impacts on soil erosion rates: a review. J Soil Water Conserv 59:43–50
Nietsch SL, Arnold JG, Kiniry JR, Williams JR (2001) SWAT: soil and water assessment tool theoretical documentation. USDA, ARS, Temple, TX
Neitsch SL, Arnold JG, Kiniry JR, Williams JR, King KW (2005) Soil and water assessment tool: theoretical documentation, version 2005, Texas, USA
O'Neal MR, Nearing MA, Vining RC, Southworth J, Pfeifer (2005) Climate change impacts on soil erosion in Midwest United States with changes in crop management. Catena 61:165–184
Parry ML (2007) Climate change 2007: impacts, adaptation and vulnerability: working group I contribution to the fourth assessment report of the IPCC. Cambridge University Press, Cambridge
Pruski FF, Nearing MA (2002a) Climate-induced changes in erosion during the 21st century for eight US locations. Water Resour Res 38:34-1–34-11
Pruski FF, Nearing MA (2002b) Runoff and soil-loss responses to changes in precipitation: a computer simulation study. J Soil Water Conserv 57:7–16
Santhi C, Arnold JG, Williams JR, Dugas WA, Srinivasan R, Hauk LM (2001) Validation of the SWAT model on a large RWER basin with point and non-point sources. JAWRA J Am Water Resour Assoc 37:1169–1188
Santos H, Idinoba M, Imbach P (2008) Climate scenarios: what we need to know and how to generate them. CIFOR working paper

Schulze RE (1997) Impacts of global climate change in a hydrologically vulnerable region: challenges to South African hydrologists. Prog Phys Geogr 21:113–136

Setegn SG, Srinivasan R, Dargahi B, Melesse A (2009a) Spatial delineation of soil erosion prone areas: application of SWAT and MCE approaches in the Lake Tana basin, Ethiopia. Hydrol Process Spec Issue: Nile Hydrol 23(26):3738–3750

Setegn SG, Srinivasan R, Melesse A, Dargahi B (2009b) SWAT model application and prediction uncertainty analysis in the Lake Tana basin, Ethiopia. Hydrol Process 24(3):357–367

Setegn SG, Dargahi B, Srinivasan R, Melesse AM (2010) Modelling of sediment yield from Anjeni gauged watershed, Ethiopia Using SWAT. JAWRA 46(3):514–526

Setegn SG, Rayner D, Melesse AM, Dargahi B, Srinivasan R (2011) Impact of climate change on the hydroclimatology of Lake Tana basin, Ethiopia. Water Resour Res 47:W04511

Setegn SG, Melesse AM, Haiduk A, Webber D, Wang X, McClain M (2014) Spatiotemporal distribution of fresh water availability in the Rio Cobre Watershed, vol 120, CATENA, Jamaica, pp 81–90

Tadege A (2001) Initial National Communication of Ethiopia to the United Nations Framework Convention on Climate Change (UNFCCC). National Meteorological Services Agency, Addis Ababa, Ethiopia

Thorpe AJ (2005) Climate change prediction: a challenging scientific problem. Institute of Physics

Von Storch H, Zorita E, Cubasch U (1993) Downscaling of global climate change estimates to regional scales: an application to Iberian rainfall in wintertime. J Clim 6(1161):1171

Wale A (2008) Hydrological balance of Lake Tana Upper Blue Nile basin, Ethiopia. In: ITC thesis, PP 159–180

Walling DE (2006) Human impact on land–ocean sediment transfer by the world's rivers. Geomorphology 79:192–216

Walling DE (2009) The impact of global change on erosion and sediment transport by rivers: current progress and future challenges, UNESCO

Wang X, Melesse AM (2005) Evaluations of thc SWAT Model's snowmelt hydrology in a Northwestern Minnesota watershed. Trans ASAE 48(4):1359–1376

Wang X, Melesse AM (2006) Effects of STATSGO and SSURGO as inputs on SWAT Model's Snowmelt simulation. J Am Water Resour Assoc 42(5):1217–1236

Wang X, Melesse AM, Yang W (2006) Influences of potential evapotranspiration estimation methods on SWAT's hydrologic simulation in a Northwestern Minnesota watershed. Trans ASAE 49(6):1755–1771

Wang X, Shang S, Yang W, Melesse AM (2008a) Simulation of an agricultural watershed using an improved curve number method in SWAT. Trans Am Soc Agric Bio Eng 51(4):1323–1339

Wang X, Yang W, Melesse AM (2008b) Using hydrologic equivalent wetland concept within SWAT to estimate streamflow in watersheds with numerous wetlands. Trans Am Soc Agric Bio Eng 51(1):55–72

Wang X, Garza J, Whitney M, Melesse AM, Yang W (2008c) Prediction of sediment source areas within watersheds as affected by soil data resolution. In: Findley PN (ed) Environmental modelling: new research, Chap 7. Nova Science Publishers, Inc., Hauppauge, pp 151–185. ISBN 978-1-60692-034-3

Wilby RL, Dawson CW (2007) Statistical downscaling model (SDSM), version 4.2. A decision support tool for the assessment of regional climate change impacts, UK

Wilby RL, Dawson CW, Barrow EM (2002) SDSM—a decision support tool for the assessment of regional climate change impacts. Environ Model Softw 17:145–157

Wilby RL, Charles SP, Zorita E, Timbal B, Whetton P, Mearns LO (2004) Guidelines for use of climate scenarios developed from statistical downscaling methods. IPCC task group on data and scenario support for impacts and climate analysis

Williams JR (1975) Sediment-yield prediction with Universal Equation using runoff energy factor. In: Present and prospective technology for predicting sediment yield and sources. US Dep Agr ARS-S40, pp 244–252

Williams JR (1995) The EPIC model. In: Singh VP (ed) Computer models of watershed hydrology. Water Resources Publications, Highlands Ranch, pp 909–1000

Wischmeier WH, Smith DD (1978) Predicting rainfall erosion losses-a guide to conservation planning. Agriculture handbook 282, USDA-ARS

Xu CY (1999) From GCMs to river flow: a review of downscaling methods and hydrologic modelling approaches. Prog Phys Geogr 23:229–249

Yitayew M, Melesse AM (2011) Critical water resources management issues in Nile River basin. In: Melesse A (ed) Nile River basin: hydrology, climate and water use, Chap 20. Springer Science Publisher, pp 401–416. doi:10.1007/978-94-007-0689-7_20

Zhang X, Srinivasan R, Hao (2007) Predicting hydrologic response to climate change in the Luohe River basin using the SWAT model. Trans ASABE 50:901–910

Chapter 29
Climate Change Impact on Stream Flow in the Upper Gilgel Abay Catchment, Blue Nile basin, Ethiopia

Anwar A. Adem, Seifu A. Tilahun, Essayas K. Ayana, Abeyou W. Worqlul, Tewodros T. Assefa, Shimelis B. Dessu and Assefa M. Melesse

Abstract According to Intergovernmental Panel on Climate Change (IPCC) future projections, precipitation and temperature will increase over eastern Africa in the coming century. This chapter presents basin-level impact of climate change on stream flow in Upper Gilgel Abay catchment, Blue Nile basin, Ethiopia, by downscaling HadCM3 global climate model (GCM) using statistical downscaling model (SDSM). IPCC recommended baseline period (1961–1990) was used for analysis of baseline scenario. For future scenario analysis time periods of the 2020s, 2050s and 2080s were used. Globally, HadCM3 model is widely applied for climate change studies and it contains A2 (medium–high emission) and B2 (medium–low emission) scenarios. The impact assessment on stream flow was done using the soil and water assessment tool (SWAT) hydrological model. The performance of

A.A. Adem (✉)
Tana Sub-Basin Office, Blue Nile Authority, Bahir Dar, Ethiopia
e-mail: unnxuss@gmail.com

S.A. Tilahun · E.K. Ayana · A.W. Worqlul · T.T. Assefa
School of Civil and Water Resources Engineering, Bahir Dar University, Bahir Dar, Ethiopia
e-mail: seifuad@yahoo.co.uk

E.K. Ayana
e-mail: ekk45@cornell.edu

A.W. Worqlul
e-mail: Abeyou_wale@yahoo.com

T.T. Assefa
e-mail: ttaffese@gmail.com

E.K. Ayana
Geospatial Data and Technology Center, Bahir Dar University, Bahir Dar, Ethiopia

S.B. Dessu
Department of Civil Engineering, Addis Ababa University, Addis Ababa, Ethiopia
e-mail: sbehilu@gmail.com

A.M. Melesse
Department of Earth and Environment, Florida International University, Miami, USA
e-mail: melessea@fiu.edu

© Springer International Publishing Switzerland 2016

A.M. Melesse and W. Abtew (eds.), *Landscape Dynamics, Soils and Hydrological Processes in Varied Climates*, Springer Geography,
DOI 10.1007/978-3-319-18787-7_29

SWAT model in simulating the stream flow was shown with a Nash–Sutcliffe Efficiency (NSE) of 0.76 and 0.78 for calibration and validation periods, respectively. Mean annual changes of precipitation and temperature (maximum and minimum) were applied to quantify these impacts. The result of downscaled precipitation and temperature reveals a systematic increase in all future time periods for both A2 and B2 scenarios. These increases in climate variables are expected to increase mean annual stream flow by 7.1, 9.7, and 10.1 % for A2 scenario and by 6.8, 7.9, and 6.4 % for B2 scenario for 2020s, 2050s, and 2080s, respectively. Future work need to consider impact of land use change on the catchment for future sustainable development plan.

Keywords Climate change · IPCC · A2 and B2 scenarios · Precipitation and temperature · Hadcm3 · SWAT · Upper Gilgel Abay · Nile basin

29.1 Introduction

The environmental crisis on natural ecosystems, agriculture and food supplies, human health, forestry, water resources and availability, energy use, and transportation directly or indirectly in many areas of the world are partly caused by global climate change (Chen et al. 2012). For instance, it has significant effects on the hydrological cycle because the distribution of water resources is very sensitive to climate change (Solomon 2007). Global climate change has also the potential to impose additional pressures on water availability and water demand (Bates et al. 2008).

Global warming is directly linked to changes in the large-scale hydrological cycle. Linkages between global warming and hydrological cycle components include increasing atmospheric water vapor content; changing precipitation patterns, intensity, and extremes; reduced snow cover and widespread melting of ice; and changes in soil moisture and runoff (Bates et al. 2008). This linkage needs to be understood and models are often used as tools to examine such linkages.

Changes in natural systems and resources, which are partly results of climate change, pose a big threat to Africa's economic growth (AMCEN 2011). In addition, water resource issues have not been adequately addressed in climate change analyses and climate policy formulations (Bates et al. 2008). One can imagine that under future climate conditions (second half of the century) with increased precipitation in the Greater Horn of Africa, direct runoff and more erosion will be initiated (Parry 2007). Additionally, soil loss due to sheet and rill erosion will increase and cause accelerated erosion and deposition in Lake Tana. Consequently, research on how climate affects stream flow and impact the sediment yield of catchment is important.

Modeling future stream flow using the projected changes of local-scale climate variables is necessary to properly plan for future climate conditions. For Ethiopia, suitable planning and sustainable utilization of water resources development based on climate change impact is very critical. In spite of the implication and significance of Lake Tana and its basin for the continued settlement of the people in the basin, only little is done in planning for the future (Gebremariam 2009). Blue Nile River basin is one of most sensitive basins to climate change and water resources variability (Kim and Kaluarachchi 2009). Nevertheless, water availability effects of climate change in water resource analysis, management, and policy formulation in Lake Tana basin have not been addressed adequately (Setegn et al. 2011).

Hydrology of the Nile River basin has been studied by various researchers. These studies encompass various areas including stream flow modeling, sediment dynamics, teleconnections and river flow, land use dynamics, climate change impact, groundwater flow modeling, hydrodynamics of Lake Tana, water allocation, and demand analysis (Melesse et al. 2009a, b, 2011a, b, 2014; Abtew et al. 2009a, b; Abtew and Melesse 2014a, b, c; Melesse 2011; Yitayew and Melesse 2011; Chebud and Melesse2013, 2009a, b; Setegn et al. 2009a, b, 2010).

Studies on the Upper Blue Nile basin (Soliman et al. 2009; Kim and Kaluarachchi 2009; Elshamy et al. 2009; Eguavoen 2009; Beyene et al. 2010; Taye et al. 2011) and Lake Tana sub-basin (Tarekegn and Tadege 2006; Setegn et al. 2011) showed a significant evidence on sensitivity of the region for climate variability and change. For instance, in Lake Tana sub-basin, some studies have been done on hydrological impact assessment of climate change, especially on stream flow. The study by Tarekegn and Tadege (2006) showed that runoff is much more seasonal in Lake Tana sub-basin. The CCCM and GFD3 GCMs predict a reduction of 18.2 and 12.6 %, respectively, in the annual runoff by 2075, while the UK89 GCM predicts wetter conditions and as a result an increase of 2.5 % in annual runoff. Setegn et al. (2011) used 15 GCMs to quantify climate change impact and showed an increase pattern in future temperature in the area for all time periods and emission scenarios. The stream flow response for this change in temperature and rainfall followed the pattern of the changes shown by the two variables.

The investigation by Elshamy et al. (2009) showed that runoff is more sensitive to precipitation changes than to potential evapotranspiration changes and that the sensitivity increases with the aridity of the catchment. A 10 % increase in rainfall yields a 29 % increase of flow, while a 10 % reduction in rainfall yields flow reduction of 19 % for Gilgel Abay catchment. Climate change impact study on stream flow of Gilgel Abay catchment has not shown any systematic increase or decrease in rainfall despite studies indicating that there will be a distinct increase in minimum and maximum temperatures in the future (Abdo et al. 2009). According to Abdo's result for the 2080s, the runoff volume in the rainy season will reduce by approximately 11.6 and 10.1 % for the A2 and B2 scenarios, respectively. While 33 % of the seasonal and annual runoff is expected to reduce when temperature increases by 2 °C and when rainfall decreases by approximately 20 %. Trend-based land use and climate change scenario (2011–2025) will increase the average annual sediment load by 29 % and stream flow by 5.2 %. This significantly affects soil erosion and stream flow compared to other scenario condition by using GCM of European Centre Hamburg Model (ECHAM5, A1B emission scenario) (Mekonnen and Tadele 2012).

The impact of climate change on water resources variability, especially on river flows, soil moisture, evapotranspiration, and groundwater flow has been studied by analyzing projected and downscaled climatic data and using hydrological models (Mango et al. 2011a, b; Behulu et al. 2014; Setegn et al. 2011; Setegn and Melesse 2014; Anwer et al. 2014; Melesse et al. 2009a, b; Dessu and Melesse 2012, 2013; Grey et al. 2013).

The main objective of this study is to quantify the impact of climate change on stream flow in Upper Gilgel Abay catchment using SWAT hydrological model. This was done by downscaling selected GCM outputs to reproduce changes of local-scale climate variables. The findings in this research could help in examining the applicability of GCMs under various geographical and physiographic settings.

29.2 Study Area and Dataset

29.2.1 Study Area

Gilgel Abay River is the largest tributary in Lake Tana sub-basin covering an area of about 1654 km^2. It originates from the highland spring of Gish-Abay town in north eastern Ethiopia. It contributes 44.2 % of water to Lake Tana (Wale 2008). Geographically, Upper Gilgel Abay catchment is a northern part the Upper Blue Nile basin and is the southern part of Lake Tana sub-basin with the latitudes and longitudes between 10° 56′ 53″–11° 21′ 58″N and 36° 49′ 29″–37° 23′ 34″E, respectively (Fig. 29.1).

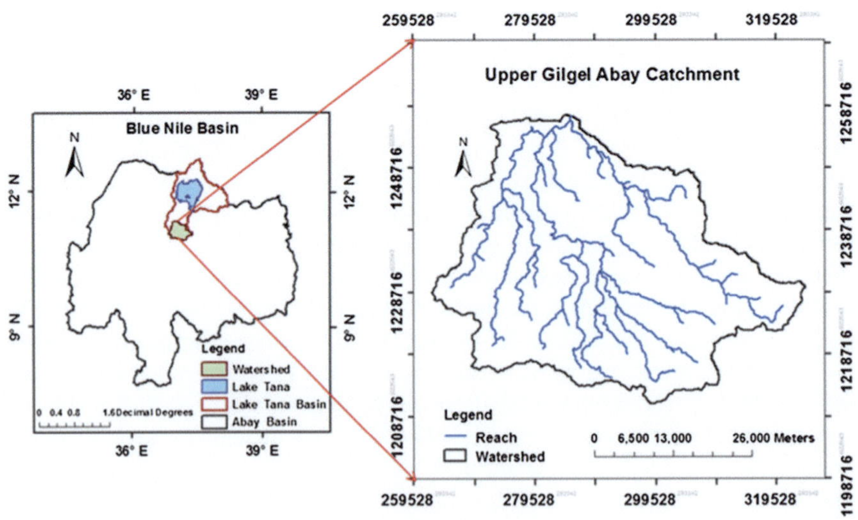

Fig. 29.1 Location of study Upper Gilgel Abay catchment

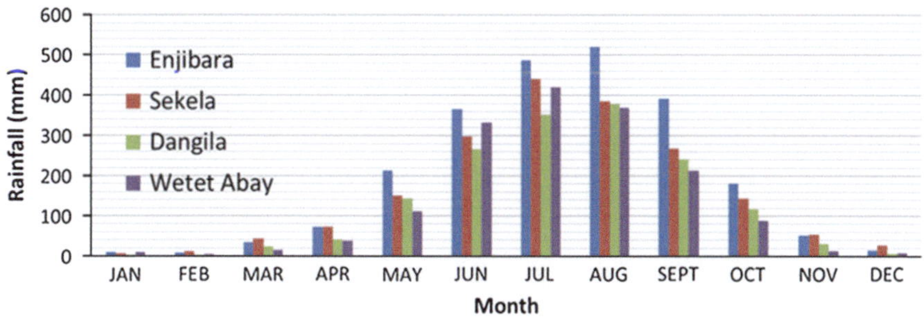

Fig. 29.2 Mean monthly rainfall in Upper Gilgel Abay catchment (1994–2008)

The elevation of Upper Gilgel Abay catchment ranges from 1890 to 3514 m above mean sea level (amsl). The highest elevation of the catchment is located in the south eastern tip. Nearly two-third of the catchment has an elevation ranges from 1890 to 2427 m amsl which extends from the center to the north tip (the outlet of the river).

Mean monthly rainfall (1994–2008) plot of Injibara, Sekela, Dangila, and Wetet Abay stations indicates the study area has one peak per year (Fig. 29.2). Therefore, Upper Gilgel Abay catchment lies in mono-modal climate class according to Ethiopian climate classification with respect to rainfall regimes (Tadege 2001). The main rainfall season of the study area is from June to September and accounts for 70–90 % of the annual rainfall (Abdo et al. 2009).

Mean annual areal rainfall of the study area varies from 1624 to 2349 mm. Majority of the study area has a mean annual rainfall between 1842 and 1986 mm. The temperature of the study area is dependent on elevation where temperature decreases with increase in elevation. For instance, mean monthly maximum temperature of Wetet Abay (1994–2008) at elevation of 1915 m amsl varies from 24.3 to 31.3 °C and Gundil at elevation of 2574 m amsl varies from 18.3 to 24.9 °C. Generally, daily variation between maximum and minimum temperature is high compared to the seasonal variation of temperature in the study area (Fig. 29.3).

29.2.2 Dataset

29.2.2.1 Meteorological Data

Meteorological variables such as rainfall, and maximum and minimum temperatures were required as a predictand to downscale the global climate GCM data to local climate variables. Ten meteorological stations with 15 years daily data were selected for this study (Fig. 29.4). To derive statistical relationships between the predictand and predictor, only Bahir Dar station data are used as this is the only station that satisfies the baseline historical data recommended by

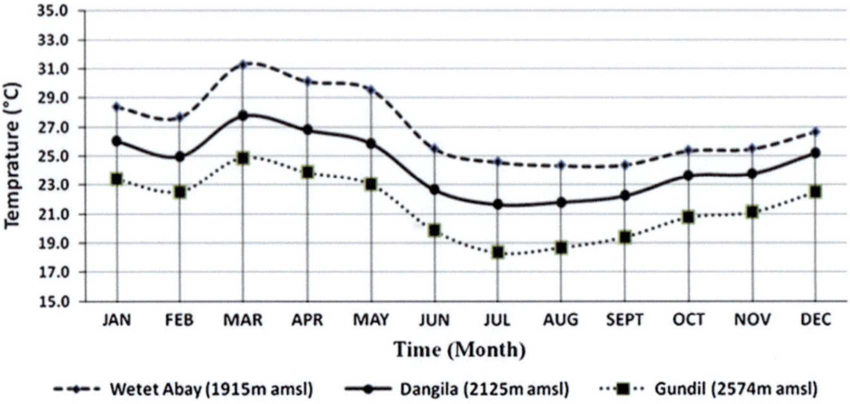

Fig. 29.3 Mean monthly temperature at three stations with different elevations (1994–2008)

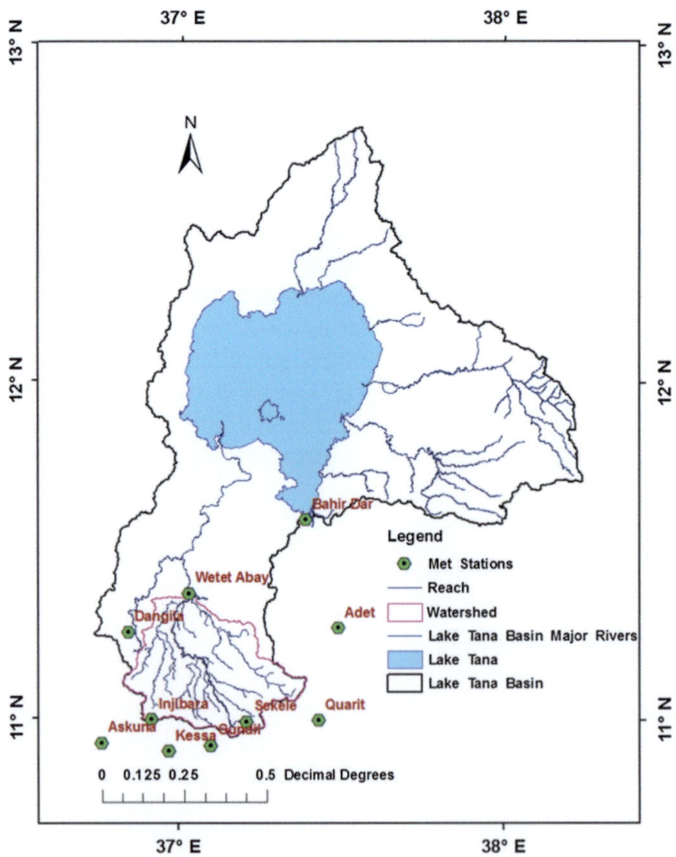

Fig. 29.4 Meteorological station locations

IPCC (i.e., 1961–1990). The meteorological variables from all the ten stations were used for stream flow modeling in SWAT.

29.2.2.2 GCM Data

GCM data are required to project and quantify the relative change of climate variables between the current and future time period. One of the global circulation models, HadCM3, was used for this study because the model is widely applied in many climate change studies and the model provides daily predictor variables which can be used for the SDSM. The predictor variables are supplied on a grid box by grid box basis, 2.5° latitude by 3.75° longitude. The location of the study area was entered and the correct grid box calculated and a zip file was downloaded (http://www.cics.uvic.ca/scenarios/sdsm/select.cgi).

To apply SDSM to GCM data, both observed predictand and GCM data should ideally be available on the same grid spacing. Individual predictor and predictand files (one variable to each file, time series data only) are denoted by the extension * .DAT (Wilby et al. 2002). The predictor represents large-scale atmospheric variables, whereas the predictand represents local surface variables such as temperature and precipitation.

29.2.2.3 Hydrological Data

Stream flow was used for calibrating and validating the SWAT model simulation. Daily stream flow data were obtained from Ministry of Water and Energy (MoWE) for Gilgel Abay River near Merawi town.

29.2.2.4 Spatial Data

Digital elevation model (DEM), land use/land cover, and soil are the three spatial data inputs required by SWAT model. DEM describes the elevation of any point in a given area at a specific spatial resolution as a digital file. DEM is one of the essential inputs required by SWAT: (1) to delineate the watershed into a number of sub-watersheds or sub-basins and (2) to analyze the drainage pattern of the watershed, slope, stream length, width of channel within the watershed. The DEM was obtained from the NASA Shuttle Radar Topographic Mission (SRTM) with a resolution of 90 m by 90 m (http://srtm.csi.cgiar.org/SELECTION/inputCoord.asp).

Land use/Land cover and soil data were collected from MoWE Meta data section, which was used during Upper Blue Nile (Abay) River master plan study by Le Bureau, central d'Etudes Pourles Equipments d'outre-Mer (BCEOM) in 1996–1998. The land use/Land cover and soil map scale used during the master plan study were 1:250,000 (Gebrekristos 2008).

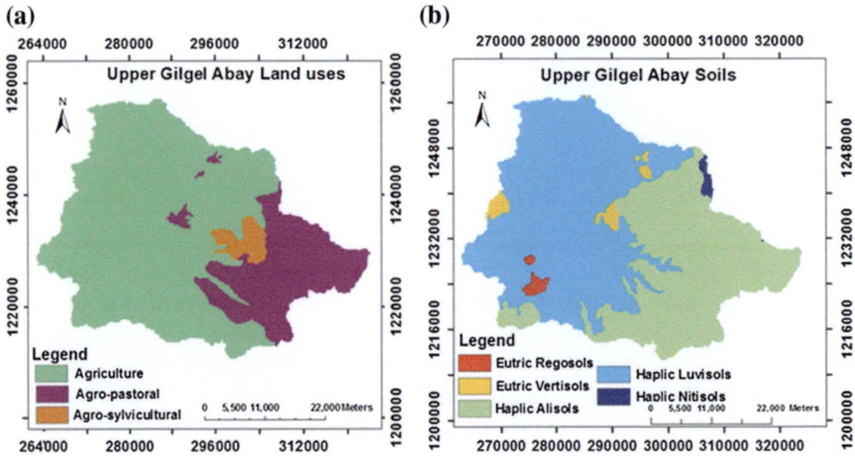

Fig. 29.5 Spatial inputs of SWAT defined **a** land use and **b** soil

As per BCEOM, there are three land uses in the catchment, agriculture (AGRL), agro-pastoral (AGRC) and agro-sylvicultural (FRSD) (Fig. 29.5a). Whereas Eutric Regosols, Eutric Vertisols, Haplic Alisols, Haplic Luvisols, and Haplic Nitisols are the major soils in the study area (Fig. 29.5b).

29.3 Methodology

Defining the climatological baseline, the IPCC Special Report on Emissions Scenarios (SRES) which is considered in HadCM3 global climate model and the downscaling technique which was applied on HadCM3 is the first and basic step for this impact assessment to be described. Stream flow modeling with SWAT is the second step in the methodology. Finally, climate change impact study using hydrological model is the subject to be discussed.

29.3.1 *Climatological Baseline*

In order to assess the implications of future changes in the environment, society, and economy on an exposure unit, it is first necessary to have information about the present-day or recent conditions as a reference point or baseline (Carter et al. 1999; McCarthy et al. 2001). Baseline information is important for: (1) characterizing the prevailing conditions under which an exposure unit functions and to which it must adapt; (2) describing average conditions, spatial and temporal variability, and anomalous events, some of which can cause significant impacts;

(3) calibrating and testing impact models across the current range of variability; (4) identifying possible ongoing trends or cycles; and (5) specifying the reference situation with which to compare future changes (Carter et al. 1999). According to the above-listed importance and advantage, this study applied the suggested IPCC climatological baseline period (1961–1990). For the study area, only Bahir Dar meteorological station fulfills IPCC baseline period because the observed data cover the range from 1961 to 1990. Therefore, this data were used as a predictand for downscaling.

29.3.2 SRES Emissions Scenarios

According to IPCC description, climate scenarios are plausible representations of the future that are consistent with assumptions about future emissions of greenhouse gases and other pollutants and with our understanding of the effect of increased atmospheric concentrations of these gases on global climate. These assumptions include future trends in energy demand, emissions of greenhouse gases, land use change as well as assumptions about the behavior of the climate system over long time scales. The IPCC–Task Group on Data and Scenario Support for Impact and Climate Assessment (IPCC-TGCIA) classified climatic scenarios into three main types based on how they are constructed (Carter et al. 2007). These are synthetic scenarios, also known as incremental scenarios, analogue scenarios, and climate model-based scenarios.

The IPCC SRES in replacing the old IPCC scenarios (IS92) identifies 40 different scenarios following four families of story lines (Santoso et al. 2008). Each story line represents a distinctly different direction for future developments such as demographic, socioeconomic, technological, and environmental developments. The four qualitative story lines yield four sets of scenarios called families (A1, A2, B1, and B2). The main characteristics of the two SRES story lines scenario families which are considered in HadCM3 global climate model are described below (Nakicenovic et al. 2000). The A2 story line and scenario family describe a very heterogeneous world. The underlying theme is self-reliance and preservation of local identities. Fertility patterns across regions converge very slowly, which results in continuously increasing global population. Economic development is primarily regionally oriented, and per capita economic growth and technological changes are more fragmented and slower than in other story lines. The B2 story line and scenario family describes a world in which the emphasis is on local solutions to economic, social, and environmental sustainability. It is a world with continuously increasing global population at a rate lower than A2, intermediate levels of economic development, and less rapid and more diverse technological change than in the B1 and A1 story lines. While the scenario is also oriented toward environmental protection and social equity, it focuses on local and regional levels.

29.3.3 Statistical Downscaling Techniques

General circulation models (GCMs) indicate that rising concentrations of greenhouse gases will have significant implications for climate at global and regional scales (Wilby and Dawson 2007). Due to their coarse spatial resolution and inability to resolve important sub-grid-scale features such as clouds and topography, GCMs are restricted in their usefulness for local impact studies by their coarse spatial resolution. GCMs depict the climate using a three-dimensional grid over the globe, typically having a horizontal resolution of between 250 and 600 km, 10–20 vertical layers in the atmosphere and sometimes as many as 30 layers in the oceans (Nakicenovic et al. 2000). Their resolution is thus quite coarse relative to the scale of exposure units in most impact assessments. Several methods have been adopted for developing regional GCM-based scenarios at the sub-grid scale, a procedure variously known as "regionalization" or "downscaling" is applied. Two different approaches to downscaling are possible which are dynamical and statistical (Hewitson and Crane 1996).

The typical application in dynamic (nested model) downscaling is to drive a regional dynamic model at mesoscale or finer resolutions with the synoptic-scale and larger-scale information from a GCM (Giorgi and Mearns 1991; Jenkins and Barron 1997). Detailed information at spatial scales down to 10–20 km and at temporal scales of hours or less may be achieved in such applications (Hewitson and Crane 1996). Such models are computationally demanding and are not easily accessible, but in the long term, this technique is likely to be the best solution and needs to be encouraged.

Statistical (empirical) downscaling is computationally efficient in comparison with dynamical downscaling. It is a practical approach for addressing current needs in the climate change research community, especially in many of the countries liable to be most sensitive to climate change impacts (Hewitson and Crane 1996). Many of the processes which control local climate, e.g., topography, vegetation, and hydrology, are not included in coarse-resolution GCMs. The development of statistical relationships between the local scales and large scales may include some of these processes implicitly. Under the broad empirical/statistical downscaling techniques, the following three major techniques have been developed. These are weather classification/typing schemes, transfer function/regression model, and stochastic weather generators methods.

Regression models are a conceptually simple means of representing linear or nonlinear relationships between local climate variables (predictands) and the large-scale atmospheric forcing (predictors) (Wilby et al. 2004). Commonly applied methods include canonical correlation analysis (CCA) (von Storch et al. 1993), artificial neural networks (ANN) which are skin to nonlinear regression (Crane and Hewitson 1998), and multiple regression (Murphy 1999).

For this particular study, a type of regression model was used which is SDSM. SDSM is widely applied in many regions of the world over a range of different climatic condition. It permits the spatial downscaling of *daily* predictor–predictand relationships using multiple linear regression techniques. The *predictor* variables provide daily information concerning the large-scale state of the atmosphere, while the *predictand* describes conditions at the site scale (CCIS 2008). SDSM is a decision support tool that facilitates the assessment of regional impacts of global warming by allowing the process of spatial scale reduction of data provided by large-scale GCMs. It is best described as a hybrid of the stochastic weather generator and regression-based methods. This is because large-scale circulation patterns and atmospheric moisture variables are used to linearly condition local-scale weather generator parameters (e.g., precipitation occurrence and intensity) (Wilby et al. 2002).

Users are allowed to simulate, through combinations of regressions and weather generators, sequences of daily climatic data for present and future periods by extracting statistical parameters from observed data series (Gagnon et al. 2005). The stochastic component of SDSM permits the generation of 100 simulations. The SDSM software reduces the task of statistically downscaling daily weather series into seven discrete steps (Wilby and Dawson 2007): (1) quality control and data transformation; (2) screening of predictor variables; (3) model calibration; (4) weather generation (using observed predictors); (5) statistical analyses; (6) graphing model output; and (7) scenario generation (using climate model predictors).

HadCM3 was downscaled using SDSM by considering the IPCC climatological baseline (1961–1990). It was calibrated (1991–1980) and validated (1981–1990) using observed historical climate variables.

29.3.4 SWAT Stream Flow Modeling Approach

SWAT is a river basin, or watershed, scale model developed by Dr. Jeff Arnold for the United States Department of Agriculture (USDA) Agricultural Research Service (ARS). It is physically based, conceptual and computationally efficient model that operates on a daily time step at basin scale (Arnold et al. 2000). It is designed to predict impact of land management practices on water, sediment, and agricultural chemical yields in large complex watersheds with varying soils, land use, and management conditions over long periods of time (Neitsch et al. 2005). There are benefits of SWAT model such as watersheds with no monitoring data (e.g., stream gauge data) can be modeled. The relative impact of alternative input data (e.g., changes in management practices, climate, and vegetation) on water quality or other variables of interest can be quantified (Neitsch et al. 2005).

SWAT uses a two-level disaggregation scheme. A preliminary sub-basin identification is carried out based on topographic criteria, followed by further discretization using land use and soil-type considerations. Areas with the same soil type and land use form a hydrological response unit (HRU), a basic computational unit assumed to be homogeneous in hydrological response to land cover change.

SWAT model has been applied for various catchment areas ranging from 0.015 km^2 (Mapfumo et al. 2003) to as large as 491,700 km^2 (Arnold et al. 2000).

Mostly water enters in the SWAT watershed system in form of precipitation. Flows and water quality parameters are routed in the model on the basis of HRU to each sub-basin and subsequently to the watershed outlet. The model reflects difference in evapotranspiration for various land use and soil type in the subdivision of watersheds. The runoff was predicted separately from each HRU and routed to obtain the total yield for the watershed. Hence, increase the accuracy and gives a better physical description of water balance. In this research, SWAT Soil Conservation Service Curve Number (SWAT-CN) method was used. Water balance is the key for the simulation of hydrology within a watershed. SWAT uses two phase for the simulation of hydrology, land phase, and routing phase.

The application of SWAT in predicting stream flow and sediment as well as evaluation of the impact of land use and climate change on the hydrology of watersheds has been documented by various studies (Dessu and Melesse 2012, 2013; Dessu et al. 2014; Wang et al. 2006, 2008a, b, c; Wang and Melesse 2005, 2006; Behulu et al. 2014, 2013; Setegn et al. 2009a, b, 2010, 2011; Setegn and Melesse 2014; Mango et al. 2011a, b; Getachew and Melesse 2012; Anwer et al. 2014; Grey et al. 2013).

29.3.4.1 Land Phase of Surface Runoff

The land phase simulation calculates the amount of sediment, water pesticide, and nutrient loading into the main channel in each basin. The land phase of the hydrological cycle is simulated by SWAT based on the water balance equation (Eq. 29.1).

$$\mathrm{SW_t} = \mathrm{SW_0} + \sum_{i=1}^{t} \left(R_{\mathrm{day}} - Q_{\mathrm{surf}} - E_{\mathrm{a}} - w_{\mathrm{seep}} - Q_{gw}\right) \tag{29.1}$$

where SW_t is the final soil water content (mm), SW_0 is the initial soil water content on day i (mm), t is the time (days), R_{day} is the amount of precipitation on day i (mm), Q_{surf} is the amount of surface runoff on day i (mm), E_{a} is the amount of evapotranspiration on day i (mm), w_{seep} is the amount of water entering the vadose zone from the soil profile on day i (mm), and Q_{gw} is the amount of return flow on day i (mm) (Neitsch et al. 2005).

ArcSWAT uses the concept of infiltration excess runoff. It assumes the runoff occurs whenever the rainfall intensity is greater than the rate of infiltration (Taffese 2012). This process is very important in areas where significant soil crusting and/or surface sealing occur during storm events, in irrigated fields, in urban areas, and more generally during very high rainfall intensity storms. To estimate surface runoff, SWAT uses two methods based on the above assumption: the Soil Conservation Service Curve Number (SCS-CN) method, SCS 1972, and Green and

Ampt infiltration method, Green and Ampt 1911. For this particular research, SCS-CN was used. This is because of the absence of sufficient hydrological and meteorological data collected at sub-daily scale.

In this method, land use and soil properties are lumped into a single parameter (White 2009). SWAT also uses Natural Resource Conservation Service (NRCS) soil classification based on the infiltration properties of the soil (Neitsch et al. 2005) into four groups (A, B, C, and D) having high, moderate, low, and very low infiltration rate, respectively. In the classification, a soil group has similar runoff potential under similar storm and cover condition. To determine CN, the model then defines antecedent moisture condition based on the Curve Number–Antecedent Moisture Condition (CN–AMC) (USDA–NRCS, 2004) based on the soil moisture content calculated by the model (Neitsch et al. 2005). The retention parameter (S) is then determined using the daily CN value (Eq. 29.2).

$$S = 25.49\left[\frac{1000}{\mathrm{CN}} - 10\right] \tag{29.2}$$

The direct runoff is determined by integrating the above-mentioned empirical model with SCS runoff equation (29.3).

$$Q_{\mathrm{surf}} = \frac{\left[P_{\mathrm{day}} - I_{\mathrm{a}}\right]^2}{P_{\mathrm{days}} - I_{\mathrm{a}} + S} \tag{29.3}$$

where Q_{surf} is the surface runoff or rainfall excess (mm), P_{day} is precipitation depth for the day (mm), S is the retention parameter (mm), and I_{a} is the initial abstraction which usually approximated as $0.2S$.

29.3.4.2 Routing Phase of Surface Runoff

The routing phase of SWAT defines the movement of water, nutrients, sediment, and pesticides through the channel network of the watershed into the outlet. Variable storage routing or Muskingum River routing can be used as routing technique for flow, sediment, and contaminant loading.

In this study, flow was routed through stream network of the watershed from upland areas to the main channel by variable storage routing which is a process analogous to HYMO (Williams and Hann 1973). Continuity equation is the concept behind storage routing (Eq. 29.4).

For a given reach,

$$\Delta V_{\mathrm{stored}} = V_{\mathrm{in}} - V_{\mathrm{out}} \tag{29.4}$$

where V_{in} is the volume of inflow during the time step (m^3), and V_{out} is the volume of outflow during the time step (m^3), and ΔV_{stored} is the change in volume of storage during the time step (m^3). The calculation can be further specified (Eq. 29.5).

$$V_{stored,2} - V_{stored,1} = \frac{\Delta t}{2} * \left\{ \left[q_{in,1} + q_{in,2} \right] - \left[q_{out,1} + q_{out,2} \right] \right\} \tag{29.5}$$

where $q_{in,1}$ is the inflow rate at the beginning of time step (m^3/s), $q_{in,2}$ is inflow rate at the end of time step (m^3/s), $q_{out,1}$ is the outflow rate at the beginning of time step (m^3/s), $q_{out,2}$ is the outflow rate at the end of time step (m^3/s), Δt is the length of the time step (s), $V_{stored,2}$ is the storage volume at the end of time step (m^3), and $V_{stored,1}$ is the storage volume at the beginning of time step (m^3). The volume of water in the channel is divided by the outflow rate to compute the travel time.

$$TT = \frac{V_{stored}}{q_{out}} = \frac{V_{stored,1}}{q_{out,1}} = \frac{V_{stored,2}}{q_{out,2}} \tag{29.6}$$

where TT is the travel time, V_{stored} is storage volume (m^3), and q_{out} is the flow rate.

29.4 SWAT Model Setup

29.4.1 Watershed Delineation

Watershed delineator tool in ArcSWAT allows the user to delineate the watershed and sub-basins using DEM. Flow direction and accumulation are the concepts behind to define the stream network of the DEM in SWAT. The monitoring point is added manually and the numbers of sub-basin are adjusted accordingly. Finally, the catchment area is delineated to be 1654 km^2 for Upper Gilgel Abay catchment, and 19 sub-basins are formed for the whole catchment.

29.4.2 HRU Analysis

HRU analysis helps to load land use map and soil map and also incorporates classification of HRU into different slope classes. The land use map as well as soil map was overlapped 100 % with the delineated watershed and 74 HRUs were formed for the study area. Spatial inputs of slope, soil, and land use were used to define the catchment. Ninety meter resolution of SRTM DEM topographic data was used for slope classification. Extracted topographic map with the catchment boundary shows an elevation range from 1890 to 3514 m amsl.

29.4.3 Weather Data Definition

Available meteorological records (1994–2008) (i.e., precipitation, minimum and maximum temperatures, relative humidity, and wind speed) and location of meteorological station are prepared based on ArcSWAT 2009 input format and integrated with the model using weather data input wizards. Bahir Dar meteorological station data were used as weather generator for this study.

29.4.4 Sensitivity Analysis

Sensitivity analysis is a simple technique for assessing the effect of uncertainty on the system performance. It is also a measure of the effect of change of one parameter on another. Sensitivity analysis was undertaken by using a built-in tool in SWAT-CUP that uses the global sensitivity design method. SWAT-CUP uses t-test to rank the most sensitive parameter that corresponds to greater change in output response. A t-test result provides a table of output having columns of t-stat and p-values. A t-stat provides a measure of sensitivity (larger in absolute values are more sensitive) and p-values determined the significance of the sensitivity. A value close to zero has more significance. To improve simulation result and thus understand the behavior of hydrological system in Upper Gilgel Abay catchment, sensitivity analyses were conducted using the 26 flow parameters.

29.4.5 Calibration and Validation

Model calibration involves modification of input parameters and comparison of predicted output with observed values until a defined objective function is achieved (James and Burges 1982). Parameters identified in sensitivity analysis that influence the simulation result significantly were used to calibrate the model. For the study reported in this chapter, the measured stream flow data of Gilgel Abay River at Merawi were used to calibrate the model by SWAT-CUP using Sequential Uncertainty Fitting version-2 (SUFI-2) algorithm (Abbaspour 2011) from a period of 1996–2004 by skipping two-year data as a warm-up period. The rest of the data were used for validation.

The calibration process using SUFI-2 algorithm was selected for this research from other two algorithms (ParaSol and GLUE) for its best final fitted parameters for Gilgel Abay, Gumara, Ribb, and Megech catchments (Setegn et al. 2009a). In SUFI-2, parameter uncertainty accounts for all sources of uncertainties such as uncertainty in driving variables (e.g., rainfall), conceptual model, parameters, and measured data.

29.5 Hydrological Models in Climate Change

Operational hydrologists use hydrological models to seek solutions to water resource problems, such as flood protection, frequency and duration of extreme hydrological events (floods, droughts), irrigation, spillway design, hydropower evaluation, or water supply design, under different climatic conditions at one location or at different locations (Xu 1999).

The advantages of hydrological models in climate change impact studies include the following: (1) models tested for different climatic/physiographic conditions, as well as models structured for use at various spatial scales and dominant process representations, are readily available; (2) GCM-derived climate perturbations (at different levels of downscaling) can be used as model input; (3) hence, a variety of responses to climate change scenarios can be modeled, including transient ones; and (4) the models can convert climate change output to relevant water resources variables related, for example, to reservoir operation, groundwater recharge, or irrigation demand (Schulze 1997).

The general procedure for estimating the impacts of hypothetical climate change on hydrological behavior is as follows (Xu 1999): (1) Determine the parameters of a hydrological model in the study catchment using current climatic inputs and observed river flows for model validation; (2) perturb the historical time series of climatic data according to some climate change scenarios; (3) simulate the hydrological characteristics of the catchment under the perturbed climate using the calibrated hydrological model; (4) compare the model simulations of the current and possible future hydrological characteristics.

Climate simulations from the GCMs could be used directly to drive hydrological models, which in turn could be used to evaluate the hydrological and water resource effects of climate change. Accordingly, methods of simple alteration of the present conditions are widely used by hydrologists. In the simple alteration method, the generation of climate scenarios consists of two steps (Xu 1999). First, estimate average annual changes in precipitation and temperature using either GCM results or historical measurements of change, or personal estimates (typically, $\Delta T = +1$, +2 and +4 °C and $\Delta P = 0$, ±10 %, ±20 %). Second, adjust the historic temperature series by adding ΔT, and for precipitation, by multiplying the values by $(1 + \Delta P/100)$.

29.6 Results and Discussion

29.6.1 *Selected Predictor Variables*

The best correlated predictor variables selected for precipitation and maximum and minimum temperatures with the corresponding month which have a strong correlation between predictands and each predictor are listed in Table 29.1.

Table 29.1 Selected large-scale predictor variables for the predictands of Bahir Dar station

Predictand	Predictor	Description	Month correlated
Precipitation	ncepr500	Relative humidity at 500 hpa	Jan, Feb, Apr, Aug, Nov, Dec
	Nceprhum	Near-surface relative humidity	Jun, Sep, Oct
Maximum temperature	ncepp5_u	500 hPa zonal velocity	Feb, Apr
	ncepp500	500 hpa geopotential height	Jan, Oct, Nov, Dec
	Nceptemp	Mean temperature at 2 m	Mar, May, Jun, Jul, Aug, Sep
Minimum temperature	ncepp500	500 hpa geopotential height	Feb, Mar, Apr, May, Jun, Jul, Aug
	Ncepshum	Surface-specific humidity	Sep, Oct, Nov, Dec
	Nceptemp	Mean temperature at 2 m	Jan

Predictand and predictor have good correlations mean the predictor has best performance to downscale the global climate variables to local-scale climate variable compared to others. For instance, relative humidity at 500 hpa and near surface had good performance to downscale precipitation rather than other predictors. Relative humidity at 500 hpa was very good predictor for the months of January, February, April, August, November, and December, whereas precipitation for months of June, September, and October was better downscaled with near-surface relative humidity.

29.6.2 Baseline Climate Analysis

Baseline scenario analysis was performed for Bahir Dar station within 30-year period from 1961 to 1990. The HadCM3 was downscaled for daily base period for two emission scenarios (A2 and B2), and some of the statistical properties of the downscaled data were compared with daily observed data. One of the criteria commonly used in evaluating the performance of any useful downscaling is whether the historical or observed condition can be replicated or not.

The downscaled baseline daily temperatures show good agreement with observed data. However, due to the conditional nature of daily precipitation, downscaled values have less concurrence with observed daily data. In conditional models, there is an intermediate process between regional forcing and local weather (e.g., local precipitation amounts depend on wet–dry/dry–day occurrence, which in turn depend on regional-scale predictors such as humidity and atmospheric pressure) (Wilby et al. 2004). Additionally, complicated nature of precipitation processes and its distribution in space and time is the other reason for the disagreement. Climate model simulation of precipitation has improved over time, but is still problematic (Bader et al. 2008) and has a larger degree of uncertainty than those for

Table 29.2 Downscaled daily precipitation, maximum and minimum temperature efficiency (R^2) relative to observed data

	R^2	
	Calibration (1961–1980)	Validation (1981–1990)
Precipitation	0.24	0.25
Maximum temperature	0.66	0.64
Minimum temperature	0.57	0.56

temperature (Thorpe 2005). This is because rainfall is highly variable in space, and so the relatively coarse spatial resolution of the current generation of climate models is not adequate to fully capture that variability. Coefficient of determination (R^2) between daily observed versus simulated (downscaled) clearly shows the difference between unconditional and conditional models for both calibration and validation (Table 29.2).

29.6.3 Future Climate Projection

Checking the efficiency of the downscaling model which can replicate the observed statistical properties for future scenarios or not helps to project daily future climate variables for the next century using the HadCM3 (A2 and B2) GCM. The projection generates 20 ensembles of daily temperature and 100 ensembles of daily precipitation variables. These ensembles are averaged out in order to consider the characteristics of all those ensembles.

The analysis was done for three 30 years of data ranges based on the recommendation of the World Meteorological Organization (WMO) for the 2020s (2011–2040), 2050s (2041–2070) and 2080s (2071–2099). The generated scenarios were considered individually for each baseline predictand as below.

29.6.3.1 Precipitation

The rainfall expected to experience a mean annul increase by 2.21, 2.23, and 1.89 % for A2 scenario at 2020s, 2050s, and 2080, respectively. This mean annual increase was repeated by B2 scenario with 2.06, 1.85, and 0.36 % at 2020s, 2050s, and 2080s, respectively. These values show that increasing in rainfall is not uniform; instead, it differs from time period to time period.

A rainfall projection of Bega (October–January) shows increasing for the two emission scenarios except 2080s of B2 scenario for the month of January whereas Belg (February–May) projection shows a increasing mean monthly rainfall for the first two months (February and March) and decreasing for the last two months (April and May) of the season for A2 scenario. In case of B2 scenario with this Belg season, rainfall is increasing on May at 2020s. The Kiremt (June–September) season of the two months (June and July) shares the rainfall decreasing of April and May at 2080s. Except September (increasing), Kiremt season more or less has nearly constant rainfall (Fig. 29.6a, b).

The projected mean monthly rainfall of this study has a similar pattern to that of the work of Abdo et al. (2009) and (deBoer 2007) using HadCM3 and ECHAM5/MPI-OM global climate models in the same catchment. These works were done on Gilgel Abay catchment and northern Ethiopian highlands. Both of the studies agreed with the mean monthly rainfall decrease in May, June, and July and increase in September, October, and November compared to the baseline period. Similarly on IPCC third assessment report of McCarthy et al. (2001), rainfall is predicted to increase in December–February and decrease June–August in parts of East Africa under intermediate warming scenarios. This IPCC report strengths this research output on increasing mean monthly rainfall from December to February and a little bit decreasing from June to August for both A2 and B2 scenarios.

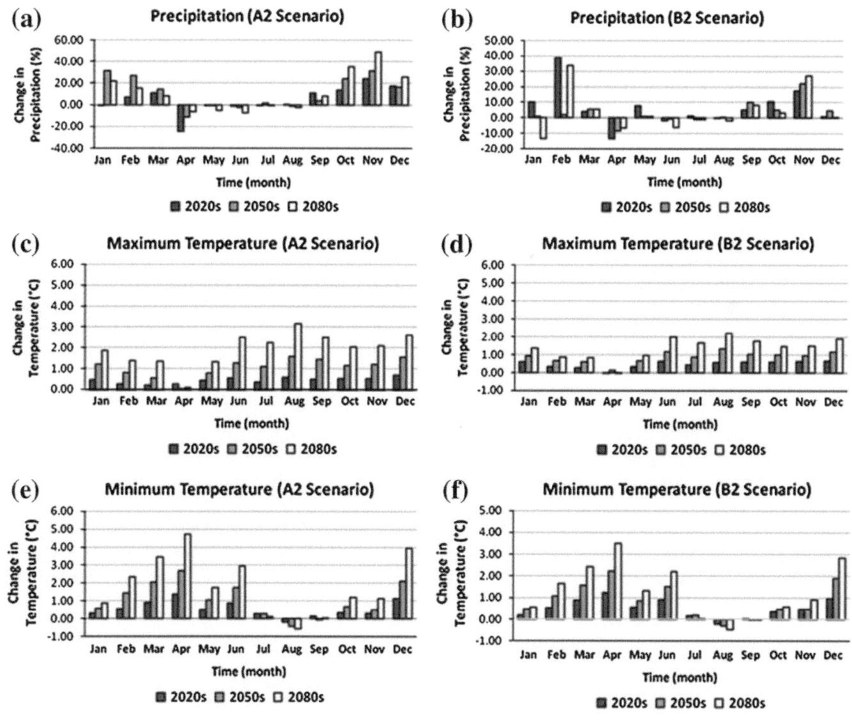

Fig. 29.6 Future changes in mean monthly precipitation for **a** A2 and **b** B2 scenarios; maximum temperature for **c** A2 and **d** B2 scenarios; minimum temperature for **e** A2 and **f** B2 scenarios from the baseline period

29.6.3.2 Maximum Temperature

The projected mean annual maximum temperature shows increasing trend for all time periods by 0.43, 1.05, and 1.92 °C for A2 scenario for 2020s, 2050s, and 2080, respectively. B2 scenario also shows an increase of mean annual maximum temperature with 0.47, 0.87, and 1.38 °C for 2020s, 2050s and 2080s, respectively. As compared to B2 scenario, A2 scenario has higher increasing trend. The increase includes all months largely for all time periods except April (Fig. 29.6c, d).

29.6.3.3 Minimum Temperature

The projected minimum temperature shows an increasing trend in all time periods. In this case, both the A2 and B2 emission scenarios predict the future minimum temperature in similar manner. For A2 scenario, mean annual minimum temperature increases by 0.55, 1.06, and 1.83 °C and for B2 scenario 0.50, 0.87, and 1.29 °C for 2020s, 2050s, and 2080s, respectively. Mean monthly variation of minimum temperature is higher than maximum temperature. For both A2 and B2 emission scenarios, minimum temperature will increase from October to June. Changes in July, August, and September are smaller with August showing a decreasing trend (Fig. 29.6e, f).

Generally, the projected minimum and maximum temperatures in all time periods is within the range projected by IPCC which reported average temperature will rise by 1.4–5.8 °C toward the end of this century. In relation to this, one can understand and link the result of maximum and minimum temperature predictions to the IPCC emission scenario story lines that increment for A2 scenario is greater than B2 scenario because A2 scenario represents a medium–high scenario which produces more carbon dioxide concentration than the B2 scenario which represent a medium–low scenario. The ranges of projected changes of precipitation and maximum and minimum temperatures from the present condition in Kiremt/wet season are shown in Table 29.3.

Table 29.3 Ranges of projected changes of precipitation, maximum and minimum temperature from the present condition in wet season

Scenarios	Time period	Precipitation (%)	Maximum temperature (°C)	Minimum temperature (°C)
A2	2020s	−1.5 to 10.8	0.34 to 0.57	−0.2 to 0.87
	2050s	−2.3 to 3.9	1.1 to 1.59	−0.41 to 2.74
	2080s	−7.1 to 8.0	2.23 to 3.15	−0.54 to 2.94
B2	2020s	−1.8 to 5.0	0.43 to 0.63	−0.23 to 0.9
	2050s	−0.8 to 10.1	0.86 to 1.34	−0.31 to 1.53
	2080s	−5.9 to 8.1	1.67 to 2.21	−0.47 to 2.21

29.7 Hydrological Model Calibration and Validation

Impact assessments on stream flow and sediment yield were done by soil and water assessment tool (SWAT) hydrological model. Results of mean annual changes of precipitation and temperature (maximum and minimum) obtained from this section was applied to quantify these impacts.

As mentioned in the methodology part, flow sensitivity analysis has been carried out for 26 hydrological parameters using SWAT-CUP global sensitivity analysis. The most sensitive parameters considered for calibration were Manning's "*n*" value for the mean channel (CH_N2), initial SCS runoff curve number for moisture condition II (CN2), surface runoff lag coefficient (SURLAG), base flow alpha factor (ALPHA_BF), effective hydraulic conductivity in main channel alluvium (CH_K2), available water capacity of the soil layer (SOL_AWC), maximum canopy storage (CANMX), groundwater "revap" coefficient (GW_REVAP), and soil evaporation compensation factor (ESCO). Result of sensitivity analysis was used to conduct the calibration of SWAT. Upper Gilgel Abay stream flow gauged near Merawi town was used to calibrate the hydrological model for a period of January 1, 1996, to December 31, 2004, using SWAT-CUP SUFI-2 algorithm as described in the methodology section. The calibration was performed with the nine sensitive parameters of stream flow. After several iterations, fitted values for the nine parameters were gained (Table 29.4).

SWAT-CUP in parallel gives the best simulation of stream flow considering the above-fitted parameters. Performance of the best simulation of stream flow result which used these fitted parameter values for calibration and validation is shown in Table 29.5 and also in Fig. 29.7 for scatter plot of stream flow. Santhi et al. (2001) stated that Nash–Sutcliff efficiency (NSE) values greater than or equal to 0.50 are considered adequate for SWAT model application. Hence, it is observed that SWAT exhibited strong performance in representing the hydrological conditions of the catchment.

It can be seen from the flow hydrographs (Fig. 29.8a, b) that the simulated flows well matched the observed flows except for peak values in the calibration period and low values in the validation period for both daily and monthly time steps.

Table 29.4 Calibrated fitted values of flow-sensitive parameters

Rank	Parameter	Fitted value
1	CH_N2.rte	0.0135
2	CN2.mgt	76.79
3	SURLAG.bsn	9.685 (days)
4	ALPHA_BF.gw	0.77 (days)
5	CH_K2.rte	129.375 (mm/h)
6	SOL_AWC.sol	75.26 (mm)
7	CANMX.hru	2.345 (mm)
8	GW_REVAP.gw	−0.03
9	ESCO.hru	0.46

Table 29.5 Calibrated model simulation performance

Criteria	Calibration (1996–2004)	Validation (2005–2008)
Coefficient of determination (R^2)	0.8	0.76
Nash–Sutcliffe efficiency (NSE)	0.77	0.75
RMSE-observations standard deviation ratio (RSR)	0.48	0.5

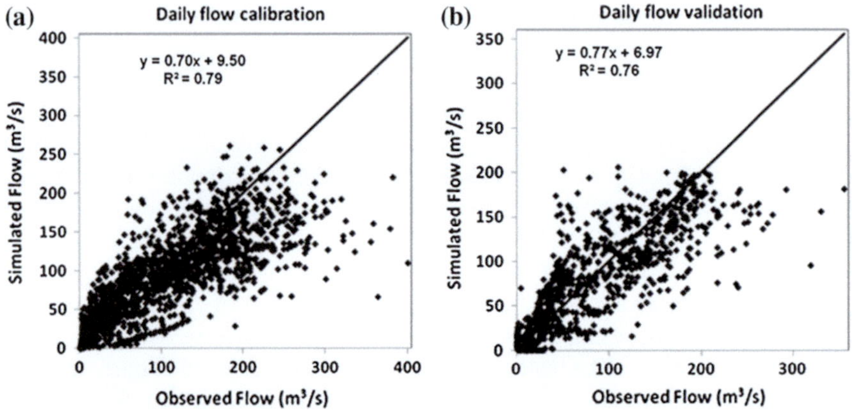

Fig. 29.7 Scatter plot of simulated versus observed flow daily **a** calibration **b** validation

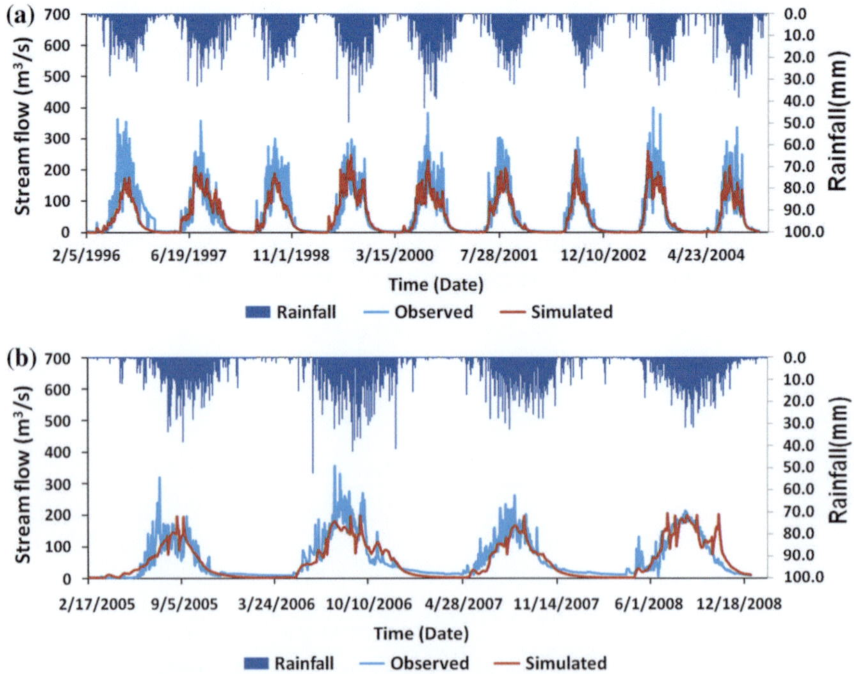

Fig. 29.8 Observed and simulated daily stream flow in comparison with areal rainfall for Upper Gilgel Abay catchment for the **a** calibration and **b** validation periods

Setegn et al. (2009b) showed that on an annual basis, 59 and 54 % of stream flow from the catchment was base flow for measured and simulated flows, respectively. In this work, it was found that 52.2 and 56.9 % of stream flow from the catchment was base flow for measured and simulated flow, respectively.

29.8 Impact of Climate Change on Stream Flow

One of the main objectives of this study was assessing the impact of climate change on stream flow over the study area. Therefore, this section of result and discussion is one of the ultimate goals of this study. For this purpose, the changes of down-scaled climate variables (absolute difference for temperature and percentage difference for precipitation) from the baseline climate were used to get 2020s, 2050s, and 2080s (each has 30 years daily climatic data including the baseline) time series data. SWAT simulation was run four times (for the baseline, 2020s, 2050s, and 2080s) by keeping constantly calibrated soil, crop, and slope parameters to quantify climate change impact only. Simulation results of stream flow for the three future time periods, 2020s, 2050s, and 2080s, were compared with the baseline period simulation.

Mean annual stream flow may increase by 7.1, 9.7, and 10.1 % for 2020s, 2050s, and 2080s, respectively, from the baseline stream flow for A2 scenario whereas for B2 scenario, the model shows an increase by 6.8, 7.9, and 6.4 % for 2020s, 2040s, and 2080s, respectively. The study area receives most of its rainfall in the wet season (June to September). Future ranges of changes of stream flow for this season is important to understand hydrological impact of climate change (Table 29.6).

Mean monthly change of stream flow clearly shows the season which may be impacted highly with climate change. The high-flow months from July to November may be impacted highly for both A2 and B2 scenarios similarly (Fig. 29.9). For instance, August stream flow is projected to increase by 4.3, 5.9, and 4.8 % on 2020s, 2050s, and 2080s time periods, respectively, for A2 scenario and by 4.1, 4.8, and 1.9 % for 2020s, 2050s, and 2080s, respectively. Percentage change of low-flow season months is higher than that of high-flow season but the volumetric changes as compared to the high-flow season is very low.

Table 29.6 Ranges of projected changes of stream flow from the present condition in the wet season

Scenarios	Time period	Change ranges of stream flow (%)
A2	2020s	4.33–8.10
	2050s	5.90–9.62
	2080s	4.80–15.07
B2	2020s	4.09–7.67
	2050s	4.76–7.60
	2080s	1.921–8.34

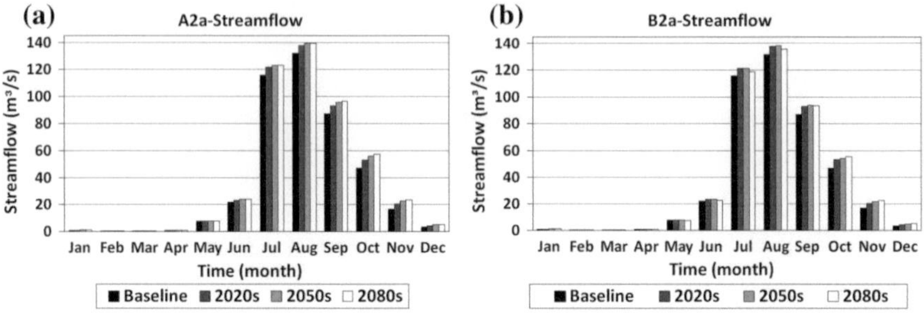

Fig. 29.9 Mean monthly flow for **a** A2 and **b** B2 scenarios

The result of this study on impact of climate change on stream flow was comparable to that of Mekonnen and Tadele (2012) who analyzed the combined effects of land use and climate change on soil erosion and stream flow in the Upper Gilgel Abay River catchment using SWAT hydrological model and European Centre Hamburg Model (ECHAM5, A1B emission scenario). The result showed that the combined effect of trend based on the land use and climate change scenario (2011–2025) will increase the average annual stream flow by 5.2 %.

29.9 Uncertainties on Studying Climate Change and Its Impact

Uncertainties in projected changes in the hydrological system arise from the following: (1) internal variability of the climate system, (2) uncertainty in future greenhouse gas and aerosol emissions, (3) the translation of these emissions into climate change by GCMs, (4) hydrological model uncertainty, (5) uncertainty from insufficient field data at all scales (Xu et al. 2005), and uncertainty of downscaling techniques (Bates et al. 2008). In general, the sources of uncertainties of climate scenarios are multiple. The climate system itself is too complex to be represented in a numerical model and contains a number of assumptions and parameterizations that each climate modeling centers approach differently. Uncertainties in climate scenarios and GCM outputs may be large (Abdo et al. 2009).

The uncertainties related to SDSM model in this work could be as follows: (1) the relationship between the predictor and predictand is achieved by only considering the data statistical condition, i.e., the model does not take into consideration the physical nature of the catchments (major drawback), (2) it requires high-quality data for model calibration; and (3) the model is highly sensitive to the choice of predictor variables and empirical transfer scheme. Parameter uncertainties and structural deficiencies in the hydrological models that are used for impact assessments are others sources of uncertainties. This is why different hydrological models may give different stream flow results for a given input (Abdo et al. 2009).

29.10 Conclusions

The use of GCMs for climate change impact assessment on stream flow requires assessment of hydrological model performance results with downscaling of the parameters. Hydrological model performance checking for a specific basin or catchment is a basic task before using GCMs for impact assessment of climate change on the hydrology and water resources. Stream flow simulation results of SWAT model gave a satisfactory performance with three model efficiency criteria after calibrating the model with observed stream flow data gauged near Merawi town.

Mean annual absolute and percentage changes of climate variables from the baseline period were used to simulate future projections of stream flow. Stream flow projections for future time periods showed that mean annual stream flow may increase by 7.1, 9.7, and 10.1 % at 2020s, 2050s, and 2080s, respectively, from the baseline period for A2 scenario, whereas for B2 scenario, it will be expected to increase by 6.8, 7.9, and 6.4 % for 2020s, 2040s, and 2080s, respectively. The variation in change of stream flow between months is high. The high-flow months from July to November may be impacted highly for both A2 and B2 scenarios similarly.

Acknowledgment We would like to acknowledge Blue Nile Water Institute (BNWI) and Tana sub-basin Organization (TaSBO) for their financial support. We also thankful to Ministry of Water and Energy (MoWE) and National Meteorology Agency (NMA)—Bahir Dar Branch Directorate for their help by providing necessary data for the study.

References

Abbaspour KC (2011) SWAT-CUP4: SWAT calibration and uncertainty programs—A user manual. p21

Abdo KS, Fiseha BM, Rientjes THM, Gieske ASM, Haile AT (2009) Assessment of climate change impacts on the hydrology of Gilgel Abay catchment in Lake Tana basin, Ethiopia. Hydrol Process 23:3661–3669

Abtew W, Melesse AM (2014a) Nile River Basin hydology. In: Melesse AM, Abtew W, Setegn S (eds) Nile River Basin: ecohydrological challenges, climate change and hydropolitics. Springer, Berlin, pp 7–22

Abtew W, Melesse AM (2014b). Climate teleconnections and water management. In: Nile River Basin. Springer International Publishing, Berlin, pp. 685-705

Abtew W, Melesse AM (2014c) Transboundary Rivers and the Nile. In: Nile River Basin. Springer International Publishing, Berlin, pp 565–579

Abtew W, Melesse AM, Desalegn T (2009a) Spatial, inter and intra-annual variability of the Blue Nile River Basin rainfall. Hydrol Process 23(21):3075–3082

Abtew W, Melesse AM, Desalegn T (2009b) El Niño Southern Oscillation link to the Blue Nile River Basin hydrology. Hydrol Process 23(26):3653–3660 (Special issue: Nile Hydrology)

AMCEN (2011) Addressing climate change challenges in Africa: A practical guide towards sustainable development, p3

Arnold JG, Muttiah RS, Srinivasan R, Allen PM (2000) Regional estimation of base flow and groundwater recharge in Upper Mississippi River basin. J Hydrol 227:21–40

Anwar A, Melesse AM, Admasu S (2014) Climate change in upper Gilgel Abay River catchment, Blue Nile Basin Ethiopia. In: Melesse AM, Abtew W, Setegn S (eds) Nile River Basin: ecohydrological challenges, climate change and hydropolitics. Springer, Berlin, pp 363–388

Bader D, Covey C, Gutowski W, Held I, Kunkel K, Miller R, Tokmakian R, Zhang M (2008) Climate models: an assessment of strengths and limitations. US Department of Energy Publications, p 8

Bates B, Kundzewicz ZW, Wu S, Palutikof J (2008) Climate change and water. Intergovernmental Panel on Climate Change (IPCC)

Behulu F., Setegn S., Melesse A.M. and Fiori A., (2013) Hydrological analysis of the Upper Tiber Basin: A Watershed Modeling Approach, Hydrological Processes, 27(16), 2339–2351

Behulu F, Setegn S, Melesse AM, Romano E, Fiori A (2014) Impact of climate change on the hydrology of Upper Tiber River Basin using bias corrected regional climate model. Water Resour Manag 1–17

Beyene T, Lettermaier DP, Kabat P (2010) Hydrologic impacts of climate change on the Nile River Basin: implications of the 2007 IPCC scenarios. Clim Change 100:433–461

Carter TR, Hulme M, Lal M (1999) Guidelines on the use of scenario data for climate impact and adaptation assessment v1

Carter TR, Hulme M, Lal M (2007) Guidelines on the use of scenario data for climate impact and adaptation assessment v2

CCIS (2008) Frequently asked questions. SDSM Background 1of 4

Chebud, Y., Melesse, A.M. (2013) Stage Level, Volume, and Time-frequency change information content of Lake Tana using Stochastic Approaches, Hydrological Processes, 27(10): 1475–1483 DOI: 10.1002/hyp.9291

Chebud YA, Melesse AM (2009a) Numerical modeling of the groundwater flow system of the Gumera Sub-Basin in Lake Tana Basin, Ethiopia. Hydrol Process 23(26):3694–3704 (Special issue: Nile Hydrology)

Chebud YA, Melesse AM (2009b) Modeling lake stage and water balance of Lake Tana, Ethiopia. Hydrol Process 23(25):3534–3544

Chen H, Xiang T, Zhou X, CY XU (2012) Impacts of climate change on the Qingjiang Watershed's runoff change trend in China. Stoch Env Res Risk Assess 26:847–858

Crane RG, Hewitson BC (1998) Doubled CO_2 precipitation changes for the susquehanna basin: down-scaling from the genesis general circulation model. Int J Climatol 18:65–76

Deboer B (2007) The impact of climate change on rainfall extremes over Northeast Africa. KNMI, Royal Netherlands Meteorological Institute, De Bilt, Netherlands p5

Dessu SB, Melesse AM (2012) Modeling the rainfall-runoff process of the Mara River Basin using SWAT. Hydrol Process 26(26):4038–4049

Dessu SB, Melesse AM (2013) Impact and uncertainties of climate change on the hydrology of the Mara River Basin. Hydrol Process 27(20):2973–2986

Dessu SB, Melesse AM, Bhat M, McClain M (2014) Assessment of water resources availability and demand in the Mara River Basin. CATENA 115:104–114

Eguavoen I (2009) The acquisition of water storage facilities in the Abay River Basin. University of Bonn, Ethiopia

Elshamy ME, Balirira R, Abdel-Gaffar E, Moges SA (2009) Investigating the climate sensitivity of different Nile sub-basins. In: 13th international water technology conference, IWTC 13 2009, Hurghada, Egypt

Gagnon SB, Singh B, Rousselle J, Roy L (2005) An application of the statistical downscaling model (SDSM) to simulate climatic data for streamflow modelling in Québec. Can Water Resour J 30:297–314

Gebrekristos ST (2008) Watershed modeling of Lake Tana basin using SWAT. MSc Thesis, ArbaMinch University, 45

Gebremariame ZH (2009) Assessment of climate change impact on the net basin supply of Lake Tana Water balance. ITC MSc thesis, ENSCHEDE, The Netherlands

Getachew HE, Melesse AM (2012) Impact of land use /land cover change on the hydrology of AngerebWatershed, Ethiopia. Int J Water Sci 1(4):1–7. doi:10.5772/56266

Giorgi F, Mearns LO (1991) Approaches to the simulation of regional climate change: a review. Rev Geophys 29:191–216

Grey OP, Webber Dale G, Setegn SG, Melesse AM (2013) Application of the soil and water assessment tool (SWAT Model) on a small tropical Island State (Great River Watershed, Jamaica) as a tool in integrated watershed and coastal zone management. Int J Trop Biol Conserv 62(3):293–305

Green WH, Ampt GA (1911) Studies on soil physics, 1, The flow of air and water through soils, J. Agric. Sci., 4(1), 1–24

Hewitson BC, Crane RG (1996) Climate downscaling: techniques and application. Climate Research, 7, 85–95

James LD, Burges SJ (1982) Selection, calibration, and testing of hydrologic models. Hydrol Model Small Watersheds, 437–472

Jenkins GS, Barron EJ (1997) Global climate model and coupled regional climate model simulations over the eastern United States: GENESIS and RegCM2 simulations. Global Planet Change 15:3–32

Kim U, Kaluarachchi JJ (2009) Climate change impacts on water resources in the Upper Blue Nile River Basin, Ethiopia1. JAWRA J Am Water Resour Assoc 45:1361–1378

Mapfumo E, Chanasyk DS, Baron VS (2003) Patterns and simulation of soil water under different grazing management systems in central Alberta. Can J Soil Sci 83:601–614

Mango L, Melesse AM, McClain ME, Gann D, Setegn SG (2011a) Land use and climate change impacts on the hydrology of the upper Mara River Basin, Kenya: results of a modeling study to support better resource management. Hydrol Earth Syst Sci 15:2245–2258. doi:10.5194/hess-15-2245-2011 (Special issue: Climate, weather and hydrology of East African Highlands)

Mango L, Melesse AM, McClain ME, Gann D, Setegn SG (2011b) Hydro-meteorology and water budget of Mara River basin, Kenya: a land use change scenarios analysis. In: Melesse A (ed) Nile River Basin: hydrology, climate and water use. Springer Science Publisher, Berlin, Chapter 2, pp 39–68. doi:10.1007/978-94-007-0689-7_2

McCarthy JJ, Canziani OF, Leary NA, Dokken DJ, White KS (2001) Climate change 2001: impacts, adaptation, and vulnerability: contribution of Working Group II to the third assessment report of the Intergovernmental Panel on Climate Change. Cambridge University Press, Cambridge

Mekonnnen K, Tadele K (2012) Analyzing the impact of land use and climate changes on soil erosion and stream flow in the Upper Gilgel Abbay Catchment, Ethiopia Ohrid, Republic of Macedonia, pp 1–13

Melesse AM (2011) Nile River Basin: hydrology, climate and water use. Springer Science & Business Media, Berlin

Melesse AM, Loukas AG, Senay G, Yitayew M (2009a) Climate change, land-cover dynamics and ecohydrology of the Nile River Basin. Hydrol Process 23(26):3651–3652

Melesse A, Abtew W, Desalegne T, Wang X (2009b) Low and high flow analysis and wavelet application for characterization of the Blue Nile River System. Hydrol Process 24(3):241–252

Melesse A, Abtew W, Setegn S, Dessalegne T (2011a) Hydrological variability and climate of the Upper Blue Nile River Basin. In: Melesse A (ed) Nile River Basin: hydrology, climate and water use. Springer Science Publisher, Berlin, Chapter 1, pp 3–37. doi:10.1007/978-94-007-0689-7_1

Melesse A, Abtew W, Setegn SG (2014) Nile River Basin: ecohydrological challenges, climate change and hydropolitics. Springer Science & Business Media, Berlin

Mohammed, H., Alamirew, T., Assen, M., Melesse, A.M 2015. Modeling of sediment yield in Maybar gauged watershed using SWAT, northeast Ethiopia, CATENA, 127, 191–205
Murphy J (1999) An evaluation of statistical and dynamical techniques for downscaling local climate. J Clim 12:2256–2284
Nakicenovic N, Alcamo J, Davis G, De Vries B, Fenhann J, Gaffin S, Gregory K, Grubler A, Jung TY, Kram T (2000) Special report on emissions scenarios: a special report of working group III of the Intergovernmental Panel on Climate Change. Pacific Northwest National Laboratory, Richland, WA (US), Environmental Molecular Sciences Laboratory (US)
Neitsch SL, Arnold JG, Kiniry JR, Williams JR, Williams JR, King KW (2005) Soil and water assessment tool: theoretical documentation, version 2005. Texas, USA
Parry ML (2007) Climate change 2007: impacts, adaptation and vulnerability: working group I contribution to the 4th assessment report of the IPCC. Cambridge University Press, Cambridge
Santhi C, Arnold JG, Williams JR, Dugas WA, Srinivasan R, Hauck LM (2001) Validation of SWAT model on a large RWER basin with point and nonpoint sources. JAWRA J Am Water Resour Assoc 37:1169–1188
Santoso H, Idinoba M, Imbach P (2008) Climate scenarios: what we need to know and how to generate them. CIFOR Working Paper
Schulze RE (1997) Impacts of global climate change in a hydrologically vulnerable region: challenges to South African hydrologists. Prog Phys Geogr 21:113–136
Setegn S, Melesse, AM (2014) Climate change impact on water resources and adaptation strategies in the Blue Nile River Basin, In: Melesse AM, Abtew W, Setegn S (eds) Nile River Basin: ecohydrological challenges, climate change and hydropolitics. Springer, Berlin, pp 389–420
Setegn SG, Srinivasan R, Dargahi B, Melesse AM (2009a) Spatial delineation of soil erosion prone areas: application of SWAT and MCE approaches in the Lake Tana Basin, Ethiopia. Hydrol Process 23(26):3738–3750 (Special issue: Nile Hydrology)
Setegn SG, Srinivasan R, Melesse AM, Dargahi B (2009b) SWAT model application and prediction uncertainty analysis in the Lake Tana Basin, Ethiopia. Hydrol Process 24(3):357–367
Setegn SG, Bijan Dargahi B, Srinivasan R, Melesse AM (2010) Modelling of sediment yield from Anjeni Gauged Watershed, Ethiopia Using SWAT. JAWRA 46(3):514–526
Setegn SG, Rayner D, Melesse AM, Dargahi B, Srinivasan R (2011) Impact of climate change on the hydroclimatology of Lake Tana Basin, Ethiopia. Water Resour Res 47:W04511
Soliman ESA, Sayed MAA, Jeuland M (2009) Impact assessment of future climate change for the Blue Nile basin using a RCM nested in a GCM. Nile Basin Water Eng Sci Mag 2:15–30
Solomon S (2007) Climate change 2007-the physical science basis: Working group I contribution to the fourth assessment report of the IPCC. Cambridge University Press, Cambridge
Tadege A (2001) Initial national communication of Ethiopia to the United Nations framework convention on climate change (UNFCCC) National Meteorological Services Agency. Addis Ababa, Ethiopia
Tafffese T (2012) Physically based rainfall: runoff modelling in the northern Ethiopian highlands: The case of Mizewa watershed. MSc Thesis, Bahir Dar University
Tarekegn D, Tadege A (2006) Assessing the impact of climate change on the water resources of the Lake Tana sub-basin using the WATBAL model. Discuss. Pap, 30
Taye MT, Ntegeka V, Ogiramoi NP, Williams P (2011) Assessment of climate change impact on hydrological extremes in two source regions of the Nile River Basin. Hydrol Earth Syst Sci 15:209–222
Thorpe AJ (2005) Climate change prediction: a challenging scientific problem. Institute of Physics
Von Storch H, Zorita E, Cubasch U (1993) Downscaling of global climate change estimates to regional scales: an application to Iberian rainfall in wintertime. J Clim 6:1161–1171
Wale A (2008) Hydrological balance of Lake Tana Upper Blue Nile Basin, Ethiopia. ITC Thesis 2008:159–180
Wang X, Melesse AM (2005) Evaluations of the SWAT model's snowmelt hydrology in a Northwestern Minnesota Watershed. Trans ASAE 48(4):1359–1376
Wang X, Melesse AM (2006) Effects of STATSGO and SSURGO as inputs on SWAT model's snowmelt simulation. J Am Water Resour Assoc 42(5):1217–1236

Wang X, Melesse AM, Yang W (2006) Influences of potential evapotranspiration estimation methods on SWAT's hydrologic simulation in a Northwestern Minnesota Watershed. Trans ASAE 49(6):1755–1771

Wang X, Shang S, Yang W, Melesse AM (2008a) Simulation of an agricultural watershed using an improved curve number method in SWAT. Trans Am Soc Agri Bio Eng 51(4):1323–1339

Wang X, Yang W, Melesse AM (2008b) Using hydrologic equivalent wetland concept within SWAT to estimate streamflow in watersheds with numerous wetlands. Trans Am Soc Agri Bio Eng 51(1):55–72

Wang X, Garza J, Whitney M, Melesse AM, Yang W (2008c) Prediction of sediment source areas within watersheds as affected by soil data resolution. In: Paul NF (ed) Environmental modelling: new research. Nova Science Publishers, Inc., Hauppauge, Ch. 7, pp 151–185. ISBN: 978-1-60692-034-3

White ER, Easton ZM, Fuka DR, Steenhuts TS (2009) SWAT-WB theoretical documentation. Soil and Water Lab, department of biological and Environmental engineering, Cornell University, Ithaca, NY

Wilby RL, Dawson CW (2007) Statistical Downscaling Model (SDSM), Version 4.2, A decision support tool for the assessment of regional climate change impacts. United Kingdom

Wilby RL, Dawson CW, Barrow EM (2002) SDSM—a decision support tool for the assessment of regional climate change impacts. Environ Model Softw 17:145–157

Wilby RL, Charles SP, Zorita E, Timbal B, Whetton P, Mearns LO (2004) Guidelines for use of climate scenarios developed from statistical downscaling methods. IPCC task group on data and scenario support for impacts and climate analysis

Williams JR, Hann RW (1973) HYMO: problem-oriented language for hydrologic modeling-User's manual. USDA. ARS-S-9, 45

Xu CY (1999) From GCMs to river flow: a review of downscaling methods and hydrologic modelling approaches. Prog Phys Geogr 23:229–249

Xu CY, Widen E, Halldin S (2005) Modelling hydrological consequences of climate change—progress and challenges. Adv Atmos Sci 22:789–797

Yitayew M, Melesse AM (2011) Critical water resources management issues in Nile River Basin. In: Melesse A (ed) Nile River Basin: hydrology, climate and water use. Springer Science Publisher, Berlin, Chapter 20, pp 401–416. doi:10.1007/978-94-007-0689-7_20'

Chapter 30
Climate Change Impact Assessment on Groundwater Recharge of the Upper Tiber Basin (Central Italy)

Fiseha Behulu, Assefa M. Melesse and Aldo Fiori

Abstract This chapter discusses the effects of climate change on groundwater recharge of the upper Tiber River basin (UTRB) in central Italy. The chapter also provides summaries and overview of climate change studies over the Italian territory. A calibrated and validated watershed model, soil and water assessment tool (SWAT), was forced by the climate model outputs obtained from three dynamically downscaled regional climate models (RCMs) to evaluate the recharge characteristics of the basin. During calibration, the watershed model was highly sensitivity to groundwater flow parameters. The climate change analysis from the three RCMs indicated that by the end of twenty-first century, rainfall will decrease up to 40 % in the dry period and there will be an increase in temperature that could reach as high as 3–5 °C. Moreover, the bias-corrected RCM outputs showed different results among models. Such variation of result calls for further investigation of model uncertainty. Groundwater recharge shows a decreasing trend as a response to changes in rainfall. However, as the timing of both precipitation and recharge is critical for future groundwater development in the basin, further analysis through different climate models, downscaling approach and groundwater modeling needs to be taken into account.

Keywords Tiber basin · Climate change · Regional climate models · SWAT · Groundwater recharge · Central Italy

---

F. Behulu (✉) · A. Fiori
Department of Civil Engineering, Roma Tre University, Rome, Italy
e-mail: fishbehulu@gmail.com

A. Fiori
e-mail: aldo.fiori@uniroma3.it

A.M. Melesse
Department of Earth and Environment, Florida International University, Modesto A. Maidique Campus, Mimi, FL 33199, USA
e-mail: melessea@fiu.edu

© Springer International Publishing Switzerland 2016
A.M. Melesse and W. Abtew (eds.), *Landscape Dynamics, Soils and Hydrological Processes in Varied Climates*, Springer Geography,
DOI 10.1007/978-3-319-18787-7_30

30.1 Introduction

In the last three decades, numerous studies have been conducted focusing on the investigation of regional climate change or variability impacts on water resources. However, these climate impact studies in the past focused more on surface water resources. Despite its importance for drinking water across the world and its vital role in maintaining ecological value of many areas, there have been few studies conducted in relation to climate change impact on groundwater. This is because of the visibility, accessibility, and more obvious recognition of climate effects on surface water than on groundwater. However, after the IPCC's fourth assessment report (AR4), there were numbers of studies emerging with due emphasis on aquifer recharge and groundwater storage using climate change predictions from general circulation models (GCMs) or regional climate models (RCMs). The studies conducted by Loaiciga (2003), Allen et al. (2004, 2010), Chen et al. (2002, 2004), Scibek and Allen (2006), Toews and Allen (2009), and Roosmalen et al. (2007, 2009) are among few of the comprehensive works related to climate change and groundwater. The recent draft report of IPCC's fifth assessment report (AR5) has also cited the attention given to groundwater resources. The review papers by Green et al. (2011), Taylor et al. (2013), and Kløve et al. (2014) are the most recent comprehensive review works related to climate change impact on groundwater resources.

Most studies used outputs from GCMs through statistical downscaling except the work of Roosmalen et al. (2007) who used outputs of RCM through bias correction in European region. Moreover, all the studies have used the estimation of recharge using surface hydrologic model and further applied simulation of transient or steady-state groundwater conditions. The work by Allen et al. (2010) used four GCMs (namely CGCM3.1, ECHAM5, PCM1, and CM2.1) to compare recharge simulation with the historical time period. They have shown that there are both relative increases and decreases, where by in the 2080s, the range of model prediction spans −10.5 to +23.2 % relative to historical recharge. Scibek and Allen (2006) developed a methodology for linking climate models and groundwater models to investigate future impacts of climate change on groundwater resources in an unconfined aquifer, situated near Grand Forks in south-central British Columbia, Canada. They found that the effect of spatial distribution of recharge on groundwater levels, compared to that of a single uniform recharge zone, is much larger than that of temporal variation in recharge. This was compared to a mean annual recharge. Similarly, Woldeamlak et al. (2007) modeled the effects of climate change on the groundwater systems in the GroteNete catchment in Belgium using a physically distributed water balance model and a finite difference groundwater model for different scenarios (wet, cold, and dry). Their result showed that the wet and dry scenarios are more representative for the study area.

The consideration of different scenarios has also an impact on future groundwater availability in a given region. Loaiciga et al. (2000) observed that the effect of climate change on a groundwater system in Texas resulted in a reduction of the aquifer's groundwater resources under climate scenarios with a doubling of atmospheric CO_2 concentration. They pointed out that CO_2 doubling is a common assumption in such impact studies. Under the same condition, Roosmalen et al. (2007) compared the effects of future climate change on groundwater recharge, storage, and discharge to streams for two regions in Denmark. They demonstrated the importance of using site-specific models that capture the physical characteristics of the area by applying the same climate change scenarios at two hydrologically and geologically different areas. Later, Roosmalen et al. (2009) investigated the sensitivity of groundwater system to different climate change scenarios with 2.2 and 3.2 °C increases in temperature for B2 and A2 scenarios, respectively, as compared to the base period (1961–1990). Their comparative assessment showed that groundwater systems tend to respond more slowly to variability in climatic conditions than do surface water systems. Due to this fact, assessments and models of groundwater resources are commonly based on long-term average climatic conditions, as annual average recharge, and potentially underestimate the importance of variations from the norm (Allen et al. 2010). Further, the IPCC (2007) report states that climate change affects groundwater recharge rates, renewable groundwater resources, and depths of groundwater tables. However, knowledge of current recharge and levels in both developed and developing countries is poor, and there has been still very little research on the future impact of climate change on groundwater, or groundwater–surface water interactions.

The impact of climate change on water resources variability, especially on river flows, soil moisture, evapotranspiration, and groundwater flow, has been studied by analyzing projected and downscaled climatic data and using hydrological models (Mango et al. 2011a, b; Behulu et al. 2014; Assefa et al. 2014; Melesse et al. 2009, 2011; Dessu and Melesse 2013; Grey et al. 2013).

30.2 Overview of Climate Change Studies in Italy

Italy lies at the center of the Mediterranean region, which has been identified as one of the most sensitive areas to GHG-induced global warming (Giorgi 2006; IPCC 2007). The Italian territory is expected to be susceptible to climate change. Like any other part of the world, this in turn will have considerable impacts on various sectors including water resources, agriculture, and tourism.

Despite its importance, studies related to projected impacts of climate change in Italy are limited. This is due to the complex and fine-scale variability in topography, coastlines, and vegetation cover. The north–south elongated shape of the country also experiences different climatic behavior. The northern most regions that consist the Alps are characterized in cold climate. The central part of the territory experiences temperate condition, and the southernmost part (Sardinia and Sicily) are known for their semiarid and hot climate. The major mountain systems, the Alpine chains in the north and the Apennines extending from north to south along the entire Italian Peninsula also modulate the climate conditions of the entire nation.

Hence, accurate characterization and its representation in climate models are reported to be very difficult (Coppola and Giorgi 2010).

Majority of the studies on climate change issues focused more on the precipitation and temperature characteristics as they determine the hydrologic cycle and other natural phenomena. These studies pointed out significant variations in the climate of Italy exit since early twentieth to the end of twenty-first century. Coppola and Giorgi (2010) have conducted a comprehensive assessment of precipitation and temperature projections over Italy using 19 recent GCMs from CMIP3 and 10 RCMs from PRUDENCE simulations. They indicated that both precipitation and temperature have seasonally varying signals. Precipitation is found to be substantially decreasing over the entire peninsula in the summer as much as −40 % and to the lesser extent in spring and fall seasons. Their result also indicates summer precipitation will tend to increase in the north, whereas it shows transitional signal over central Italy and decreases over southern Italy. Inter-annual variability of precipitation tends to increase in all seasons, but temperature variability increases only in the summer. Both minimum and maximum temperatures are expected to show warming up to several degrees in all the seasons, with maximum in summer and minimum in winter. Through analysis of different types of extremes, three main results for the Italian Peninsula have been reported (1) an increase in frequency, intensity, and duration of heat waves; (2) an increase in drought occurrence as measured by the maximum length of dry spells; and (3) an increase of intense precipitation events over northern Italy in winter and a decrease over central and southern Italy in summer (Beniston et al. 2007). Kjellstrom et al. (2007) analyzed daily maximum and minimum temperature extremes under warmer scenario conditions and found increase of extremes greater than the increase in mean temperature. Similarly, Vergni and Todisco (2011) used observed dataset in central Italy and found the rate of change in the minimum temperature greater than the maximum that will result in reduction of daily temperature range.

As the global- and regional-scale climate model-based assessments are not sufficient to understand the extent of change at local scale, various authors have explored trends of precipitation and temperature in Italy based on indices derived from observation of long time series data (Moonen et al. 2002; Brunetti et al. 2001, 2002, 2004, 2006; Di Matteo and Dragoni 2006; Colombo et al. 2007; Todisco and Vergni 2008; Fatichi et al. 2009; Matzneller et al. 2010; D'Agostino et al. 2010; Vergni and Todisco 2011; Romano and Preziosi 2012). Unlike precipitation, most of the studies agree in general increase in trends of temperature depending on the site and data analysis. However, some authors reported different results on the precipitation patterns of Italy as summarized by (Romano and Preziosi 2012). For example, temperatures in the annual series have a positive trend of 1 °C per century at national level with a systematic increase in winter droughts (Brunetti et al. 2006). Increasing temperature trend of 0.4 °C per 100 years for northern Italy and 0.7 °C per 100 years for southern Italy is reported (Brunetti et al. 2004). Colombo et al. (2007) divided the available ground observation stations into mountain, continental, and coastal areas and described a positive trend in temperature, mainly for the mountain stations (1980–2000). According to Brunetti et al. (2006), precipitation shows a decreasing tendency in the whole of Italy over the last two centuries. On yearly basis, a negative trend is evident for northern and southern Italy, with −47 mm per 100 years and −104 mm per 100 years, respectively (Brunetti et al. 2004). Colombo et al. (2007) found that stations in the mountains of Italy have recorded significant increase of precipitation events during autumn and winter, but for the rest of the Italian territory, precipitation is reduced during early springs.

The high spatiotemporal variability of climate over the entire peninsula has triggered site-specific studies to evaluate the local-scale climate variability on the hydrologic behavior. For example, the Tiber River Basin (TRB) Authority has reported that most severe climate change scenarios for central Italy where a 6–8 °C increase in temperature is forecasted by 2080. A decreasing trend in rainfall throughout the year is projected with most notably between October and April with a drop by as much as 50 %. These predictions are partially confirmed by measurements and summarized as follows: (i) From 1920 to 1938, there has been a modest increase in mean annual precipitation (from 914.2 to 923.3 mm) with an average increase of about 0.48 mm per year; (ii) From 1938 to 2003, there has been a decrease (from 923.3 to 806.9 mm) with an average decrease of 1.79 mm per year; and (iii) from 2003 onwards, the rate of decrease amounts to 3.65 mm per year. Based on the data collected between 1952 and 2007, the authority has reported that there is a consistent trend of gradual decrease in annual precipitation, mainly in winter decreasing up to 30 %, and rising temperature.

Romano and Preziosi (2012) analyzed daily time series data of rainfall over the period of 1920–2010 in the Tiber River Basin, central Italy. They analyzed the precipitation patterns through standardized indices and showed significant decrease in annual precipitation (−8 %) over the entire TRB with decrease in winter precipitation of about −16 %. They stated that such reduction is related to decrease in the number of rainy days. A 1.27 m^3 s^{-1} reduction in annual river discharge was also reported to be related to the reduction in precipitation (Romano et al. 2011). In the same basin, Brocca et al. (2011) showed preliminary investigation on the climate change effects on the flood frequency in selected sub-basins of the UTRB. They used HadCM3 and their result showed different responses to climate change for the selected sub-basins. The A2 emission scenario is more critical in short term (2020s) than the B2 scenario for which an increase in maximum discharge reaches up to 78 %. Another comprehensive work on the analysis of change in precipitation regime in central Italy was the work of Fatichi and Caporali (2009). They used 40 indices to evaluate change in precipitation pattern and trend detection. Contrary to other studies in the area, they presented that there is no evidence for non-stationary. They presumed that the complexity of the climate in central Italy, the presence of numerous feedbacks, might distort or remove the consequences of global warming on the precipitation regime. Bartolini et al. (2012) argued that Mediterranean warming is especially due to changes in the summer season. They supported their findings with observed data from Tuscany, central Italy. Their result highlighted a positive trend for mean temperature of about 0.9 °C per 50 years with a slightly more pronounced increase in maximum temperature.

Some authors have also made thorough analysis based on single-station observations using various indices and comparison of results with different sites. For example, in Bologna-Cadriano area, analysis in the second half of the twentieth century (1952–2007) showed an increase in mean annual temperature of 1.2 °C and significant increasing trend in reference evapotranspiration. No clear sign of a decrease in precipitation, but maximum groundwater table level dropped by 42 cm (Matzneller et al. 2010). In order to evaluate climate change risk on agriculture, Moonen et al. (2002) used a station from Pisa with time series data of 122 years. Their results indicated that there is a shift toward more extremely low rainfall events but negligible effect on agriculture and drought risk. From trend analysis over the same period, they indicated that no significant changes in soil water surplus or deficit on annual basis.

From hydrological study point of view, Burlando and Rosso (2002) evaluated effects of transient climate change impacts on runoff variability in the Arno River, central Italy. They showed that a reduction of the annual maxima daily flows is expected for the Arno basin. For all the scenarios used in their analysis, a general reduction of water availability was also expected, but it does not necessarily mean a reduction in total discharge in the river rather a different distribution in time and space that could substantially constrain the effective availability of water for use. D'Agostino et al. (2010) used distributed catchment-scale model to study the impact of land use and climate change in Apulia region, Candelaro catchment, southern Italy. They forced the hydrologic model by climate scenario that showed rainfall reduction of 5–10 % during winter and 15–20 % during summer, while temperatures are expected to increase between 1.25 and 1.5 °C during winter and 1.5 and 1.75 °C during summer. Their result showed that by 2050, groundwater recharge in the catchment would decrease by 21–31 % and stream flows by 16–23 %. Di Matteo and Dragoni (2006) also evaluated the effect of climate change on the water resources of Firenzuola Lake in Umbria region, central Italy. Their result indicated that the yield of the basin is likely to decrease.

30.3 Hydrometeorology of the Upper Tiber Basin

30.3.1 Location and General Characteristics

TRB is the largest river basin in Central Apennines District of Italy. Geographically, the basin is located between 40.5°N to 43°N latitudes and 10.5°E to 13°E longitudes covering an area of about 17,500 km^2 occupying roughly 5 % of the Italian territory. The basin crosses six administrative regions and twelve provinces. Almost 90 % of the basin lies in the regions of Umbria and Lazio, and the remaining 10 % falls within the regions of Emilia Romagna, Tuscany, Marche, and Abruzzo. The area under each region is shown in parentheses in the legend of Fig. 30.1. Including the oldest city of Rome, major cities such as Perugia, Terni, and Rieti are located within the basin (http://www.abtevere.it/node/379).

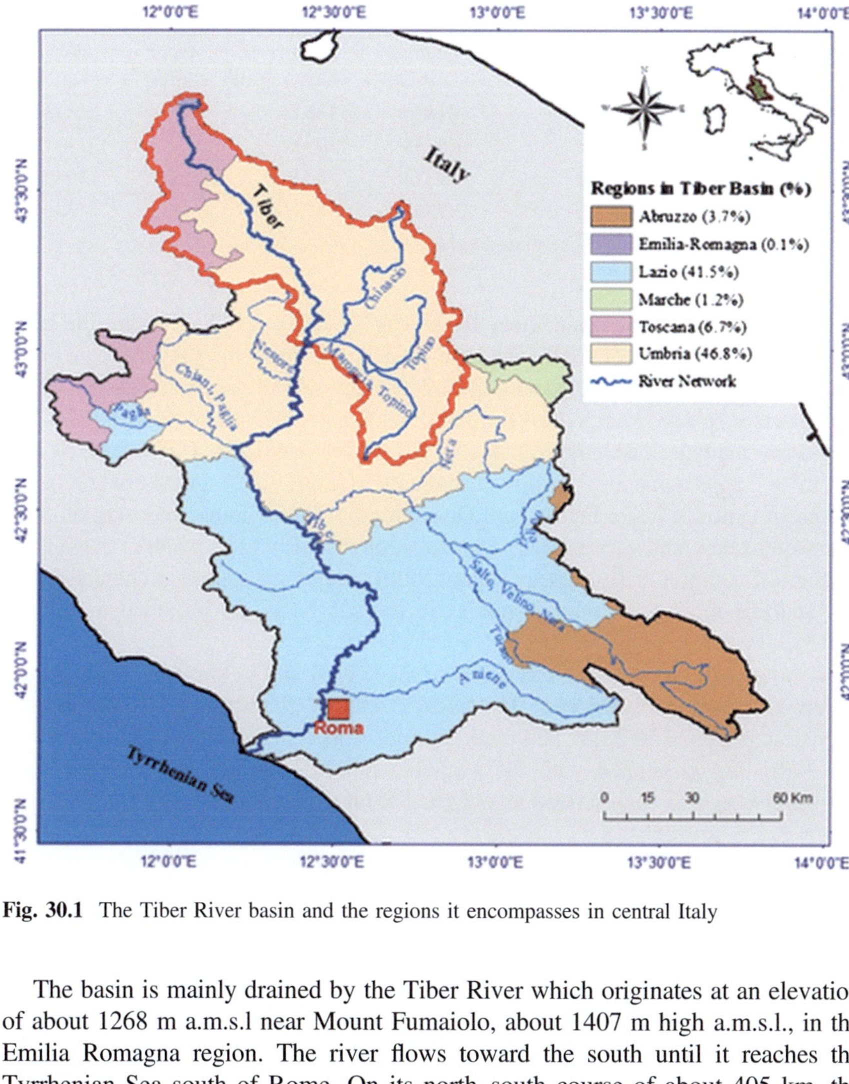

Fig. 30.1 The Tiber River basin and the regions it encompasses in central Italy

The basin is mainly drained by the Tiber River which originates at an elevation of about 1268 m a.m.s.l near Mount Fumaiolo, about 1407 m high a.m.s.l., in the Emilia Romagna region. The river flows toward the south until it reaches the Tyrrhenian Sea south of Rome. On its north–south course of about 405 km, the Tiber River receives flows from different tributaries of small to medium river systems. The main contributing river systems include the Chiani–Paglia and Nestore River systems from the west; the Chiascio–Topino, Salto–Turano–Velino–Nera, and Aniene river systems from the east. In addition to the river systems, there are small natural lakes in the basin, including Lake Trasimeno (122.7 km^2), Lake Piediluco (1.7 km^2), Lake Vico (12.3 km^2), and Lake Albano (6.0 km^2). Figure 30.1 shows the location of the TRB and the case study area (red boundary) selected for the hydrological simulation. The upper Tiber River basin (UTRB) is part of the TRB that covers an area of 4145 km^2, about 20 % of the TRB, with its outlet at Ponte Nuovo. The elevation of the catchment ranges from 145 to 1560 m above sea

Table 30.1 Summary of major physiographic characteristics of the UTRB

Sub-basins	Outlet elevation (m a.s.l.)	Drainage area (km^2)	Main river length (km)	Main river slope (%)	Basin lag (h)[a]
UTRB at Ponte Nuovo	165.0	4145.0	136.0	0.23	18–22
Tiber at Ponte Felcino	197.0	2033.0	109.6	0.27	14–17
Tiber at Santa Lucia	265.0	932.0	63.9	0.52	10–13
Chiascio at Rosciano	171.0	1956.0	89.6	0.32	13–15

[a]Information obtained from the literature review

level. In addition to the main Tiber River, the Chiascio and Topino are the main tributaries that drain the UTRB. The hydrological regime of the river systems in the TRB commonly receives its maximum inflow from rainfall generated runoff which characterizes spring (March–May) and autumn (September–November) floods. The long-term average discharge of the Tiber River into the Tyrrhenian Sea is 225 $m^3 s^{-1}$ approximately 7 billion m^3 annually (Cesari 2010). In the context of the European Union's Water Framework Directive, the TRB belongs to ecoregion 3 for rivers and lakes and ecoregion 6 for transitional and coastal waters—Annex XI Directive 2000/60/CE (European Union 2000). Major hydrological characteristics and analysis related to water resources in the sub-basin are provided in Behulu (2013).

Summary of the main characteristics of the UTRB and its sub-basins at selected outlets is shown in Table 30.1. The basin is mainly characterized by the longest river flow path of 136.2 km, and main channel slope ranging from 0.23 to 2.85 %. The basin lag associated with the available rainfall and runoff events for each sub-basin was also summarized as obtained from different reports.

30.3.2 Air Temperature

Relatively, small amount of time series data for minimum and maximum temperatures (T_{min} and T_{max}) exists in the basin with many missing values and unevenly distributed in space and time. Seventy-eight gauging station was provided by the Hydrographic Services of Umbria and Lazio regions. The summary of the stations and their data availability is given in Behulu (2013). Four stations with relatively complete time series data were selected to explore the inter-annual surface temperature variation in the UTRB for the period of 1961–1990. These stations are Assisi, Gubbio, Perugia, and Spoleto.

The inter-annual variability of daily T_{max} and T_{min} is shown in Fig. 30.2a, b respectively, with the 95 % confidence of the mean values. The patterns of both daily minimum and daily maximum temperature are similar. From Fig. 30.2a, it is clear that the highest daily maximum temperature is observed in July and August which are the dry period of the region. The lowest value of maximum temperature is recorded in the months of January and February. Like that of the daily maximum

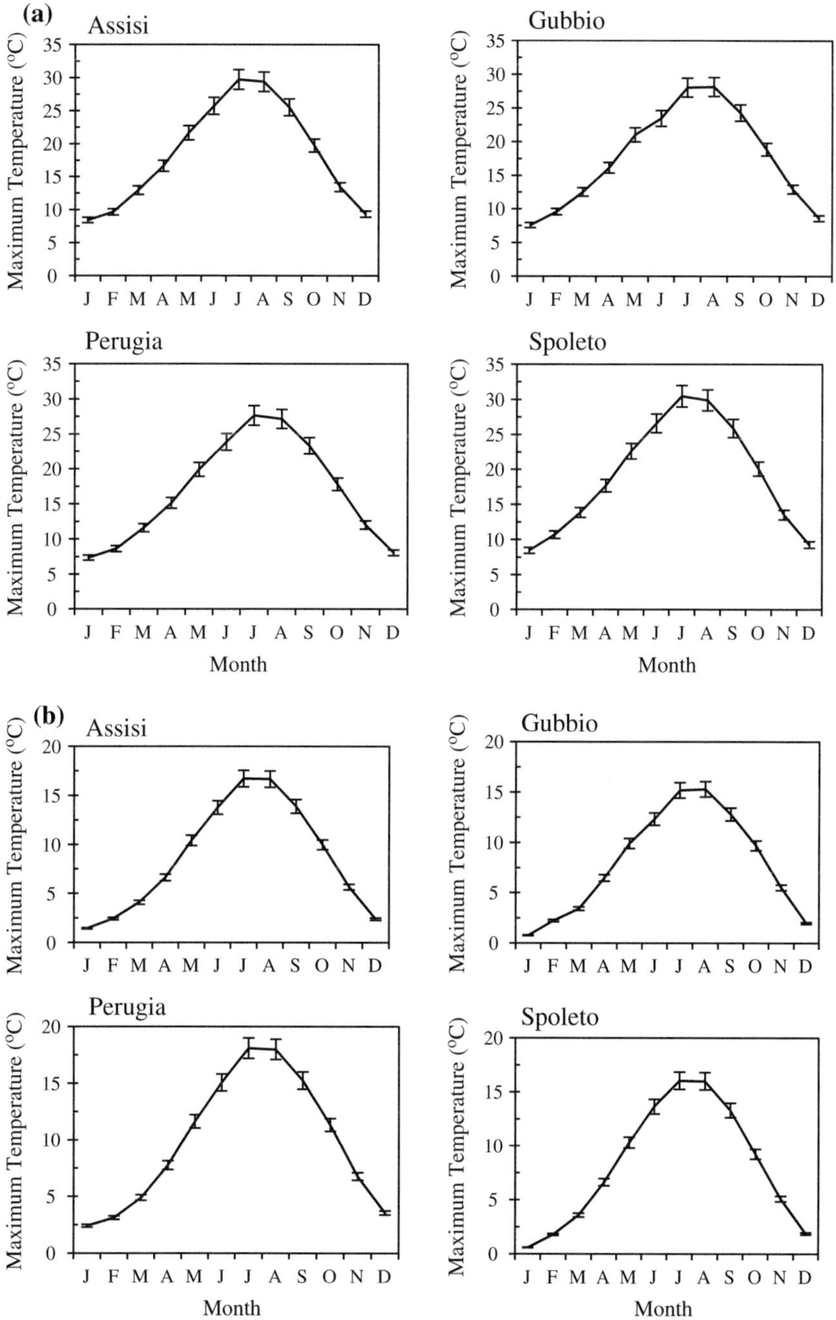

Fig. 30.2 **a** Inter-annual variability of daily maximum temperature at selected stations in the UTRB (January 1, 1961–December 31, 1990). **b** Inter-annual variability of daily minimum temperature at selected stations in the UTRB (January 1, 1961–December 31, 1990)

temperature, higher values of daily minimum temperatures are recorded in July and August at all the stations. The lowest value is observed in the month of January. Also, the temperature values show dependence on elevation of the area. For the dry season (JJA), daily T_{max} shows a decrease of 5 °C per 100 m elevation and the daily T_{min} shows 8 °C for every 100 m of elevation.

30.3.3 Rainfall

Daily time series rainfall data were obtained from Regional Hydrographic Services of Umbria. More than 80 rainfall stations were available in the basin but with several missing values. For the period 1961–1990, rainfall stations with a long data record, 70 % or above, were selected. Out of the available stations, only 52 stations satisfied this criterion. The selected stations with their mean annual value and corresponding daily missing percentage out of the 30 years of record are summarized in Behulu (2013).

Rainfall in the UTRB shows high temporal and spatial variability. The mean annual rainfall data at the stations with complete records were summarized, and then, spatial interpolation is performed over the entire basin (Fig. 30.3). Ordinary kriging interpolation with exponential variogram is used to show the spatial variation of rainfall for the period of 1961–1990 based on the selected gauging stations. The mean annual rainfall in the UTRB ranges between 790 and 1338 mm, which shows large spatial variability with a maximum rainfall as large as 1.8 times the minimum rainfall. Based on the interpolation method used, the mean annual rainfall for the period is estimated at 980 mm.

Temporal variation of 30-year mean annual rainfall from selected stations is shown in Fig. 30.4 as a ratio of annual deviation from the 30-year mean and the 30-year mean. Years with above-average rainfall have positive values, and drier years have negative values. The mean annual rainfall in the basin showed a decreasing trend similar to most climate change-related studies in the area (Coppola and Giorgi 2010; Behulu et al. 2013, 2014).

30.3.4 River Flow

Observed river flow data from Santa Lucia, Ponte Felcino, Ponte Nuovo, and Petrignano di Assisi were analyzed. The first three gauging stations are located along the Tiber River, and only the station at Petrignano di Assisi is along Chiascio River which is one of the tributaries of the upper Tiber River (Fig. 30.5). Relatively, the Ponte Nuovo station which is the basin outlet has long record of time series data as compared to the other three gauging stations. For the calibration of the hydrologic model used in this study, the flow data at Ponte Nuovo are used and the other stations which are located in the upstream are used for validation.

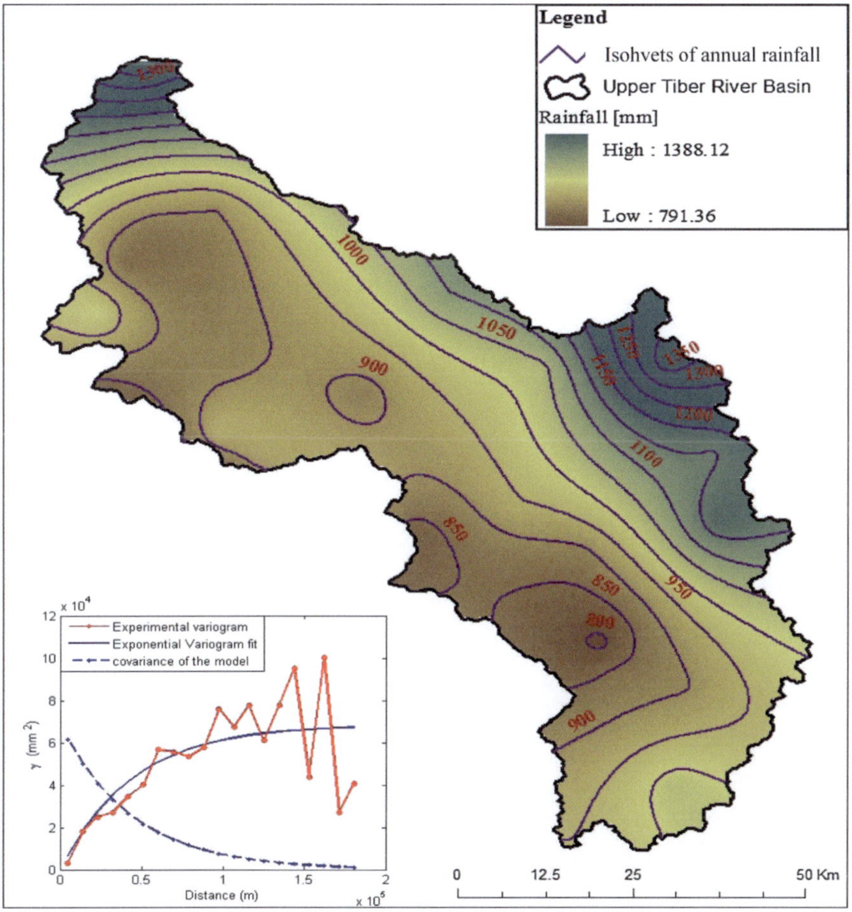

Fig. 30.3 Mean annual rainfall variation in the upper Tiber Basin (variogram model used in the ordinary kriging shown in the *lower left corner*)

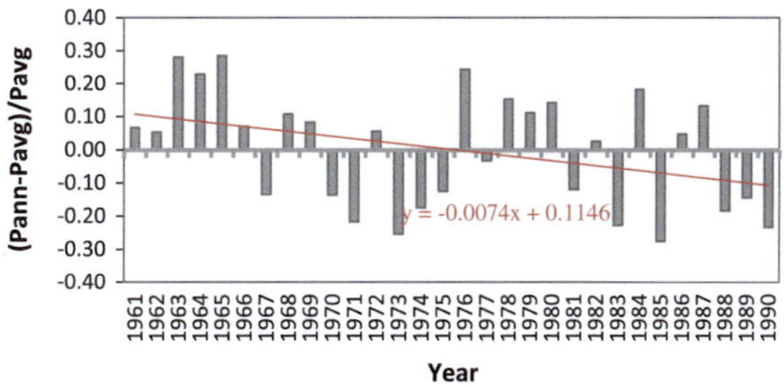

Fig. 30.4 Mean annual precipitation anomaly for the study area (1961–1990)

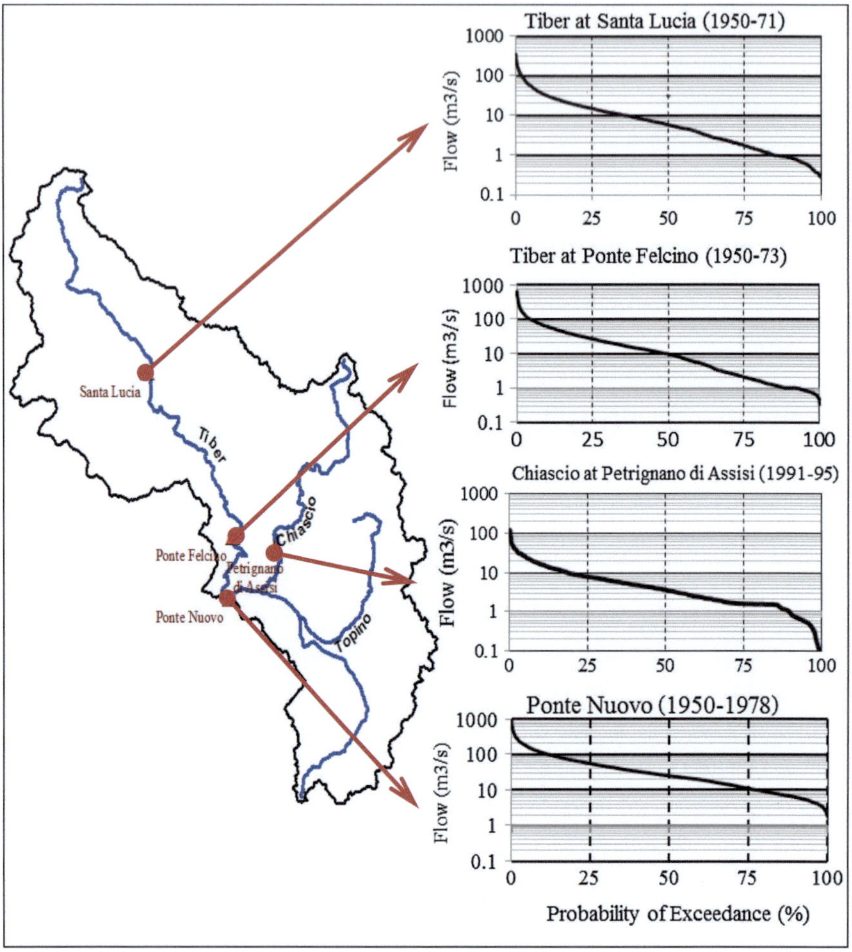

Fig. 30.5 Flow gauging stations and their corresponding frequency of flow

Figure 30.5 shows the frequency of flow based on the average daily flow recorded at the four gauging stations along the Tiber River and its tributary in the UTRB. Figure 30.5 shows that the maximum daily flow at all the recorded station is above 100 $m^3\ s^{-1}$ and the minimum daily flow varies depending on the location of the station. The minimum daily flow at Ponte Nuovo is above 1 $m^3\ s^{-1}$, as the frequency curve ends at the 100 % probability of exceedance. However, 10 % of the flows at the other upstream stations show daily flow less than 1 $m^3\ s^{-1}$. At downstream, 50 % of the flow at Ponte Nuovo shows daily average flow larger than 20 $m^3\ s^{-1}$, whereas the other upstream three stations show a corresponding value of 10 $m^3\ s^{-1}$.

30.4 Recharge Estimation Through Watershed Modeling

Groundwater recharge generally refers to the component of hydrologic cycle that reaches the groundwater table. Direct measurement of recharge is unfortunately not possible as the processes vary from place to place and time to time. There is no hard and fast rule to follow for estimation of recharge in any area because a method developed for a given locality will not give a reliable result when used in another locality. Quite often, groundwater recharge is estimated as a fraction of rainfall. However, estimation based on hydrologic model and hydrograph separation methods is commonly practiced by surface water hydrologists. In this chapter, a watershed model was used for estimation of groundwater recharge. A study by Luo et al. (2012) can be considered as good example of watershed model application for groundwater recharge estimation.

30.4.1 The Soil and Water Assessment Tool (SWAT) Model

Among the large number of watershed models (Singh and Woolhiser 2002), the soil and water assessment tool (SWAT) model (Arnold et al. 1998) was selected to simulate hydrological components in the UTRB. The model was selected due to three main reasons: (i) It is widely used in different part of the world even under scarce data condition, (ii) it is freely available with detailed documentation and good reviews (Gassman et al. 2007; Neitsch et al. 2009), and (iii) its GIS interface simplifies the spatial data handling and catchment delineation in the modeling process.

SWAT is a physically based, a spatially distributed, and continuous time watershed model developed to predict the impact of land management practices on water, sediment, and agricultural chemical yields in large complex watersheds with varying soils, land use, and management conditions over long periods of time (Arnold et al. 1998; Neitsch et al. 2009). It is actively supported by the US Department of Agriculture (USDA), Agricultural Research Service (ARS) at the Grassland, and Soil and Water Research Laboratory in Temple, Texas, USA (Neitsch et al. 2005). As a physically based model, SWAT uses hydrologic response units (HRUs) to describe spatial heterogeneity in land cover and soil types within a watershed. The HRU is therefore the smallest spatial unit for rainfall–runoff calculations which is a lumped land area within a sub-watershed comprising of unique land cover, soil, slope, and management combinations.

SWAT has eight major components: hydrology, weather, sedimentation, soil temperature, plant growth, nutrients, pesticides, and land management. For all types of problem studied with SWAT, water balance is the driving force behind everything that happens in the watershed (Neitsch et al. 2009). The general water balance equation in SWAT is given in Eq. (30.1). Within the HRU, water balance is represented by four storage volumes: snow, soil profile (0–2 m), shallow aquifer (typically 2–20 m), and deep aquifer (≥20 m). The model simulates relevant hydrologic processes such as surface runoff, evapotranspiration, infiltration, percolation, shallow aquifer and deep aquifer flow and storage, and channel routing (Arnold and Allen 1996). The simulation of the hydrologic processes within the watershed can be done in four subsystems: surface soil, intermediate zone, shallow and deep aquifers, and channel flow. Stream flow in the main channel is determined by three sources: surface runoff, lateral flow, and base-flow from shallow aquifers. In general, the hydrology of a watershed can be separated into two major components as the land phase and the routing phase of the hydrologic cycle. The land phase of the hydrologic cycle controls the amount of water, sediment, nutrient, and pesticide loadings to the main channel in each sub-basin. The routing phase of the hydrologic cycle controls the movement of water, sediments, and chemicals through the channel network of the watershed to the outlet. Detail description of each phase is given in the theoretical description documents of SWAT model (Neitsch et al. 2009).

$$\mathrm{SW_t} = \mathrm{SW_o} + \sum_{i=1}^{t} (R_{\mathrm{day}} - Q_{\mathrm{surf}} - E_{\mathrm{a}} - W_{\mathrm{seep}} - Q_{\mathrm{gw}} \tag{30.1}$$

where $\mathrm{SW_t}$ is the final soil water content (mm), $\mathrm{SW_o}$ is the initial soil water content on day i (mm), t is time (days), R_{day} is the amount of precipitation on day i (mm), Q_{surf} is the amount of surface runoff on day i (mm), E_{a} is the amount of evapotranspiration on day i (mm), W_{seep} is the amount of water entering the vadose zone from the soil profile on day i (mm), and Q_{gw} is the amount of return flow on day i (mm).

SWAT's application for understanding hydrological processes and estimate fluxes of water, sediment, and pollutants is getting greater attention in recent years. The application of SWAT in predicting stream flow and sediment as well as evaluation of the impact of land use and climate change on the hydrology of watersheds has been documented by various studies (Dessu and Melesse 2012, 2013; Dessu et al. 2014; Wang et al. 2006, 2008a, b, c; Wang and Melesse 2005, 2006; Behulu et al. 2012, 2013, 2014; Mango et al. 2011a, b; Getachew and Melesse 2012; Assefa et al. 2014; Grey et al. 2013; Mohammed et al. 2015).

30.4.2 Groundwater in SWAT Model

SWAT differentiates the underground storage into two portions, shallow aquifer and deep aquifer, as shown in Fig. 30.6. The shallow aquifer receives recharge from the unsaturated soil profile percolation. An exponential decay weighting function is utilized to account for the time delay in aquifer recharge once the water exits the soil profile (Neitsch et al. 2009). The delay function accommodates situations where the recharge from the soil zone to the aquifer is not instantaneous, 1 day or less. The recharge to aquifer on a given day is calculated by Eq. 30.2.

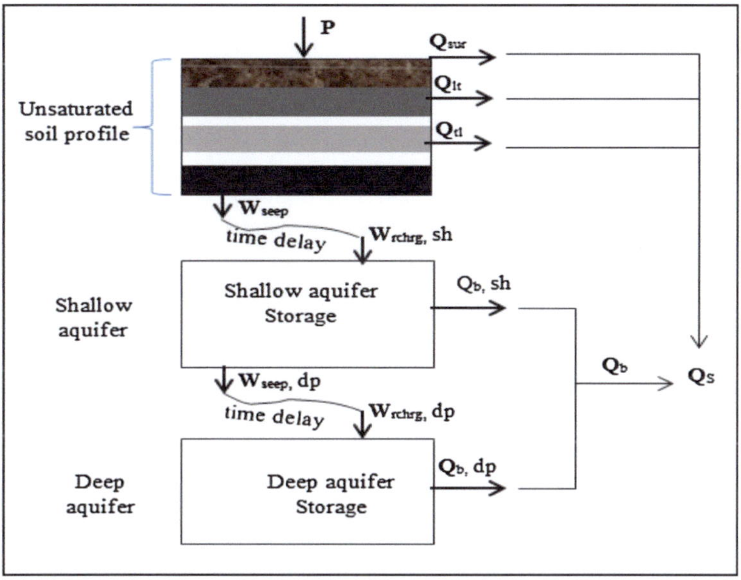

Fig. 30.6 Conceptual representation of hydrologic processes in SWAT (Luo et al. 2012)

$$W_{\mathrm{rchrg},i} = \left[1 - \exp\left(\frac{-1}{\delta_{\mathrm{gw,sh}}}\right)\right] * W_{\mathrm{seep}} + \exp\left(\frac{-1}{\delta_{\mathrm{gw,sh}}}\right) * W_{\mathrm{rchrg},i-1} \tag{30.2}$$

where W_{rchrg} is the amount of recharge entering the aquifers (mm day^{-1}), $\delta_{\mathrm{gw,sh}}$ is the delay time of the overlying geologic formations (days), and W_{seep} is the total amount of water exiting the bottom of the soil profile (mm day^{-1}); subscriptions "seep" indicates seepage water exiting bottom of the unsaturated soil profile, "rchrg" indicates recharge, *i* is the sequential number of days, and "sh" indicates the shallow aquifer storage.

A fraction of the total daily recharge can be routed to the deep aquifer. The amount of water diverted from the shallow aquifer due to percolation to the deep aquifer on a given day is given by Eq. 30.3.

$$W_{\mathrm{seep,dp},i} = \beta_{\mathrm{dp}} * W_{\mathrm{rchrg},i} \tag{30.3}$$

where β_{dp} is a coefficient of shallow aquifer percolation to deep aquifer and subscription "dp" indicates deep aquifer. The amount of recharge entering the shallow and deep aquifers is given by Eqs. 30.4a and 30.4b, respectively.

$$\mathrm{W}_{\mathrm{rchrg,sh},i} = W_{\mathrm{rchrg},i} - W_{\mathrm{seep,dp},i} \tag{30.4a}$$

$$W_{\mathrm{rchrg,dp},i} = W_{\mathrm{rchrg,dp},i-1} * \left[1 - \exp\left(\frac{-1}{\delta_{\mathrm{gw,dp}}}\right)\right] + W_{\mathrm{seep,dp},i} * \left[1 - \exp\left(\frac{-1}{\delta_{\mathrm{gw,dp}}}\right)\right] \quad (30.4b)$$

where W_{rchrg} is the amount of recharge entering the aquifer (mm day^{-1}); subscripts "sh" and "dp" indicates the shallow and deep aquifers; $\delta_{\mathrm{gw,dp}}$ is the delay time or drainage time of the deep aquifer geologic formations (days); and $W_{\mathrm{seep,dp}}$ is the total amount of water exiting the bottom of the shallow aquifer (mm day^{-1}).

In the SWAT simulation process, if only the shallow reservoir is used to generate the baseflow, the parameter for the recharge to deep aquifer is disabled. Otherwise, the parameter β_{dp} is determined through calibration. Other parameters to be calibrated for baseflow modeling in SWAT are the delay time (δ_{gw}) and the recession constants (α_{gw}).

In addition to SWAT, the use of digital filters to estimate baseflow contribution of a watershed is suggested by some authors (Arnold and Allen 1999; Luo et al. 2012). A public domain automatic baseflow separation program was used in this study in order to evaluate the groundwater contribution of the watershed (Arnold and Allen 1999). The program is available at http://swat.tamu.edu/software/baseflow-filter-program/ accessed October 12, 2014. The filter uses Eqs. 30.5 and 30.6.

$$Q_{\mathrm{sf},i} = \lambda Q_{\mathrm{sf},i-1} + \frac{1+\lambda}{2}\left(Q_{\mathrm{s},i} - Q_{\mathrm{s},i-1}\right) \quad (30.5)$$

where Q_{sf} is the filtered surface runoff (quick response) at time step i and Q_{s} is the original stream flow (surface runoff), and λ is the filter parameter. Baseflow is then calculated by Eq. 30.6.

$$Q_{\mathrm{b},i} = Q_{\mathrm{s},i} - Q_{\mathrm{sf},i} \quad (30.6)$$

30.5 Climate Change Scenario and Bias Correction of RCMs

30.5.1 The Climate Scenarios and RCMs

For the assessment of climate change impact on hydrological processes, simulated climate dataset from Prediction of Regional scenarios and Uncertainties for Defining European Climate change risks and Effects (PRUDENCE) archive were used. The climate scenarios utilized during the preparation of the AR4 (IPCC 2000) were used. These scenarios are alternative images of how the future might unfold and are an appropriate tool with which to analyze how driving forces may influence future emission outcomes. All PRUDENCE datasets contain simulations of 30 years (1961–1990) for control period and future period (2071–2100) with A2 and B2 emission scenarios. The A2 scenarios are close to the high end of the range (CO_2 concentration of about 850 ppm by 2100), and the B2 scenarios are close toward the lower end (CO_2 concentration of about 620 ppm by 2100). The three RCMs used in this study have a spatial resolution of 50 km with their lateral boundary forcing fields from HadAM3H (Buonomo et al. 2007). HadAM3H is an atmospheric global model developed at the Hadley Center, the resolution of which is considered high in its class and is about 15 km. Details on the RCMs are given in Jacob et al. (2007) and Christensen et al. (2007), and short summaries of the selected models are given as follows.

30.5.1.1 RegCM

RegCM is a RCM built by International Center for Theoretical Physics (ICTP) (Giorgi et al. 1993a, b). The dynamical core of the RegCM is equivalent to the hydrostatic version of the mesoscale model MM5 of NCAR Pennsylvania State University. Surface processes are handled via the biosphere-atmosphere transfer scheme (BATS), while there are special schemes for precipitation and convection. Energy transfers involving radiation are computed with the radiation package of the NCAR community climate model. The model has a grid resolution of 50–70 km with lambert conformal conical projection covering Europe with 119 × 98 grid boxes.

30.5.1.2 PROMES

The PROMES regional climate model is developed in Universidad Complutense de Madrid (UCM) (Arribas et al. 2003). This is the climate version of the PROMES model. It is a hydrostatic and primitive equation model. Prognostic variables are potential temperature, surface pressure, horizontal wind components, specific humidity, cloud, and rainwater. PROMES runs at 50-km resolution with lambert conformal conical projection covering Europe in 112 × 96 grid boxes.

30.5.1.3 RCAO

RCAO is the Swedish Meteorological and Hydrological Institute (SMHI) Rossby Centre regional Atmosphere–Ocean model (Jones et al. 2004). It incorporates a regional atmospheric (RCA) and a regional ocean model (RCO), both developed in the Rossby Centre, and a river routine based on the HBV hydrological model. The RCA model has its roots in the limited area model HIRLAM and it is run in the resolution range of 10–70 km and with 24–60 vertical levels. Variables are temperature, horizontal wind components, specific humidity, cloud water, turbulent kinetic energy, surface pressure, soil temperature, and water content. The RCO model is based on the OCCAM version of the Bryan–Cox–Semtner primitive equation ocean model with a free surface. The model covers the European window over 106 × 102 grid boxes with its South Pole rotation of 25°E and 32°S.

30.5.2 Bias Correction of RCMs

The dynamically downscaled RCMs are considered to have robust application capabilities to reproduce local climate. However, studies have showed that all RCMs are susceptible to inherent systematic biases due to their imperfect conceptualization, discretization, and spatial averaging within grid cells (Graham et al. 2007; Teutschbein and Seibert 2012). Hence, bias correction is considered to be the minimum requirement in using RCM for impact assessment.

Several bias correction methods have been developed since the development of climate models and their application in hydrologic impact studies. A comprehensive review of currently available methods in hydrology is given in Teutschbein and Seibert (2012) and Themeßl et al. (2010). In this study, a simple bias correction method called "delta change or scaling" is used to prepare climate inputs for the hydrologic model (Lenderink et al. 2007). The method is commonly applied to transfer the signal of climate change derived from a climate model simulation to an observed database. The general bias correction procedure for temperature is presented by Eqs. 30.7a and 30.7b.

$$T_{\text{corrected}}(i,j) = T_{\text{scen}}(i,j) + \Delta T(j) \tag{30.7a}$$

$$\Delta T(j) = \bar{T}_{\text{obs}}(j) - \bar{T}_{\text{ctrl}}(j) \tag{30.7b}$$

where $T_{\text{corrected}}$ is the bias-corrected temperature input for the hydrological model during the scenario simulation; T_{scen} is the simulated temperature in the scenario period; (i, j) is the ith day of jth month; and ΔT (j) is the change in temperature between the mean monthly observation (obs) and mean monthly control period temperature, where $\bar{T}_{\text{obs}}$ is the mean monthly observed temperature for the month of j and $\bar{T}_{\text{ctrl}}$ is mean monthly temperature form month j in the control or reference period (usually taken as 30 years). The indices *ctrl* stand for control period (1960–1990) and *scen*, scenario period (2070–2100). The general bias correction procedure for precipitation is presented by Eqs. 30.8a and 30.8b.

$$P_{\text{corrected}}(i,j) = P_{\text{scen}}(i,j) * \Delta P(j) \tag{30.8a}$$

$$\Delta P(j) = \frac{\bar{P}_{\text{obs}}(j)}{\bar{P}_{\text{ctrl}}(j)} \tag{30.8b}$$

where $P_{\text{corrected}}$ is precipitation input for the hydrological scenario simulation; P_{obs} is the observed precipitation in the historical period at each station; (i, j) is the ith day of *j*th month; $\Delta P\ (j)$ is the change in precipitation calculated using as a ratio; $\bar{P}_{obs}(j)$ is the mean observed precipitation for the month of *j*; and $P_{ctrl}(j)$ is mean rainfall for month *j* for the control or reference period (1961–1990). The indices *obs* and *ctrl* stand for the observed and control period (1961–1990), respectively.

30.6 Results and Discussion

30.6.1 Calibration and Validation of SWAT Model

In the SWAT model, there are twenty-six hydrological parameters that control the different hydrologic processes in the study under question. The choice of those parameters depends on the processes to be modeled, the method applied, and the location of the watershed. In the present study, eighteen out of the twenty-six parameters were evaluated and the top ten parameters were selected based on their sensitivity. Detailed explanation on the model data inputs, calibration, and validation procedures including sensitivity analysis for the parameters of SWAT model is given in Behulu et al. (2013). Table 30.2 shows the selected parameters with their final values in the modeling processes.

Using the identified parameters, the behavior of the basin in terms of response to stream flow at the outlet was successfully evaluated. The observed stream flow data over the time period of January 1, 1961 to December 31, 1978 from the Ponte

Table 30.2 Parameters used for SWAT simulation

Parameter	Description	Model process	Rank	Variation range	Fitted value
ALPHA_BF	Baseflow recession constant days	Groundwater	1	0–1	0.055[a]
	SCS runoff curve number for moisture				
CN2	Condition II	Runoff	2	±20 %	−9[b]
CH_N2	Channel	Channel flow	3	0–1	−0.004[c]
	Effective hydraulic conductivity in main				
CH_K2	Channel alluvium (mm h^{-1})	Channel flow	4	0–150	15.0[a]
	Threshold water level in the shallow aquifer				
GWQMN	For return flow to occur	Groundwater	5	0–5000	350[c]
RCHRG_DP	Aquifer percolation coefficient	Groundwater	6	0–1	0.01[a]
SURLAG	Surface runoff lag coefficient	Runoff	7	0–2	1.00[a]
ESCO	Soil evaporation compensation factor	Evaporation	8	0–1	0.85[a]
SLOPE	Average slope steepness (mm^{-1})	Geomorphology	9	±20 %	–
SOL_Z	Soil depth	Soil water	10	20 %	–

[a]Default values are replaced by this value (absolute change)
[b]Default values are multiplied by this percentage (relative change)
[c]Default values are increased by this value (absolute change)

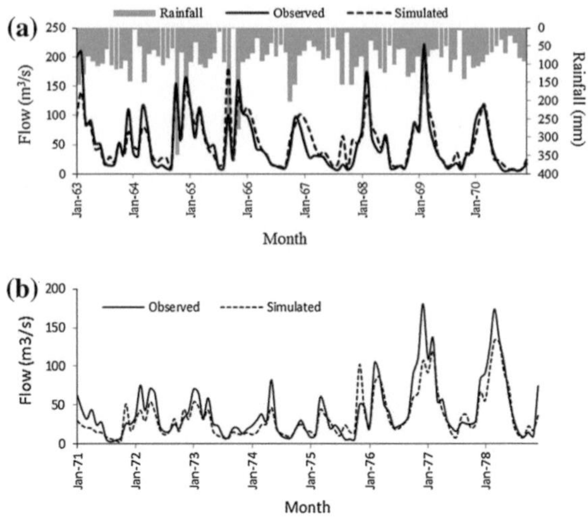

Fig. 30.7 **a** Calibration results for monthly flow at Ponte Nuovo (1963–1970). **b** Validation results for monthly flow at Ponte Nuovo (1971–1978)

Nuovo outlet was used for both calibration and validation. Moreover, independent data from Santa Lucia, Ponte Felcino, and Petrignano Di Assisi gauging stations were used for validation purposes. All the performance indicators showed acceptable limits as recommended by Moriasi et al. (2007). The model performance during the calibration and validation period is shown in Fig. 30.7a, b, respectively.

For all the stations in the study area, a simple bias correction to both temperature and precipitation data was applied for the control period (1962–1990). The corrections were made based on monthly basis, and each of the monthly corrections was applied to the corresponding daily data of each month. The same correction procedure was applied to all stations, but only Assisi station results are presented.

Figure 30.8a shows the result of biases at Assisi from two different models. Relatively larger bias was observed for precipitation data from RegCM than RCAO. Temperature bias from the two models is different with RCAO showing small biases. RegCM shows higher biases for both maximum and minimum temperatures (Fig. 30.8b).

30.6.2 Impact of Climate Change on Recharge in the UTRB

The change in baseflow contribution at the basin outlet was determined for both the A2 and B2 scenarios in the basin. The digital filter for baseflow separation presented by Arnold and Allen (1999) was used to determine the baseflow both from the simulated flow data and for the control period (1961–1990) and the scenario period (2070–2099). Assuming the simulated flows based on observed dataset is "true" values, the change in the baseflow was calculated, and the result is shown in Fig. 30.9a, b. In both scenarios, the results show that the baseflow contribution tends to decrease. However, evaluation of the river flow at the basin outlet and

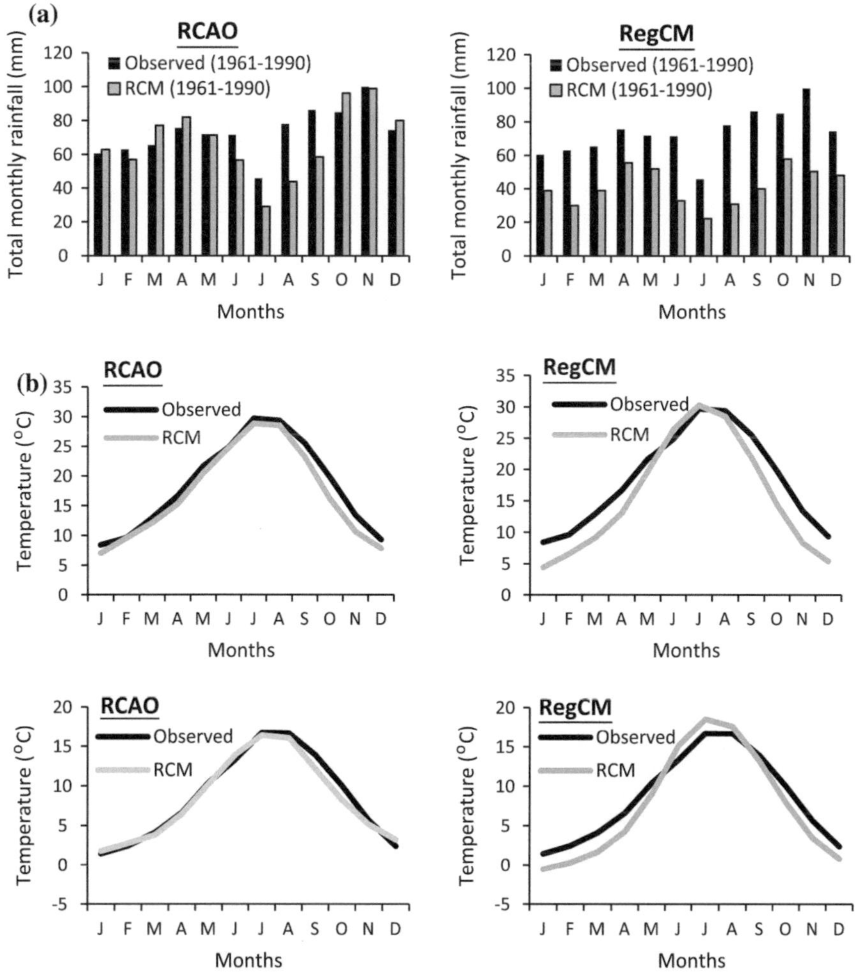

Fig. 30.8 **a** Precipitation bias at Assisi station from regional climate models RCAO and RegCM for 1961–1990. **b** Temperature bias at Assisi station from regional climate models RCAO and RegCM for 1961–1990 (T_{max} *upper panel*; T_{min} *lower panel*)

water balance analysis showed that the PROMES model gave in different results, especially in the wet season. On the other hand, the recharge condition in the basin showed different results for A2 and B2 scenarios, with the former showing a pronounced decrease in all the RCMs.

The spatial distribution of average relative change in recharge is shown in Fig. 30.10. In case of B2 scenario, the basin responded to an increase in groundwater recharge that reaches as high as 35 %, whereas in case of A2 scenario, the decrease could reach as low as −20 %. From these results, accepting limitations of the methods used for bias correction and simulation capability of the watershed

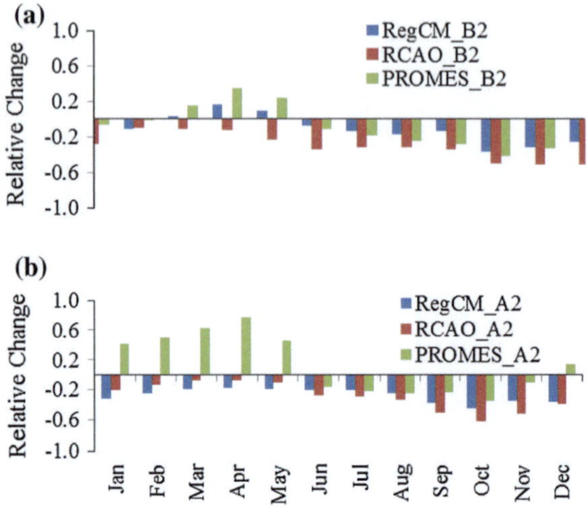

Fig. 30.9 **a** Monthly average change in baseflow contribution under B2 scenario from the three models. **b** Monthly average change in baseflow contribution under A2 scenario from the three models

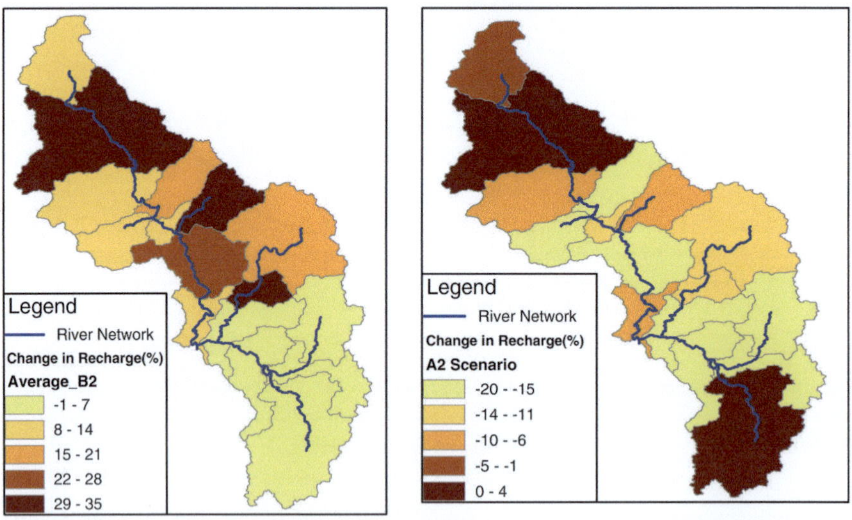

Fig. 30.10 Spatial distribution of average relative changes in recharge over the UTRB for B2 and A2 scenarios

model, it can be inferred that choice of scenario has impact on basin hydrology. However, in this study, the variation with recharge is in agreement with the change in precipitation and temperature. It is also good to note that precipitation and temperature are the determinant variables in SWAT simulation as compared to the other model inputs.

30.7 Summary

The current state-of-the-art climate change in groundwater was reviewed considering the progressive works of various research papers. There is a relatively higher concern of impact assessment on groundwater recharge since the IPCC's AR4 came out. The overall hydrological characteristics of the Italian territory under climate change and variability issues were also summarized in this chapter. Most of the studies in the Italian context are based on observed dataset that shows agreement with the increase in temperature and decreasing trend in precipitation. Being located in the hot spot of climate change, the north–south elongated topography of the territory experiences range of climatic behavior.

The general hydrometeorological characteristics of the UTRB was investigated using its relative location, air temperature, rainfall, and river flow data obtained from Hydrographic Services of Umbria region. The basin is characterized by Mediterranean climate having low permeability and mean annual rainfall of 790–1300 mm. The rainy seasons are mostly in autumn to spring seasons.

The UTRB is also used as a case study area to evaluate the effects of climate change on groundwater recharge through calibrated and validated watershed model, SWAT. The model is forced by the precipitation and temperature data obtained from dynamically downscaled and bias-corrected regional climate model outputs. The groundwater recharge shows a decreasing trend as a response to changes in rainfall. However, as the timing of both precipitation and recharge is critical for future groundwater development in the basin, further analysis through different climate models, downscaling approach, and groundwater modeling needs to be taken into account.

The climate change analysis from the three RCMs indicated that by the end of twenty-first century, rainfall will decrease up to 40 % in the dry period and there will be an increase in temperature that could reach as high as 3–5 °C. Moreover, the three bias-corrected RCM (RCAO, PROMES and RegCM) outputs showed different results. Such variation of result calls for further investigation of model uncertainty. These uncertainties can be due to one or more of the four sources: (i) uncertainty due to climate scenarios; (ii) choice of the deriving GCM; (iii) transfer of large-scale climatology to regional-scale climatology, downscaling method; and (iv) hydrologic model input variables. In this case, the three RCMs were used with the same boundary condition from HadAM3H under A2 and B2 emission scenarios which resulted in different predictions of the recharge component. Also, the result of the case study is based on the assumption that the present model characteristics remain unchanged in the future. This may not be the case. Also, scenarios can only assist in climate change analysis, including climate modeling and the assessment of impacts, adaptation, and mitigation to indicate the path of change. However, the possibility that any single emission path will occur as described in a scenario is highly uncertain. Such issues are expected to be addressed in the forthcoming IPCC's report.

Acknowledgments The authors would like to acknowledge the Hydrographic Service of Umbria region for the hydrometeorological data of the study area. The authors also thank the Food and Agricultural Organization (FAO) for the providing soil data free of charge.

References

Allen DM, Mackie DC, Wei M (2004) Groundwater and climate change: a sensitivity analysis for the Grand Forks aquifer, southern British Columbia, Canada. Hydrogeol J 12(3):270–290

Allen DM, Cannon AJ, Toews MW, Scibek J (2010) Variability in simulated recharge using different GCMs. Water Resour Res 46(10)

Arnold JG, Allen PM (1996) Estimating hydrologic budgets for three Illinois watersheds. J Hydrol 176(1–4):57–77

Arnold JG, Allen PM (1999) Automated methods for estimating baseflow and groundwater recharge from streamflow records. JAWRA J Am Water Resour Assoc 35(2):411–424

Arnold JG, Srinivasan R, Muttiah RS, Williams JR (1998) Large area hydrologic modeling and assessment—part 1: model development. J Am Water Resour Assoc 34(1):73–89

Arribas AA, Gallardo CG, Gaertner MG, Castro MC (2003) Sensitivity of the Iberian Peninsula climate to a land degradation. Clim Dyn 20(5):477–489

Assefa A, Melesse AM, Admasu S (2014) Climate change in upper Gilgel Abay River catchment, Blue Nile basin Ethiopia. In: Melesse AM, Abtew W, Setegn S (eds) Nile River basin: ecohydrological challenges, climate change and hydropolitics, pp 363–388

Bartolini G, di Stefano V, Maracchi G, Orlandini S (2012) Mediterranean warming is especially due to summer season evidences from Tuscany (central Italy). Theoret Appl Climatol 107(1–2):279–295

Behulu F (2013) Modelling the hydrologic behavior of the upper Tiber River basin under climate change scenarios. PhD dissertation

Behulu F, Melesse AM, Romano E, Volpi E, Fiori A (2012) Statistical downscaling of precipitation and temperature for the Upper Tiber Basin in Central Italy. Int J Water Sci 1(3)

Behulu F, Setegn S, Melesse AM, Fiori A (2013) Hydrological analysis of the Upper Tiber basin: a watershed modeling approach. Hydrol Process 27(16):2339–2351

Behulu F, Setegn S, Melesse AM, Romano E, Fiori A (2014) Impact of climate change on the hydrology of upper Tiber River basin using bias corrected regional climate model. Water Resour Manage 28:1327–1343

Beniston M, Stephenson D, Christiansen O, Ferro C, Frei C, Goyette S, Halsnaes K, Holt T, Jylha K, Koffi B, Palutikof J, Scholl R, Semmler T, Woth K (2007) Future extreme events in European climate: an exploration of regional climate model projections. Clim Change 81:71–95

Brocca L, Camici S, Tarpanelli A, Melone F, Moramarco T (2011) Analysis of climate change effects on floods frequency through a continuous hydrological modelling. In: Baba A et al (eds) Climate change and its effects on water resources NATO science for peace and security series C: environmental security, vol 3, pp 97–104

Brunetti M, Colacino M, Maugeri M, Nanni T (2001) Trends in the daily intensity of precipitation in Italy from 1951 to 1996. Int J Climatol 21(3):299–316

Brunetti M, Colacino M, Nanni T, Navarra A (2002) Droughts and extreme events in regional daily Italian precipitation series. Int J Climatol 22:543–558

Brunetti M, Buffoni L, Mangianti F, Maugeri M, Nanni T (2004) Temperature, precipitation and extreme events during the last century in Italy. Global Planet Change 40(1–2):141–149

Brunetti M, Maugeri M, Monti F, Nanni T (2006) Temperature and precipitation variability in Italy in the last two centuries from homogenised instrumental time series. Int J Climatol 26:345–381

Buonomo E, Jones R, Huntingford C, Hannaford J (2007) On the robustness of changes in extreme precipitation over Europe from two high resolution climate change simulations. Q J Royal Meteorol Soc 133(622):65–81
Burlando P, Rosso R (2002) Effects of transient climate change on basin hydrology. 2. Impact on runoff variability in the Arno River, Central Italy. Hydrol Process 16:1177–1199
Cesari G (2010) Il bacino del Tevere, il suo ambiente idrico e l'impatto antropico. http://www.museoenergia.it/museo.php?stanza=12&ppost=962. Accessed 20 Dec 2014
Chen Z, Grasby SE, Osadetz KG (2002) Predicting annual groundwater levels from climatic variables: an emperical model. J Hydrol 260:102–117
Chen Z, Grasby SE, Osadetz KG (2004) Relation between climate variability and groundwater levels in the upper carbonate aquifer, southern Manitoba, Canada. J Hydrol 290:43–62
Christensen JH, Carter TR, Rummukainena M, Amanatidis G (2007) Evaluating the performance and utility of regional climate models: the PRUDENCE project. Clim Change 81:1–6
Colombo T, Pelino V, Vergari S, Cristofanelli P, Bonasoni P (2007) Study of temperature and precipitation variations in Italy based on surface instrumental observations. Glob Planet Change 57:308–318
Coppola E, Giorgi F (2010) An assessment of temperature and precipitation change projections over Italy from recent global and regional climate model simulations. Int J Climatol 30(1):11–32
D'Agostino DR, Trisorio LG, Lamaddalena N, Ragab R (2010) Assessing the results of scenarios of climate and land use changes on the hydrology of an Italian catchment: modelling study. Hydrol Process 24(19):2693–2704
Dessu SB, Melesse AM (2012) Modeling the rainfall-runoff process of the Mara River basin using SWAT. Hydrol Process 26(26):4038–4049
Dessu SB, Melesse AM (2013) Impact and uncertainties of climate change on the hydrology of the Mara River basin. Hydrol Process 27(20):2973–2986
Dessu SB, Melesse AM, Bhat M, McClain M (2014) Assessment of water resources availability and demand in the Mara River basin. CATENA 115:104–114
Di Matteo L, Dragoni W (2006) Climate change and water resources in limestone and mountain areas: the case of Firenzuola Lake (Umbria, Italy). In: Proceedings of the 8th conference on limestone hydrogeology 2006, Neuchâtel, Switzerland
European Union (2000) Directive 2000/60/EC of the European parliament and of the. Official J Eur Communities L327(2):72–73
Fatichi S, Caporali E (2009) A comprehensive analysis of changes in precipitation regime in Tuscany. Int J Climatol 29(13):1883–1893
Gassman PW, Reyes MR, Green CH, Arnold JG (2007) The soil and water assessment tool: historical development, applications, and future research directions. Trans ASABE 50 (4):1211–1250
Getachew HE, Melesse AM (2012) Impact of land use /land cover change on the hydrology of Angereb watershed, Ethiopia. Int J Water Sci 1,4:1–7. doi:10.5772/56266
Giorgi F (2006) Climate change hot-spots. Geophys Res Lett 33:L08707, doi:10.1029/2006GL025734
Giorgi F, Marinucci MR, Bates GT (1993a) Development of a second-generation regional climate model (RegCM2). Part I: boundary- layer and radiative transfer processes. Mon Weather Rev 21:2794–2813
Giorgi F, Marinucci MR, Bates GT (1993b) Development of a second-generation regional climate model (RegCM2). Part II: convective processes and assimilation of lateral boundary conditions. Mon Weather Rev 121:2814–2832
Graham LP, Hagemann S, Jaun S, Beniston M (2007) On interpreting hydrological change from regional climate models. Clim Change 81:97–122
Green TR, Taniguchi M, Kooi H, Gurda JJ, Allen DM, Hiscock KM, Treidel H, Aureli A (2011) Beneath the surface of global change: impacts of climate change on groundwater. J Hydrol 405:532–560
Grey OP, Dale F St, Webber G, Setegn Shimelis G, Melesse Assefa M (2013) Application of the soil and water assessment tool (SWAT Model) on a small tropical Island State (Great River

Watershed, Jamaica) as a tool in integrated watershed and coastal zone management. Int J Trop Biol Conserv 62(3):293–305
IPCC (2000) Special report on emission scenarios (SRES): a special report of working group II of the intergovernmental pannel on climate change. Cambridge University Press, Cambridge
IPCC (2007) Climate change 2007: the physical science basis. IPCC-Report AR4(WG-I)
Jacob D, BÃ¤rring L, Christensen O, Christensen J, de Castro M, DÃ©quÃ© M, Giorgi F, Hagemann S, Hirschi M, Jones R, KjellstrÃ¶m E, Lenderink G, Rockel B, SÃ¡nchez E, SchÃ¤r C, Seneviratne S, Somot S, van Ulden A, van den Hurk B (2007) An inter-comparison of regional climate models for Europe: model performance in present-day climate. Clim Change 81(0):31–52
Jones CG, Willén U, Ullerstig A, Hansson U (2004) The Rossby centre regional atmospheric climate model part I. Model climatology and performance for the present climate over Europe. Ambio 33(4–5):199–210
Kjellstrom E, Barring L, Jacob D, Jones R, Lenderink G, Schar C (2007) Modeling daily temperature extremes: recent climate and future changes over Europe. Clim Change 81:249–265
Kløve B, Ala-Aho P, Bertrand G, Gurdak JJ, Kupfersberger H, Kværner Muotka T, Mykrä H, Preda E, Rossi P, Uvo CB, Velasco E, Pulido-Velazquez M (2014) Climate change impacts on groundwater and dependent ecosystems. J Hydrol 518:250–266
Lenderink G, Buishand A, Van Deursen W (2007) Estimates of future discharges of the river Rhine using two scenario methodologies: direct versus delta approach. Hydrol Earth Syst Sci 11(3):1145–1159
Loaiciga HA (2003) Climate change and groundwater. Ann Assoc Am Geogr 93(1):30–41
Loaiciga HA, Maidment DR, Valdes JB (2000) Climate-change impacts in a regional karst aquifer, Texas, USA. J Hydrol 227(1–4):173–194
Luo Y, Arnold J, Allen P, Chen X (2012) Baseflow simulation using SWAT model in an inland river basin in Tianshan Mountains, Northwest China. Hydrol Earth Syst Sci 16:1259–1267
Mango L, Melesse AM, McClain ME, Gann D, Setegn SG (2011a) Land use and climate change impacts on the hydrology of the upper Mara River Basin, Kenya: results of a modeling study to support better resource management. Spec Issue Clim Weather Hydrol East Afr Highlands Hydrol Earth Syst Sci 15:2245–2258. doi:10.5194/hess-15-2245-2011
Mango L, Melesse AM, McClain ME, Gann D, Setegn SG (2011b) Hydro-meteorology and water budget of Mara River basin, Kenya: a land use change scenarios analysis. In: Melesse A (ed) Nile River basin: hydrology, climate and water use (chapter 2). Springer Science Publisher, Berlin, pp 39–68. doi:10.1007/978-94-007-0689-7_2
Matzneller P, VenturaF Gaspari N, Pisa PR (2010) Analysis of climatic trends in data from the agrometeorological station of Bologna-Cadriano, Italy (1952–2007). Clim Change 100(3–4):717–731
Melesse A, Loukas Athanasios G, Senay Gabriel, Yitayew Muluneh (2009) Climate change, land-cover dynamics and ecohydrology of the Nile River Basin, Hydrological Processes. Spec Issue Nile Hydrol 23(26):3651–3652
Melesse A, Bekele S, McCornick P (2011) Hydrology of the Niles in the face of land-use and climate dynamics. In: Melesse A (ed) Nile River basin: hydrology, climate and water use. Springer Science Publisher, Berlin, pp vii–xvii. doi:10.1007/978-94-007-0689-7
Mohammed H, Alamirew T, Assen M, Melesse AM (2015) Modeling of sediment yield in Maybar gauged watershed using SWAT, northeast Ethiopia. CATENA 127:191–205
Moonen AC, Ercoli L, Mariotti M, Masoni A (2002) Climate change in Italy indicated by agrometeorological indices over 122 years. Agric For Meteorol 111:13–27
Moriasi DN, Arnold JG, Van Liew MW, Bingner RL, Harmel RD, Veith TL (2007) Model evaluation guidelines for systematic quantification of accuracy in watershed simulations. Trans ASABE 50(3):885–900
Neitsch SL, Arnol JG, Kiniry JR, Williams JR (2005) Soil and water assessment tool, theoretical documentation version 2005
Neitsch SL, Arnold JG, Kiniry JR, Williams JR (2009) Soil and water assessment tool, theoretical documentation version 2009

Romano E, Preziosi E (2012) Precipitation pattern analysis in the Tiber River Basin (Central Italy) using standardized indices. Int J Climatol 33(7):1791792
Romano E, Petrangeli AB, Preziosi E (2011) Spatial and time analysis of rainfall in the Tiber River basin (Central Italy) in relation to discharge measurements (1920–2010). Proc Environ Sci 7:258–263
Roosmalen VL, Christensen BSB, Sonnenborg TO (2007) Regional differences in climate change impacts on groundwater and stream discharge in Denmark. Vadoze Zone J 6(3):554–571 (special issue)
Roosmalen VL, Sonnenborg OT, Jensen HK (2009) Impact of climate and land use change on the hydrology of a large-scale agricultural catchment. Water Resour Res 45(7)
Scibek J, Allen DM (2006) Modeled impacts of predicted climate change on recharge and groundwater levels. Water Resour Res 42(W11405):1–18
Singh VP, Woolhiser DA (2002) Mathematical modeling of watershed hydrology. J Hydrol Eng 7 (4):270–292
Taylor RG, Scanlon B, Döll P, Rodell M, van Beek R, Wada Y, Longuevergne L, Leblanc M, Famiglietti JS, Edmunds M, Konikow L, Green TR, Chen J, Taniguchi M, Bierkens MFP, MacDonald A, Fan Y, Maxwell RM, Yechieli Y, Gurdak J, Allen D, Shamsudduha M, Hiscock K, Yeh PJF, Treidel H (2013) Groundwater and climate change. Nat Clim Change 3:322–329
Teutschbein CA, Seibert J (2012) Bias correction of regional climate model simulations for hydrological climate-change impact studies: review and evaluation of different methods. J Hydrol (456–457):12–29
Themeßl MJ, Gobiet AA, Leuprecht A (2010) Empirical-statistical downscaling and error correction of daily precipitation from regional climate models. Int J Climatol 31(10):1530–1544
Todisco F, Vergni L (2008) Climatic changes in Central Italy and their potential effects on corn water consumption. Agric For Meteorol 148(1):1–11
Toews MW, Allen DM (2009) Simulated response of groundwater to predicted recharge in a semi-arid region using a scenario of modelled climate change. Environ Res Lett 4(035003):19
Vergni L, Todisco F (2011) Spatio-temporal variability of precipitation, temperature and agricultural drought indices in Central Italy. Agric For Meteorol 151(3):301–313
Wang X, Melesse AM (2005) Evaluations of the SWAT model's snowmelt hydrology in a Northwestern Minnesota Watershed. Trans ASAE 48(4):1359–1376
Wang X, Melesse AM (2006) Effects of STATSGO and SSURGO as inputs on SWAT model's snowmelt simulation. J Am Water Resour Assoc 42(5):1217–1236
Wang X, Melesse AM, Yang W (2006) Influences of potential evapotranspiration estimation methods on SWAT's hydrologic simulation in a Northwestern Minnesota watershed. Trans ASAE 49(6):1755–1771
Wang X, Shang S, Yang W, Melesse AM (2008a) Simulation of an agricultural watershed using an improved curve number method in SWAT. Trans Am Soc Agri Bio Eng 51(4):1323–1339
Wang X, Yang W, Melesse AM (2008b) Using hydrologic equivalent wetland concept within SWAT to estimate streamflow in watersheds with numerous wetlands. Trans Am Soc Agri Bio Eng 51(1):55–72
Wang X, Garza J, Whitney M, Melesse AM, Yang W (2008c) Prediction of sediment source areas within watersheds as affected by soil data resolution, In: Findley PN (ed) Environmental modelling: new research, chap 7. Nova Science Publishers, Inc., Hauppauge, pp 151–185 (ISBN: 978-1-60692-034-3)
Woldeamlak ST, Batelaan O, De Smedt F (2007) Effects of climate change on the groundwater system in the Grote-Nete catchment, Belgium. Hydrogeol J 15:891–901

Chapter 31
Estimation of Climate Change Impacts on Water Resources in the Great River Watershed, Jamaica

Orville P. Grey, St. Dale F.G. Webber, Shimelis Setegn and Assefa M. Melesse

Abstract The Great River Watershed, located in north-west Jamaica, is recognized as a critical watershed for development, particularly in the areas of housing, tourism, agriculture, and mining. Additionally, it is a source for sediment and other nutrient loading to the coastal environment. Island states frequently make decisions based on watershed impacts. This requires an in-depth understanding and analysis of factors such as water resources, land use/cover, and climate change impact among others at the watershed level. An integrated modelling framework has been proposed for this watershed that includes modelling of hydrological parameters using the Soil and Water Assessment Tool (SWAT) and projected changes in temperature and rainfall on water resources. The simulated model performance statistics for high flow discharge yielded a Nash–Sutcliffe Efficiency (NSE) value of 0.68 and a coefficient of determination (R^2) value of 0.70 suggesting good measured and simulated (calibrated) discharge correlation. Projected climate change will likely have huge impacts on available water in stream flow. Similarly, the potential impacts to soil water content, evapotranspiration, and surface runoff are also shown to be highly sensitive and important. The calibrated SWAT model can and should be used for further analysis of the effect of climate change as well as other different management scenarios for the remaining 25 watersheds in Jamaica as well as other island states in the Caribbean.

Keywords Soil and water assessment tool · Climate change · Great River Watershed · Jamaica

O.P. Grey (✉) · St. D.F.G. Webber
Centre for Marine Sciences, University of the West Indies, Mona, Kingston 7, Jamaica
e-mail: opgjunior@gmail.com

S. Setegn
Applied Research Center (ARC) and Department of Environmental and Occupational Health, Florida International University, Miami, USA

A.M. Melesse
Department of Earth and Environment, Florida International University, Miami, USA

© Springer International Publishing Switzerland 2016

A.M. Melesse and W. Abtew (eds.), *Landscape Dynamics, Soils and Hydrological Processes in Varied Climates*, Springer Geography,
DOI 10.1007/978-3-319-18787-7_31

31.1 Introduction

The Fourth Assessment Report of the Intergovernmental Panel on Climate Change (IPCC 2007a) projects a bleak future for water resources availability in the Caribbean regions. The report suggests that decreases in mean annual precipitation (in some cases by as much as 20 %) are likely in the regions of the sub-tropics (Cashman et al. 2010; IPCC 2007b). Unlike other regions, major fluctuations in precipitation tend to significantly influence low-latitude tropical island watersheds. This influence is exposed when one considers the impacts of severe precipitation events such as hurricanes and tropical storms, and is further impacted by infrequent drought episodes (Smith et al. 2003).

The impact of climate change on water resources variability, especially on river flows, soil moisture, evapotranspiration, and groundwater flow, has been studied by analysing projected and downscaled climatic data and using hydrological models (Mango et al. 2011a, b; Behulu et al. 2014; Setegn et al. 2011, 2014; Setegn and Melesse 2014; Assefa et al. 2014; Melesse et al. 2009, 2011; Dessu and Melesse 2013; Dessu et al. 2014).

Arnell (2004) suggested that many Caribbean islands are likely to become increasingly water-stressed in the future, as a result of climate change irrespective of the climate scenario employed based on recent projections from a macro-scale hydrological model using the IPCC emission scenarios. A significant reduction of rainfall is projected by the "Providing Regional Climates for Impacts Studies (PRECIS)" model in the latitudinal band encompassing the Caribbean Sea, Central America, and part of the western half of northern South America. In this zone, the reductions are in the range of 10–50 %. The patterns of change in annual rainfall indicate a general future trend to drier conditions in various areas of the Caribbean sub-region such as Jamaica (Chen et al. 2008; ECLAC 2010; Taylor et al. 2007).

Jamaica's freshwater resources are derived from surface sources (rivers and streams), underground sources (wells and springs), and through rainwater harvesting. Groundwater resource satisfies for approximately 80 % of the water demand and represents approximately 84 % of the island's exploitable water (Climate Studies Group Mona 2012). The surface water resources have seasonal variability in flow which can in part be linked to changes in rainfall. Groundwater is also directly recharged by rainfall and indirectly from rivers and streams. Water resources are important to Jamaica's economy. Water supports important sectors including tourism and recreation, mining, food and beverage processing, irrigated agriculture, and manufacturing. The major users of water are irrigated agriculture (33 %), residential water use (21 %), and the environment (39 %). Manufacturing, hotels, and mining use less than 7 % of the annual available water (Government of Jamaica 2011; Climate Studies Group Mona 2012).

This project was initiated to develop and apply the Soil Water Assessment Tool (SWAT) to evaluate the ability of the model to predict stream flow and consequently the impacts of climate change on specific hydrological parameters (e.g., stream flow, evapotranspiration, and surface runoff); thereby testing the applicability of SWAT on a tropical island watershed. In this study, calibration and validation of the hydrological component of SWAT was conducted using weather data for a 9-year period (1998–2006) focusing on a critically important watershed in fairly good health. The Great River Watershed (GRW) on the islands north coast was selected for this investigation.

31.2 Materials and Methodology

31.2.1 Study Area

The Great River Watershed is located in the north-western section of the island (Fig. 31.1). It is one of the 26 watersheds in the island and forms the western boundary of the parish of St James with Westmoreland and Hanover. The Great River is approximately 74 km long and encompasses drainage areas of these three parishes as well as the parish of St. Elizabeth to the south with an area of 327.27 km^2. The Great River has five major tributaries as follows: Brown's River, Sevens River, Quashies River, Lambs River, and Roaring River. It flows in a north–north-westerly direction from the community of Pisgah at an approximate elevation of 430 m above mean sea level in the parish of St. Elizabeth to the Caribbean Sea at

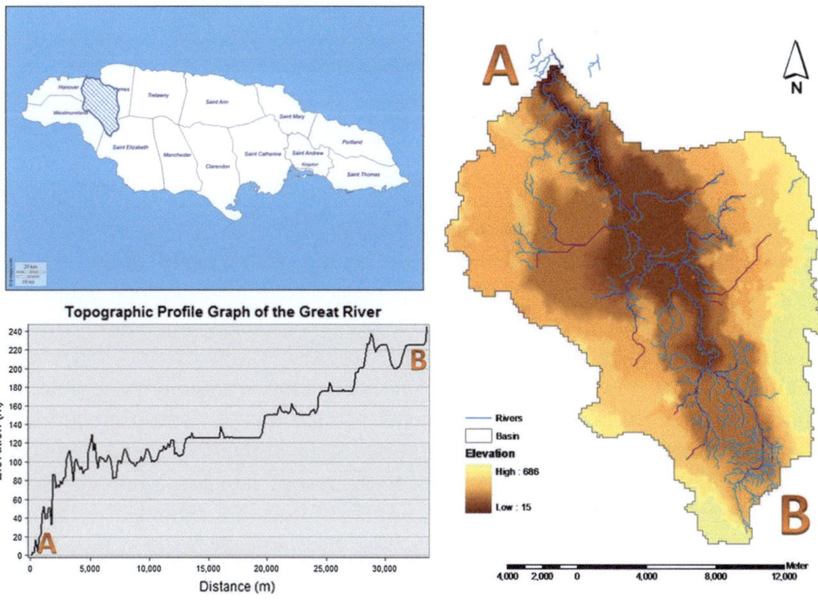

Fig. 31.1 Great River watershed, Jamaica (Hayman 2001)

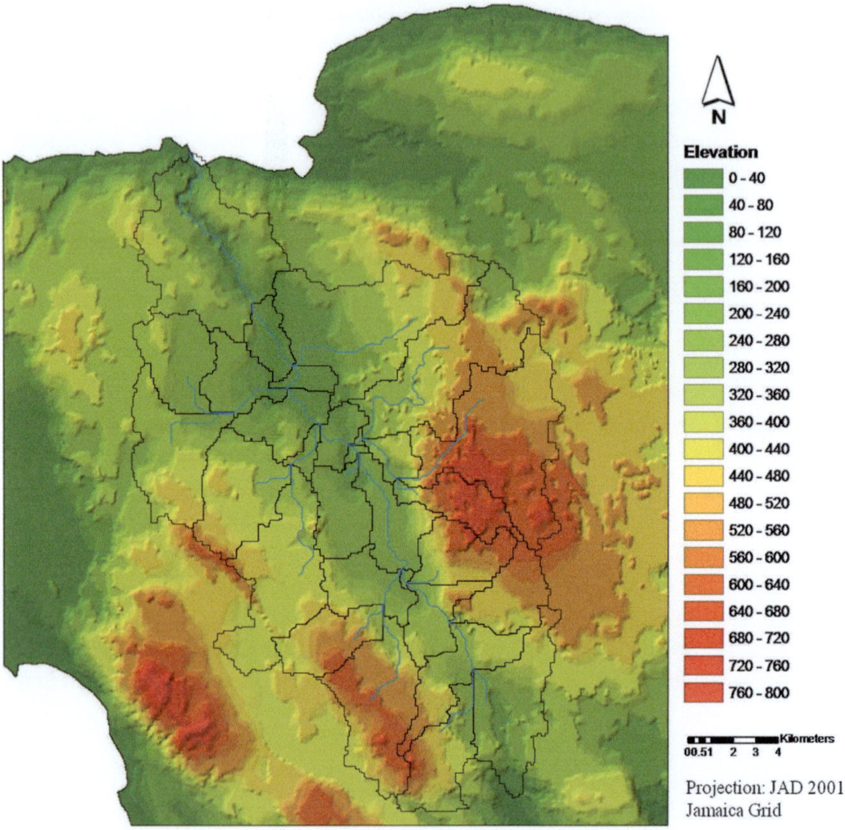

Fig. 31.2 Topographic map of the Great River watershed

Great River Bay on the north coast (ARD 2003; Hayman 2001). Figure 31.2 shows the elevation of the watershed.

The hydrostratigraphy is influenced by the geology (lithography) of the area. The dominant hydrostratigraphic unit within the watershed is the limestone aquifer followed by the basal and limestone aquiclude (Fig. 31.3). The aquifers have sufficient permeability and storage to support economic, exploitable groundwater yield whereas the aquicludes have low permeability and do not support economic water yield. The tributaries of these rivers have steep gradients, and stream flow is reduced during the dry season. Runoff from the slopes above the bed of the tributaries is rapid, and flow is of short duration (Hayman 2001). Similar to the rest of Jamaica, a bimodal annual rainfall pattern is observed in the watershed. Both peaks are exhibited in the rainy season that starts in May and ends in November, the first peak being in June and the second in September and October. The latter is the most dominant peak. Regular monitoring of water quality within the watershed is not routinely carried out, and as such, monitoring data are inconsistent (Greenaway 2004; Hayman 2001).

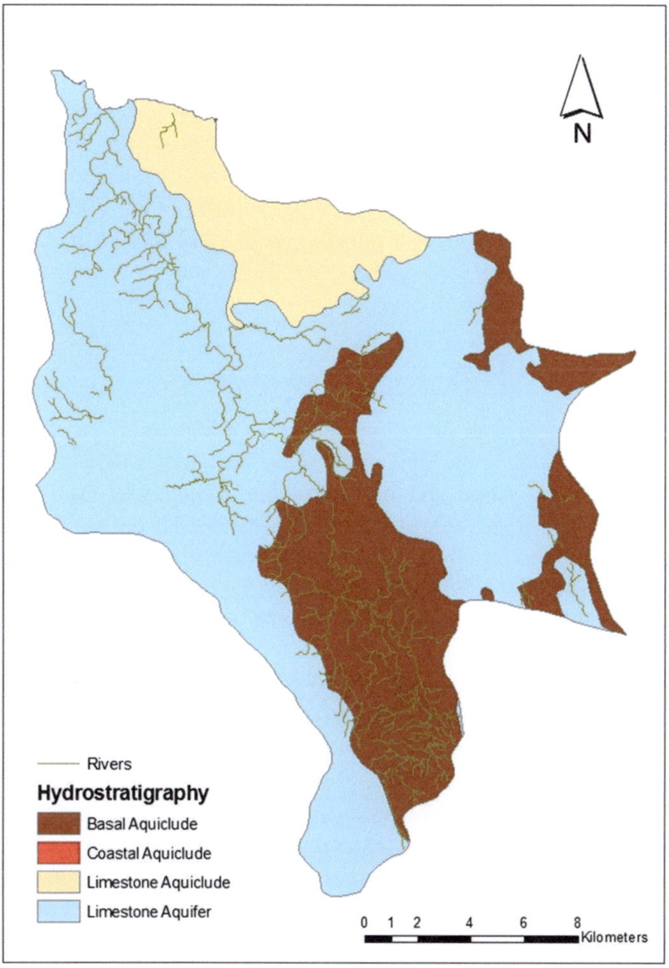

Fig. 31.3 Hydrostratigraphic units within the Great River watershed

31.2.2 *Soil and Water Assessment Tool (SWAT)*

The SWAT model (Arnold et al. 1998) is a physically-based, continuous simulation model developed for a watershed assessment of short- and long-term hydrology and water quality. The model requires extensive input data, which can be aided by GIS data and interface (Di Luzio et al. 2002). ArcSWAT interface for the SWAT 2005 model was used for this study. SWAT allows for the discretization of a watershed by dividing it into multiple sub-watersheds, which can then be further subdivided into hydrological response units (HRUs) that consist of homogeneous land use, management, and soil characteristics (Neitsch et al. 2005). ArcGIS was used to calculate basic hydrological information for the model (i.e., surface slope and water flow paths), the position, and the size of the HRUs, and provide the necessary files that are used by the SWAT model (Winchell et al. 2007, 2009).

The application of SWAT in predicting stream flow and sediment as well as evaluation of the impact of land use and climate change on the hydrology of watersheds has been documented by various studies (Dessu and Melesse 2012, 2013; Wang et al. 2006, 2008a, b; Wang and Melesse 2005, 2006; Behulu et al. 2013, 2014; Setegn et al. 2009a, b, 2014; Mango et al. 2011a, b; Getachew and Melesse 2012; Assefa et al. 2014; Grey et al. 2013; and Mohammed et al. 2015).

31.2.2.1 Model Input Data

A 56-m resolution digital elevation model (DEM), supplied by the Mona Geo-Informatics Institute at the University of the West Indies (Mona, Jamaica), was used to determine the slope and flow direction, which was used to determine sub-basin outlets and areas contributing discharge to the outlets. Land-use/Land cover data were supplied by the Forestry Department of Jamaica. This dataset was created in 1998. The land-use/land cover map was reclassified to match the SWAT Land Cover database using a lookup table with attributive fields containing the codes/categories (Table 31.1).

Weather data (rainfall and temperature) were supplied by the Meteorological Service of Jamaica for the period 1998–2006, the longest continuous dataset with the most meteorological stations (4). Stream network and soil data were supplied by the Water Resources Authority of Jamaica (WRA). Stream gauge was used as the outlet to the stream network. Downscaled projections from Climate Studies Group at UWI Mona were also used. This dataset represented the absolute change in temperature and percentage change in precipitation relative to the simulated baseline. The applied scenarios were A2, B1, and B2 SRES scenarios based on the dataset available at the time of the research for 2015s, 2030s, and 2050s. Additional data layers were supplied by the Natural Environment and Planning Agency (NEPA). All digital datasets were projected to the Lambert Conformal Conic Projection, and the projected coordinate system used was the JAD 2001 Jamaica Grid.

31.2.2.2 Watershed Delineation

The modelling process generated 30 HRUs that represent the entire GRW. Calibration of the model was performed by comparing the simulated discharge with the monitoring (measured) discharge data. The measured data was divided into two time periods covering the period 1998–2006. The data represented information from four of eight meteorological stations. Only four meteorological stations had continuous daily data for the time period under investigation.

Calibration of the discharge was achieved by adjusting the input parameters. The top 10 parameters were selected based on ranking achieved from the sensitivity

Table 31.1 Definition of land-use/cover types based on LANDSAT TM 1998 (1:100,000 scale)

Original land use	Original definition	Definition of land use	[a]SWAT definition
Disturbed broadleaf forest (2° Forest)	SF	Disturbed broadleaf forest with broadleaf trees at least 5 m tall and species indicators of disturbance such as *Cecropia peltata* (trumpet tree)	FRST
Buildings and other infrastructure	BA	Buildings and other constructed features such as airstrips and quarries.	URBN
Secondary forest and fields	SC	>50 % Disturbed broadleaf forest; >25 % fields	FRSD
Open dry forest—tall (Woodland/Savannah)	WL	Open natural woodland or forest with trees at least 5 m tall and crowns not in contact, in drier part of Jamaica with species indicators such as *Bursera simaruba* (red birch)	FRST
Fields and secondary forest	CS	>50 % fields; >25 % disturbed broadleaf forest	AGRL
Fields (herbaceous crops, fallow, cultivated vegetables)	FC	Herbaceous crops, fallow, cultivated grass/legumes	AGRR
Bamboo and secondary forest	BF	>50 % bamboo; >25 % disturbed broadleaf forest	FRST
Bamboo and fields	BC	>50 % bamboo; >25 % fields	RNGB
Closed broadleaf forest (1° forest)	PF	Closed primary forest with broadleaf trees at least 5 m tall and crowns interlocking, with minimal human disturbance	FRSE

[a]Original land use was converted to the most appropriate SWAT definition which resulted in more than one original definition matching a SWAT definition

analysis conducted. Adjusting the selected parameters allowed for a better match of measured and simulated discharges. The most sensitive parameters were used to calibrate the model for the GRW. The first-year data (1998) were used as start-up/warm-up in the calibration process and were therefore not included in the final model simulations. No formal optimization procedure during calibration exists, and therefore, subjective decisions were generally made in calibrating the model (Santhi et al. 2006).

31.2.3 Model Calibration and Validation

The weather data used for driving the hydrological balance were daily precipitation and air temperature for the period 1998–2006. The weather data used represent the Meteorological Stations at Cacoon Castle, Catadupa, Montpellier, and Mount Horeb.

The simulated daily stream flows were manually calibrated and validated against observed flow measured at Lethe on the Great River (1998–2006). The gauge station is located closer to the mouth of the Great River and represents the only functional gauging station in the watershed. The split-sample approach was utilized for calibration and validation. The period 1998–2002 was selected for calibration with the first year used as a model start-up. The period 2002–2006 was selected for validation. For determination of model efficiency, we used the coefficient of determination (R^2) and the Nash–Sutcliffe model efficiency (NSE) (Nash and Sutcliffe 1970).

The Nash–Sutcliffe efficiency (NSE) is a normalized statistic that determines the relative magnitude of the residual variance ("noise") compared to the measured data variance ("information") (Nash and Sutcliffe 1970). NSE indicates how well the plot of observed versus simulated data fits the 1:1 line. Nash–Sutcliffe efficiencies range from $-\infty$ to 1. Essentially, the closer to 1, the more accurate the model is. NSE of 1 corresponds to a perfect match of modelled to the observed data. NSE of zero indicates that the model predictions are as accurate as the mean of the observed data and values of $-\infty < \text{NSE} < 0$ indicate that the observed mean is better predictor than the model.

31.2.4 Climate Change Projections

This study used the absolute change in temperature and percentage change in precipitation projections relative to the simulated baseline under the A2, B1, and B2 SRES scenarios. No data were readily available for the regional climate model (RCM) for the B1 scenario projections; hence, the B2 scenario projections were substituted. In all cases, model projections are for a 10-year period centred on 2015, 2030, and 2050. The monthly projections were used to make adjustment in the SWAT model to match monthly RCM and global circulation model (GCM) data with observed data during simulation period. These changes were made by editing the sub-basin database and adjusting the measurements in the weather adjustment dialogue box for rainfall and temperature, then rerun SWAT for each scenario.

The RCM and GCM data for precipitation and temperature were obtained from the Climate Change Group at University of West Indies (Mona Campus). The RCM data were based on type (2) downscaling and subsequent climate change projections for the Caribbean region using version 1.3 of the Hadley Centre HadRM3P model, which is a RCM within the PRECIS model (Campbell et al. 2010). PRECIS model data used in this study used 50-km resolution largely due to the unavailability of resolutions of up to 25 km at the time of the study.

The GCM data for precipitation and temperature were obtained from the Climate Studies Group Mona at University of West Indies (Mona Campus). Data from three GCMs were used: the ECHAM 5/MPI-OM developed by the Max Planck Institute for Meteorology in Germany, HadCM3 developed by the Hadley Centre in the

United Kingdom, and the MRI-CGCM2 developed by the Meteorological Research Institute in Japan.

The data were analysed for model performance as well as comparison of four parameters of the hydrological cycle, and evaluating their potential impact to the watershed considering environmental and socioeconomic concerns. The parameters examined were stream flow, soil water content, evapotranspiration, and surface water runoff. The performance of the calibrated model was used as the basis to examine various aspects of the hydrological cycle.

31.3 Results

31.3.1 Monthly Stream Flow

The SWAT model performance for calibration is acceptable. The mean measured monthly flow was 10.18 m^3/s (std. dev. = 9.94), slightly lower than the mean simulated monthly flow of 10.25 m^3/s (std. dev. = 7.06). The mean measured monthly flow was 11.50 m^3/s (std. dev. = 13.32), slightly higher than the mean simulated monthly flow of 10.66 m^3/s (std. dev. = 7.72). Figure 31.4 shows comparisons of the observed and modelled flow for the calibration and validation phases.

The model performance statistics for discharge yielded a NSE value of 0.68 and an R^2 value of 0.70, suggesting fairly good correlation between measured and simulated (calibrated) discharge. The validation period similarly demonstrated good model performance. The NSE value of 0.61 and R^2 value of 0.67 also suggest a fairly good correlation.

31.3.2 Climate Change Projections

The impact of climate change was evaluated using the validated model. Despite the fact that land-use change is expected, this condition was kept constant. As such, the results are only indicative assuming climate change impacts with all other parameters kept constant. The three GCMs are referred to in the text as ECH, HAD, and MRI. The calibrated stream flow was used as the basis for analysis.

31.3.3 Impact of Climate Change on Stream Flow

The predicted changes for the three time periods are not as clear-cut as expected. Three A2 scenarios (PRECIS, ECH, and HAD) predicted a reduction in mean

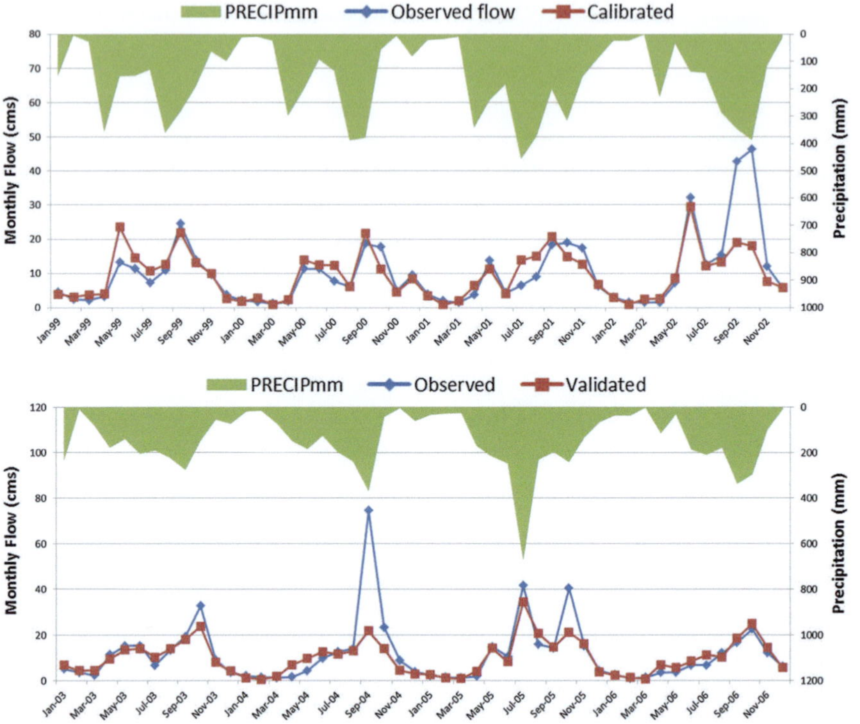

Fig. 31.4 Comparison of observed and modelled flow for calibration (*top*) and validation (*bottom*) phases

annual stream flow for all time periods. The GCM-based MRI-A2 scenario was the only A2 scenario to predict increases over the baseline with the exception of the 2030 time period. The projections observed indicated a general trend of reductions in stream flow for subsequent time periods.

The RCM-based PRECIS suggest stream flow will be highest during the 2030s. This trend is fairly similar to the ECH-based scenario with mean annual stream flow at or less than the current baseline. The HAD scenarios suggest a consistent decrease in stream flow in each time period, despite being marginally higher than the baseline in the 2015s and 2030s. The 2050 time period is projected to have the lowest average annual stream flow of all time periods and may be a cause for concern. The RCM-based scenarios also provide variability with stream flow for all three time periods. The MRI-A2 scenarios were the only scenarios to predict an increase in stream flow for the 2050s above baseline (Fig. 31.5).

The B1/B2 scenarios recorded more variability than the A2 scenarios. All scenarios predicted both increases and decreases against the baseline. Unlike the A2 scenarios, PRECIS-B2 predicted an increase in stream flow for the 2030s (Fig. 31.5). The HAD-B1 scenarios, similar to the A2 scenarios, predicted a consistent decrease in stream flow. The ECH-B1 scenario predicted stream flow fairly

Fig. 31.5 Average annual stream flow predictions for all time periods and scenarios. **a** A2 scenarios. **b** B1/B2 scenarios

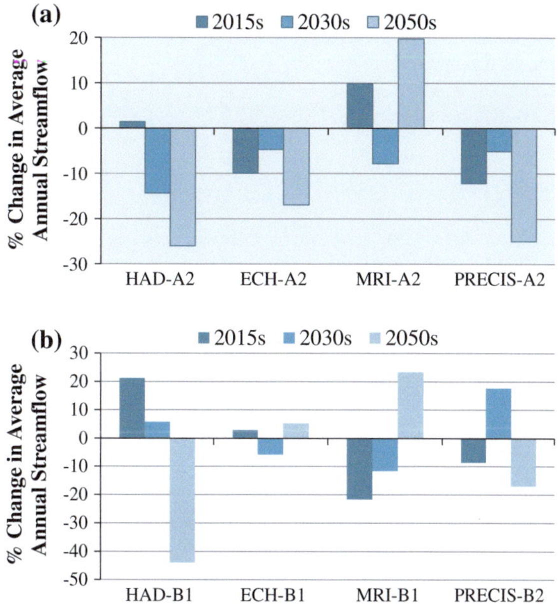

similar to the baseline. This trend is similar to the A2 scenario. The MRI-B1 scenario projected a consistent increase in stream flow, although the 2015s and 2030s were lower than the baseline.

31.3.4 Impact Climate Change on Soil Water Content

The data show a general reduction in soil water content across all time periods modelled, irrespective of scenario. The change in soil water content projected reflects a maximum increase of 3 % to a maximum decrease of 10 %. ECH predicted a progressive decrease in soil water with increasing time period, the only model to do so consistently (Fig. 31.6).

Both the HAD-A2 and ECH-A2 scenarios consistently predicted decrease in soil water content of up to 6 %. Decreases were recorded for the PRECIS-A2 scenarios for all time periods except 2030s, but were not greater than 4 % (Fig. 31.6). The projected increase was marginal at less than 1 %. The MRI-A2 scenario also reflected decreases for most time periods of up to 9 %. Minor increase was observed for the 2015s, but no more than 1 % but gradually increased up to 7 % for the 2050s. No scenario increased in the 2050s, but a decreasing range between 4 and 7 %. The general trend of decreasing soil water content was also observed with the B1/B2 scenarios. Two of three GCMs (HAD-B1 and ECH-B1) predicted a decreasing trend similar to the A2 scenarios. All scenarios predicted decreases for at least 2 time periods. No scenario increased in the 2050s, but a decreasing range of 4–9 %.

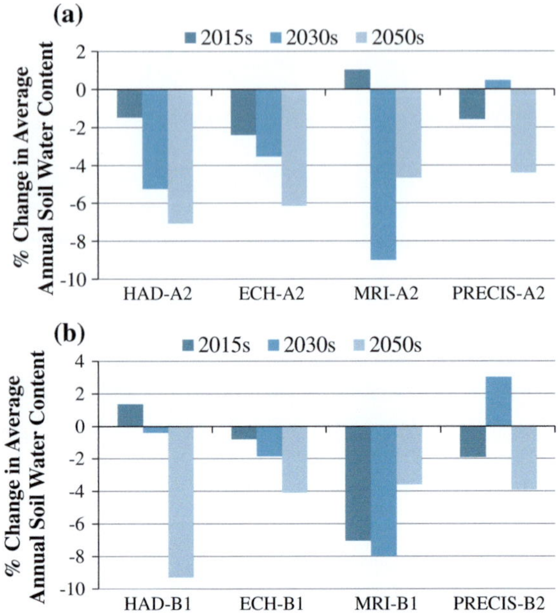

Fig. 31.6 Average annual soil water content predictions for all time periods and scenarios. **a** A2 scenarios. **b** B1/B2 scenarios

31.3.5 Impact of Climate Change on Actual Evapotranspiration (AET)

Unlike the other parameters, there was greater variability observed in the projected actual evaporation. The 2030s is projected to have an overall marginal increase in AET of up to 3 % with the regional model (Fig. 31.7). However, this scenario is not repeated by the GCM-based scenarios. Reductions in AET by the GCM scenarios suggest average annual AET may consistently decrease by as much as 7 %. Marginal increases were recorded for the 2015s whereas decreases were observed for other time periods. The greatest decrease was observed for the 2030s (Fig. 31.7).

The changes were more obvious for the B1/B2 scenario, which is based on a lower overall population growth and reduced land-use changes. The ECH-B1 scenario predicted a similar AET when compared with the baseline, not same for the A2 scenario. The MR1-B1 was the only scenario to predict consistent decrease when compared with the baseline. The most clear-cut trend was a general decrease in AET for the 2050s.

31.3.6 Impact Climate Change on Surface Runoff

The A2 scenarios project a slightly lower reduction range when compared with the B1/B2 scenarios for surface runoff due to the projected reduced precipitation

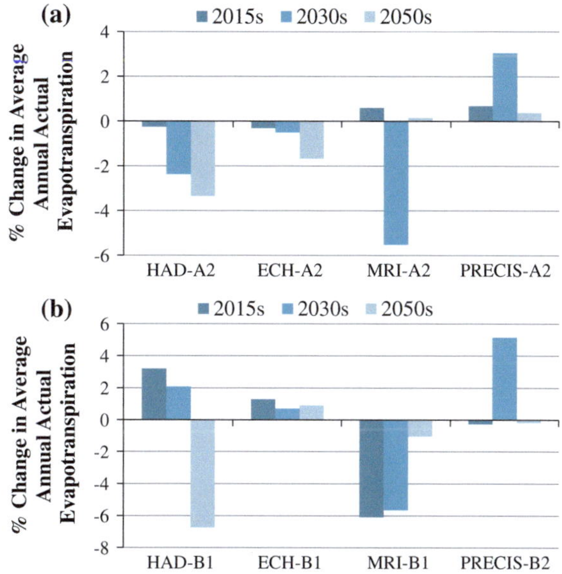

Fig. 31.7 Average annual evapotranspiration predictions for all time periods and scenarios. **a** A2 scenarios. **b** B1/B2 scenarios

projected as time periods increase (Fig. 31.8). Predicted decreases of up to 40 % were observed with the A2 scenarios. The MRI-A2 scenario was the only scenario to project consistent increases in mean annual surface runoff of up to 44 %. The ECH-A2 and the regional PRECIS-A2 scenario both project a consistent decreasing trend for annual surface runoff of between 5 and 30 %.

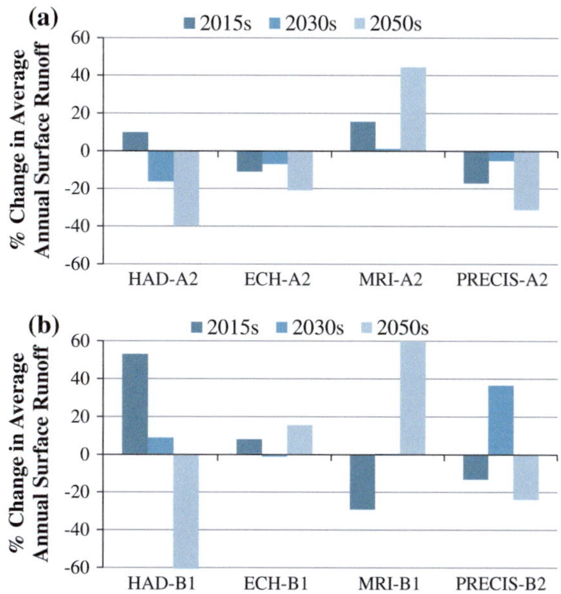

Fig. 31.8 Average annual surface runoff predictions for all time periods and scenarios. **a** A2 scenarios. **b** B1/B2 scenarios

The B1/B2 scenarios showed greater variability with each scenario predicting at least one time period with consistent reductions. These scenarios predicted a range of −64 to 61 %. Each time period recorded an even split in terms of number of scenarios showing reductions or increases. The 2030s recorded the narrowest range for surface runoff of −1 to 37 %. Consistent decreases were observed for the HAD-B1 scenario whereas the MRI-B1 scenario predicted the opposite.

31.4 Discussion

The availability of water resources is considered critical in the Caribbean islands. According to the IPCC (2007a), water resources are very critical in the low limestone islands of the eastern Caribbean, and it can be considered similarly important in larger Caribbean islands such as Jamaica that depends on surface flow from rivers for potable water, with many water takeoffs in close proximity to the river mouths. The islands volcanic bedrock accounts for no more than a third of Jamaica's rivers, and as such the water resources are extremely important in this environment since groundwater is not a significant resource due to geology. The remaining seventy per cent occurs on limestone bedrock, which tends to be karst.

Much of the rainfall is strongly associated with the genesis and passage of easterly waves, tropical depressions, and storms. Thus, changes in the occurrence of these heavy rainfall events will certainly impact the water supply of many Caribbean islands (IPCC 2007a). Droughts are considered to be more frequent in the Caribbean during El Niño years, whereas conditions tend to be wetter in La Niña years (IPCC 2007a).

The climate change scenario data used in this research are based on researched possible variations in greenhouse gas emissions. These climate change models predicted the likely climate change responses under varying scenarios globally due to the net effect of mixing of carbon dioxide. More recently, significant work has been done at the local and regional levels where it is expected that these changes will have a greater measureable impact compared with global possibilities. The two scenarios used, A2 and B1/B2, though different are not extremes and are the two most researched and peer-reviewed.

The A2 scenario predicts higher carbon dioxide emissions when compared with the B1/B2 scenario. Additionally, it is based on having a much larger population with a greater consumption of energy resources with a much lower technology application rate. This greater population will also demand greater land-use changes. Scarcity of resources is therefore greater for the A2 scenario.

Stream flow is largely dependent on the amount of rainfall that occurs across the entire watershed and is also strongly influenced by the evapotranspiration rates. Therefore, it is safe to suggest that rainfall and temperature will significantly affect stream flow. Climate change therefore has a significant potential impact on water availability for small island states. The data suggest mean stream flow (annual and monthly) will be reduced marginally in the 2015s. This suggests water availability will be marginally affected throughout the GRW, assuming no significant increase in population or socioeconomic developments.

Assuming no large-scale increases in population within the watershed, and no significant conversion of land use, it is quite likely water availability within the watershed will be adequate to meet the needs especially under B1 scenarios. The management of the watershed will be critical after the 2030s to ensure the population needs at that time are adequate.

The RCM-based data projections used in this scenario suggest a reduction in mean stream flow of 12 % on an annual basis. This potential change can be considered a significant potential impact on water availability, as well as impacts on agriculture, tourism destinations, and ecotour attractions that utilize the stream for economic benefit. All A2 scenarios simulated a decrease in mean annual stream flow for the 2030s. Despite the greater spatial resolution of the RCM scenario, the decreases in simulated stream flow are fairly similar to two of three GCM scenarios simulated.

The trend is also similar for the 2050s and will continue the trend of adverse water availability bearing in mind the great likelihood of an increase in population and development within the watershed. The increases suggested by the MRI scenarios, however, would improve water availability to support the potential increase in population and infrastructural need through development. Reduction in stream flow for the 2050s was much more than predicted for the 2030s time period.

Significant backyard farming takes place within the watershed. Additionally, important economic crops such as coffee, citrus, coconuts, and pineapples are grown on a moderate scale. Without adequate soil moisture, irrigation systems may have to be put in place or ramped up and this demand will likely heavily impact on the availability of water for other uses, particularly potable water. The agricultural impacts will most likely be observed during growing and reproductive phases. Whereas grass can make better use of soil moisture through a greater density of roots per unit area, vegetables (a major local cash crop) with much lower root density will potentially be impacted.

Soil moisture content is invariably linked with climate change. Much of the water that is absorbed through the roots of plants is lost through transpiration; this transfer of moisture as vapours additionally aids in the transfer of heat into the atmosphere. The potential reduction in soil moisture content will likely impact heavily on precipitation contributions at the local level; drier soils contribute little or no water vapour, thus reducing the potential for recharge through ad hoc local closed-loop systems. Drier soils, in the context of this tropical island watershed, may result in lowering of specific humidity and temperatures at the local level. As such, there is significant importance that should be given to the potential reduction of soil moisture content in the upper soil profile.

Most scenarios predicted a reduction in mean annual soil water content. With the exception of the MRI-B1 scenario, predicted soil water content will be marginally changed in the 2015s and 2030s. This suggests the soil moisture annual budget is not expected to be adversely affected throughout the GRW. The ability for vegetated areas to remain fairly moist should continue and the potential for bushfires or drought conditions may remain similar to current conditions during this period.

The PRECIS-A2 scenario simulated an increase of approximately 75 %, slightly greater than that simulated for the 2030s monthly.

All A2 scenarios for the 2050s projected a decreasing trend in mean wet season soil water content, albeit minor similar to the 2030s. Two of three GCM scenarios suggest average annual soil water content will consistently decrease up to the 2030s by as much as 10 %. Decreases of that magnitude are significant due to the potential impact to be expected seasonally or in a particular month. The impact on soil water content annually is not expected to be very significant, but may be seasonally or more noticeably where monthly changes are considered.

Projected actual evapotranspiration (AET) is a major component of water balance. In the Caribbean, evapotranspiration rates are consistently high due to the annually high solar radiation and temperatures; precipitation loss through AET is significant. Evapotranspiration is dependent on several weather variables, chief among them being: wind, solar radiation, humidity, and temperature. It is also influenced by plant variables such as plant and leaf size, and density. The results suggest that the annual budget for AET may be affected marginally, with the greatest possible change being a 6 % reduction in AET. The seasonal and monthly changes have greater significance than the annual change in AET. The A2 scenarios generally forecast a reduction in AET with the exception of the RCM-based scenarios where an increasing trend was observed in the 2030s. The loss in AET cannot be considered significant within the context of the GRW. Accurate calculations of AET are very important to understand its impact on water availability and impact on soils and nutrients within watersheds (Lu et al. 2003). However, the direct measurement of AET on a watershed scale is generally considered difficult if not impossible to measure. Hence, a modelling approach is generally considered (Lu et al. 2003). A general trend is observed for a reduction in AET from the baseline.

According to the WRA, surface water runoff largely occurs on the outcrops of basement rocks and also the interior valley alluviums. Environmentally, it is a leading source of pollution to rivers and streams due to its ability to bring entrained pollutants on surfaces (e.g., urban surfaces and agricultural soil) into streams and rivers (Brian and Cleveland 2010). In most instances, it experiences little or no filtering. The hydrological component may provide significant water resources to rivers and streams. This component may be important for water availability and treatment, as well as potential for flooding impacts.

Reduction in precipitation may have adverse impacts on surface flow particularly where soils' absorptive capacities are good and landscapes such as urban influences are reduced within watersheds. Increased surface runoff also has significance for impact on erosion within watersheds (Brian and Cleveland 2010). Agricultural systems are likely to suffer where irrigation is not provided and systems rely largely on precipitation. In the Caribbean, much of the surface water contributions to stream flow terminate in the sea at the terminal watershed outfall. Integrated coastal zone management becomes very important not only for water availability but also for the management of water resources for the coastal environment (Barker and McGregor 1995; FAO 2001). This becomes very important when consideration is given to the potential impact on tourism-related coastal resources.

These projections are based on current land use. Much of the GRW is covered with vegetation with human populations along transportation routes (mainly roads), which are close to rivers in many instances. Several unregulated settlements, also known as squatter settlements, have emerged over the years, and the trend appears to favour an increase in such until adequate housing solutions are provided as called for in Jamaica's most recent national planning document, the Vision 2030 Jamaica–National Development Plan.

Surface runoff has resulted in significant economic losses particularly in relation to floods, erosion, and damage to infrastructure through adverse weather in the past decade, much of which has been blamed on climate change. The more built-up an area is the more likely that surface runoff will be an issue. There are significant clay soils within the GRW, and this translates to a lower infiltration rate. Any changes to the land use of the watershed towards an increase in infrastructural developments such as mining (limestone), residential, commercial, as well as roads may have a significant impact on future surface runoff.

The results suggest that the annual budget for surface runoff will be adversely affected and may be exacerbated depending on changes in land use/land cover. The greatest possible increase in surface runoff by 61 % was predicted by the HAD-B1 scenario while the greatest reduction was similar. An increase of those magnitudes will no doubt increase water availability, but depending on how that increase is delivered spatially, it could bring a significant environmental and socioeconomic impact. Such increases bring with it the potential for flooding and transport of pollutants to sensitive areas and ecosystems such as flood plains and coral reefs. The reduced surface runoff potential may translate to reduced flooding incidences but may have impact on water availability and sustenance of critical habitats.

The impact on surface runoff is very important. Although the contributions to stream flow are being considered, it should be noted that increases in surface runoff not reaching the stream flow may have significant impact elsewhere in the GRW, such as flooding in urban and residential areas. Public infrastructure may also be affected with changes to surface runoff. Surface runoff for the RCM-based model showed a significant increase in mean annual surface runoff of 37 %. Although this suggests increased water availability, it may also signify increasing negative impacts to coastal ecosystems and infrastructure as well increased cost to treat water due to sedimentation and pollutant load levels.

The Great River is very important to the supply of potable water in north-western Jamaica. Additionally, this river empties into the westernmost boundary of a Marine Park (Montego Bay Marine Park). The north coast is covered with coral reefs. The supply of pollutants to the coastal zone may have significant impact of these and other coastal ecosystems in the region of the mouth of river.

31.5 Conclusion

The SWAT model was used successfully to calibrate and validate using the 9-year (1998–2006) weather and hydrological data in the Great River Watershed. The measured and simulated flows agree fairly well, and this is indicated by the confidence placed in the modelling through the model efficiencies calculated. Although peak flows were slightly underestimated for some peak flows, this did not negate the confidence that can be placed in the model.

The NSE statistic showed that the data had a goodness of fit greater than 60 %. This indicates that the model predictions are very good in replicating the observed stream flow data. The NSE values indicated the model is particularly sensitive to low flows, but it still performs fairly well to peak flows. The R^2 of greater than 0.67 suggests the data simulated are well correlated.

The fluctuations in annual and seasonal stream flows and other hydrological parameters due to climate change were also significant. The data generated by the UWI Climate Change Group using a RCM and three global climate model (GCM) outlined potential impacts to water availability in the future. The model predicts a general trend for a reduction in available water through reduction in stream flow. If this were to occur, the potential increase in population and urbanization and increase in tourism would significantly strain the water supply.

There are also significant potential for increase in surface water in the future. The economic impact on development infrastructure, people, and livestock is significant during periods of sustained moderate to heavy precipitation. The impacts of these have been felt in the watershed to a great extent. All efforts must be taken now to examine our development strategy for the watershed to ensure the sustainability of one of Jamaica's critical watersheds. The efforts to address squatting issues on marginal lands must be tackled as a matter of priority; otherwise, we risk the possibility of a significant reduction in the forested component and a subsequent increase in adverse impacts on the main river and its tributaries.

Although this research puts forward impact to watershed based on changes produced by RCMs and GCMs on rainfall and temperature, it should be highlighted that it is very much theoretical. There is still large uncertainty in predicting future impacts due to climate and land-use changes. The pace at which technology will be developed to arrest the increasing threat of climate change is still unknown.

References

Anwar A, Melesse AM, Admasu S (2014) Climate change in upper Gilgel Abay River Catchment, Blue Nile Basin Ethiopia. In: Melesse AM, Abtew W, Setegn S (eds.) Nile River Basin: ecohydrological challenges, climate change and hydropolitics, pp 363–388

ARD (2003) strategic plan for sustainable development of the great river watershed. Ridge to reef watershed project. ARD, Burlington

Arnell NW (2004) Climate change and global water resources: SRES emissions and socio-economic scenarios. Glob Environ Change 14(1):31–52

Arnold JG, Srinivasan R, Muttiah RS, Allen PM (1998) Continental scale simulation of the hydrologic balance. J Am Water Resour Assoc 35(5):1037–1051. doi:10.1111/j.1752-1688.1999.tb04192.x

Barker D, McGregor DFM (1995) Environment and development in the Caribbean: geographical perspectives. UWI Press, Kingston

Behulu F, Setegn S, Melesse AM, Fiori A (2013) Hydrological analysis of the upper Tiber Basin: a watershed modeling approach. Hydrol Process 27(16):2339–2351

Behulu F, Setegn S, Melesse AM, Romano E, Fiori A (2014) Impact of climate change on the hydrology of upper Tiber River Basin using bias corrected regional climate model. Water Resour Manag 28(5):1327–1343. doi: doi: 10.1007/s11269-014-0546-x

Brian L, Cleveland C (2010) Surface runoff. In: Cleveland C (ed.) Environmental information coalition, national council for science and the environment. Encyclopedia of Earth, Washington, DC

Campbell JD, Taylor MA, Stephenson TS, Watson RA, Whyte FS (2010) Future climate of the caribbean from a regional climate model. Int J Climatol 31(12):1866–1878. doi:10.1002/joc.2200

Cashman A, Nurse L, John C (2010) Climate change in the Caribbean: the water management implications. J Environ Dev 19(1):42–67. doi:10.1177/1070496509347088

Chen A, Taylor M, Centella A, Farrell D (2008) Climate trends and scenarios for climate change in the insular Caribbean. Report of working group I, climate change and biodiversity in the insular Caribbean. Technical report 381, CANARI, Trinidad

Climate Studies Group Mona (2012) State of the Jamaican climate 2012: information for resilience building (full report). Produced for the Planning Institute of Jamaica (PIOJ), Kingston

Dessu SB, Melesse AM (2012) Modeling the rainfall-runoff process of the Mara River Basin using SWAT. Hydrol Process 26(26):4038–4049

Dessu SB, Melesse AM (2013) Impact and uncertainties of climate change on the hydrology of the Mara River Basin. Hydrol Process 27(20):2973–2986

Dessu SB, Melesse AM, Bhat M, McClain M (2014) Assessment of water resources availability and demand in the Mara River Basin. CATENA 115:104–114

Di Luzio M, Srinivasa R, Arnold JG (2002) Integration of watershed tools and SWAT model into BASINS. J Am Water Resour Assoc 38(4):1127–1141. doi:10.1111/j.1752-1688.2002.tb05551.x

ECLAC (2010) Regional climate modeling in the Caribbean. Port of Spain, ECLAC subregional headquarters for the Caribbean

FAO (2001) Land resources information systems in the Caribbean. In: Proceedings of a sub-regional workshop held in Bridgetown Barbados (World Soil Resources Reports). FAO, Rome, 2–4 Oct 2000

Getachew HE, Melesse AM (2012) Impact of land use /land cover change on the hydrology of Angereb watershed, Ethiopia. Int J Water Sci 1(4):1–7. doi:10.5772/56266

Government of Jamaica (2011) The second national communication of Jamaica to the United Nations framework convention climate change. National Meteorological Service Jamaica, Jamaica, 409 p

Greenaway A (2004) Water quality of the Great River watershed, St. James/Hanover/Westmoreland. National environment and planning agency and the United States agency for international development, Kingston

Grey OP, Webber DFG, Setegn SG, Melesse AM (2013) Application of the Soil and water assessment tool (SWAT model) on a small tropical Island State (Great River Watershed, Jamaica) as a tool in integrated watershed and Coastal Zone management. Int J Trop Biol Conserv 62(3):293–305

Hayman A (2001) Rapid rural appraisal of the Great River watershed: ridge to reef watershed project. National environment and planning agency and the United States agency for international development, Burlington

IPCC (2007a) Climate change 2007: synthesis report. contribution of working groups I, II and III to the fourth assessment report of the intergovernmental panel on climate change. Geneva

IPCC (2007b) Climate change 2007: the physical science basis—summary for policymakers. IPCC WGI fourth assessment report. Intergovernmental panel on climate change, Paris

Lu J, Sun G, McNulty SG, Amatya DM (2003) Modeling actual evapotranspiration from forested watersheds across the southeastern United States. J Am Water Resour Assoc 39(4):887–896

Mango L, Melesse AM, McClain ME, Gann D, Setegn SG (2011a) Land use and climate change impacts on the hydrology of the upper Mara River Basin, Kenya: results of a modeling study to support better resource management, special issue: climate, weather and hydrology of East African Highlands. Hydrol Earth Syst Sci 15:2245–2258. doi:10.5194/hess-15-2245-2011

Mango L, Melesse AM, McClain ME, Gann D, Setegn SG (2011b) Hydro-meteorology and water budget of Mara River basin, Kenya: a land use change scenarios analysis, In: Melesse A (ed.) Nile River Basin: hydrology, climate and water use, chapter 2, pp 39–68. Springer, Berlin. doi:10.1007/978-94-007-0689-7_2

Melesse AM, Loukas AG, Senay G, Yitayew M (2009) Climate change, land-cover dynamics and ecohydrology of the Nile River Basin. Hydrol Process Spec Issue Nile Hydrol 23(26):3651–3652

Melesse AM, Bekele S, McCornick P (2011) Hydrology of the Niles in the face of land-use and climate dynamics. In Melesse A (ed.) Nile River Basin: hydrology, climate and water use, pp vii-xvii. Springer, Berlin. doi:10.1007/978-94-007-0689-7

Mohammed H, Alamirew T, Assen M, Melesse AM (2015) Modeling of sediment yield in Maybar gauged watershed using SWAT, northeast Ethiopia. CATENA 127:191–205

Nash JE, Sutcliffe JV (1970) River flow forecasting through conceptual models part I—a discussion of principles. J Hydrol 10(3):282–290

Neitsch SL, Arnold JG, Kiniry JR, Williams JR, King KW (2005) Soil and water assessment tool. Theoretical documentation, version 2005. Blackland research center, Texas agricultural experiment station, Temple

Santhi C, Srinivasan R, Arnold JG, Williams JR (2006) A Modeling approach to evaluate the impacts of water quality management plans implemented in a watershed in Texas. Environ Model Softw 21(8):1141–1157

Setegn S, Melesse AM (2014) Climate change impact on water resources and adaptation strategies in the Blue Nile River Basin. In: Melesse AM, Abtew W, Setegn S (eds.) Nile River Basin: ecohydrological challenges, climate change and hydropolitics, pp 389–420

Setegn SG, Srinivasan R, Dargahi B, Melesse AM (2009a) Spatial delineation of soil erosion prone areas: application of SWAT and MCE approaches in the Lake Tana Basin, Ethiopia. Hydrol Process 23(26):3738–3750

Setegn SG, Srinivasan R, Melesse AM, Dargahi B (2009b) SWAT model application and prediction uncertainty analysis in the Lake Tana Basin. Ethiop Hydrol Process 24(3):357–367

Setegn S, Rayner D, Melesse AM, Dargahi B, Srinivasan R (2011) Impact of climate change on the hydro-climatology of Lake Tana basin. Ethiop Water Resour Res 47(W04511):13. doi:10.1029/2010WR009248

Setegn SG, Melesse AM, Haiduk A, Webber D, Wang X, McClain M (2014) Spatiotemporal distribution of fresh water availability in the Rio Cobre watershed. Jam CATENA 120:81–90

Smith GC, Covich AP, Brasher AMD (2003) An ecological perspective on the biodiversity of tropical island streams. BioScience 53(11):1048–1051. doi:10.1641/0006-3568(2003)053[1048:aepotb]2.0.co;2

Taylor MA, Centella A, Charlery J, Borrajero I, Bezanilla A, Campbell J, Rivero R, Stephenson TS, Whyte FS, Watson RA (2007) Glimpses of the future: a briefing from the PRECIS Caribbean climate change project. Caribbean Community Climate Change Centre, Belmopan

Wang X, Melesse AM (2005) Evaluations of the SWAT Model's snowmelt hydrology in a northwestern Minnesota watershed. Trans ASAE 48(4):1359–1376

Wang X, Melesse AM (2006) Effects of STATSGO and SSURGO as inputs on SWAT Model's snowmelt simulation. J Am Water Resour Assoc 42(5):1217–1236

Wang X, Melesse AM, Yang W (2006) Influences of potential evapotranspiration estimation methods on SWAT's hydrologic simulation in a northwestern Minnesota watershed. Trans ASAE 49(6):1755–1771

Wang X, Shang S, Yang W, Melesse AM (2008a) Simulation of an agricultural watershed using an improved curve number method in SWAT. Trans Am Soc Agri Bio Eng 51(4):1323–1339

Wang X, Yang W, Melesse AM (2008b) Using hydrologic equivalent wetland concept within SWAT to estimate streamflow in watersheds with numerous wetlands. Tans Am Soc Agri Bio Eng 51(1):55–72

Winchell M, Srinivasan R, di Luzio M, Arnold J (2007) ArcSWAT interface for SWAT 2005. User's guide. Blackland research center, Texas agricultural experiment station, Temple

Winchell M, Srinivasan R, di Luzio M, Arnold J (2009) ArcSWAT 2.3.4 interface for SWAT2005: User's guide. Blackland research center, Texas agricultural experiment station, Temple

Part VII
Water and Watershed Management

Chapter 32
Koga Irrigation Scheme Water Quality Assessment, Relation to Streamflow and Implication on Crop Yield

Degarege Fentie Densaw, Essayas K. Ayana and Temesgen Enku

Abstract The Koga irrigation scheme is the first attempt of eleven major proposed large-scale irrigation schemes in Lake Tana sub-basin in the Upper Blue Nile Basin. The schemes are to provide water for small holder farmers. The completion of Koga scheme poses various challenges and provides important lessons in implementing and operating large-scale irrigation projects. With anticipated increased agricultural productivity, one of the foreseeable challenges will be the continuous assessment of the water quality. This study assesses the quality status of Koga River by analysing samples collected from planned sampling locations. Two sampling campaigns were undertaken, and 23 physical and chemical water quality parameters were measured using standard methods. On the basis of FAO irrigation water quality guideline, Koga irrigation water is well within limits of salinity, sodicity and toxicity hazard except above the threshold value of boron. Among the dominant crops cultivated potato, barley and wheat production are susceptible to yield reduction. A detailed investigation is required to quantify the extent of yield reduction incurred due to higher level of boron. Multivariate statistical technique is applied for the evaluation of relationship between streamflow and various water quality parameters.

Keywords Koga irrigation scheme · Koga Dam · Koga River · Irrigation water quality · Upper Blue Nile Basin

---

D.F. Densaw · E.K. Ayana (✉) · T. Enku
School of Civil and Water Resources Engineering,
Bahir Dar Institute of Technology, Bahir Dar, Ethiopia
e-mail: ekk45@cornell.edu

D.F. Densaw
e-mail: degaregefentie@yahoo.com

T. Enku
e-mail: temesgenku@gmail.com

E.K. Ayana
Department of Ecology, Evolution and Environmental Biology,
Columbia University, New York, USA

E.K. Ayana
GeoSpatial Data and Technology Center (GDTC), Bahir Dar University, Bahir Dar, Ethiopia

© Springer International Publishing Switzerland 2016

A.M. Melesse and W. Abtew (eds.), *Landscape Dynamics, Soils and Hydrological Processes in Varied Climates*, Springer Geography,
DOI 10.1007/978-3-319-18787-7_32

32.1 Introduction

Crop production consumes a significant proportion of available water resources in the world (Viala 2008). With the increase in world population, the demand for food will undoubtedly increase. In subsistence farming economies where the productivity is considerably low, this will translate into land use change and subsequently decline in ecosystem services (Marx 2011). Studies indicate that climate change will further strain the availability of water resources mostly in areas where adequate water infrastructures are absent. In Ethiopia, where the study is conducted, agriculture employs about 84 % of the country's population and accounts for nearly half of the gross domestic product (GDP) (EPA 1997). High agricultural productivity can be achieved by applying appropriate irrigation management systems through intensive cropping systems to alleviate problems of drought and limitations posed by land shortage. Accordingly, the Government of Ethiopia gives emphasis to the development of the irrigation sub-sector by assisting and supporting farmers to improve irrigation management practices and the promotion of modern irrigation systems on small-scale (less than 200 ha), medium-scale (200–3000 ha) and large-scale (over 3000 ha) schemes (OCDE 2001). According to the report on marketing assessment and cost recovery study in Koga carried out in February 2013, Ethiopian irrigation potential is estimated at 3.7 million hectare, but only some 197,000 ha has been developed (GIRDC 2013).

Expanding small-scale irrigation is challenging in the highlands due to the very low streamflow during the dry season, and the only way to irrigate large areas of peasant land during the dry season would be to construct storage reservoirs (OCDE 2001). The Koga irrigation scheme is the first attempt by the Government of Ethiopia to develop a large-scale irrigation (7000 ha) and supposed to be managed and funded farmers (Marx 2011). Koga irrigation scheme is intended to reduce poverty through substantial increase in household income and improve food security and nutritional practices thereby promoting human and economic development in the region. As irrigation infrastructure gradually stretched to the project extent, more land came under irrigation. Documents at the site office indicate that the irrigated area increased substantially from 700 ha in 2009 to more than 5100 ha in 2012 (GIRDC 2013).

Irrigation water quality can affect crop productivity. Irrigation water quality affects crop yield and soil physical conditions and can affect fertility, irrigation system performance, longevity, and irrigation demand. Therefore, knowledge of irrigation water quality is critical to understanding what management changes are necessary for long-term productivity (Bauder et al. 2007).

Quality of surface water is mainly controlled by geological structure and mineralogy of the watersheds, the chemical reactions that take place within the watershed as well as the type of land uses and anthropogenic activities (Alexakis 2008, 2011). Irrigation waters contain salts of varying concentration. Most of the applied water is removed by evapotranspiration leaving the salt behind. This leads to an increase in salt concentration of soils and hence additional water requirement to leach (Rhoades 1972). Extensive literature is available on water quality monitoring and assessment for irrigation uses (Saeedi et al. 2010; Aghazadeh and Mogaddam 2011; Palma et al. 2010; Suthar et al. 2010; Pazand and Hezarkhani 2012).

The objective of this study is to assess the suitability of water quality of Koga Dam with respect to FAO's irrigation water quality standards and widely used crop types. In irrigation water evaluation, emphases are placed on the chemical and physical characteristics of the water and only rarely are any other factors considered important (Ayalneh 2004). The analysis is thus confined to chemical and physical characteristics as these induce major water quality-related problems in irrigation agriculture such as salinity, water infiltration rate alteration and specific ion toxicity. In this study, field samples were collected and analysed in the laboratory, but existing water quality data were also included to expand the data for evaluating temporal variation of the water quality parameters.

32.2 Materials and Methods

32.2.1 Study Area

The Koga irrigation and watershed management project is located south of Lake Tana in the Upper Blue Nile Basin of Ethiopia (Lat. 11° 10′N–11° 25′N, Long. 37° 02′E–37° 17′E), Fig. 32.1. The Koga River that drains the 250 km^2 watershed is a tributary of the Gilgel Abay River in the headwaters of the Blue Nile basin which flows into Lake Tana. With the elevation between 1900 and 3200 m, the watershed is subject to the inter-tropical convergence zone and a single rainy season that begins in June and lasts through September resulting in 70 % of the river flow. The high rainfall variability renders high vulnerability among small-scale farmers who rely on rain-fed agriculture. According to Ministry of Water, Irrigation and Energy (MoWIE), there is 80 % probability of at least 1245 mm rainfall in a year. Koga River minimum flow occurs in April and in May, just before the onset of the rainy season. The average discharge based on 44 years of record (1959–2002) is 4.78 $m^3 s^{-1}$. The most important environmental problems observed in the area are soil erosion, deforestation, and poor land use and management. The rate of soil loss in the furthest upstream portions of the watershed exceeds the soil formation rate, in part because of severe deforestation in the 1970s and 1980s (Tilahun 2009).

The Koga River is a tributary of the Blue Nile River draining watersheds of various levels of degradation. Hydrology of the Nile River Basin has been studied by various researchers. These studies encompass various areas including streamflow modelling, sediment dynamics, teleconnections and river flow, land use dynamics, climate change impact, groundwater flow modelling, hydrodynamics of Lake Tana, water allocation and demand analysis (Melesse et al. 2009a, b, 2011, 2014; Abtew et al. 2009a, b; Abtew and Melesse 2014a, b, c; Yitayew and Melesse 2011; Chebud and Melesse 2009a, b, 2013; Setegn et al. 2009a, b, 2010; Melesse 2011).

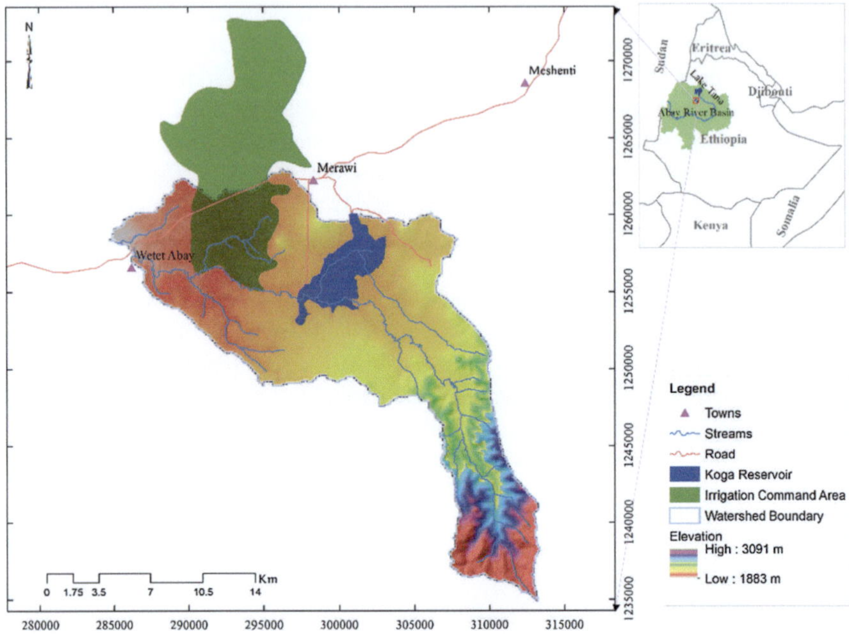

Fig. 32.1 The Koga watershed, reservoir and irrigation command area in the Lake Tana sub-basin

32.2.2 Data and Analysis

Site visits were made to Koga watershed and command areas to assess the existing situations in the area including the experiences of the farmers in rain-fed as well as irrigated agriculture. Water samples were collected at three sampling sites in April 2014 and at two sampling sites in May 2014 (Fig. 32.2). Sites recommended for future monitoring are used in collecting the samples so as to ensure data continuity. The sampling sites are selected based on their representativeness, safety and security, availability of stable streambed and suitability to avoid disturbing influences (BRL 2009). Grab samples were collected using plastic bottles rinsed with the water to be sampled. Koga River water quality test results compiled during the project design and measurements made by other researchers (Eriksson 2012) and project-related data were retrieved from MoWIE and Koga Irrigation and Watershed Management Office (KIWMO).

All samples were labelled with sampling date, location, time and other pertinent information. Samples were collected in April 2014 at river entry to the reservoir (river mouth), station 1; in the reservoir nearby the intake tower, station 2; and at a secondary irrigation canal, station 3. On a second campaign conducted in May 2014, samples were collected at two sites: at river entry to the reservoir and nearby the intake tower. Water release to the canals was interrupted, and hence, samples could not be collected from the irrigation canal. Twenty three water quality

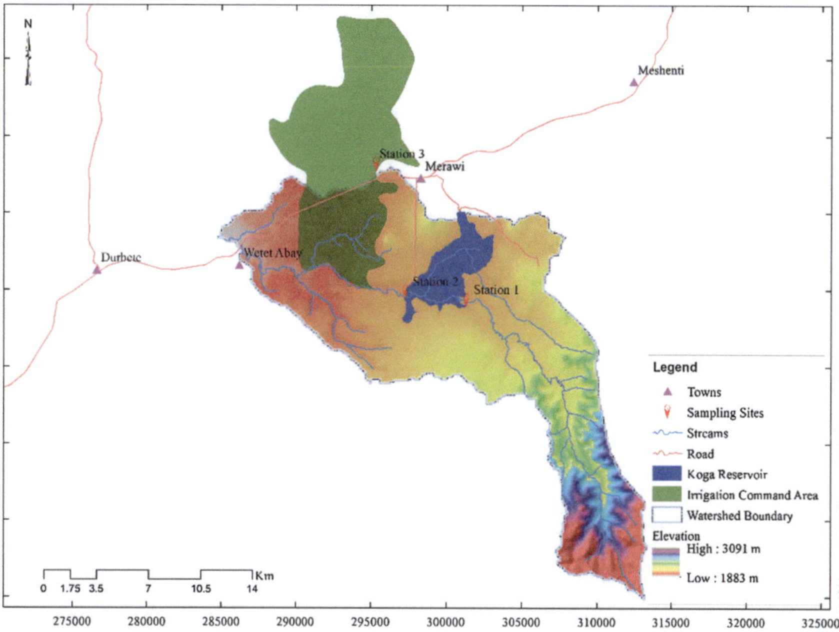

Fig. 32.2 Water quality sampling locations in Koga watershed

parameters were measured from samples collected in April 2012 and May 2014. Sampling, preservation and transportation of the water samples to the laboratory were as per standard methods. Sample analysis in the laboratory followed the standard protocols (APPA 2005).

According to survey carried out in May 2007, farmers' preferences of crops from the most to the least to be grown were onion, potato, tomato, pepper, barley, fruits, wheat, maize, sugarcane, cabbage and beetroot (Horecha 2008). Nevertheless, production statistics for the year 2012 showed that potato covered 42 %, barely 20 % and wheat 31 % of the total irrigated area. On the basis of available literature, the implication of current water quality on the yield is described (Rhoades 1972; Bauder et al. 2007; Fipps 2003a; Wilcox 1955). The impact of the water quality on the sustainability of Koga irrigation scheme is also explained. Statistical correlation technique is used to assess correlation between outflows to the irrigation command areas with physico-chemical parameters of Koga irrigation water.

32.3 Results and Discussion

The water quality analysis result from the April and May 2014 sampling is provided in Table 32.1a, where parameter concentrations and applicable limits for irrigation water are shown. Table 32.1b shows data from previous sampling and analysis by

Table 32.1 Koga River water quality. a April and May 2014. b 1995, 2011 and 2012

(a)

Parameter	Limit	May 2014		April 2014		
		River mouth	Intake	River mouth	Intake	Canal
Temperature (°C)		25.4	25.2	28	23.5	22.5
pH	6.5–8.4	7.71	7.97	8.55	8.22	8.17
DO (mg L^{-1})				9.85	7.5	7.85
EC (μS cm^{-1})	<750	40	87	249	135.8	124.9
TDS (mg L^{-1})	500	27	58	148.2	87.5	75
Turbidity (NTU)		855	309	74.1	311	228.5
Alkalinity (mg L^{-1})		30	45	157.5	62.5	67.5
TSS (mg L^{-1})				21.5	30.5	25.5
Na (mg L^{-1})		3.6	3.6	7.5	3.5	3.7
SAR	<10	0.15	0.12	0.35	0.11	0.18
Chloride (mg L^{-1})	70	0	1	28	24	24.5
NH_3 (mg L^{-1})				0.16	0.09	0.115
NO_2 (mg L^{-1})				0.025	0.12	0.035
NO_3 (mg L^{-1})	45			0.66	1.7	1.13
SO_4 (mg L^{-1})	250			3	9.5	10.5
PO_4 (mg L^{-1})				1.4	0.135	0.15
Boron (mg L^{-1})		1.7	1.85	0.8	1.75	1.8
Mg (mg L^{-1})		20	28	11	34	42.5
Ca (mg L^{-1})		12	23	16.5	18	18.5
K (mg L^{-1})				3.2	0.7	0
TH (mg L^{-1})				497	186	222
Fe (mg L^{-1})	5			0	0.05	0.02
Mn (mg L^{-1})	0.2			0.003	0.007	0.005

(b)

Parameter	Limit	Eriksson (canal)	Tana Sub-basin Office (sampling year and month)						MoWIE
		2012/4	2011/12	2012/01	2012/3	2012/6	2012/7	2012/8	1995
Temperature (°C)	6.5–8.4	24.7	21.4	14.9	20	16.7	19.3	19.3	20.1
pH		7.3	7.9	7.79	7.96	6.99	7.58	7.78	6.33
DO (mg L^{-1})	<750				4.44	3.9	3.63	4.02	8.8
EC (μS cm^{-1})	500	275	100	110	140	38	72	64	122.1
TDS (mg L^{-1})			50	50	89	24	46	41	61.3
Turbidity (NTU)		90	58.8	56.3	57	110	1002	1002	
Alkalinity (mg L^{-1})		79	117	130	130	285	430	355	
TSS (mg L^{-1})									
Na (mg L^{-1})	<10								
SAR	70								1.59
Chloride (mg L^{-1})									
NH_3 (mg L^{-1})		0.04	0.13	0.05	0.71	3.7	0.71	0.51	
NO_2 (mg L^{-1})	45	0.005	0.03	0.03	0.03	1.47	0.35	0.29	

(continued)

Table 32.1 (continued)

(b)									
Parameter	Limit	Eriksson (canal)	Tana Sub-basin Office (sampling year and month)						MoWIE
		2012/4	2011/12	2012/01	2012/3	2012/6	2012/7	2012/8	1995
NO_3 (mg L^{-1})	250		0.35	0.18	0.27	0.85	0.85	1	
SO_4 (mg L^{-1})			22	21	22	83	200	185	0
PO_4 (mg L^{-1})		0.15	0.67	0.3	0.42	10	2.13	1.82	
Boron (mg L^{-1})		2.2							1.3
Mg (mg L^{-1})									
Ca (mg L^{-1})									
K (mg L^{-1})									
TH (mg L^{-1})	5		95	108	130	500	750	420	
Fe (mg L^{-1})	0.2		2.28	1	2.14	5	3.6	2.8	0.5
Mn (mg L^{-1})									

Eriksson (2012), by Tana Sub-basin Office and by MoWIE with data collected in 1995, 2011 and 2012. Table 32.2 shows all the data and comparison of each parameter observation with FAO irrigation water quality recommended standards.

32.3.1 *Salinity Hazard*

Salinity level with regard to irrigation is very important as increase in salinity increases the osmotic potential of the soil water. The energy required by plants to uptake water increases with increased osmotic potential resulting in increased respiration and decline in yield (Akinyemi and Souley 2014). FAO irrigation water quality guidelines recommend EC_w be less than 250 µS cm^{-1} and TDS below 450 mg L^{-1} (Ayers and Westcot 1985), Table 32.2. Water with EC less than 250 µmhos cm^{-1} is considered excellent, from 251–750 µmhos cm^{-1} is good, from 751–2000 µmhos cm^{-1} is permissible and that with greater than 2000 µmhos cm^{-1} is unsuitable for irrigation (Fipps 2003b). Koga irrigation water mean values of salinity indicators, EC_w and TDS (Fig. 32.3), on the basis of all available data, are 120 µmhos cm^{-1} and 63 mg L^{-1} and 127 µmhos cm^{-1} and 79 mg L^{-1} based on samples in this study. µmhos cm^{-1} is equivalent to µS cm^{-1}.

32.3.2 *Sodicity Hazard*

Sodic waters are classified in relation to irrigation based on the ranges of SAR (sodium absorption ratio) values. Water with SAR ranging from 0 to 3 is considered

Table 32.2 Comparison of Koga irrigation water quality with FAO standards

Parameter	1995	Dec	Jan	Mar	Jun	Jul	Aug	Apr	Apr	May	FAO threshold	Remark
		2011	2012	2012	2012	2012	2012	2012	2014	2014		
EC_w (μS cm^{-1})	122.6	100	110	140	38	72	64	275	124.9	87	<250	Ok
Na (mg L^{-1})									3.7	3.6		
SAR	1.59								0.18	0.12	<12	Ok
Chloride as Cl (mg L^{-1})	2	–	–	–	–	–	–	–	15	0	<70	Ok
Boron (mg L^{-1})	1.3	–	–	–	–	–	–	2.2	1.8	1.85	<0.75	Not Ok
pH	6.33	7.9	7.79	7.96	6.99	7.58	7.78	7.3	8.17	7.97	6.5–8.4	Ok
Turbidity (NTU)	–	58.8	56.3	57	110	1002	1002	90	228.5	309	NRL	
Alkalinity (mg L^{-1})	–	117	130	130	285	430	355	79	67.5	105	NRL	
TDS (mg L^{-1})	61.3	50	50	89	24	46	41	–	240	58	<450	Ok
TSS (mg L^{-1})									255		NRL	
DO (mg L^{-1})	8.8	–	–	4.44	3.9	3.63	4.02	–	7.85		NRL	
SO_4 (mg L^{-1})	0	22	21	22	83	200	185	–	10.5		<90	Partly Ok
PO_4 (mg L^{-1})	–	0.67	0.3	0.42	10	2.13	1.82	0.15	0.15		<0.4	Ok
NH_3 (mg L^{-1})	–	0.13	0.05	0.71	3.7	0.71	0.51	0.04	0.12		NRL	
NO_2 (mg L^{-1})	–	0.03	0.03	0.03	1.47	0.35	0.29	0.005	0.035		NRL	
NO_3 (mg L^{-1})	–	0.35	0.18	0.27	0.85	0.85	1	–	1.13		<50	Ok
Ca (mg L^{-1})								96	18.5	23	<100	Ok
Mg (mg L^{-1})									42.5	28		Ok
K (mg L^{-1})									0		<20	Ok
Fe (mg L^{-1})	0.5	2.28	1	2.14	5	3.6	2.8	–	0.02		<5	Ok

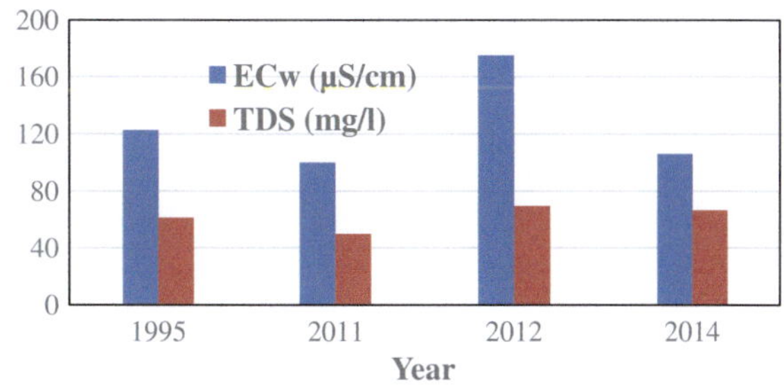

Fig. 32.3 Salinity indicators for Koga River irrigation water

good and with greater than 9 considered unsuitable for irrigation purpose (Joshi 2009). Sodium, when present in the soil in exchangeable form, causes adverse change in soil structure. An excess of sodium will cause clay particles to disperse instead of aggregate and in some cases also to swell. This results in a soil with low porosity, poor permeability and poor aeration when wet, large clods separated by deep cracks when dry. At higher concentration (>40 mg L^{-1}), sodium has also the ability to disperse soil. Dispersion of soil results in reduced rates of infiltration of water and air into the soil. When such soils dry up, crusting occurs making tilling difficult and thus adversely interfering with seed germination and seedling emergence (Jensen 1980). Overall mean sodicity for Koga irrigation water as determined from the whole data is SAR value of 0.63 which is much below the threshold of 12 recommended by FAO guideline (Table 32.2). In the present study, SAR was found to be 0.18 at the irrigation canal. The low value of SAR for Koga irrigation water can be categorized as water of excellent category. It can be used for irrigation on almost all soil types with little danger of the development of harmful levels of exchangeable sodium.

32.3.3 Toxicity

Toxicity indicators with reference to the FAO guideline are all well below the threshold values with Cl (5.7 vs. 70), SO_4 (15 vs. 90) and NO_3 (0.5 vs. 50) mg L^{-1}. However, observed boron concentration in the overall data (1.79 mg L^{-1}) as well as the measurements made for this study (1.83 mg L^{-1}) indicates higher level of concentration as compared to the 0.75 mg L^{-1} FAO threshold (Fig. 32.4, Table 32.2). Figure 32.4 also shows concentration of nutrients (PO_4, NH_3, NO_2, NO_3 and boron).

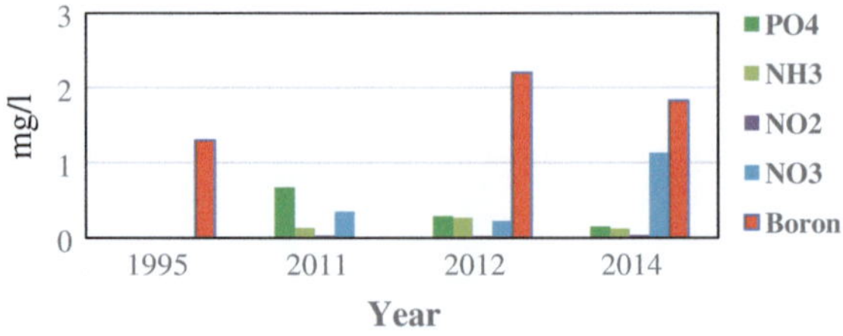

Fig. 32.4 Nutrient concentration, toxicity indicator and boron concentration

32.4 Implication on Crop Yield and Irrigation Sustainability

Boron is essential in low amounts, but toxic at higher concentrations and affects both yield and germination process. Boron toxicity can occur at concentrations less than 1.0 mg L^{-1} on sensitive crops. The literatures indicate that boron tends to occur in association with saline conditions and the yield response of plants to increasing boron concentrations is similar to their response to increasing salinity. The yield remains unaffected until a threshold concentration is reached, after which yield declines linearly (Forestry 1996). Ashagre et al. (2014) reported a decrease in germination percentage and rate, shoot and root lengths, shoot and root fresh and dry weights, root number, root–shoot ratio and seedling vigour index beyond 0.25 mg L^{-1}. Chauhan and Powar (1978) found in a micro-plot experiment that

Table 32.3 Major crops in Koga area and sensitivity to water quality parameters

Crop type	Parameter				Remark
	Salinity EC_w (dS m^{-1})	Na (SAR)	Chloride	Boron (mg L^{-1})	
Potato	1.1		Moderately sensitive	1–2	Allowable limit (Ayers and Westcot 1985)
Wheat	4	Semi-sensitive (10.1–18)	Highly tolerant	0.75–1	
Barley	5.3	Tolerant (>26.1)	Highly tolerant	0.75–1	
Maize	1.1	Sensitive (<10)	Moderately tolerant	2–4	
Onion	0.8	Semi-tolerant (18.1–26)	sensitive	0.5–0.75	
Pepper	1		Moderately sensitive	1–2	
Cabbage	1.2		Moderately sensitive	2–4	
Tomato	1.7	Semi-sensitive (10.1–18)	Moderately sensitive	4–6	

Table 32.4 Correlation between streamflow and water quality parameter concentration

Parameter	Sampling months					r	p-value	Remark
	Dec, 2011	Jan, 2012	Mar, 2012	Apr, 2012	Apr, 2014			
Q ($m^3 s^{-1}$)	5.5	4.5	3	3	4			
EC (μS/cm)	100	110	140		124.9	−0.98	0.02	Q versus EC
pH	7.9	7.79	7.96	7.3	8.17	0.35	0.56	Q versus pH
Turbidity (NTU)	90	58.8	57	90	228.5	0.06	0.93	Q versus Tur
Alkalinity (mg L^{-1})	117	130	130	79	67.5	0.25	0.68	Q versus Alkalinity
NH_3 (mg L^{-1})	0.13	0.05	0.71	0.04	0.12	−0.44	0.46	Q versus NH_3
PO_4 (mg L^{-1})	0.67	0.3	0.42	0.15	0.15	0.63	0.25	Q versus PO_4
NO_2 (mg L^{-1})	0.03	0.03	0.03	0.005	0.035	0.49	0.4	Q versus NO_2
NO_3 (mg L^{-1})	0.35	0.18	0.27	–	1.13	−0.1	0.9	Q versus NO_3
SO_4 (mg L^{-1})	22	21	22	–	10.5	0.15	0.85	Q versus SO_4
Fe (mg L^{-1})	2.28	1	2.14	–	0.02	0.13	0.87	Q versus Fe

1.5 mg L^{-1} boron in the irrigation water was toxic for wheat sensitivity. Major crops produced in Koga area and tolerance to sodium, chloride and boron are shown in Table 32.3.

32.5 Water Quality Parameters Relation to Streamflow

Correlation between streamflows and water quality parameters were evaluated to examine concentration changes with change in flow. It was found that only electric conductivity (salinity) has a statistically significant negative correlation with flow ($r = -0.98$, $p = 0.02$). As shown in Table 32.4, the listed parameters have no significant relationship with streamflow.

32.6 Conclusions

Water quality parameters monitoring indicated that Koga irrigation water quality is generally suitable for irrigation purpose except the relatively high content of boron. The pH, EC (salinity), sodium (SAR), chloride, calcium, magnesium, potassium, nitrogen, phosphorous, sulphate, iron and manganese contents of Koga irrigation water are all within the acceptable range based on irrigation water quality standard of FAO. Koga River water has no major water quality problem with respect to irrigation in the last 19 years based on Koga irrigation water quality test results obtained by previous studies and this study. The boron content of Koga irrigation water is relatively high, which is in the range of 0.7–3 mg L^{-1} (slight to moderate degree of restriction on using this water) as set by FAO. Referring to the major irrigated crops in the study area, wheat production will be affected by the higher boron concentration. The exact impact on yield should be investigated in further detail. Crops such as onion, barley and garlic are also highly sensitive to boron according to FAO's sensitivity ranges, and hence, increased care in the selection of crop and management alternatives is strongly recommended. Increased frequency of sampling should show temporal variation and source identification of boron in the Koga watershed.

References

Abtew W, Melesse AM (2014a) Nile river basin hydology, In: Melesse AM, Abtew W, Setegn S (ed) Nile river basin: ecohydrological challenges, climate change and hydropolitics, pp 7–22

Abtew W, Melesse AM (2014b) Climate teleconnections and water management. In Nile river basin. Springer International Publishing, Heidelberg, pp 685–705

Abtew W, Melesse AM (2014c) Transboundary rivers and the Nile. In Nile River Basin. Springer International Publishing. Heidelberg, pp 565–579

Abtew W, Melesse A, Desalegn T (2009a) Spatial, inter and intra-annual variability of the blue Nile River basin rainfall. Hydrol Process 23(21):3075–3082

Abtew W, Melesse A, Desalegn T (2009b) El Niño southern oscillation link to the blue Nile River basin hydrology. Hydrol Process, Special Issue: Nile Hydrol 23(26):3653–3660

Aghazadeh N, Mogaddam A (2011) Investigation of hydrochemical characteristics of groundwater in the Harzandat aquifer, Northwest of Iran. Environ Monit Assess 176(1–4):183–195

Akinyemi J, Souley S (2014) Monitoring the quality of some sources of irrigation water in different parts of Ogun State, Nigeria. IERI Procedia 9:123–128

Alexakis D (2008) Geochemistry of stream sediments as a tool for assessing contamination by arsenic, chromium and other toxic elements: East Attica region. Greece Eur Water 21(22):57–72

Alexakis D (2011) Diagnosis of stream sediment quality and assessment of toxic element contamination sources in East Attica. Greece Environ Earth Sci 63(6):1369–1383

Appa A (2005) Standard methods for the examination of water and wastewater. American Public Health Association, Washington DC

Ashagre H, Hamza IA, Fita U, Nedesa W (2014) Influence of boron on seed germination and seedling growth of wheat (Triticum aestivum L.). Afr J Plant Sci 8(2):133–139

Ayalneh W (2004) Socio-economic and environmental impact assessment of community based small scale irrigation in the upper Awash basin, a case study of four community based irrigation scheme Addis Ababa University, Addis Ababa

Ayers R, Westcot D (1985) Water quality for agriculture. FAO Irrigation and drainage paper 29 Rev. 1. Food and Agriculture Organization of the United Nations 174

Bauder TA, Waskom RM, Davis JG, Sutherland PL (2007) Irrigation water quality criteria. Colorado State University Cooperative Extension

Brl I (2009) Water information & knowledge management project, component 2—strengthening water quality data generation and management, B2—water quality guidelines. Ministry of Water Resources, Addis Ababa
Chauhan R, Powar S (1978) Tolerance of wheat and pea to boron in irrigation water. Plant Soil 50 (1–3):145–149
Chebud YA, Melesse AM (2009a) Numerical modeling of the groundwater flow system of the Gumera sub-basin in Lake Tana basin. Ethiop Hydrol Process 23(26):3694–3704
Chebud YA, Melesse AM (2009b) Modeling lake stage and water balance of Lake Tana. Ethiop Hydrol Process 23(25):3534–3544
Chebud Y, Melesse AM (2013) Stage level, volume, and time-frequency change information content of Lake Tana using stochastic approaches. Hydrol Process 27(10):1475–1483. doi:10.1002/hyp.9291
EPA (1997) Environmental policy. Environmental Protection Authority of FDRE, Addis Ababa
Eriksson S (2012) Water quality in the Koga irrigation project, Ethiopia: a snapshot of general quality parameters
Fipps G (2003a) Irrigation water quality standards and salinity management strategies. Available electronically from http://www.hdlhandlenet/1969.1:87829
Fipps G (2003b) Irrigation water quality standards and salinity management strategies
Department of Water Affairs and Forestry (1996) South African water quality guidelines, agricultural use, vol 4
GIRDC (2013) Marketing assessment and cost recovery study of Koga irrigation and watershed management project, socio-economic aspects and agricultural development options. Ministry of Water Resources, Addis Ababa
Horecha D (2008) Cost recovery study in Koga irrigation and watershed management project. Ministry of Water Resources, Addis Ababa
Jensen ME (1980) Design and operation of farm irrigation systems. Monograph Series-American Society of Agricultural Engineers (USA), (3)
Joshi DM, Kumar A, Agrawal N (2009) Assessment of the irrigation water quality of river Ganga in Haridwar district. Rasayan J. Chem 2(2):285–292
Marx S (2011) Large-scale irrigation in the blue Nile basin: chances and obstacles in implementing farmers' self-management: a case study of the Koga irrigation and watershed management project in Amhara Region. Ethiopia, IWMI, February
Melesse AM (2011) Nile river basin: hydrology, climate and water use. Springer Science & Business Media, Heidelberg
Melesse AM, Loukas Athanasios G, Senay Gabriel, Yitayew Muluneh (2009a) Climate change, land-cover dynamics and ecohydrology of the Nile river basin. Hydrol Process 23(26):3651–3652
Melesse AM, Abtew W, Desalegne T, Wang X (2009b) Low and high flow analysis and wavelet application for characterization of the blue Nile river system. Hydrol Process 24(3):241–252
Melesse AM, Abtew W, Setegn S, Dessalegne T (2011) Hydrological variability and climate of the upper blue Nile river basin In: Melesse A (ed) Nile river basin: hydrology, climate and water use. Springer Science Publisher Chapter 1, pp 3–37, doi:10.1007/978-94-007-0689-7_1
Melesse A, Abtew W, Setegn SG (2014) Nile river basin: ecohydrological challenges, climate change and hydropolitics. Springer Science & Business Media, Heidelberg
OCDE (2001) Koga irrigation and watershed management project appraisal report. African Development Fund, Addis Ababa
Palma P, Alvarenga P, Palma VL, Fernandes RM, Soares AM, Barbosa IR (2010) Assessment of anthropogenic sources of water pollution using multivariate statistical techniques: a case study of the Alqueva's reservoir. Portugal Environ Monit Assess 165(1–4):539–552
Pazand K, Hezarkhani A (2012) Investigation of hydrochemical characteristics of groundwater in the Bukan basin. Northwest of Iran Appl Water Sci 2(4):309–315
Rhoades J (1972) Quality of water for irrigation. Soil Sci 113(4):277–284
Saeedi M, Abessi O, Sharifi F, Meraji H (2010) Development of groundwater quality index. Environ Monit Assess 163(1–4):327–335

Setegn SG, Srinivasan R, Dargahi B, Melesse AM (2009a) Spatial delineation of soil erosion prone areas: application of SWAT and MCE approaches in the Lake Tana Basin. Ethiopia. Hydrol Process 23(26):3738–3750

Setegn SG, Srinivasan R, Melesse AM, Dargahi B (2009b) SWAT model application and prediction uncertainty analysis in the Lake Tana Basin, Ethiopia. Hydrol Process 24(3):357–367

Setegn SG, Bijan Dargahi B, Srinivasan R, Melesse AM (2010) Modelling of sediment yield from Anjeni gauged watershed, Ethiopia using SWAT. JAWRA 46(3):514–526

Suthar S, Sharma J, Chabukdhara M, Nema AK (2010) Water quality assessment of river Hindon at Ghaziabad, India: impact of industrial and urban wastewater. Environ Monit Assess 165(1–4):103–112

Tilahun H (2009) Payment for environmental service to enhance resource use efficiency and labor force participation in managing and maintaining irrigation infrastructure, the case of upper blue Nile basin. Bahir Dar University, Bahir Dar

Viala E (2008) Water for food, water for life a comprehensive assessment of water management in agriculture. Irrigat Drain Syst 22(1):127–129

Wilcox L (1955) Classification and use of irrigation waters

Yitayew M, Melesse AM (2011) Critical water resources management issues in Nile river basin. In: Melesse A (ed) Nile river basin: hydrology, climate and water use. Springer Science Publisher, Chapter 20, pp 401–416. doi:10.1007/978-94-007-0689-7_20

Chapter 33
Managing Wicked Environmental Problems as Complex Social-Ecological Systems: The Promise of Adaptive Governance

Kofi Akamani, Eric J. Holzmueller and John W. Groninger

Abstract Lessons learned from the failure of rational-comprehensive planning approaches to the management of water and other natural resources have led to a rethinking of the nature of resource management problems. The concept of wicked problems is increasingly used to describe problems that lack consensus in terms of their definition and solutions. In line with the ongoing search for appropriate conceptual frameworks and institutions for managing wicked problems, this chapter argues for a reconceptualization of wicked problems as complex social-ecological systems. Such a reconceptualization calls for a resilience-based approach to managing wicked problems through adaptive governance institutions. Adaptive governance provides a multi-level institutional framework for managing the uncertainties and conflicts that characterize wicked environmental problems. This argument is illustrated using the case of the Cache River watershed in southern Illinois, USA.

Keywords Wicked problems · Adaptive management · Analytic deliberation · Ecosystem management · Rational-comprehensive planning · Restoration · Water resources management

33.1 Introduction

Planning is a systematic process by which societal problems are identified and solved to attain desirable futures, such as enhanced human well-being and ecosystem health. One of the major theories in planning that has influenced the field of natural resource

K. Akamani (✉) · E.J. Holzmueller · J.W. Groninger
Department of Forestry, Southern Illinois University, 1205 Lincoln Drive, Agriculture Building, MC 4411, Carbondale, IL 62901, USA
e-mail: k.akamani@siu.edu

E.J. Holzmueller
e-mail: eholzmue@siu.edu

J.W. Groninger
e-mail: groninge@siu.edu

© Springer International Publishing Switzerland 2016

A.M. Melesse and W. Abtew (eds.), *Landscape Dynamics, Soils and Hydrological Processes in Varied Climates*, Springer Geography,
DOI 10.1007/978-3-319-18787-7_33

management is rational-comprehensive planning theory (Lachapelle et al. 2003; Plummer and Fennell 2009). With its roots in economic assumptions about human behavior (Brooks 2003) and positivist assumptions about science (Dalton 1986), rational-comprehensive planning entails the application of the scientific method to decision-making with the aim of deriving the most efficient means of enhancing the public welfare (Rittel and Webber 1973). Here, the planner is a value-neutral technician whose role is to apply the scientific method in undertaking a comprehensive and objective analysis of planning problems without political interference.

Rational-comprehensive planning has been critiqued as a utopian model of how decisions should be made in an ideal world. According to Forester (1984), the model is appropriate in situations where planning problems are well-defined, the actor has knowledge on the full range of alternative solutions and their consequences, baseline information exists on the context of the problem, information exists on the values and preferences of citizens, and there is unlimited time, skills, and resources. Unfortunately, these ideal conditions rarely exist in the real world that is characterized by time and resource constraints, political volatility, and so forth (Lachapelle et al. 2003). Besides the constraints in knowledge and capacity (Etzioni 1967), the centralized expert-driven approach of rational-comprehensive planning has been critiqued as undemocratic and inequitable (Davidoff 1965). Rational-comprehensive planning has also been critiqued for its outdated assumptions about the world as existing in a near-equilibrium state and composed of linear relationships that can be understood, predicted, and controlled (Plummer and Fennell 2009). Importantly, rational-comprehensive planning assumes the existence of a unitary public interest, and as such is poorly suited for addressing conflicts inherent in the diversity of goals pursued by an increasingly diverse society (Lane 2001).

Recently, planning problems are increasingly being conceptualized as wicked problems that cannot be adequately solved using the rational-comprehensive planning model. Wicked problems are characterized by a lack of consensus and clarity in their definition and potential solutions (Batie 2008; Balint et al. 2011). According to Ritchey (2013), wicked problems are "messy, devious, and they fight back when you try to deal with them" (p. 1). In their seminal article, Rittel and Webber (1973) offered a useful way to describe wicked problems by opposing them to the "tame" problems that rational-comprehensive planning is designed to solve under ideal conditions. The authors identified ten defining attributes of wicked problems (Table 33.1). The concept of wicked problems has been used to describe the most intractable problems in a range of resource management arenas, including water resources policy (Gunderson and Light 2006), forest planning (Allen and Gould 1986), and recreation planning (Brooks and Champ 2006). Research attention is increasingly being given to the search for conceptual approaches and institutional mechanisms for managing wicked environmental problems (Gollagher and Hartz-Karp 2013; Light et al. 2013; Ritchey 2013).

In line with these efforts, this chapter proposes a new way of analyzing and addressing wicked problems using insights from social-ecological resilience and adaptive governance. Wicked problems, just as social-ecological systems, exhibit

Table 33.1 Ten defining attributes of wicked problems according to Rittel and Webber (1973)

1. No definitive formulation of a wicked problem
2. Wicked problems have no stopping rule
3. Solutions to wicked problems are not true or false, but good or bad
4. No immediate and no ultimate test of a solution to a wicked problem
5. Every solution to a wicked problem is a "one-shot operation"
6. No established set of potential solutions
7. Every wicked problem is unique
8. Every wicked problem can be considered a symptom of another problem
9. Causes of a wicked problem can be explained in numerous ways; the choice of explanation determines the nature of the solution to the problem
10. In solving wicked problems, the planner has no right to be wrong

the attributes of complex adaptive systems, such as emergence, scale-sensitivity, heterogeneity, surprise, and path-dependence. A rethinking of wicked problems as complex social-ecological systems creates new opportunities for understanding and managing wicked problems using the established literature on resilience and adaptive governance. The key attributes of adaptive governance include the pursuit of adaptive and integrated goals, utilization of diverse sources of knowledge, reliance on a polycentric institutional structure, and an analytic-deliberation decision-making process that is informed by communicative action theory (Dietz et al. 2003; Folke et al. 2005; Gunderson and Light 2006; Akamani 2014).

The next section of this chapter briefly examines the shared attributes in the literature on social-ecological systems and wicked problems and argues for the need to analyze wicked problems from the lens of social-ecological systems. The subsequent section examines the key features of adaptive governance and how they can contribute to managing wicked problems. Next, the case of the Cache River watershed is presented as an illustration of a wicked water resources management problem. Concluding remarks are offered in the final section of the chapter.

33.2 Rethinking Wicked Problems as Complex Social-Ecological Systems

Consistent with the attributes of rational-comprehensive planning, conventional resource management regimes were characterized by the dominance of centralized institutions, strong reliance on experts trained in reductionist science, and an emphasis on the maximization of output from selected components of ecosystems (Nelson et al. 2008b; Armitage et al. 2009). A common underlying assumption in the conventional approach to resource management is that natural and social systems are distinct from each other and that natural ecosystems are predictable and controllable (Folke et al. 2002; Berkes 2003). However, there is growing realization that this approach is inadequate for dealing with the complexity and uncertainty entailed in current resource management challenges (Westley et al. 2011). In recognition of the flawed belief in human ability to control ecosystems, Ludwig et al. (1993) stated that "it is more appropriate to think of resources as managing humans than the converse" (p. 17). Similarly, Holling (1993) has noted that the limitations in human ability to understand ecosystems, stating that "there is an inherent unknowability as well as unpredictability concerning these evolving managed ecosystems and the societies with which they are linked" (p. 554).

Since the publication of Holling's (1973) seminal paper, a rich body of the literature has accumulated on the dynamics of human–environment interactions, resulting in radical changes in the old assumptions that underpinned conventional resource management. It is now widely recognized that the study of social and ecological systems as distinct from each other is no longer reasonable (Redman et al. 2004). The concept of social-ecological systems has now emerged to describe the dynamic inter-dependence and coevolution between social and ecological systems (Folke 2007; Folke et al. 2011). A closely related concept to social-ecological system is the "human ecosystem" which has been defined by Machlis et al. (1997) as "a coherent system of biophysical and social factors capable of adaptation and sustainability over time" (p. 351). Significant research efforts have gone into understanding the characteristics of social-ecological systems. These studies have shown that social-ecological systems could be characterized using the attributes of complex adaptive systems, such as nonlinearity, thresholds, uncertainty, emergence, scale, self-organization, and path-dependence (Berkes 2004; Folke et al. 2004; Liu et al. 2007). Costanza (1996) defines systems as "groups of interacting, interdependent parts linked together by exchanges of energy, information, and matter" (p. 981). Simple systems are systems in which the relationships among the components are predominantly linear (Newell 2001). As such, they are easily understood and predicted. On the other hand, complex systems comprise a large number of interacting and interdependent components that are connected through nonlinear relationships (Newell 2001). It is the contention of this chapter that both wicked problems and social-ecological systems share the attributes of complex systems, such as scale, uncertainty, and path-dependence (Table 33.2).

First, wicked problems and social-ecological systems are characterized by scale-sensitivity as well as cross-scale (e.g., spatial, temporal, and jurisdictional) and cross-level (e.g., community, regional, and national) interactions. Holling's (2001) panarchy theory posits that social-ecological systems are composed of a hierarchically nested set of adaptive cycles that are engaged in cross-scale interactions with one another. Adaptive cycles represent the four-phase cycles of growth, conservation, release, and reorganization entailed in the dynamics of social-ecological systems (Walker et al. 2006). The dynamic interactions among smaller and faster adaptive cycles at lower levels and larger and slower adaptive cycles at higher levels account for the balance between change and stability in

Table 33.2 Similarities of wicked problems and social-ecological systems

1. Sensitive to scale
2. Cross-scale interactions
3. Nonlinear interactions
4. Uncertainty and surprise are common
5. Path-dependency—legacy effects of past actions

social-ecological systems (Holling 2004). While the various levels in a social-ecological system are interlinked, each level operates under relatively unique principles, thus making one-size-fits-all solutions inappropriate (Berkes 2004). The multi-scalar nature of social-ecological systems also implies that there can be multiple correct perspectives about the system rather than a single correct perspective (Berkes 2003). Given the interactions across the various levels, a focus on a single level of analysis cannot yield adequate understanding without analyzing processes occurring below and above the chosen level (Walker et al. 2006). Similar to the social-ecological systems, wicked problems tend to be intricately interconnected across multiple levels and scales. As such, every wicked problem appears to be a symptom of another problem (Allen and Gould 1986). There is no single correct level at which wicked problems can be analyzed and addressed, and each chosen level of analysis has trade-offs (Rittel and Webber 1973). While wicked problems have shared attributes, each problem has its own uniqueness that makes it difficult to devise generalizable solutions (Nie 2003). Wicked problems also tend to have multiple explanations about their causes. The choice of definition and the appropriateness of interventions depend on the perspectives or world views of the policy participants rather than the use of some objective criteria (Allen and Gould 1986).

Second, both wicked problems and social-ecological systems are characterized by nonlinear relationships that give rise to uncertainties and surprise. Complex systems cannot be adequately understood and predicted using reductionist techniques because of the nonlinear relationships among the components and the emergent patterns of behavior (Costanza 1996; Gunderson 2010; Allen and Gunderson 2011). The inadequacy of understanding of complex systems increases the likelihood of surprise in human–environment interactions (Holling 1993; Liu et al. 2007). From the perspective of panarchy theory, uncertainties and surprise in social-ecological systems are inherent in the dynamics of the adaptive cycles at each level, as well as the interactions across the various levels and scales of the panarchy. Adaptive cycles are dynamic rather than static (Benson and Garmestani 2011). The dynamics of each adaptive cycle is characterized by long periods of slow and gradual change that lead to resource accumulation (from growth to conservation), followed by short periods of abrupt change (release and reorganization) that allow for system renewal (Holling 2001). The first trajectory, known as the "front loop," is relatively predictable, whereas the second trajectory, known as the "back loop," is marked by high levels of uncertainty and unpredictability. Uncertainties are also associated with the interactions among adaptive cycles at various levels of the panarchy. While larger and slower adaptive cycles condition the dynamics of small and faster adaptive cycles, changes occurring at lower levels also have the potential to trigger cascading effects at higher levels of the panarchy (Holling 2001; Walker et al. 2006). Similar to social-ecological systems, wicked problems are marked by uncertainties and surprise. There is no room for sufficient understanding and exhaustive formulation of wicked problems because of the endless chain of interconnectedness among their components (Rittel and Webber 1973; Batie 2008). Consequently, decision-makers have no objective criteria for verifying that all possible solutions to wicked problems have been considered in the decision-making process. Furthermore, there are no immediate or ultimate tests of solutions to wicked problems due to uncertainties about the consequences of chosen solutions (Rittel and Webber 1973; Gollagher and Hartz-Karp 2013). Thus, interventions on wicked problems will always occur in the face of incomplete knowledge about the problem and its potential solutions which makes surprise likely.

Third, both wicked problems and social-ecological systems illustrate the importance of system history and context in shaping current conditions and future trajectories. Path-dependency suggests that historical and contextual factors constitute important explanatory variables in the dynamics of social-ecological systems (Berkes 2007). Past decision and actions create incentives and constraints that influence subsequent options (Nelson et al. 2007; Heinmiller 2009). Similar to social-ecological systems, path-dependency is an attribute of wicked problems. The literature on wicked problems recognizes that every attempted solution on wicked problems constitute a context that shapes the future. According to Rittel and Webber (1973: 163), "every implemented solution is consequential. It leaves 'traces' that cannot be undone….And every attempt to reverse a decision or to correct for the undesired consequences poses another set of wicked problems." Given the difficulty of reversing the adverse effects of past actions, decision-makers have no right to be wrong in attempts to manage wicked problems in spite of the ambiguity about their causes and solutions.

Given the shared attributes between wicked problems and social-ecological systems as complex systems, this chapter argues for the adoption of a social-ecological system approach to analyzing wicked problems. A reconceptualization of wicked problems as social-ecological systems calls for a resilience-based approach to managing wicked problems. The sustainable management of complex social-ecological systems requires building their resilience. The resilience concept was originally used by Holling (1973) to refer to the amount of disturbance an ecosystem could absorb before shifting into a different state. Following this meaning, Walker et al. (2006) define resilience as "the capacity of a system to experience shocks while retaining essentially the same function, structure, feedbacks, and therefore identity" (p. 2). In coupled social-ecological systems, the resilience concept has three interrelated meanings: the magnitude of disturbance a system can absorb while remaining in a given state; the degree to which a system is capable of self-organization; and the degree to which a system can build and increase its capacity for learning and adapting to change (Folke et al. 2002). Social-ecological responses to drivers of change could be categorized into coping, adaptability, and transformability. Coping refers to short-term or emergency responses to drivers of change, usually undertaken at the level of the individual or lower spatial scales (Berkes and Jolly 2001). Adaptability is the capacity of social-ecological systems to learn and adjust to external drivers and internal processes while maintaining their current state (Folke et al. 2010). In contrast to coping and adaptation, transformability refers to the capacity of actors in a social-ecological system to create a fundamentally new system when the existing system becomes untenable (Walker et al. 2004). Social-ecological transformation may occur through planned or purposeful collective action in response to undesirable situations (Walker et al. 2006; Ernstson et al. 2010) or through processes occurring at higher levels (Folke et al. 2010). The resilience concept can be viewed positively or negatively (Gunderson and Light 2006). Hence, the outcomes of coping, adaptability, or transformation could be either positive or negative. Building resilience in social-ecological systems requires flexible, multi-level institutions that emphasize learning (Folke et al. 2002).

33.3 Managing Wicked Problems Using Adaptive Governance

The previous section argued for a reconceptualization of wicked problems as complex social-ecological systems in view of their shared attributes, and the need to pursue a resilience-based management of wicked problems. This section examines the concept of adaptive governance, one of the promising institutional mechanisms for building resilience in social-ecological systems. Adaptive governance regimes provide multi-level institutional frameworks within which various actors and institutions are connected across multiple scales in a continuous decision-making and implementation process aimed at building the capacity of the social-ecological system for managing conflicts and responding to change (Dietz et al. 2003; Walker et al. 2006; Walker 2012).

A key component of adaptive governance is adaptive management (Gunderson and Light 2006). Adaptive management represents one of the early attempts to apply the resilience concept to resource management. Adaptive management is based on the assumption that resource management will always occur in a context of incomplete knowledge about the dynamics of social-ecological systems, thus giving rise to uncertainties and unpredictability (Allen et al. 2011). Adaptive management seeks to increase knowledge and reduce uncertainty through an iterative decision-making and implementation process that prioritizes learning (Allen and Gunderson 2011). Adaptive management is ultimately aimed at building the resilience of social-ecological systems to enhance their sustainability in the face of uncertainty (Folke et al. 2002). While adaptive management has received widespread attention in the resource management literature, failures have often resulted from its application (Walters 2007). The inadequate consideration of the social and institutional context of resource management has repeatedly been noted as a major source of weakness in the adaptive management concept (McLain and Lee 1996; Gunderson and Light 2006). As Allen and Gunderson (2011) have cautioned, adaptive management may not be an appropriate response to dealing with wicked resource management problems.

Adaptive governance addresses the social and institutional context that is required to facilitate adaptive management (Gunderson and Light 2006) as well as adaptive co-management, an institutional mechanism that combines the learning attribute of adaptive management and the vertical and horizontal linkages of co-management (Olsson et al. 2006). Olsson et al. (2007) have noted that "adaptive governance conveys the difficulty of control, the need to proceed in the face of substantial uncertainty, and the importance of dealing with diversity and reconciling conflict among people and groups who differ in values, interests, perspectives, power, and the kinds of information they bring to situations" (p. 2). Thus, adaptive governance is not only necessary in facilitating adaptive management, but it is a useful mechanism for conflict management and social-ecological transformation (Olsson et al. 2006; Walker 2012). Given these attributes, adaptive governance is critical for promoting ecosystem-based resource management, particularly in times of abrupt change (Folke et al. 2005, 2011). As such, adaptive governance seems promising as an institutional mechanism within which wicked environmental problems could be addressed. Among the factors that have been found to contribute to the emergence and sustenance of adaptive governance systems are the presence of windows of opportunity, leadership, economic incentives, enabling legislation and policies, bridging organizations, and appropriate networks (Gunderson and Light 2006; Folke 2007; Folke et al. 2011). In the ensuing section of this chapter, the key attributes of adaptive governance are described and their relevance for managing wicked environmental problems is discussed.

33.3.1 Adaptive, Integrated Resource Management Goals

Conventional approaches to resource management often emphasized efficiency and predictability in the supply of products from selected components of ecosystems (Holling and Meffe 1996; Folke et al. 2005). The reduction in diversity and variability in ecosystems through these output-maximizing approaches has led to the erosion of resilience and an increase in vulnerabilities (Holling and Meffe 1996; Holling 2012). Also, the broader arena of the development and the environment has traditionally treated social, economic, and environmental issues as distinct from each other, and often focused on one to the neglect of the others. The inadequacy and adverse consequences of this sector-specific approach to resource management have led to a shift toward ecosystem management (Folke et al. 2011). Ecosystem management seeks to balance social, economic, and ecological goals with the aim of sustaining both the social and ecological system (Endter-Wada et al. 1998; Endter-Wada and Blahna 2011). Ecosystem management is a new approach to sustainability that is informed by an awareness of the dynamic inter-dependence between social and ecological systems across multiple spatial and temporal scales and the need to build the resilience of the coupled social-ecological system (Folke et al. 2011). Success in the implementation of ecosystem management requires an adaptive governance framework that promotes integrated resource management goals across the various sectors, as well as learning to deal with uncertainty through adaptive management (Folke et al. 2005; Gunderson and Light 2006).

Through the pursuit of policies that integrate concerns across various sectors, adaptive governance could potentially contribute to managing wicked problems. As has been noted earlier, wicked problems tend to be intricately interconnected with one another; hence, each problem represents a symptom of other problems. Adopting a sector-specific approach to wicked problems is likely to lead to partial solutions and unintended consequences. The use of an integrated approach to solving wicked problems through adaptive governance is more likely to account for the multiple causal relationships among wicked problems. The use of holistic resource management goals is also more likely to accommodate the diverse perspectives of participants in the resource management process, thereby reducing the potential for conflict. Importantly, the centrality of adaptive management within adaptive governance allows for managing wicked problems through an ongoing process of learning and adaptation. These insights are captured in a report by the Australian Public Service Commission (2007: 11) which noted that "The handling of wicked problems requires holistic rather than linear thinking. This is thinking capable of grasping the big picture, including the interrelationships between the full range of causal factors and policy objectives."

33.3.2 Integrated Knowledge Systems

As part of the influence of enlightenment thinking, the role of science in conventional resource management has been to serve as an instrument for the domination of nature (Cortner and Moote 1999). The type of science that has been useful for this purpose is positivist and reductionist science with its emphasis on quantifiable and generalizable knowledge about objective realities (Cortner and Moote 1999; Nelson et al. 2008b). The dominance of experts in the field of resource management trained in the reductionist scientific method has proven non-participatory, too expensive in most instances, and not always trusted by local stakeholders (Moller et al. 2004; Nelson et al. 2008b; Raymond et al. 2010). Importantly, reductionism and the increasing specialization and isolation among disciplines are inadequate for understanding the complexity of social-ecological systems (Costanza 1996; Nelson et al. 2008b; Armitage et al. 2009). The reductionist scientific method assumes reality to be composed of a series of isolated parts, and that the whole can be understood by an analysis of the parts and their relationships (Costanza 1996). However, the nonlinear relationships, emergent properties, cross-scale interactions, uncertainties, and surprise in social-ecological systems make them inherently unknowable and unpredictable using reductionist methods (Nelson et al. 2008b). A more promising way to understand such complex systems is the use of multiple sources and types of knowledge. Adaptive governance emphasizes the use of diverse sources of knowledge as a way of gaining a holistic understanding on the dynamics of social-ecological systems (Nelson et al. 2008b; Bark et al. 2012). In broadening the scope of knowledge beyond traditional reductionist science, the adaptive governance literature emphasizes various forms of collaboration among disciplines (multidisciplinary, interdisciplinary, and transdisciplinary), recognition of local and traditional knowledge, and the promotion of social learning that occurs through social interaction processes (Folke et al. 2005; Nelson et al. 2008b).

The pursuit of learning through diverse sources of knowledge in adaptive governance frameworks is particularly promising in managing wicked environmental problems. As complex systems, wicked problems are characterized by uncertainty. The multiple causal chains inhibit a full understanding of wicked problems, and outcomes of potential solutions are never clear or predictable. Wicked problems and their solutions tend to depend on the scale of analysis and the perspective of the observer (Rittel and Webber 1973). All these attributes of wicked problems necessitate the use of diverse sources of knowledge. As Weber and Khademian (2008: 337) have suggested, “Any effort to effectively manage a wicked problem will require efforts to draw on a broad range of knowledge, to develop a new base of knowledge to address the complexities of the wicked problem and to serve as a premise for co-operation.” Similarly, the Australian Public Service Commission’s (2007) report emphasizes that managing wicked problems requires developing a shared understanding of the multiple dimensions of the problem from the diverse perspectives of stakeholders, as well as providing mechanisms for the sharing of experience on wicked problems. The report also highlights the need for a multi-disciplinary approach to generating the skills and knowledge needed for dealing with wicked problems.

33.3.3 Polycentric Institutional Structure

The institutional structures for adaptive governance radically depart from the centralized, command-and-control mechanisms of conventional resource management (Dietz et al. 2003). Centralized institutions have been criticized for lacking the flexibility to respond to social-ecological change (Armitage et al. 2009). Such institutions also tend to lead to mismatch between policy interventions and the scale at which problems occur (Olsson et al. 2007). Adaptive governance responds to these shortfalls by relying on a polycentric institutional structure, composed of nested semi-autonomous institutions with multiple centers of decision-making and implementation authority (Olsson et al. 2007; Ostrom 2010). The nesting of institutions across different jurisdictional scales enhances the fit between institutional interventions and the scales at which problems occur (Olsson et al. 2007). Through the nested institutional structure, adaptive governance also facilitates the devolution of authority to lower levels, thereby enhancing participation and accountability (Folke et al. 2005; Akamani and Wilson 2011). Within the nested system, new rules could be tested at lower levels which enhance opportunities for experimentation and learning (Ostrom 1999; Folke et al. 2011). Through the polycentric institutional structure, adaptive governance utilizes different types of institutions, both formal and informal, from the public and private sectors, as well as community-based institutions (Dietz et al. 2003; Nelson et al. 2008b; Rijke et al. 2013). This diversity of institutions in the adaptive governance framework enhances the resilience of the system by promoting response diversity (Olsson et al. 2007) and reducing the likelihood of failure associated with reliance on a single type of institution (Dietz et al. 2003; Berkes 2007; Akamani and Wilson 2011).

The unique nature of wicked problems requires interventions that are specific to the scale and context of each problem. Yet, wicked problems also tend to be interconnected across various levels and scales with no level being the best point of intervention (Rittel and Webber 1973). Adaptive governance provides a framework for dealing with wicked problems through its multi-level framework which allows for devising scale-sensitive and context-specific interventions to problems at each level while accounting for their implications on other levels. The multi-level framework allows for devising solutions to wicked problems at multiple levels, thereby enhancing the fit between policy interventions and the scale at which wicked problems occur. Also, given that wicked problems are characterized by a high level of uncertainty regarding their causes and solutions, the use of different types of institutions within the adaptive governance framework enhances response diversity and reduces the likelihood of failure. Consistent with the attributes of adaptive governance, the Australian Public Service Commission's (2007) report calls for a devolved approach to managing wicked problems that promote collaboration among the public sector, private sector, and communities as a means of enhancing flexibility and accountability. The report further highlights the need for government to create awareness and encourage collaboration, as well as the need for involvement of non-governmental organizations in building community capacity for managing wicked problems through bottom-up processes.

33.3.4 Analytic-Deliberation Process

The decision-making process entailed in adaptive governance differs from that of the rational-comprehensive planning approaches of conventional resource management. As has been noted previously, the rational-comprehensive model aspires to advance the common interest by treating the decision-making process as a scientific and technical exercise that is devoid of values and politics. However, differences in values and interests, as well as knowledge and power of policy participants, make conflict an inherent component of the resource management process. An alternative to rational-comprehensive planning is communicative planning which is concerned with "how collective deliberation can accommodate different forms of knowledge and styles of reasoning to promote social learning and yield creative agreements" (Goldstein 2009: 2). The role of the planner includes the facilitation of dialog, moderation of debate, and mediation of negotiation among parties involved in conflict situations (Forester 2009).

Consistent with the turn toward communicative planning, adaptive governance recognizes that conflict is an integral part of decision-making and that well-managed conflicts could provide opportunities for learning and adaptation (Dietz et al. 2003). Adaptive governance relies on an analytic-deliberation process, one that combines technical analysis with public deliberation (Webler and Tuler 1999; Dietz et al. 2003). Dietz et al. (2003) interpret analytic deliberation as "Well-structured dialogue involving scientists, resource users, and interested public, and informed by analysis of key information about environmental and human-environment systems" (p. 1910). The use of analytic deliberation provides opportunities for interaction among various policy participants, including scientists, decision-makers, and ordinary citizens throughout the various stages of the decision-making and implementation process (Tuler and Webler 1999). These interaction processes enhance the consideration and reconciliation of the diverse values held by participants, thereby enhancing the management of conflicts (Dietz et al. 2003; Ballint et al. 2011). Analytic deliberation also enhances the integration of different types of knowledge, thereby leading to enhanced understanding about the social-ecological system (Dietz and Stern 1998). Through analytic deliberation, adaptive governance meets the requirements of good governance such as participation, accountability, transparency, and legitimacy (Akamani and Wilson 2011) and also promotes the building of trust and social capital for enhancing collective responses to social-ecological change (Dietz et al. 2003).

Through the use of an analytic-deliberation process, adaptive governance offers a promising framework for managing wicked problems. The definition of a wicked problem is in the mind of the beholder (Allen and Gould 1986). As such, different participants see wicked problems differently. Similarly, solutions to wicked problems are not true or false based on some objective criteria, but rather are judged as good or bad based on the value aspirations of participants (Rittel and Webber 1973). Decision-making processes that are driven by experts or advocates are inadequate for dealing with such problems (Carcasson 2013). Managing wicked problems requires decision-making processes that promote adequate representation and meaningful participation among the full range of stakeholders (Ballint et al. 2011; Gollager and Hartz-Karp 2013). Opportunities for communication and interaction among stakeholders are essential for enhancing understanding about wicked problems as well as building trust for collective action among stakeholders (Brooks and Champ 2006; Australian Public Service Commission 2007). Analytic deliberation can provide opportunities for interaction among participants, through which shared meanings could be developed, thereby reducing the conflicting perspectives on wicked problems. The opportunities for knowledge integration and information sharing afforded through analytic deliberation also seem useful in managing the uncertainties entailed in the definition and solution of wicked problems.

33.4 The Cache River Watershed: An Illustration of a Wicked Problem

The Cache River watershed (Fig. 33.1) occupies an area of 1944 km^2 in southern Illinois. Although relatively small in size, the watershed is characterized by a high level of biophysical, sociocultural, and administrative complexity. Located at the intersection of four physiographic regions and influenced by two climatic zones (Duram et al. 2004), the watershed has a unique and diverse ecosystem that supports diverse plant and animal species including the neotropical song birds and the migratory waterfowl. The watershed is also home to 1000-year-old cypress and water tupelo trees (Kraft and Penberthy 2000). The wetlands in the watershed have been designated as Wetlands of International Importance by the Ramsar Convention since 1994, and the watershed currently has more than 50 threatened and endangered species (Davenport et al. 2010). Administratively, the watershed cuts across five county boundaries and different land ownership types. The Cache River watershed is also characterized by sociocultural diversity in terms of the composition of the population and the types of land use (Adams et al. 2005).

The last 200 years have seen alterations in the hydrology and land use of the watershed through the interaction of various drivers of change, including demography, economics, technology, and policy implementation (Duram et al. 2004). According to Duram et al. (2004), the arrival of European settlers in the early 1800s began the process of conversion of forests into human settlements and farmlands. By the 1870s, large-scale logging had emerged due to expanded markets created by the construction of railroads. Equipped with technological advances and fueled by economic motivations, the 1900s brought about further modification of the watershed through logging, conversion of forests into farmlands, and various engineering projects aimed at draining of wetlands as well as the diversion, dredging and channelization of sections of the Cache River (Nelson et al. 2008a, 2009). One of the major engineering projects that has been described as "a severe wound in the southern Illinois landscape" (Cache River Watershed Planning Committee 1995: 4) is the construction of the Post Creek Cutoff in 1915 which divided the Cache River into two: the lower and upper Cache River. Although the Post Creek Cutoff facilitated the draining of wetlands for agriculture, it also radically modified the hydrology of the river (Kraft and Penberthy 2000). This project also led to a decline in the logging industry, resulting in the loss of timber-related jobs and subsequent population loss in the 1930s and 1940s (Duram et al. 2004). The 1940s and 1950s saw further modification of the Cache River and its tributaries through engineering projects aimed at straightening and channelizing parts of the river (Kraft and Penberthy 2000). With more than 223 miles of channelized streams existing in the Cache River watershed, various ecological consequences emerged, including erosion and sedimentation (Cache River Ecosystem Partnership 1999).

Consistent with the growing influence of the environmental movement, the period since the 1970s has seen increasing prioritization of ecological values in the Cache River watershed. Beginning with the acquisition of wetlands in the Cache

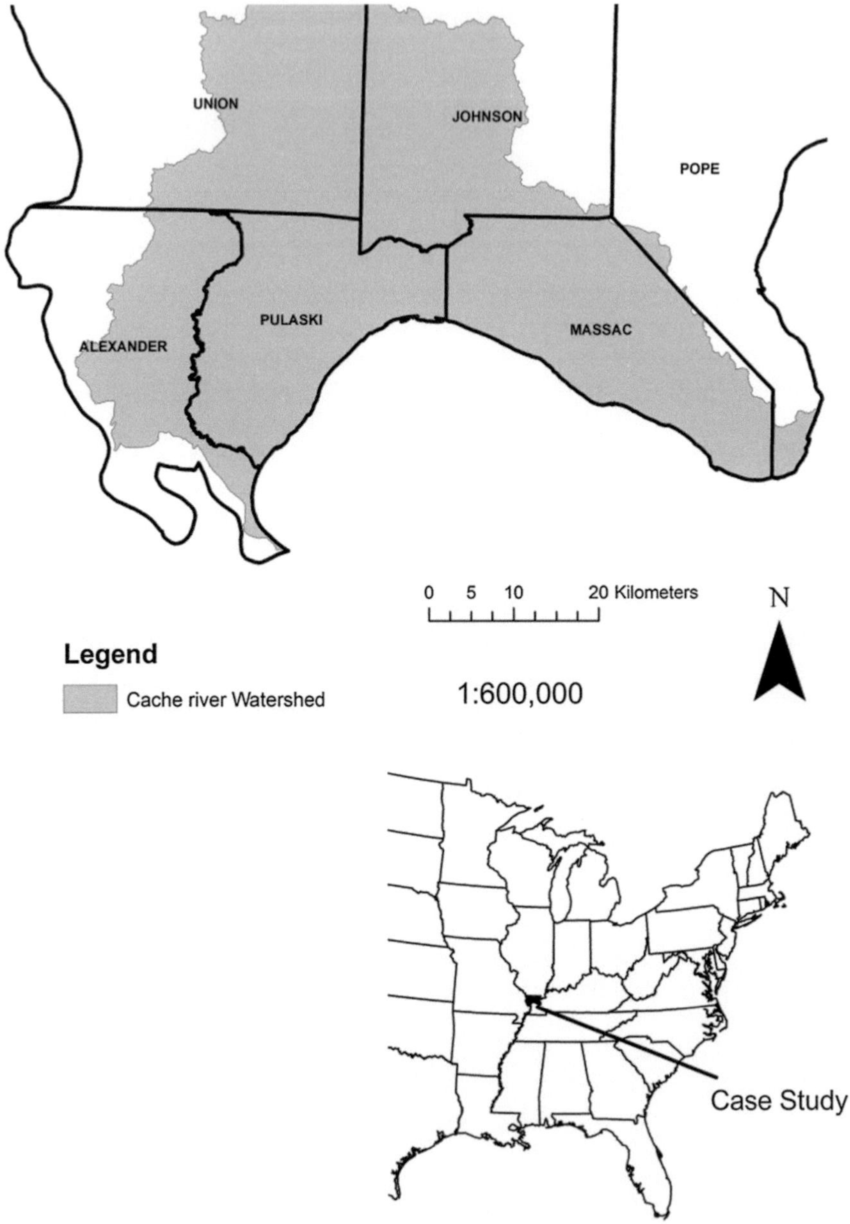

Fig. 33.1 Cache River watershed located in the midwestern United States in southern Illinois

River watershed by the Illinois Department of Natural Resources (IDNR) and The Nature Conservancy (TNC) for protection and restoration in the 1970s, a collaborative network of state, federal, and private organizations has emerged to address the ecological challenges in the watershed. In 1991, the Joint Venture Partnership (JVP) was formed to protect and restore 60,000 acres of wetlands along the Cache River, and ultimately to reconnect the lower and upper Cache Rivers. Specific goals of the JVP include forest and wetland habitat restoration, reduction of sedimentation and streambank/bed erosion, and a managed reconnection of the upper and lower segments of the Cache River (IDNR 2006; Davenport et al. 2010; Behnken 2013). Membership of the JVP currently comprises TNC, IDNR, the US Fish and Wildlife Service (FWS), Ducks Unlimited, and the Natural Resource Conservation Service (NRCS). Between 1993 and 1995, TNC and NRCS led the preparation of the Cache River Watershed Resource Plan (Resource Plan) with funding from the Environmental Protection Agency. The plan has been used by members of the JVP to acquire funding for various restoration projects in the watershed (Adams et al. 2005). Also, the implementation of various federal policies related to the Farm Bill in the 1980s and 1990s, such as the Conservation Reserve Program, has contributed to reducing the soil erosion problem and increasing forested acreage in the watershed (Adams et al. 2005). In spite of the obvious progress in achieving regionally popular restoration objectives, various social and ecological problems in the Cache River watershed still persist.

This brief description of the Cache River watershed illustrates issues of scale, uncertainty, and history that characterize wicked problems. First, the Cache River watershed is marked by uncertainty and surprise in the interaction among the various components of the biophysical and social system. For instance, the construction of the Post Creek Cutoff in 1915 yielded short-term economic benefits in terms of availability of agricultural land. However, it also led to long-term unintended social and ecological consequences. Current efforts in the restoration of the watershed and reconnection of the lower and upper Cache Rivers are also faced with uncertainties about the outcomes of such interventions (Davenport et al. 2010). Second, the current and future conditions in the watershed are also conditioned by the path-dependent effect of past interventions. For instance, past modifications of the watershed to promote agriculture have given rise to a large agricultural constituency whose interests need to be considered in future management options in the watershed. The watershed also has a social history of conflicts and mistrust that continue to shape planning efforts (Kraft and Penberthy 2000; Adams et al. 2005). Finally, differences in perspectives exist within and among the various stakeholder groups in the Cache River watershed, such as communities and the external organizations that constitute the JVP. A study on community members' perspectives on wetlands in the Cache River watershed revealed that community members associate ecological, cultural, and economic meanings with the landscape (Davenport et al. 2010). Differences also exist among the organizations involved in the restoration of the watershed. Such differences constitute potential sources of conflict in the definition and solution of problems in the watershed.

While the Cache River watershed clearly exhibits the attributes of wicked problems, ongoing institutional mechanisms for managing the watershed do not appear adequate in dealing with the complexity and wickedness of the watershed.

These institutional issues are discussed below with a focus on the preparation and implementation of the Resource Plan.

Managing wicked problems using adaptive governance calls for an institutional structure that engages different types of actors at multiple levels, as well as a decision-making process that combines scientific analysis with stakeholder deliberation in order to devise socially acceptable solutions that are informed by sound knowledge. A careful analysis of the institutional structures and processes for the preparation and implementation of the Resource Plan in the Cache River watershed indicated limited opportunities for stakeholder participation in the planning process. The 25 member Planning Committee, that was composed of large-scale commercial farmers and a handful of environmentalists (Kraft and Penberthy 2000), has been critiqued for not adequately representing the full range of local stakeholder interests in the watershed (Adams et al. 2005).

Beside the inadequate representation, the process for preparing the Resource Plan did not allow for adequate consideration of the diverse values and knowledge of local stakeholders. The process for the preparation of the plan entailed monthly meetings among members of the Planning Committee, field trips to wetlands and farms by the Planning Committee, and four public meetings held at the beginning and the end of the nine-step planning process to solicit stakeholder comments and to announce the outcome of the planning process. Additionally, a telephone survey was undertaken to identify resource concerns of residents in the watershed (Cache River Watershed Resource Planning Committee 1995). In all, the process offered limited opportunities for stakeholder interaction and deliberation since face-to-face interaction occurred only through the meetings that were held at the beginning and the end of the planning process. Although the telephone survey revealed major differences in perspectives between farm and non-farm residents, as well as differences between residents of the watershed and the Planning Committee that was supposed to represent them, no explicit mechanisms were employed to address these conflicts (Kraft and Penberth 2000). As such, the planning process failed to realize the expected benefits of analytic deliberation, which include enhanced social capital, legitimacy, accountability, participation, and transparency.

Using adaptive governance to manage wicked problems requires adaptive and integrated resource management goals that are informed by complexity and resilience thinking. The content of the Resource Plan has received criticism for its narrow focus on ecological goals, failure to prioritize learning, and neglect of the socioeconomic concerns in the watershed (Lant 2003; Akamani 2014). The plan emphasized nine resource management concerns ranging from soil erosion to protection of private property. However, these goals focus mainly on ecological and hydrologic issues in the watershed while socioeconomic concerns are neglected. This outcome has been attributed to the dominance of the Technical Committee, composed of 15 experts representing various organizations that shaped the focus of the plan (Adams et al. 2005). While the plan has contributed to improving upon ecological conditions in the watershed, it may have also contributed to the erosion of trust and social capital among stakeholders (Lant 2003; Davenport et al. 2010).

The planning goals also reflect a lack of appreciation of the complexity and wickedness of the problems in the watershed. For example, common goals of hydrologic reconnection and habitat restoration have alternative outcomes due to the differing mandates and priorities of JVP parties. For example, flooding regimes that optimize waterfowl habit may also damage or kill mature trees and prevent regeneration of native forests. Agencies also differ in their approaches to vegetation restoration: IDNR and TNC tree planting plans emphasize restoring native tree species to sites where they historically were found (Ruzicka et al. 2010). In contrast, NRCS and FWS support moist soil units and/or food plots and advocate large-scale planting of small-seeded oaks, reflecting prevailing game wildlife conservation priorities of many stakeholders (Groninger 2005; Behnken 2013). Beyond the multitude of stakeholder views, agencies themselves are sometimes bound by policies and command structures that restrict their flexibility and responsiveness within an adaptive governance framework.

33.5 Conclusion

In view of the shortfalls associated with the rational-comprehensive approach to planning and resource management, the concept of wicked problems, a class of problems that lack consensus in their definition and solution, is receiving attention among academics and policy-makers. Yet, pathways to managing such problems are unclear. Borrowing from the field of social-ecological systems research, this chapter has argued for a rethinking of wicked problems as complex social-ecological systems, and the need to manage wicked problems using adaptive governance institutions. An adaptive governance approach to managing wicked problems calls for adaptive, integrated resource management goals, integrated knowledge systems, polycentric institutional structures, and analytic-deliberation processes.

Analysis of the Cache River watershed has shown that the interactions between the complex social system and complex biophysical system are characterized by surprise and uncertainties, multiple conflicting perspectives, and path-dependency that are typical of wicked problems. Yet, policy responses to managing the watershed do not appear to recognize this complexity and wickedness. Partly due to constraints by the missions and policies outside their control, land management agencies involved in the ongoing restoration efforts in the watershed since the 1970s have not adequately prioritized the need for learning to live with complexity and wickedness. Institutional structures and processes for decision-making in the watershed also do not offer adequate opportunities for representation and participation by local stakeholders. As such, planning processes are largely expert-driven and management options constrained by individual agency missions, leading to the prioritization of the ecological concerns of powerful actors. Furthermore, land management is entrusted to state or national agencies lacking mechanisms to truly integrate local values and knowledge. Although the restoration efforts have led to some significant improvements in the ecological system as some of the least controversial actions were undertaken, the wickedness of the watershed remains unmanaged. Dealing with wicked problems in the Cache River watershed will require policies and incentives to enable land management agencies and local stakeholders to embrace complexity and wickedness, building local institutional capacity, and investment in conflict management skills.

In all, the case of the Cache River watershed illustrates a broader problem in current resource management—the difficulty of breaking away from the dominance of the rational-comprehensive approach of conventional resource management and transitioning toward an integrated adaptive governance approach that supports ecosystem management (Lachapelle et al. 2003; Folke et al. 2011). In the context of water resources management, Pahl-Wostl et al. (2009) have noted that although the shortfalls of the conventional command-and-control paradigm of water management is widely known, and the need for adaptive, integrated, and participatory approaches to water management has been discussed for decades, actual change has been slow due to the path-dependent effects of the older regimes. Thus, the process of transition toward adaptive governance is itself a wicked problem that requires research and policy focus. Allen and Gould (1986) have stated that "People are what make problems wicked, and people are the ones who can solve them" (p. 23). Greater consideration of the human dimensions of land and water resources management using relevant theoretical perspectives from the social sciences will be needed to enhance understanding of the social and institutional dynamics entailed in the transition toward adaptive management and adaptive governance (Pahl-Wostl 2007). Such a transition is necessary for the successful management of wicked environmental problems.

References

Adams J, Kraft S, Ruhl JB et al (2005) Watershed planning: pseudo-democracy and its alternatives—the case of the Cache River watershed, Illinois. Agric Hum Values 22:327–338

Akamani K (2014) Challenges in the transition toward adaptive water governance. In: Imtaez MA (ed) Water conservation: practices, challenges and future implications. Nova Science Publishers, Inc., Hauppauge

Akamani K, Wilson IP (2011) Toward the adaptive governance of transboundary water resources. Conserv Lett 4:409–416

Allen GM, Gould EM (1986) Complexity, wickedness, and public forests. J Forest 84:20–24

Allen CR, Gunderson LH (2011) Pathology and failure in the design and implementation of adaptive management. J Environ Manage 92:1379–1384

Allen CR, Fonteine JJ, Pope KL et al (2011) Adaptive management for a turbulent future. J Environ Manage 92:1339–1345

Armitage DR, Plummer R, Berkes R et al (2009) Adaptive co-management for social-ecological complexity. Front Ecol Environ 7:95–102

Australian Public Service Commission (2007) Tackling wicked problems: a public policy perspective. Commonwealth of Australia

Balint PJ, Stewart RE, Desai A et al (2011) Wicked environmental problems: managing uncertainty and conflict. Island Press, London

Bark RH, Garrick DE, Robinson CJ et al (2012) Adaptive basin governance and the prospects for meeting indigenous water claims. Environ Sci Policy 19:169–177
Batie SS (2008) Wicked problems and applied economics. Am J Agric Econ 5:1176–1191
Behnken JA (2013) Natural resource management knows no bounds: a case study of the Cache River joint venture partnership. Unpublished M.S. Thesis, Department of Forestry, Southern Illinois University, Carbondale
Benson MH, Garmestani AS (2011) Embracing panarchy, building resilience and integrating adaptive management through a rebirth of the national environmental policy act. J Environ Manage 92:1420–1427
Berkes F (2003) Alternatives to conventional management: lessons from small-scale fisheries. Environments 31:5–19
Berkes F (2004) Rethinking community-based conservation. Conserv Biol 18:621–630
Berkes F (2007) Community-based conservation in a globalized world. PNAS 104:15188–15193
Berkes F, Jolly D (2001) Adapting to climate change: social-ecological resilience in a Canadian western Arctic community. Conservation Ecology 5:18. Available at http://www.consecol.org/vol5/iss2/art18. Accessed 20 Aug 2006
Brooks MP (2003) Planning theory for practitioners. Planners Press, Washington DC
Brooks JJ, Champ PA (2006) Understanding the wicked nature of "unmanaged recreation" in Colorado's front range. Environ Manage 38:784–798
Cache River Ecosystem Partnership (1999) The Cache River watershed strategic resource plan: a supplement to the 1995 Cache River watershed resource plan. Cache River Ecosystem Partnership
Cache River Watershed Resource Planning Committee (1995) Resource plan for the Cache River watershed: Union, Johnson, Massac, Alexander and Pulaski counties Illinois. Cache River Watershed Resource Planning Committee, Grand Chain
Carcasson M (2013) Tackling wicked problems through deliberative engagement. Colorado Municipalities (October 2013) 9–13
Cortner HJ, Moote MA (1999) The politics of ecosystem management. Island Press, Washington DC
Costanza R (1996) Ecological economics: reintegrating the study of humans and nature. Ecol Appl 6:978–990
Dalton LC (1986) Why the rational paradigm persists—The resistance of professional education and practice to alternative forms of planning. J Plann Educ Res 5:147–153
Davenport MA, Bridges CA, Mangun JC et al (2010) Building local community commitment to wetlands restoration: a case study of the Cache River watershed in southern Illinois, USA. Environ Manage 45:711–722
Davidoff P (1965) Advocacy and pluralism in planning. J Am Inst Planners 31:331–338
Dietz T, Stern PC (1998) Science, values, and biodiversity. Bioscience 48:441–444
Dietz T, Ostrom E, Stern P (2003) The struggle to govern the commons. Science 302:1907–1912
Duram L, Bathgate J, Ray C (2004) A local example of land-use change: southern Illinois—1807, 1938, and 1993. The Prof Geogr 56:127–140
Endter-Wada J, Blahna DJ (2011) Linkages to public land framework: toward embedding humans in ecosystem analysis using "inside-out social assessment". Ecol Appl 21:3254–3271
Endter-Wada J, Blahna DJ, Krannich R et al (1998) A framework for understanding social science contributions to ecosystem management. Ecol Appl 8:891–904
Ernstson H, van der Leeuw S, Redman C et al (2010) Urban transitions: on urban resilience and human-dominated ecosystems. Ambio 39:531–545
Etzioni A (1967) Mixed-scanning: a "third" approach to decision-making. Public Adm Rev 27:385–392
Folke C (2007) Social-ecological systems and adaptive governance of the commons. Ecol Restor 22:14–15
Folke C, Carpenter SR, Elmqvist T et al (2002) Resilience and sustainable development: building adaptive capacity in a world of transformations. Ambio 31:437–440

Folke C, Carpenter S, Walker B et al (2004) Regime shifts, resilience, and biodiversity in ecosystem management. Annu Rev Ecol Evol Syst 35:557–581
Folke C, Hahn T, Olsson P et al (2005) Adaptive governance of social-ecological systems. Annu Rev Environ Resour 30:441–473
Folke C, Carpenter SR, Walker B et al (2010) Resilience thinking: integrating resilience, adaptability and transformability. Ecology and Society 15:20. Available at http://www.ecologyandsociety.org/vol15/iss4/art20/. Accessed 8 Dec 2012
Folke C, Jansson A, Rockstrom J et al (2011) Reconnecting to the biosphere. Ambio 40:719–738
Forester J (1984) Bounded rationality and the politics of muddling through. Public Adm Rev 44:23–31
Forester J (2009) Dealing with differences: dramas of mediating public disputes. Oxford University Press, Oxford
Goldstein BE (2009) Resilience to surprises through communicative planning. Ecology and society 14: 33. Available at http://www.ecologyandsociety.org/vol14/iss2/art33. Accessed 8 Dec 2012
Gollagher M, Hartz-Karp J (2013) The roles of deliberative collaborative governance in achieving sustainable cities. Sustainability 5:2343–2366
Groninger JW (2005) Increasing the impact of bottomland hardwood afforestation. J Forest 103:184–188
Gunderson L (2010) Ecological and human community resilience in response to natural disasters. Ecology and society 15: 18. Available at http://www.ecologyandsociety.org/vol15/iss2/art18/. Accessed 8 Dec 2012
Gunderson L, Light SS (2006) Adaptive management and adaptive governance in the everglades ecosystem. Policy Sci 39:323–334
Heinmiller TB (2009) Path dependency and collective action in common pool governance. Int J Commons 3:131–147
Holling CS (1973) Resilience of ecological systems. Annu Rev Ecol Syst 4:1–23
Holling CS (1993) Investing in research for sustainability. Ecol Appl 3:552–555
Holling CS (2001) Understanding the complexity of economic, ecological, and social systems. Ecosystems 4:390–405
Holling CS (2004) From complex regions to complex worlds. Ecology and society 9:11. Available at http://www.ecologyandsociety.org/vol9/iss1/art11/. Accessed 10 Aug 2010
Holling CS (2012) Response to "panarchy and the law." Ecology and society 17: 37. Available at doi:10.5751/ES-05402-170437. Accessed 15 Dec 2012
Holling CS, Meffe GK (1996) Command and control and the pathology of natural resource management. Conserv Biol 10:328–337
Illinois Department of Natural Resources (2006) The cache river wetlands joint venture: a watershed scale restoration project. Available at http://dnr.state.il.us/lands/landmgt/parks/r5/ccr/ccr_jointventureaccomplishdoc.pdf. Accessed 22 Feb 2012
Kraft S, Penberthy J (2000) Conservation policy for the future: what lessons have we learned from watershed planning and research. J Soil Water Conserv 55:327–333
Lachapelle PR, McCool SF, Patterson ME (2003) Barriers to effective natural resource planning in a "messy" world. Soc Nat Resour 16:473–490
Lane MB (2001) Affirming new directions in planning theory: comanagement of protected areas. Soc Nat Resour 14:657–671
Lant C (2003) Watershed governance in the United States: the challenges ahead. Water Resour Update 126:21–28
Light S, Medema W, Adamowski J (2013) Exploring collaborative adaptive management of water resources. J Sustain Dev 6:31–46
Liu J, Dietz T, Carpenter SR et al (2007) Complexity of coupled human and natural systems. Science 317:1513–1516
Ludwig D, Hilborn R, Walters C (1993) Uncertainty, resource exploitation, and conservation: lessons from history. Science 260:17–18

Machlis GE, Force JE, Burch WR Jr (1997) The human ecosystem Part I: the human ecosystem as an organizing concept in ecosystem management. Soc Nat Resour 10:347–367
McLain RJ, Lee RG (1996) Adaptive management: promises and pitfalls. Environ Manage 20:437–448
Moller H, Berkes F, Lyver PO et al (2004) Combining science and traditional ecological knowledge: monitoring populations for co-management. Ecology and society 9: 2. Available at http://www.ecologyandsociety.org/vol9/iss3/art2
Nelson DR, Adger WN, Brown K (2007) Adaptation to environmental change: contributions of a resilience framework. Annu Rev Environ Resour 32:395–419
Nelson JL, Ruffner CM, Groninger JW et al (2008a) Drainage and agriculture impacts on fire frequency in a forested wetland. Can J For Res 38:2932–2941
Nelson R, Howden M, Smith MS (2008b) Using adaptive governance to rethink the way science supports Australian drought policy. Environ Sci Policy 11:588–601
Nelson JL, Groninger JW, Ruffner CM et al (2009) Past land use, disturbance regime change, and vegetation response in a southern Illinois bottomland conservation area. J Torrey Bot Soc 136:242–256
Newell WH (2001) A theory of interdisciplinary studies. Issues Interdisc Stud 19:1–25
Nie M (2003) Drivers of natural resource-based political conflict. Policy Sci 36:307–341
Olsson P, Gunderson LH, Carpenter SR et al (2006) Shooting the rapids: navigating transitions to adaptive governance of social-ecological systems. Ecology and society 11:18. Available at http://www.ecologyandsociety.org/vol11/iss1/art18. Accessed 4 June 2009
Olsson P, Folke C, Galaz V et al (2007) Enhancing the fit through adaptive co-management: creating and maintaining bridging functions for matching scales in the Kristianstads Vettenrike Biosphere Reserve, Sweden. Ecology and society 12: 28. Available at http://www.ecologyandsociety.org/vol12/iss1/art28. Accessed 4 June 2009
Ostrom E (1999) Coping with the tragedies of the commons. Annu Rev Polit Sci 2:493–535
Ostrom E (2010) Polycentric systems for coping with collective action and global environmental change. Glob Environ Change 20:550–557
Pahl-Wostl C (2007) Transitions towards adaptive management of water facing climate change and global change. Water Resour Manage 21:49–62
Pahl-Wostl C, Sendzimir J, Jeffrey P (2009) Resources management in transition. Ecology and society 14:46. Available at http://www.ecologyandsociety.org/vol14/iss1/art46/. Accessed 10 June 2010
Plummer R, Fennell DA (2009) Managing protected areas for sustainable tourism: prospects for adaptive co-management. J Sustain Tourism 17:149–168
Raymond CM, Fazey I, Reed M et al (2010) Integrating local and scientific knowledge for environmental management. J Environ Manage 91:1766–1777
Redman CL, Grove MJ, Kuby LH (2004) Integrating social science into the long term ecological research (LTER) network: social dimensions of ecological change and ecological dimensions of social change. Ecosystems 7:161–171
Rijke J, Farrelly M, Brown R, Zavenbergen C (2013) Configuring transformative governance to enhance resilient urban water systems. Environ Sci Policy 25:62–72
Ritchey T (2013) Wicked problems: modelling social messes with morphological analysis. Acta Morphologica Generalis 2:1–8
Rittel HW, Webber MM (1973) Dilemmas in a general theory of planning. Policy Sci 4:155–169
Ruzicka KJ, Groninger JW, Zaczek JJ (2010) Deer browsing, forest edge effects, and vegetation dynamics following bottomland forest restoration. Restor Ecol 18:702–710
Tuler S, Webler T (1999) Designing an analytic deliberative process for environmental health policy making in the US nuclear weapons complex. Risk: Health Saf Environ 10:65–87
Walker BH (2012) A commentary on "resilience and water governance: adaptive governance in the Columbia River Basin." Ecology and society 17: 29. Available at doi:10.5751/es-05422-170429. Accessed 25 Feb 2013

Walker B, Holling CS, Carpenter S, Kinzig A (2004) Resilience, adaptability and transformability in social-ecological systems. Ecology and society. 9:5. Available at http://www.ecologyandsociety.org/vol9/iss2/art5/. Accessed 4 June 2009

Walker B, Gunderson L, Kinzig A et al (2006) A handful of heuristics and some propositions for understanding resilience in social-ecological systems. Ecology and society. 11:1. Available at http://www.ecologyandsociety.org/vol11/iss/art13/. Accessed 15 May 2007

Walters CJ (2007) Is adaptive management helping to solve fisheries problems? Ambio 36:304–307

Weber EP, Khademian AM (2008) Wicked problems, knowledge challenges, and collaborative capacity builders in network settings. Public Adm Rev 68:334–349

Webler T, Tuler S (1999) Integrating technical analysis with deliberation in regional watershed management planning: applying the national research council approach. Policy Stud J 27:530–543

Westley F, Olsson P, Folke C et al (2011) Tipping toward sustainability: emerging pathways of transformation. Ambio 40:762–780

Chapter 34
Evaluation of thc Effects of Water Harvesting on Downstream Water Availability Using SWAT

Ayalkibet M. Seka, Adane A. Awass, Assefa M. Melesse, Gebiaw T. Ayele and Solomon S. Demissie

Abstract There are many water-related human interventions that modify the natural hydrological systems of watersheds. Rainwater harvesting (RWH) is such an intervention that involves harnessing of water. Water harvesting used in upstream prevents surface runoff to downstream impacting biodiversity and ecosystems. The main objective of the study is to assess the effects of water harvesting technologies on downstream water availability in Alaba District, Ethiopia. The Soil and Water Assessment Tool (SWAT) model, cost-benefit ratio, and optimal control approach (OCA) were used to analyze the hydrological, socioeconomic impact and trade-offs on water availability of the community, respectively. The downstream impact of increasing water consumption in the upstream rainfed areas of the Bilate and Shala watershed is simulated using the semi-distributed SWAT model. The two land use scenarios tested at sub-basin levels for the conventional land use represent the current land use practice (Agri-CON) and infield rainwater harvesting (IRWH) for

A.M. Seka (✉) · A.A. Awass
Water Resources Research Center & Department of Water Resource & Irrigation Engineering, Institute of Technology, Arba Minch University, P.O. Box 21
Arba Minch, Ethiopia
e-mail: aabmdb.seka4@gmail.com

A.A. Awass
e-mail: adaneabe@gmail.com

A.M. Melesse
Department of Earth and Environment, Florida International University, Miami, USA

G.T. Ayele
Faculty of Civil and Water Resource Engineering, Bahir Dar Institute of Technology, Bahir Dar University, P.O. Box 252, Bahir Dar, Ethiopia
e-mail: gebeyaw21@gmail.com

G.T. Ayele
Department of Hydraulic and Water Resources Engineering, Blue Nile Water Institute, Bahir Dar University, P.O. Box 252, Bahir Dar, Ethiopia

S.S. Demissie
Department of Civil and Environmental Engineering, University of California, Los Angeles, Los Angeles 90095, CA, USA

© Springer International Publishing Switzerland 2016

A.M. Melesse and W. Abtew (eds.), *Landscape Dynamics, Soils and Hydrological Processes in Varied Climates*, Springer Geography,
DOI 10.1007/978-3-319-18787-7_34

improving soil water availability through RWH land use scenario. The simulated water balance results showed that the highest peak mean monthly direct flow was obtained from Agri-CON land use (12.71 m^3/ha), followed by Agri-IRWH land use (11.5 m^3/ha). The Agri-IRWH scenario reduced direct flow by 10 % compared to Agri-CON and more groundwater flow contributed by Agri-IRWH (190 m^3/ha) than Agri-CON (125 m^3/ha). The overall result suggests that the water yield of the area may not be negatively affected by the Agri-IRWH land use scenario. The technology in the district benefited positively having an average cost-benefit ratio of 4.2. Storage tanks, series of check dams, and gravel filled dams are alternative solutions for water harvesting.

Keywords Water harvesting · SWAT · Land use scenario · Water availability · Ethiopia

34.1 Introduction

Many water-related human interventions such as water storage, diversion, regulation, distribution, application, pollution, purification, and other associated acts that modify the natural water systems do alter the hydrologic regimes of watersheds. The common effect of all of these is their impact on the downstream (Ngigi 2003), requiring a holistic approach of a river basin scale analysis and management. This approach should enhance the common understanding of the impacts of the different activities on the overall productivity of water and sustainability of natural resource use.

In Ethiopia, agriculture, mainly commercial farming, is the economic backbone of the society in the region. The rural small-scale farmers are currently being introduced to improved agricultural practices through improved surface water (rainwater) management. The hydrological balance of any river basin is directly and indirectly influenced by the spatial and temporal distribution of land use/land cover changes. Changes in land cover can modify crucial hydrological processes, such as evapotranspiration and ground water recharge. Since the demand for water and land is closely related to socioeconomic development, it is necessary to analyze and model both the physical environment and the human dimension of these processes.

Rainwater harvesting (RWH), which involves harnessing of water in the upstream watershed and is designed for "on-site" gains, may have hydrological impacts on downstream water availability (Ngigi 2003). Increased water consumption at upstream level is an issue of concern for downstream water availability, but it is generally assumed that there are overall gains and synergies by maximizing the efficient use of rainwater at farm level (Rockstrom et al. 2002). However, expansion of RWH practices could have unintended hydrological consequences on river basin water resources and may have negative implications on downstream water availability to sustain hydro-ecological and ecosystem services (De Winnaar and Jewitt 2010).

In general, increasing the residence time of water in a watershed through RWH may have positive environmental as well as hydrological implications/impacts downstream (Rockstrom et al. 2002). However, it may also result in uninformed decisions by policy makers. For instance, Irrigation Department in India ordered the destruction of community RWH structures, fearing that it would threaten the supply of irrigation water to downstream users. Therefore, there is a need for further research and understanding on the possible impact of wider expansion of RWH technologies.

Water harvesting systems have been successfully utilized by people in some parts of the world where water shortage exists. In India, RWH is used to recharge groundwater wells (Stiefel et al. 2007, 2008, 2009). Although the application of water harvesting techniques is potentially high, their practice is limited in Ethiopia. In order to meet the water demand for various purposes, sustainable systems of water harvesting and managing should be developed. Local approaches and indigenous experiences have to be encouraged and be applied easily at both village and household levels. The effort to collect water for completing the need and the use of water in efficient ways is becoming very urgent. Water harvesting techniques are promoted to be introduced to community for handling the water scarcity and disaster due to flooding. Collected water from direct rainfall and runoff will be very valuable for covering the needs. Water harvesting may also increase recharge of the groundwater leading to increase of groundwater storage. This is the reason why water harvesting techniques are important for sustaining water resource management (Hidayat et al. 2008).

Understanding and predicting the impact of surface processes on water resource requires long-term historical reconstructions and projections into the future of land use changes at local and regional scales. Understanding of the causes of land use change has moved from simplistic representations to a much more profound understanding that involves situation-specific interactions among a large number of factors at different spatial and temporal scales (Lambin et al. 2003).

Land use/land cover dynamics is an important landscape process capable of altering the fluxes of water, sediment, contaminants, and energy. Mainly caused by human, impact of land use on water resource availability is high. Degraded watersheds tend to accelerate overland flow reducing soil moisture and base flow recharge and increase sediment detachment and transport. Various studies used land cover mapping tools and methods to understand land use changes, inventory of forest, and natural resources as well as understand the changes in the hydrologic behavior of watersheds (Getachew and Melesse 2012; Mango et al. 2011a, b; Wondie et al. 2011, 2012; Melesse and Jordan 2002, 2003; Melesse et al. 2008; Mohammed et al. 2013).

Water harvesting has now become one of the most important pillars of food security strategy of Ethiopian government. Over the years, the Ethiopian government together with nongovernmental organizations has been involved in the development of RWH to enhance water availability for crop production, and drinking water for humans and livestock. Although many organizations are involved in water-related activities, the capacity of the community and government is still low to take advantage of the potential that RWH offers in justifying the effects of water scarcity.

As a result of long history of agriculture and high population in Alaba District, Ethiopia, vegetation cover is very low. Consequently, erosion hazards in the sloppy areas are enormous. Huge gullies are observed toward the southern end of the district, where soils are totally removed beyond recovery. This is believed to have been aggravated due to the easily detachable nature of the soil, and even though there were some efforts of soil and water conservation (SWC) over the last twenty years, these efforts were limited.

In addition to this, upstream land use change could also bring about a significant on-site effect on water resources and off-site effect on downstream users. It may increase or decrease discharge of water at the outlet of a watershed. For instance, the introduction of RWH in the upper stream of Bilate and Shala watershed in Alaba District may result in a major impact on the downstream water users and the ecological balance of the system, if allowed to expand unchecked. The decline in rainfall due to the drought reduces appreciably the groundwater and water table recharge (IPMS 2005). With all stated problems, it is reasonable to assess the effect of existing water harvesting technology on downstream water availability of Alaba District to determine their success and sustainability.

The overall objective of the study is to assess the effect of water harvesting structure on the downstream water availability and to find the viable options of surface water harvesting with specific objectives to (1) identify hydrologic modeling factors that are responsible for the downstream effect, (2) conduct a cost-benefit analysis (CBA) and modeling impacts of RWH technologies considering economic growth across different scale, (3) evaluate trade-off on upstream and downstream water availability of the community under water harvesting influence and environmental externality, and (4) suggest an alternative and economically viable methods of surface water harvesting for crop production, domestic uses, and livestock.

34.2 Study Area, Dataset, and Methods

34.2.1 Study Area

Alaba District, Ethiopia, is located at 7° 17′N latitude and 38° 06′E longitudes (Fig. 34.1). The Bilate River basin is among the major sub-basins that are part of the Abaya–Chamo basin, which is the sub-basin of the Rift Valley Lake Basin. It drains the northern watershed of the Lake Abaya–Chamo drainage basin. The watershed with an area of 3450.6 km^2 is located roughly between 37° 47′ 14″ to 38° 20′ 14″E and 6° 33′ 18″ to 8° 6′ 57″N latitudes and longitudes, respectively. With high spatial and temporal variation of rainfall, the total annual rainfall in the basin ranges between 1280 and 1339 mm in the upper sub-basin, 1061 and 1516 mm in the middle sub-basin, and 769 and 956 mm in the lower sub-basins of the study area.

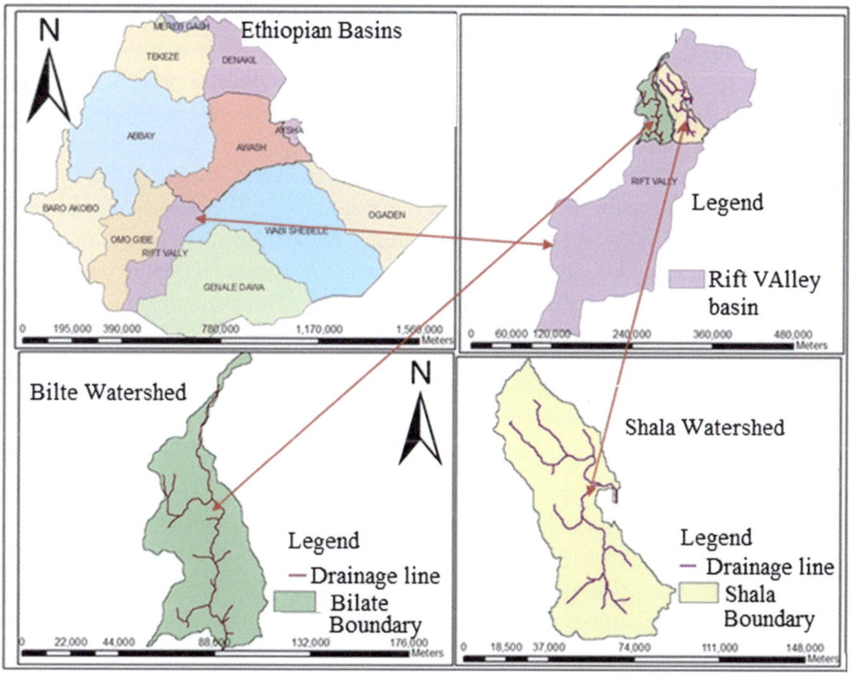

Fig. 34.1 Location of the study area (Bilate and Shala watershed)

34.2.2 Dataset

A comprehensive review of existing data about soil, climate, topographic map, land use/land cover change pattern, agricultural practice, and socioeconomic data has been made at the beginning stage of the study. The general method for this study is divided into three main parts, namely pre-field work, field work, and post-field work.

The topographic map of the study area was scanned. The satellite images were geo-referenced using the topographic map of the study area. Preliminary image interpretation, band selection, layer stacking, and image preprocessing were done in this phase. A collection of different secondary data such as rainfall and temperature data, soil map, digital elevation model (DEM), and unsupervised classification of images before ground truth work were done in the first phase.

There are three basic datasets needed for the modeling work. These are the meteorological data (rainfall, temperature, relative humidity, and sunshine hours), the hydrological (stream flow) data, and watershed physiographic data. The Soil Water Assessment Tool (SWAT) requires meteorological and hydrological input data in a daily time step including rainfall, temperature, relative humidity, sunshine hours, and monthly flow data at the sub-basin outlet.

Monthly cumulative rainfall, mean maximum and minimum temperatures, mean relative humidity, and mean sunshine hours data were compiled for all available stations in the basin. Most of the meteorological stations for which data were collected are located inside the basin and some are located outside. Record lengths of these stations vary from a few years to more than 25 years. Taking into account the length of record, continuity of data, parallel period of observation, and the distribution of stations in the sub-basin, nine meteorological stations were selected for the study. The distribution of these stations within the sub-basin is not even. Hence, the Thiessen polygon method was used to estimate the areal rainfall and minimize the error introduced by spatial variability. The method is also used to estimate temperature, relative humidity, and sunshine hours.

34.2.3 Methods

The methodology of this study is summarized in a form of a flowchart as shown in Fig. 34.2. The flowchart depicts the steps followed in carrying out the study.

Land cover classification: Landsat satellite imageries were used to identify changes in LULC distribution in the Bilate and Shala watersheds over a 32-year

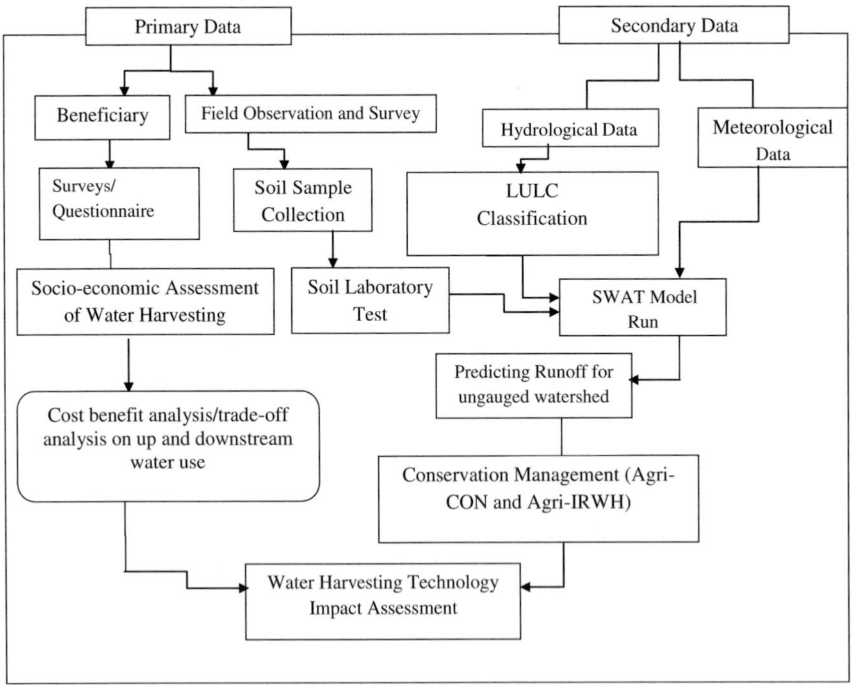

Fig. 34.2 Flowchart of general methodology

period from 1973 to 2005. Landsat Multispectral Scanner (MSS), Thematic Mapper (TM), and Enhanced Thematic Mapper Plus (ETM+) sensors were selected to represent the land cover conditions in the years 1973, 1984, and 2005, respectively. The Landsat MSS imagery is in four channels (2 visible, 2 near-infrared) at 57-m resolution. Landsat TM imagery provides seven multispectral channels (3 visible, 1 near-infrared, 2 mid-infrared, 1 thermal infrared) at 30-m resolution (120-m resolution for the thermal infrared band). The ETM+ adds an extra 15-m resolution panchromatic band and improved resolution for the thermal infrared map of 60-m resolution.

Since the image had different file format, all images were imported in the tagged file formats (TIF). The acquisition dates, sensor, path/row, and resolution of the images are shown in Table 34.1.

Earth Resources Data Analysis System (ERDAS): Imagine that an image processing software was used to perform image analysis and image classification (ERDAS 2005). This analysis was done to provide information on LULC trends and for generation of land cover scenario.

Unsupervised classification: Unsupervised classification used to cluster pixels in a dataset into classes based on statistics only. These classes are spectral classes, and their identity is not initially known, until they are compared with some reference data. This method calculates class means evenly distributed in the data space and then iteratively clusters the remaining pixels using minimum distance techniques. Each iteration recalculates means and reclassifies pixels with respect to the new means. All pixels classified to the nearest class unless a standard deviation or distance threshold is specified. If some pixels do not meet the selected criteria, they are unclassified. This process continues until the number of pixels in each class changes by less than the selected pixel change threshold or the maximum number of iterations is reached. The final clusters are used to classify the image with classifiers such as the minimum distance or maximum likelihood. The output, however, requires post-classification operations to make the results more meaningful. Image classification was first done by unsupervised classification using the Iterative Self-Organizing Data Analysis Technique (ISODATA) algorithm. This produced a map with 10 classes which was then displayed over color composite images, and classes assigned to a specific land cover category.

Supervised classification: This method used to cluster pixels in a dataset into classes corresponding to user-defined area of interest (AOIs) or training classes which are selected as representative areas to be mapped in the output. Supervised classification done using maximum likelihood algorithm. Training classes were defined prior to performing supervised classification. For the supervised

Table 34.1 Profile of satellite image

Image	Acquisition date	Sensor	Landsat mission	Resolution (m)	Number of bands
1973	26695	MSS	1	80	4
1984	31038	TM	5	30	7

classification, the ground control points collected in the field were used as the training sample set. Supporting information was obtained from field observation of land use land cover and interviews with local elder people. The sample set was created using the band combination of 4, 3, 2 (image of 1984 and 2005) and 4, 2, 1 (image of 1973 for visual interpretation of the image in their true color). Figure 34.3 shows the flowchart for the land cover mapping.

Soil type identification: The importance of soil properties stems from the important role they play in hydrological modeling. The use of models for the prediction of runoff and impacts of LULC depends heavily on detailed data on soil physical properties and the understanding of these data. However, such data are unavailable in most part of Ethiopia including the study watershed. The only information available in the study area is the FAO soil map (1:1,000,000).

To enable soil data collection, a random sampling method was used to include all the land use and land cover classes and all sub-watersheds that were delineated. Totally, 40 representative sites were identified throughout the watershed and a soil database was established through an intensive data collection from each of the sample sites.

Soil types in the study area and laboratory analysis of soil samples for particle size distribution (sand, silt and clay content), organic carbon, cation exchange capacity, pH, bulk density, and saturated hydraulic conductivity were analyzed.

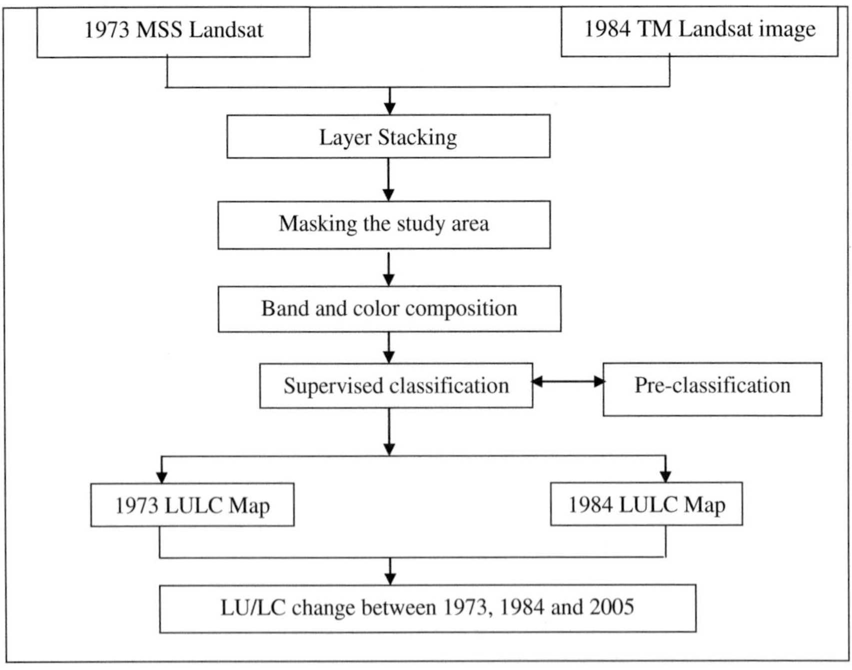

Fig. 34.3 Flowchart methodology for analysis of LULC (Larsson 2002)

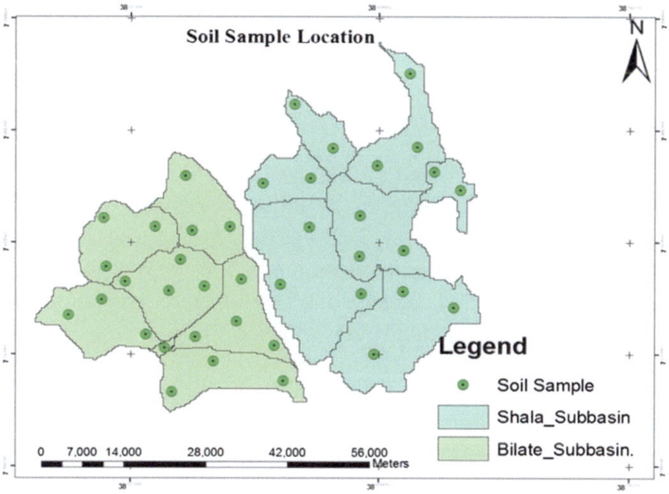

Fig. 34.4 Soil sampling locations

Figure 34.4 indicates that the locations of soil sample points taken for analysis in the two watersheds. The figure clearly shows that from Bilate watershed, 22 soil samples and from Shala watershed, 18 soil samples were collected.

Analysis of soil data at Bilate and Shala watersheds: Though the major aim of the soil data collection was to prepare the input data for the SWAT model, analyses were performed to get more information on the distribution of soil properties and to identify the variation in soil types. The spatial distribution of the different soil types and their properties by soil layers and sampling sites in the study area were prepared. Accordingly, the soils were classified into different textural classes using the textural triangle using the computer program for soil textural classification developed by Gericke et al. (2004).

34.2.4 Model Input Data Preparation and Processing

The study used SWAT for simulating flows. The application of SWAT in predicting stream flow and sediment as well as evaluation of the impact of land use and climate change on the hydrology of watersheds has been documented by various studies (Dessu and Melesse 2012, 2013; Dessu et al. 2014; Wang et al. 2006, 2008a, b, c; Wang and Melesse 2005, 2006; Behulu et al. 2013, 2014; Setegn et al. 2009a, b, 2010, 2011, 2014; Mango and Melesse 2011a, b; Getachew and Melesse 2012; Assefa et al. 2014; Grey et al. 2013; Mohammed et al. 2015).

The spatially distributed data needed for the ArcSWAT interface include the DEM, soil, and land use data. Data on weather and flow are used for the prediction of water balance and calibration purposes.

Land use scenario

An assessment of the impact of land use change on the water balances of Bilate and Shala watersheds was also undertaken by creating logically perceived land use scenarios using ArcGIS. In this study, land use data for the year 2005 were used as a benchmark against which different land use scenarios were compared. Two land use scenarios were considered, Scenario-1 (Agri-CON): conventional land use which represents the current land use practice in the area and Scenario-2 (Agri-IRWH): infield RWH based on the work of Hensley et al. (2000) which was aimed at improving the precipitation use efficiency (PUE).

With the use of the Agri-IRWH technique, runoff and soil loss from the cropland were reduced to zero (Hensley et al. 2000). It is also reported that use of mulch in the basins reduced evaporation significantly, contributing to the increase in yield, on average 30–50 percent, compared to production under conventional tillage. On the other hand, it has been shown by several hydrological studies at watershed level that upstream shifts in water-flow partitioning may result in complex and unexpected downstream effects, both negative and positive, in terms of water quantity and quality.

Runoff for Ungauged Watersheds

Runoff for ungauged watersheds is estimated by three procedures that are of different complexity. The first procedure applies a reorganization procedure where a regional model (Merz and Bloschl 2004) is established between watershed characteristics and model parameters that in turn are used for ungauged watershed modeling. A second procedure simply transfers model parameters from neighboring or nearby watersheds to ungauged watersheds to allow for runoff simulation. In the third procedure, parameters of gauged watersheds are transferred to ungauged watersheds by simple comparison of watershed size. In all three procedures, the SWAT model is selected for the simulation of watershed runoff.

34.2.5 Socioeconomic Analysis of Water Harvesting Technology

Field survey and data collection: The ultimate purpose of the field survey was to collect qualitative and quantitative information to help better understand, explain, and interpret the land use and land cover change. Hence, understanding trends in resource dynamics required historical information, which can be achieved using qualitative and quantitative data collected through interview and group discussion with selected informants believed to have a good understanding of the issues of interest.

A questionnaire covering a wide range of topics relevant to the central issue of interest was developed. It was also pretested so as to evaluate the understandability of the questions and modifications were made accordingly. Besides this, field observation was made to have better information about the nature of the various land use and land cover classes prevailing in the area. Bilate watershed is composed of 22,294 households, and Shala has 35,719 households, out of which 59 in Bilate and 41 in Shala that use water harvesting technology were interviewed in 2013 to produce a comprehensive baseline survey of socioeconomic conditions in the watersheds. Some of the findings from this survey are presented in this chapter.

Economic evaluation of water harvesting: Cost-benefit analysis (CBA) study of a community water harvesting system aims to inform the economic returns of agricultural, livestock, and domestic water utilization in Alaba and contribute to a larger effort in solving irrigation water scarcity. This section will present the current water harvesting technology, discuss preceding economic valuations of water for crop production (irrigation water) and domestic purpose, and present the method and result of economic valuation of water harvesting in Alaba. Water management alternatives and technologies for water harvesting and collection in all three case studies were evaluated using CBA based on social evaluation, according to regulations for use of public funds of the government and NGOs. All projects and their technical alternatives were evaluated using both financial and social assessments. Financial assessments only considered the financial returns of the investment. Social assessments, however, considered financial, economic, and social returns of the investment.

The CBA was used to assess the water harvesting technology in Alaba for comparison of cost and benefit that accrues from a specific project. A CBA involves calculating and comparing all of the costs and benefits, which are expressed in monetary terms. The comparison of expected costs and benefits can help to inform decision makers about the likely efficiency of an adaptation investment. CBA provides a basis for prioritizing possible adaptation measures.

The bottom line for choosing an adaptation option is the comparison of the monetized elements of costs and benefits. The costs and benefits need to be discounted to properly calculate their present value. Adaptation planners can choose between three indicators of whether their options are efficient:

In this study, both economic and computational approaches were applied to analyze trade-offs between different water uses under water harvesting techniques. The study focuses on the economic conditions under which multi-purpose water use is optimized and the social, environmental and management conditions under which it can evolve and enhance the performance and resilience of the coupled systems in the face of increasing variability in water availability, as expected. A total of 20 households, which were selected using a stratified sampling technique, have been surveyed. From each stratum, 6 farming associations were selected randomly.

Optimization of the water use trade-off (OCA): The OCA was used to determine the optimal system level allocation of the available water resources to the two competing sectors under equilibrium conditions. Returns from irrigated agriculture are modeled as the net returns from the use of the water for the production of a crop.

The returns from the domestic are the net revenues from selling the domestic animal at market price.

Data analysis: Data were analyzed using descriptive statistics. Qualitative data were used to specify contexts of the study and enrich information generated from quantitative data analysis. The household survey statistics was used for socioeconomic analysis and trade-off between upstream and downstream community that use water harvesting technology. To assess the economic viability of water harvesting ponds constructed in Alaba by the community, government and international NGO were considered. This project collected a large amount of socioeconomic and physical data from the community in the area. Data collection activities involved the active participation of the local self-governments and members of the local community.

Detailed information was collected from the owners of each WHS including dates of construction, costs of construction, water holding capacity, number of acres irrigated, and current status (i.e., working/not working).This illustrated the advantages of participatory study and involving the local community in the mapping process. Land use maps, of the district, were created through the supervised classification.

34.3 Results and Discussion

34.3.1 Hydrological Modeling

Results of sensitivity analysis of stream flow at the gauging point for a period of six years which includes both calibration periods (January 1, 1997, to December 31, 2002) and setup period (January 1, 1995, to December 31, 1996) have shown that 27 parameters are considered in the analysis.

Model calibration: The default SWAT simulation showed a weak hydrography in matching the simulated and observed stream flows. Manual and automatic calibrations for average annual water balance and stream flow were conducted. Once the model was calibrated for average annual conditions, it was also calibrated for monthly and daily simulation. For this study, both average annual and monthly calibrations were done. During this step, flow was simulated for eight years from January 1, 1995, to December 31, 2002. Some of the initial/default and finally adjusted parameter values are shown in Table 34.2.

Results of the calibration analysis revealed an R^2 (coefficient of determination) of 0.75. The other measure used is percent difference/percent bias (PBIAS). PBIAS measures the average tendency of the simulated data to be larger or smaller than their observed counterparts (Gupta et al. 1999). Figures 34.5 and 34.6 show comparisons of the simulated and observed flows for the calibration phase.

The optimal value of PBIAS is 0.0, with low-magnitude values indicating accurate model simulation. Positive values indicate model underestimation bias,

Table 34.2 Initial and adjusted manual calibrated flow parameter values at the outlet of Bilate (Bilate Tena) watershed

Parameters	Description	Range	Initial (default) values	Adjusted parameter values
CN2	Curve number	±25 %	Default*	22 %
Esco	Soil evaporation compensation factor	0–1	0.246	0.1
Revapmn	Threshold depth of water in the shallow aquifer required for "revap" or percolation to deep aquifer to occur	0–500	0.217	0.8
Gw_Revap	Groundwater revap coefficient	0.02–0.2	0.02	0.17
Gwqmn	Threshold depth of water in the shallow aquifer required for return flow to occur	0–5000	0.318	200

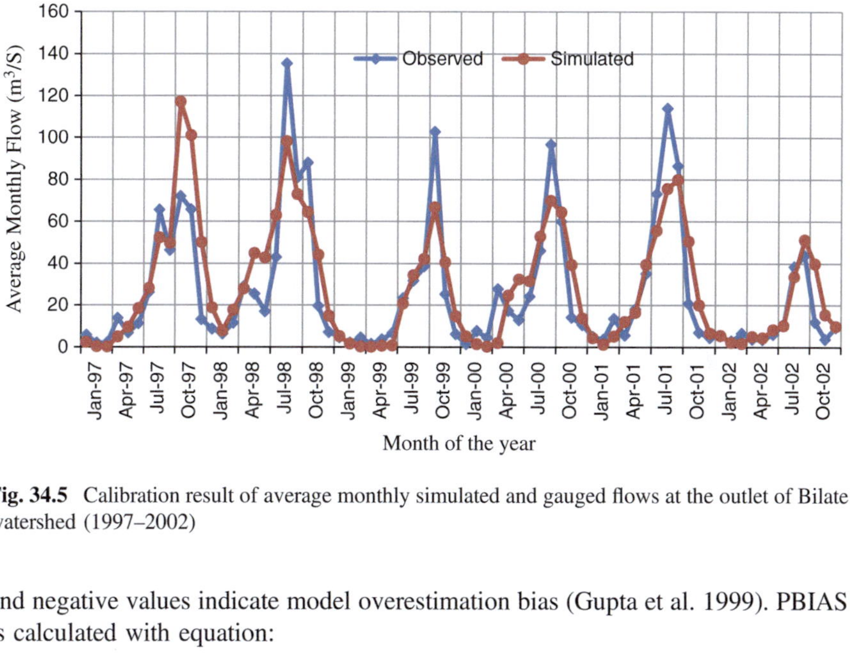

Fig. 34.5 Calibration result of average monthly simulated and gauged flows at the outlet of Bilate watershed (1997–2002)

and negative values indicate model overestimation bias (Gupta et al. 1999). PBIAS is calculated with equation:

$$\text{PBIAS} = \frac{\sum_{i=1}^{n}\left(Y_i^{\text{Obs}} - Y_i^{\text{sim}}\right) * (100)}{\sum_{\text{i}=1}^{\text{n}}\left(Y_i^{\text{Obs}}\right)} \tag{34.1}$$

where Y_i^{Obs} = observed flow data and Y_i^{Sim} = model simulated flow data.

The systematic and unsystematic root-mean-square errors (RMSEs and RMSEu) are also minimal. The ratio of the unsystematic root-mean-square error (RMSEu) to the root-mean-square error provided a value of 0.85, indicating good correlation between the observed and simulated flows and that the error is not possibly of a

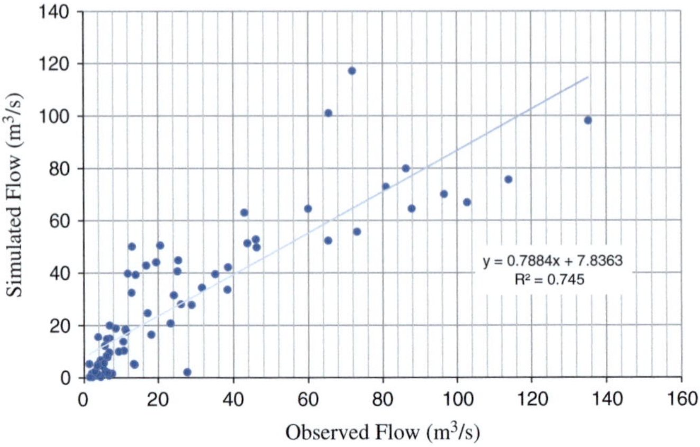

Fig. 34.6 Scatter plot of monthly simulated versus gauged flow at Bilate gauged station for calibration period (1997–2002)

systematic nature (Welderufael et al. 2009). RMSE is calculated with observed and simulated flows.

$$\mathrm{RMSE} = \left(\sqrt{\sum_{i=1}^{n} \left(Y_i^{\mathrm{Obs}} - Y_i^{\mathrm{Sim}} \right)^2} \right) \quad (34.2)$$

where, Y_i^{Obs} = observed flow data and Y_i^{Sim} = model simulated flow data.

The Nash and Sutcliffe efficiency (NSE) (Nash and Sutcliff 1970) revealed a value of 0.74 for the monthly stream flow calibration, describing a satisfactory correlation between the observed and simulated monthly stream discharges.

Nash and Sutcliffe efficiency can be calculated as:

$$\mathrm{NSF} = 1 - \left[\frac{\sum_i^n \left(Y_i^{\mathrm{Obs}} - Y_i^{\mathrm{Sim}} \right)^2}{\sum_i^n \left(Y_i^{\mathrm{Obs}} - Y_i^{\mathrm{Mean}} \right)^2} \right] \quad (34.3)$$

where Y_i^{Obs} = observed flow data and Y_i^{Sim} = model simulated flow data, and Y^{mean} = mean of observed flow data (Table 34.3).

Model validation: Validation was carried out for four years from January 1, 2003, to December 31, 2005, period without further adjustment of parameters during the calibration period (Figs. 34.7 and 34.8). The correlation coefficient

Table 34.3 Calibration statistics of average monthly simulated and gauged flows at the outlet of Bilate watershed

Period	Total flow (m^3/s)		Average flow (m^3/s)		R^2	NSE
	Gauged	Simulated	Gauged	Simulated		
1997–2002	1906.03	2066.85	26.505	28.713	0.75	0.74

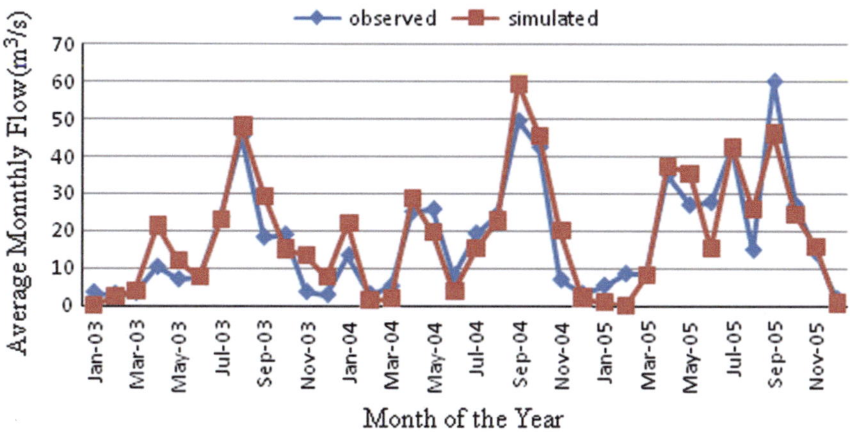

Fig. 34.7 Validation result of average monthly simulated and gauged flows at the outlet of Bilate watershed (2003–2005)

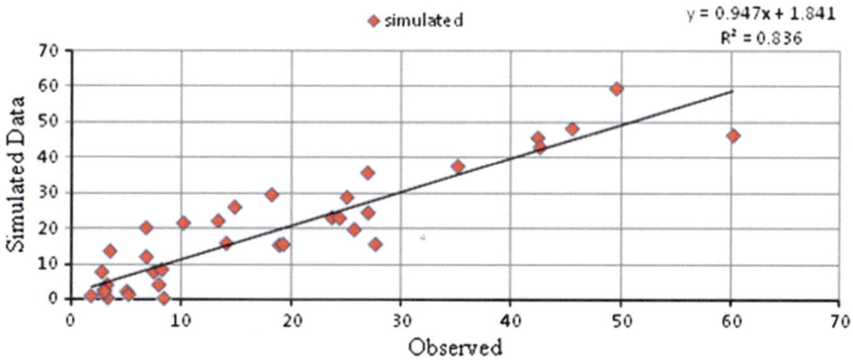

Fig. 34.8 Scatter plot of monthly simulated versus gauged flow at Bilate gauged station for the validation period (2003–2005)

during the validation period was R^2 = 0.836 and the Nash–Sutcliffe efficiency was (NSE = 0.80).The calibration values show a good agreement between observed and simulated flow of the watershed.

Annual water balance: The standard land phase hydrologic parameters used in SWAT were considered for annual water balance. For this study, average annual water balance in the watershed for both calibration and validation period was done (Table 34.4). The result of simulation showed that the largest portion of the average annual precipitation (66.63 %) falling in the watershed is lost through transmission losses. This value indicates that there is high sensitivity of transmission losses to any change than any other hydrologic parameter governing the water balance of sub-watersheds (Table 34.5).

Table 34.4 Validation statistics of average monthly simulated and gauged flows at the outlet of Bilate watershed

Period	Total flow (m^3/s)		Average flow (m^3/s)		R^2	NSE
	Gauged	Simulated	Gauged	Simulated		
2003–2005	405.05	540.04	16.842	22.5	0.836	0.801

Table 34.5 Average annual water balance (mm) for both calibration and validation based on the land use scenario on Bilate watershed

Hydrological parameter	Calibration period (1997–2002)	Validation period (2003–2005)	Weighted average	% of yearly precipitation
Precipitation	1047.9	1085.7	1066.8	
Surface runoff	122.68	123.57	123.12	11.54
Revap/shallow aquifer recharges	105.07	83.35	94.21	8.83
Deep aquifer recharges	132.84	139.15	135.99	12.74
Total aquifer recharge	11.94	11.06	11.5	1.077
Total water yield	238.76	221.17	229.965	21.55
Percolation out of soil	225.91	204.95	215.43	20.19
Actual evapotranspiration	235.84	217.91	226.875	21.26
Transmission losses	682.5	739.3	710.9	66.63

34.3.2 Modeling Factors Responsible for the Downstream Effect

In this study, the model simulated several parameters among which the direct flow (DIRQ), the base flow or the ground water flow (GWQ), water yield (WY), and evapotranspiration (ET) were selected for evaluation purposes. The effect of land use on hydrologic responses of the study watersheds is also evaluated. Figure 34.9 shows the variation of the water yield for the three land use datasets at the outlet of Bilate and Shala watersheds.

It was shown that there was an average decrease of WY by about 42.76 % over a period of 6 years (1997–2002) when land use data of 1973 were compared to the land use data with 1984 and also decrease of WY by about 52.6 % between 1984 and 2005.

From Figs. 34.10 and 34.11, the highest annual surface average runoff 252 mm and 403 mm was generated in Bilate and Shala watersheds, respectively, in 1973.

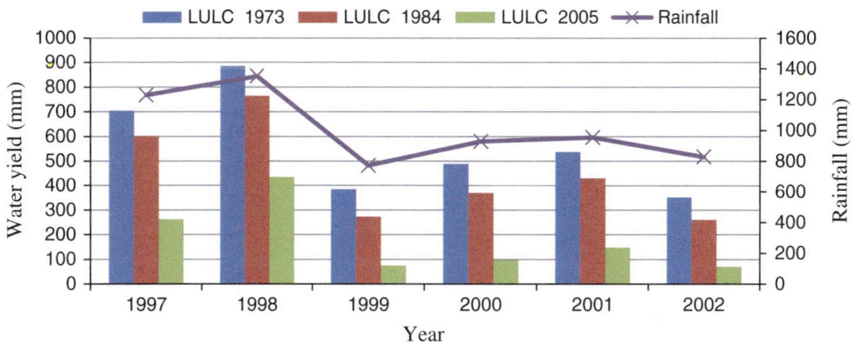

Fig. 34.9 Comparison of the simulated average annual water yields for the period of 1997–2002 based on the three land use datasets on Bilate and Shala watershed

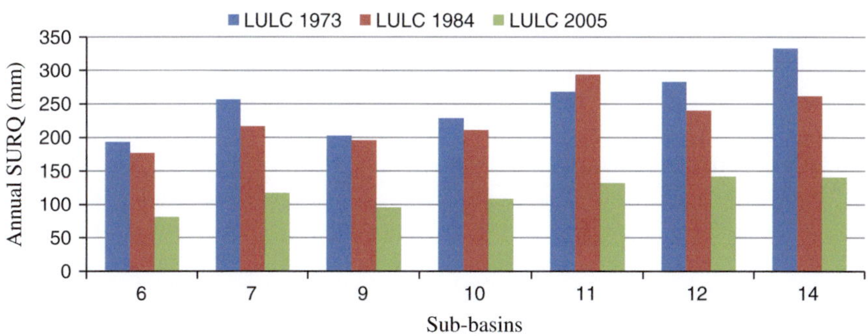

Fig. 34.10 Surface runoff simulated for selected sub-basins (6, 7, 9, 10, 11, 12, and 14) of Bilate watershed for LULC 1973, 1984, and 2005

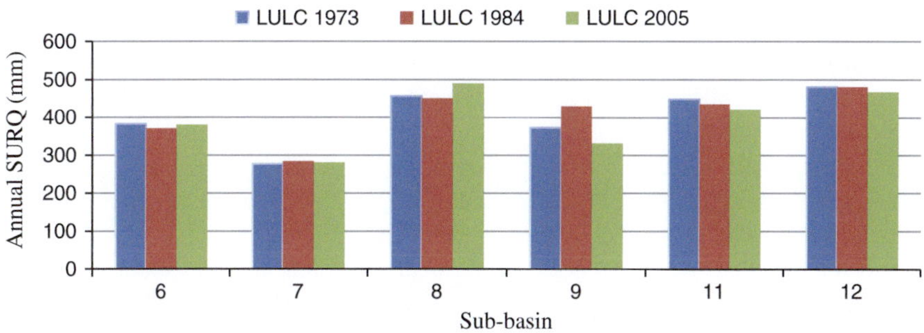

Fig. 34.11 Surface runoff simulated for selected sub-basins (6, 7, 8, 9, 11, and 12) of Shala watershed for LULC 1973, 1984, and 2005

The surface runoff of the selected sub-basins of the two watersheds increased by 22 and 28 % of 1984 and 2005, respectively, compared with LULC 1973.

34.3.3 Land Use Scenario

Ex situ management options: There is a number of water harvesting structures currently in place in Bilate and Shala watersheds, in the form of community ponds. In the near future, more ex situ management options are expected to be put in place by the community. In order to assess the impact of these structures on the downstream hydrology of the watershed, especially on surface runoff and groundwater recharge, two treatment levels were selected. In the first scenario, no reservoirs were modeled, and in the second case, all current and projected structures were incorporated into the SWAT simulations as reservoirs, from the first year of simulations. The types of structures being modeled, along with their effective capacity, are summarized in Tables 34.6, 34.7, 34.8, and 34.9.

With regard to evapotranspiration (ET), there was no significant difference in the total annual amount. The ET from Agri-CON and Agri-IRWH land use followed the same pattern due to the same type of crop (maize) considered in both cases. The annual ET in Agri-IRWH showed 20 and 30 m^3/ha more water loss than Agri-CON and LULC 2005 land use scenarios, respectively. The results suggest that the water yield may not be adversely affected by the Agri-IRWH land use scenario despite its design for surface runoff abstraction. It is expected that this result will assist in taking a practical measure regarding water resource management in general and a strategic allocation and use of water in particular.

Generally, it was found that RWH structures, as well as in-situ water conservation measures, have the potential to greatly improve water availability for agricultural production throughout the year by increasing groundwater recharge and increasing soil water storage, respectively. However, there is always a break point, where upstream WH technology also reduces sediment flows, and thus has a positive impact, and also reduce land degradation in upstream areas.

The economic benefits of this improved water harvesting technology have two main components: (1) productivity gained and (2) the value of any time saving that result from the installation of the new water resources. The economic value of time saving to households will vary greatly depending on local labor market conditions and economic opportunities.

This chapter details the time-saving benefits by multiplying (1) an estimate of the time saving associated with the initial quantity of water consumed; (2) an estimate of the market wage for labor (ranging from ETB 32 to ETB 40/day); and (3) a parameter less than one to denote the ratio of the value of time spent collecting water to the market wage (ranging from 0.1 to 0.5). The estimates of the time saving assume that the time required to collect water from the improved source varies from 0.1 to 0.5 h/20 L water. Thus, it is possible in the model for a positive time costs to households using the new source instead of time saving. With the optimization

Table 34.6 Components of direct flow of the different land use scenarios

Year	PRECIP (mm)	LULC 2005			Agri-IRWH			Agri-CON		
		SURQ (m^3/ha)	LAT_Q (m^3/ha)	DIRQ (m^3/ha)	SURQ (m^3/ha)	LAT_Q (m^3/ha)	DIRQ (m^3/ha)	SURQ (m^3/ha)	LAT_Q (m^3/ha)	DIRQ (m^3/ha)
1997	63.61	214.7	35.8	250.5	161	37.1	198.1	240.6	37	277.6
1998	126.69	194	31.8	225.8	135.9	33.8	169.8	219.2	33.7	252.9
1999	99.19	191.4	36.1	227.6	153.2	36.3	189.5	220	37	257
2000	116.4	264.5	48.3	312.9	225.1	48.3	273.5	295.9	48.3	344.2
2001	95.72	254.5	46.7	301.2	200.7	47.3	248	277.1	46.9	324
2002	131	200.6	32.9	233.5	140.52	35.04	175.6	226.6	34.84	26.15
Mean	105.435	219.95	38.6	258.6	169.4	39.64	209.1	246.6	39.6	247.0

Table 34.7 Simulated annual groundwater on the different land use scenario

Year	PRECIP (mm)	Annual ground water (m^3/ha)		
		Agri-IRWH	Agri_CON	LULC 2005
1997	63.61	339.1	272.5	127.3
1998	126.69	437.7	351.9	193.8
1999	99.19	257.4	202.6	104.7
2000	116.40	337.4	276.5	146.2
2001	95.72	383.5	306.4	169.6
2002	131.00	218.8	175.8	92.4
Mean	105.44	329.0	264.3	139.0

Table 34.8 Simulated annual deep water percolation on the different land use scenarios

Year	PRECIP (mm)	Annual deep percolation (m^3/ha)		
		Agri-IRWH	LULC 2005	Agri-CON
1997	63.61	393.6	332.3	314.1
1998	126.69	437.8	369.8	354.5
1999	99.19	269.7	247.1	214.3
2000	116.40	381.8	328.7	313.1
2001	95.72	390.9	338.2	307.9
2002	131.00	240.9	238.5	196.8
Mean	105.44	352.5	309.1	283.4

Table 34.9 Simulated annual evapotranspiration for the different land use scenario

Year	PRECIP (mm)	Evapotranspiration (m^3/ha)		
		Agri-IRWH	Agri_CON	LULC 2005
1997	63.61	463.0	475.0	567.0
1998	126.69	575.0	492.0	481.0
1999	99.19	430.0	443.0	418.0
2000	116.40	415.0	428.0	409.0
2001	95.72	455.0	472.0	435.0
2002	131.00	413.0	423.0	411.0
Mean	105.43	458.5	455.5	453.5

approach, delivering water for domestic purposes is not optimal since the value of the water per demand harvested is less than the amount of water needed for domestic purpose. Taking the advantage of the geological suitability of the area (a floodplain with alluvial soils of few meters deep), other alternatives such as storage tanks, sand filled dams, and check dams for artificial recharge need to be given attention.

34.4 Conclusions and Recommendations

34.4.1 Conclusions

The purpose of this study reported in this chapter was to assess the impact of WH technology on a downstream water availability of communities. Field survey and social survey have been used for the assessment.

The SWAT hydrological model was used to assess the impact of two possible land use scenarios, in addition to the 1973, 1984, and 2005 actual land use data, on the water resources of the Bilate and Shala watersheds. The 2005 actual land use scenario was used as the baseline scenario to assess the different water balance components of Bilate and Shala watersheds against the two new scenarios constructed in the study.

The two new possible land use scenarios are (1) agriculture on the 0–3 % slope of 2005 land use changed to maize crop land using conventional tillage (Agri-CON) and (2) LULC of 2005 on the 0–3 % slope of land use changed to maize crop land using IRWH (Agri-IRWH). The model was manually calibrated from January 1997 to December 2002, and Nash and Sutcliffe efficiency (NSE) of 0.74 and R^2 of 0.75 were achieved. The results of the water balance simulation by comparing land uses for the years 1973, 1984, and 2005 over an 8-year period (1995–2002) showed that land use of 1984 gave 42.76 % lower water yield (WY) compared to land use of 1973. Similarly, land use of 2005 gave 52.6 % lower WY compared to land use of 1984. The scenario analysis results confirmed that there was improvement of water infiltration into the soil by Agri-IRWH and LULC 2005 land uses. Both resulted in higher base flow than Agri-CON land use. Both also demonstrated high deep water percolation with a significant difference ($t_{stat} > t_{crit}$) in annual amounts compared to Agri-CON. The results of the scenario analysis also indicate that conventional agricultural land use type generated the highest direct flow compared to the ones subjugated by LULC 2005 or IRWH land use types. The Agri-IRWH showed 11 % higher base flow compared to the Agri-CON land use scenario.

Overall, the results suggest that the WY of Bilate and Shala watersheds will not be adversely affected by the Agri-IRWH land use scenario despite its design for surface runoff abstraction. It is expected that this result will assist in taking a proactive measure regarding water resource management in general and a strategic allocation and use of water in particular.

Water harvesting is one option that increases the amount of water per unit cropping area, reduces drought impact, and enables the use of runoff beneficially. The investigative process and the results of the CBA of communities under water harvesting technology have been extremely informative. The system results have shown that rainwater is the major water source that satisfies the agricultural and domestic water needs of the study area. The CBA results have demonstrated that the average estimated benefits from all household groups are positive having cost-benefit ratio of greater than one (BC ratio 4.3). The CBA result provides a good starting point to investigate the economic valuations of clean water in study area.

In most cases, water management strategies improve rainfed agriculture and can result in trade-offs with downstream water users and ecosystems. However, depending on the choice of management strategy, these trade-offs can be minimized. An increase in agricultural yield in areas where the productivity is presently very low results in a relatively large improvement in water productivity, compared with yield improvements in areas with higher yields.

Storage tanks seem easily adaptable due to the existing indigenous practices, which can enable us to overcome people's uncertainty of the effectiveness of SWC. A series of check dams in association with small ponds can also be cheaper, but reliability for how long they can provide the water needs further study.

Finally, before construction of any of the structures, it is important that the amount of flood within a given watershed, sedimentation processes in the waterways, and the socioeconomic and sociocultural aspects in the area should be studied. Irrigation development requires basic investment for irrigation structures that enables long-term development (Admasu 1996). Pastoralists/agro-pastoralists living in the floodplains could not be encouraged to consider crop production as an alternative under the existing water harvesting practices unless well-designed diversion structures appropriate to the area are invested.

34.4.2 Recommendations

The result of this study clearly shows that water productivity, social considerations, and economics are important when decisions are made to implement WH technology. This also implies that downstream irrigation farming should be scaled down in favor of the application of the WH technology at the upstream level.

It had been indicated that most of the assessed RWH ponds are technically suitable. Therefore, governmental and nongovernmental organizations replicate community-based RWH projects with community partial cost coverage principle to achieve various global and national goals and objectives, such as those of the recent "Development and Transformation" plan of the Ethiopian Government.

Awareness creation for water professionals, policy makers, international financiers, and donors through workshops, meetings, and publications such as brochures and the like is very important works to be undertaken to improve the promotion of RWH in the effort made to improve the low-level safe water supply of the country. Finally, to ensure the sustainability and safety of the ponds, strong extension system and responsible institutions provide training on the use and operation of ponds.

References

Admasu G (1996) Assessment of crop production and vulnerability to famine: case study Wollo in 1994. In: Andah K and Sannoh S (eds) Water resources management in drought prone areas. Proceduring of international workshop, 16th–18th May 2012, Bahir Dar University, Italian Cultural Institution, Antwerpen, Belgium. Univesiteitsplein 1:B-2610

Assefa A, Melesse AM, Admasu S (2014) Climate change in Upper Gilgel Abay River catchment, Blue Nile basin Ethiopia. In: Melesse AM, Abtew W, Setegn S (eds) Nile River basin: ecohydrological challenges, climate change and hydropolitics, pp 363–388

Behulu F, Setegn S, Melesse AM, Fiori A (2013) Hydrological analysis of the Upper Tiber Basin: a watershed modeling approach. Hydrol Process 27(16):2339–2351

Behulu F, Setegn S, Melesse AM, Romano E, Fiori A (2014) Impact of climate change on the hydrology of Upper Tiber River basin using bias corrected regional climate model. Water Resour Manage 28:1327-1343.

De Winnaar G, Jewitt G (2010) Ecohydrological implications of runoff harvesting in the head waters of the Thukela River basin, South Africa. Phys Chem Earth Parts A/B/C 35:634–642

Dessu SB, Melesse AM (2012) Modeling the rainfall-runoff process of the Mara River basin using SWAT. Hydrol Process 26(26):4038–4049

Dessu SB, Melesse AM (2013) Impact and uncertainties of climate change on the hydrology of the Mara River basin. Hydrol Process 27(20):2973–2986

Dessu SB, Melesse AM, Bhat M, McClain M (2014) Assessment of water resources availability and demand in the Mara River basin. CATENA 115:104–114

ERDAS (Earth Resources Data Analysis System) (2005)User manual of ERDAS IMAGINE: Lieca geosystems, Norcross, GA 30092-2500, USA

Gericke O, Pretorius J, Wagenaarl ED, Loyd C (2004) Hydrological modelling of river basins Using HSPF model. In: Proceedings of the 2004 Water Institute of Southern Africa (WISA) biennial conference, Cape Town. ISBN:1-920-01728-3

Getachew HE, Melesse AM (2012) Impact of land use/land cover change on the hydrology of Angereb Watershed, Ethiopia. Int J Water Sci 1, 4:1–7. doi:10.5772/56266

Grey OP, Webber DG, Setegn SG, Melesse AM (2013) Application of the soil and water assessment tool (SWAT Model) on a small tropical Island State (Great River Watershed, Jamaica) as a tool in Integrated Watershed and coastal zone management. Int J Trop Biol Conserv 62(3):293–305

Gupta HV, Sorooshian S, Yapo PO (1999) Status of automatic calibration for hydrologic models: comparison with multilevel expert calibration. J Hydrologic Eng 4(2):135–143

Hensley M, Botha JJ, Anderson JJ, Van Staden PP, Du Toit A (2000) Optimizing rainfall use efficiency for developing farmers with limited access to irrigation water. Water Research Commission, Pretoria, South Africa

Hidayat Y, Sinukaban N, Pawitan H, Tarigan SD (2008) Impact of rainforest conversion on surface runoff and soil erosion in nopu upper watershed of Central Sulawesi. J Trop Soils 13 (1):59–65

IPMS (Improving Productivity and Market Success) (2005) Information resources portal, Ethiopia, 15 July 2005, pp 9–13. Accessed 12 March 2009

Lambin F, Geist HJ, Lepers E (2003) Dynamics of land use and land cover change in tropical regions. Annu Rev Environ Resour 28:205–241

Larsson H (2002) Analysis of variations in land covers between 1972 and 1990, Kassala Province, Eastern Sudan, using Landsat MSS data. Int J Remote Sens 23(2):325–333

Mango L, Melesse AM, McClain ME, Gann D, Setegn SG (2011a) Land use and climate change impacts on the hydrology of the upper Mara River Basin, Kenya: results of a modeling study to support better resource management. Special issue: Climate, weather and hydrology of East African Highlands. Hydrol Earth Syst Sci 15:2245–2258. doi:10.5194/hess-15-2245-2011

Mango L, Melesse AM, McClain ME, Gann D, Setegn SG (2011b) Hydro-meteorology and water budget of Mara River basin, Kenya: a land use change scenarios analysis. In: Melesse A (ed) Nile River basin: hydrology, climate and water use, Chap 2. Springer Science Publisher, pp 39–68. doi:10.1007/978-94-007-0689-7_2

Melesse AM, Jordan JD (2002) A comparison of fuzzy vs. augmented-ISODATA classification algorithm for cloud and cloud-shadow discrimination in landsat imagery. Photogram Eng Remote Sens 68(9):905–911

Melesse AM, Jordan JD (2003) Spatially distributed watershed mapping and modeling: land cover and microclimate mapping using landsat imagery, Part 1. J Spat Hydrol (e-journal) 3(2)

Melesse AM, Weng Q, Thenkabail P, Senay G (2008) Remote sensing sensors and applications in environmental resources mapping and modeling. Spec Issue: Remote Sens Nat Resour Environ Sens 7:3209–3241

Merz M, Bloschl G (2004) Regionalization of watershed model parameter. J Hydrol 287:95–123

Mohammed H, Alamirew A, Assen M, Melesse AM (2013) Spatiotemporal mapping of land cover in Lake Hardibo drainage basin, Northeast Ethiopia: 1957–2007. In: Water conservation: practices, challenges and future implications. Nova Publishers, Hauppauge, pp 147–164

Mohammed H, Alamirew T, Assen M, Melesse AM (2015) Modeling of sediment yield in Maybar gauged watershed using SWAT, northeast Ethiopia. CATENA 127:191–205

Nash JE, Sutcliffe JV (1970) River flow forecasting through con-ceptual models: Part I. A discussion of principles. J Hydrol 10:282–290

Ngigi SN (2003) Rain water harvesting for improved food security. Promoting technologies in the Greater Horn of Africa. Greater Horn of Africa Rainwater Partnership (GHARP), Kenya Rainwater Association (KRA)

Rockstrom J, Barron J, Fox P (2002) Rainwater management for increased productivity among small holder farmers in drought prone environment. Phys Chem Earth 27:949–959

Setegn SG, Srinivasan R, Dargahi B, Melesse AM (2009a) Spatial delineation of soil erosion prone areas: application of SWAT and MCE approaches in the Lake Tana basin, Ethiopia. Hydrol Process 23(26):3738–3750

Setegn SG, Srinivasan R, Melesse AM, Dargahi B (2009b) SWAT model application and prediction uncertainty analysis in the Lake Tana Basin, Ethiopia. Hydrol Process 24(3):357–367

Setegn SG, Dargahi B, Srinivasan R, Melesse AM (2010) Modelling of sediment yield from Anjeni Gauged watershed, Ethiopia using SWAT. J Am Water Resour Assoc 46(3):514–526

Setegn S, Rayner D, Melesse AM, Dargahi B, Srinivasan R (2011) Impact of climate change on the hydro-climatology of Lake Tana basin, Ethiopia. Water Resour Res 47, W04511:13. doi:10.1029/2010WR009248

Setegn SG, Melesse AM, Haiduk A, Webber D, Wang X, McClain M (2014) Spatiotemporal distribution of fresh water availability in the Rio Cobre Watershed, Jamaica. CATENA 120:81–90

Stiefel J, Melesse AM, McClain M, Price R (2007) Rainwater Harvesting in Rajasthan, India: recharge estimation using tracers. 13th international rainwater catchment systems conference and 5th international water sensitive urban design conference, 21–23 Aug 2007, Sydney, Australia, 8 pp (CD proceeding)

Stiefel JM, Melesse AM, McClain ME, Price RM, Chauhan NK (2008) The impact of artificial recharge from rainwater harvesting structures on the groundwater of nearby wells in rural Rajasthan. In: International groundwater conference on groundwater dynamics and global change, Jaipur, India, 11–14 March 2008

Stiefel J, Melesse AM, McClain M, Price RM, Anderson AP, Chauhan NK (2009) Effects of rainwater harvesting induced artificial recharge on the groundwater of wells in Rajasthan, India. Hydrogeol J 17(8):2061–2073

Wang X, Melesse AM (2005) Evaluations of the SWAT model's snowmelt hydrology in a northwestern minnesota watershed. Trans ASAE 48(4):1359–1376

Wang X, Melesse AM (2006) Effects of STATSGO and SSURGO as inputs on SWAT model's snowmelt simulation. J Am Water Resour Assoc 42(5):1217–1236

Wang X, Melesse AM, Yang W (2006) Influences of potential evapotranspiration estimation methods on SWAT's hydrologic simulation in a northwestern Minnesota watershed. Trans ASAE 49(6):1755–1771

Wang X, Shang S, Yang W, Melesse AM (2008a) Simulation of an agricultural watershed using an improved curve number method in SWAT. Trans Am Soc Agri Bio Eng 51(4):1323–1339

Wang X, Yang W, Melesse AM (2008b) Using hydrologic equivalent wetland concept within SWAT to estimate streamflow in watersheds with numerous wetlands. Trans Am Soc Agri Bio Eng 51(1):55–72

Wang X, Garza J, Whitney M, Melesse AM, Yang W (2008c). Prediction of sediment source areas within watersheds as affected by soil data Resolution. In: Findley PN (ed) Environmental modelling: new research, Chap 7. Nova Science Publishers, Inc., Hauppauge, pp 151–185. ISBN 978-1-60692-034-3

Welderufael WA, Le Roux PAL, Hensley M (2009) Quantifying rainfall-runoff relationships on the Melkassa hypo calcic regosol ecotope in Ethiopia. Water SA 35(5):634–648

Wondie M, Schneider W, Melesse AM, Teketay D (2011) Spatial and temporal land cover changes in the Simen Mountains National Park, a world heritage site in Northwestern Ethiopia. Rem Sens 3:752–766. doi:10.3390/rs3040752

Wondie M, Schneider W, Melesse AM, Teketay D (2012) Relationship among environmental variables and land cover in the Simen Mountains National Park, a world heritage site in Northern Ethiopia. Int J Rem Sens Appl (IJRSA) 2(2):36–43

Chapter 35
Evaluation of Technical Standards of Physical Soil and Water Conservation Practices and Their Role in Soil Loss Reduction: The Case of Debre Mewi Watershed, North-west Ethiopia

Getachew Engdayehu, Getachew Fisseha, Mulatie Mekonnen and Assefa M. Melesse

Abstract Soil and water conservation (SWC) measures had been implemented for the last four to five decades to reduce the severe soil erosion problem in north-west Ethiopia. A study was conducted at Debre Mewi watershed, north-west Ethiopia, Upper Blue Nile Basin, to evaluate the technical standard of the implemented SWC measures and their role in reducing soil loss using the RUSLE model, farmers' interviews, transect walk, field observation and direct measurements. Soil bund, fanya juu, stone check dam, gully reshaping and artificial waterways were evaluated based on vertical interval and width between bunds, gradient and length of bunds, embankment height and compaction and size of artificial waterway parameters. As a result, most of the gully treatment designs and waterway layouts fail to meet the standard. Most of the SWC structures constructed in the study area didn't meet vertical interval, bund length, bund gradient, embankment compaction and height standards, which reduced the effectiveness of the measures in reducing soil erosion. SWC measures technical standards deficient construction and management was

G. Engdayehu
Amhara National Regional State, Bureau of Agriculture, Natural Resource Conservation and Management Department, Bahir Dar, Ethiopia
e-mail: ztseatgeta@yahoo.com

G. Fisseha
Faculty of Agriculture, Bahir Dar University, Bahir Dar, Ethiopia
e-mail: fisseha.getachew@yahoo.com

M. Mekonnen (✉)
Amhara National Regional State, Bureau of Agriculture, GIS and ICT Department, Amhara, Ethiopia
e-mail: mulatiemekonneng@gmail.com

A.M. Melesse
Department of Earth & Environment, AHC 5-390 Florida International University, 11200 SW 8th Street, Miami, USA
e-mail: melessea@fiu.edu

© Springer International Publishing Switzerland 2016

A.M. Melesse and W. Abtew (eds.), *Landscape Dynamics, Soils and Hydrological Processes in Varied Climates*, Springer Geography,
DOI 10.1007/978-3-319-18787-7_35

found to be an important cause of soil erosion. Although based on modelling, soil erosion after SWC applications is reduced, but is still higher than the tolerable soil loss rate. Increased effort in guideline improvement and implementing SWC measures based on their technical standards is required to manage soil erosion.

Keywords Debre Mewi watershed · Soil and water conservation measures · RUSLE · Soil loss · Erosion control structure standards · Soil erosion · Nile Basin

35.1 Introduction

Land degradation in the form of soil erosion is a serious global problem (Pimentel et al. 1995; Bekele and Holden 1999). Globally, about 80 % of current degradation of agricultural land is caused by soil erosion (Angima et al. 2003; Morgan 2005). However, wide spatial disparity exists on the extent and depth of the problem (Stocking and Murnaghan 2000; FAO 2004). Human activity triggers the problem (Bewket and Sterk 2002). It is associated with rapid population growth, inadequate attention to the basic natural resources, intense land cultivation, uncontrolled grazing and deforestation (Bekele and Holden 1999; Girma 2001; Bewket 2003; Mekonnen and Melesse 2011).

Soil erosion creates severe limitations to sustainable agricultural land use as it reduces farm soil productivity and causes food insecurity (Sonneveld and keyzer 2003; Beshah 2003; Awdenegest and Holden 2007). It frustrates economic development efforts, especially in developing countries like Ethiopia where there is heavy dependence on the land with low external input farming systems and subsistence production (Bekele and Holden 2001). In the highlands of Ethiopia, the degradation of agricultural land creates a serious threat to food production (Bekele 2001; Bewkete and Sterk 2002).

Ethiopia has a total surface area of 112 million ha of which 60 million ha is estimated to be agriculturally productive (Bedadi 2004; Makombe et al. 2007). However, about 27 million ha is significantly eroded, 14 million ha is seriously eroded, and 2 million ha has reached the point of no return with an estimated total loss of 2 billion m^3 of topsoil per year (Bewket and Sterk 2002; Bedadi 2004). Visible erosion features such as rills and gullies indicate seriously affected hot spot areas by water erosion (Haile et al. 2006; Mekonnen and Melesse 2011). Soil erosion is the most threatening ecological degradation observed in Ethiopia affecting the precious soil resources which are the basis of agricultural production and food for the people (Hurni 1995; Herweg 1996; Carrucci 2000). Water erosion (sheet and rill) is the most important erosion process which in the mid-1980s significantly eroded about 27 million ha (almost 50 %) of the highland area. Fourteen million ha is seriously eroded, and over 2 million ha is eroded beyond reclamation with an average annual soil loss from cultivated lands of 100 t ha^{-1} y^{-1} (FAO 1986). On the other hand, the annual soil formation of the Ethiopia highlands is estimated to be between 2 and 22 t ha^{-1} y^{-1} and varies with geologic and climatic conditions, topographical set-up and agricultural practices (Hurni 1983).

It is recently, in the past four decades, that the Ethiopian government recognized the impact of soil erosion and put considerable efforts to reduce the problem and stop further degradation (Bekele and Holden 1999; Bewket and Sterk 2002; Amsalu and Graaff 2004). As a result, huge areas were covered with SWC structures and billions of trees were planted through mass mobilization (Admassie 2000; Girma 2001). In spite of the rapidly growing awareness of soil erosion processes and increased conservation methods' application in the past few decades, erosion remains widespread.

Soil erosion is the main threat in the study area, Debre Mewi watershed. Reported rate of soil loss is 36 t ha^{-1} yr^{-1} (Derebe 2009) and 12.5–50 t ha^{-1} yr^{-1} (Mekonnen and Melesse 2011). Farmers in the watershed also responded that soil erosion and fertility decline as their major problem (SWHISA 2007). Major active gullies and rills are steel observed in most land uses, specifically in farmlands. The major causes for such soil erosion in general and formation of active gullies in particular are overgrazing above field carrying capacity, very poor attention and misuse of land for centuries, improper discharge of run-off from roads, absence of integrated watershed management, land use change and fragile nature of the soil which make it susceptible for erosion (SWHISA 2007; Fisseha et al. 2011; Mekonnen and Melesse 2011).

Poor construction and lack of regular maintenance of SWC measures are also important issues. To maximize the effectiveness of any SWC measure, regular maintenance is required to avoid structural failure (Zhang et al. 2010; Mekonnen et al. 2014a). When a drainage ditch behind the SWC measure is filled with sediment, its sediment trapping efficacy quickly decreases with additional pressure being placed on the structure leading to possible structural failure (Mekonnen et al. 2014a). Inadequately, localized structure alignment can concentrate run-off at one location which creates large erosive potential and can generate major problems further down the landscape (Nyssen et al. 2007).

Hence, this study was conducted at Debre Mewi watershed to assess and evaluate the technical standards of soil and water conservation (SWC) measures against the technical guideline of MoARD (2005). The study objectives were (1) assessing the appropriateness of selected SWC measures based on agro-ecological zone, slope, soil type and land use; (2) evaluating the physical SWC measures implemented against the recommended design and layout; (3) checking the acceptance of SWC measures by farmers and follow-up responsibility to ensure sustainability; and (4) assessing the amount of soil loss difference before and after SWC implementation using the RUSLE model. To achieve the above objectives, the following research questions were designed.

1. Are SWC measures selected based on agro-ecology, slope and land use?
2. To what extent the implemented SWC measures met the technical guideline of MoARD (2005)?
3. Do farmers accept the implemented SWC measures and take management responsibility for the sustainability of the outputs?
4. How much soil loss is decreased because of SWC implementation?

To carry out this study, surveying and measurement materials such as global positioning system (GPS), tape meter, pegs, gauged pole, clinometer, line level, 1:50,000 topo map, digital camera, analysis software and Arc GIS version 9.3 were used. Satelite imagery and digital elevation model (DEM) were used for land use / cover mapping and to generate slope. Data were collected through semi-structured interviews, transect walks, field observation and field measurements. The final questionnaire was modified after pretesting the draft questionnaire specially to capture the farmers' knowledge and attitude using local language. The questionnaire included issues of household and farm characteristics, perceptions on SWC measures and preferred conservation activities by the farmers and why. Also included were major technical gaps according to farmers' view and farmers' traditional conservation knowledge and practice.

Transect walk with the planning team was made to cross-check the information on soils, slopes and bio-physical activities and status of previously constructed SWC structures. Focus group discussions with the watershed planning team were carried out, and informal interviews with development agents and Woreda (District) experts were conducted. In addition, the researchers made a field visit of the entire watershed during 2013 mass mobilization for SWC works to capture farmers' attitude on SWC structures. Secondary data such as reports and official documents were used for cross-checking and supporting data analysis.

35.2 Description of the Study Site

Debre Mewi watershed is located in Amhara region, north-west Ethiopia, Blue Nile Basin (Fig. 35.1). It covers about 498 ha and is located between 11°20′10″ and 11°21′58″N latitude and 37°24′07″ and 37°25′55″E longitude. The watershed is located about 500 km from the capital of Ethiopia, Addis Ababa, and 30 km south of Lake Tana and Bahir Dar town, the capital of Amhara Regional State.

35.2.1 Climate

There is a meteorological station at Adet town, which is the nearest to the study area. Based on 10-year rainfall data analysis, the area is characterized by a unimodal rainfall pattern with peaks in July and August with annual average of 1230 mm.

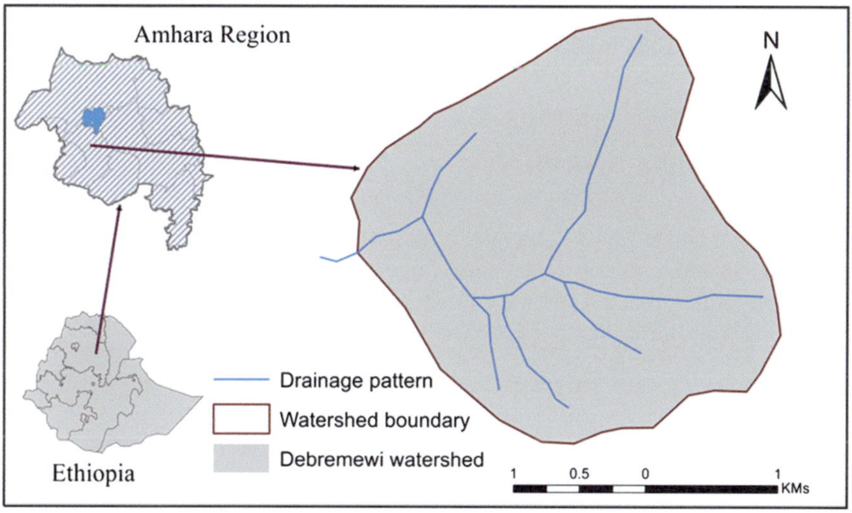

Fig. 35.1 Location map of the Debre Mewi watershed

About 80–90 % of the rainfall falls in the main rainy season (Kiremt), which starts in June and extends to August/September. It is in this season that the major agricultural activities, such as ploughing, sowing seeds and weeding, are performed. The dry months are between November and March locally known as Bega. The annual temperature ranges from 8.75 °C to 25.7 °C. Thus, the watershed is under the category of moist (sub-humid) Woina Dega agro-ecology zone.

35.2.2 Topography

The slope or gradient of the watershed ranges up to 50 % slope. Elevation ranges between 2200 and 2366 m.a.s.l. The eastern and north-eastern part of the watershed along the main road is nearly gentle slope followed by steep, moderately steep and gently sloping to its outlet.

35.2.3 Soils

The soils of the study area include Eutric Vertisols (33.28 %, 150 ha), Nitosols (24.83 %, 132 ha), Pellic Vertisols (19.55 %, 103 ha), Eutric Cambisols (8.29 %, 42 ha), Eutric Fluvisols (7.43 %, 40 ha) and Eutric Aquic Vertisols (6.62 %, 31 ha) (Mekonnen and Melesse 2011; Fisseha et al. 2011). The Nitosols and Fluvisols occupy the higher altitudes. Pellic Vertisols and Cambisols are found in gently undulating lands, while the other soil types occupy the lower altitudes covered by black and fluvial soils highly susceptible for rill and gully erosion.

35.2.4 Farming System

Agriculture in Debre Mewi watershed is rain-fed with a subsistence mixing farming system. Land and livestock are the most important livelihood assets. Tef (Eragrostis *tef*), finger millet, maize (Zea *mays*) and wheat (Triticum *vulgare*) are the major crops cultivated during the rainy season.

35.3 Soil and Water Conservation Measures

Soil and water conservation measures are more effective when integrated at catchment scale (Mekonnen et al. 2014a). The isolated or single case conservation approach so far observed from field experience has not satisfied the social, economic and environmental needs. For instance, the construction of physical SWC measures on farmland or placing check dams inside gullies or planting tree spices was considered as conservation intervention. However, this approach is understood as an individual approach that cannot solve the problem effectively, and an integrated approach should be used to solve the problem sustainably in a combined and integrated way (Mekonnen et al. 2014a).

Information on SWC methods and techniques (InfoTech's) are action-oriented summaries for the different SWC measures commonly applied in various parts of the country (MoARD 2005). The guidelines are prepared to help development agents and various experts at Woreda (District) level with minimum practical information on work norms and technical standards. The technical standards include soil conservation, water harvesting and some basic community infrastructures like feeder roads. The main purpose is to guide field staff to apply the measures following technical standards applicable to local agro-ecological conditions of soils, slopes, vegetation and rainfall patterns. The guideline summarizes the proposed interventions, providing information on key design features of the measures and their implementation requirements. The main contents of the InfoTech's include period/phases for technology implementation, objectives/remarks, suitability, ecology and adaptability based upon local knowledge, main land use, technical preparedness, potential to increase productivity, sustainable productivity and environmental protection and also includes required surveying tools, minimum technical standards, layout and vertical intervals, work norm, integration opportunities, planning and implementation arrangements, management requirements, limitations and institutional responsibility (MoARD 2005).

35.3.1 Agro-Ecological Zone for Soil and Water Conservation

Agro-ecological zone is one important variable in selecting and implementing SWC measures. Classifying area into dry, moist and wet helps in understanding what types of water harvesting and SWC structure could be used. What is designed for dry area may not serve sufficiently for wet and moist areas. To avoid such technical mistakes, identification of the agro-climatic zone of the working area is important. The simplest method by which an agro-climate of an area can be identified is by using rainfall, altitude and temperature. According to MoA (2001), areas with annual rainfall less than 900 mm, 900–1,400 mm and greater than 1400 mm are considered as dry, moist and wet agro-climatic zones, respectively. In Ethiopia, for natural resource development and conservation purpose, SWC methods are selected based on agro-ecological zones (Hurni 1995; MoARD 2005; BoA 2011).

35.3.2 Sound Soil and Water Conservation Principles

Effective implementations of SWC measures need to be guided through basic principles. According to Hurni (1995), MoARD (2005), BoA (2011) and Mekonnen et al. (2014a): (1) SWC measures should start from hilltop and proceed downhill to safely handle and protect concentrated surface run-off (ridge to valley approach); (2) SWC measures should be simple to understand and easy to implement by land owners. Labour and cost demand has to be maintained at optimum level to meet its objective and achieve the maximum benefit; (3) SWC measures should get acceptance by the implementers for its sustainability (socially accepted and sustainable); (4) SWC structures should be located at appropriate area in the land escape, properly designed and constructed based on accepted techniques, (proper location, design, layout and construction); (5) most conservation structures constructed on cultivated lands may take 10 % of farmland out of production which may create hardship on farmers to accept physical structures. Therefore, integrating the structures with productive practices to maximize benefit and to get acceptance by land owners is important; (6) recognizing the role of women in SWC activities and ensuring their active participation in decision-making; (7) build on traditional practices, local knowledge and experience as far as possible; and (8) developing appropriate SWC strategies involving farmers in identifying priorities, analysing problems, devising solutions, encouraging group work and strengthen local institutions for this purpose can be taken as key principles.

35.3.3 *Description of Selected Soil and Water Conservation Measures*

35.3.3.1 Traditional SWC Measures

Assessing both traditional and improved SWC measures is important in understanding farmers' way of thinking about the interventions (Hudson 1992). To prevent soil erosion, farmers use a number of traditional SWC measures which include application of manure, traditional cut-off drains, plantation of trees, stone bunds, planting local grass species along farm boundaries, leaving crop residues on the field and fallowing the farmland. Such locally available methods have not been given attention by development agents, researchers, soil and water conservationists and government staff (IFAD 1992). For effective soil erosion control, locally available and accepted practices should be considered.

35.3.3.2 Farm Bunds

Farm bunds are physical SWC measures used to reduce surface run-off velocity, trap water and sediment and thus control soil erosion. Bunds are simply embankment like structures made of soil and/or stone with basin at its upper side and constructed across the slope. Generally, it is used on cultivated lands with slopes above 3 % based on the soil type, intensity and amount of rainfall. In wet and moist regions and where the permeability of the soil is low, graded channel terraces are used with the purpose of draining run-off from the arable land at non-erosive velocity to a site where it can safely be discharged. A fanya juu is a soil bund with a basin at its lower side. It reduces velocity of overland flow and consequently soil erosion. Stone bunds are bunds without water retention basin and constructed in areas where stones are available.

35.3.3.3 Degraded Hillside Measures

Area closure with few rows of hillside terrace, trench or percolation trench is the most suitable method to enhance the rehabilitation process of hillside degraded areas. Different moisture harvesting and soil conservation methods can be applied together with area closure. The selection of SWC structures depend on the existing condition of the degraded areas specially land use, vegetation cover and soil condition of the area. The method is suitable for medium and high rainfall areas. It is more suitable on less degraded hillside with good soils and having some vegetation cover, such as grass and bushes (Nedassa et al. 2005).

Hillside terrace is a soil conservation and water harvesting structure constructed on steep and degraded hillside along the contour and commonly used for tree planting and run-off and erosion control. It is suitable for all agro-ecological zones, in steep degraded hillsides with shallow soils and slopes ranging from 20 to 50 %. The minimum standard for hillside terrace is height of stone riser 50–75 cm, foundation 30 cm by 30 cm, width of terrace 1–1.5 m, trench width 50 cm and depth 50 cm. The spacing between two consecutive hillside terraces depends on the plantation objectives, type of tree seedling to be planted and workability of the area and generally ranges from 1 to 2 m vertical interval, or 2–5 m horizontal distance.

35.3.3.4 Percolation Trench

Percolation trench is a type of physical measure used to manage storm water run-off to prevent flooding and downstream sedimentation and to improve water quality in an adjacent river, stream or lake. It is a shallow excavated trench that is designed to infiltrate storm water through permeable soils into the groundwater aquifer. Trenches collect and store considerable amount of run-off water and protect cultivated fields located downstream from flood and erosion. It is suitable in all agro-ecology from arid to high rainfall areas and on soils greater than 50 cm depth. It is mostly implemented on degraded hillsides above croplands to harvest rainwater and protect cropland from flood hazards. The minimum standard is >50 cm width, >50 cm depth and mostly 6–10 m length as the main purpose is to collect and store rainfall water.

35.3.3.5 Gully Treatment Measures

Check dam is an obstruction wall across the bottom of a gully or a small stream for the purpose of reducing the velocity of run-off and preventing the deepening and widening of a gully channel. Check dam can be wooden, lose stone, gabion, plastic or sand filled sack (Carucci 2000). Check dams or sediment storage dams trap large amounts of sediment eroded and transported from the upper catchment as well as within the gully itself and reduce downstream sedimentations with sediment trapping efficacy of 67–74 % (Mekonnen et al., 2014b). Areas occupied by sediments behind a check dam may be planted with crops, trees/shrubs and grasses for providing additional income for farmers or to increase the sediment trapping efficacy of the structures (Mekonnen et al. 2014b). A spillway is provided to control the run-off from overtopping and to prevent the deepening of gullies below the structure. Check dams also serve to recharge groundwater and give rise to springs at downstream sites.

Criteria that should be considered in the design of check dams include proper side keys, adequate spillway for safe water disposal, an apron of non-erodible material to dissipate the energy of water passing through the spillway with a length of 1.5 times the height of the check dam and proper spacing between successive dams. The size of catchment area and peak flow should be considered. The minimum technical specifications of check dams include side key 0.7–1 m, bottom key 0.5 m, height 1–1.5 m excluding foundation and base width 1.5–3.5 m (Carucci 2000). The spacing of check dams can also be determined by using an empirical equation (Eq. 35.1).

$$S = \frac{1.2H}{G} \tag{35.1}$$

where S is spacing in m, H is the effective height of the check dam, and G is the gully gradient in decimal (MoARD 2005).

35.3.3.6 Artificial Waterway

A waterway is a drainage channel used to discharge excess run-off from constructed, graded SWC measure or farmlands to rivers or lakes. Waterways should be constructed one year before the graded SWC measure. The channel can have gradient up to a maximum of 1 % with embankment height of 60 cm after compaction. The base width ranges from 1 to 1.2 m in stable soils and 1.2–1.5 m in unstable soils. It is a mistake to design waterway structures to discharge run-off if there is no suitable outlet such as a natural waterway, artificial waterway or grassed or forested areas (Carucci 2000; MoARD 2005).

35.4 Soil and Water Conservation Efforts and Its Acceptance

35.4.1 Soil and Water Conservation Efforts in Amhara Region

After analysing the intrinsic and extrinsic risks of soil, land and water resource degradation, the region launched public mobilization in 2010 to implement SWC measures. This effort is very much strengthened and taken as a priority objective with the political leaders and experts conducting SWC campaigns every year (BoA 2013). In those campaigns, it has been observed that huge amount of resources (labour and finance) have been mobilized and large tract of land has been covered. At regional level, before the commencement of the SWC activities through public mobilization, different preparatory works were conducted such as labour force organization, public awareness creation, capacity building and local resource inventory. Compared with previous years, the past three year's community mobilization strategies were effective in motivating collective action and involving all active labour forces (elders, women, men and youth) that are expected to participate in catchment development activities as planned. The community mobilization arrangements have allowed all concerned bodies to play their parts according to their ability. The active work forces both men and women who were engaged in physical hard work (digging, excavation, stone transport), while the elders look after children and animals as well as they guard villages from theft.

Table 35.1 Area coverage by soil and water conservation measures in Amhara region in 2010–2013

No.	SWC measures	Unit	Area covered with SWC measures			
			2010/11	2011/12	2012/13	Total
1	Steep slope area closure	ha	425,818	267,920	267,256	960,994
2	Hillside terrace	ha	116,619	147,876	139,990	404,485
3	Moisture harvesting structures	m/no	54	50	26.4	130.4
4	Farmland terrace	ha	458,508	621,849	630,909	1,711,266
5	Terrace maintenance	ha	299,060	339,383	232,135	870,578
6	Biological measures on terraces	ha	175,211	481,391	144,867	801,469
7	Drainage structures	km	16,002	19,827	14,606	50,435
8	Seedling raising	m/no	1.53	1.7	1.975	5.205
9	Seedling planting	m/no	1.1	1.3	1.62	4.02
10	Plantation	ha	155,330	224,647	317,677	697,654
11	Gully rehabilitation	ha	6044	18,953	9971	34,968
12	Participant day	million no	107	198	205	511

Source BoA (2013)

The initiative is well taken by the farming community and is becoming a common objective. In most cases, the campaign gives due emphasis for SWC measures which are to be constructed on communal areas and on those measures which serve considerable number of community members. In this regard, the attention given for gully rehabilitation and drainage management structures is encouraging. The regional government also gives due attention for drainage management structures such as cut-off drains, waterways and graded bunds, particularly in the highland areas where there is high rainfall and the drainage management structures have paramount importance (BoA 2013). Major achievements since 2010 are presented in Table 35.1.

35.4.2 Soil and Water Conservation Efforts in Debre Mewi

Soil and water conservation in the study area started recently (2010). More than 300 ha of the watershed is covered with physical and biological SWC measures by water and land resource project and BoA programs. Traditional ditches (locally called feses or shina) were indigenous practices used by 75 % of the surveyed farmers to drain excess run-off from their farms. Such traditional ditches have little value compared to other conservation structures like bunds, because they are mostly constructed at higher gradient and cause gully formation and accelerate sediment transport. Gully rehabilitation through biological measures with community participation was fruitful and opened the eye of the community in gully rehabilitation possibilities (Fisseha et al. 2011).

35.4.3 Acceptance and Management Responsibility of SWC Measures

Farmers' decisions on the acceptance and sustainability of SWC measures are largely determined by the awareness of the problems and perceived benefits of conservation (Amsalu and Graaff 2004). In the top-down approach, soil conservation methods are selected on the basis of technical criteria rather than required costs and benefits associated with their adoption. With such top-down planning, the target beneficiaries are largely passive recipients resulting in a lack of enthusiasm for implementation, follow-up and maintenance of the measures. Participation, where it has occurred, has typically been a case of the professionals gathering data, analysing it, preparing plans and then asking the local community if they agree before requesting mobilization of local resources as labour to implement these plans (Haile et al. 2006).

The Amhara Regional State Council has issued a proclamation that enables the community to rehabilitate, protect and utilize the watershed that has been already rehabilitated and going to be rehabilitated. The objective of the proclamation is to bring about sustainable and active users' participation in the watershed by the efforts of the community, establish suitable condition of the rehabilitated natural resource, use, administer and protect natural resources. Finally, promote capacity for management, use and protection of their rehabilitated resource in accordance with their regulations, establishing their own organization and associations of watershed users in Amhara region (CANRS 2013).

35.5 Evaluation of SWC Measure Technical Standards

The approach was designed to integrate the technical standards with knowledge, skill, attitude and preference of farmers on SWC activities applied in a watershed. SWC guidelines which have been used as field guidelines for development agents and experts in Ethiopia were developed by Hurni (1995) and later modified by MoARD (2005), Table 35.2. The guideline has detailed description and options of methods for different agro-ecologies and land use classifications. Moreover, it presents general layout and design specifications for each type of method. On different slope categories, 108 farm bunds were evaluated for bottom and upper width, height, vertical interval, spacing between bunds, length of bunds and other layout characteristics. Availability of proper discharge structures at the end of graded bunds has been assessed and evaluated against the minimum technical standards.

For gully treatment and rehabilitation measures, application of gully treatment principles, selection of appropriate method, selection of construction materials for

Table 35.2 Vertical interval (VI) and bund spacing or horizontal distance (Hd)

Slope (%)	Soil depth >75 cm		Soil depth 50–75 cm		Soil depth 25–50 cm	
	VI (m)	Hd (m)	VI (m)	Hd (m)	VI (m)	Hd (m)
3	1	33				
4	1	25				
5	1	20	0.7	15	0.5	10
6	1	17	0.7	12	0.6	10
7	1	14	0.8	12	0.7	10
8	1	12	0.8	10	0.7	9
9	1	11	0.9	10	0.8	9
10	1	10	0.9	9	0.8	8
11	1.1	10	1	9	0.9	8
12	1.1	9	1	8	0.9	8
13	1.2	9	1.1	8	1	8
14	1.2	8	1.1	8	1	7
15	1.2	8	1.1	7	1	7
16	1.3	8	1.1	7	1	6
17	1.3	8	1.2	7	1.1	6
18	1.3	7	1.2	7	1.1	6
19	1.3	7	1.2	6	1.1	6
20	1.4	7	1.2	6	1.1	6

Source MoARD (2005)

the gully size and run-off amount and application of gully reshaping were assessed. Height, side key, apron and space between check dams were the parameters measured and evaluated against the guideline. For this purpose, measurements were taken from 50 % of the treated gullies in the watershed.

In addition to the graded farm bund structures, small amount of artificial waterways that receive run-off from graded bunds were constructed in the watershed. Determining the amount of run-off generated is a prerequisite to determine the capacity and shape of an artificial waterway. The run-off amount was determined with the rational method. Rational Method is the oldest and the most applied hydrological model of predicting peak rate run-off (Eq. 35.2).

$$Q = \frac{\mathrm{CIA}}{360} \quad (35.2)$$

where Q is run-off ($\mathrm{m}^3\ \mathrm{s}^{-1}$), C is run-off coefficient (0–1), I is rainfall intensity (mm hr^{-1}), and A is area of catchment (ha).

Once the run-off is roughly estimated, the size of artificial waterway is evaluated whether it can adequately accommodate the estimated run-off. Major design parameters evaluated include channel gradient, channel shape, dimensions (depth, bottom and top width) and sequence of implementation with the graded structures and type of lining material.

35.6 Estimation of Soil Loss by RUSLE Model

A number of studies used laboratory, field scales and modelling studies to understand soil erosion and sediment dynamics in various regions (Defersha and Melesse 2011, 2012; Defersha et al. 2011, 2012; Maalim and Melesse 2013; Maalim et al. 2013; Setegn et al. 2009, 2010; Melesse et al. 2011; Msagahaa et al. 2014; Wang et al. 2008; Mekonnen and Melesse 2011; Mohammed et al. 2015). In this study, annual soil loss was assessed using the revised universal soil loss equation (Eq. 35.3), which was developed by Wischmeir and Smith (1978) and adapted for Ethiopian condition by Hurni (1985) and used by Mekonnen and Melesse (2011) for the study area.

$$E = R * K * L * S * C * P \tag{35.3}$$

where E is estimated soil loss (t ha^{-1} yr^{-1}), R is rainfall erosivity factor, K is soil erodibility factor, L is slope length factor, S is slope gradient factor, C is land cover factor, and P is management practice factor.

The R-factor is defined as the product of kinetic energy and the maximum 30-min intensity and shows the erosivity of rainfall events (Wischmeier and Smith 1978). However, in this study, to determine the value of the R-factor, the average annual historical rainfall event (10 years) was used from Adet meteorological station, which is located at 10 km from the watershed. Then, the R value is corrected to the mean annual rainfall of the watershed using the R-correlation established for Ethiopia condition Hurni (1985). The annual R-factor for the average annual rainfall (1230 mm) for the watershed is 682 mm. The soil erodibility (K), slope length (L), slope gradient (S) and C and P-factors of RUSLE for the entire watershed based on landforms, slope, soil colour, land use, rainfall and conservation practice are shown in Table 35.3.

Table 35.3 RUSLE parameters and estimated values

Land form	Slope	Area (ha)	Collected data for USLE parameters in each field					
			Soil colour	Slope gradient	Slope length	Cover	Conservation practice	Rainfall (mm)
Undulating plains	2–9	35	Black	5.5	28	Teff	Farm bunds	1230
		112	Brown	5.5	28	Teff and millet	Farm bunds	1230
Rolling land	9–15	42	Black	11.04	19	Teff	Farm bunds	1230
		101	Red and brown	11.04	19	Maize and teff	Farm bunds	1230
Hill to rolling	15–30	97.63	Light brown	17.4	19.5	Maize and millet	Farm bunds	1230

To determine the value of the *K*-factor, soil type and colour were considered (Hurni 1985; Mekonnen and Melesse 2011). For soils having different colour in the same land use and landform, the *K*-factor was taken separately and the amount of soil loss was calculated separately. As an example, the *K*-factor for undulating plains for black and brown soils was 0.15 and 0.2, respectively, and the same approach was used in determining the value of the soil colour for the other landforms in the watershed.

Slope length, the distance between two terraces, was measured using tape meter, and the slope gradient factor was recorded using clinometers on different landform and land uses and extrapolated based on Hurni (1985) for Ethiopian condition. Type of land use/cover was identified separately for each land unit. This was done by classifying the watershed into similar land uses and land forms. For variations in land cover with specific land unit or landform, the *C*-factor was obtained using weighted value of the different land cover.

The soil loss values can be used for determining critical area of soil erosion in the watershed by comparing with soil loss tolerance. If the soil loss is less than or equal to the soil loss tolerance, the soil loss is acceptable. The maximum soil loss tolerance for tropical regions is 25 t ha^{-1} y^{-1} (Ringo 1999). A commonly used soil loss tolerance rate is 5–12 t ha^{-1} yr^{-1} for shallow to deep soils (Lal 1994). The tolerance value for tropical soils has not yet been formulated at international level, but Hurni (1985) and Hudson (1995) established annual soil loss tolerance limits that vary between 0.2 and 11 t ha^{-1} y^{-1} as cited in Ringo (1999).

35.7 Results and Discussion

35.7.1 Household and Farm Characteristics

The total households living in Debre Mewi watershed is 88 % male headed (88 %) and 12 % female headed. The total population is 1034 of which 513 (49.6 %) were male and 521 (50.4) female, 36 % were below 15 years, 6.2 % were above 64 and the rest were in the active labour force age (16–64 years). The age of sampled farmers ranged from 23 to 75. From the total respondents, 5 % were female. Over half of the respondents (59 %) were illiterate, and the rest can read and write with basic education and religious schooling. The mean land holding size was about 1.09 ha. There was a significant variation in the size of land holdings. Among the sampled households, the majority (48 %) have from 0.75 to 1 ha, 17 % hold from 1.1 to 1.75 ha, 27 % have less than 0.75 ha, and only 8 % have more than 2 ha. One household has up to six plots of farmlands within their holdings. This fragmentation has its own negative effects to implement SWC measures.

Livestock are the most essential possession of the people in the study area. Livestock are not only used to produce meat and milk, but also used as a store of wealth, an insurance or a saving mechanism that fulfil urgent need for cash.

The farmer can sell his livestock at the local markets whenever he is in need of urgent cash and often to fulfil social obligations. Cattle are 885 (37 %), sheep 470 (20 %), goats 80 (3 %) and mule and donkeys 138 (6 %), and the rest were different animals. Possession of oxen determines the capacity to farm. Nearly 13 % of the households did not have an ox; some 29 % owned only a single ox; 52 % have a pair of oxen; 5 % have 3 oxen; and 1 % have 4 oxen. Farmers have no other option to produce crop without a pair of oxen. Farmers without oxen usually rent their lands for crop sharing or for money. A pair of oxen is needed to plough fields. Those who possess one ox have used pairing (locally called kenja), alternate use of the pair of oxen by two farmers. Pairing is also practiced with those that have two oxen.

The major source of fodder in the watershed is crop residue and hay collected during harvesting time. In the rainy season (kiremt), the livestock stay on the upland and rocky communal lands during the day. Edible weed species from the crop fields are also sources of livestock feed during the rainy season. During the dry season (bega), crop residues (mainly tef straw) and grass collected during October and December are the main sources of feed.

35.7.2 Traditional Soil and Water Conservation Measures

Traditional SWC practices give us an understanding of farmers' way of dealing with soil erosion. Traditionally, farmers commonly use traditional drainage ditches, farm boundaries as barriers and roads as waterways, and contour ploughing and few practices of stone terraces/bunds to protect their farmlands from soil erosion. Out of the surveyed farmers, 67 % used traditional ditches to drain excess run-off and keep fallow and stabilized farm boundary as erosion control method. Traditional ditches (locally called feses) are indigenous practices widely used to drain excess run-off. These are micro-channels constructed on cultivated fields to remove excess water from cultivated fields. However, they all know that such traditional ditches cause run-off concentration and sometimes gully formation in addition to the source of conflict with neighbouring farmers for flooding.

35.7.3 Fertilizers and Soil Erosion Control

Ninety five per cent of farmers in the watershed use commercial fertilizer on their farm. From these, 39 % commonly use compost and animal manure in addition to commercial fertilizer and 56 % use compost with fertilizer at least in one piece of land. However, lack of knowledge, shortage of crop residues, green leaf shortage for compost preparation and lack of commitment by farmers and development agents limit the quality and quantity of compost. Although farmers believe that commercial fertilizer, compost and manure increase production, they have limited knowledge on its fertility restoration and soil erosion control. Crop residue on the farm has vital role in improving the physical and chemical properties of soil and erosion control. After harvest, crop residue is collected for animal feed, roof thatching, fuel and other purposes followed by free grazing which severely reduces residue left on the field.

35.8 Technical Evaluation of Soil and Water Conservation Measures

35.8.1 Farmers' Technical Assessment

Farmers were asked to evaluate the introduced SWC technologies with their own criteria. The criteria selected by the farmers to evaluate the technologies include suitability for farming, reduction of farmland and effectiveness of soil conservation. Farmers raised criticism on fanya juu terrace constructed on gentle slope (0.5 % gradient) for the following reasons. Since the ditch is below the embankment and the gradient was not enough to drain out the excess run-off, the run-off retained by the embankment drained to their farm and water logging occurred resulting in yield reduction. Due to this reason, some farmers were forced to break the embankment to drain the collected run-off water. On the other hand, since the nature of the soil in most farmland where fanya juu terrace was implemented was fragile to carry the concentrated run-off resulted in accidental breakage of the embankment and causing rill and gully formation and crop damage where poorly constructed. Some respondents criticized the exaggerated width of the embankment and narrow space between bunds since it reduces their small farmland. Little economic incentive from bunds was also an issue raised by most farmers. The other evaluation by the farmers was the construction of very long and many graded bunds without appropriate waterways.

Pigeon pea, elephant grass and Sesbania sesban were introduced species to stabilize the farm bunds and to compensate the reduction in cropping growing area due to the area occupied by the bunds. But the productivity of the pigeon pea was very poor as mentioned by all farmers. The main reason for the poor performance of the pigeon pea may be poor variety selection which does not fit to the local agro-ecology.

Farmers during personal interview and group discussion were asked to recommend what should be done to improve the technical standard and effectiveness of SWC measures. They suggested farmers' full participation in the selection of best method for their farmland, experience sharing from areas having such practice, better grass species for embankment stabilization and economic return from the bunds.

35.8.2 SWC Evaluation Against Design Specifications

35.8.2.1 Vertical Interval (VI) and Length of Bunds

The spacing between existing terraces ranged from 8 to 50 m, with an average of 23 m on 3–19 % slope lands. As shown in Table 35.4, 78 % did not meet the minimum technical standards of vertical interval. Similar vertical interval and bund spacing were used at different slopes. Ninety two per cent of the bunds constructed on slopes greater than 9 % (51 out of 55 bunds) did not meet the technical standard. Some wider spacing has reduced the effectiveness of bunds since the concentrated run-off causes breakage through weak terrace segments.

35.8.2.2 Gradient and Length of Bunds

According to the guideline, recommended bund gradient ranges from 0.5 to 1 % (Hurni 1985; MoARD 2005). In this study, it was found that less than 1 % followed the technical standard and 7 % of the deficiencies were due to lack of skill during

Table 35.4 Recommended and existing vertical interval and horizontal distance

Slope (%)	Recommended for soil depth >75 cm		Measured terraces	Existing average		Number of terraces VI dismantled
	VI (m)	HD (m)		VI (m)	HD (m)	
3	1	33	6	1	32.6	1
4	1	25	11	1.2	30	4
5	1	20	16	1.46	30	10
6	1	17	3	1	18	no
7	1	14	5	1.44	20	2
8	1	12	12	2.3	28	10
9	1	11	6	1.7	18	6
10	1	10	21	1.78	17.8	21
11	1.1	10	7	2.24	20.3	5
12	1.1	9	4	2.2	18	3
13	1.2	9	3	1.95	15	2
14	1.2	8	3	2.52	18	3
15	1.2	8	4	2.8	14	4
16	1.3	8	2	2.95	18.5	2
17	1.3	8				
18	1.3	7	5	3.6	20	5
19	1.3	7				
Total			108			78

Fig. 35.2 fanya juu terraces unable to drain out the excess water (*left*) and soil buried under gravel on top (*right*)

layout. Overflow and embankment breakage were phenomena observed in farms treated by fanya juu. According to Haile et al. (2006) and MoARD (2005), fanya juus retain small amounts of run-off above the embankment and drain excess run-off through the ditch causing overflow and topping the embankment with down slope drainage problems (Fig. 35.2 left). In some areas, there is also risk of overtopping and breakage during high run-off in fragile soils. The guideline recommends placing embankment. Figure 35.2 (right) shows where the soil is buried below the gravel unlike the guideline.

The average length of bund was 124 m with a maximum of 360 m and a minimum of 30 m. Out of 108 farm bunds, 76 bunds (70 %) have length above 80 m, which is outside the technical recommendation (max 50–80 m). The major determinant factor to determine the length of terrace is the availability of natural waterway or suitable boundary between farms to construct artificial waterway. This is because farmers did not want to construct waterway on their plot since it takes out large area of land. As a result, there were exceptional bunds having length up to 360 m. The guideline recommends to increase channel cross section towards the end because of water concentration (from 25 cm depth and 50 cm width to 50 cm depth and 100 cm width), but fixed channel dimension throughout the length was observed.

35.8.2.3 Embankment Compaction and Height

Assessment of embankment height of soil bund and fanya juu terraces showed that bunds constructed in 2013 have an approximate average height of 42 cm, which meets 70 % of the technical recommendation. But such terraces shrink gradually due to poor compaction. Lack of compaction is creating quality problem on SWC measures (Fig. 35.3).

Fig. 35.3 Poorly compacted soil bund (*left*) and bund construction (*right*)

35.8.2.4 Embankment Width, Channel Depth and Bund Width

The bottom width of all bunds was within the range of the technical standard, but only 81 and 83 % of the top width for fanya juu and soil bund, respectively, met the technical standard (Table 35.5). In the study area, bunds had about 40 cm depth and 43 cm width channel, which meet 85 and 90 % of the standard, respectively.

35.8.2.5 Gully Treatment

Proper land management and slope treatment measures such as retention and infiltration ditches above the gully area are required to reduce the rate and amount of surface run-off. Such measures can be constructed before starting gully treatment check dam construction. In the study area, to retain run-off from rangelands and forest lands on the upper part of the gully, different SWC measures were put in place (Fig. 35.4). But all of these structures were done after the construction of different wooden, gabion and lose stone check dams which did not follow the guideline. These physical structures were built after the start of the rainfall season which led to structural failures. The area closure practices above the main gully contributed to have dense cover on the communal land and decrease run-off amount.

Table 35.5 Design specification and observed measured average bund width and height after compaction

Type of measure	Design specification (m)			Measurement result (m)		
	Height after compaction	Base width	Top width	Height after compaction	Base width	Top width
fanya juu	0.6	1–1.5	0.3–0.5	0.42	1.39	0.62
Soil bund	0.6	1–1.5	0.3–0.5	0.42	1.4	0.6

Fig. 35.4 Upper gully treatment measures

Brushwood check dams (single and double) are temporary structures for small gully heads and should not be used for large concentrated run-off but can be used where run-off is less than 1 $m^3 s^{-1}$ (Adugna 2010; MoARD 2005), and were used in areas where there is 2.4, 1.4 $m^3 s^{-1}$ run-off generated from the upper catchment and 1 $m^3 s^{-1}$ run-off diverted from roads to the gully. Therefore, it is advisable to construct SWC measures to reduce excess run-off above the structures and to tackle the problem effectively. Brushwood and soil-filled sack check dams were constructed in the study area. But, in the nearby sites, plenty of stones were available which can be used to construct stone check dams which are stronger than sac and brush check dams. It is not advisable to build soil-filled sack and wooden check dams which are damaged by water erosion easily (Fig. 35.5).

Gully wall reshaping is cutting off steep slopes of active gully flanks into gentle slope (minimum at 45 % slope), up to two-third of the total depth of the gully and constructing small trenches along contours for revegetation. Gully reshaping has been done in the watershed with gullies of 65 % slope which are far higher than the standards as shown in Fig. 35.6 (left). It has promise to reduce severe gully erosion

Fig. 35.5 Failed wooden and soil-filled sack check dams

Fig. 35.6 Gully reshaping with steep slope (*left*) and stone use attempt to control gully erosion (*right*)

problem and help stabilization. Small stones that are not angular, which cannot connect easily, have been used for lose stone check dam construction and this coupled with low skill of stone pilling by the farmers had resulted in check dam failures (Fig. 35.6 right).

Spacing between measured loose stone check dams was not far from the technical standard. It was wide so that the top of the lower check dam was not level with the foot of the upper check dam. The average height of the loose stone check dams was 67 cm which is acceptable since depth of the gully is below 1 m. Out of the evaluated 14 lose stone check dams, 50 % have no side keys, 27 % have 19 cm which is the minimum technical standard, and only 35 % of the check dams have apron. Loose stone check dams should have 1.5 to 3.5 m base width but what was observed was 1.16 m base width on average (below the minimum technical standard).

35.8.2.6 Artificial Waterway

An artificial waterway is a drainage feature that can receive and dispose excess run-off from cut-off drains and graded terraces to the natural watercourse. The dimension of a waterway depends on the expected discharge calculated in $m^3\ s^{-1}$. Based on estimated run-off, the recommended dimension (depth, bottom width and top width) of the artificial waterways was found as 0.35, 0.42 and 0.55 m depth; 1.1, 0.7 and 2.5 m bottom width; 1.9, 2.2 and 3.6 m top width, for the studied waterways. This is within the range of technical standards except the top width of one artificial waterway.

Depending on the amount of discharge expected, slope and type of soil in which the artificial waterway was constructed, it is possible to use different materials for stabilization such as grass and stone. The width can be reduced if stone is combined with grass. If land is scarce and/or steep, narrow waterways lined by stone can be used. Waterways running diagonally across the slope are not recommended because if they are breached or overtopped serious damage can occur. Out of the three waterways evaluated, two of them were not done along the slope because they were constructed following the boundary between two farm ownerships. As a result,

Fig. 35.7 Fragile gully without treatment used as waterway

breach of the artificial waterway occurred and run-off overtopped and caused damage to croplands. The third was not properly lined with stone or planted with grasses.

Grass-lined waterways should be laid one year ahead of graded bunds or cut-off drains. As soon as construction is completed, the waterway should be lined by planting a suitable spreading grass or with stone or a combination of grass and stone. Every effort should be made to avoid discharging run-off into the waterway until the grass is established. In the study area, unfortunately all waterways were constructed after the graded bunds have been done. Gullies that are fragile and susceptible for erosion have been also used as waterways without any gully treatment (Fig. 35.7).

Administrative boundary between districts (Bahir Dar zuria and Yilmana Densa) is also used as a waterway. Two waterways have been constructed side by side along the boundary due to land ownership issues (Fig. 35.8) and causing productive soil loss from both sides. This is because of lack of an integrated watershed management approach.

35.8.3 Management Responsibility and Sustainability

35.8.3.1 Farmers' Acceptance of Introduced SWC Measures

The introduced SWC measures were widely acknowledged and accepted as effective measures against soil erosion. The results of the field survey showed that almost 100 % of the farmers recognized soil erosion as a problem and are aware of

Fig. 35.8 Two waterways constructed side by side along two districts boundary due to land ownership issues and lack of integrated water management

the need for soil conservation. Most of the respondents (90 %) confirmed that erosion has been decreased in areas where SWC measures have been taken. As a result, able-bodied farmers contributed from 20 to 40 days free labour for SWC conservation during the campaign from January to February, 2011. Gradual observations of the positive effects and benefits of previous conservation activities, specially gully rehabilitation results, contributed to the acceptance of SWC measures by farmers. Almost all of the farmers reported that SWC measures were very helpful for erosion control and fertility restoration. All farmers accepted soil bund with ditch since it drains the excess run-off safely to a waterway and reducing overtopping.

Farmers responded positively to the effectiveness of SWC methods application on their farm. Large amount of sediment that would have left their fields was trapped by the structures and crops near the bunds were better. All farmers replied that they like the benefits of SWC measures but are not willing to accept control grazing due to livestock feed shortage. A common complaint was cattle reproduction became low due to cows getting weak when feeding area is limited. Cut and feed is labour-intensive making it impossible to feed and water their large number of livestock. They are not willing to reduce livestock size because they need them for ploughing, trampling teff and millet fields, for planting, trashing crops during harvest and to transport their product from field to their homes and to market. This study results are different from Derebe (2009) work on constructing terraces or bunds on small-sized farmlands.

35.8.3.2 Soil and Water Conservation Measures Management Responsibility

During group discussion, farmers confirmed that they have already developed a by-law based on the proclamation number 204/2013 rectified by the Council of the Amhara National Regional State (CANRS 2013). The proclamation states that everybody should participate in community asset management and natural resource conservation to bring about sustainability and active users' participation in the watersheds. They agreed that individual farmers were responsible to keep and maintain SWC measures constructed on individual farms. Area closure and SWC structures on communal lands were the responsibility of the whole community by the coordination of the planning team. For the management of the whole watershed, they were in the process of establishing formal association.

35.8.4 Soil Loss

Different soil loss values were obtained on different landforms due to differences in model inputs (Table 35.3). Soil loss on the undulating plain landform was 7.88 t ha^{-1} yr^{-1} with annual rainfall 1230 mm (R = 682), black soil colour (K = 0.15), 28 m average slope length (L = 1.4), 5.5 % slope gradient (S = 0.44), teff land cover (C = 0.2) and soil bund conservation practice (P = 0.5). On similar landforms but different cover (teff and millet) and black soil colour, soil loss was found to be 8.4 t ha^{-1} yr^{-1}. On rolling landforms (8–15 % slope), soil loss was 19.97 and 23.97 t ha^{-1} y^{-1} for brown and black soils, respectively. The higher soil loss was obtained on hill to rolling land form (15–30 % slope), which was 26.58 t ha^{-1} y^{-1}. The average model estimated annual soil loss from farmlands with different landforms treated by SWC measures was estimated as 17 t ha^{-1} y^{-1}. In general, soil loss increased as slope steepness increased in the watershed.

Different rates of soil loss before the implementation of intensive SWC practice have been reported as 39 t ha^{-1} y^{-1} (Derebe 2009) and 12.5–50 t ha^{-1} y^{-1} (Mekonnen and Melesse 2011). In this study, it was found loss after implementation of SWC was 17 t ha^{-1} yr^{-1} which was reduced by 22 t ha^{-1} yr^{-1} compared to estimates by Derebe (2009) but is in the lower range of estimates by Mekonnen and Assefa (2011) for preconservation losses. Since the three studies were conducted in the same watershed with similar conditions, the reduction in soil loss should be attributed to the implemented SWC measures. In general, soil loss was found to be above the tolerable soil loss rate (0.2–12 t ha^{-1} y^{-1}) estimated for Ethiopian condition (Hurni 1985; Ringo 1999). This implies that there is a need to implement more SWC measures following the standard and supported with sound integrated watershed management approaches to reduce soil loss in Debre Mewi watershed. Table 35.6 shows predicted soil loss for various model parameters.

Table 35.6 RUSLE parameters and estimated values

Estimated values for each parameter						Predicted soil loss t ha^{-1} yr^{-1}
Erodibility	Slope gradient (S)	Slope length (L)	Cover (C)	Conservation practice (P)	Rainfall (mm)	
0.15	0.44	1.4	0.25	0.5	682	7.88
0.2	0.44	1.4	0.2	0.5	682	8.4
0.15	1.1	1.42	0.25	0.5	682	19.97
0.225	1.1	1.42	0.2	0.5	682	23.97
0.2	1.856	1.4	0.15	0.5	682	26.58
Average soil loss						17.36

35.9 Conclusion and Recommendation

Physical soil and water conservation measures implemented in Debre Mewi watershed were identified and technically evaluated with farmers' criteria and survey. The contribution of SWC for erosion control was assessed using the RUSLE model. From the study, it is concluded that SWC technologies were selected and implemented based on the agro-ecological zone and land use requirement. Soil fertility management is widely applied by farmers to increase productivity, but they have limited knowledge on fertilizers contribution on erosion control. Some problems were observed in applying physical SWC measures on slopes less than 3 % with fanya juu structures on very fragile soils. Even though the community widely acknowledged and accepted SWC measures as effective erosion control, land owners complained that they were not involved in the selection of the type of SWC method implemented on their farmland. This violates the SWC extension approach of encouraging participation.

Soil and water conservation structures done according to the standard can overcome erosion problem. But to achieve the intended objective and for sustainability of the practice, technical standards of SWC need to be followed. In some areas, there is very narrow bund spacing that competes with farmable land and creates difficulty for ploughing. Also, wider spacing of bunds causes high run-off concentration and damage farmland. Inappropriate drainage management activities and poorly selected, designed and constructed gully treatment activities aggravate soil erosion and land degradation. There are several cases where various standards are not met.

The by-law ratified by watershed community association clearly assigned management responsibility of SWC measures and is found to be excellent tool to ensure sustainability. Still, farmers do not understand enough the win-win opportunity between controlled grazing and soil and water conservation.

Soil erosion assessment result using the RUSLE model indicated the annual soil loss decreased due to the implementation of SWC measures in the watershed. But the erosion rate is still far higher than the tolerance limit. Application of compost, manure, crop residue after crop harvesting, improved plough other than the traditional plough "maresha", improved soil fertility, increased infiltration and decrease in surface run-off contribute to reduction in soil erosion.

Based on the findings of the study and frequent field observations, the following recommendations are forwarded. Farmers should be consulted to participate in the selection of SWC measures before implementation on their farms. It is better to treat vertisol farmland that have gentle slope with only grass strip and other biological measures. If fanya juu terrace is to be built on gentle slopes, the embankment should be stone faced and gradient should be 1 %. Inter-terrace/inter-structural management practices such as ploughing in graded contours, strip cropping, cover crop, lye farming and planting in rows along graded contours can be applied.

Gully treatment should be started from upper catchment, and before selecting the type of structure, the nature of the gully should be known and the amount of run-off should be estimated. The type of material for gully treatment, and design parameters like, side key, apron, spillway and gully reshaping should be selected following guidelines.

The technical guideline prepared by MoARD (2005) and used as a standard for SWC in Ethiopia needs to be updated. It should include compass, GIS and GPS technologies to select appropriate SWC measures. The standard revision should include design of gully treatment structures for water harvesting and standard for bund length. Further studies should be made to evaluate SWC failures based on the technical standards to reduce erosion.

References

Admassie Y (2000) Twenty years to nowhere: property rights, land management and conservation in Ethiopia. Red Sea Press, Lawrenceville, New Jersey

Adugna B (2010) Drainage management and control: training manual prepared for zonal and woreda experts. Bahirdar, Ethiopia

Amsalu A, Graaff JD (2004) Farmers' views of soil erosion problems and their conservation knowledge at Beressa watershed, central highlands of Ethiopia. Agric Hum Values 23:99–108

Angima SD, Stott DE, O'Neill MK, Ong CK, Weesies GA (2003) Soil erosion prediction using RUSLE for central Kenya highland conditions. Agriculture Ecosystem and Environment, pp 295–308

Awdenegest M, Holden NM (2007) farmers' perceptions of soil erosion and soil fertility loss in southern Ethiopia. Land Degrad Dev 18(5):543–554

Bedadi B (2004) Evaluation of soil erosion in the Haregie Region of Ethiopia using soil loss models, rainfall simulation and field trials, Doctorial thesis, Pretoria University

Bekele A (2001) Status and dynamics of natural resources in Ethiopia. In: Taye A (ed) Food security through sustainable land use: population, environment and rural development issues for sustainable livelihoods in Ethiopia. NOVIB Partners Forum on Sustainable Land Use, Addis Ababa, Ethiopia, pp 165–184

Bekele S, Holden ST (1999) Soil erosion and smallholders' conservation decisions in the highlands of Ethiopia. World Dev 27(4):739–752

Bekele S, Holden ST (2001) Farm-level benefits to investments for mitigating land degradation: empirical evidence from Ethiopia. Environ Dev Econ 6:335–358

Beshah T (2003) Understanding farmers: explaining soil and water conservation in Konso, Wolaita and Wello, Ethiopia. Tropical Resource Management Papers, vol 41. Wageningen University, 245 p

Bewket W (2003) Towards integrated watershed management in highlands of Ethiopia: The Chemoga watershed case study. Tropical resource management papers no 44. Wageningen Agricultural University, The Netherlands

Bewkete W, Sterk G (2002) Farmers' participation in soil and water conservation activities in the Chemoga watershed, Ethiopia. Land Degrad Develop 13:189–200

BOA (2011) Soil and water conservation, field manual. Bahir Dar, Ethiopia

BOA (2013) Growth and transformation plan achievement report. Bahir Dar, Ethiopia

CANRS (2013) Administration and use of watersheds rehabilitated and being rehabilitated with community participation proclamation. Proclamation no. 204/2013. Proclamation issued to determine the administration and use of watersheds rehabilitated and being rehabilitated by community participation. Bahir Dar, Ethiopia

Carucci V (2000) Guidelines on water harvesting and soil conservation for moisture deficit areas in Ethiopia: the productive use of water and soil, manual for trainers, first draft. Ministry of Agriculture, Addis Ababap, p 383

Defersha MB, Melesse AM (2011) Field-scale investigation of the effect of land use on sediment yield and surface runoff using runoff plot data and models in the Mara River basin, Kenya. CATENA 89:54–64

Defersha MB, Melesse AM (2012) Effect of rainfall intensity, slope and antecedent moisture content on sediment concentration and sediment enrichment ratio. CATENA 90:47–52

Defersha MB, Quraishi S, Melesse AM (2011) Interrill erosion, runoff and sediment size distribution as affected by slope steepness and antecedent moisture content. Hydrol Proc 7 (4):6447–6489

Defersha MB, Melesse AM, McClain M (2012) Watershed scale application of WEPP and EROSION 3D models for assessment of potential sediment source areas and runoff flux in the Mara River Basin, Kenya. CATENA 95:63–72

Derebe A (2009) Assessment of land upland erosion process and farmers perception of land conservation in Debre watershed, near Lake Tana, Ethiopia. Msc thesis. Bahirdar, Ethiopia

FAO (1986) Highlands reclamation study, Final report. Addis Ababa, Ethiopia

FAO (2004) Methodological framework for land degradation assessment in dry-lands. ftp:fao.org/agl/agll/lada/LADA-Methframwk-simple.pdf. Accessed 12 June 2013

Fisseha G, Gebrekidan H, Yitaferu B, Bedadi B (2011) Analysis of land use/land cover changes in the Debre-Mewi watershed at the upper catchment of the Blue Nile Basin, Northwest Ethiopia

Girma T (2001) Land degradation: a challenge to Ethiopia. Environ Manage J Int livestock research institute 27:815–824

Haile M, Herweg K, Stillhardt B (2006) Sustainable land management—a new approach to soil and water conservation in Ethiopia. Mekele University, Ethiopia, pp 79–89

Herweg K 1996. Field manual for assessment of current erosion damage. Soil conservation research programme (SCRP), Ethiopia and centre for development and environment (CDE). University of Berne, Switzerland

Hudson N (1992) Soil conservation, 2nd edn. Batsford Limited, London

Hudson N (1995) Soil conservation, 3rd edn. Batsford Limited, London

Hurni H (1983) Ethiopian highlands reclamation study. Soil formation rates in Ethiopia, Working paper no 2. Addis Ababa, Ethiopia

Hurni H (1985) Erosion-productivity systems in Ethiopia. Proceedings of the fourth international conference on soil conservation. Venezuela

Hurni H (1995) Guidelines for development agents on soil conservation in Ethiopia. Watershed development and land use department. MORDA, Addis Ababa

IFAD (1992) Soil and water conservation in Sub-Saharan Africa Towards sustainable production by the rural poor. Centre for Development Cooperation Services, Free University, Amsterdam

Lal RE (1994) Soil erosion research methods. St. Lucie Press. Soil and Water Conservation Society, Delray Beach

Maalim FK, Melesse AM (2013) Modeling the impacts of subsurface drainage systems on runoff and sediment yield in the Le Sueur watershed, Minnesota. Hydrol Sci J 58(3):1–17

Maalim FK, Melesse AM, Belmont P, Gran K (2013) Modeling the impact of land use changes on runoff and sediment yield in the Le Sueur watershed Minnesota using GeoWEPP. Catena 107:35–45

Makombe G, Kelemwork D, Dejene A (2007) A comparative analysis of rain fed and irrigated agricultural production in Ethiopia. Irrig Drainage Syst 21:35–44

Mekonnen M, Melesse AM (2011) Soil erosion mapping and hotspot area identification using gis and remote sensing in Northwest Ethiopian highlands, near lake Tana. Nile River Basin; hydrology, climate & water use. Springer, Berlin, pp 207–224

Mekonnen M, Keesstra SD, Baartman JEM, Stroosnijder L, Maroulis J (2014a) Soil conservation through sediment trapping: a review. Land Degrad Dev. doi:10.1002/ldr.2308

Mekonnen M, Keesstra SD, Baartman JEM, Coen J. Ritsema (2014b) Sediment storage dam: a structural gully erosion control and sediment trapping measure, northern Ethiopia. Geophysical Research Abstracts, vol 16, EGU2014-114, 2014. EGU General Assembly 2014

Melesse AM, Ahmad S, McClain M, Wang X, Lim H (2011) Sediment load prediction in large rivers: ANN approach. Agric Water Manage 98:855–886

MoA (2001) A guide to watershed management, Ethiopia

MoARD (2005) Community based participatory watershed development: a guideline. Ministry of Agriculture and Rural development, Addis Ababa, Ethiopia

Mohammed H, Alamirew T, Assen M, Melesse AM (2015) Modeling of sediment yield in Maybar gauged watershed using SWAT, northeast Ethiopia. CATENA 127:191–205

Morgan RPC (2005) Soil erosion and conservation, 3rd edition. Blackwell Publishing, National Soil Resources Institute, Cranfield University, London

Msagahaa J, Ndomba PM Melesse AM (2014) Modeling sediment dynamics: effect of land use, topography and land management. In: Melesse AM, Abtew W, Setegn S (eds) Nile River Basin: ecohydrological challenges, climate change and hydropolitics, pp 165–192

Nedessa B, Ali J, Nyborg I (2005) Exploring Ecological and socio-economic issues for the improvement of area enclosure management. A case study from Ethiopia DCG report no 38. Addis Ababa, Ethiopia

Nyssen J, Descheemaeker K, Haregeweyn N, Haile M, Deckers J, Poesen J (2007) Lessons learnt from 10 years research on SWC in Tigray. Tigray livelihood papers no.7. Zala-Daget Project, Mekelle University, K.U.Leuven, Tigray BOARD. ISBN 978-90-8826-027-8: 53

Pimentel D, Harvey C, Resosudarmo P, Sinclair K, Kurz D, McNair M, Crist S, Shpritz L, Fitton L, Saffouri R, Blair R (1995) Environmental and economic costs of soil erosion and conservation benefits. Science 267:1117–1123

Ringo D (1999) Assessment of erosion in the Turasha catchment in the lake Naivasha area, Kenya. Unpublished, MSc Enschede

Setegn SG, Srinivasan R, Dargahi B, Melesse AM (2009) Spatial delineation of soil erosion prone areas: application of SWAT and MCE approaches in the Lake Tana Basin, Ethiopia. Hydrol Process Spec Issue Nile Hydrol 23(26):3738–3750

Setegn SG, Dargahi B, Srinivasan R, Melesse AM (2010) Modelling of sediment yield from Anjeni Gauged Watershed, Ethiopia using SWAT. JAWRA 46(3):514–526

Sonneveld B, Keyzer M (2003) Land under pressure: soil conservation concerns and opportunities for Ethiopia. Land Degrad Develop 14:5–23

Stocking M, Murnaghan N (2000) Land degradation guidelines for field assessment. http://archive.unu.edu/env/plec/l-degrade/index-toc.html. Accessed 10 May 2013

SWHISA (2007) Participatory rural appraisal report of Debre Mewi watershed. IRG-09. Bahir Dar, Ethiopia

Wang X, Garza J, Whitney M, Melesse AM, Yang W (2008) Prediction of sediment source areas within watersheds as affected by soil data resolution. In: Findley PN (ed) Environmental modelling: new research, Chap. 7. Nova Science Publishers, Inc., Hauppauge, pp 151–185. ISBN: 978-1-60692-034-3

Wischmeir WH, Smith D (1978) Predicting rainfall erosion losses. A guide to conservation planning. USDA. Hand book no 357, Washington, DC

Zhang SH, Liu SX, Mo XG, Shu C, Zheng CL, Hou B (2010) Impacts of precipitation variation and soil and water conservation measures on runoff and sediment yield in the Wuding River Basin, middle reaches of the Yellow River. J Beijing For Univ 32:161–168

Chapter 36
Can Watershed Models Aid in Determining Historic Lake Sediment Concentrations in Data-Scarce Areas?

Essayas K. Ayana and Tammo S. Steenhuis

Abstract In most developing countries, there is a lack of historical sediment data. In Ethiopia, over the past 40 years, sediment concentrations in rivers are periodically measured for determining sediment rating curves. But, no such historical data are available for lakes such as Lake Tana in Upper Blue Nile River basin, greatly hampering the ability to track the trend of environmental pollution. The objective of this study is to determine whether models can be used to estimate the historical record of lake sediment concentrations. In this study, a relationship between reflectance and TSS was developed and 10-year TSS time series data were generated from MODIS/Terra images. The 10-year TSS time series data were used to calibrate and validate an erosion model. The soil and water assessment tool variable source area (SWAT-VSA) model was selected because it has performed well in the Lake Tana basin. The result showed that at monthly timescale, TSS at the river mouth can be replicated with Nash–Sutcliffe efficiency (NS) of 0.34 for calibration and 0.21 for validation periods. Percent bias (PBIAS) and ratio of root mean square error (RSR) to the standard deviation of measured data are all within range. Given the inaccessibility and costliness to measure TSS at river mouths to a lake, the results found here are considered best estimates for suspended sediment budget in water bodies of the basin. However, direct measurement or improved models should yield better results.

E.K. Ayana · T.S. Steenhuis
School of Civil and Water Resources Engineering, Bahir Dar Institute of Technology, Bahir Dar, Ethiopia
e-mail: tss1@cornell.edu

E.K. Ayana
Department of Ecology Evolution and Environmental Biology, Columbia University, New York, USA

E.K. Ayana (✉)
The Nature Conservancy, Arlington, VA, USA
e-mail: eka2115@columbia.edu; essayas.ayana@tnc.org

T.S. Steenhuis
Department of Biological and Environmental Engineering, Cornell University, Ithaca, NY, USA

© Springer International Publishing Switzerland 2016

A.M. Melesse and W. Abtew (eds.), *Landscape Dynamics, Soils and Hydrological Processes in Varied Climates*, Springer Geography,
DOI 10.1007/978-3-319-18787-7_36

Keywords TSS · MODIS · Lake Tana · East Africa · SWAT · Upper Blue Nile River basin

36.1 Introduction

The ability to numerically model water resource systems has progressed enormously with advances in computational power and the understanding of processes at finer scale (Silberstein 2006). However, water resource data collection at varying scales is expensive, so that modelers often tend to conceptualize processes based on simplified views of nature (Dozier 1992) or match the observed data even if the underlying premises are unrealistic (Kirchner 2006). In addition, collection of water resource data, especially in developing countries, is characterized by inadequate monitoring, gaps in observations, a decline in the number of stations, chronic underfunding, and differences in processing and quality control (Vörösmarty et al. 2001; Harvey and Grabs 2003). Our ability today to monitor extreme events with ground-based systems is less than it was 45 years ago (Macauley and Vukovich 2005).

Space-borne remote sensing has become a potential data source to model land and water resource systems. Various remotely sensed image-based tools are also developed to measure turbidity (Chen et al. 2007, 2009; Shen et al. 2010), suspended sediment concentration (Jiang et al. 2009; Nechad et al. 2010; Onderka et al. 2011), chlorophyll a (Fiorani et al. 2006; Wang et al. 2010), phytoplankton (Kwiatkowska and McClain 2009), cyanobacterial blooms (Kutser et al. 2006), and other physical water quality parameters (Liu et al. 2003; Hu et al. 2004). Using these data in hydrologic modeling requires an understanding of the potentials and limitations of the data sets.

A number of studies used laboratory, field scales, and modeling studies to understand soil erosion and sediment dynamics in various regions (Defersha and Melesse 2011, 2012; Defersha et al. 2011, 2012; Maalim and Melesse 2013; Maalim et al. 2013; Melesse et al. 2011; Msagahaa et al. 2014; Wang et al. 2008; Mekonnen and Melesse 2011; Mohamed et al. 2015).

Modeling sediment concentration in freshwater lakes in developing countries is hampered by lack of historical sediment concentration data. There is an existing knowledge base with respect to stream discharge and sediment modeling using SWAT (Conway 2000; Kebede et al. 2006, 2011; Tarekegn and Tadege 2006; Chebud and Melesse 2009a, b, 2013; Wale et al. 2009; Setegn et al. 2010, 2011; White et al. 2011; Dile et al. 2013; Yasir et al. 2014; Mohamed et al. 2015). Improved hydrologic models using the saturation excess runoff processes that are dominant in wet humid areas such as Ethiopia have also been developed for predicting runoff and erosion (Easton et al. 2008; Steenhuis et al. 2009; Tilahun et al. 2012).

None of the modeling studies simulated sediment concentration referred as total suspended solids (TSS), for lakes. Recently, Kaba et al. (2014) generated a ten-year TSS time series data from remotely sensed MODIS/Terra version 5 images for Lake Tana at the Gumera River mouth. This data set was used to test whether a calibrated SWAT-VSA model is able to predict the TSS concentration in Lake Tana (Easton et al. 2008). The results in this study provides scientific basis for testing the suitability of using calibrated watershed models for predicting sediment concentration time series for freshwater lakes in developing countries where historic data are scarce.

36.2 Materials and Methods

36.2.1 Study Area

The Gumera catchment drains an area of about 1280 km^2 (Fig. 36.1). The watershed drains into Lake Tana, a freshwater lake and source of the Blue Nile River. Agriculture being a dominant activity in the area represents 96 % of the watershed, and only 4 % is forested.

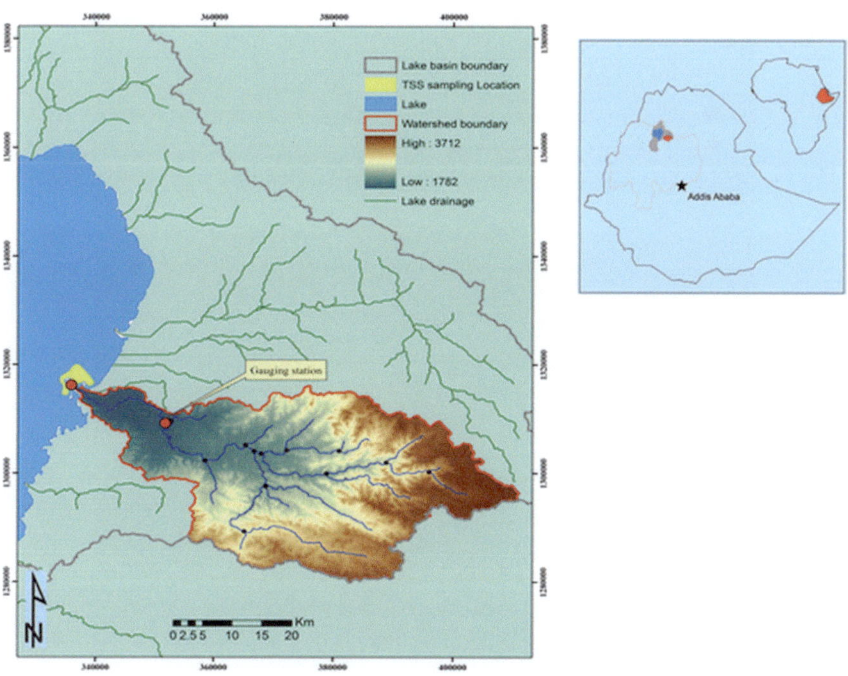

Fig. 36.1 Lake Tana stream network and gauging station in the Gumera watershed

Elevation of the Gumera watershed ranges from 1792 to 3712 m. The watershed climate and vegetation are characteristic of a sub-humid zone with a high diurnal temperature variation between daytime extremes of 30 °C to night lows of 6 °C. Rainfall may reach up to 2000 mm per year with rainy season from May to October with July and August being the wettest (Vijverberg et al. 2009). Gumera River, 75 km long, drains into Lake Tana. The average discharge over a 33-year period is 34.4 $m^3 s^{-1}$. Minimum in this period was 0.04 $m^3 s^{-1}$, and the maximum was 406 $m^3 s^{-1}$ (Ministry of Water and Energy, FDRE).

36.2.2 Establishing the TSS–Reflectance (ρ) Relationship

Three campaigns (November 27, 2010, May 13, 2011, and November 7, 2011) were carried out to collect water samples during the satellite overpass time over Lake Tana near the mouth of the Gumera River during the rainy season. Samples were collected along transects parallel to the shore during each overpass (Fig. 36.2).

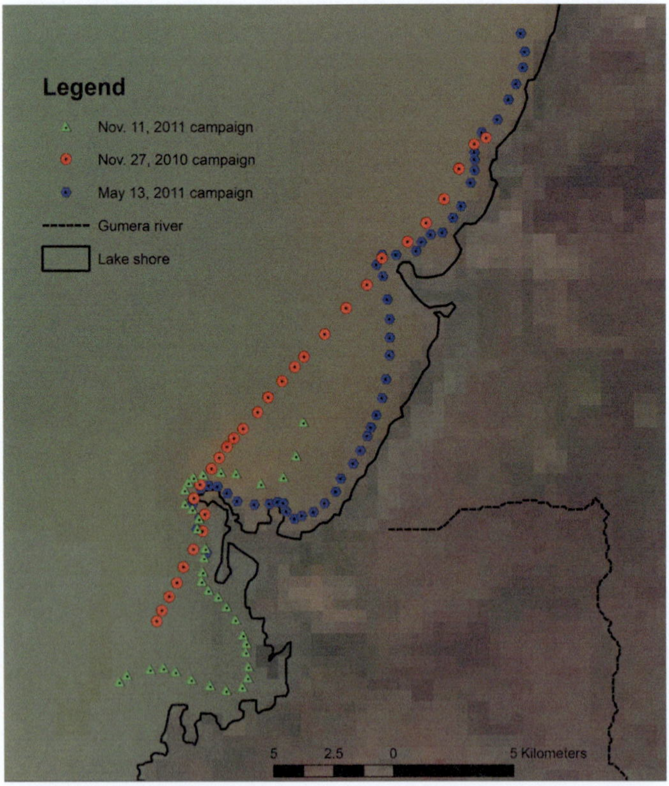

Fig. 36.2 Sampling locations for the three sampling campaigns near the mouth of the Gumera River

At each location along the sampling path, bulk water samples were collected from the upper 0.2 m of the water column in a 750-ml container for turbidity and TSS analysis. GPS coordinates of sampled locations were also recorded. Measurements of total suspended solids were taken in laboratory by drawing 10 ml aliquot from a well-mixed container, centrifuging for ten minutes at 4000 rpm, pouring off supernatant, separating and drying the retained solids, and weighing.

MODIS red and NIR images corresponding to the field water sampling dates (November 27, 2010, May 13, 2011, and November 7, 2011) were downloaded. The red (620–670 nm) and NIR (841–876 nm) bands labeled "MOD09GQ" are available on a nearly daily basis at 250-m spatial resolution. These images are sensitive for turbid water applications (Hu et al. 2004). A number of previous studies have successfully used MODIS 250-m images to establish a reflectance–TSS, reflectance–turbidity, and reflectance–Secchi depth relationships (Miller and McKee 2004; Dall'Olmo et al. 2005; Kutser et al. 2006; Chen et al. 2007; Petus et al. 2010). TSS, turbidity, and Secchi depth values reflect the amount of particulate in water.

In order to obtain the relationship between the MODIS reflectance and TSS at the river mouth, a multiple regression analysis was performed on various combinations of red and near-infrared red (NIR) bands (Table 36.1). Normalized ratios (NIR to red), band sum, and band difference are used along with single-band regression. Samples from the first two campaigns of November 27, 2010, and May 13, 2011, were used to establish the relation (i.e., calibration), and the third sample collected on November 7, 2011, was used to validate the relation. The goodness of fit of the model was evaluated based on the resulting coefficient of determination (R^2). Adjusted R^2 is also calculated for each regression to test whether an improvement in the R^2 is due to the inclusion of a band to the regression model or a random chance. For the validation step, the accuracy of predicted TSS was assessed using root mean square error (RMSE). Figure 36.3 shows the best-fit relationship between NIR and turbidity. Turbidity is a viable surrogate for TSS.

The calibrated regression model between TSS and reflectance in the red (ρ_{NIR}) and NIR (ρ_{NIR}) bands is shown in order of decreasing R^2, Table 36.1. A single-band regression best-fit equation for turbidity is shown in Eq. 36.1 (Fig. 36.3).

$$\mathrm{TSS} = 2371\rho_{\mathrm{NIR}} - 62.8 \qquad (36.1)$$

Table 36.1 Result of multiple regression analysis

TSS (N = 54)				
Band combination	R^2	Adjusted R^2	Standard error	Significance F
NIR	**0.96**	**0.95**	10.77	0.000
NIR/Red	0.89	0.88	16.86	0.000
NIR + Red	0.76	0.76	24.34	0.000
Red − NIR	0.86	0.85	18.96	0.000
$\frac{\mathrm{Red-NIR}}{\mathrm{NIR+Red}}$	0.88	0.88	16.92	0.000

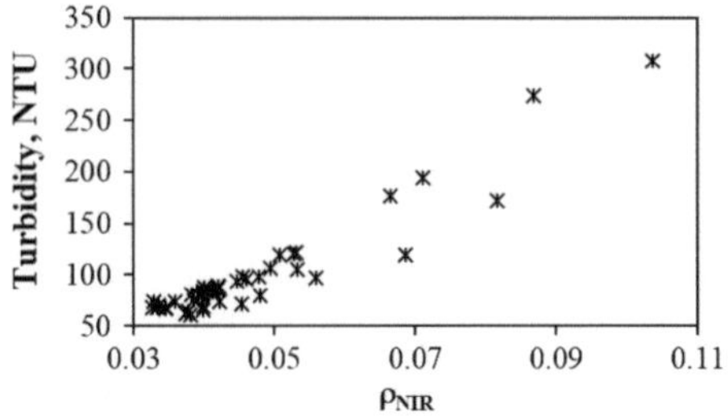

Fig. 36.3 NIR–TSS relationship

where TSS is in mg L^{-1} ($n = 54$ and $p < 0.001$) and P_{NIR} is reflectance measurement in NIR band obtained from MODIS image.

36.2.3 SWAT Model Setup

Detailed description of the SWAT model is provided in the literature (Neitsch et al. 2005; Arnold et al. 2007). Easton et al. (2008) reconceptualized SWAT for mountainous areas by using the topographic wetness index in combination with land use to define the HRU. In SWAT-VSA, HRUs are defined using topographic wetness index in combination with land use. In this way, the saturation excess runoff from variable source areas which is the dominant process in Ethiopian highlands is incorporated into SWAT (Steenhuis et al. 2009). Water balance is simulated by SWAT using Eq. 36.2.

$$\mathrm{SW}_t = \mathrm{SW}_o + \sum_{t=1}^{n} \left(R_{\mathrm{day}} - Q_{\mathrm{surf}} - E_{\mathrm{a}} - W_{\mathrm{seep}} - Q_{\mathrm{gw}} \right) \tag{36.2}$$

where SW_t and SW_o are final and initial soil water contents at time t and t_o in mm, respectively, R_{day} is precipitation on day i in mm, Q_{surf} is surface runoff on day i in mm, E_{a} is evapotranspiration on day i in mm, W_{seep} is percolation on day i in mm, and Q_{gw} is return flow on day i in mm.

Sediment transport processes are simulated via soil erosion and sediment transport from the hillslopes of the catchment and the sediment processes in the stream channel (Neitsch et al. 2005). The sediment yield from a HRU is calculated using the modified universal equations (MUSLE) which depends on the rainfall runoff energy to entrain and transport sediment (Williams and Singh 1995), Eq. 36.3.

$$\mathrm{Sed} = 11.8\left(Q_{\mathrm{surf}} \cdot q_{\mathrm{peak}} \cdot \mathrm{area}_{\mathrm{hru}}\right)^{0.56} \cdot K_{\mathrm{USLE}} \cdot C_{\mathrm{USLE}} \cdot P_{\mathrm{USLE{\cdot}LS_{USLE}}} \cdot \mathrm{CFRG} \tag{36.3}$$

where Sed is the sediment yield on a given day (metric ton); Q_{surf} is the surface runoff volume (mm ha^{-1}); q_{peak} is the peak runoff rate ($\mathrm{m}^3\ \mathrm{s}^{-1}$); $\mathrm{area}_{\mathrm{hru}}$ is the area of

HRU (ha); K_{USLE} is the soil erodibility factor [0.013 metric ton m^2 h (m^3-metric ton cm)$^{-1}$]; C_{USLE} is the cover and management factor; P_{USLE} is the support practice factor; LS_{USLE} is the topographic factor; and CFRG is the coarse fragment factor. More detailed descriptions of the model can be found in Neitsche et al. (2005).

The model setup involved five steps of data preparation, sub-basin discretization and HRU definition, sensitivity analysis, calibration, and validation. The spatial data required in SWAT are the digital elevation model (DEM), soil, and land use data.

A 30-m by 30-m resolution DEM is used to delineate the watershed, analyze the drainage patterns of the land surface terrain and generate the topographic wetness index. Sub-basin parameters are derived from the DEM. The soil data were acquired from the new Harmonized World Soil Database (HWSD). The land use map was obtained from the Abay Basin master plan document (BCEOM 1999). The weather variables' data were acquired from Ethiopian National Meteorological Agency (NMA) for stations located within and near the watershed. In order to fill gaps in some of the data, the weather generator file created by White et al. (2011) was used. Daily river discharge for Gumera River is available since 1976. Daily discharge data from January 2000 to December 2009 were used to calibrate and validate the model. This time span was selected for its overlap to 10-year lake sediment concentration data generated from MODIS images (Kaba et al. 2014).

The sensitivity analysis tool in SWAT is used in ranking parameters based on their influence in governing flow or sediment. SWAT calibration uncertainty program (SWAT-CUP), linked to ArcSWAT, was used to perform uncertainty analysis and calibrate the model. The SWAT-CUP program includes five calibration routines (SUFI-2, ParaSol, GLUE, MCMC, and PSO). Previous detailed studies have shown sequential uncertainty fitting (SUFI-2) program that performs better for Gumera watershed (Setegn et al. 2008).

Calibration of monthly flow and TSS was performed from 2000 to 2006 with the first year as a warm-up period, and the validation period was 2007–2009. The water balance was calibrated first followed by the TSS time series data obtained from the time series generation process. The model parameters were checked for maintaining their physical meaning (i.e., whether they are within the specified limits). The performance of the simulation was evaluated using the Nash–Sutcliffe coefficient of efficiency (Nash and Sutcliffe 1970), the *p*-factor, and *r*-factor (Rouholahnejad et al. 2012). In addition, percent bias (PBIAS) and ratio of the root mean square error (RSR) to the standard deviation of measured data were used to evaluate the model output (van Griensven et al. 2012). The Nash–Sutcliffe coefficient of efficiency is expressed by Eq. 36.4.

$$E_{NS} = \frac{\sum_{t=1}^{n} (Q_{i,s} - Q_{i,m})^2}{\sum_{t=1}^{n} Q_{i,m} - \bar{Q}_m)^2} \tag{36.4}$$

where $Q_{i,s}$ is simulated quantity (flow or TSS), $Q_{i,m}$ is measured quantity, and $\bar{Q}_m$ is mean of the measured quantity.

For *p*-factor, the 95 percent prediction uncertainty (95PPU) is calculated at the 2.5 and 97.5 % levels of the cumulative distribution of an output variable obtained through Latin hypercube sampling. The average distance $\bar{d}_{\mathrm{m}}$ between the upper and the lower 95PPU is used to calculate the *r*-factor expressed by Eq. 36.5 (Abbaspour 2008):

$$r\text{-factor} = \frac{\bar{\mathrm{d}}x}{\sigma x} \tag{36.5}$$

Uncertainty is an inherent characteristic of hydrologic models. These uncertainties should be properly addressed and quantified for the models to be usable in decision making. In SUFI-2, a measure, *p*-factor is used to quantify the degree to which all uncertainties are accounted. The *p*-factor is the percentage of measured data bracketed by the 95 % prediction uncertainty (95PPU) (Abbaspour 2008).

36.3 Results and Discussion

Reflectance–TSS relationship has been developed (Eq. 36.1) and depicted in Fig. 36.3. Figure 36.4a shows observed TSS and TSS estimated by Eq. 36.1. Figure 36.4b shows residuals of the estimated and measured TSS.

36.3.1 Generating TSS Time Series data

The 10-year time span (2000–2009) NIR images of Gumera River entering Lake Tana were downloaded via USGS MODIS Reprojection Tool Web Interface (MRTWeb). Cloud-contaminated images were excluded, and the images were masked with the water sampling location polygon. This location is consistently

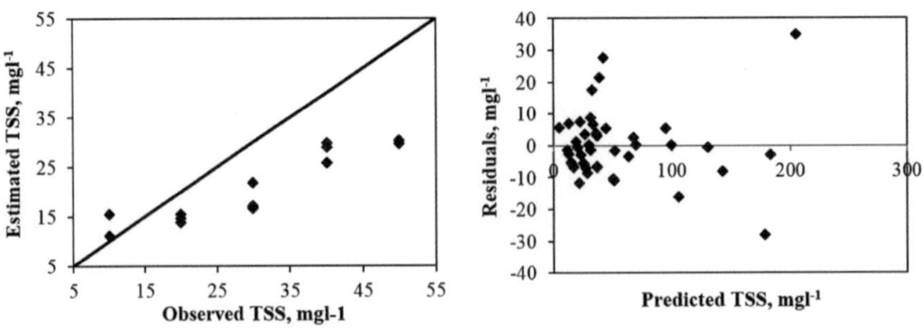

Fig. 36.4 Estimated TSS versus observed TSS (*left*); validation result for Eq. 36.1 using data collected on November 7, 2011 (*right*)

more turbid in the images from other times during the rainy season (Fig. 36.2). A mean monthly reflectance raster is created using cell statistic operation in ArcGIS. In each mean reflectance image, the pixel with the largest reflectance is identified using the Getis-Ord Gi* statistic data mining technique (Getis and Ord 2010). The pixel with the highest z score for the Getis-Ord Gi* statistic is identified for each mean reflectance raster at Gumera River mouth. Using monthly reflectance from 2000 to 2009, TSS estimates were generated using Eq. 36.1 demonstrating the application of estimating TSS from reflectance at sites where there are no measurements of TSS (Fig. 36.5).

36.3.2 Application of the SWAT Model

Daily river discharge for Gumera River is available since 1976, but sediment data generated from MODIS images are only available from January 2000 to December 2009. The calibration and validation of the SWAT model to simulate flow and sediment was performed for the flow and sediment overlap period (2000–2009). Calibration of monthly flow and TSS was performed from 2000 to 2006 with the first year as a warm-up period, and the validation period was 2007–2009.

The sensitivity analysis tool in SWAT is used in ranking parameters based on their influence in governing flow or sediment. Table 36.2 shows the results of the sensitivity analysis for the SWAT-VSA monthly simulation of flow and sediment. For flow simulation, the sensitivity parameters CN2 and ALPHA_BF are ranked first and second, respectively. As expected, the channel erodibility factor (CH_EROD) and the sediment transport coefficient (SPCON) were the most sensitive parameters governing TSS in the lake at the river mouth (Table 36.2).

NSE (Eq. 36.4), *r*_factor (Eq. 36.5), R^2, and *p*_factor are computed for the objective functions in SUFI-2. The Nash–Sutcliffe objective function yields the best result (Table 36.3). Other statistical variables were used for comparative judgment.

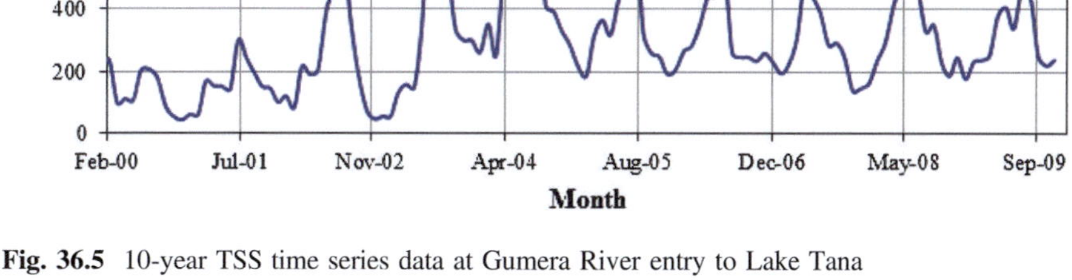

Fig. 36.5 10-year TSS time series data at Gumera River entry to Lake Tana

Table 36.2 Parameter sensitivity in SWAT-VSA

	Fitted value	Range	Rank
Flow parameters			
CN2	0.074	±0.25	1
ALPHA_BF	0.89	0–1	2
SURLAG	2.8	1–10	3
REVAPMN	471.75	0–500	5
USLE_C	−0.08	±0.25	11
Sediment parameters			
CH_ERODE	0.24	0–0.6	4
SPCON	0.005	0.001–0.01	6

Table 36.3 Monthly statistical coefficients for discharge and sediment calibration and validation

Variable	Period	R^2	E_{NS}	PBIAS	RSR
Discharge	Calibration	0.79	0.75	21.4	0.5
	Validation	0.8	0.64	21.7	0.6
TSS	Calibration	0.39	0.34	−9.4	0.8
	Validation	0.32	0.21	−2.2	0.8

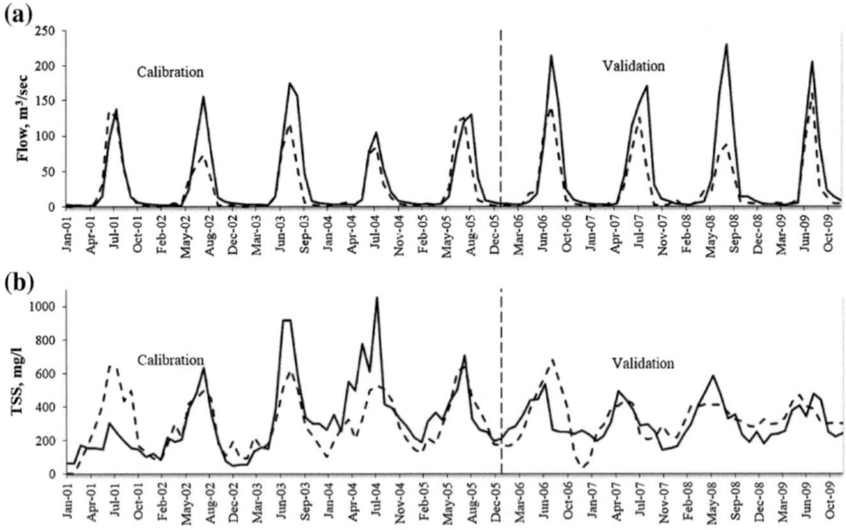

Fig. 36.6 Monthly calibration and validation: **a** monthly observed flow (*solid line*), simulated (*dashed*), and **b** monthly TSS derived from MODIS images (*solid line*), simulated (*dashed*)

Results of the time series SWAT model calibration and validation for flow and sediment are shown in Fig. 36.6a, b, respectively. Figure 36.7a, b shows graphic statistical fit of observed and simulated flows and sediment, respectively.

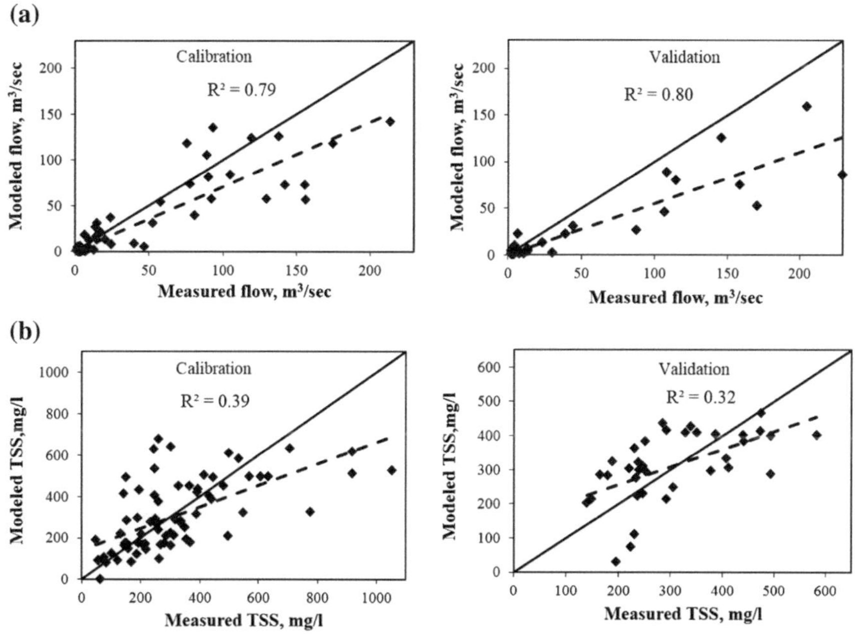

Fig. 36.7 Comparison between measured and simulated: **a** flow and **b** TSS (2001–2009)

Flow was calibrated at the gauging station. Like any water quality model, SWAT must first accurately simulate the hydrologic processes before it can realistically predict pollutants. In this application, the model simulated flow very well (R^2 of 0.79 and 0.80) for calibration and validation periods, respectively. The simulation for TSS did not yield as high R^2 as for flow simulation (0.39 for calibration and 0.32 for validation).

In evaluating simulation results, van Griensven et al. (2012) recommended three criteria: fitness to observations, fitness to reality, and fitness to purpose. Fitness to observations refers to the difference between the observed and simulated values. Fitness to reality evaluates how well a model represents the physical process while maintaining parameters within their meaningful range, and fitness to purpose accounts for how well certain watershed characteristics which the model output is needed to address are taken into consideration.

Based on the criteria for "fitness to observations," models are considered fit if NSE > 0.5 and RSR ≤ 0.7, and if PBIAS is ±25 % and ±55 % for flow and sediment, respectively, for a monthly time step (van Griensven et al. 2012). Moriasi et al. (2007), however, indicated that NSE values between 0 and 1 are generally viewed as acceptable. The RSR and the PBIAS criteria are satisfied. Simulated flow satisfies the entire criterion for fitness to purpose including the dry season flow. PBIAS values tend to vary more among different autocalibration methods, during dry and wet years (Moriasi et al. 2007). The wet season flow is especially important as it carries the major proportion of the TSS into the lake. The PBIAS for flow indicates slight underestimation bias which will eventually degrade the TSS simulation outcome.

With respect to the model "fitness to reality," the parameter values are all within the recommended range (Table 36.2). The average sediment yield is greater than 10 tons ha^{-1} which is in accordance with the values found in other studies (Hurni 1988; Hawando 1997; Tebebu et al. 2010). The shift in TSS after the 2002–2003 droughts is reasonably well represented both with the simulated and with the MODIS-derived time series.

The model "fitness to purpose" is the most important criteria for assessing the usability of the simulated TSS time series data. The modest NSE of 0.34 for calibration and 0.21 for validation and R^2 of 0.39 and 0.32 (Table 36.3) show that the predictions are more accurate than taking an average value, and thus, watershed processes are taken into account, but could be improved. Thus, the model is reasonably fit for its purpose because it provides a method to estimate TSS concentrations using watershed characteristics for a changing climate and for the past when measurements are not available. At the same time, these results show that the model should be improved to better represent the lake processes.

36.4 Conclusion

In this study, the usability of MODIS image-generated TSS time series data is evaluated to estimate TSS input into Lake Tana from Gumera River. An equation is developed where TSS can be estimated from reflectance. Estimates of TSS from MODIS-derived reflectance are shown for ten years. A second approach to estimate sediment input from Gumera River was tested using the SWAT hydrologic model. The model was calibrated and validated with a modest performance. The simulation over a period of 10 years (2000–2009) allowed an estimation of the annual average input of TSS into Lake Tana. Given the complicated sediment transport processes that are not fully understood, the data mining techniques applied in constructing the TSS time series data gave modest results. Harmel et al. (2006) noted that the uncertainty in using manual single point random time grab sampling could be in excess of 50 % and −5.3 to 4.4 % due to the method used in sample analysis. The inability to incorporate the major landscape element (i.e., the floodplain) as well as the lake sediment processes to the model results in reduced accuracy in the model output according to van Griensven et al. (2012). Thus, there is a clear need to develop better simulation models for estimation of lake sediment concentration; however, until these are available, the current model can be used to estimate TSS concentrations using watershed characteristics for a changing climate and for the past when measurements are not available.

References

Abbaspour K (2008) SWAT-CUP2: SWAT calibration and uncertainty programs—a user manual. Department of Systems Analysis, Swiss Federal Institute of Aquatic Science and Technology, Dübendorf

Arnold JG, Srinivasan R, Muttiah RS, Williams J (2007) Large area hydrologic modeling and assessment part I: model development1. JAWRA J Am Water Resour Assoc 34:73–89

BCEOM (1999) Abay River basin integrated master plan, main report. Ministry of Water Resources, Addis Ababa

Chebud YA, Melesse AM (2009a) Modelling lake stage and water balance of Lake Tana, Ethiopia. Hydrol Process 23:3534–3544

Chebud YA, Melesse MA (2009b) Numerical modelling of the groundwater flow system of the Gumera sub basin in Lake Tana basin, Ethiopia. Hydrol Process 23:3534–3544

Chebud Y, Melesse AM (2013) Stage level, volume, and time-frequency change information content of Lake Tana using stochastic approaches. Hydrol Process 27(10):1475–1483. doi:10.1002/hyp.9291

Chen Z, Hu C, Muller-Karger F (2007) Monitoring turbidity in Tampa Bay using MODIS/Aqua 250-m imagery. Remote Sens Environ 109:207–220

Chen S, Fang L, Zhang L, Huang W (2009) Remote sensing of turbidity in seawater intrusion reaches of Pearl River Estuary—a case study in Modaomen water way, China. Estuar Coast Shelf Sci 82:119–127

Conway D (2000) The climate and hydrology of the Upper Blue Nile River. Geograph J 166:49–62

Dall'Olmo G, Gitelson AA, Rundquist DC, Leavitt B, Barrow T, Holz JC (2005) Assessing the potential of SeaWiFS and MODIS for estimating chlorophyll concentration in turbid productive waters using red and near-infrared bands. Remote Sens Environ 96:176–187

Defersha MB, Melesse AM (2011) Field-scale investigation of the effect of land use on sediment yield and surface runoff using runoff plot data and models in the Mara River basin, Kenya. CATENA 89:54–64

Defersha MB, Melesse AM (2012) Effect of rainfall intensity, slope and antecedent moisture content on sediment concentration and sediment enrichment ratio. CATENA 90:47–52

Defersha MB, Quraishi S, Melesse AM (2011) Interrill erosion, runoff and sediment size distribution as affected by slope steepness and antecedent moisture content. Hydrol Proc 7 (4):6447–6489

Defersha MB, Melesse AM, McClain M (2012) Watershed scale application of WEPP and EROSION 3D models for assessment of potential sediment source areas and runoff flux in the Mara River basin, Kenya. CATENA 95:63–72

Dile YT, Karlberg L, Temesgen M, Rockström J (2013) The role of water harvesting to achieve sustainable agricultural intensification and resilience against water related shocks in sub-Saharan Africa. Agric Ecosyst Environ 181:69–79

Dozier J (1992) Opportunities to improve hydrologic data. Rev Geophys 30:315–331

Easton ZM, Fuka DR, Walter MT, Cowan DM, Schneiderman EM, Steenhuis TS (2008) Re-conceptualizing the soil and water assessment tool (SWAT) model to predict runoff from variable source areas. J Hydrol 348:279–291

Fiorani L, Okladnikov IG, Palucci A (2006) First algorithm for chlorophyll—a retrieval from MODIS—Terra imagery of Sun—induced fluorescence in the Southern Ocean. Int J Remote Sens 27:3615–3622

Getis A, Ord JK (2010) The analysis of spatial association by use of distance statistics. Persp Spatial Data Anal 127–145

Harmel R, Cooper R, Slade R, Haney R, Arnold J (2006) Cumulative uncertainty in measured streamflow and water quality data for small watersheds. Trans Am Soc Agric Eng 49 (3):689701

Harvey K, Grabs W (2003) WMO report of the GCOS, pp 18–20

Hawando T (1997) Desertification in Ethiopian highlands. RALA Rep 200:75–86

Hu C, Chen Z, Clayton TD, Swarzenski P, Brock JC, Muller-Karger FE (2004) Assessment of estuarine water-quality indicators using MODIS medium-resolution bands: Initial results from Tampa Bay, FL. Remote Sens Environ 93:423–441

Hurni H (1988) Degradation and conservation of the resources in the Ethiopian highlands. Mt Res Dev 123–130

Jiang X, Tang J, Zhang M, Ma R, Ding J (2009) Application of MODIS data in monitoring suspended sediment of Taihu Lake, China. Chin J Oceanol Limnol 27:614–620

Kaba E, Philpot W, Steenhuis T (2014) Evaluating suitability of MODIS-Terra images for reproducing historic sediment concentrations in water bodies: Lake Tana, Ethiopia. Int J Appl Earth Obs Geoinf 26:286–297

Kebede S, Travi Y, Alemayehu T, Marc V (2006) Water balance of Lake Tana and its sensitivity to fluctuations in rainfall, Blue Nile basin, Ethiopia. J Hydrol 316:233–247

Kebede S, Admasu G, Travi Y (2011) Estimating ungauged catchment flows from Lake Tana floodplains, Ethiopia: an isotope hydrological approach. Isot Environ Health Stud 47:71–86

Kirchner JW (2006) Getting the right answers for the right reasons: linking measurements, analyses, and models to advance the science of hydrology. Water Resour Res 42:W03S04

Kwiatkowska E, McClain C (2009) Evaluation of SeaWiFS, MODIS Terra and MODIS aqua coverage for studies of phytoplankton diurnal variability. Int J Remote Sens 30:6441–6459

Liu Y, Islam MA, Gao J (2003) Quantification of shallow water quality parameters by means of remote sensing. Prog Phys Geogr 27:24–43

Maalim FK, Melesse AM (2013) Modeling the impacts of subsurface drainage systems on Runoff and Sediment Yield in the Le Sueur Watershed, Minnesota. Hydrol Sci J 58(3):1–17

Maalim FK, Melesse AM, Belmont P, Gran K (2013) Modeling the impact of land use changes on runoff and sediment yield in the Le Sueur Watershed, Minnesota using GeoWEPP. Catena 107:35–45

Macauley MK, Vukovich FM (2005) Earth science remote sensing data: contributions to natural resources policymaking. Resources for the Future, Washington, DC

Mekonnen M, Melesse A (2011) Soil erosion mapping and hotspot area identification using GIS and remote sensing in northwest Ethiopian highlands, near Lake Tana. In: Melesse A (ed) Nile River basin: hydrology, climate and water use, Chap 10. Springer Science Publisher, pp 207–224. doi:10.1007/978-94-007-0689-7_10

Melesse AM, Ahmad S, McClain M, Wang X, Lim H (2011) Sediment load prediction in large rivers: ANN approach. Agric Water Manage 98:855–886

Miller RL, McKee BA (2004) Using MODIS Terra 250 m imagery to map concentrations of total suspended matter in coastal waters. Remote Sens Environ 93:259–266

Mohammed H, Alamirew T, Assen M, Melesse AM (2015) Modeling of sediment yield in Maybar gauged watershed using SWAT, northeast Ethiopia. CATENA 127:191–205

Moriasi D, Arnold J, Van Liew M, Bingner R, Harmel R, Veith T (2007) Model evaluation guidelines for systematic quantification of accuracy in watershed simulations. Trans ASABE 50:885–900

Msagahaa J, Ndomba PM, Melesse AM (2014) Modeling sediment dynamics: effect of land use, topography and land management. In: Melesse AM, Abtew W, Setegn S (eds) Nile River basin: ecohydrological challenges, climate change and hydropolitics, pp 165–192

Nash JE, Sutcliffe J (1970) River flow forecasting through conceptual models part I—a discussion of principles. J Hydrol 10:282–290

Nechad B, Ruddick K, Park Y (2010) Calibration and validation of a generic multisensor algorithm for mapping of total suspended matter in turbid waters. Remote Sens Environ 114:854–866

Neitsch S, Arnold J, Kiniry J, Williams J, King K (2005) Soil and water assessment tool: theoretical documentation, version 2005, Texas, USA

Onderka M, Rodny M, Veliskova Y (2011) Suspended particulate matter concentrations retrieved from self-calibrated multispectral satellite imagery. J Hydrol Hydromech 59:251–261

Petus C, Chust G, Gohin F, Doxaran D, Froidefond JM, Sagarminaga Y (2010) Estimating turbidity and total suspended matter in the Adour River plume (South Bay of Biscay) using MODIS 250-m imagery. Cont Shelf Res 30:379–392

Rouholahnejad E, Abbaspour K, Vejdani M, Srinivasan R, Schulin R, Lehmann A (2012) A parallelization framework for calibration of hydrological models. Environ Model Softw

Setegn SG, Srinivasan R, Dargahi B (2008) Hydrological modelling in the lake Tana Basin, Ethiopia using SWAT model. Open Hydrol J 2:49–62

Setegn SG, Srinivasan R, Melesse AM, Dargahi B (2010) SWAT model application and prediction uncertainty analysis in the Lake Tana Basin, Ethiopia. Hydrol Process 24:357–367

Setegn SG, Rayner D, Melesse AM, Dargahi B, Srinivasan R (2011) Impact of climate change on the hydroclimatology of Lake Tana basin, Ethiopia. Water Resour Res 47:W04511

Shen F, Salama MHDS, Zhou YX, Li JF, Su Z, Kuang DB (2010) Remote-sensing reflectance characteristics of highly turbid estuarine waters—a comparative experiment of the Yangtze River and the Yellow River. Int J Remote Sens 31:265–2639

Silberstein R (2006) Hydrological models are so good, do we still need data? Environ Model Softw 21:1340–1352

Steenhuis TS, Collick AS, Easton ZM, Leggesse ES, Bayabil HK, White ED, Awulachew SB, Adgo E, Ahmed AA (2009) Predicting discharge and sediment for the Abay (Blue Nile) with a simple model. Hydrol Process 23:3728–3737

Kutser T, Metsamaa L, Vahtmae E, Strombec, N (2006) Suitability of MODIS 250 m resolution band data for quantitative mapping of cyanobacterial blooms, pp 318–328

Tarekegn D, Tadege A (2006) Assessing the impact of climate change on the water resources of the Lake Tana sub-basin using the WATBAL model. Discuss Pap 30

Tebebu T, Abiy A, Zegeye A, Dahlke H, Easton Z, Tilahun S, Collick A, Kidnau S, Moges S, Dadgari F (2010) Surface and subsurface flow effect on permanent gully formation and upland erosion near Lake Tana in the Northern Highlands of Ethiopia. Hydrol Earth Syst Sci 14:2207–2217

Tilahun S, Guzman C, Zegeye A, Engda T, Collick A, Rimmer A, Steenhuis T (2012) An efficient semi-distributed hillslope erosion model for the sub humid Ethiopian highlands. Hydrol Earth Syst Sci Discuss 9:2121–2155

van Griensven A, Ndomba P, Yalew S, Kilonzo F (2012) Critical review of SWAT applications in the upper Nile basin countries. Hydrol Earth Syst Sci 16:3371–3381

Vijverberg J, Sibbing FA, Dejen E (2009) Lake Tana: source of the Blue Nile. The Nile 163–192

Vörösmarty C, Askew A, Grabs W, Barry R, Birkett C, Döll P, Goodison B, Hall A, Jenne R, Kitaev L (2001) Global water data: a newly endangered species. EOS Trans Am Geophys Union 82:54–58

Wale A, Rientjes T, Gieske A, Getachew H (2009) Ungauged catchment contributions to Lake Tana's water balance. Hydrol Process 23:3682–3693

Wang X, Garza J, Whitney M, Melesse AM, Yang W (2008) Prediction of sediment source areas within watersheds as affected by soil data resolution. In: Findley PN (ed) Environmental modelling: new research, Chap 7. Nova Science Publishers, Inc., Hauppauge, pp 151–185. ISBN: 978-1-60692-034-3

Wang H, Hladik C, Huang W, Milla K, Edmiston L, Harwell M, Schalles J (2010) Detecting the spatial and temporal variability of chlorophyll-a concentration and total suspended solids in Apalachicola Bay, Florida using MODIS imagery. Int J Remote Sens 31:439–453

White ED, Easton ZM, Fuka DR, Collick AS, Adgo E, McCartney M, Awulachew SB, Selassie YG, Steenhuis TS (2011) Development and application of a physically based landscape water balance in the SWAT model. Hydrol Process 25:915–925

Williams JR, Singh V (1995) The EPIC model. In: Computer models of watershed hydrology, pp 909–1000

Yasir S, Crosato A, Mohamed YA, Abdalla SH, Wright NG (2014) Sediment balances in the Blue Nile River basin. Int J Sedim Res 29:316–328

Index

© Springer International Publishing Switzerland 2016

A.M. Melesse and W. Abtew (eds.), *Landscape Dynamics, Soils and Hydrological Processes in Varied Climates*, Springer Geography,
DOI 10.1007/978-3-319-18787-7

MIX
Papier aus verantwortungsvollen Quellen
Paper from responsible sources
FSC® C105338

If you have any concerns about our products,
you can contact us on
ProductSafety@springernature.com

In case Publisher is established outside the EU,
the EU authorized representative is:
Springer Nature Customer Service Center GmbH
Europaplatz 3, 69115 Heidelberg, Germany

Printed by Libri Plureos GmbH
in Hamburg, Germany